W9-AVM-005

Drug Stability

DRUGS AND THE PHARMACEUTICAL SCIENCES

DRUGS AND THE PHARMACEUTICAL SCIENCES

A Series of Textbooks and Monographs

edited by

James Swarbrick
AAI, Inc.
Wilmington, North Carolina

1. Pharmacokinetics, *Milo Gibaldi and Donald Perrier*
2. Good Manufacturing Practices for Pharmaceuticals: A Plan for Total Quality Control, *Sidney H. Willig, Murray M. Tuckerman, and William S. Hitchings IV*
3. Microencapsulation, *edited by J. R. Nixon*
4. Drug Metabolism: Chemical and Biochemical Aspects, *Bernard Testa and Peter Jenner*
5. New Drugs: Discovery and Development, *edited by Alan A. Rubin*
6. Sustained and Controlled Release Drug Delivery Systems, *edited by Joseph R. Robinson*
7. Modern Pharmaceutics, *edited by Gilbert S. Banker and Christopher T. Rhodes*
8. Prescription Drugs in Short Supply: Case Histories, *Michael A. Schwartz*
9. Activated Charcoal: Antidotal and Other Medical Uses, *David O. Cooney*
10. Concepts in Drug Metabolism (in two parts), *edited by Peter Jenner and Bernard Testa*
11. Pharmaceutical Analysis: Modern Methods (in two parts), *edited by James W. Munson*
12. Techniques of Solubilization of Drugs, *edited by Samuel H. Yalkowsky*
13. Orphan Drugs, *edited by Fred E. Karch*
14. Novel Drug Delivery Systems: Fundamentals, Developmental Concepts, Biomedical Assessments, *Yie W. Chien*
15. Pharmacokinetics: Second Edition, Revised and Expanded, *Milo Gibaldi and Donald Perrier*
16. Good Manufacturing Practices for Pharmaceuticals: A Plan for Total Quality Control, Second Edition, Revised and Expanded, *Sidney H. Willig, Murray M. Tuckerman, and William S. Hitchings IV*
17. Formulation of Veterinary Dosage Forms, *edited by Jack Blodinger*
18. Dermatological Formulations: Percutaneous Absorption, *Brian W. Barry*
19. The Clinical Research Process in the Pharmaceutical Industry, *edited by Gary M. Matoren*
20. Microencapsulation and Related Drug Processes, *Patrick B. Deasy*
21. Drugs and Nutrients: The Interactive Effects, *edited by Daphne A. Roe and T. Colin Campbell*
22. Biotechnology of Industrial Antibiotics, *Erick J. Vandamme*
23. Pharmaceutical Process Validation, *edited by Bernard T. Loftus and Robert A. Nash*

24. Anticancer and Interferon Agents: Synthesis and Properties, *edited by Raphael M. Ottenbrite and George B. Butler*
25. Pharmaceutical Statistics: Practical and Clinical Applications, *Sanford Bolton*
26. Drug Dynamics for Analytical, Clinical, and Biological Chemists, *Benjamin J. Gudzinowicz, Burrows T. Younkin, Jr., and Michael J. Gudzinowicz*
27. Modern Analysis of Antibiotics, *edited by Adjoran Aszalos*
28. Solubility and Related Properties, *Kenneth C. James*
29. Controlled Drug Delivery: Fundamentals and Applications, Second Edition, Revised and Expanded, *edited by Joseph R. Robinson and Vincent H. Lee*
30. New Drug Approval Process: Clinical and Regulatory Management, *edited by Richard A. Guarino*
31. Transdermal Controlled Systemic Medications, *edited by Yie W. Chien*
32. Drug Delivery Devices: Fundamentals and Applications, *edited by Praveen Tyle*
33. Pharmacokinetics: Regulatory • Industrial • Academic Perspectives, *edited by Peter G. Welling and Francis L. S. Tse*
34. Clinical Drug Trials and Tribulations, *edited by Allen E. Cato*
35. Transdermal Drug Delivery: Developmental Issues and Research Initiatives, *edited by Jonathan Hadgraft and Richard H. Guy*
36. Aqueous Polymeric Coatings for Pharmaceutical Dosage Forms, *edited by James W. McGinity*
37. Pharmaceutical Pelletization Technology, *edited by Isaac Ghebre-Sellassie*
38. Good Laboratory Practice Regulations, *edited by Allen F. Hirsch*
39. Nasal Systemic Drug Delivery, *Yie W. Chien, Kenneth S. E. Su, and Shyi-Feu Chang*
40. Modern Pharmaceutics: Second Edition, Revised and Expanded, *edited by Gilbert S. Banker and Christopher T. Rhodes*
41. Specialized Drug Delivery Systems: Manufacturing and Production Technology, *edited by Praveen Tyle*
42. Topical Drug Delivery Formulations, *edited by David W. Osborne and Anton H. Amann*
43. Drug Stability: Principles and Practices, *Jens T. Carstensen*
44. Pharmaceutical Statistics: Practical and Clinical Applications, Second Edition, Revised and Expanded, *Sanford Bolton*
45. Biodegradable Polymers as Drug Delivery Systems, *edited by Mark Chasin and Robert Langer*
46. Preclinical Drug Disposition: A Laboratory Handbook, *Francis L. S. Tse and James J. Jaffe*
47. HPLC in the Pharmaceutical Industry, *edited by Godwin W. Fong and Stanley K. Lam*
48. Pharmaceutical Bioequivalence, *edited by Peter G. Welling, Francis L. S. Tse, and Shrikant V. Dinghe*
49. Pharmaceutical Dissolution Testing, *Umesh V. Banakar*
50. Novel Drug Delivery Systems: Second Edition, Revised and Expanded, *Yie W. Chien*
51. Managing the Clinical Drug Development Process, *David M. Cocchetto and Ronald V. Nardi*
52. Good Manufacturing Practices for Pharmaceuticals: A Plan for Total Quality Control, Third Edition, *edited by Sidney H. Willig and James R. Stoker*
53. Prodrugs: Topical and Ocular Drug Delivery, *edited by Kenneth B. Sloan*
54. Pharmaceutical Inhalation Aerosol Technology, *edited by Anthony J. Hickey*

55. Radiopharmaceuticals: Chemistry and Pharmacology, *edited by Adrian D. Nunn*
56. New Drug Approval Process: Second Edition, Revised and Expanded, *edited by Richard A. Guarino*
57. Pharmaceutical Process Validation: Second Edition, Revised and Expanded, *edited by Ira R. Berry and Robert A. Nash*
58. Ophthalmic Drug Delivery Systems, *edited by Ashim K. Mitra*
59. Pharmaceutical Skin Penetration Enhancement, *edited by Kenneth A. Walters and Jonathan Hadgraft*
60. Colonic Drug Absorption and Metabolism, *edited by Peter R. Bieck*
61. Pharmaceutical Particulate Carriers: Therapeutic Applications, *edited by Alain Rolland*
62. Drug Permeation Enhancement: Theory and Applications, *edited by Dean S. Hsieh*
63. Glycopeptide Antibiotics, *edited by Ramakrishnan Nagarajan*
64. Achieving Sterility in Medical and Pharmaceutical Products, *Nigel A. Halls*
65. Multiparticulate Oral Drug Delivery, *edited by Isaac Ghebre-Sellassie*
66. Colloidal Drug Delivery Systems, *edited by Jörg Kreuter*
67. Pharmacokinetics: Regulatory • Industrial • Academic Perspectives, Second Edition, *edited by Peter G. Welling and Francis L. S. Tse*
68. Drug Stability: Principles and Practices, Second Edition, Revised and Expanded, *Jens T. Carstensen*
69. Good Laboratory Practice Regulations: Second Edition, Revised and Expanded, *edited by Sandy Weinberg*
70. Physical Characterization of Pharmaceutical Solids, *edited by Harry G. Brittain*
71. Pharmaceutical Powder Compaction Technology, *edited by Göran Alderborn and Christer Nyström*
72. Modern Pharmaceutics: Third Edition, Revised and Expanded, *edited by Gilbert S. Banker and Christopher T. Rhodes*
73. Microencapsulation: Methods and Industrial Applications, *edited by Simon Benita*
74. Oral Mucosal Drug Delivery, *edited by Michael J. Rathbone*
75. Clinical Research in Pharmaceutical Development, *edited by Barry Bleidt and Michael Montagne*
76. The Drug Development Process: Increasing Efficiency and Cost-Effectiveness, *edited by Peter G. Welling, Louis Lasagna, and Umesh V. Banakar*
77. Microparticulate Systems for the Delivery of Proteins and Vaccines, *edited by Smadar Cohen and Howard Bernstein*
78. Good Manufacturing Practices for Pharmaceuticals: A Plan for Total Quality Control, Fourth Edition, Revised and Expanded, *Sidney H. Willig and James R. Stoker*
79. Aqueous Polymeric Coatings for Pharmaceutical Dosage Forms: Second Edition, Revised and Expanded, *edited by James W. McGinity*
80. Pharmaceutical Statistics: Practical and Clinical Applications, Third Edition, *Sanford Bolton*
81. Handbook of Pharmaceutical Granulation Technology, *edited by Dilip M. Parikh*
82. Biotechnology of Antibiotics: Second Edition, Revised and Expanded, *edited by William R. Strohl*

83. Mechanisms of Transdermal Drug Delivery, *edited by Russell O. Potts and Richard H. Guy*
84. Pharmaceutical Enzymes, *edited by Albert Lauwers and Simon Scharpé*
85. Development of Biopharmaceutical Parenteral Dosage Forms, *edited by John A. Bontempo*
86. Pharmaceutical Project Management, *edited by Tony Kennedy*
87. Drug Products for Clinical Trials: An International Guide to Formulation • Production • Quality Control, *edited by Donald C. Monkhouse and Christopher T. Rhodes*
88. Development and Formulation of Veterinary Dosage Forms: Second Edition, Revised and Expanded, *edited by Gregory E. Hardee and J. Desmond Baggot*
89. Receptor-Based Drug Design, *edited by Paul Leff*
90. Automation and Validation of Information in Pharmaceutical Processing, *edited by Joseph F. deSpautz*
91. Dermal Absorption and Toxicity Assessment, *edited by Michael S. Roberts and Kenneth A. Walters*
92. Pharmaceutical Experimental Design, *Gareth A. Lewis, Didier Mathieu, and Roger Phan-Tan-Luu*
93. Preparing for FDA Pre-Approval Inspections, *edited by Martin D. Hynes IIIv*
94. Pharmaceutical Excipients: Characterization by IR, Raman, and NMR Spectroscopy, *David E. Bugay and W. Paul Findlay*
95. Polymorphism in Pharmaceutical Solids, *edited by Harry G. Brittain*
96. Freeze-Drying/Lyophilization of Pharmaceutical and Biological Products, *edited by Louis Rey and Joan C. May*
97. Percutaneous Absorption: Drugs—Cosmetics—Mechanisms—Methodology, Third Edition, Revised and Expanded, *edited by Robert L. Bronaugh and Howard I. Maibach*
98. Bioadhesive Drug Delivery Systems: Fundamentals, Novel Approaches, and Development, *edited by Edith Mathiowitz, Donald E. Chickering III, and Claus-Michael Lehr*
99. Protein Formulation and Delivery, *edited by Eugene J. McNally*
100. New Drug Approval Process: Third Edition: The Global Challenge, *edited by Richard A. Guarino*
101. Peptide and Protein Drug Analysis, *edited by Ronald E. Reid*
102. Transport Processes in Pharmaceutical Systems, *edited by Gordon Amidon, Ping I. Lee, and Elizabeth M. Topp*
103. Excipient Toxicity and Safety, *edited by Myra L. Weiner and Lois A. Kotkoskie*
104. The Clinical Audit in Pharmaceutical Development, *edited by Michael R. Hamrell*
105. Pharmaceutical Emulsions and Suspensions, *edited by Francoise Nielloud and Gilberte Marti-Mestres*
106. Oral Drug Absorption: Prediction and Assessment, *edited by Jennifer B. Dressman and Hans Lennernäs*
107. Drug Stability: Principles and Practices, Third Edition, Revised and Expanded, *edited by Jens T. Carstensen and C. T. Rhodes*

ADDITIONAL VOLUMES IN PREPARATION

Containment in the Pharmaceutical Industry, *edited by James Wood*

Good Manufacturing Practices for Pharmaceuticals: Fifth Edition, Revised and Expanded, *Sidney H. Willig*

Advanced Pharmaceutical Solids, *Jens T. Carstensen*

Drug Stability

Principles and Practices

Third Edition, Revised and Expanded

edited by

Jens T. Carstensen
Madison, Wisconsin

C. T. Rhodes
University of Rhode Island
Kingston, Rhode Island

MARCEL DEKKER, INC. NEW YORK • BASEL

ISBN: 0-8247-0376-6

This book is printed on acid-free paper.

Headquarters
Marcel Dekker, Inc.
270 Madison Avenue, New York, NY 10016
tel: 212-696-9000; fax: 212-685-4540

Eastern Hemisphere Distribution
Marcel Dekker AG
Hutgasse 4, Postfach 812, CH-4001 Basel, Switzerland
tel: 41-61-261-8482; fax: 41-61-261-8896

World Wide Web
http://www.dekker.com

The publisher offers discounts on this book when ordered in bulk quantities. For more information, write to Special Sales/Professional Marketing at the headquarters address above.

Current printing (last digit):
10 9 8 7 6 5 4 3 2 1

PRINTED IN THE UNITED STATES OF AMERICA

The first and second editions of Drug Stability *were authored by Jens T. Carstensen.*

Preface

The considerable success achieved by the first and second editions of *Drug Stability* has led to the need for a third edition. Although this edition is firmly based on the foundations laid by JTC, it has undergone a radical change in that additional chapter authors have contributed to the book and CTR is now assisting in the editorial duties.

In the period of almost five years that has elapsed since the publication of the second edition, there have indeed been significant developments in the science, technology, and regulatory aspects of pharmaceutical stability testing. However, innovation has not occurred at a uniform rate in all the disciplines of relevance to our field. There have been no fundamental changes in the equations that govern chemical reaction; our understanding of the mechanisms of oxidation has matured but the underlying theory is still basically unaltered. Important new technologies for stability testing (especially for macromolecules) have been introduced, however, and have influenced the approaches used by pharmaceutical scientists in laboratory stability studies. It is probably in the regulatory arena that the most substantial developments have occurred. The several ICH Guidelines on stability topics and the subsequent FDA document have exerted, and will continue to exert, considerable influence on approaches used by pharmaceutical scientists in stability work. In particular, all the effects of the lengthy FDA document are still not clear.

Thus, in making our plans for this third edition, we realized that although the discussion of some topics was in relatively little need of modification from the second edition, others needed radical modification and additional coverage. Further, we felt it appropriate to introduce some topics not previously discussed in *Drug Stability* to any significant extent.

We hope that the third edition not only continues the tradition established in the first and second editions, but also blends in new topics and opinions so that this book can continue to be regarded as a reliable text covering many, if not all, subjects that relate to drug stability.

Jens T. Carstensen
C. T. Rhodes

Contents

Contributors

Jens T. Carstensen, Ph.D. Madison, Wisconsin

D. A. Dean Consultant, Beeston, Nottingham, England

Mary D. DiBiase, Ph.D. Biogen, Cambridge, Massachusetts

Wolfgang Grimm, Ph.D. Biberach, Germany

Donald D. Hong, Ph.D. Pharmaceutical Consultant, Raleigh, North Carolina

Mary K. Kottke, Ph.D. Cubist Pharmaceuticals, Inc., Cambridge, Massachusetts

Brian R. Matthews, Ph.D. Alcon Laboratories (U.K.) Limited, Hemel Hempstead, Hertfordshire, England

C. T. Rhodes, Ph.D. Department of Applied Pharmaceutical Sciences, University of Rhode Island, Kingston, Rhode Island

Mumtaz Shah, Ph.D. Trigen Laboratories, Salisbury, Maryland

Shri C. Valvani, Ph.D. Pharmaceutical Consultant, Kalamazoo, Michigan

Drug Stability

1

Introductory Overview

C. T. RHODES

University of Rhode Island, Kingston, Rhode Island

1. STABILITY IS AN ESSENTIAL QUALITY ATTRIBUTE FOR DRUG PRODUCTS

> "There never was anything by the wit of man so well devized or so sure established which hath not in the continuance of time become corrupted . . ."
>
> Thomas Cranmer

Everything made by human hands—from the sublime Parthenon to the trivial milkshake—is subject to decay. Pharmaceuticals are no exception to this general statement. If there is any functionally relevant quality attribute of a drug product that changes with time, evaluation of this change falls within the purview of the pharmaceutical scientists and regulators who quantify drug product stability and shelf life.

The rate at which drug products degrade varies dramatically. Some radiopharmaceuticals must be used within a day or so. Other products may, if properly stored and packaged, retain integrity for a decade or more, although in many jurisdictions the maximum shelf life that a regulatory agency will approve for a drug product is five years. (This restriction is hardly an onerous one, since even for a product with a five-year shelf life it is probable that over 95% of the product will be sold and used within thirty months of manufacture, providing all involved in the distribution process obey the first law of warehousing: FIFO—first in, first out.)

Since the evaluation of the stability of a drug product is highly specialized and esoteric in nature, reliance on the patient's suck-it-and-see organoleptic evaluation is of distinctly limited value. Thus governments in many parts of the world—most importantly in Western Europe, North America, and Japan—have, because of concerns about drug product safety, efficacy, and quality, found it appropriate to require some form of stability testing for drug products. However, it must be recognized that even before governments became active in this area many reputable companies were already giving attention to drug product stability and developing their own in-house approaches. The increasing intervention by regulatory agencies such as the FDA (U.S. Federal Food and Drug Administration) and the HPB (Canadian Health Protection Branch) stimulated standard approaches to stability testing in those parts of the world subject to their control (1). More recently, the process of globalization and harmonization has stimulated the development of world-wide standards. (This topic is further considered in Chapter 18.) It is now well accepted that stability is an essential property of drug products; thus the assignment of a shelf life is a routine regulatory requirement.

2. POTENTIAL ADVERSE EFFECTS OF INSTABILITY IN PHARMACEUTICAL PRODUCTS

There is a variety of mechanisms by which drug products may degrade, and thus a quite wide range of adverse effects that can occur.

2.1. Loss of Active

Obviously, loss of drug is of major importance in the stability studies of many pharmaceutical products. Unfortunately, one sometimes gets the impression that some regard this as the only adverse effect of drug product stability, This is, of course, not true, and for some products loss of active is not the critical variable that determines shelf life. However, it is certainly true that for many products loss of potency is of major importance. In general, we regard any product that contains less than 90% of label claim of drug as being of unacceptable quality. Therefore, for many drug products, determination of the time that elapses before the drug content no longer exceeds 90% (when the product is stored in conformance to label instructions) is an essential element in determining shelf life (2).

The essence of the conventional way of determining shelf life from loss of active is as follows. The potency of product stored at the appropriate temperature (25°C for products to be labelled "Store at Controlled Room Temperature") is determined as a function of time and the best straight line of potency as a function of time determined by least squares regression analysis (Fig. 1). Of course, because of analytical and sampling error there will normally be some scatter of the experimentally determined data points around the mean regression line. Thus in order to have a high comfort level about the shelf life that we will assign to the product, we use conventional

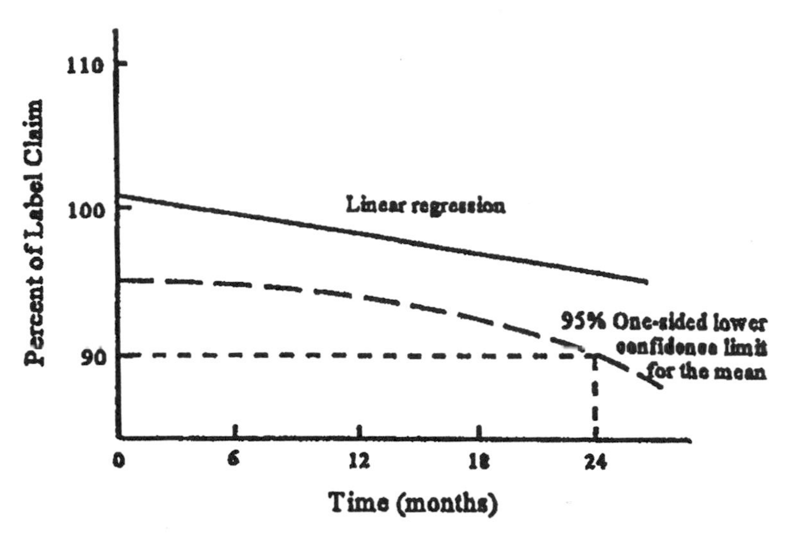

Figure 1 Least squares regression analysis.

statistical methods to calculate the 90% confidence zone of the regression line. This means that there is a 90% probability that the true regression line (of label claim potency as a function of time) is within the zone. Thus there is only a 5% chance that the true line is really below the 90% zone. It is apparent that any shelf life derived from the intersection of the lower 90% confidence bound and the 90% potency value has a 95% confidence level. In other words, there is only a 5% chance that our estimate of the shelf life will be too high. In fact, there is an additional safety margin built into most estimates of shelf life in that the period of time determined as described above (termed by A. J. Smith the *conformance period*) is usually significantly greater than the period defined as the shelf life. Suppose that for three separate pharmaceutical products we obtained 95% confidence estimates of the conformance periods of 13.2, 26.1, and 39.4 months. We would probably assign shelf lives of 12, 24, and 36 months to the three products. The difference between the conformance period and the assigned shelf life provides an extra stability reserve.

Those conversant with kinetic theory as outlined in Chapter 2 may wonder if the order of reaction affects the form of graph obtained as shown in Fig. 1. It might be felt that if the degradation process is governed by first-order kinetics, the plot would yield a curve rather than a straight line. In fact, since this type of plot only covers potency values from 100% down to 90%, both first and zero kinetic processes appear to be essentially linear. Yes, there are some degradation processes (3) that do produce nonlinear plots, but fortunately these are remarkably rare.

Shelf life values are normally assigned to a product rather than a batch. In order for such a practice to be legitimate, there must be reliable data that shows that, for at least three batches of the product, there is no significant difference in the slopes or the intercepts of the types of plot shown in Fig. 1. In those instances, where this level of batch similarity cannot be demonstrated, it might be possible to assign a shelf life based on a worst-case scenario (i.e., the shelf life is based on the worst-case batch). In this situation, we probably need more than three batches so that we can be comfortable that we have indeed identified the worst-case batch.

Most pharmaceutical products are characterized by only one shelf life. However, in some cases a product may have two. For example, a freeze-dried (lyophilized) protein product may have one shelf life, say two years, for the product stored in the dry condition and a second shelf life, say two days, for the product when it has been reconstituted with the appropriate vehicle and is ready for injection.

2.2. Increase in Concentration of Active

For some products, loss of vehicle can result in an increase in the concentration of active drug. For example, some lidocaine gels exhibit this behavior. Perfusion bags sometimes allow solvent to escape and evaporate so that the product within the bag shows an increase in concentration. This behavior is rare, but for such situations a horizontal mirror image of Fig. 1 should be used to estimate shelf life. In such a case, the point where the upper 90% confidence bound intersects that 110% potency value will define the conformance period.

2.3. Alteration in Bioavailability

Bioavailability and bioequivalence of drug products is a subject of great importance to those concerned with drug product quality (4). If the rate or extent of absorption that characterizes a product changes on storage, then this is, of course, a stability problem. In particular, if dissolution test data shows significant changes with time, there should be concern about possible clinically relevant modification of bioavailability or bioequivalence.

Examination of lists of products that have been subject to recalls shows that a number of products subsequent to release onto the market have been shown to fail relevant dissolution tests. Such failure has resulted in a Class II or Class III recall.* Case hardening of the surface of tablets or pellicle formation with hard gelatin capsules have been involved in this type of problem (4,5). It is thus most appropriate to include dissolution (or other release tests) in stability evaluations of pharmaceutical products.

2.4. Loss of Content Uniformity

Suspensions are the drug delivery systems most likely to show a loss of content uniformity as a function of time. For such systems, determination of ease of redispersion or sedimentation volume may therefore be included in a stability protocol (6).

2.5. Decline of Microbiological Status

In the comparatively recent past, say 15 years ago, it was relatively common practice to give attention to the microbiological status of pharmaceutical products only if they were designed for administration by the parenteral or ophthalmic routes. The situation has now changed, and though we do not expect all pharmaceutical

* FDA classifies recalls as I (most serious), potentially with serious, possibly with fatal, consequences; II (quite serious), may impair therapeutic response; and III (least serious), unlikely to have substantial adverse effect on therapy.

products to be sterile (that is, totally devoid of all forms of life both vegetative and sporing), we do have concern about the microbiological status of all drug delivery systems. We have concern about the extent of the total bioburden, and we also have a specific interest in excluding pathogens.

Basically, there are two possible ways in which the microbiological status of a pharmaceutical product can change significantly with time. First, microorganisms present in the product at the time of manufacture may reproduce and thus increase the number of viable organisms. Thus a product that, when assayed for total bioburden at the time of manufacture, is within limits may, when tested after say 6 months storage, exceed the maximum permitted bioburden. Second, if package integrity is compromised during distribution or storage, it is possible that the microbiological status of the product may be adversely affected as a result of the ingress of microorganisms.

In order to reduce or eliminate the first type of microbiological problem, attention should be given to the quality of the raw materials and the nature of the manufacturing facility and its operation. Certain raw materials that are often the source of microorganisms (both pathogenic and nonpathogenic) are of natural origin (e.g., corn starch, lecithin), and thus there should be particularly careful monitoring of the microbiological status of such raw materials. In terms of excluding contamination during manufacture, such factors as positive pressure air flow, equipment design, personnel training, and clear SOPs (standard operating procedures) all have roles to play.

2.6. Loss of Pharmaceutical Elegance and Patient Acceptability

It is not easy to define pharmaceutical elegance. It includes any aspect of the product that might suggest that the product is somehow substandard or variable. For example, some drugs that contain amino functional groups, when made into direct compression tablets that contain spray-dried lactose, may show some slight yellow or brown speckling on the surface of the tablet. The speckling is caused by the interaction of the drug with a minor component in the lactose, which results in the formation of a chromatophore that absorbs strongly in the visual part of the spectrum. Analysis of the speckled tablets might reveal no loss of potency or change in dissolution, but of course no reputable manufacturer will market tablets that look as though they are suffering from measles.

Pharmaceutical elegance does not mean that all drug products are expected to look and taste nice. Indeed, for some patients, particularly those of the older generation, the reverse is sometimes true. Some patients seem to believe that in order to be effective a product should have an unpleasant smell and taste. No, the important point about drug products is that attributes such as appearance, taste, and smell should be reproducible and not show significant batch-to-batch variation.

Also of relevance to pharmaceutical elegance is the ease (or lack thereof) of patient use and any change in patient acceptability. For example, if a topical product exhibits a change in play time or skin drag, this may not necessarily directly affect inherent safety or efficacy, but it may well impair the likelihood of the patient accepting the product or using it appropriately. Similarly, apparently trivial matters, such as a label showing some loss of adhesion at the extreme corners, may not reasonably be said to modify the essential safety and efficacy of the product. However, it may well engender doubts in the patient's mind about the quality of the product and thus adversely affect patient compliance.

2.7. Formation of Toxic Degradation Products

If a drug degrades to a molecular species that is toxic, there must be special attention given to the quantity of such a species found in the product during its shelf life. The classic example often quoted in this regard is the formation of epianhydrotetracycline from tetracycline, although the evidence in support of the alleged toxicity in this example may not be overwhelming. However, with protein drugs it is quite likely that even perturbation of molecular structure in a domain well removed from that responsible for therapeutic activity may result in serious toxic potential. Thus it does seem quite likely that this aspect of stability testing may become more important in the future. Concerns about potential levels of toxic degradation products are probably of considerable importance in the present reluctance by regulatory authorities to approve stability overages for new products.

2.8. Loss of Package Integrity

Change in package integrity during storage or distribution can be a stability problem that may require careful monitoring. For example, if a plastic screw cap loses back-off-torque, the possibility of chemical or microbiological hazard may be significantly increased. Thus when there is reason to believe that such a problem might exist, it is appropriate to use specific package integrity tests in the stability test protocol.

2.9. Reduction of Label Quality

The label of a drug product must be regarded as an essential element of the product. It provides information on identity, use, and safety. Thus if any aspect of the label deteriorates with time, this can be a serious stability problem. For example, if the plasticizer in a plastic bottle migrates into the label and causes the ink to run and thus adversely affects legibility, this is a major problem.

2.10. Modification of Any Factor of Functional Relevance

If there is any time-dependent change of any functionally relevant attribute of a drug product that adversely affects safety, efficacy, or patient acceptability or ease of use, monitoring such a change will be within the purview of stability evaluation. For example, when some transdermal patches were first introduced into use in the United States, a problem of adhesion aging was observed. Freshly manufactured patches showed excellent skin adhesion. However, in some instances products stored at room temperature for a period of weeks or months showed a loss of adhesion. Thus in use the patches had a tendency to fall off the patient's skin. For products subject to this adhesion aging, it would obviously be important to include quantification of adhesion potential in any protocol designed to evaluate stability.

3. THE GAMUT OF STABILITY CONCERNS

3.1. Bulk Drug Substance and Excipients

For the manufacturer of pharmaceutical products, such as compressed tablets or ophthalmic solutions, the first stability type concern will be specifications for the raw materials from which the drug delivery system is fabricated (i.e., drug substance excipients, water) and the extent, if any, to which such materials may degrade in the period between purchase and use.

The seriousness of this type of concern has abated to a considerable extent in the many companies that have adopted the just-in-time philosophy of drug product manufacture. This method requires that the time for which raw materials are stored at the facility where the drug delivery system is to be manufactured is kept to a minimum—often only a week or less. This procedure means that whereas in the past it was common to have raw materials that had been purchased as long as two years before use, it is now uncommon in many companies to see any raw material that has been held more than six weeks. Although the just-in-time method of pharmaceutical manufacturing was introduced primarily to reduce the amount of capital tied up in raw materials, it has also had the effect of reducing stability problems with raw materials. However, there are still some raw materials such as proteins (e.g., lecithin) for which stability is still a real concern and for which in-house testing by the pharmaceutical manufacturer may well be prudent.

In general, pharmacopoeias such as USP have performed well in providing standards for drug substances, and the tests described in the compendia can often be of great value in assuring the quality of drugs to be used in manufacture of drug delivery systems.

The situation concerning standards for excipients is less satisfactory. It is, unfortunately, only comparatively recently that pharmaceutical scientists have given substantial attention to standards for excipients that have national or international recognition. The *Handbook of Pharmaceutical Excipients* (6) has, however, become widely used by pharmaceutical scientists (6) and has attained quasi-official status. Also of value in providing information about excipients is the *Handbook of Pharmaceutical Additives* (7). This book provides data on over 6,000 materials and can be a useful source of data for possible standards for excipients and might be used as the basis for a vendor's Certificate of Analysis.

One of the problems that may face a manufacturer of a dosage form is that a vendor of an excipient may be unwilling to provide comprehensive data on the material because of the fear that the information could be exploited by a competitor. Fortunately, there is a rather clever regulatory procedure to getting around this problem. The manufacturer of the excipient establishes a Drug Master File at the FDA. This document contains comprehensive data on such topics as impurities, toxicology, and stability. Its contents are made available to FDA reviewers when a pharmaceutical manufacturer refers to the particular excipient in a request concerning a formulation that will contain the excipient.

3.2. Research and Development Formulations

A common practice in the pharmaceutical industry is to evaluate a number of formulations for such critical attributes as stability. As time goes by, those formulations that are shown to be unsatisfactory are rejected, and thus the number of potentially viable formulations is reduced.

Unfortunately, the formulators in charge of research and development (R&D) projects do not always inform the quality control chemists responsible for stability testing that certain test formulations are no longer potential candidates to be marketed. This results in a waste of time and money.

There is no universally accepted procedure for evaluating stability of R&D formulations. Often, however, accelerated testing and/or comparisons with similar products already on the market can be useful. When a formulation has been finally selected, the official guideline should be followed in obtaining appropriate stability data so that market approval can be obtained.

3.3. Clinical Trials Materials

Chapter 15 gives specific attention to this important matter. The book *Drug Products for Clinical Trials* (8) may also be of value to readers who are especially interested in this topic.

3.4. Marketed Product

A very substantial part of the effort of a stability testing program is devoted to producing data that will convince a regulatory authority that a particular formulation, process, and shelf life for a new or reformulated product are acceptable. Also, once a product is marketed, the manufacturer will wish to generate reliable stability data that will provide assurance that the marketed product continues to justify the shelf life that has been assigned. It is these two linked areas that are a major focus of this book.

The process of finalizing the formulation and process intended for the marketed product should be completed as early as possible during clinical trials. The FDA has issued specific requirements about preapproval inspections that it conducts before an NDA (New Drug Application) or similar marketing approval document is accepted (9). The Agency has also issued (9) useful information concerning its requirements regarding scale-up (11).

3.5. Reformulation, Change of Manufacturing Site, Troubleshooting, Complaints

It would be naive to think that once regulatory approval has been obtained and a product is on the market, the work of a stability group becomes routine and mundane. Usually the stability of the product is subject to intermittent new studies for a variety of reasons. For example, a decision may be made to modify one or more formulation or process variables, so that, depending on the significance of such changes, stability studies of varying complexity may be necessary.

Also, if it is decided to change the site at which the product is manufactured, regulatory agencies may also require additional data. The new FDA Guidelines (10) gives specific attention to this point and should, of course, be carefully followed by those subject to FDA jurisdiction.

Even if no change of formulation, process, or site of manufacture is contemplated, there may be other reasons for additional stability studies. Unfortunately, it is not unknown for a new product that we believe to have been fully validated with respect to all quality attributes (including stability) to exhibit unexpected stability problems. These problems may progressively develop in a most insidious way, affecting all batches or, in some instances, only some batches intermittently. In either event, troubleshooting directed at identifying the cause and then taking appropriate remedial action is necessary.

Similarly, if complaints from patients, health professionals, or others involve stability problems, it is obviously important that stability group personnel should be involved in the evaluation of the problem and be consulted when it is decided if remedial action is required.

3.6. Product in the Channel of Distribution

It is not sufficient to restrict our concerns about drug product stability to the quality of the pristine, freshly manufactured material that we regard with justifiable pride as it waits in our warehouse for distribution after it has been cleared from quarantine by our QC/QA (Quality Control/Quality Assurance) department. Of course, it is normal to store some stability samples in our stability storage areas (retained samples). However, the evaluation of samples that have been stored under the utopian conditions in the manufacturer's stability storage areas is of limited value. Samples retained for stability testing are not dropped off the back of a truck; they are not left on a loading dock in the blazing sun; nor are they left in the freezing cold. Thus it is somewhat unrealistic to expect retained stability samples to reflect accurately the stability status range of products that are in the channel of distribution. As is discussed in Chapter 18, there is now increased concern about the stability status of products in the channel of distribution.

3.7. Product Under the Control of the Patient

There is good reason to believe that, in many instances, the conditions under which patients store their drug products is far removed from optimal. At one time some regulatory authorities were considering the possibility of requiring shelf lives that could be guaranteed right through to the time when the patient used the last dose of the product. It is now probably generally appreciated that this idea is not feasible. It certainly is, however, most appropriate that pharmacists should take time and trouble to counsel patients on the appropriate ways to store drug products.

3.8. *In Vivo* Stability

The final stability concern is the degradation of the drug *in vivo*. In particular, the hydrolysis of drug at the low pH conditions of the stomach can be particularly serious. The traditional answer to this particular problem is to enteric coat a tablet. In the past, the materials used as enteric coats were not always effective. The polymers now available for enteric coating are much more reliable (11).

4. REASONS FOR STABILITY TESTING

4.1. Our Concerns for Patients' Welfare

Obviously, our primary reason for stability testing should be our concern for the well-being of the patients who will use our products. Sometimes in the mad rush to comply with other requirements, this important fundamental may be discounted or forgotten. Indeed, sometimes one gains the impression that in some quarters stability is regarded as having little clinical relevance. Certainly, if a product that does not degrade to toxic decomposition products and that is not characterized by a narrow therapeutic ratio is present on the market at only 85% of label claim, one would not expect patients to be dropping dead in the streets because of this deficiency instability. However, this is not to say that stability problems can never have serious clinical consequences. For example, in the early 1980s a packaging stability problem with nitroglycerin tablets unfortunately resulted in some nitroglycerin tablets being available in the Midwest with potency values of less than 10% of label claim. Since nitroglycerin is used for the emergency treatment of a most serious cardiac condition, angina, there is unfortunately strong cause for concern that some patients may have died as a result of this stability problem.

Even if death is not likely because of stability problems with a particular drug product, the inconvenience, discomfort, and cost associated with the use of product that is subpotent or exhibits an unacceptably wide range of potencies may be a serious problem needing radical remedial response. For example, concerns about possible potency problem with L-thyroxene products were of considerable importance in stimulating the FDA to require, in August 1997, that all human L-thyroxene products that were on the U.S. market at that time could only remain on the market until August 2000, unless new regulatory NDAs (New Drug Applications) or ANDAs (Abbreviated New Drug Applications) were approved (12).

4.2. To Protect the Reputation of the Producer

We should all be jealous for the reputation that the stability of our pharmaceutical products—compounded or manufactured—enjoys. Thus a most important reason for conducting a stability testing program is to assure ourselves that our products will indeed retain fitness for use with respect to all functionally relevant attributes for as long as they are on the market.

4.3. Requirements of Regulatory Agencies

In many parts of the world, there are legal requirements that certain types of stability tests, as required by regulatory agencies, must be performed (13). Obviously, the law must be obeyed. However, it is wrong to abdicate from all scientific judgement and only conduct those stability tests that a regulatory agency is perceived as requiring. Indeed, there are occasions when any manufacturer with a true dedication to quality will perform stability tests that are over and above those required by regulation.

4.4. To Provide a Database That May Be of Value in the Formulation of Other Products

Data obtained in the stability evaluation of product X in 1999 may prove to be of value when, in 2003, we start developing product Y. There may be occasions, although they are probably rare, when it will be worthwhile to continue stability testing on an R&D formulation that we know will never be marketed just because we are interested in the stability of a new excipient that we have included in the formulation.

5. MODES OF DEGRADATION

5.1. Chemical

Chemical degradation (solvolysis, oxidation, etc.) is common and is described in subsequent chapters of this book. Our knowledge of kinetics can be of material assistance in dealing with chemical degradation.

5.2. Physical

Physical degradation can be caused by a range of factors (e.g., impact, vibration, abrasion, and temperature fluctuations such as freezing, thawing, or shearing). Physical testing is described in Chapter 10.

Unfortunately, in many instances, our knowledge of the exact mechanisms involved in physical degradation is incomplete. It is also unfortunate that a number of the physical test methods that could be used in evaluation of physical stability (e.g., tablet friability, tablet impact resistance, suspension redispersibility, or injection syringeability) are still nonofficial and variable. It is noteworthy that it was not until 1997 that official standardized test methods for the quantification of bulk and tap density were introduced into the USP, although such tests have value in helping to evaluate compressibility.

5.3. Biological (Especially Microbiological)

In North America, Japan, and Western Europe it is microbiological factors that are most likely to be involved in biological stability problems. However, in some parts of the world rats, roaches, ants, and other nonmicrobiological organisms can be responsible for stability problems.

5.4. Limitations of This Classification

Useful though the above tripartite classification of degradation mechanisms may be, there is a danger that its use may overcompartmentalize our approach to drug product stability. This can be dangerous. In fact, many stability problems involve more than one mechanism. For example, insufficient antioxidant in a rubber condom may result in oxidation of the device by a chemical mode. However, the effect that may be detected is loss of tensile strength, which is, of course, a physical parameter.

6. THE ESSENTIAL ELEMENTS OF A HIGH-QUALITY, COST-EFFECTIVE STABILITY PROGRAM

6.1. Commitment of the Organization to Quality

It is essential that throughout the organization responsible for the development and production of pharmaceutical products there be a *true* commitment to quality.

Process

ing ampoules under nitrogen for solutions that are susceptible to oxidation is one
mple of a processing method that can improve stability. Of course, the most
mon process variable that is adjusted to control stability is selection of the pack-
components and materials, and readers specifically interested in this topic are
rred to Dixie Dean's chapter.

. Formulation

literature is replete with accounts of proven and potential methods of improving
duct stability and hence shelf life. All that is provided in this section are some
eral concepts that can, if appropriate, be explored in more detail.

In the past, stability overages,* which allowed a relatively easy method to
rove shelf life, were quite common. Indeed, there are many drug products on
market in different parts of the world that contain a stability overage of up
10% of label claim. However, a number of regulatory agencies, including FDA,
now showing much more reluctance than previously to approve such overages
drug products.† This reluctance to approve the use of stability overages probably
ms from a number of causes.

First, there is concern about the possible increase in toxicity that might
company the use of a stability overage. If a product for which compendial potency
its are 90–110% is released onto the market at 100% of label claim, then the
ximum amount of any degradation product that could be present in the product
until the expiration date is 10%. However, if the product is released at 110%
label claim, then it is conceivable that in some instances there could be up to
% of degradation product. If the degradation product, or part thereof, is toxic,
e of a stability overage has *doubled* the potential hazard to which a patient is
posed.

Second, if stability overages are allowed, then the range of potencies to which a
tient may be subjected is increased. For example, suppose that a patient who has a
peat prescription for drug X (which is known to have a relatively low range of
ceptable therapeutic blood levels) finishes tablets of lot A101, which has a potency
90%, and is then supplied with tablets from B103, which has a potency of 110%.
hen (even neglecting degradation of drug while the tablets are under control of
e patient and not considering content uniformity) we can see that the patient
ay experience a 20% variation in blood levels. In contrast, in the absence of a
ability overage, the maximum potency variance would only be 10%. This substan-

Overages are of three types: container, manufacturing, and stability. A *container* overage is added to
llow for the fact that it is not possible in some cases to remove all the contents from a container. Thus
mpoules labeled 1.0 mL are normally filled with 1.1 mL. A *manufacturing* overage is added when it
known that relatively small and reproducible amounts of active are always lost during the manufac-
uring process although we are using modern equipment and facilities and well-trained staff. A manu-
acturing overage is, of course, dissipated by the time final product testing is completed.

Vitamin products, which are classified by the FDA as food supplements (unless they are administered by
he oral route or supplied under a doctor's prescription), still have substantial overages—sometimes up to
00% of label claim.

Absent this commitment, it is likely that a stability program will be regarded simply as a burdensome, nonproductive expense. If such an attitude pervades top or middle management—although such attitudes are rarely expressed directly in writing but rather transmitted by a nod and a wink—it is quite possible that the stability testing group will be starved of essential equipment and personnel. One sometimes visits companies where stability testing is two or three months, or even more, behind schedule. Top managers say they cannot understand how the problem has developed since, as everyone in the company knows, their personal dedication to product quality is second only to their commitment to God, the flag, and the family. Unfortunately, such individuals sometimes "talk big but spend little."

6.2. Firm Grasp of Underlying Scientific Theory

It should hardly need to be stated that stability testing of pharmaceuticals requires in-depth education in the science of pharmaceutical formulation, evaluation, analysis, and statistics. Unfortunately, there still are companies where personnel with such education are lacking.

6.3. Up-to-Date Knowledge of All Relevant Policies of Regulatory Agencies and Applicable Pharmacopoeial Standards

Official regulations and standard test methods continue to evolve. Thus it is important that at least one person in every company be charged with the responsibility of keeping up-to-date files on data from the FDA, the USP, or such other entities as may be relevant that impinge on any aspect of the design, execution, or interpretation of stability tests. Perusal of Pharmacopoeial Forum (*PF*), the journal in which the USP provides trailer-type information about possible new or modified test methods or monographs, should be mandatory in all companies for which USP standards may be of relevance.

6.4. Effective Communication Between R&D, Production, QC/QA, Complaints, and Regulatory Affairs

In order to have a successful stability testing program, it is important that there be clear, effective, and rapid communication between all the various organizational entities in a company that can provide useful input into the stability program.

6.5. A General Understanding of the Limitations of the Analytical Methods Used in the Stability Testing Program

Everyone with any degree of responsibility for decisions about a stability program—not just those performing the tests in the laboratory—should have a general understanding of the parameters that characterize the test methods used in stability testing (accuracy, precision, sensitivity, reproducibility, transferability, etc.). We do not require that everyone be expert at say HPLC (high-performance liquid chromatography) or ELISA (enzyme-linked imunosorbest assay), but we should expect that the decision makers be aware of the salient characteristics of the test methods, the results of which are used in decisions about stability.

6.6. Careful Monitoring of the Stability Budget

It is surprising that some companies have no stability budget. It is even more surprising to find that there are scientists designing stability protocols who select test method A instead of test method B (both of which might be technically satisfactory but of significantly different cost), who have knowledge of or interest in the relative costs of the two tests.

It is not easy to devise a mechanism for evaluating a stability budget such that we can be quite certain that we have accounted for all monies spent on the program. However, even though the budget that we estimate may be relative, rather than absolute, it still can be of substantial value.

6.7. Managerial Skills to Coordinate and Optimize the Program

The capstone of a high-quality cost-effective stability program must be managerial skills that nurture and coordinate the personal and professional skills of all involved with the program.

7. CONFORMANCE PERIODS, SHELF LIVES, AND EXPIRATION DATES

The conformance period of drug product is defined by the most vulnerable time-dependent quality attribute. As has already been indicated in Sec. 2.1 of this chapter, loss of potency is, for many products, the critical parameter. In those cases where some other attribute is more vulnerable, it will be that property that defines the conformance period. The same general approach as that shown in Fig. 1 should be followed; however, instead of plotting potency as a percentage of label claim on the y-axis, one plots the appropriate critical stability parameter. The conformance period is then determined from the intersection of the lowest (or highest) acceptable value of the parameter and the 95% confidence bound of the regression line. In the rather rare event that there are two stability attributes of about the same criticality, then both should be quantified and the lower conformance period used as the basis for the assignment of the shelf life of the product.

As has been previously indicated, the shelf life assigned to a product is equal to, or less than, the conformance and is usually a convenient round number (e.g., 7 days, 1 month, 1 year, 18 months, or 2, 3, or 5 years).

The expiration (or expiry) date placed on the label of any given batch indicates the date at which the shelf life ends for the batch. Thus if the product is stored in accordance with label instructions, it is expected that the product will retain fitness for use up to that date. With the exception of products that have very short shelf lives, it is conventional in many parts of the world to give only the month and year of the expiration date. It is expected that for such dates, e.g., May '03, the product should remain of acceptable quality until the *end* of the stated month.

When products have a 5-year shelf life, the practice of only giving expiration dates for the months of January or July seems to be becoming more common. This practice simplifies stock control, since there are fewer dates to deal with. This approach is used as follows: Suppose we have a product that has a five-year shelf life, and we manufacture batches of the product in February, April, June, August, and November of 2002. The first three batches would be dated January '07; the last two would be dated July '07.

Obviously, if a product is not stored in accordance w
expiration date cannot necessarily be relied on.

8. SOME POSSIBLE STRATEGIES TO IMPROVE SHEL

It is fortunate that many drug substances and products ar
with little difficulty, we can justify a shelf life of 3 years o
are drug substances that are very much more liable to d
require much skill and hard work to develop a product with
mercially acceptable. Since this book is not focused on form
tion only outlines some of the general approaches that
efforts to improve shelf life.

8.1. Sampling and Analytical

Examination of Fig. 1 reveals that the more scatter that we
the wider the 90% zone of confidence will be. If we were so
experimental points that all fitted exactly on the regressio
upper and lower 90% confidence bounds would also be on
line; thus our estimate of the conformance period wou
intersection of the regression line and the 90% potency line
substantially extend the shelf life that we could legitimate
it is impossible to obtain such perfect data that the 90%
no width whatsoever. However, anything that we can do to
will improve our shelf life.

There are two main causes for the fact that stability plots,
Fig. 1, show scatter, *viz.*, sampling error and analytical error.
do to reduce either or both of these errors will improve o
our having made any change to the formulation or process u

It is not often easy to see how sampling error could be redu
of near-infrared spectroscopy for single-tablet assay (see Cha
known, individual tablets throughout the shelf life testing pe
the data so obtained at each time point, might be a practicab
error due to content uniformity variation (14). Perhaps reductio
is one of the incentives that we have in making sure that all
on time.

In terms of analytical error, if we can improve precision an
will slim the 90% confidence envelope and improve our shelf life.
been shown that the extra cost of a more sophisticated assay ma
improvement in shelf life that results.

8.2. Statistical

If testing of samples is continued beyond the point at which
reached the 90% confidence of the label claim value, we move th
of the 90% confidence zone to later times and thus improve o
valid statistical approach was specifically mentioned in the 19
Guidelines.

Absent this commitment, it is likely that a stability program will be regarded simply as a burdensome, nonproductive expense. If such an attitude pervades top or middle management—although such attitudes are rarely expressed directly in writing but rather transmitted by a nod and a wink—it is quite possible that the stability testing group will be starved of essential equipment and personnel. One sometimes visits companies where stability testing is two or three months, or even more, behind schedule. Top managers say they cannot understand how the problem has developed since, as everyone in the company knows, their personal dedication to product quality is second only to their commitment to God, the flag, and the family. Unfortunately, such individuals sometimes "talk big but spend little."

6.2. Firm Grasp of Underlying Scientific Theory

It should hardly need to be stated that stability testing of pharmaceuticals requires in-depth education in the science of pharmaceutical formulation, evaluation, analysis, and statistics. Unfortunately, there still are companies where personnel with such education are lacking.

6.3. Up-to-Date Knowledge of All Relevant Policies of Regulatory Agencies and Applicable Pharmacopoeial Standards

Official regulations and standard test methods continue to evolve. Thus it is important that at least one person in every company be charged with the responsibility of keeping up-to-date files on data from the FDA, the USP, or such other entities as may be relevant that impinge on any aspect of the design, execution, or interpretation of stability tests. Perusal of Pharmacopoeial Forum (*PF*), the journal in which the USP provides trailer-type information about possible new or modified test methods or monographs, should be mandatory in all companies for which USP standards may be of relevance.

6.4. Effective Communication Between R&D, Production, QC/QA, Complaints, and Regulatory Affairs

In order to have a successful stability testing program, it is important that there be clear, effective, and rapid communication between all the various organizational entities in a company that can provide useful input into the stability program.

6.5. A General Understanding of the Limitations of the Analytical Methods Used in the Stability Testing Program

Everyone with any degree of responsibility for decisions about a stability program—not just those performing the tests in the laboratory—should have a general understanding of the parameters that characterize the test methods used in stability testing (accuracy, precision, sensitivity, reproducibility, transferability, etc.). We do not require that everyone be expert at say HPLC (high-performance liquid chromatography) or ELISA (enzyme-linked imunosorbest assay), but we should expect that the decision makers be aware of the salient characteristics of the test methods, the results of which are used in decisions about stability.

6.6. Careful Monitoring of the Stability Budget

It is surprising that some companies have no stability budget. It is even more surprising to find that there are scientists designing stability protocols who select test method A instead of test method B (both of which might be technically satisfactory but of significantly different cost), who have knowledge of or interest in the relative costs of the two tests.

It is not easy to devise a mechanism for evaluating a stability budget such that we can be quite certain that we have accounted for all monies spent on the program. However, even though the budget that we estimate may be relative, rather than absolute, it still can be of substantial value.

6.7. Managerial Skills to Coordinate and Optimize the Program

The capstone of a high-quality cost-effective stability program must be managerial skills that nurture and coordinate the personal and professional skills of all involved with the program.

7. CONFORMANCE PERIODS, SHELF LIVES, AND EXPIRATION DATES

The conformance period of drug product is defined by the most vulnerable time-dependent quality attribute. As has already been indicated in Sec. 2.1 of this chapter, loss of potency is, for many products, the critical parameter. In those cases where some other attribute is more vulnerable, it will be that property that defines the conformance period. The same general approach as that shown in Fig. 1 should be followed; however, instead of plotting potency as a percentage of label claim on the *y*-axis, one plots the appropriate critical stability parameter. The conformance period is then determined from the intersection of the lowest (or highest) acceptable value of the parameter and the 95% confidence bound of the regression line. In the rather rare event that there are two stability attributes of about the same criticality, then both should be quantified and the lower conformance period used as the basis for the assignment of the shelf life of the product.

As has been previously indicated, the shelf life assigned to a product is equal to, or less than, the conformance and is usually a convenient round number (e.g., 7 days, 1 month, 1 year, 18 months, or 2, 3, or 5 years).

The expiration (or expiry) date placed on the label of any given batch indicates the date at which the shelf life ends for the batch. Thus if the product is stored in accordance with label instructions, it is expected that the product will retain fitness for use up to that date. With the exception of products that have very short shelf lives, it is conventional in many parts of the world to give only the month and year of the expiration date. It is expected that for such dates, e.g., May ’03, the product should remain of acceptable quality until the *end* of the stated month.

When products have a 5-year shelf life, the practice of only giving expiration dates for the months of January or July seems to be becoming more common. This practice simplifies stock control, since there are fewer dates to deal with. This approach is used as follows: Suppose we have a product that has a five-year shelf life, and we manufacture batches of the product in February, April, June, August, and November of 2002. The first three batches would be dated January ’07; the last two would be dated July ’07.

Obviously, if a product is not stored in accordance with label instructions, the expiration date cannot necessarily be relied on.

8. SOME POSSIBLE STRATEGIES TO IMPROVE SHELF LIFE

It is fortunate that many drug substances and products are inherently stable; thus, with little difficulty, we can justify a shelf life of 3 years or more. However, there are drug substances that are very much more liable to degradation, and it may require much skill and hard work to develop a product with a shelf life that is commercially acceptable. Since this book is not focused on formulation *per se*, this section only outlines some of the general approaches that might be considered in efforts to improve shelf life.

8.1. Sampling and Analytical

Examination of Fig. 1 reveals that the more scatter that we have on a stability plot the wider the 90% zone of confidence will be. If we were somehow able to obtain experimental points that all fitted exactly on the regression line, then both the upper and lower 90% confidence bounds would also be on the mean regression line; thus our estimate of the conformance period would be given by the intersection of the regression line and the 90% potency line. Clearly, this would substantially extend the shelf life that we could legitimately claim. Of course, it is impossible to obtain such perfect data that the 90% confidence zone has no width whatsoever. However, anything that we can do to reduce its thickness will improve our shelf life.

There are two main causes for the fact that stability plots, such as that shown in Fig. 1, show scatter, *viz.*, sampling error and analytical error. Anything that we can do to reduce either or both of these errors will improve our shelf life without our having made any change to the formulation or process used for our product.

It is not often easy to see how sampling error could be reduced. Possibly the use of near-infrared spectroscopy for single-tablet assay (see Chapter 18) of the same known, individual tablets throughout the shelf life testing period, and averaging the data so obtained at each time point, might be a practicable method to reduce error due to content uniformity variation (14). Perhaps reduction of sampling error is one of the incentives that we have in making sure that all samples are tested on time.

In terms of analytical error, if we can improve precision and reproducibility we will slim the 90% confidence envelope and improve our shelf life. In some cases, it has been shown that the extra cost of a more sophisticated assay may be justified by the improvement in shelf life that results.

8.2. Statistical

If testing of samples is continued beyond the point at which degradation has reached the 90% confidence of the label claim value, we move the narrow "waist" of the 90% confidence zone to later times and thus improve our shelf life. This valid statistical approach was specifically mentioned in the 1984 FDA Stability Guidelines.

8.3. Process

Filling ampoules under nitrogen for solutions that are susceptible to oxidation is one example of a processing method that can improve stability. Of course, the most common process variable that is adjusted to control stability is selection of the package components and materials, and readers specifically interested in this topic are referred to Dixie Dean's chapter.

8.4. Formulation

The literature is replete with accounts of proven and potential methods of improving product stability and hence shelf life. All that is provided in this section are some general concepts that can, if appropriate, be explored in more detail.

In the past, stability overages,* which allowed a relatively easy method to improve shelf life, were quite common. Indeed, there are many drug products on the market in different parts of the world that contain a stability overage of up to 10% of label claim. However, a number of regulatory agencies, including FDA, are now showing much more reluctance than previously to approve such overages for drug products.† This reluctance to approve the use of stability overages probably stems from a number of causes.

First, there is concern about the possible increase in toxicity that might accompany the use of a stability overage. If a product for which compendial potency limits are 90–110% is released onto the market at 100% of label claim, then the maximum amount of any degradation product that could be present in the product up until the expiration date is 10%. However, if the product is released at 110% of label claim, then it is conceivable that in some instances there could be up to 20% of degradation product. If the degradation product, or part thereof, is toxic, use of a stability overage has *doubled* the potential hazard to which a patient is exposed.

Second, if stability overages are allowed, then the range of potencies to which a patient may be subjected is increased. For example, suppose that a patient who has a repeat prescription for drug X (which is known to have a relatively low range of acceptable therapeutic blood levels) finishes tablets of lot A101, which has a potency of 90%, and is then supplied with tablets from B103, which has a potency of 110%. Then (even neglecting degradation of drug while the tablets are under control of the patient and not considering content uniformity) we can see that the patient may experience a 20% variation in blood levels. In contrast, in the absence of a stability overage, the maximum potency variance would only be 10%. This substan-

---

* Overages are of three types: container, manufacturing, and stability. A *container* overage is added to allow for the fact that it is not possible in some cases to remove all the contents from a container. Thus ampoules labeled 1.0 mL are normally filled with 1.1 mL. A *manufacturing* overage is added when it is known that relatively small and reproducible amounts of active are always lost during the manufacturing process although we are using modern equipment and facilities and well-trained staff. A manufacturing overage is, of course, dissipated by the time final product testing is completed.

† Vitamin products, which are classified by the FDA as food supplements (unless they are administered by the oral route or supplied under a doctor's prescription), still have substantial overages—sometimes up to 100% of label claim.

tial potency variation could lead to sub- or supratherapeutic blood levels and perhaps the need to retitrate the patient.

Third, and perhaps most important, there is a perception in some quarters that use of a stability overage is a cop-out that represents an easy Band-Aid approach to formulation that is quite unacceptable in modern pharmaceutical technology. It is thought that a more thorough investigation of the problem and a willingness to devote appropriate resources of time, personnel, and money might well allow the problem to be solved by other, more conventional, formulation approaches that do not require a stability overage.

Formulation approaches to reduce the problem of hydrolysis of drugs in solution have generally been of rather limited success. Recently, complexation of drugs with cyclodextrins has attracted considerable interest (15). Such complexes may show improved resistance to hydrolysis, faster dissolution, and better bioavailability. Of course, since most hydrolysis reactions are catalyzed by hydronium and hydroxyl ions, pH control might appear to have great value as a formulation approach to reducing hydrolysis. In practice, however, this approach has had rather limited success. For drugs liable to hydrolysis that are formulated into tablets, the use of a coating may be of value in improving stability.

In contract to hydrolysis, degradation by oxidation can often be successfully controlled by formulation approaches. There is a range of chelating agents and both oil- and water-soluble antioxidants that are used in products in various parts of the world. When a product contains an antioxidant, it is normal to monitor the amount (or concentration) of antioxidant as part of stability studies. In theory, it would be acceptable if all the antioxidant were used by the end of the shelf life period. In practice, most of us would feel rather uncomfortable if we did not have, say, 25% remaining at the end of the shelf life.

Antimicrobial preservatives, such as sodium benzoate, are commonly added to many pharmaceutical products. The amount (or concentration) of such components should be monitored during stability studies. Although chemical assay for antimicrobial preservatives may be acceptable at most time points, the testing performed at the last time point should be by a microbiological challenge test, such as that specified in the USP.

Perhaps the area where formulation approaches are particularly important in controlling stability problems is the field of protein drugs, an area of ever-increasing importance. Dr. Kottke and Dr. DiBiase give this topic specific attention in their chapter in this book.

REFERENCES

1. FDA Guidance for Industry. Draft Stability Testing of Drug Substances and Drug Products, 1998.
2. B. Kommanaboyina, C. T. Rhodes. Drug Devel. Indus. Pharm. 25, in press. (1999). 2.
3. C. M. Won, Pharm. Res. 9:131–137, 1992.
4. J. T. Carstensen, C. T. Rhodes. Drug Devel. Indus. Pharm. 19:2709–2714, 1993.
5. S. E. Tabibi, C. T. Rhodes. In: Modern Pharmaceutics. 3 ed. G. S. Banker, C. T. Rhodes, eds. New York: Marcel Dekker, 1995.
6. The Handbook of Pharmaceutical Excipients, 2nd ed. American Pharmaceutical Association and the Royal Pharmaceutical Association of Great Britain, 1996.

7. M. Ash, I. Ash. Handbook of Pharmaceutical Additives. Gower, Croft Road, England, 1995.
8. Donald C. Monkhouse and C. T. Rhodes, eds. Drug Products for Clinical Trials. New York: Marcel Dekker, 1998.
9. Martin D. Hynes, III. ed. Preparing for FDA Pre-Approval Inspections. New York: Marcel Dekker, 1998.
10. FDA. SUPAC Guidelines, 1995.
11. C. T. Rhodes, S. C. Porter. Drug Devel. Indus. Pharm. 24:1139–1153, 1998.
12. C. T. Rhodes. Clin. Res. Drug Reg. Affairs 15, 180–185 1998.
12. ICH Harmonized Tripartite Guideline for Stability Testing of New Drug-Substances and Products, 23 September, 1994 (ICH QIA).
14. K. M. Morisseau, C. T. Rhodes. Pharm. Tech (Tabletting Yearbook) 6–12, 1997.
15. M. D. Dhanaraju, K. Senthil Kumaran, T. Baskaran, M. Sree Rama Moorthy. Drug Develop. Indus. Pharm. 24:583–587, 1998.

2

Solution Kinetics

JENS T. CARSTENSEN

Madison, Wisconsin

Stability is not synonymous with chemical kinetics, yet most of the rate-limiting phenomena are either associated with chemical reactions or are describable by some equation system that bears a resemblance to those encountered in chemical kinetics. It is, therefore, of importance to lay the proper kinetic foundation before discussing the actual phenomena encountered in dosage forms.

These fundamental principles are most conveniently described by solution kinetics. The simpler a system is, the easier it is to make it reproducible, and it is therefore not surprising that the largest number of pharmaceutical publications on the subject of kinetics deal with solution systems. Furthermore, the more dilute a system is, the more it will adhere to ideal laws, and hence the largest number of publications to be found deal with dilute systems. There are obviously pharmaceutical dosage forms that are solutions, viz. oral, parenteral, nasal, ophthalmic, and otic solutions. Of these, it is only the parenteral and ophthalmic solutions that are chemically fairly simple, i.e. contain only a few number of components. These are systems that would behave similarly to the patterns described in, for instance, the chemical literature. In oral solutions, there are many ingredients (sweeteners, solubilizers, etc.), so that, here, one would expect definite vehicle effects and interaction possibilities.

The Stability Guidelines make certain requirements on basic stability that are best elucidated (or only elucidated) through solution kinetics: First of all it is necessary to develop a stability-indicating assay. This is defined in lines 111 of the 1987 Guidelines as "Quantitative analytical methods that are based on the characteristic structural, chemical, or biological properties of each active ingredient of a drug product and that will distinguish each active ingredient from its degradation products so that the active ingredient content can be accurately measured." The 1993 ICH Guidelines state,

> Analytical test procedures should be fully validated and the assays should be stability-indicating. The need or the extent of replication will depend on the results of validation studies (194–196).
>
> The focus may instead be on assuring the specificity of the assay ... of identified degradants as indicators of the extent of degradation via particular mechanisms (386–389).

This means that the assay must be capable of detecting quantitatively the amount of parent drug present, and identify, and to some degree quantitate, the decomposition products. Lines 265–277 of the 1987 Guidelines state, "When degradation products are detected, the following information about them should be submitted when available:

(a) Identity and chemical structure,
(b) cross-reference to any available information about biological effect and significance at the concentrations likely to be encountered,
(c) procedure for isolation and purification,
(d) mechanism of formation, including order of reaction ...,
(e) physical and chemical properties,
(f) specifications and directions for testing for their presence at the levels or concentrations expected to be present."

Lines 141–144 further state that "the stability-indicating methodology should be validated by the manufacturer (and the accuracy and precision established) and described in sufficient detail to permit validation by FDA laboratories."

In developing stability-indicating assay methodology, it is customary to deliberately decompose the drug in solution, so as to challenge the assay and insure its capability of separating the parent drug from decomposition products. It is obvious, also, that it is desired to establish the kinetic order of the decomposition.

1. THE ORDER OF A REACTION

The order of a reaction will be defined below, but in essence it determines how the degradation data are treated. That it is important to establish the order of a reaction is evident in that the 1993 ICH Guidelines specifically state,

> The nature of any degradation relationship will determine the need for transformation of the data for linear regression analysis. Usually the relationship can be represented by a linear, quadratic or cubic function on an arithmetic or logarithmic scale. Statistical methods should be employed to test the goodness of fit of the data on all batches and combined batches (where appropriate) to the assumed degradation line or curve (138–143).

They also state (in respect to mass balance),

> This concept is a useful scientific guide for evaluation data but it is not achievable in all circumstances. The focus may instead be on assuring the routes of degradation, and the use, if necessary, of identified degradants as indicators of the extent of degradation via particular mechanisms (385–389).

Although the presentation modes outlined in this quotation are not (at least not in the case of the quadratic or cubic functions) of scientific bent, it is obvious that efforts must be made, before the formal stability program is started, to establish the order of the reaction.

Establishing the order is, furthermore, of financial importance, because the establishing of expiration periods (which will be discussed later) depends, to some degree, on the investigator's capability of extrapolating the concentration of drug beyond the last time point of testing. The 1993 ICH Guidelines further state,

> Limited extrapolation of the real time data beyond the observed range to extend expiration dating at approval time, particularly where the accelerated data supports this, may be undertaken. However, this assumes that the same degradation relationship will continue to apply beyond the observed data and hence the use of extrapolation must be justified in each application in terms of what is known about the mechanism of degradation, the goodness of fit of any mathematical model, batch size, existence of supportive data, etc. (149–155).

The longest possible expiration period is, of course, economically desirable, and many of the efforts of the stability programs of pharmaceutical companies are geared towards lengthening this period. As for definition of the order or a reaction, if

$$A + B \rightarrow E \tag{2.1}$$

then the reaction rate is given by

$$\frac{dC}{dt} = -k_{(n+m)}[A]^n[B]^m \tag{2.2}$$

where C is the concentration of the species being studied, brackets denote concentrations of A and B, and k denotes a rate constant, then the reaction is said to be of the order $n+m$. The rate constant, in this writing, will most often carry the subscript denoting the order of the reaction whenever reaction orders are discussed and being distinguished. (A notable exception is the notation in the section dealing with pH profiles). The most important orders of interest in the pharmaceutical sciences are integral orders, i.e. those in which the sum of n and m is 0, 1, or 2. (Orders of higher than two are rare.)

As alluded to above, knowledge of the order of a reaction is of great importance in stability determination of drug substances, in particular in solution. The problem is frequently to judge whether the concentration-time profiles are linear (zero order) or curved (first or other order). When large amounts of data are at hand (e.g., at different temperature, where the order does not depend on temperature), then a data-consolidation technique described by Carstensen and Franchini (1994), Carstensen (1997), and Franchini and Carstensen (1994, 1999). The technique has later been used by Shalaev et al. (1997) in the study of solid-state methyl transfer reactions. For instance, the data in Fig. 1, when plotted linearly, give *fairly* good plots, but since there are only a few points it is difficult to say with reasonable certainty whether the data are, indeed, linear or curve-linear.

A *fractional life* is the length of time it takes for a product or drug substance to decrease to the level indicated by the fraction: the half-life, t_{50}, of a substance is the length of time it takes to decrease the content of active compound to 50% of its value. If, in Fig. 1, a given fractional life (e.g., t_{90} as shown in the figure) is read

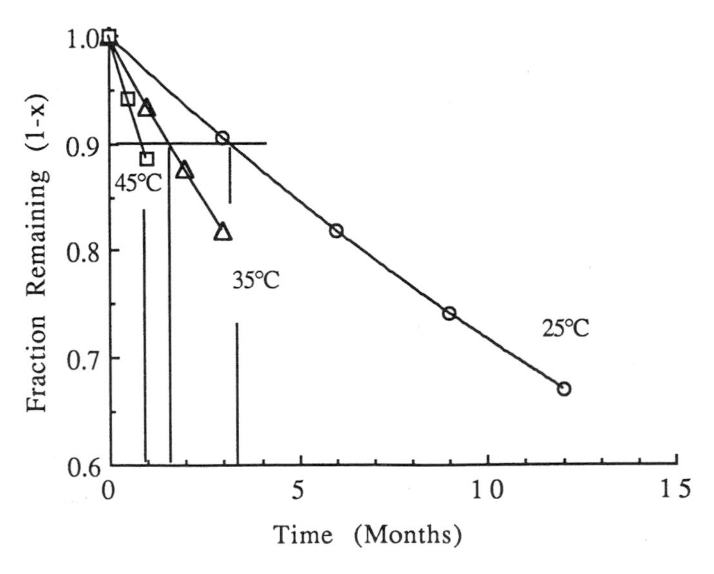

Fig. 1 Example of sparse data at three temperatures, such as is often encountered in early development of drug products.

Table 2.1 Example of Reduced Data Treatment in Kinetics

Time (Months)	Temperature (t_{90})	Fraction retained	Reduced time
0	25°C (3.5)	1.000	1.000
3		0.905	0.905
6		0.819	0.819
9		0.741	0.741
12		0.670	0.670
0	37°C (1.7)	1.000	1.000
1		0.935	0.935
2		0.875	0.875
3		0.800	0.800
0	45°C (0.95)	1.000	1.000
0.5		0.942	0.942
1		0.887	0.887

off the graphs, then the data may be consolidated. The t_{90} values shown are 45°C: 0.95; 35°C: 1.7; and 25°C: 3.5. The data in the first column in Table 1 are then reduced to the fourth column, and this is used as abscissa and the fraction retained (regardless of temperature) is used as ordinate in column 3, and the data are shown in Fig. 2.

The method allows better extrapolation tolerances, since the number of points is larger than for the individual temperatures. If, as in the unstable shown in Fig. 1, the t_{90} value is 3.5 months, then extrapolation could be made to $24/3.5 = 6.7$ half-lives, and the estimated potency after 24 months could be estimated with better precision than if only the 25°C data had been used.

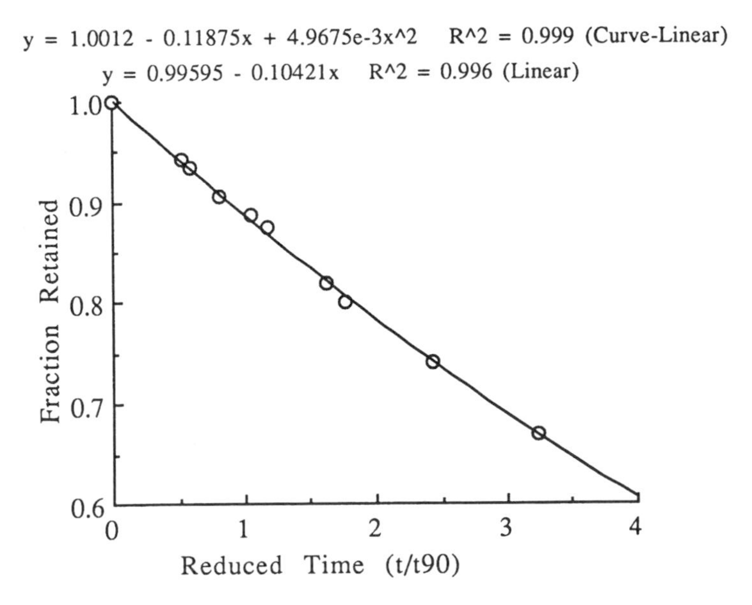

Fig. 2 Data from Table 1 plotted by reduced time treatment.

The fractional life should be *read off the graph* (Fig. 1) since at that point the reaction order is not known. Once the data are plotted, it is possible to execute curve fitting and estimate the best fit. In the case cited, the number of points is probably still too small to make a decision, but the indication is that the data are first order. Including data from even higher temperatures will help in this respect, but it is necessary that the order of reaction not change at the higher temperatures.

It has been mentioned that above 85% it is difficult to distinguish between different reaction orders. Li et al. (1998) report on apparent first-order plots for oxidation of a 4-[2-(2-amino-4-oxo-4,6,7,8-tetrahydro-3*H*-pyrimidino[5,4-6][1,4] thiazin-6-yl)-(S)-ethyl-2,5-thenoyl-L-glutamic acid. It is to be noted (from their Fig. 6) that there is definite downward curvature in the plots, and that they probably are S-shaped, as is discussed further in the chapter on oxidation. Here again, using a method of fractional times will help in deciding on which orders are plausible, or if a certain order of reaction can be ruled out.

It should, finally, be mentioned that Mälkki-Lame and Valkeile (1988) have described a method for transforming regression curves to the determination of reaction order of given situations in stability studies, using the Box–Cox technique and the Link function transformations.

2. THE ZERO-ORDER REACTION

There are not many truly zero-order reactions in the pharmaceutical field. It will be shown at a later point that there are several types of reactions that will appear to be zero order, i.e., are pseudo-zero order. The equation for zero-order reactions is

$$\frac{dC}{dt} = -k_0 \tag{2.3}$$

where C is concentration, t is time, and k_0 is the zero order rate constant.

It is seen that the unit of k is concentration units per time unit, e.g. molar per second. The integrated form of Eq. (2.3) is

$$C = C_0 - k_0 t \tag{2.4}$$

or

$$C_0[1 - a] = kt_a \tag{2.5}$$

where a is the fraction remaining at time t_a.

A quantity often utilized is the half-time, $t_{1/2}$, which is given by

$$t_{1/2} = \frac{C_0}{2k_0} \tag{2.6}$$

It is noted that this is dependent on the initial concentration.

Zero-order data may be graphed on plain Cartesian graph paper, using concentration as ordinate and time as abscissa. An example (Higuchi and Rheinstein, 1959) is shown in Fig. 1.

3. FIRST-ORDER REACTIONS

In this case Eq. (2.2) takes the form

$$\frac{dC}{dt} = -k_1 C \tag{2.7}$$

which integrates to

$$\ln\left[\frac{C}{C_0}\right] = -k_1 t \tag{2.8}$$

It is noted that the *a*-fractional life is given by

$$\ln[a] = -k_1 t_a \tag{2.9}$$

The most common of the *a*-lives is the half-life and the t_{90} (i.e., the point where 90% of the original concentration is left), which adhere to Eq. (2.9) by the equations

$$k_1 t_{/2} = -0.693 \tag{2.10}$$

$$k_1 t_{0.9} = -0.105 \tag{2.11}$$

Fig. 3 and Table 2 show an example of a straight first-order reaction.

In stability situations it is required to monitor both the disappearance of the drug and the appearance of decomposition product(s). In most cases there is more than one decomposition product (and simple cases of this will be treated below). In the simplest case there is only one decomposition product. There are cases of this, e.g., aspirin in simple systems (Carstensen et al., 1985; Carstensen and Attarchi, 1988a, 1988b) decomposes, by a pseudo-first-order reaction, in a simple fashion, i.e. to salicylic acid and acetic acid. In such cases, if the assay of the decomposition product is fairly good, the decomposition can be monitored best by monitoring

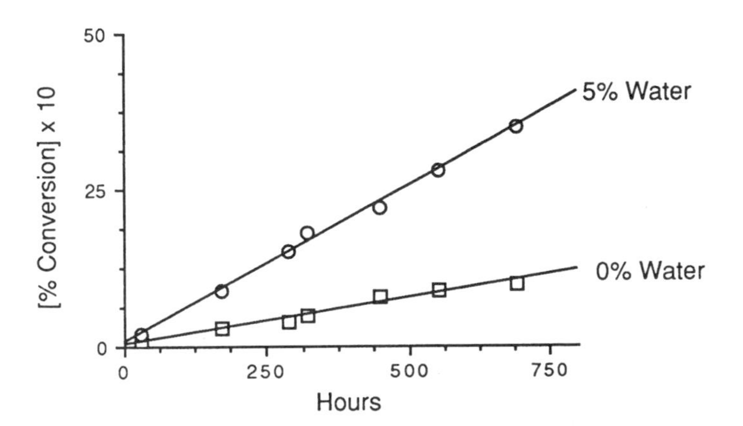

Fig. 3 Decomposition of vitamin A acetate (to anhydrovitamin A). The least squares fits are (with 5% water) $y = 0.48 + 0.015x$; (without water) $y = 0.75 + 0.05x$. (Figure constructed from data published by Higuchi and Rheinstein, 1959.)

Table 2.2 Decomposition of Decarboxymoxalactam

Time (Min)	% Retained	ln[% Retained]	% Decomposed
0	100	4.61	0
10	78	4.36	32
20	50	3.91	50
30	38	3.64	62
40	27	3.30	73
50	17	2.83	83

Source: Reconstructed from data published by Hashimoto et al. (1984).

the appearance rate of the decomposition product, which should follow the reaction

$$[B] = A_0[1 - \exp(-k_1 t)] \tag{2.12}$$

An example of this is shown in Table 2.

It should be noted that whenever this approach is taken, it is mandatory still to monitor the content of parent drug, because mass balance should persist throughout the reaction period. (If the molar quantities do not sum up to A_0 (within experimental error), then either the reaction is not simple $A \rightarrow B$, or the analytical procedure fails in aged samples). The easiest way of plotting the data in Table 2 is obviously to subtract each [B] figure from A_0 and plot is as [A]. One might then argue that one might simply plot the experimental value of [A]. But for fairly stable systems, the values of [A] may not differ (decrease) much and may be masked by experimental error. The percentage change in [B], however, is substantial, as seen in the table, and plotting becomes more meaningful (see Fig. 4).

There are many reported instances of first-order reactions in solution. For instance, Jordan (1998) has shown that timolol and propanolol decompose by straight first-order kinetics at pH 7.4. Aso et al. (1997) have shown that aqueous solutions of cephalothin decompose by first-order kinetics and that they follow

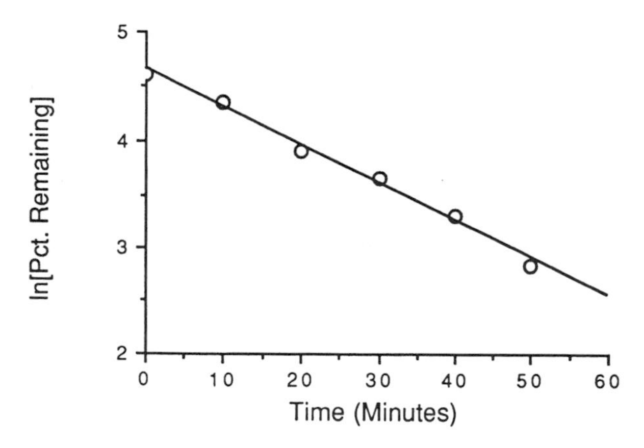

Fig. 4 Plot of first-order data from Table 1. (Graph constructed from data by Hashimoto et al., 1984.)

an Arrhenius equation. Hammad and Müller (1998) have found clonazepam to degrade first-order in phosphate buffers at pH 7.4 and to adhere to an Arrhenius equation.

Heat conduction *microcalorimetry* has been used as a method to evaluate stability and excipient stability by a series of researchers Angerg et al., (1988, 1990, 1993). Hansen et al. (1989), and Wilson et al. (1995) have described the general method and results interpretation. Oliyai and Lindenbaum (1991) have studied the decomposition of ampicillin in solution by means of microcalorimetry.

4. FIRST-ORDER REACTIONS WITH MORE THAN ONE END PRODUCT

The considerations above have assumed that the scheme is simply a reaction of type A → B, but often there is more than one decomposition product.

4.1. Consecutive Reactions of The First Order

The 1993 ICH Guidelines state that mass balance (or material balance) is

> The process of adding together the assay value and levels of degradation products to see how closely these add up to 100 per cent of the initial value, with due consideration of the margin of analytical precision (382–384).

It is possible that the primary decomposition product itself is not stable, and in such cases the reaction scheme is

$$A \underset{k_1}{\rightarrow} B \underset{k_2}{\rightarrow} C \tag{2.13}$$

In other words, there will be more than one decomposition product. If all the products can be identified and quantitated, then it follows that *the number of moles of A, B, and C should always add up to the initial number of moles of A*. It is noted that it is the number of moles that must add up. Addition on a weight basis would be futile if there is a substantial difference between the molecular weights of the drug and the products. The guidelines recognize that it can be difficult, at times, to ascertain mass balance, partly due to analytical precision.

More often it is "unknowns" that cause the problem. If C were not identified, for instance, and was detected as a peak in a HPLC chromatogram, then its "content" is often stated as the area under the peak, using the drug as the unit of measure. But if, for instance, a UV detector is used, and C is lacking the amount of chromophores that A possesses, then the area under the C peak may grossly underestimate the amount of C.

An example of this is chlorbenzodiazepine, which hydrolyzes to the lactam form, and then further to the benzophenone (Carstensen et al., 1971). In fact in this reaction, for some of the benzodiazepines, C can progress further with the formation of the carbostyril and the acridone derivative, and some of the steps are associated with equilibrium conditions.

The rate equations governing scheme (2.13) are

$$\frac{d[A]}{dt} = -k_1[A] \tag{2.14}$$

$$\frac{d[\text{B}]}{dt} = -k_2[\text{B}] + k_1[\text{A}] \tag{2.15}$$

and

$$\frac{d[\text{C}]}{dt} = k_2[\text{B}] \tag{2.16}$$

These simultaneous differential equations may be solved by conventional means and yield the following results:

$$[\text{A}] = A_0\, e^{-k_1 t} \tag{2.17}$$

$$[\text{B}] = A_0 \frac{k_1}{k_2 - k_1}(e^{-k_1 t} - e^{-k_2 t}) \tag{2.18}$$

and

$$[\text{C}] = A_0 - [\text{A}] - [\text{B}] \tag{2.19}$$

It is noted that the above expressions refer to molar quantities. An example of consecutive reactions is shown in Table 3 and Fig. 5. The table shows $C = 100 - [\text{A}] - [\text{B}]$. It often happens that one of the decomposition products is difficult to assay for, and in such a case, it may be obtained by difference, provided that mass balance is checked occasionally, e.g., in the early stages and at the end. Of course, there are reactions that have a multitude of end products, and in such cases it is conventional to assume that if e.g. a HPLC peak is less than 0.5% then it is considered negligible. This may be dangerous, because (especially if it is a constant wavelength peak), the actual molar content of the product(s) in the peak may be more than 0.5% (in which case mass balance would be lost).

To ascertain that the unidentified products are not toxic (and since they are unidentified, specific toxicity cannot be checked), it is conventional, as well, in such cases, to degrade a sample considerably and check its toxicity. It is worthwhile,

Table 2.3 Photolysis of Cefotaxime

Time (hours)	Cefotaxime A (% moles)	Anti-isomer, B (% moles)	*C* by difference[a]
0	100	0	0
0.25	82	10	8
0.5	70	15	15
0.75	55	18	27
1	43	19	38
1.5	28	18	54
2	20	15	65
3	10	10	80
4	5	5	90

[a] Column 3 not reported by Lerner et al. (1988). C may be more than one product.
Source: Constructed from data published by Lerner et al. (1988).

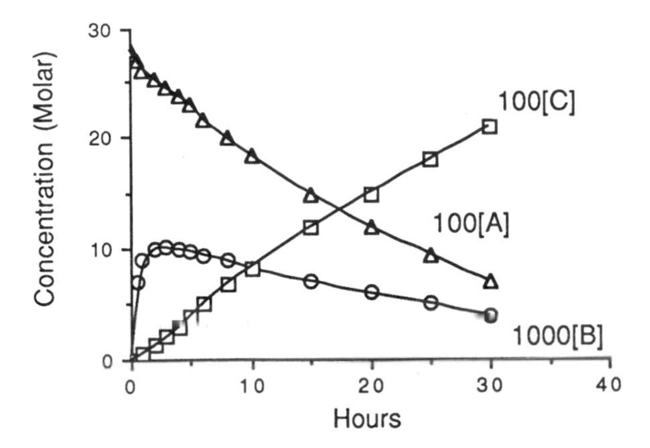

Fig. 5 The A → B part of cefatoxime photolysis. C_{total} has been obtained as 100-[A]-[B], but this does not account for the possibility of other reactions of C. (Graph constructed from data by Lerner et al., 1988.)

however, also to check the toxicity at intermediate points, because C might be toxic, but degrade into nontoxic products, and the toxicity of a partly degraded sample might be worse than that of a fully degraded sample).

The A → B → C reaction is rather common; for instance, it has been reported by Misra et al. (1993).

There continues to be, in present literature, reports of this type of reaction; for instance, Archontaki et al. (1998) reported on the decomposition of nordazepam and showed typical A–B–C plots with the A degradation being first order, the B profile having a maximum, and the C profile having the typical upswing. Burke et al. (1997) reported on the decomposition of theo-m-GLA and found it to be biexponential.

Buur and Bundgaard (1984) and Beal et al. (1993, 1997) reported that the hydrolyses of 3-acetyl- and 3-propionyl-5-FU were biexponential and found that an initial equilibrium of 3-acyl-5FU with O^2-acyl-5FU, which then hydrolyzed to 5-FU, explained this.

4.2. Parallel Reactions

If A can decompose into two species, B and C, then the reactions may be represented by:

$$A \rightarrow B \qquad \text{(rate constant } k_1) \tag{2.20}$$

and

$$A \rightarrow C \qquad \text{(rate constant } k_2) \tag{2.21}$$

The rate equation is

$$\frac{dA}{dt} = -k_1[\text{A}] - k_2[\text{A}] = -(k_1 + k_2)[\text{A}] \tag{2.22}$$

Table 2.4 Parallel Reactions (5-azacytosine decomposition)

Time (hours)	5-azacycytosine ($\times 10^4$ Molar)	5-azouracil ($\times 10^4$ Molar)	Nonchromophoric compounds ($\times 10^4$ *M*)
0	1.65	0	0
0.5	1.4	0.13	0.3
1	1.18	0.20	0.5
1.5	1.00	0.25	0.6
2	0.85	0.27	0.67
2.5	0.72	0.29	0.72

Source. Table constructed from data by Notari and de Young (1975).

which integrates to

$$\ln\left[\frac{A}{A_0}\right] = -[k_1 + k_2]t \tag{2.23}$$

At any given time the (molar) ratio of formation of B and C is given by

$$\frac{[\mathrm{B}]}{[\mathrm{C}]} = \frac{k_1}{k_2} \tag{2.24}$$

An example of this is shown in Table 4 and Fig. 6.

Other examples are those of Visconti et al. (1984), who have studied the degradation profile of cadralazine in aqueous solution. The reaction consists of four parallel reactions. Fabre et al. (1984) have shown that 3-acetoxymethylcephalosporin, cefotaxime sodium salt, in aqueous solution, decomposes by the scheme

Cefotaxime —(k_2)⟶lactone —(k_3)⟶products

Cefotaxime —(k_2)⟶products

i.e., a combination of a parallel and a consecutive reaction.

5. EQUILIBRIA

Frequently a reaction will proceed and level off. In such cases there is often an equilibrium:

$$\mathrm{A} \longleftrightarrow \mathrm{B} \tag{2.25}$$

with an equilibrium constant, K, given by

$$\frac{[\mathrm{B}]}{[\mathrm{A}]} = K \tag{2.26}$$

Denoting the forward rate constant $k_>$ and the backwards rate constant $k_<$ it follows that when equilibrium has been achieved (at $t = \infty$), the amount going to the right in the reaction must equal the amount going to the left, i.e.

$$k_>[\mathrm{A}] = k_<[\mathrm{B}] \tag{2.27}$$

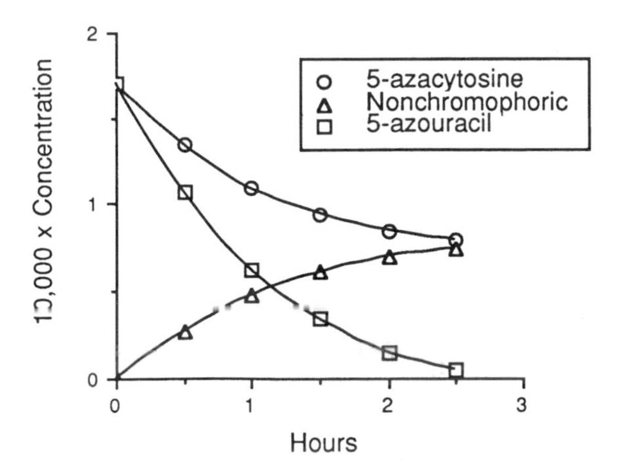

Fig. 6 Example of parallel reactions: decomposition of azacytosine. (Graph constructed from data published by Notari and deYoung, 1975.)

or

$$\frac{k_>}{k_<} = \frac{[\mathrm{B}]}{[\mathrm{A}]} = K \tag{2.28}$$

Denoting by A_∞ the infinity concentration of A (and hence by $A_0 - A_\infty$ the infinity concentration of B), Eq. (2.27) may be written

$$k_>[A_\infty] = k_<[A_0 - A_\infty] \tag{2.29}$$

which may be rewritten

$$A_\infty = \frac{A_0 k_<}{k_< + k_>} \tag{2.30}$$

The rate equation for Eq. (2.25) is

$$\begin{aligned} -\frac{dA}{dt} &= k_>[\mathrm{A}] + k_<[\mathrm{B}] = k_>[\mathrm{A}] + k_<[A_0 - A] \\ &= k_>[\mathrm{A}] + k_<[A_0] - k_<[\mathrm{A}] \\ &= [k_> + k_<]\{A - A_\infty\} \end{aligned} \tag{2.31}$$

which integrates to

$$\ln[A - A_\infty] = [k_> + k_<]t + \ln[A_0 - A_\infty] \tag{2.32}$$

or

$$\ln\left[\frac{A - A_\infty}{A_0 - A_\infty}\right] = -[k_> + k_>]t \tag{2.33}$$

The work regarding the hydrolysis of hydrocortisone butyrate by Yip et al. (1983) is an example of this type of decomposition combined with an $\mathrm{A} \rightarrow \mathrm{B} \rightarrow \mathrm{C}$

Table 2.5 Decomposition of Progabide in pH 1.75 Buffer

Time (min)	Concentration C (Molar)	ln[C-0.0055]
0	0.00896	−5.666
7	0.0076	−6.166
13	0.00700	−6.502
19	0.00660	−6.812
25	0.00612	−7.386
31	0.00600	−7.601

Source: Table constructed from data published by Farraj et al. (1988)

reaction, with the equilibrium occurring between A and B. Ghebre-Sellassie et al. (1984) have described the epimerization of benzylpenicilloic acid in alkaline media and shown it to be an equilibrium between 5R,6R-benzylpenicilloic acid with penamaldic acid (enamine).

Table 5 and Fig. 7 show data by Farraj et al. (1988) showing a leveling effect. If the data in Table 4 represent a simple equilibrium, then the equilibrium level could be obtained by iteration, by assuming different values for the equilibrium level and choosing the one giving best linearity in the form

$$\ln[C_\infty - C] = -kt + \ln[C_\infty - C_0] \tag{2.34}$$

In this case $C = 0.0055$ and the data are plotted in this fashion in Fig. 8, but it should be underscored that they have simply been used as an example. The alkaline hydrolysis of chlorambucil (Owen and Stewart, 1979) is another example of an equilibrium situation.

Beal et al. (1993) tested the hydrolysis of 3-acetyl-5-fluorouracil and showed equilibrium kinetics, as did Prankerd et al. (1992) in the case of rifampicin.

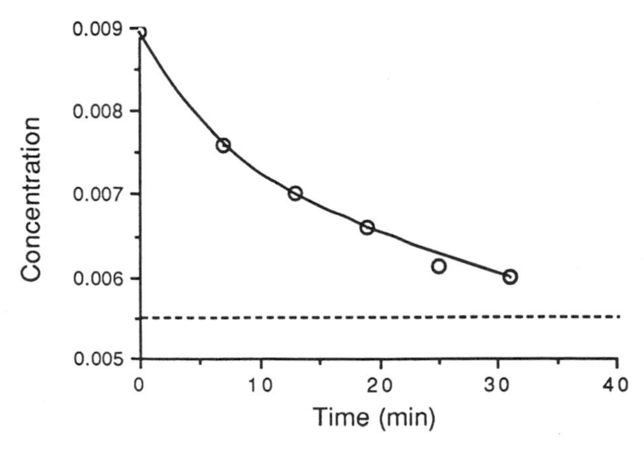

Fig. 7 Decomposition of progabide in pH 1.75 buffer. (Graph constructed from data reported by Farraj et al., 1988.)

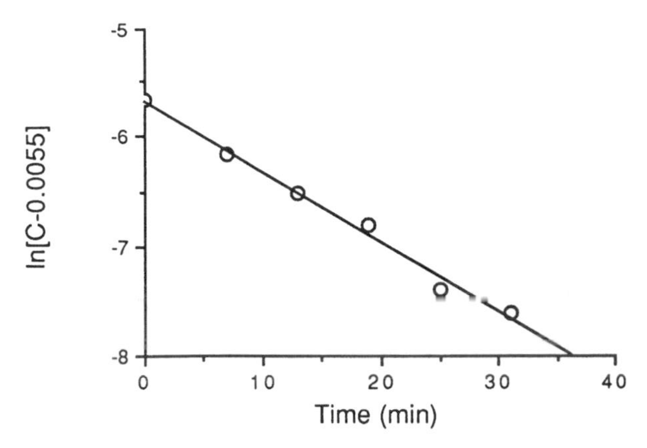

Fig. 8 Data from Fig. 6 treated using a niveau level of $C = 0.0055$. Least squares fit is $y = -0.764 = 0.0127x$ ($R = 0.97$). (Graph constructed from data reported by Farraj et al., 1988.)

5.1. Steady-State Situations

If a situation occurs where $A \rightarrow B \rightarrow C$ and the latter is fast, the kinetics can be simplified by assuming that [B] is "at steady state" throughout the time course. This, obviously cannot be true at the onset. The equations governing this situation are

$$\frac{d[A]}{dt} = -k_1[A] \tag{2.35}$$

$$\frac{d[B]}{dt} = k_1[A] - k_2[B] = 0 \tag{2.36}$$

$$\frac{d[C]}{dt} = k_2[B] \tag{2.37}$$

where the steady state has been imposed by setting the expression in Eq. (2.36) equal to zero. Hence

$$[A] = A_0 e^{-k_1 t} \tag{2.38}$$

and since it follows from Eq. (2.36) that

$$[B] = \frac{k_1[A]}{k_2} \tag{2.39}$$

then Eqs. (2.38) and (2.39) inserted into Eq. (2.37) give

$$\frac{d[C]}{dt} = k_1 e^{-k_1 t} \tag{2.40}$$

which integrates to

$$C = C_\infty[1 - e^{-k_1 t}] \tag{2.41}$$

i.e., the reaction occurs as if B were not in the picture at all.

In general, if there is such a fast step in the first step of a complex reaction, it is not incorrect to consider it an A–B–C reaction.

The same arguments can be made in the case of A → B → C → D reaction where the B → C was much more rapid than the others, and in such a case it would be justified to think of this as an A → C → D reaction. To be more exacting, a steady-state approach would probably be better. It should be pointed out however, that the steady-state approach is a fundamental approximation, and if it is used, then the reasonableness of the approximation should always be checked.

The steady-state approach is often used, particularly in Michaelis–Menten type kinetics. Here, as an example, let us consider the situation, often occurring, that *many* low level decomposition products are encountered. There are different regulatory views on this, one being that no more than 1% of a product may be formed for it to be considered a minor decomposition product. The situation is hazy, at best, at all times, because often the compounds are unknown. In such cases the "amount" of the decomposition product in the small peak is estimated by the ratio of its area to that of the main peak. But if the decomposition product has a different λ_{max} then this estimate is incorrect, and this is likely to occur if, for instance, in an HPLC setup a single-wavelength UV detector is used.

A well documented and elucidated example is the case reported by Vilanova et al. (1994), who showed in alkaline hydrolysis of cefotaxime the presence of deacetylcefotaxime, the 7-epimer of cefotaxime, the 7-epimer of deacetylcefotaxime, the exocyclic methylene compound, and examine compounds. With such an array of decomposition products, it is important to establish the major products, and treat, in approximation, the decomposition in this light. In the simplest case, cefotaxime shows an A–B–C and A–D–E reaction, with two B curves and three C curves.

Another case that serves as such an example is the case of relaxin oxidation by hydrogen peroxide reported by Nguyen et al. (1993) shown in Fig. 9 where there are two intermediates (B and C) showing maxima and a final product, D, showing the monotonically increasing pattern.

6. PSEUDO-ZERO-ORDER REACTIONS

When only small amounts of decomposition occur, it is difficult to distinguish between zero- and first-order reactions. This is because for small values of x (< 0.15)

$$\ln[1 - x] \approx -x \tag{2.42}$$

where x is the fraction decomposed. If the initial amount of drug substance is A_0, then the fraction decomposed is

$$\frac{A_0 - A}{A_0} = 1 - \frac{A}{A_0} = x \tag{2.43}$$

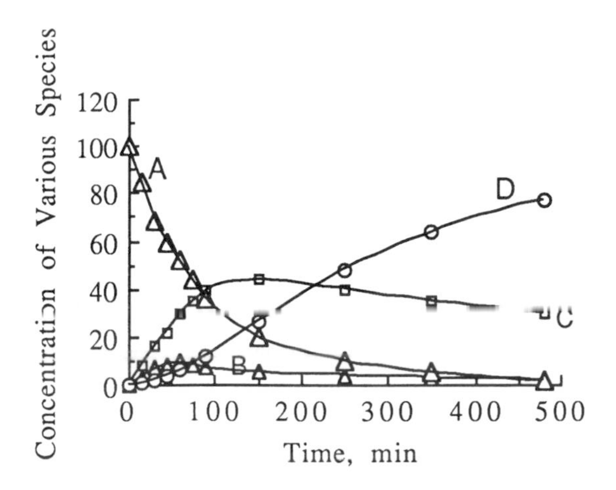

Fig. 9 Decomposition of an A → B; A → C; B → C → D reaction. (Graph constructed from data published by Nguyen et al., 1993.)

or

$$\frac{A}{A_0} = 1 - x \tag{2.44}$$

A first order reaction would require that

$$\ln \frac{A}{A_0} = -kt \tag{2.45}$$

but this may, via Eq. (2.37), be written

$$\ln[1 - x] \approx -x = -kt \tag{2.46}$$

or

$$x = 1 - \frac{A}{A_0} = kt \tag{2.47}$$

which may be written

$$A = A_0 - A_0 kt \tag{2.48}$$

i.e. a zero-order reaction. Since it actually was a first-order reaction [Eq. (2.45)] such a situation is referred to as a pseudo-zero-order reaction.

A set of data is shown in Table 6, treated in zero-order fashion in Fig. 10A and in first-order fashion in Fig. 10B; and it is seen that the fits are comparable. The least squares fit data are shown in Table 6.

It is noted that different time intervals are used for the different temperatures, and it is one of the tasks, before starting studies at higher temperatures, to establish what the time intervals should be. There is no sense in e.g. testing at 3, 6, and 9 months at 55°C, if all the drug is lost after 3 months' storage.

Table 2.6 Assays for an Arrhenius Study

	Potency (°C)				
Months	15	25	37	45	55
0	100	100	100	100	100
3		99.5	98.5	96.5	91
6		99	97	93	83
9		98.5	95.5		
12	99.5	98.5			
18		97.5			
24	99.0	96.5			
36	98.5				
48	98.0				
k_0 %/mo	0.042	0.0017	0.05	1.17	3
ln[k_0]	−3.17	−1.77	−0.69	0.16	1.10

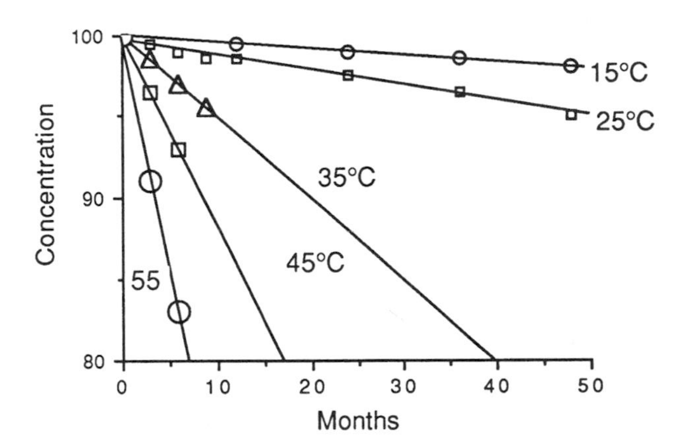

Fig. 10A Data from Table 5 treated by zero-order kinetics.

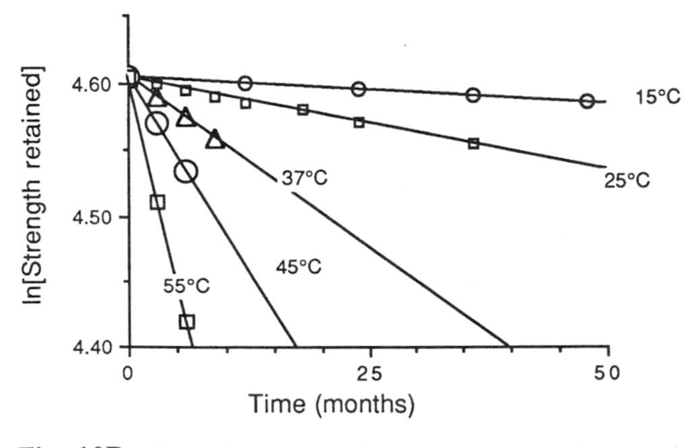

Fig. 10B Data from Table 5 treated by first-order kinetics.

7. THE ARRHENIUS EQUATION

Rate constants are, of course, a function of temperature, and the data shown in Table 5 are graphed in Figs. 10A and 10B.

If rapid results are desired for a given product, it is at times a practice to store it at elevated temperatures. The purpose of this is to force sufficiently large degrees of decomposition in a short time, so that they may be assessed with accuracy. The data in Table 6 are artificially precise, and with a bit of assay error, the 25°C data would not show a discernible loss after 6 months. Is it possible to get some idea of what the loss would actually be, and what it would be after 24 months, without having to wait too long? To get an answer to this (an estimate, not a precise answer) is one of the reasons that Arrhenius plotting is carried out for drug products. The method is actually quite precise in solution systems.

The temperature dependence of a chemical reaction (as long as it is the rate-determining rate constant that is being treated) follows the so-called Arrhenius equation given by

$$\ln[k] = -\frac{E_a}{RT} + \ln[Z] \tag{2.49}$$

or its antilogarithmic form,

$$k = Z\exp\left[-\frac{E_a}{RT}\right] \tag{2.50}$$

where E_a is the activation energy, R is the gas constant, and T is the absolute temperature (°K) obtained by adding 273.15° to the degrees Celcius (Centigrade).

Often the variable $1000/T$ is used (because the numbers then are between 2 and 4 rather than between 0.002 and 0.004 and hence are easier to handle). The slope of a plot according to Eq. (2.42) is still E_a/R, but E_a will now be in kCal (rather than in cal) per degree per mole.

An example of this type of treatment of the first-order data in Fig. 10A is shown in Table 7.

A similar table may be constructed for zero-order treatment, and the graphical presentation is shown in Fig. 10A. When the rate constants are plotted according to Eq. (2.49), then Fig. 11 emerges.

Table 2.7 Least Squares Fit Parameters from Fig. 10B

Temperature °C	Temperature °K	k (mo^{-1})	$1000/T$	$\ln[k]$
15	288.15	0.00042	3.473	−7.783
25	298.15	0.0014	3.356	−6.571
37	400.15	0.0052	3.224	−5.259
45	408.15	0.0118	3.145	−4.440
55	418.15	0.031	3.049	−3.471

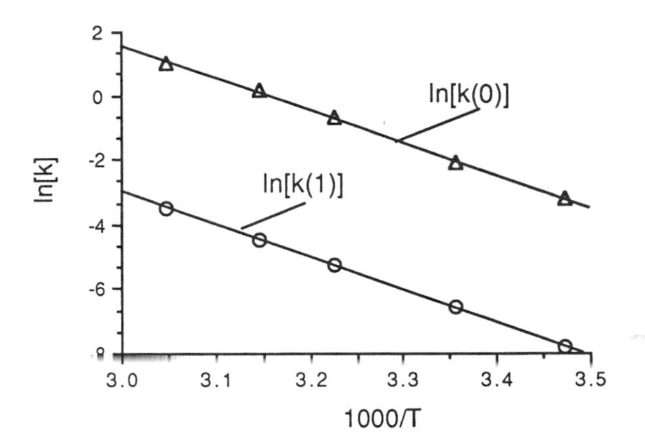

Fig. 11 Data from Figs. 10 A and B treated by the Arrhenius equation.

The Arrhenius equations for the two types of plotting are

$$\ln[k_0] = 31.63 - 10.03(1000/T) \tag{2.51}$$

$$\ln[k_1] = 27.631 - 10.2(1000/T) \tag{2.52}$$

It is noted that both plotting modes give good results and about the same activation energy. Using a value of $R = 1.99$ cal/degree-mole gives an activation energy of about 20 kCal/mole. *This is quite a common activation energy for many reactions.*

If only the data at the three high temperatures had been present, they could have been plotted as in Fig. 11 and extrapolated to 25°C (where $1000/T = 3.356$). Inserting this into Eq. (2.49) gives

$$[k_{25}] = 27.631 - 10.2 \times 3.356 = -6.601 \text{ or } [k_{25}] = 0.00136$$

which is close to the value in Table 7.

One can now construct a curve, as shown in Fig. 12, which is an extrapolated curve. The data points from Table 6 are shown for comparison. In general, extrapolations are not that good, but in solution systems they frequently approximate the curve that, after time has elapsed, is the actual curve. A case in point is flurogestone decomposition at pH 7.3 reported by Kabadi et al. (1984).

Arrhenius plotting can also be carried out by using t_{90} data. Since $k_1 t_{90} = -0.105$ it follows that

$$\ln[k_1 t_{90}] = \ln[t_{90}] + \ln[k_1] = \ln[t_{90}] - \frac{E_a}{RT} + \ln[Z] = -0.105$$

or

$$\ln[t_{90}] = \frac{E_a}{RT} - \ln[Z] - 0.105 \tag{2.53}$$

i.e. a plot of $\ln[t_{90}]$ versus $1000/T$ will be linear and the slope will be E_a/R, where E_a will be in kCal (rather than in cal) per degree per mole.

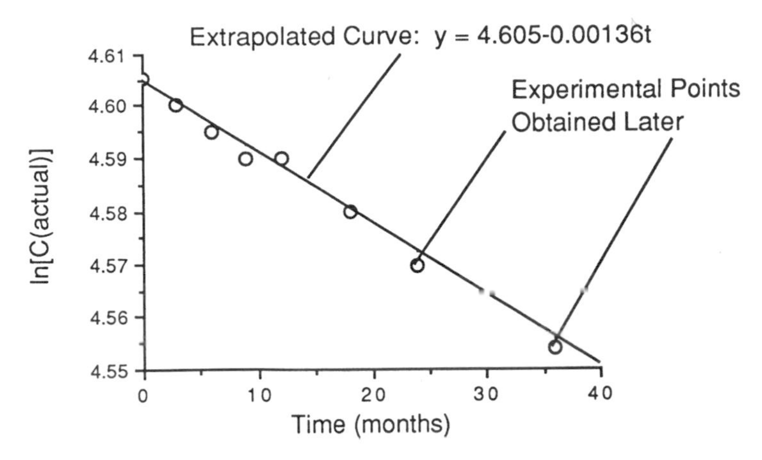

Fig. 12 Extrapolated decomposition curve of a solution obtained (e.g., in January 1993 after three months accelerated data) compared with the actual data accumulated in time and plotted 33 months later (i.e., in October 1995).

Table 2.8 Data from Table 6 Plotted by the t_{90} Method

Temperature	$1000/T$	k_1 (mo^{-1})	t_{90} (months)a	$\ln[t_{90}]$
55°C	3.049	0.031	3.40	1.233
45°C	3.145	0.0118	8.93	2.189
37°C	3.224	0.0052	20.26	3.009

[a] Calculated from k_1 values from the equation $kt_{90} = -0.1054$.

Table 8 shows the data from Tables 6 and 7 plotted by the t_{90} method. The data are shown in Fig. 13. The least squares fit equation in Fig. 13 is $\ln[t_{90}] = -29.695 + (10.142/T)$, and it is noted that the slope (activation energy) is the same as in Fig. 11 (where the slope is 10.153). The small difference lies in rounding off errors in calculating t_{90} from the k_1 values.

The t_{90} value at room temperature is determined to be 4.072 [Eq. (2.53)]. If plotted on semilogarithmic paper, the value will emerge directly, and this type of plotting is more easily understood by those not familiar with kinetics than plots of the type in Fig. 11.

7.1. Cyclic Testing

One advantage of testing at higher temperatures is that it is possible to construct decomposition profiles at nonconstant temperature. It should be pointed out that extended room temperature is defined in the USP, 1990, as between 15 and 30°C but is now 15–25°C. Stability studies are usually carried out isothermally, e.g., as mentioned, at 25°C. The rationale for testing temperatures will be discussed at a later point in the book; suffice it to say at this point that a product in the marketplace will never experience an isothermal shelf history.

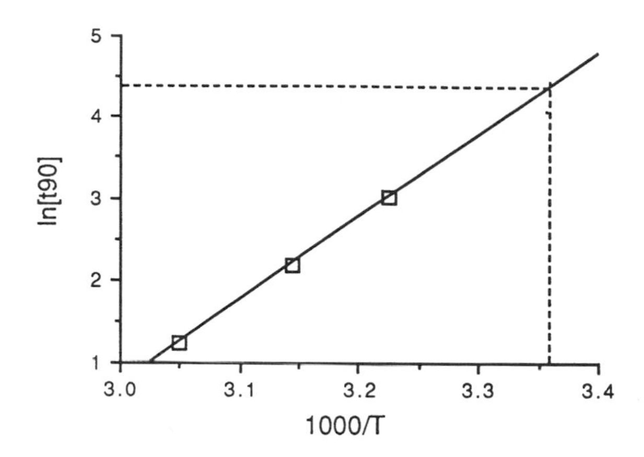

Fig. 13 Arrhenius plotting using t_{90}.

At best there will be daily fluctuations. This concept was investigated by Carstensen and Rhodes (1986), in the following fashion. Suppose a product is stored at 25°C with daily fluctuations of ±5°C. This means that the dependence of T on time, t (in days, if the cycle is one day), is given by

$$T = T_1 + T_2 \sin(2\pi t) \tag{2.54}$$

This is shown in Fig. 12.

Assume, as well, that the reaction is zero order, i.e.

$$C = C_0 - kt \tag{2.55}$$

Introducing Eqs. (2.50) and (2.54) into this and applying the situation to a differential time element dt, gives

$$dC = \left[Z \exp\left[-\frac{E}{R\{T_1 + T_2 \sin(2\pi t)\}} \right] dt \right] \tag{2.56}$$

so that to obtain the concentration after a time period t of cyclic storage, where $0 < t < 1$ day, is given by

$$C = C_0 - \int_0^t Z \exp\left(-\frac{E}{(R\{T_1 + T_2 \sin(2\Pi t)\}} \right) \tag{2.57}$$

These types of integrals can be solved by computer programs. The loss in e.g. one year will be 360 times that after one day, and so on.

The authors calculated the loss after 3 years of storage, using different activation energies, and using k_{25} (isothermal) = 0.01%, and using a daily cycle with a fluctuation of ±5°C. They arrived at the results in Table 9.

It is seen (Table 9, Fig. 14) that as long as the activation energy is less than 22 kCal per mole, the percent increase in the amount lost after a given storage period is less than 10%. For example, if a dosage form lost 5% after 3 years storage at static room temperature, it would lose 5.5% after 3 years of cyclic room temperature. This point will be of importance in the following section.

Table 2.9 Cyclic Versus Constant 25°C Data. $k_{25} = 0.01\%$ Per Day

E, kCal per mole	Loss after 3 years	Percent increase in loss
10	11.16	1.9
15	11.38	3.9
20	11.77	7.5
25	12.27	12.1
30	12.89	17.7

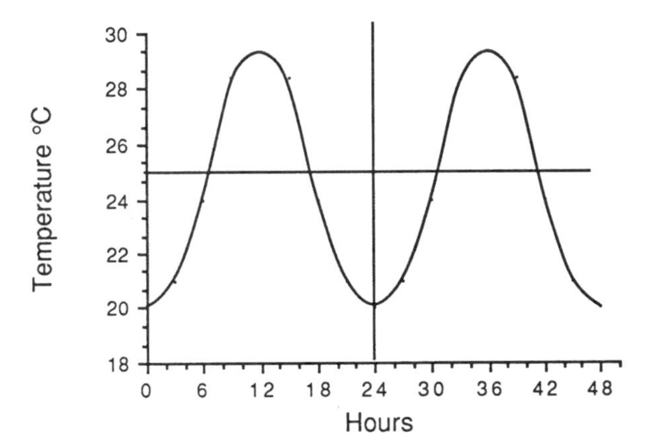

Fig. 14 Daily temperature fluctuations according to Eq. (2.54).

It is noted that any type of cycle can be used, e.g. seasonal cycles could be used as well. The problem of cyclic testing (for chemical stability) is raised from time to time. But to decide on a cycle is difficult, and it is much more rational to use the data from accelerated studies to produce the desired profile.

7.2. Nonisothermal Kinetics

It is possible, rather than studying a reaction at a fixed temperature (isothermally), to vary the temperature in a given fashion, and fit the data to Eq. (2.50).

For a zero-order reaction we can write

$$\frac{C_0}{C} = kt = \left[Z \exp\left(-\frac{E_a}{RT}\right)\right] \cdot t \tag{2.58}$$

We may allow T to vary in a given manner, e.g., in the simplest case as

$$\frac{1}{T} = a - bt \tag{2.59}$$

where a and b are the constants that we input into a programmable temperature

Table 2.10 Alkaline Decomposition of Riboflavin by Isothermal and by Several Nonisothermal Programs

Temperature program	Rate constant at 25°C	Activation energy (kcal/mole)	Reference
Isothermal	0.016	19.2	Cole and Leadbeater (1966)
Linear up	0.014	20.1	Guttman (1962)
Linear up	0.016	20.3	Rosenberg et al. (1984)
Log up	0.018	17.9	Madsen et al. (1974)
Log up[a]	0.015	20.9	Rosenberg et al. (1984)
Log down	0.015	18.9	Rosenberg et al. (1984)

[a] Triplicate experiments.
Source: Table constructed from data published by Rosenberg et al. (1984).

bath. By inserting Eq. (2.59) into Eq. (2.58), the concentration profile in time becomes

$$\frac{C_0}{C} = Z \exp\left[-\frac{E_a}{R}(a - bt)\right] \tag{2.60}$$

or, logarithmically,

$$\begin{aligned} \ln\left[\frac{C_0}{C}\right] &= \ln[Z] - \frac{E_a}{R}(a - bt) \\ &= \left[\ln\{Z\} - \frac{aE_a}{R}\right] + \left(\frac{bE_a}{R}\right)t \end{aligned} \tag{2.61}$$

This gives rise to a linear plot when $\ln[C_0/C]$ is plotted versus t, and (since a and b are the constants for the program we have chosen for our temperature bath), the slope (divided by b/R) will give us the value of E_a; and now $\ln[Z]$ can be obtained from the intercept.

The same procedure can be used for first-order reactions, although they are somewhat more complicated. In such a case a computer program is best, and curves can be generated to match the curve obtained experimentally.

Table 10 shows some of the investigations that have been carried out in the last 20 years, using this procedure.

7.3. Kinetic Mean Temperature

Since stability studies carried out in an industrial setting are isothermal, there has been a fair amount of discussion over the last 2 decades as to what the actual temperature of the study ought to be. Up to 1993, the FDA required stability studies to be carried out at 30°C for the approval of expiration periods. In contrast, the European community would consider the United States as an area where 25°C would be the appropriate temperature to require for isothermal testing.

The resolution of the problem in Europe came from by Futscher and Schumacher (1972), who established the climate zones, and Wolfgang Grimm (1985,

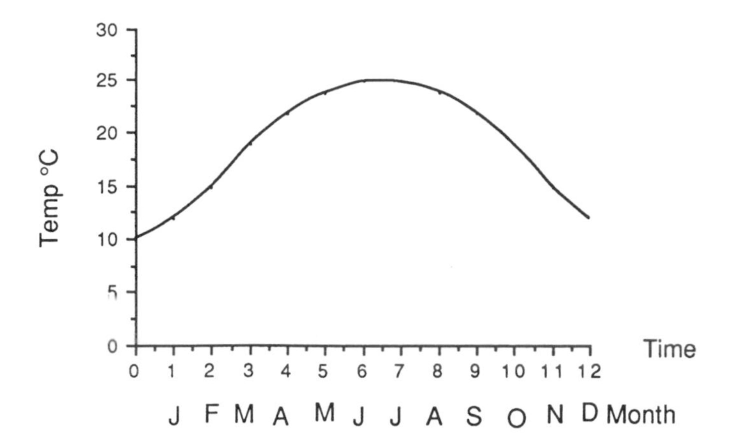

Fig. 15 Hypothetical temperature profile a drug might experience in noncontrolled conditions in a temperate climate.

1986), who employed the concept of virtual temperature (Haynes, 1971) to systematize them. The zones are as in the table.

Temperate	21°C (45% RH)
Mediterranean	25°C (60% RH)
Tropical moist	30°C (70% RH)
Desert	30°C (35% RH)

The temperatures listed are not the average yearly temperatures in these areas but rather the kinetic mean temperatures (KMTs). These are obtained by the following considerations. What is important is not what the temperature is, but how much drug is lost.

The problem in the United States is that certain areas (for instance Arizona) are in the desert zone and certain areas (e.g., Miami, the Keys) are in the tropical zone. If a product is shipped to any of these areas it may experience temperatures or relative humidities other than those of Zone II.

Consider a drug product that experiences the temperature profile in Fig. 15. If a sample experiences this temperature profile, then the rate constant will first increase from a low in January to above the "average" in the summer, then fall back to the "average" rate constant right after the summer, and decrease below the "average" as the fall months set in. The actual concentration profile is the curve shown in Fig. 16. If the average temperature had been used, isothermally, then the line in the figure would have resulted. It is seen that if the average temperature from Fig. 15 had been used, then the fraction of drug retained after 12 months' storage would have been higher than in the case of the nonisothermal storage.

There is a temperature, which, had the drug been stored at that temperature isothermaly, would have caused a drop exactly equal to that brought about by the nonisothermal storage. This is shown in Fig. 17. The question is, What is that temperature?

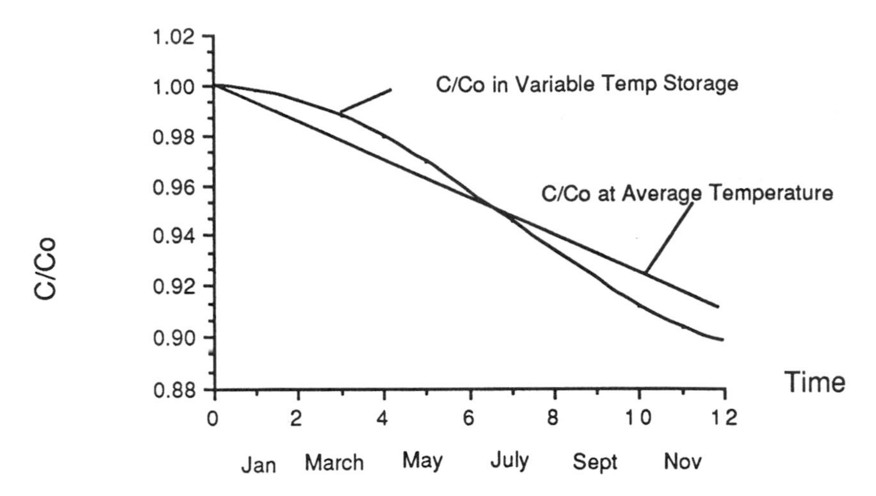

Fig. 16 Nonisothermal temperature loss using the temperature profile in Fig. 15, compared with loss encountered by isothermal storage at the average temperature from Fig. 15.

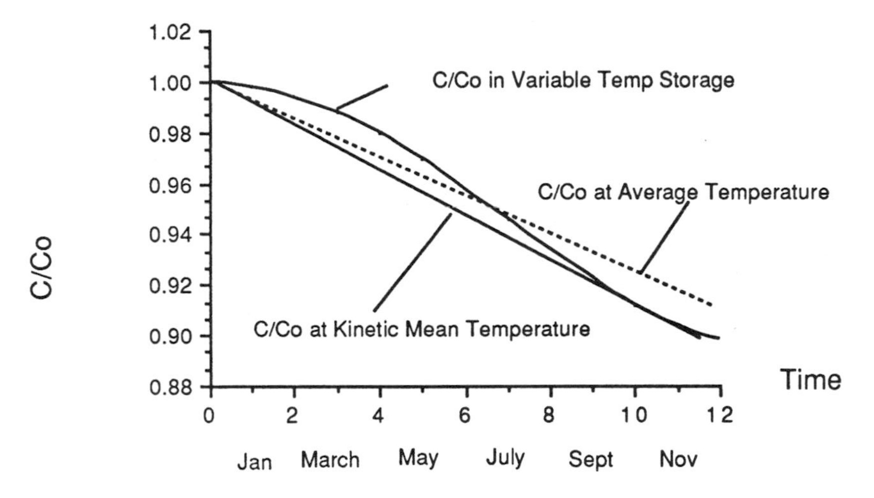

Fig. 17 Definition of kinetic mean temperature as the temperature at which the loss, isothermally, equals that of the variable temperature storage.

So for this approach it is necessary to know (a) the absolute temperature (T°K) profile as a function of time, t:

$$T = f(t) \tag{2.62}$$

and (b) the activation energy.

If we do, we may calculate the amount retained after, say, 12 months as:

$$\left[\frac{C}{C_0}\right] = Z\int_0^{12} \exp\left[-\frac{E_a}{R \cdot f(t)}\right] dt \tag{2.63}$$

The kinetic mean temperature (T_{kmt}) is then defined as that temperature at which isothermal storage (at T_{kmt}) would have caused the same drop in potency, i.e.,

$$Z \exp\left\{-\frac{E_a}{[R \cdot (T_{kmt})]}\right\} \cdot 12 = Z \int_0^{12} \exp\left[-\frac{E_a}{\{R \cdot f(t)\}}\right] dt \tag{2.64}$$

It is noted that Z cancels out, so that knowledge of the pre-exponential factor is not necessary. Hence the kinetic mean temperature *for the reaction in question* is given by

$$\exp\left\{-\frac{E_a}{R \cdot T_{kmt}}\right\} 12 = \int_0^{12} \exp\left(-\frac{E_a}{\{R \cdot f(t)\}}\right) dt \tag{2.65}$$

The approach of Grimm for a harmonized, global approach was to assume that (as has been reported, e.g., by Kennon, 1964) that on the average

$$\frac{E_a}{R} = 10\,\text{kCal/mol} \tag{2.66}$$

T_{kmt} with this restriction is called the kinetic mean temperature (KMT), and the only requirement for its calculation is knowledge of the temperature profile of storage. Grimm (1985, 1986) calculated the KMT for a series of cities and countries and arrived at the climate zones listed above.

As a result of an FDA advisory board meeting (6/25/1993, Silver Springs, MD) and the following ICH Guidelines, the agency has adopted this policy for the United States as well. This is shored up by the so-called Prescription Drug Marketing Act (PDMA), which requires that all warehouses, distribution centers, transportation chains, and pharmacies retain KMTs of less than 25°C. That is, the average measured temperature should be below 25°C and only "spurious excursions" to 30°C are allowed.

It is noted that the average measured temperature is always (except in truly isothermal situations) lower than the KMT. In actual *isothermal* storage, the KMT is equal to the temperature of the storage.

7.4. Eyring Plots

Recalling the thermodynamic relationship

$$\Delta G = \Delta H - T\Delta S \tag{2.67}$$

the Arrhenius equation can be written

$$k = Z \exp\left[\left\{\left(\frac{\Delta S¶}{R}\right)\right\} - \left\{\left(\frac{\Delta H¶}{RT}\right)\right\}\right] \tag{2.68}$$

Eyring has shown that

$$Z = \frac{\kappa T}{h} \tag{2.69}$$

where κ is Boltzmann's constant,

$$\kappa = 1.38 \cdot 10^{-16}\,\text{erg/deg} \tag{2.70}$$

and h is Planck's constant,

$$h = 6.63 \cdot 10^{-27}\,\text{erg s} \tag{2.71}$$

hence

$$\frac{\kappa}{h} = \left(\frac{1.38}{6.63}\right) \cdot 10^{11} = 2.08 \cdot 10^{10} \tag{2.72}$$

or

$$\ln\left[\frac{\kappa}{h}\right] = 23.76 \tag{2.73}$$

Introducing this into Eq. (2.68) then gives

$$k = \left[\frac{\kappa T}{h}\right] \exp\left[\left\{\left(\frac{\Delta S\P}{R}\right)\right\} - \left\{\frac{\Delta H\P}{RT}\right\}\right] \tag{2.74}$$

or

$$\ln\left[\frac{k}{T}\right] = 23.8 + \left[\left\{\left(\frac{\Delta S\P}{R}\right)\right\} - \left\{\left(\frac{\Delta H\P}{RT}\right)\right\}\right. \tag{2.75}$$

where the superscript refers to the transition state (Laidler, 1965). Kearny et al. (1993) employed these principles for interconversion kinetics and equilibrium of the lactone and hydroxyacid forms or the HMG-CoA reductase inhibitor, CI-981 = [R-(R*, R*)]-2-(4-fluorophenyl)-b,d-dihydroxy-5-(1-methylethyl)-3-phenyl-4-[(phenylamino)carbonyl]-1H-pyrrole-1-heptanoic acid.

The equation they arrived at was

$$\ln\left[\frac{k}{T}\right] = 31.95 + \frac{\Delta S^*}{1.987} - \frac{\Delta H^*}{1.987T} \tag{2.76}$$

8. SECOND-ORDER REACTIONS

If a drug substance A reacts with a second substance B according to a common scheme:

$$\text{A} + \text{B} \rightarrow \text{C} \tag{2.77}$$

then the rate equation is

$$\frac{d[\text{A}]}{dt} = -k_2[\text{A}][\text{B}] \tag{2.78}$$

If the initial concentration of A present is a, the initial concentration of B present is b, and the amount of C formed at time t is x, then at time t

$$[A] = a - x \tag{2.79}$$

and

$$[B] = b - x \tag{2.80}$$

Inserting this in (2.61) gives

$$-\frac{dx}{dt} = -k_2(a - x)(b - x) \tag{2.81}$$

which integrates to

$$\ln\left[\frac{b(a - x)}{a(b - x)}\right] = (a - b)k_2 t \tag{2.82}$$

or

$$\ln\left[\frac{(a - x)}{(b - x)}\right] = (a - b)k_2 t + \ln\left[\frac{a}{b}\right] \tag{2.83}$$

It is noted that the term on the left-hand side of the equation is dimensionless, so that the right-hand side is so as well. Hence the concentrations, a, b, and x may be in any concentration unit, and the unit for k will be such that the right-hand side is dimensionless, i.e., [time concentration]$^{-1}$.

To establish the general shape of the curves of a, b, and x, and second-order rate curve has been generated using $A(0) = a = 100$ and $B(0) = b = 70$ and $k_2 = 0.0001$. This curve is shown in Fig. 18.

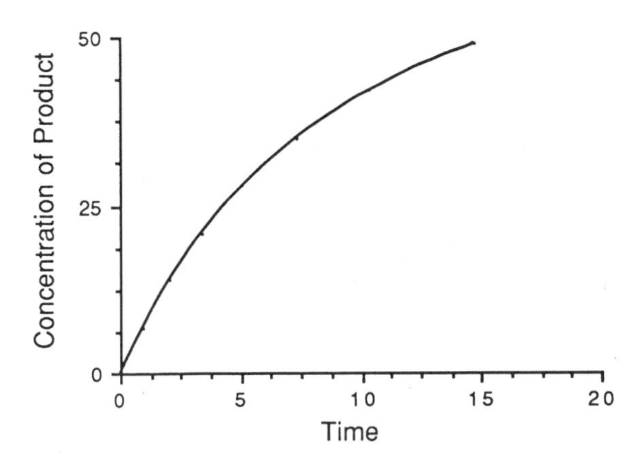

Fig. 18 Generated second-order reaction curve. Concentration of reactant (decomposition product) is plotted versus time. $a = 100$, $b = 70$, and $k_2 = 0.0001$.

Table 2.11 Second Order Data Where $a = A(0) = 100$, Assuming Values of $b = B(0) = 60$ and 70

Time	x	$[B] = 60 - x$	$[B] = 70 - x$	$[A] = a - x$
0	0	60	70	100
2	14	46	56	86
4	24	36	46	76
6	31	29	39	69
8	37	33	43	63
10	42	18	28	58
12	47	13	23	53

A set of second-order date are shown in Table 11, and it is assumed that A, with an initial concentration of 100, reacts with an unknown substance, B, i.e. that b is unknown.

a and b could be, for instance in mg/mL and k could be in units of mL(month · mg). The number of reacted concentration units (x) is shown graphically in Fig. 18 as a function of time. (i.e., the ordinate could be in mg/mL and the abscissa in months).

If the task were to study the reaction between A and B, then the experimental design would simply be to carry out the reaction with varying amounts of A and B and plot these according to Eq. (2.82) or (2.83). The problem in stability work, however, is quite different. In general the investigator follows the concentration of active drug, say A, and does not know that an interaction is taking place. (It is not possible, a priori, to imagine all the possible interactions that may occur.)

An experience of the author's is a parenteral, aqueous diluent that contained maleic acid and propylene glycol. In general, one expects that an ester will hydrolyze, but rarely that it will be created by the opposite reaction in an aqueous system. In this case, however, there was a formation of both the mono and diester of propylene glycol and maleic acid. This was eventually deduced, but it was not thought of originally. With good assay methodology, this may be discovered early, but not necessarily, because, if it is not suspected, the assay may not account for the possibility of the interaction, and the reaction product peak can hide under a reactant peak, if HPLC is used.

An example is shown in Table 12, where concentration is monitored versus time for progabide in pH 1.75 buffer. This decomposition has been reported by Farraj et al. (1988).

The data in Table 12 are the same as those used to create Figs. 8 and 9. Suppose, first of all, that it were not suspected that the decomposition of the drug A was due to interaction. Then the logical procedure would be to plot the data first order (semilogarithmically), and, in this vein, the data from Fig. 18 are plotted in Fig. 19. Note that the terms $(a - x)$ and $(b - x)$ are the concentrations of unreacted (undecomposed, intact) A and B, respectively, but that since b is not known, then neither is x nor $(a - x)$ nor $(b - x)$.

Table 2.12 Decomposition of Progabide in pH 1.75 Buffer as a Function of Time

Time (min)	Concentration C	[C-0.0055]	ln[C-0.0055]
0	0.00896	0.00343	−4.715
7	0.00760	0.00205	−4.880
13	0.00700	0.00245	−4.962
19	0.00660	0.00105	−5.021
25	0.00612	0.00062	−5.096
31	0.00600	0.00045	−5.116

Source: Data taken off graph published by Farraj et al. (1988). Table numbers are subject to the magnitude of error associated with reading off graphs. The data are used in a pedagogical sense.

The manner in which this is first approached is by iteration. Several values for the niveau value, C_{inf}, are selected and plotted by the equation

$$\ln[(C - C_{\text{inf}})] = -kt + \ln[(C_0 - C_{\text{inf}})] \tag{2.84}$$

The plot using $C_{\text{inf}} = 0.0055$ is shown in Fig. 8.

This type of process is usually carried out by computer by inserting different values of C_{inf} into Eq. (2.84) and selecting the "best" value by the criterion that the produced line gave the smallest value of the sum of squares (s^2_{yx}, to be discussed in Chapter 11, "Statistical Aspects").

It is seen in Fig. 19 that there is a "trend in the data," i.e. that the plain first-order plot is not linear. It is not just that the correlation coefficient is rather small* but it is the fact that the points collect below the line in the middle region.

If indeed this were an interaction, then one might guess that there would be a concentration where there would be a leveling off (from Fig. 5 an infinity value $C_{\text{inf}} = 0.0055$ would seem reasonable). If this were a second-order reaction, then this concentration, with concentrations expressed in molarities, would be the value of $a - b$. It is known, now, that the value of b in the above example is $C_0 - C_{\text{inf}} = 0.00896 - 0.00555 = 0.00341$ molar. It is not known *what* B is, but the concentration of it is known. This can now be used to plot the data by Eq. (2.82) or (2.83) as has been done in Fig. 20.

The intercept is not quite in line with what it should be, since $\ln[0.00896/0.00346] = 1.24$, but it is sufficiently close to assume that the treatment is in the right direction.

As mentioned, computer programs are the easiest way of handling second-order reactions, and a program for data treatment of second-order kinetic data is shown here.

---

* The correlation coefficient is close to 1.00 for very linear data. 0.97 is not too good a correlation coefficient for six data points.

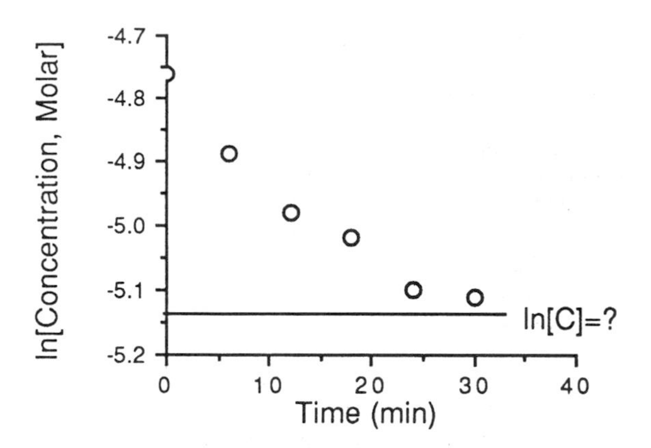

Fig. 19 Data from Table 11 plotted semilogarithmically. The least squares fit is $y = -4.764 - 0.0127x$ ($R = 0.97$).

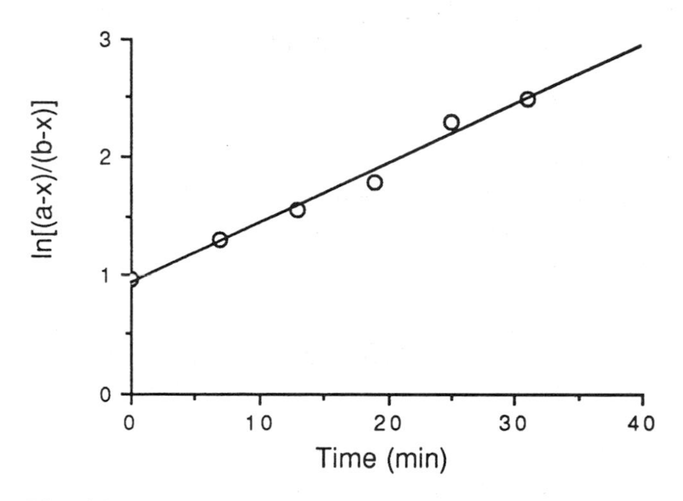

Fig. 20 The data in Table 4 treated by Eq. (2.82) using $a = 0.00896$ and $b = 0.00896 - 0.0055 = 0.00346$. The least squares fit is $y = 0.92 + 0.051x$.

```
100 INPUT "INITIAL DRUG CONCENTRATION=";W1
105 INPUT "INITIAL REACTANT CONCENTRATION=";W2
110 INPUT "NUMBER OF POINTS=";N1
115 PRINT "PRINT DATA IN 400 BLOCK AS: DATA,CONC REACTANT"
120 PRINT
125 PRINT "TIME";SPC(6);"CONC.DECOMP.";SPC(6);
    "LN{B(A-X)/(A(B-X))}"
130 PRINT "--------------------------------------"
200 READ B,A
205 X=A
210 W3=W1-B
215 W4=W2-B
```

```
220 W5=W2/W1
225 W6=W5*W3/W4
230 Y=LOG(W6)
235 X1=X1+X
240 X2=X2+(X^2)
245 Y1=Y1+Y
250 Y2=Y2+(Y2^2)
255 Z1=Z1+(X*Y)
260 N2=N2+1
265 PRINT X;SPC(5);Y
270 IF N2=N1 GOTO 700
300 GOTO 200
400 DATA 1.32,6
405 DATA 2.93,12
410 DATA 4.96,18
415 DATA 7.59,24
420 DATA 11.2,30
425 DATA 16.5,36
430 DATA 25.8,42
435 DATA 51/3.48
700 Z2=X2-((X1^2)/N2)
710 Z3=Y2-(Y1^2/N2)
720 Z4=Z1-(X1*Y1/N2)
730 Z5=Z4/Z2
740 Z6=(Y1=(Z5*X1))/N2
750 Z7=(Z4^2)/(Z3*Z2)
760 78=Z7^(0.5)
770 PRINT
800 PRINT "SLOPE=";Z5
810 PRINT
820 PRINT "INTERCEPT=";ZL6
830 PRINT
840 PRINT "CORRELATION COEFFICIENT=";Z8
850 PRINT
860 PRINT "S^2=";Z9
870 END
```

Second-order reactions, per se, are less frequent in the literature, because often the reactant is unknown in stability situations. Ohta et al. (1998) found that oxidized glutathione (GSSG) interacted with *trans*- and *cis*-diamminedichloroplatinum (DDP) by way of the S–S bond. The reaction was pseudo-first order and the rate constants linearly related to *cis*-DDP concentrations. If one considers the reaction

$$\text{GSSG} + \text{DDP} \rightarrow \text{products}$$

then, if the GSSG concentration is high, the reaction *is* second order but will appear first order and, indeed, the pseudo-first order rate constants will be linear in DDP concentration.

8.1. Equal Initial Concentration. Reciprocal Plots

In the case where $a=b$, the rate equation becomes

$$\frac{d(a-x)}{dt} = -k_2(a-x)^2 \tag{2.85}$$

This may be written

$$\frac{d(a-x)}{(a-x)^2} = -k_2 t \tag{2.86}$$

which integrates to

$$\left[\frac{1}{(a-x)}\right] = \left[\frac{1}{a}\right] + k_2 t \tag{2.87}$$

It should be noted in the foregoing whether one of the two components A and B is much less than the other. If $[B] \gg [A]$, then the rate equation becomes

$$\frac{dA}{dt} = -k_2[A][B] \approx -k_{ps}[A] \tag{2.88}$$

where

$$k_{ps} = k_2[B] \tag{2.89}$$

k_{ps} is a pseudo-first-order rate constant, so that in this case the decomposition will appear first order in the predominant species. Particularly where B is water, then $k_{ps} = k_2 \cdot 55.5$ since 55.5 is the molarity of water in water, and since it is assumed that $[B] \gg [A]$, so that the final concentration of B is

$$[B] - [A] \approx [B] \tag{2.90}$$

8.2. Pseudo-First-Order Reactions

If $b > a$, then the integration of Eq. (2.81) gives

$$\ln\left[\frac{a(b-x)}{b(a-x)}\right] = -(b-a)k_2 t \tag{2.91}$$

If, however, $b \gg a$, then (2.78) becomes

$$\frac{d[A]}{dt} = -\{k_2[B]\}[A] \approx -k_1'[A] \tag{2.92}$$

where

$$k_2[B] = k_1' \tag{2.93}$$

is almost constant and is denoted the pseudo-first-order rate constant. It is noted that this will have the unit of time $^{-1}$ as it should. Most reactions actually are of this type. A common case in point is that of a hydrolysis. The molarity of pure water is

$1000/18 = 55.5$. An example is the case of hydrolysis of aspirin (MW 180) (Edwards, 1950):

$$(CH_3COO)C_6H_4COOH + H_2O \rightarrow CH_3COOH + HOC_6H_4COOH \qquad (2.94)$$

If the solution of aspirin were, e.g., 0.1 molar at the onset, then if the reaction went to completion, the final water concentration would be 55.4 molar, i.e., would hardly change at all. Hence most hydrolyses are pseudo first order.

The literature is replete with examples of pseudo-first-order hydrolysis. Lazarevski et al. (1978) reported the kinetics of the acid catalyzed hydrolysis of erythromycin oxime and erythromycylamine to be pseudo first order. Sternson and Shaffer (1978) reported the hydrolysis of digoxin in acid solutions to be pseudo first order, although the reaction scheme was complex.

Hydrolysis is usually thought of as being undesirable. However, in the case of prodrugs, it is a built-in feature: prodrugs are utilized if the parent drug is poorly soluble (i.e., making a soluble derivative) or poorly bioavailable. In the proper biological environment, it should decompose into the active species (e.g., hydrolyze in the bloodstream after absorption). The point is that A is a derivative of an active species which is not absorbed well, and that A itself is well adsorbed. It must then (unless it too is active) decompose into the active species in the bloodstream so that the action is indistinguishable from that of the parent compound. Vaira et al. (1984) have described the kinetics, pH profile, and Arrhenius plots of derivatives of 3-[hydroxymethyl]-phenytoin esters. It was found that various amino groups containing acyl esters of 3-(hydroxymethyl)phenytoin were potential orally useful products and that the disodium phosphate ester of 3-[hydroxy methyl]-phenytoin appeared to be a good parenteral form of the drug.

9. COMPLICATED HYDROLYSIS SCHEMES

Frequently it is assumed that since a reaction is a hydrolysis, then, the scheme should be pseudo first order. Although this may always be true in the initial stages of a decomposition, caution should be observed in extending the conclusions. An example is the work by Teraoka et al. (1993) dealing with hydrolysis at the S-bond of ranitidine in acetate buffer and in water. Their findings are shown in Fig. 19. When such occurrences take place, more sophisticated mechanisms must be thought of (e.g., Franchini and Carstensen, 1994).

Teraoka et al. (1993) in the above case established that a set of parallel reactions (hydrolysis, two rearrangements) occurred, i.e., denoting ranitidine by I and other products by higher roman numerals, the scheme proposed was

$$\text{I} \rightarrow \text{III}$$

$$\text{I} \rightarrow \text{II}$$

$$\text{I} \rightarrow \text{IV} + H_2O \text{ (ring closure)}$$

$$\text{II} + \text{I} \rightarrow \text{condensation}$$

Of these the first three will be first order (parallel), and hence, often in stability work, a first order assumption is not unreasonable (Teraoka et al., 1993) (see Fig. 21).

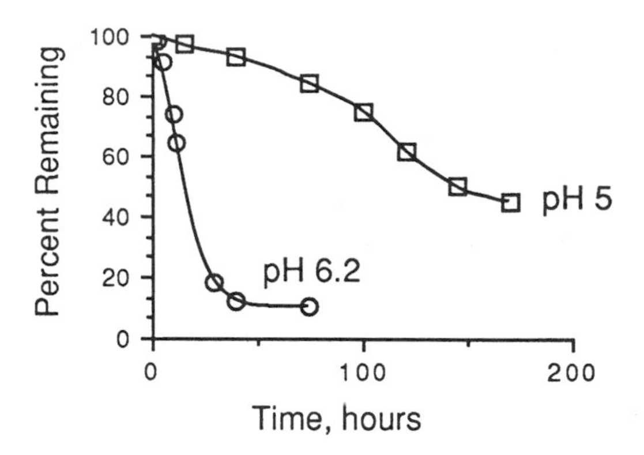

Fig. 21 Ranitidine stability in acetate buffers (0.1 M) of ionic strength 0.1. (Graph constructed from data published by Teraoka et al., 1993.)

REFERENCES

Angerg, M., Nyström, C., Castensson, S. (1988). Acta Pharm. Suec. 25:307.

Angerg, M., Nyström, C., Castensson, S. (1990). Int. J. Pharm. 61:66.

Angerg, M., Nyström, C., Castensson, S. (1993). Int. J. Pharm. 90:19.

Archontaki, H. A., Gikas, E. E., Panderi, I. E. Ovszikoglou, P. M. (1998). Int. J. Pharm. 167:69.

Aso, Y, Sufang, T., Yoshka, S., Kojima, S. (1997). Drug Stability 1:237.

Beal, H. D. Pranderd, R. J., Todaro, L. J., Sloan, K. B. (1993). Pharm. Res. 10:905.

Beal, H. D., Prankerd, R. J., Sloan, K. B. (1997). Drug Dev. Ind. Pharm. 23:517.

Burke, M., Redden, P. R., Douglas, J.-A., Dick, A., Horrobin, D. F. (1997). Int. J. Pharm. 157:81.

Buur, A., Bundgaard, H. (1984). Int. J. Pharm. 21:349.

Carstensen, J. T. (1997). Modeling and Data Treatment in the Pharmaceutical Sciences, Lancaster, PA: Technomic Publishing Corp.

Carstensen, J. T., Attarchi, F. (1988a). J. Pharm. Sci. 77:314.

Carstensen, J. T., Attarchi, F. (1988b). J. Pharm. Sci. 77:318.

Carstensen, J. T., Franchini M. K. (1994). Pharm. Res. 11:S-26.

Carstensen, J. T., Rhodes, C. T. (1986). Drug Dev. Ind. Pharm. 12:1219.

Carstensen, J. T., Su, K. S. E., Maddrell, P., Johnson, J. B., Newmark, H. N. (1971). Bull. Parenteral. Drug Assoc. 25:193.

Carstensen, J. T., Attarchi, F., Hou, X.-P. (1985). J. Pharm. Sci. 74:741.

Cole, B. R., Leadbeater, L. (1966). J. Pharm. Pharmacol. 18:101.

Edwards, L. J. (1950). Trans. Faraday Soc. 46.

Fabre, H., Eddine, N. H., Berge, G. (1984). J. Pharm. Sci. 73:611.

Farraj, N. F., Davis, S. S., Parr, G. D., Stevens, H. N. E. (1988). Pharm. Research 5:226.

Franchini, M. K. (1995). Ph.D. thesis, University of Wisconsin, p. 158.

Franchini, M. Carstensen, J. T. (1994). Int. J. Pharm. 111:1 53.

Franchini, M. K., Carstensen, J. T. (1999). Pharm. Dev. Tech. 4:257.

Futscher, N., Schumacher, P. (1972). Pharm. Ind. 34:47.

Ghebre-Sellassie, I., Hem, S. L., Knevel, A. M. (1984). J. Pharm. Sci. 73:125.

Grimm, W. (1985). Drugs Made in Germany, 28:196.

Grimm, W. (1986) Drugs Made in Germany, 29:39.

Guttman, D. E. (1962). J. Pharm. Sci. 51:811.

Hammad, M. A., Müller, B. W. (1998). Int. J. Pharm. 169:55.

Hansen, L. D., Lewis, E. A., Eatough, D. J., Bergstrom, R. G., DeGraft, K., Johnson, D. (1989). Pharm. Res. 6:20.
Hashimoto, N., Tasaki, T., Tanaka, H., (1984). J. Pharm. Sci. 73:369.
Haynes, J. D. (1971). J. Pharm. Sci. 60:927.
Higuchi, T., Rheinstein, J. A. (1959). J. Am. Pharm. Assoc., Sci. Ed. 48:155.
Jordan, C. G. M. (1998). J. Pharm. Sci. 87:880.
Kabadi, M. B., Valia, K. H., Chien, Y. W. (1984). J. Pharm. Sci. 73:1461.
Kearney, A. S., Crawford, L. F., Mehta, S. C., Radebaugh, G. W. (1993). Pharm. Res. 10:1461.
Kennon, L. (1964). J. Pharm. Sci. 53:1067.
Laidler, K. J. (1965). Chemical Kinetics, New York: McGraw-Hill, p. 68.
Lazarevski, T., Radobolja, G., Djokic, S. (1978). J. Pharm. Sci. 67:1031.
Lerner, D. A., Bonneford, G., Fabre, H., Mandrou, B., deBuochberg, M. S. (1988). J. Pharm. Sci. 77:699.
Li, S., Wang, W., Chu, J., Zamansky, I., Tyle, P., Rowlings, C. (1998). Int. J. Pharm. 167:49.
Madsen, B. W., Anderson, R. A., Herbeison Evans, D., Sneddon, W. (1974). J. Pharm. Sci. 63:777.
Mälkki-Lame, L., Valkeila, E. (1988). Int. J. Pharm. 161:29.
Misra, P. K., Haq, W., Katti, S. B., Mathur, K. B., Raghubir, R., Patnaik, G. K., Dhawan, B. N. (1993). Pharm. Res. 10:660.
Nguyen, T. H., Burnier, J., Meng, W. (1993). Pharm. Res. 10:1563.
Notari, R. E., DeYoung, J. L. (1975). J. Pharm. Sci. 64:1148.
Ohta, N., Inagaki, K., Muta, H., Yotsuyanagi, T., Matsuo, T. (1998). Int. J. Pharm. 161:15.
Oliyai, R., Lindenbaum, S. (1991). Int. J. Pharm. 73:33.
Owen, W. R. Stewart, P. J. (1979). J. Pharm. Sci. 68:992.
Prankerd, R., Walters, J., Parnes, J. (1992). Int. J. Pharm. 78:59.
Rosenberg, L. S., Pelland, D. W., Black, G. D., Aunet, D. L., Hostetler, C. K. (1984). J. Pharm. Sci. 73:1279.
Shalaev, E. Y., Shalaeva, M., Byrn, S. R., Zografi, G. (1997). Int. J. Pharm. 152:75.
Sternson, L. A., Shaffer, R. D. (1978), J. Pharm. Sci. 67:327.
Teraoka, R. Otsuka, M., Matsuda, Y. (1993). J. Pharm. Sci. 82:601.
Varia, S. A., Schuller, S. Stella, V. J. (1984). J. Pharm. Sci. 73:1074.
Vilanova, B., Munoz, F., Donoso, J., Frau, J., Blanco, F. G. (1994). J. Pharm. Sci. 83:322.
Visconti, M., Citerio, L., Borsa, M., Pifferi, G. (1984). J. Pharm. Sci. 73:1812.
Yip, Y. H. W., Po, A. L. W., Irwin, W. J. (1983). J. Pharm. Sci. 72:776.

3

Kinetic pH Profiles

JENS T. CARSTENSEN

Madison, Wisconsin

One of the tasks of stability scientists, particularly in the preformulation stage, is to establish the effect of pH on the stability of the drug. To this end it is worthwhile to describe a couple of concepts in wide use, concepts that are often misused.

These concepts are closely tied to the effect of pH on reaction kinetics. It is, therefore, important to ascertain a "constant" pH during the reaction, and there are two means of doing this: (a) using a pH stat and (b) employing buffers.

In the former case, the "salt" concentration changes only very little, and one assumes (maybe not justifiably) that the determination is carried out "in water." In the latter case, there is the complication that the buffer species (and there are always at least two for the buffer to perform its function) catalyzes one or more of the reactions, so that steps must be taken to "eliminate" this effect.

The second point is that the mere presence of an electrolyte (NaCl for instance) can affect the rate constant and the pK values in question and that, therefore, a complete kinetic study must also address this point.

The important variables in a kinetic study of a drug substance in solution are therefore

Buffer effects
Ionic strength effects
pH
pK of drug substance

1. IONIC STRENGTH

The ionic strength, μ, is defined as

$$\mu = 0.5\Sigma m_i Z_i^2 \quad (3.1)$$

where m_i is the molality of the ith ion and Z_i is the charge of the ith ion. A word of caution is in order.

A "salt" is considered, in this text, to be the solid state of a strong electrolyte that, in solution, is completely dissociated, e.g., calcium chloride:

$$CaCl_2 \rightarrow Ca^{++} + 2Cl^- \quad (3.2)$$

For such a 2 : 1 electrolyte, if the molality of the "salt" were m molal, the molality of the chloride ion would be $2m$ molal, so that these figures inserted into Eqs. (3.1) and (3.2) give

$$\mu = 0.5[(m \times 2^2) + (2m \times 1^2)] = 3\,\mathrm{M} \quad (3.3)$$

and this is true for a 2 : 1 and 1 : 2 strong electrolyte in general. For a 1 : 1 electrolyte, $\mu = m$.

It should be added at this point that many investigations deal with concentrations, where, in fact activities are the important variable.

On the surface, it should therefore be a simple matter to calculate the ionic strength of a preparation. When, however, the case is one of weak electrolytes (acetic acid, acetic buffers), the problem becomes more complicated.

The degree of dissociation of an acid can for instance be calculated via the acid equilibrium constant, as will be covered shortly. This leads to a degree of dissociation, α, so that the ionic part of the solution contributed by the weak electrolyte is, e.g., in the case of a simple acid (HA),

$$\mu = m\alpha \quad (3.4)$$

The problem with determining the degree of dissociation will be dealt with below. The activity, a_A, of a compound is related to its concentration [A] by the activity coefficient f_A (Moore, 1963):

$$a_A = [A] f_A \quad (3.5)$$

It will be shown later that the ionic strength can have a definite effect on stability of drugs. van Maanen et al. (1999) have demonstrated a kinetic salt effect in the aquous solution decomposition of thiotepa.

1.1. The Debye–Hückel and Subsequent Limiting Laws

The activity coefficient, f, of a compound is the ratio between activity and concentration, e.g., for a compound HA,

$$a_{HA} = f_{HA}[HA] \quad (3.6)$$

where [HA] is the concentration of HA in molality. Often, particularly in more dilute solution, there is a tendency to replace molality with molarity, but it is a good habit to do work in units of molality, in particular because molalities are not temperature dependent; a one molal solution at 25°C is still one molal at 50°C, and this statement cannot be made for a one molal solution.

The Debye–Hückel limiting law states that

$$-\log f_z = z^2 A \frac{\sqrt{\mu}}{(\varepsilon T)^{3/2}} \tag{3.7}$$

where z is charge, ε is dielectric constant, T is absolute temperature and log is a base 10-logarithm (Brønsted, 1943). For *aqueous solutions* ($\varepsilon = 80$) at room temperature (25°C), the factor

$$\frac{A}{(\varepsilon T)^{3/2}} \approx 0.5 \tag{3.8}$$

and this factor is often used indiscriminately at other temperatures, and sometimes in nonaqueous media, so

$$-\log f_z = \{z\}^2 0.5\sqrt{\mu} \tag{3.9}$$

This is true at very high dilution. In slightly more concentrated solutions, the equation takes the form

$$-\log f_z = \{z\}^2 0.5 \frac{\sqrt{\mu}}{1 + \beta\sqrt{\mu}} \tag{3.10}$$

where the factor β is related to the ionic radius of the electrolyte and is often set to be one, i.e.

$$-\log f_z = z^2 0.5 \frac{\sqrt{\mu}}{1 + \sqrt{\mu}} \tag{3.11}$$

1.2. pK_a, pK_b, and pK

These three concepts are frequently bandied about and described indiscriminately in literature, and definitions of these (Brønsted, 1943) are quite old, but the laws of thermodynamics, we hope, are not functions of time. (The notation H^+ rather than H_3O^+ will be used in the following.)

The ionization constant of an acid, based on the equilibrium

$$HA \Leftrightarrow H^+ + A^-$$

is

$$K_{(a)} = \frac{[H^+][A^-]}{[HA]} \tag{3.12}$$

where brackets denote *concentrations*. The notation of Brønsted is used here. The term "operational pK" is used and it will be written $K_{(a)}$. It is most often simply denoted K. It will be seen in the following, that simply using the term K can be misleading, since then it would be difficult to distinguish between the same concept for a base.

The $pK_{(a)}$ is defined as the negative of the base-ten logarithm of K, i.e.

$$pK_{(a)} = -\log K_{(a)} \tag{3.13}$$

The *thermodynamic* equilibrium constant is defined as

$$K_a = \frac{a_{H+}a_{A-}}{a_{HA}} \tag{3.14}$$

i.e. there is no parenthesis in the subscript. (Actually the Brønsted notation was squared brackets, but the term K_a, thusly written, is so commonplace that it is used here without brackets). It follows that

$$pK_a = -\log K_a \tag{3.15}$$

Since activities are connected with concentrations by way of activity coefficients, f, e.g.,

$$a_{H+} = f_{H+}[H+] \tag{3.16}$$

it follows that

$$\begin{aligned} K_a &= \left\{\frac{f_{H+}f_{A-}}{f_{HA}}\right\}\frac{[H^+][A^-]}{[HA]} \\ &= \left\{\frac{f_{H+}f_{A-}}{f_{HA}}\right\}K_{(a)} \end{aligned} \tag{3.17}$$

If the charge of HA is Z_{HA} (i.e., would be 0 for acetic acid, but would be 1 for an amine hydrochloride), then that of A^-, is

$$Z_{A-} = Z_{HA} - 1 \tag{3.18}$$

and since that of the hydrogen ion is $+1$ it follows in general that

$$\begin{aligned} \log K_a &= \log K_{(a)} + \log\left\{\frac{f_{H+}f_{A-}}{f_{HA}}\right\} = -pK_{(a)} - 0.5[1^2 + Z_{A-}^2 - Z_{HA}^2]\surd\mu \\ &= -pK_{(a)} - 0.5[1^2 + (Z_{A-})^2 - (Z_{A-} + 1)^2]\surd\mu \end{aligned} \tag{3.19}$$

or

$$\begin{aligned} pK_a &= pK_{(a)} + 0.5(1 + Z_{A-}^2 - Z_{A-}^2 - 1 - 2Z_{A-})\surd\mu \\ &= pK_{(a)} - Z_{A-}\surd\mu \end{aligned} \tag{3.20}$$

This is identical to the formula quoted by Brønsted (1943) where $\log K_{(a)}$ rather than $pK_{(a)}$ is used, hence there is a plus, not a minus, in front of Z_{A^-}. It is noted that it is the charge of the conjugate *base*, which is the Z value, that must be used.

It is noted that the $pK_{(a)}$ is the value that must be used in a particular system with a particular ionic strength. It is hence a function of the system, and therefore, if tabulated, would have to be referred to in relation to that specific system. In contrast to this the pK_a (and pK_b) are universal quantities that may be tabulated.

To obtain the operational $pK_{(a)}$, the value for the pK_a and the value of $Z_{A^-}\sqrt{\mu}$ are inserted in Eq. (3.20). To obtain $\sqrt{\mu}$, when a weak acid is in question, the Henderson–Hasselbach equation, cited in the following section, is employed. This allows for calculation of the degree of dissociation, α. As will be seen below, this may require a further iteration, since the new value of $pK_{(a)}$ will give a different degree of dissociation, and hence a different ionic strength, than first employed in the equation.

pK values can be obtained spectrophotometrically. For instance, Sing et al. (1999) have used UV spectrophotometric methods to determine the pK_a of nimesulide.

1.3. pH and the Henderson–Hasselbach Equation

The hydrogen concentration of a solution is usually described by its base-ten logarithm and is then referred to as the pH (Sørensen, 1909a,b, Moore, 1963). That is,

$$pH = -\log_{10}[H^+] \tag{3.21}$$

where $[H^+]$ is the hydrogen ion *concentration* (not activity). Bates (1964), however, distinguished between

$$p_CH = -\log_{10}[H^+] \tag{3.22}$$

and

$$p_aH = -\log_{10}[a_{H+}] \tag{3.23}$$

In this writing, the definition in (3.21) will be used, so that care should be taken in the comparison of expressions derived by the second definition of pH. With the definition in Eq. (3.21) the equilibrium equation in Eq. (3.11) can be written

$$pH = pK_{(a)} + \frac{\log[A^-]}{[HA]} \tag{3.24}$$

This assumes that the activity coefficients are unity. It is usually the total concentration, C_0, of "acid" that is known, e.g., one talks about a solution that is one molal in acetic acid. Hence

$$C_0 = [A^-] + [HA] \tag{3.25}$$

and this inserted into Eq. (3.24) now gives

$$pH = pK_{(a)} + \frac{\log\{C_0 - [HA]\}}{[HA]} \tag{3.26}$$

Hence knowing the "concentration" C_0 and the $pK_{(a)}$ it is possible to calculate the amount of $[A^-]$ and [HA] at any pH.

Water is a special case:

$$H_2O \rightarrow H^+ + OH^- \tag{3.27}$$

so the concentration of $[H^+]$ and $[OH^-]$ are each 10^{-7}, at pH 7, and the ionization product of water is given by

$$K_w = [H^+][OH^-] = 10^{-14} \tag{3.28}$$

pK_b is the quantity for a base corresponding to pK_a. For a base the reaction is (Martin et al., 1993):

$$B^- + H_2O \rightarrow HB + OH^- \tag{3.29}$$

and it follows that

$$K_{(b)} = \frac{[HB][OH^-]}{[B^-]} \tag{3.30}$$

Multiplying and dividing Eq. (3.30) by Eq. (3.28) and employing $K_w = [H^+][OH^-]$ now gives

$$K_{(b)} = \frac{K_w}{K_{(a)}} \tag{3.31}$$

The relation of $K_{(b)}$ to K_b is completely analogous to the relation of $K_{(a)}$ to K_a discussed above.

Knowledge of pK_a and K_b values can be used roughly to estimate the pH of solutions of the drug substance. An example is shown here.

Example 3.1.

Vecuronium bromide (Merck Index) has a pK_b of 9. The molecular weight is 637, and if a solution is 0.9%, what would be the pH of such a solution?

Answer.

The solution would be $9/637 = 0.014$ molar. In this case the equilibrium is

$$RH^+ + H_2O \rightarrow RH_2^{++} + OH^- \tag{3.32}$$

To calculate the pH of a 0.9% solution of the quaternary salt in aqueous solution, it is first necessary to establish what the $pK_{(B)}$ is. It is hence necessary to calculate the ionic strength. To do this one would have to know how much of the vecuronium was present as monopositive and dipositive ion. To do this one would have to know the $pK_{(B)}$, i.e., one would have to know the quantity one seeks. The problem can be solved by continuous iteration.

As a first guess, one would use the pK_b value and assume it to be the $pK_{(b)}$ value. The pH is then given by (Martin et al., 1993):

$$[OH^-] = \{K_b C\}^{1/2} = (10^{-9} \times 0.014)^{1/2} = 3.7 \times 10^{-6} = 10^{0.57-6} = 10^{-5.43} \tag{3.33}$$

which gives

$$pOH = 5.43 \tag{3.34}$$

or

$$pH = 14 - 5.43 = 8.57 \tag{3.35}$$

The Martin reference is a basic text and does not go into ionic strength effects, but it is an excellent source for solving problems such as the one stated in Example 3.1.

It follows from the above that (a) at the pK the concentration of two of the species are equal; (b) moving up or down one unit makes the ratio 1 : 10, i.e., there is only 10% of one of the species; (c) moving up or down two units makes the ratio 1 : 100, and one may disconsider the more dilute species in question. Figure 1 is an example of this, phosphate pK_a values being 2.12, 7.21, and 12.67.

It is a general practice to use polybasic acids (such as phosphoric acid, Fig. 3) as buffers, since they exhibit a buffer effect over a considerable pH range. The buffer capacity, ϕ, is defined as the amount of acid (δAcid) of base (δBase) it would take to change the pH (δpH) by one unit, or on a more differential scale,

$$\phi = \frac{\delta \, \text{acid}}{\delta \, \text{pH}} \tag{3.36}$$

or a similar expression for the base.

In kinetic work, there are several factors that are of importance: the concentration of a given species (which for instance in the case of an acid or a base can be calculated from the Henderson–Hasselbach equation), its rate constant, and of course the pH.

Ong et al. (1964) and Avdeef et al. (1993) have reported that the equation by Yasuda–Shedlovski (Shedlovski, 1956) given below works well. It is akin to the Henderson–Hasselbach equation:

$$pK_a = pH - \log\left[\frac{\beta}{1-\beta}\right] - \log\left[\frac{s\gamma A}{\gamma HA}\right] \tag{3.37–3.39}$$

Here $s\gamma A$ and γHA are the activity coefficients of protolytic pair, and β is the fraction converted into more basic species.

Fallavena and Schapoval (1997) have employed potentiometric titrations in methanol–water mixtures to determined the pK_a of nimesulide and found the equation to work well. They determined the pK of the compound in various concentrations of methanol and to obtain the pK_a in water they plot the reciprocal of the dielectric constant of the methanol–water mixture, a so-called Yasuda–Shedlovski plot (Shedlovski, 1956). This is shown in Fig. 1 for nimesulide.

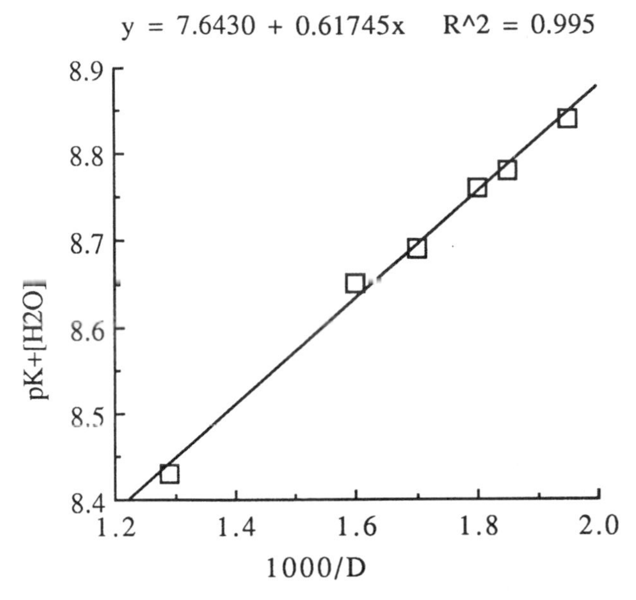

Fig. 1 Yasuda–Shedlovsky plot. (Constructed from data published by Fallavena and Schapoval, 1997, for nimesulide.)

The question of pK values in hydroorganic mixture is of interest, analytically, since such mixtures are often used in liquid chromatography solvents. Barbosa et al. (1997) determined the pK values of fluoroquinolones in acetonitrile and found them to be fairly linear in mole fraction, although pK versus reciprocal dielectric constant also appeared to be linear for norfloxacin, fleroxacin, and flumequine.

The apparent pK values of a compound is a function of the dielectric constant of the medium. As an example, Brandl and Magill (1997) have reported on the pK values of ketorolac in isopropanol. Fig. 2 shows the effect of the medium on the apparent pK.

1.4. Ionic Strength of Mixtures of Weak and Strong Electrolytes

In calculating the ionic strength for a monobasic acid (for instance acetic acid), the ionic strength is equal to the concentration, C, times the degree of dissociation, α. Since $Z_{A^-} = -1$, Eq. (3.20) becomes

$$pK_a = pK_{(a)} + \sqrt{m\alpha} \tag{3.40}$$

as quoted, e.g., by Maron and Prutton (1965).

A practical problem that arises is often that a solution is made with buffer components (e.g., a mixture of NaH_2PO_4 and Na_2HPO_4). This is supposed to give a desired pH (e.g., 7.5), but when the solution is made up, the pH might be only 7.4. It is then adjusted to 7.5 by adding a solution of KOH (see Fig. 3).

Although it is possible to calculate the ion concentrations hypothetically, these calculations become cumbersome, and it is a good practice to *burette* in KOH (or other base) of known molality, so that its exact volume is known. The pH is known, the concentrations [HA] and [A^-] are known from the Henderson–Hasselbach

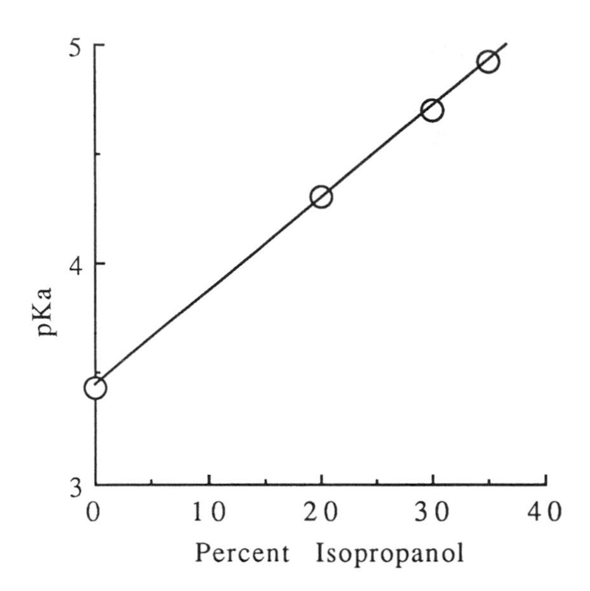

Fig. 2 pK of ketorolac versus percent isopropanol. (Figure constructed from data published by Brandl and Magill, 1997.)

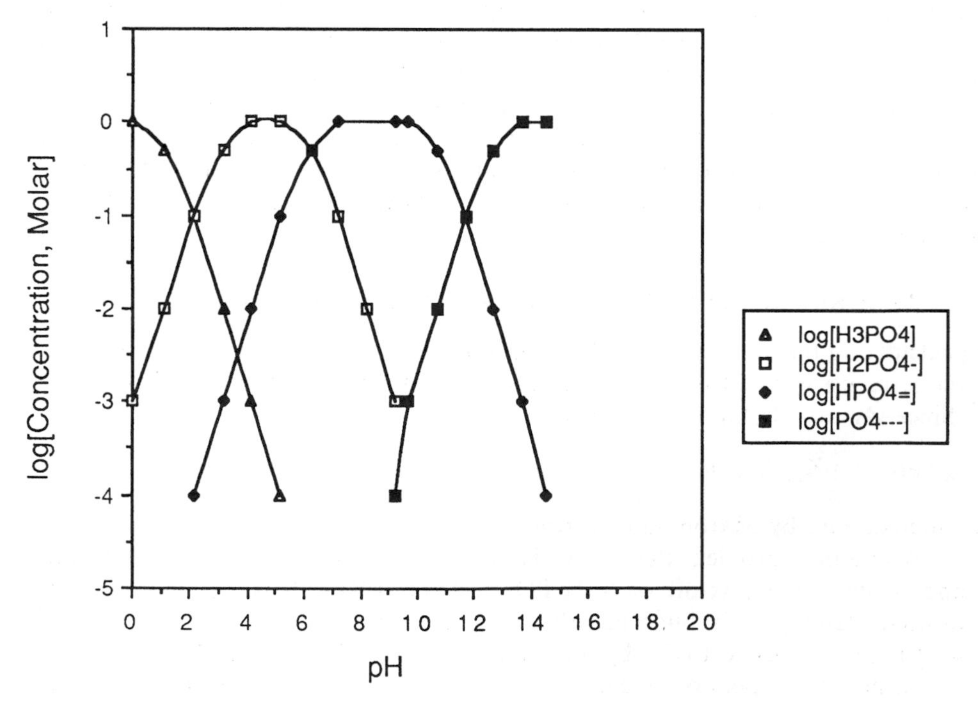

Fig. 3 Concentrations of species of phosphate ions as a function of pH. Overall concentration is one molar. An example of an application of this to a drug is that of van der Howen et al. (1994).

equation, the concentration $[K^+]$ is known, and $[OH^-]$ or $[H^+]$ is known from the pH. Hence, the ionic strength can easily be calculated. As mentioned, failing to know the exact quantity of base added makes the calculation much more difficult.

Once the approximate ionic strength (μ_1) is known, then a more accurate estimate of the $pK_{(a)}$ for the actual situation can be calculated. This necessitates a new value of α, and in essence the process should be continued until consecutive, iterated values converge. More than one iteration, however, would result in very small changes and is rarely done.

2. GENERAL AND SPECIFIC BASE CATALYSIS. BUFFER EFFECT

When a drug A hydrolyzes in aqueous solution, the reaction can be catalyzed by various ionic species, and/or simply be noncatalytic. This is schematized below. The approach is simplified (not distinguishing reaction mechanism, e.g., nucleophilic or electrophilic attack catalysis):

Noncatalyzed hydrolysis (rate constant k_0^*)

$$A + H_2O \rightarrow \text{products} \tag{3.41}$$

Specific acid catalysis (rate constant k_+^*)

$$A + H^+ + H_2O \rightarrow \text{products} \tag{3.42}$$

Specific base catalysis (rate constant k_-^*)

$$A + OH^- \rightarrow \text{products} \tag{3.43}$$

HQ (from the HQ/Q^--buffer) catalysis (rate constant k_{HQ}^*)

$$A + HQ \rightarrow \text{products} \tag{3.44}$$

and Q^- (from the HQ/Q^- buffer) catalysis (rate constant $k_{Q^-}^*$)

$$A + Q \rightarrow \text{products} \tag{3.45}$$

The rate with which the reaction takes place is (or may be) a function of the acidity of the medium in which it occurs. If an experiment is to be carried out at a given pH, then this can be achieved either by using a pH stat, and titrate acid or alkaline decomposition products, so as to maintain a constant pH, or a buffer (HQ) may be added. A buffer, by its very nature, may affect the rate, and hence the overall rate of the reaction can be written as

$$\begin{aligned} \frac{d[A]}{dt} = &- [A] \cdot [H_2O]k_0^* + k_+^*[H^+] \cdot [H_2O] + k_-^*[OH^-] \cdot [H_2O] \\ &+ k_{HQ}^*\{[HQ] \cdot [H_2O] + k_Q^*[Q^-]\} \cdot [H_2O] \end{aligned} \tag{3.46}$$

where, for convenience, in this context, the hydronium ion activity is equated with concentration. It is noted that the term $[H_2O]$ occurs in all the terms, so that it can be merged with the k values to give

$$\frac{d[A]}{dt} = -[A]\{k_0 + k_+[H^+] + k_-[OH] + k_{HQ}[HQ] + k_Q[Q^-]\} \tag{3.47}$$

As mentioned earlier, the symbol for the hydronium ion has been dropped in favor of $[H^+]$, and this will be the convention henceforth. It follows that the k values are 55.5 times the k^* values (since $[H_2O] = 55.5$ in dilute solutions). It is noted that this is only true for dilute solutions, and a section dealing with concentrated solutions will be presented later.

It is seen that at a given buffer concentration (and therefore at given hydrogen and hydroxy-ion concentrations, the reaction expression is

$$\frac{dA}{dt} = -k'[A] \tag{3.48}$$

where k' is a pseudo-first-order rate constant given by

$$k' = k_0 + k_+[H^+] + k_-[OH^-] + k_{HQ}[HQ] + k_Q[Q^-] \tag{3.49}$$

Eq. (3.49) predicts a first- or pseudo-first-order reaction, and Eq. (3.46) predicts the rate constant to be a function of several factors. It is conventional first to eliminate the effect of the buffer. This is done by performing the experiments at several buffer concentrations at several pH values (i.e., at several ratios of [HQ] to [Q]).

An example of this is the work by Maulding et al. (1975) on the hydrolysis of chlordiazepoxide. Their data are shown in Figs. 4 and 5. Buffer effects will be discussed in a later section of this chapter, and in general the observed rate constants, k_{obs}, are linear in buffer concentration. However, it should be pointed out that they are not always linear over the entire range, as shown e.g. in Fig. 5.

Figure 5 shows the hydrolysis rate of chlordiazepoxide in presence of equimolar amounts of phosphoric acid and dihydrogen phosphate ion at 79.5°C, in various buffer concentrations. It is seen that when the data are back-extrapolated to a zero concentration, the buffer-free rate constant is 0.9 h^{-1}. This extrapolated rate constant will be denoted k in the following. (It should be noted that Maulding et al. found that at phosphate concentrations higher than 0.1 M, linearity was lost,

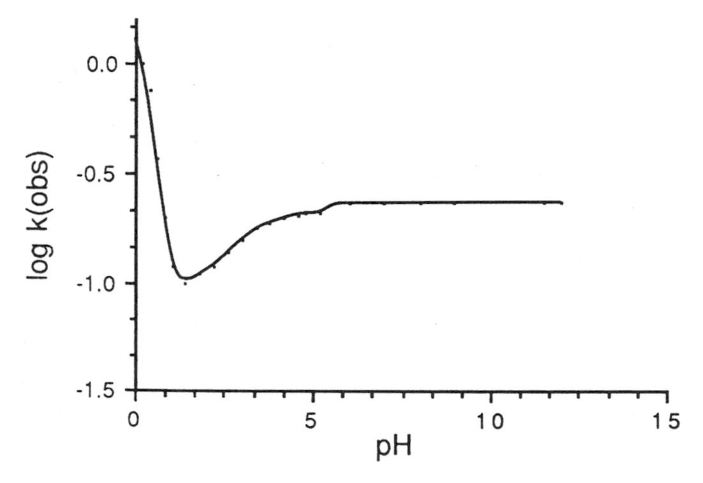

Fig. 4 pH profile of chlordiazepoxide. (Graph constructed from data by Maulding et al., 1975.)

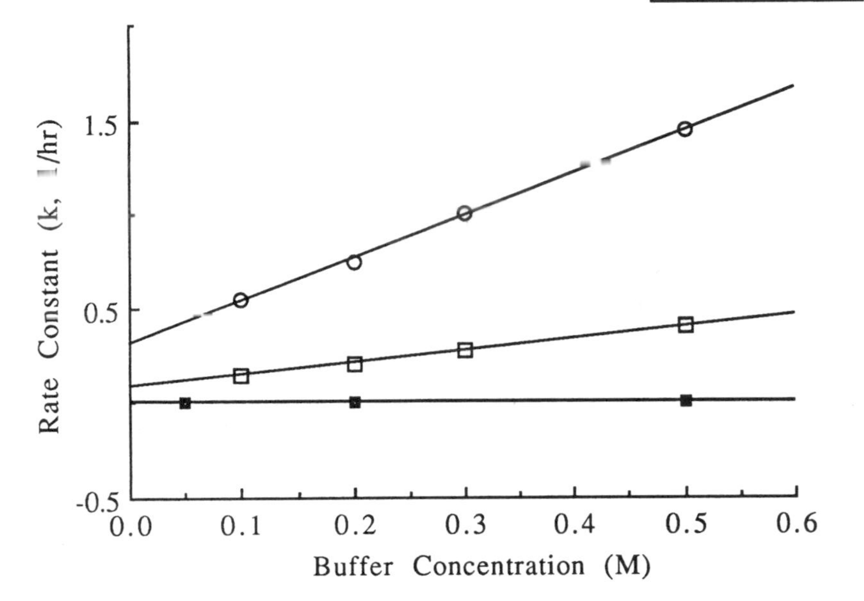

Fig. 5 Linear buffer concentration plot (panipenem). (Graph constructed from data published by Ito et al., 1997.)

indicating a shift in rate-determining step.) In other words, this buffer-free rate constant, k, is given by

$$k = k_0 + k_+[H^+] + k_-[OH^-] \tag{3.50}$$

The effect of the phosphate-to-phosphoric-acid ratio on the rate constant, in the case of linearity, can be easily calculated. If one simply considers two ratios (i.e., two pH values), then the equations for the pseudo-first-order rate constants would be

$$k(pH_1) = k_1 + k_{HQ}[HB]_1 + k_Q[B^-]_1 \tag{3.51}$$

and

$$k(pH_2) = k_2 + k_{HQ}[HB]_2 + k_Q[B^-]_2 \tag{3.52}$$

Plots of k_{obs} at a given pH (not $\log_{10}[k]$) should therefore be linear in buffer concentration, as shown in Fig. 5. For instance Hou and Poole (1969a,b) showed the traditional buffer effects in the hydrolysis of ampicillin in solution at different pH values, which enabled them to calculate the catalytic effect of the different buffer species. Ito et al. (1997) determined the buffer concentration effect on the hydrolysis of panipenem.

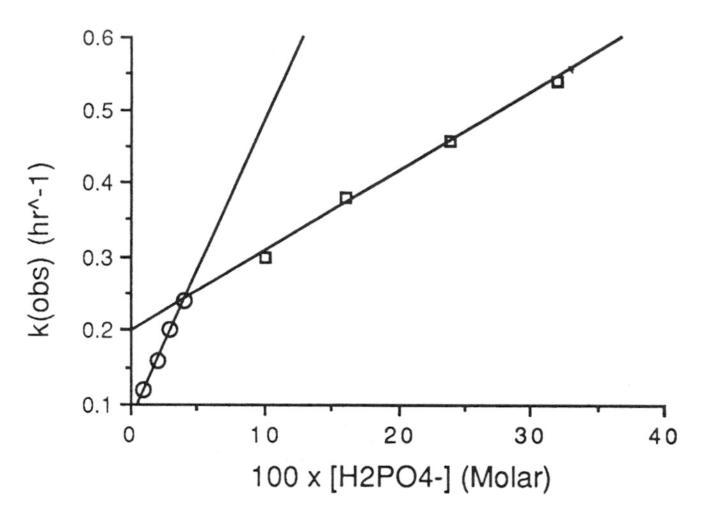

Fig. 6 Effect of buffer ratio HQ/Q- of a phosphate buffer on chlordiazepoxide decomposition. (Graph constructed from data published by Maulding et al., 1975.)

The k values at the two pH values are found from the y intercepts at the two pH values, and there are now two equations with two unknowns (k_{HQ} and k_Q), so these can be calculated. Carstensen et al. (1992) have described an abbreviated method for an experimental design that will allow such calculations with a minimum of experimentation.

At times the effect is more complicated (Fig. 6), such as showing nicks in the curve, i.e., possessing two line segments. This would imply that the basic equations (3.51) and (3.52) do not apply. This could, for instance, be due to aggregation or micellarization of the drug substance, which might be effected by buffer concentrations. The aggregates may have different degradation rate constants.

When the base-ten logarithms of the k values are plotted versus pH, then the so-called kinetic pH profiles result. The data from Maulding et al. (1975) are plotted in this fashion in Fig. 2.

When the pH is sufficiently low, then

$$k_{obs} = k_{+}[H^{+}] \tag{3.53}$$

or base-ten logarithmically,

$$\log[k_{obs}] = \log[k_{+}] - pH \tag{3.54}$$

Such a plot should be linear in pH with a slope of -1. Figure 4 shows that this is so in the low-pH region.

At high pH, only the last term in k_{obs} is important, and

$$k_{obs} = k_{-}[OH^{-}] = \frac{k_{-}K_w}{[H^{+}]} \tag{3.55}$$

or base-ten logarithmically,

$$\log[k_{obs}] = \log[k_{+}] - 14 + pH \tag{3.56}$$

so that a plot of log k_{obs} versus pH should be linear with a slope of +1 in the high-pH region.

For intermediate pH values, there can be equal weight of the terms $k_+[H^+]$ and $k_-[OH^-]$, or the term k_0 may predominate. In such a case there will be a plateau where $k_0 \gg k_+[H^+]$ and $k_-[OH^-]$.

2.1 Is the Optimum pH the Same as the pK?

There is a thought that occurs from time to time as to whether, in a simple scheme as above, the optimum pH would equal that of the pK_a of the compound. The answer to this question is no, and the failure of the question is exemplified for a situation where

$$HA + H^+ \rightarrow \text{products} \tag{3.57}$$

and

$$A^- + OH^- \rightarrow \text{products} \tag{3.58}$$

We here have two straight lines as shown in Fig. 7. The equations for the lines are

$$k_{obs} = k_+[H^+][HA] \tag{3.59}$$

or

$$\log[k_{obs}] = -pH + \log[k_+] + \log[HA] \tag{3.60}$$

where at low pH the concentration of HA is approximately equal to the analytically determined concentration of drug, and at high pH,

$$k_{obs} = k_-[OH^-][A^-] \tag{3.61}$$

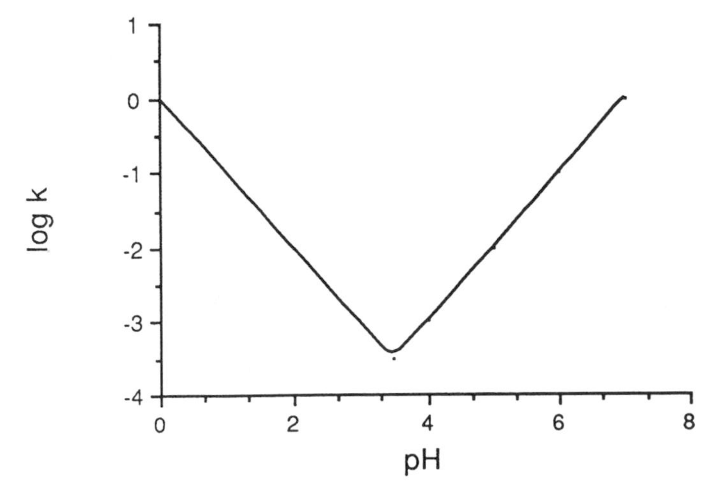

Fig. 7 A V-shaped, achematic pH profile. An example of this is the work by King et al. (1992). In reality, there is a gradual transition at the minimum, not a sharp break.

or

$$\log[k_{\mathrm{obs}}] = 14 - \mathrm{pH} + \log[k_-] + \log[\mathrm{A}^-] \tag{3.62}$$

where at low pH the concentration of A^- is approximately equal to the analytically determined concentration of drug (see Fig. 7). The intersection of the two lines with Eqs. (3.60) and (3.61) occurs when

$$k_+[\mathrm{H}^+][\mathrm{HA}] = k_-[\mathrm{OH}^-][\mathrm{A}^-] \tag{3.63}$$

Introducing K_{w} and $K_{(\mathrm{a})}$ now gives

$$k_+[\mathrm{H}^+][\mathrm{HA}] = k_-\left\{\frac{10^{-14}}{[\mathrm{H}^+]}\right\}\left\{\frac{K_{(\mathrm{a})}[\mathrm{HA}]}{[\mathrm{H}^+]}\right\} \tag{3.64}$$

or

$$[\mathrm{H}^+] = 10^{-5}\left\{\frac{10K_{(\mathrm{a})}\mathrm{k}_-}{\mathrm{k}_+}\right\}^{1/3} \tag{3.65}$$

and this is not solely a function of $K_{(\mathrm{a})}$. If, indeed, the maximum stability occurs at $\mathrm{pH} = \mathrm{pK}_{(\mathrm{a})}$, then this would be quite fortuitous.

There are many different types of pH profiles, and some of these will be discussed in detail in the following.

2.2. Temperature Effects

At each given pH, each of the rate constants (k_{i}) will adhere to the Arrhenius equation (Fig. 8). If there is a domain where one rate expression (k_{q}) is predominant, then

$$k_{\mathrm{obs}} = k_{\mathrm{q}}\phi(q) \tag{3.66}$$

where $\phi(q)$ can be unity or $[\mathrm{H}^+]$ or $[\mathrm{OH}^-]$. In such regions the Arrhenius equation can be expected to hold, but the activation energies for the various k_{q} terms may differ.

In regions of pH where more than one term is predominant then, denoting these by k_{q1} and k_{q2} it follows that

$$k_{\mathrm{obs}} = k_{\mathrm{q1}}\phi(q1) + k_{\mathrm{q2}}\phi(q2) \tag{3.67}$$

so that

$$\begin{aligned}\log[k_{\mathrm{obs}} &= \log\{k_{\mathrm{q1}}\phi(q1) + k_{\mathrm{q2}}\phi(q2)\\ &= \log\left\{Z_1\exp\left(-\frac{E_1}{RT}\right)\phi(q1) + Z_1\exp\left(-\frac{E_1}{RT}\right)\phi(q1)\right\}\end{aligned} \tag{3.68}$$

and unless

$$Z_1E_1\phi(q1) \approx Z_2E_1\phi(q1) \tag{3.69}$$

is the same over the region studied, a linear Arrhenius plot is not to be expected.

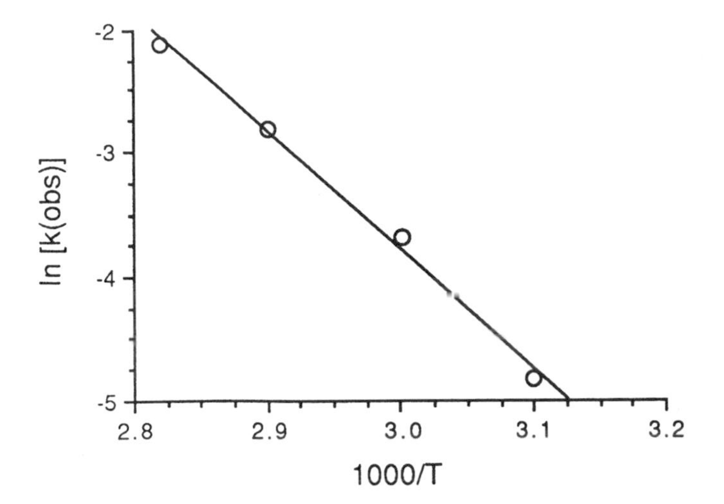

Fig. 8 Arrhenius plot of aqueous decomposition of chlordiazepoxide at pH 11. The least-squares fit of the line is $\ln[k] = 25.05 - 9.61\ (1000/T)$ ($R = 1.00$) showing an activation energy of 19.2 kCal/mol. This may be compared with the activation energy of 20 kCal/mol used in the concept of KMT. (Figure constructed from data published by Maulding et al., 1975.)

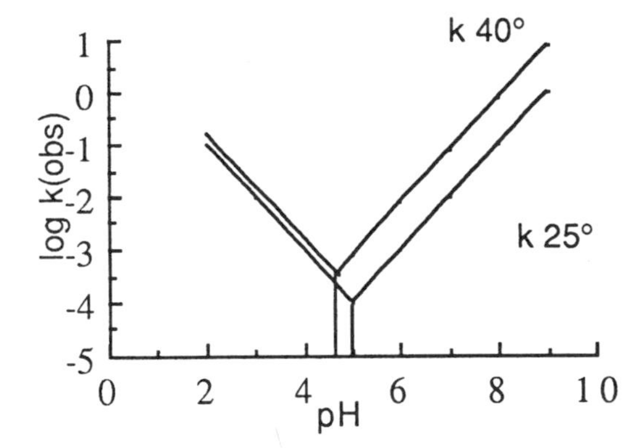

Fig. 9 Effect of temperature on optimum pH. In the figure, the activation energies are such that the value of k_+ at 40°C is 1.58 times that at 25°C, and for k_- the value at 40°C is 7.9 times that at 25°C. The pH shift is quite obvious.

Furthermore, it is not to be expected that the "optimum pH" will be the same at all temperatures. This is demonstrated in Fig. 9.

2.3. Acid or Base Catalysis

It should be pointed out that when reporting kinetic pH profiles, it is most instructive to drop out the buffer effect terms. This is often not done, at the loss of some insight into the problems at hand. There are many examples of pH profiles, e.g., Ravin et al. (1964) reported on kinetic pH profiles of carbuterol.

For pH profiles in general, obviously there are always, in buffer systems, competing reactions, since there is general acid or base catalysis (buffer catalysis) or specific acid or base catalysis ([H+] or [OH−]). If k simply denotes the bimolecular rate constant for the particular pathway (i.e., either general or specific), and if p is the number of nonequivalent protons, dissociable on the conjugate acid bade, and q the amount of sites of the general base that can accept a proton, then the relation

$$\log\left[\frac{k}{q}\right] = \log GB + \beta\left[\mathrm{pK_a} + \log\left(\frac{p}{q}\right)\right] \tag{3.70}$$

is denoted the Brønsted relation. β is the Brønsted coefficient and GB is a constant. As mentioned, $\mathrm{pH_a}$ is the negative logarithm of the conjugate acid dissociation constant (and for hydroxide attack, for instance, is $\mathrm{pK_w} - \mathrm{pH}$).

Most treatises deal with acid catalysis. Kirsch and Notari (1984) have dealt with the theoretical basis for the detection of general base catalysis in the presence of predominating hydroxide catalysis. This is difficult for two reasons. Firstly, the pH range were the stability is the least is also the range where the specific catalysts (i.e., [OH]) predominate over buffer catalysts. Most often buffer catalysis is completely overlooked in this region.

As a general statement, it can seen from the above that when the degradation of a drug is accelerated by general and specific base catalysis, then the pseudo-first-order rate constant for loss of drug, k_{obs}, is given by

$$k_{obs} = k^0_{OH}[\mathrm{OH}]\left\{\frac{f_d f_{OH}}{f^{¥}}\right\} + k^0_B[\mathrm{B}]\left\{\frac{f_d f_B}{f^{¥}}\right\} \tag{3.71}$$

where k^0_{OH} is the rate constant (bimolecular) for specific base attack at zero ionic strength, k^0_B is the rate constant (bimolecular) for general base attack, and the terms in square brackets are hydroxyl and specific base concentrations. The f terms are activity coefficients, and ¥ denotes transition state.

General base cannot be distinguished from specific hydroxide and general acid catalyses. The latter is defined as

$$k_{obs(gb)} = \left[\frac{k^0_{cat}K_a}{K_w}\right][\mathrm{HB}][\mathrm{OH}]\left\{\frac{f_d f_{HB} f_{OH}}{f_{¥}}\right\} \tag{3.72}$$

where K_a and K_w denote buffer and water dissociation constants, and where HB denotes the acid conjugate to the base. If this is divided by k_{obs} then the fraction, F_{gb}, due to general base reaction is obtained. Introducing the Brønsted relation into this then gives

$$F_{gb} = \frac{K_a[\mathrm{HB}][(p/q)K_w/(K_a a_{H_2O})]^{\beta}}{q(K_a[\mathrm{HB}][(p/q)K_w/(K_a a_{H_2O})]^{\beta}) + K_w \cdot \{f'_{¥}/f_{HB}f_{¥}\}} \tag{3.73}$$

This term as a function of the $\mathrm{pK_a}$ values of a series of buffers has been calculated by Kirsch and Notari (1984). When the Brønsted coefficient is close to unity, the equation simplifies to

$$F_{gb} = \frac{[HB]}{(1/p) + [HB]} \tag{3.74}$$

Kirsch and Notari (1984) tested the equation using the conversion in carbonate buffers at 60°C and unit ionic strength of ancitabine prodrug. The study was carried out at unit ionic strength. Their data showed excellent correlation between observed values of F_{gb} and those calculated.

3. pH PROFILES FOR THE SITUATION WHERE $HA + H^+$ AND $A^- + H^+ \rightarrow$ PRODUCTS

In the following, an acid will be denoted HA and a base B. In the former case there is a possibility that HA decomposes and A^- does not, and the possibility that both decompose. (H_2A^+ and $A^=$ are also prone to decomposition in certain cases as will be seen in the following). A similar situation exists with a base, B, where both BH+ and B (and in cases BH_2^{++} and B^-) may decompose, others where only one decomposes. There is, basically no difference between the HA and the B terminology, but it is maintained in the following for convenience. Equations derived for one will apply to the other as well. In the immediately following text the species HA will be considered.

If an acid is only hydrogen-ion catalyzed, then

$$HA + H^+ + (H_2O) \rightarrow \text{products} \tag{3.75}$$

and

$$A^- + H^+ + (H_2O) \rightarrow \text{products} \tag{3.76}$$

The quantity monitored, analytically, is the drug concentration, C, given by

$$C = [HA] + [A^-] \tag{3.77}$$

where [HA] and $[A^-]$ are related to one another by the acid equilibrium constant $K_{(a)}$, of the compound:

$$K_{(a)} = \frac{[A^-][H^+]}{[HA]} \tag{3.78}$$

or

$$[HA] = \frac{[A^-][H^+]}{K_{(a)}} \tag{3.79}$$

From this it follows that

$$[A^-] = C - [HA] = C - \frac{[A^-][H^+]}{K_{(a)}} \tag{3.80}$$

from which

$$[A^-] = C\frac{K_{(a)}}{[H^+] + K_{(a)}} \tag{3.81}$$

and

$$[HA] = C - [A^-] = C\frac{[H^+]}{[H^+] + K_{(a)}} \tag{3.82}$$

These expressions are introduced into the rate equation:

$$\begin{aligned} \text{Rate} &= -\frac{dC}{dt} = -(k_{1+})[HA][H^+] - (k_{2+})[A^-][H^+] \\ &= [H^+](k_{1+})C\frac{[H^+]}{[H^+] + K_{(a)}} + (k_{2+})C[H^+]\frac{K_{(a)}}{[H^+] + K_{(a)}} \\ &= -k_{obs}C \end{aligned} \tag{3.83}$$

Hence the reaction is apparent first order with an observed rate constant, k_{obs}, of

$$k_{obs} = \frac{k_{1+}[H^+]^2}{[H^+] + K_{(a)}} + \frac{k_{2+}K_{(a)}[H^+]}{[H^+] + K_{(a)}} \tag{3.84}$$

when plotted in this fashion, curvature will be followed with the appropriate fitting of parameter values (k_{1+}, k_{2+}, and K_a). The latter is usually known, independently.

The last term in Eq. (3.84) can be written

$$\frac{k_{2+}K_{(a)}}{1 + K_{(a)}/[H^+]} \tag{3.85}$$

At low $[H^+]$ this term vanishes, and the first term becomes

$$k_{obs} = \frac{k_{1+}[H^+]^2}{[H^+] + K_{(a)}} \tag{3.86}$$

i.e., when $[H^+] \ll K_{(a)}$,

$$k_{obs} \approx k_{1+}[H^+] \tag{3.87}$$

as expected, and a similar argument will show that at high $[H^+]$ expression becomes

$$k_{obs} = k_{2+}[H^+] \tag{3.88}$$

The logarithmic forms of Eqs. (3.87) and (3.88) are

$$\log[k_{obs}] = \log[k_{1+}] + \log[H^+] = \log[k_{1+}] - \text{pH} \tag{3.89}$$

and

$$\log[k_{obs}] = \log[k_{2+}] + \log[H^+] = \log[k_{2+}] - \text{pH} \tag{3.90}$$

so that a plot of $\log[k_{2+}]$ versus pH will be two line segments (with different intercepts) both with a slope of minus one. This is the type profile which later (in Fig. 8) is denoted type JDEN.

A similar situation where [HA] is only hydroxyl ion catalyzed will give profile of type AHJK (Fig. 11).

Of course, when more species are involved, then the kinetic expressions can become quite involved (e.g., the work on decomposition of alprazolam by Cho et al., 1983) and frequently include several terms of the type

$$k_1 K \frac{1}{K_+[H^+]} \tag{3.91}$$

3.1. Specific Acid/Base Catalysis of One Species (HA + H^+ and HA + OH^- → Products)

Another instance is the specific acid/base catalysis of a compound present as only one species. The reaction stated would be

$$HA + H^+ + (H_2O) \rightarrow \text{products} \tag{3.92}$$

and

$$HA + OH^- + (H_2O) \rightarrow \text{products} \tag{3.93}$$

As shown in Eq. (3.82),

$$[HA] = C - [A^-] = C\frac{[H^+]}{[H^+] + K_{(a)}} \tag{3.94}$$

The rate equation is

$$-\frac{dC}{dt} = k_+[HA][H^+] + k_-[HA][OH^-] \tag{3.95}$$

Introducing Eq. (3.82) and the expression $[OH^-] = K_w/[H^+]$ into Eq. (3.95) gives

$$-\frac{dC}{dt} = k_+ C\left\{\frac{[H^+]^2}{[H^+] + K_{(a)}} + \frac{k_- K_w C}{[H^+] + K_{(a)}}\right\} \tag{3.96}$$

i.e., it is first order in C with an apparent rate constant of

$$k_{obs} = \frac{k_+[H^+]^2}{[H^+] + K_{(a)}} + \frac{k_- K_w}{[H^+] + K_{(a)}} \tag{3.97}$$

where K_w is water's ionization constant.

At high pH values, $[H^+] \ll K_a$ and the last term will dominate, i.e., $[H^+]^2/([H^+] + K_{(a)}) = [H^+]/(1 + K_{(a)}/[H^+])$ is small because $1 + K_{(a)}/[H^+]$ is large and $[H^+]$ is not large. The last term then is

$$\frac{k_- K_w}{[H^+] + K_{(a)}} \approx \frac{k_- K_w}{K_{(a)}} \tag{3.98}$$

i.e.,

$$\log[k_{obs}] = \log\left[\frac{k_- K_w}{K_a}\right] \tag{3.99}$$

i.e., is constant.

At low pH (high $[H^+]$) the last term in Eq. (3.96) vanishes and the equation becomes

$$k_{obs} = \left[\frac{k_+[H^+]^2}{[H^+] + K_{(a)}}\right] \approx k_+[H^+] \tag{3.100}$$

or

$$\log[k_{obs}] = \log[k_+] + \log[H^+] = \log[k_+] - \text{pH} \tag{3.101}$$

and at high pH (low $[H^+]$) the first term in Eq. (3.96) vanishes and it becomes

$$k_{obs} = \frac{k_- K_w}{[H^+] + K_{(a)}} \approx \frac{k_- K_w}{K_{(a)}} \tag{3.102}$$

i.e., is at a plateau, since the right-hand side is a constant.

At intermediate pH values the last term will predominate, i.e.,

$$k_{obs} = \frac{k_- K_w}{[H^+] + K_{(a)}} \approx \frac{k_- K_w}{[H^+]} \tag{3.103}$$

i.e.,

$$\log[k_{obs}] = \log[k_- K_w] - \log[H^+] = \log[k_- K_w] + \text{pH} \tag{3.104}$$

i.e., the pH profile will have slope -1 at low pH, plus slope 1 at intermediate slope, and be horizontal at high pH. This corresponds to a profile denoted ABCDE at a later point (Fig. 11).

An example of this is pralidoxine (Ellin, 1958; Ellin et al., 1964; Ellin and Wills, 1964).

4. HYDROLYSES OF ACIDIC DRUGS AT A GIVEN pH

Most decompositions in solution are, as mentioned, hydrolyzes. If the drug substance is an acid, then in the final product the pH will be fixed. The situation is complicated by the number of species in solution and by the fact that the pH is usually controlled (kept constant) by the addition of buffers.

The effect of pH will be discussed in the subsequent section; suffice it to say here that in cases where both undissociated acid (HA) and anion (A^-) are fairly soluble, the following holds:

$$HA + H_2O \rightarrow E \tag{3.105}$$

with rate constant k, and

$$A^- + H_2O \rightarrow E' \tag{3.106}$$

with rate constant k'. E and E′ stand, globally, for decomposition product. The quantity measured analytically is usually the sum, C, of the concentrations of free and ionized acid, i.e.,

$$C = [HA] + [A^-] \tag{3.107}$$

The rate equations for disappearance of overall concentration, C, remaining at time t, is given by

$$\frac{dC}{dt} = -k[HA] - k'[A^-] \tag{3.108}$$

The equilibrium for the acid is

$$HA \longleftrightarrow A^- + H^+ \tag{3.109}$$

The equilibrium constant is denoted $K_{(a)}$, so

$$A^- = K_{(a)}[HA][H^+] \tag{3.110}$$

Equations (3.107), (3.108), and (3.110), if the pH is kept constant, contain four unknowns (including dC/dt) and can therefore be reduced to one equation with two unknowns. This reduction can be done in such a fashion that the unknowns are C and dC/dt, and so a differential equation results, which allows solving for C.

Inserting Eq. (3.110) into Eq. (3.107) gives

$$C = [HA] + \frac{K[HA]}{[H^+]} \tag{3.111}$$

or

$$[HA] = \frac{C}{1 + K/[H^+]} \tag{3.112}$$

Inserting Eq. (3.111) into (3.107) then gives

$$[A^-] = C - \frac{C}{1 + K/[H^+]} \tag{3.113}$$

Equations (3.111) and (3.113) are now inserted into Eq. (3.108) to give

$$-\frac{dC}{dt} = k\frac{C}{1 + K_{(a)}/[H^+]} + k'\left[C - \frac{C}{1 + K_{(a)}/[H^+]}\right] \tag{3.114}$$

which may be rearranged to read

$$\frac{dC}{C} = -k_{obs}C \tag{3.115}$$

where

$$k_{obs} = k' + \frac{k_-k'}{1 + K_{(a)}/[H^+]} \tag{3.116}$$

Equation (3.115) integrates to

$$\ln\frac{C}{C_0} = -k_{obs}t \tag{3.117}$$

which of course is first order. Changing the pH will have the effect shown in Eq. (3.116).

5. A SIMPLIFIED APPROACH TO pH PROFILES

It stands to reason that the stability of a drug substance in solution depends on its molecular environment. One of the most important macroscopic parameters of this nature is the pH, and a large volume of literature exists on this subject, in particular on the effect of pH in the acid range. A simplified approach is presented first.

In the case of hydrolysis of a nondissociating drug the decomposition is given by

$$A + H_2O \rightarrow E \tag{3.118}$$

where E, again, is a general symbol for decomposition products. The rate of the reaction is (or may be) a function of the acidity of the medium in which it occurs. If an experiment is carried out at a given pH, then this can be achieved by using a pH stat, or a buffer (HB) may be added. A buffer, by its very nature, can affect the rate, and hence the overall rate of the reaction can be written

$$\frac{d[A]}{dt} = -[A] \cdot \{k_0 + k_+[H^+] + k_-[OH^-] + k_{HB}[HB] + k_B[B^-]\} \cdot [H_2O] \tag{3.119}$$

where, for convenience, in this context, the hydrogen ion activity is equated with concentration. The subscripts to the rate constants, k, in Eq. (3.119) refer to, in the order given, the hydrolysis per se, the hydrogen ion catalyzed, the hydroxyl ion catalyzed, the buffer acid catalyzed, and the buffer anion catalyzed rate constants.

It is seen that at a given buffer concentration (and therefore at a given hydrogen and hydroxy-ion concentration) the reaction expression is

$$\frac{dA}{dt} = k[A] \tag{3.120}$$

where k is a pseudo-first-order rate constant given by

$$k = \{k_0 + k + [H^+]\} + k_-[OH^-] + k_{hb}[HB] + k_b[B^-]\} \tag{3.121}$$

Equation (3.120) predicts a first- or pseudo-first-order reaction, and Eq. (3.121) predicts the rate constant to be a function of several factors. It is conventional first to eliminate the effect of the buffer. This is done by performing the experiments at several buffer concentrations at several pH values (i.e., at several ratios of [HB] to [B]). An example of this is the work by Maulding et al. (1975) dealing with the hydrolysis of chlordiazeopxide cited earlier. Their data are shown in Fig. 4.

6. EFFECT OF BUFFER

To attain a constant pH, the most common approach is the use of buffers HB/B^-, where HB denotes an acid. Since the buffer may, as seen above, affect the kinetics, it is customary to carry out experiments where the buffer concentration is varied.

As shown in Fig. 10, this type of plot is usually linear (at low concentration of buffers), and the procedure generally used is to extrapolate the line to zero concentration to obtain the "buffer-free" rate constant, k^*.

This is done at two pH values (pH_1 and pH_2) obtained by the same buffer system. The k^* values are found from the intercepts of the straight lines at the two pH values, and the slopes yield two equations with two unknowns (k_{hb} and k_b), so that these can be calculated.

When the Briggsian (i.e. base-ten) logarithms of the k^* values are plotted versus pH, then the so-called kinetic pH profiles result. The data from Maulding et al. (1975) are plotted in this fashion in Fig. 4. k_{obs}, at given pH values, may or may not follow an Arrhenius equation. A case where they do so is shown in Fig. 8.

There are many different types of pH profiles. As has been seen in some cases, when the pH is sufficiently low, then

$$k^* \approx k_+[H^+] \tag{3.122}$$

or, logarithmically,

$$\log_{10}[k^*] = \log_{10}[k_+] - pH \tag{3.123}$$

Such a plot should be linear in pH with a slope of -1.

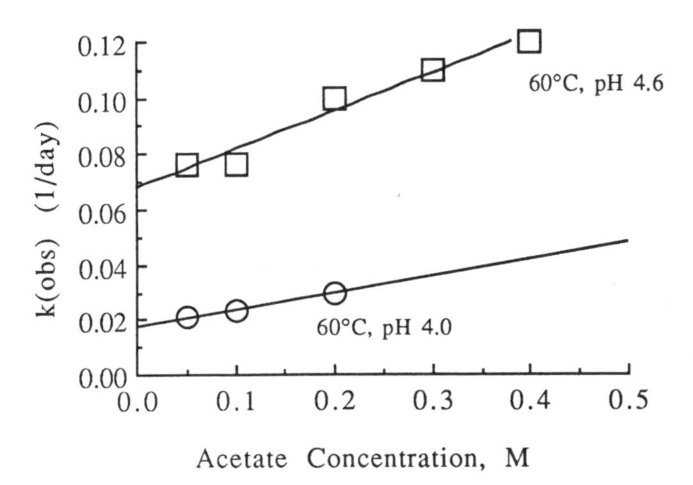

Fig. 10 Graph constructed from data published by Fassberg and Stella (1992) dealing with the hydrolysis of campothectin. Similar examples are those of Maulding et al. (1975) and of Ito et al. (1997).

At high pH, only the last term in k^* is of importance, and

$$k^* \approx k_-[OH^-] = \frac{k_- 10^{-14}}{[H^+]} \tag{3.124}$$

or

$$\log_{10}[k^*] = \log_{10}[k_-] + pH - 14 \tag{3.125}$$

i.e., a plot of log k^* versus pH should be linear with a slope of +1. An example of this is the hydrolysis of 3-quinuclidinyl benzilate in dilute aqueous solution reported by Hull et al. (1979). For pH values in between there can be an equal weight of the two terms $k^+[H^+]$ and $k^-[OH^-[$, or, when there is a plateau, the term k_0 may dominate.

It should be pointed out that when reporting kinetic pH profiles, it is most instructive to drop out the buffer effect terms. This is often not done in practice, at the loss of some insight into the problems at hand.

There are many examples of the above, for instance the kinetic pH profiles of carbuterol reported on by Ravin et al. (1964). Hou and Poole (1969a,b) showed the traditional buffer effects in the hydrolysis of ampicillin in solution at different pH values, which enabled them to calculate the catalytic effect of the different buffer species.

7. GENERAL AND SPECIFIC ACID AND BASE CATALYSIS

For pH profiles in general, there are always, in buffer systems, competing reactions, since there is general acid or base catalysis (buffer catalysis) or specific acid or base catalysis ($[H^+]$ or $[OH^-]$).

In this case, denoting by HA and A the nonionic and ionic form of the drug substance, presuming it to be an acid, the acid equilibrium would be (omitting, for convenience, water and using hydrogen rather than hydronium ion, and using concentrations):

$$HA \longleftrightarrow H^+ + A^- \tag{3.126}$$

with equilibrium constant

$$K_{(a)} = \frac{[H^+][A^-]}{[HA]} \tag{3.127}$$

The two species hydrolyze with pseudo-first-order rate constants k_{HA} and k_A, i.e., using the notation $D = d/dt$,

$$D[HA] = -k_{HA}[HA] \tag{3.128}$$

and

$$D[A^-] = -k_A[A^-] \tag{3.129}$$

The analytically measured concentration is C, given by

$$C = [HA] + [A^-] \tag{3.130}$$

Rearranging Eq. (3.127) gives

$$[A^-] = \frac{K_{(a)}[HA]}{[H^+]} \tag{3.131}$$

Introducing this into Eq. (3.130) gives

$$C = [HA] + \frac{K_{(a)}[HA]}{[H^+]} = [HA]\left\{1 + \frac{K_{(a)}}{[H^+]}\right\} \tag{3.132}$$

hence

$$[HA] = \frac{C}{1 + K_{(a)}/[H^+]} \tag{3.133}$$

The sum of Eqs. (3.128) and (3.133) is

$$DC = -k_{HA}[HA] - k_A[A^-] = -k_{HA}[HA] - k_A\{C - [HA]\} \tag{3.134}$$

$$DC = (k_{HA} - k_A)[HA] - k_A C \tag{3.135}$$

where Eq. (3.130) has been used for the next-to-last step. Introducing Eq. (3.134) into Eq. (3.135) now gives

$$\begin{aligned} DC &= \frac{k_{HA} - k_A C}{1 + K_{(a)}/[H^+]} - k_A C \\ &= -Q \cdot C = \frac{dC}{dt} \end{aligned} \tag{3.136}$$

where

$$Q = \frac{k_{HA} - k_A}{1 + K_{(a)}/[H^+]} - k_A \tag{3.137}$$

The above is the simplest of all treatments of the situation described in the heading. A good example of this is the work by Longhi et al. (1994). If the rate constants k_{HA} and k_A are functions of $[H^+]$, $[OH^-]$, and buffer concentration, then the expressions becomes quite complicated. Often some of these are of importance, others not, and in these cases it there will always be one term of the type $k/\{1 + K/[H+]\}$.

For instance for the decomposition of moxalactam Hashimoto et al. (1984, 1985) arrived at the equation

$$\log k = k_+[H^+] + k_0 + k_d\frac{1}{(1 + [H^+]/K} + k_-[OH^-] \tag{3.138}$$

where k_0 and k_d represent the first-order rate constants for the water catalyzed degradation of dianionic and trianionic moxalactam. This curve will give the usual -1 slope in the acid (in this case below pH 2) region, and a $+1$ slope in the alkaline (in this case above pH 8) region.

When the processes become more complicated, then the kinetic expressions can become quite involved (e.g., the work on decomposition of alprazolam by Cho et al., 1983).

8. THE GENERAL PROCEDURE

To arrive at an expression for the "buffer-free" observed rate constant as a function of pH, the general procedure is as follows:

1. To write up all the species in question. These in general consist of buffer (HB) and drug (HA) species.
2. To write up all the equilibrium constants existing between various ionic forms of the same species.
3. To express the analytical concentration in the form of the species present.
4. To write up all possible reactions ($HA \rightarrow$product, $HA + H^+ \rightarrow$ product, $HA + OH^- \rightarrow$ product, and so on).
5. To eliminate those reactions that are not pertinent. It will be seen later that in the absence of a kinetic salt effect, reactions of two charged species (e.g. $A^- + H^+ \rightarrow$ product) are unlikely.
6. To write the rate constants for the various drug species.
7. This should produce n equations with $n+1$ unknowns. $1/dt$ is usually denoted D and treated as a constant, and the equations are solved for each species. These can be reduced in traditional fashion to one equation with two unknowns (dC/dt and C), and this is the differential equation that is sought.

When the equation for the measured species as a function of pH is arrived at (and it can be quite complicated), then there will be M rate constants, and $M+K$ data points (i.e., K degrees of freedom).

The data are then fitted by nonlinear programs in a computer, or curves are generated with different values of the M rate constants until one set of values is found that fits the data.

These curves are frequently called "theoretical" curves, but they are indeed but best fits. It is quite possible that there is some reaction with a rate constant that is not taken into account, or conversely that there is one more reaction accounted for than necessary.

The shape of the profile can vary and will have the shape of one of the line segments in Fig. 11. For instance the profile shown in Fig. 4 would be of the type ABCDE.

The various schemes associated with different profile shapes will be discussed in the following. Van der Houwen et al. (1988, 1991, 1994, 1997) have analyzed the general procedure and called to our attention flaws in many interpretations on the subject. They have summarized the possible situations by considering the situations where 0, 1, 2, 3, and 4 protolytic equilibria are at hand. In their treatise as well as here, the subscripts 0, 1, 2, etc. are used as terminology for the successive deprotonation, e.g., subscript 0 is used for the fully protonated species. For an amine, $RNH^+ \leftrightarrow RH + H^+$ would have an equilibrium constant of K_0 and superscripts H, S, and OH are used for hydrogen ion, solvent, and hydroxyl ion catalysis, so that for the fully charged amine referred to its rate constant with hydrogen ion would be designated k_0^{H}. Van der Howen et al. (1997) take into account kinetically indistinguishable situations, and for this purpose they introduce the terms

$$M_0 = k_0^{\mathrm{H}}[\mathrm{H}^+] \tag{3.139}$$

$$M_1 = k_0^{\mathrm{S}}[\mathrm{H}^+] + k_1^{\mathrm{H}}[\mathrm{H}^+] \tag{3.140}$$

$$M_2 = k_0^{\mathrm{OH}}K_{\mathrm{w}} + k_1^{\mathrm{S}} + k_2^{\mathrm{H}}K_{\mathrm{a1}}K_{\mathrm{a2}} \tag{3.141}$$

$$M_3 = k_1^{\mathrm{OH}}K_{\mathrm{w}}K_{\mathrm{a1}} + k_2^{\mathrm{S}}K_{\mathrm{a1}}K_{\mathrm{a2}} + k_3^{\mathrm{H}}K_{\mathrm{a1}}K_{\mathrm{a2}}K_{\mathrm{a3}} \tag{3.142}$$

$$M_4 = k_2^{\mathrm{OH}}K_{\mathrm{w}}K_{\mathrm{a1}}K_{\mathrm{a2}} + k_3^{\mathrm{S}}K_{\mathrm{a1}}K_{\mathrm{a2}}K_{\mathrm{a3}} \tag{3.143}$$

$$M_5 = k_3^{\mathrm{OH}}K_{\mathrm{w}}K_{\mathrm{a1}}K_{\mathrm{a2}} \tag{3.144}$$

$$M_6 = k_4^{\mathrm{OH}}K_{\mathrm{w}}K_{\mathrm{a1}}K_{\mathrm{a2}}K_{\mathrm{a4}} \tag{3.145}$$

and the applicable equations are listed in Scheme I.

$$k_{\mathrm{obs}} = M_0[\mathrm{H}^+] + M_1 + \frac{M_2}{[\mathrm{H}^+]} \tag{3.146}$$

$$k_{\mathrm{obs}} = \frac{M_0[\mathrm{H}^+] + M_1 + \{M_2/[\mathrm{H}^+]\} + \{M_3/[\mathrm{H}^+]^2\}}{1 + \{K_{\mathrm{a}}/[\mathrm{H}^+]\}} \tag{3.147}$$

$$k_{\mathrm{obs}} = \frac{M_0[\mathrm{H}^+] + M_1 + \{M_2/[\mathrm{H}^+]\} + \{M_3/[\mathrm{H}^+]^2\} + \{M_4/[\mathrm{H}^+]^3\}}{1 + \{K_{\mathrm{a1}}/[\mathrm{H}^+]\} + \{K_{\mathrm{a1}}K_{\mathrm{a2}}/[\mathrm{H}^+]^2\}} \tag{3.148}$$

$$k_{\mathrm{obs}} = \frac{M_0[\mathrm{H}^+] + M_1 + \{M_2/[\mathrm{H}^+]\} + \{M_3/[\mathrm{H}^+]^2\} + \{M_4/[\mathrm{H}^+]^3\} + \{M_5/[\mathrm{H}^+]^4\}}{1 + \{K_{\mathrm{a1}}/[\mathrm{H}^+]\} + \{K_{\mathrm{a1}}K_{\mathrm{a2}}/[\mathrm{H}^+]^2\} + \{K_{\mathrm{a1}}K_{\mathrm{a2}}K_{\mathrm{a3}}/[\mathrm{H}^+]^3\}} \tag{3.149}$$

$$k_{\mathrm{obs}} = \frac{M_0[\mathrm{H}^+] + M_1 + \{M_2/[\mathrm{H}^+]\} + \{M_3/[\mathrm{H}^+]^2\} + \{M_4/[\mathrm{H}^+]^3\} + \{M_5/[\mathrm{H}^+]^4\} + \{M_6/[\mathrm{H}^+]^5\}}{1 + \{K_{\mathrm{a1}}/[\mathrm{H}^+]\} + \{K_{\mathrm{a1}}K_{\mathrm{a2}}/[\mathrm{H}^+]^2\} + \{K_{\mathrm{a1}}K_{\mathrm{a2}}K_{\mathrm{a3}}/[\mathrm{H}^+]^3\} + \{K_{\mathrm{a1}}K_{\mathrm{a2}}K_{\mathrm{a3}}/[\mathrm{H}^+]^4\}} \tag{3.150}$$

9. pH PROFILES IN STABILITY WORK

In preliminary stability work, the existing problem is usually shortage of drug substance. Another problem that often plagues the preformulation pharmacist/chemist is that the assay may not be sufficiently developed to, e.g., label all the decomposition products. If an assay is HPLC, a decomposition product could hide under the parent peak, and kinetic plots still appear first order. (TLC should always be resorted to guard against this possibility.)

A pH profile accounting for buffer effect (i.e., extrapolating to zero buffer concentration) can be done, even with small amounts of drug, but kinetic salt effect and the finer points of reaction mechanisms cannot. It suffices, in any event, to establish the best buffer, and the optimum pH range, and to produce a fairly good pH profile. If the terminal slopes are +1 and −1, then of course the simple acid and base catalyses have been established in those regions.

The pseudo-first-order of the reaction has also been established allowing reasonable extrapolations, and the types of conclusions needed for stability work. It will also aid in the setting of expiration periods.

As work is completed, it is good to repeat the mechanistic work in more detail, and to elucidate mechanisms. After a drug is introduced into the market place, at most one year should be allowed to pass before a detailed pH profile is known (and preferably submitted for publication).

The drug in the following will be symbolized as HA (or, if it is a zwitterion, $HA^{\pm}$), its anion A^-, and its positively charged species H_2A^+. An alternate symbol presentation also used (usually when the compound is, in general, considered a base) is to denote it B and to denote the ionized species HB^+.

The patterns in Fig. 11 are drawn as line segments, but of course, in reality the "corners" are smooth. But for schematizing purposes, the lines in Fig. 11 will suffice. In areas where one term predominates in the k_{obs} expression, this is quite acceptable, but when two terms become predominant, then it oversimplifies the actual profile. In attempting more exacting expressions for these situations, the complication is that

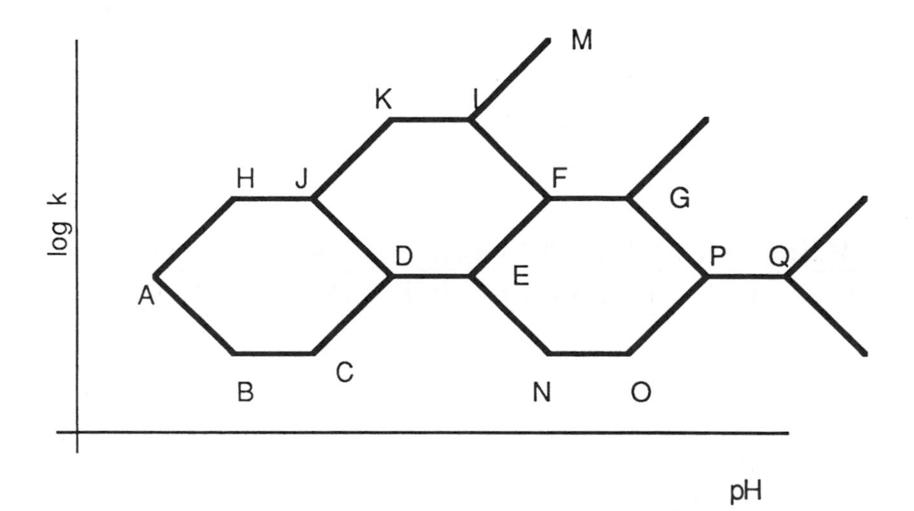

Fig. 11 Generalized pH profile polygon.

the analytically observed species, C, is not equal to either $[A^-]$ or [HA], for instance, but is the sum of these two terms.

9.1. Subtype AB

This type (Fig. 12) would most often be attributed to hydrogen-ion catalysis, i.e., in the simplest case, to

$$[HA] + [H^+] \rightarrow \text{products} \tag{3.151}$$

It is usually, when reported, part of a study over a rather narrow pH range and hence could be but part of a more complicated scheme. The necessity for only studying a compound in a narrow pH range can be dictated by many factors, such as solubility, for instance, Chloramphenicol (Higuchi et al., 1954) is of type AB and is studied from pH 0.7 to 1.7.

Digoxin (Khalil and El-Masry, 1978) is Type AB and is studied from pH 1 to 2.2. This type has also been reported for trifluoromethoxyphenyl cyanopropionamide by Brandl and Kennedy (1964) and is cited elsewhere. Hydrocortisone sodium phosphate is of this type (Marcus, 1960) and has been studied from pH 6 to 7.5. In this latter case the slope is not -1, and the rate is assumed to be

$$\text{rate} = k_1[H^+]^n[A^-] + k_2[H^+]^m[A^=] \tag{3.152}$$

Rifampin (Seydel, 1970) is of type AB and again was studied only in a narrow pH range (0.5–1.5).

Polyortho esters of 3,9-dibenzyloxy-3,9-diethyl-2,4,8,10-tetroxaspirol[3,5]undecane (Nguyen et al., 1984) are of type AB in a pH range of 3–8.

Longhi et al. (1994) have reported the pH profile of the subtype AB for isoxazolylnaphtoquinone. There is no ionic strength effect, so the equation, as seen, becomes simply $k = k[H^+]$.

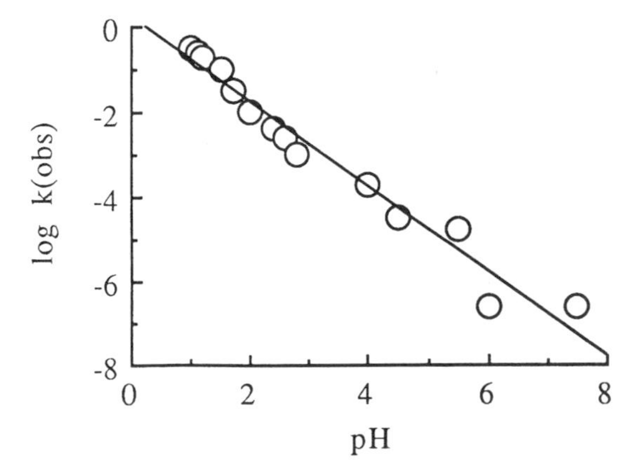

Fig. 12 (Graph constructed from data from Longhi et al. (1994).) The least squares fit is $k_{obs} = 0.23 - (1.0\text{pH})$.

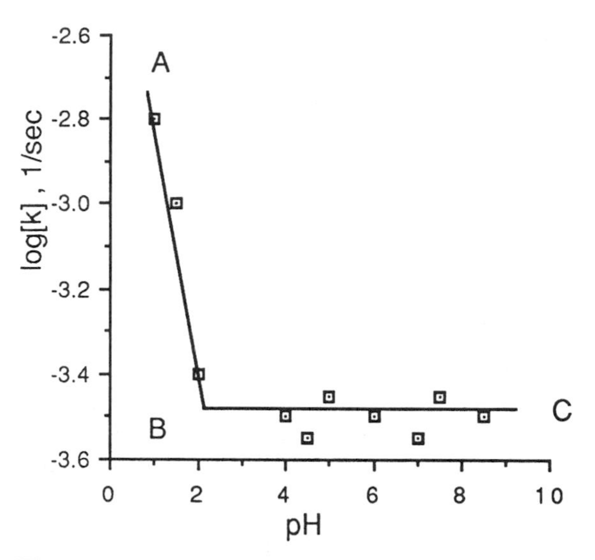

Fig. 13 k_a from dalvastin kinetics. A profile of type ABC. (Graph constructed from data published by Won, 1994.)

Type AB may simply be a part of more complicated profiles, had the study extended over a larger pH range.

9.2. Subtype ABC

A profile of type ABC (Fig. 13) is exemplified by the work published on epimerization and hydrolysis of dalvastin by Won (1994).

Lactone formation of the HMG-CoA reductase inhibitor, CI-981, which is [R-(R*,R*)]-2-(4-fluorophenyl)-β,δ-dihydroxy-5-(1-methyl-ethyl)-3-phenyl-4-[(phenylamino)carbonyl]-1H-pyrrole-1-heptanoic acid follows this scheme. The hydrolysis of this compound has been reported by Kearney et al. (1993), and is of an ABC type, with a break at pH 4.5.

This profile can often be attributed to hydrogen-ion catalysis and noncatalysis of the hydrolysis of HA, and has an observed rate constant of

$$k_{obs} = k_{+}[H+][HA] + k_1[HA] = (k_{+}[H+] + k_1)f_{HA} \tag{3.153}$$

where f_{HA} is the fraction present as HA at the pH in question. This, for instance, holds for furosemide (Cruz et al., 1979).

Promethanzine (Underberg, 1978a,b; Stavchanski et al., 1983), pH 1.5–5.5, is of this type with slope 1. This is complicated by the fact that the compound forms micelles above its critical micelle concentration.

9.3. Subtype ABCD

This type (Fig. 14), in general, is a profile where only hydrogenion and hydroxylion catalysis of one drug species plays a part. If there are horzontal parts then this is often attributable to $HA + H_2O \rightarrow$ products. This is, for instance, the case in

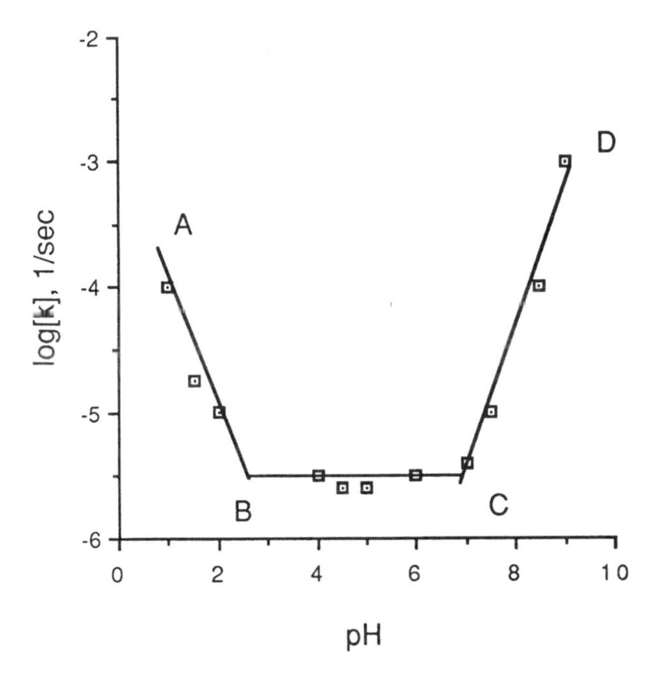

Fig. 14 Subtype ABCD. k_b from dalvastin kinetics. A profile of type ABC. (Graph constructed from data published by Won, 1994.)

meperidine and methylphenidate below. Drug substances reported to be of type ABCD include:

Acetaminophen (Koshi and Lach, 1961), pH 2–9
9-β-Arabinofuranoxyl adenine acyl migration (Anderson et al., 1985)
Amphotericin B (Hamilton-Miller, 1973), pH 4–8
Atropine (Lund and Waaler, 1968; Zvirblis et al., 1956), pH 0–10
5-Azacytidine (Notari and DeYoung 1975), pH 4–9
Benzylpenicillin (Brodersen, 1947; Finholt, et al. 1965), pH 1–11
Cefotaxime (Berge et al., 1983; Yamana and Tsuji, 1976), pH 0–10
Cephalothin (Yamana and Tsuji, 1976), pH 0–11
Cefixime (Nakami et al., 1987)
Chlorothiazide (Yamana et al., 1968), pH 3–11
Codeine sulfate (Powell, 1986), pH 0–12
Cyanocobalamine (Loy, 1952), pH 0–9
Cyclophosphamide (Hirata et al., 1967), pH 0–14
Dalvastin (Won, C. M. 1994), pH 1–10
Diazepam (Han et al., 1977), pH 0–12
Ergotamine (Kreldgaard and Kisbye, 1974), pH 2–4
Erythromycin (Amer and Takla, 1968), pH 5–10
Hydrolysis of [R-(R*, R*)]-2-(4-fluorophenyl)-β,δ-dihydroxy-5-(1-methylethyl)-3-phenyl-4-[(phenylamino)carbonyl]-1H-pyrrole-1-heptanoic acid (Kearney et al., 1993) break at pH 4.5. pH 1–9
Methylphenidate (Siegel et al., 1959) pH 1–6

Meperidine (Patel et al., 1968), pH 2–7
Methylphenidate (Siegel et al., 1959), pH 1–6
Succinyl choline chloride (Suzuki and Tanimura, 1967), pH 1–7
Sulfacetamide (Meakin et al., 1971), pH 0–13

Atropine (Lund and Waaler, 1968; Zvirblis et al., 1956) at pH below 4 attributed the decomposition to hydrogen ion catalysis, and above 6 it follows the equation

$$\text{rate} = k_{-}\{[\text{OH}^-][\text{B}]\} + k'[\text{OH}^-][\text{HB}^+] \tag{3.154}$$

In the case of phenylbutazone (Stella and Pipkin, 1976) the shape is ABCD, but the slopes are not +1 or −1. In this case there is an enolequilibrium (with equilibrium constant K^*)

$$k_{\text{obs}} = k_1 + k_2[\text{OH}^-] + \frac{k_2^*[\text{H}^+]}{([\text{H}^+] + K^*)} + \frac{k_1^* K^*}{([\text{H}^+] + K^*)} \tag{3.155}$$

Oxazepam (Han et al., 1977) is also of type ABCD, but with nonunity slopes. In this case the molecule is considered a diprotic acid (H_2A), and the fractions of diprotic, monoprotic, and uncharged species are denoted f_2, f_1, and f_0, respectively. The value for k_{obs} is then

$$\begin{aligned} k_{\text{obs}} =& k_1[\text{H}^+] \cdot f_2 + k_1'[\text{H}^+] \cdot f_1 + k_0 \cdot f_1 + k_2[\text{OH}^-] \cdot f_1 \\ &+ k_2'[\text{OH}^-] \cdot f_1 + k_2''[\text{OH}^-] \cdot f_0 \end{aligned} \tag{3.155a}$$

i.e., a combination of hydrogen or hydroxyl ion catalyzed reaction of the three protic species.

Indomethacin (Pawelczyk et al., 1979) is another case where subtype ABCD applies but where the slopes are nonunity, and this is accounted for by

$$k_{\text{obs}} = k_1[\text{HA}] + k_2[\text{H}^+][\text{HA}] + k_3[\text{A}^-] + k_2[\text{OH}^-][\text{A}^-] \tag{3.156}$$

Methicillin (Schwartz et al., 1965; Hou and Poole, 1969a,b) also shows nonunity slopes. Penethicillin (Schwartz et al., 1962) also is accounted for by this equation. Scopolamine (Smithuis, 1969; Guven and Aras, 1970) and pilocarpine (Baeschlin and Etter, 1969; Chung et al., 1970) are of type ABCD but more complicated than the remainder of the reactions quoted.

King et al. (1992) had reported the pH degradation profile of moricizine to be of type ABCD. Van der Houwen et al. (1988, 1991, 1993, 1994, 1997) reanalyzed the data and concluded that it was of type ABCDE. Fubara and Notari (1998) have shown that the degradation of cefepime in aqueous solution is of type ABCD.

9.4. Subtype ABCDE

This is type ABCD with a horizontal part at the end (Fig. 15). The simplest mechanism that this suggests is

$$HA + H^+ \rightarrow \text{products} \tag{3.157}$$

$$HA \rightarrow \text{products} \tag{3.158}$$

$$A \rightarrow \text{products} \tag{3.159}$$

$$A^- + OH^- \rightarrow \text{products} \tag{3.160}$$

Aspirin (Garrett, 1957) has a pH profile explained by this scheme. Other mechanisms accounting for the profile exist: in the case of pralidoxine (Ellin and Wills, 1964), pH 0–14, it is attributed to hydrogen and hydroxyl ion catalysis of P^+ giving rise to the equation

$$k_{\text{obs}} = \frac{k_+[H^+]^2}{[H^+] + K_a} + \frac{k_- K_w}{[H^+] + K_w} \tag{3.161}$$

At low pH where $[H^+[\gg K_a$,

$$k_{\text{obs}} = \frac{k_+[H^+]^2}{[H^+] + K_a} \approx k_{\text{obs}} = \frac{k_+[H^+]}{K_a} \tag{3.162}$$

giving rise to the minus unity slope of log k versus pH. At intermediate pH where $K_w \ll [H^+]$, it is the second term which dominates:

$$k_{\text{obs}} = k_- \frac{K_w}{[H^+] + K_w} \approx k_- \frac{K_w}{[H^+]} = k_-[OH^-] \tag{3.163}$$

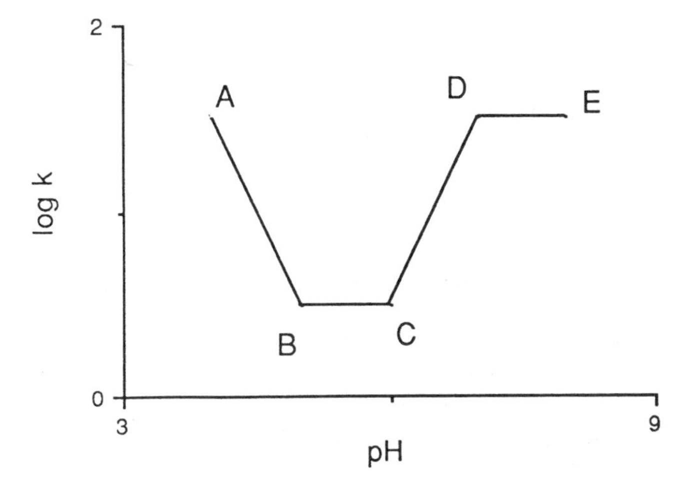

Fig. 15 Subtype ABCDE. Units on both axes are arbitrary, and the graph is intended to be an *example* only.

and hence the +1 slope of log k versus pH. At very high pH, the last term again dominates, but now $Kw/[H+] \ll k$ and the slope becomes approximately equal to k.

$$k_{obs} \approx k \tag{3.164}$$

Chlordiazepoxide (Maulding et al., 1975), pH 0 to 11, also is of this type with unity slopes in the V. Here the profile is accounted for by the reactions

$$BH^+ \rightarrow \text{products} \tag{3.165}$$

$$BH^+ + H^+ \rightarrow \text{products} \tag{3.166}$$

$$BH^+ + OH^- \rightarrow \text{products} \tag{3.167}$$

Nicotinyl-6-aminonicotinate (Wang and Patel, 1986), Nefopam hydrochloride (Tu et al., 1989), and vinpocetin (Muhammed et al., 1988) are of type ABCDE. Tu et al. (1990) have shown the pH profile of nefopam hydrochloride to be of this type.

9.5. Subtype ABCDEF

A simple explanation for a profile such as that shown in Fig. 16 is

$$HA + H^+ \rightarrow \text{products (line segment AB)} \tag{3.168}$$

$$HA \rightarrow \text{products (line segment BC)} \tag{3.169}$$

$$HA + OH^- \rightarrow \text{products (line segment CD)} \tag{3.170}$$

$$A^- \rightarrow \text{products (line segment DE)} \tag{3.171}$$

$$A^- + OH^- \rightarrow \text{products (line segment EF).} \tag{3.172}$$

Most reported reactions are of this general scheme.

Aspirin (Garrett, 1957; Edwards, 1952) is of this type. It is attributed to hydrogen ion catalyzed catalysis of the aspirin, and hydroxy ion catalyzed catalysis of the anion, and of noncatalytic hydrolysis of both aspirin and anion, so that the rate equation becomes

$$\text{rate} = k_1[HA] + k_2[HA][H^+] + k_3[A^-][OH^-] + k_4[A^-] \tag{3.173}$$

Clindamycin (Osterling, 1970; Garrett and Seyda, 1983), hydrocortisone (Bundgaard, 1977, 1980), triamcinolone (Gupta, 1983), vasopressin (Heller, 1939), and erythromycin (Amer and Takla, 1968) have profiles of the type ABCDEF.

Methyl, ethyl, and propyl parabens (Sunderland and Watts, 1984) are of this type as well, and are explained by Eqs. (3.153–3.155).

Hydrocortisone sodium succinate (Garrett, 1962) and idoxurine (Garrett et al., 1968) follow Eqs. (3.168) and (3.169) in the scheme shown above.

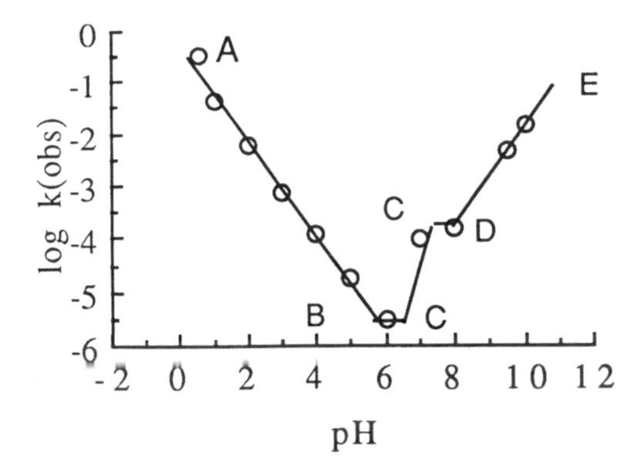

Fig. 16 Subtype ABCDEF. (Graph constructed from data published by Ismail and Simonelli, 1986, dealing with the kinetics of aspirin hydrolysis in aqueous solution of surfactants.)

Prostaglandin (Monkhouse et al., 1973), oxytetracycline (Vej-Hansen et al., 1978), and procaine (Marcus and Baron, 1959) (in the last case substituting BH^+ for HA) follow Eqs. (3.153–3.157) in the scheme above.

Van der Houwen et al. (1997) reanalyzed data published by Ugwu et al., and showed that the cyclic heptapeptide, MT=II, was of type ABCDEF.

9.6. Subtype ABCDEFG

This type (Fig. 17) could be explained by the equations above with one simple term left. This would require that there were three species that could decompose noncatalytically, one which could decompose by hydrogen ion catalysis, and one by hydroxyl ion catalysis. The profile may be explained in fewer terms, e.g., thiamin

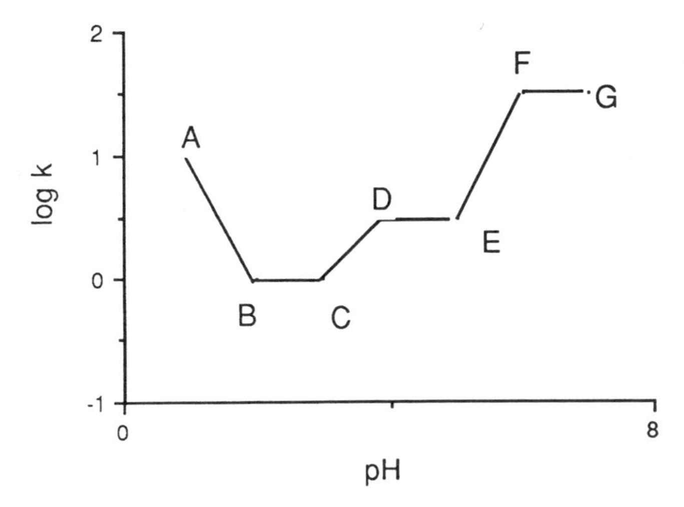

Fig. 17 Subtype ABCDEFG. Units on both axes are arbitrary, and the graph is intended to be an *example* only.

hydrochloride (Windheuser and Higuchi, 1962) is of this type and is explained by

$$BH_2^{++} + H^+ \rightarrow \text{products (line segment AB)} \tag{3.174}$$

$$BH_2^{++} \rightarrow \text{products (line segment BC)} \tag{3.175}$$

$$BH_2^{++} + OH^- \rightarrow \text{products (line segment CD)} \tag{3.176}$$

$$B^- \rightarrow \text{products (line segment DE)} \tag{3.177}$$

Ascorbic acid (Blaug and Hagratwata, 1972; Finholt et al., 1963) exhibits this type of profile, as does aztreonam (Pipkin and Davidovich, 1982; Pipkin and Barry, 1982). In the latter case it is explained by the reactions

$$H_2A^{\pm} + H^+ \rightarrow \text{products} \tag{3.178}$$

$$H_2A^{\pm} \rightarrow \text{products} \tag{3.179}$$

$$HA^- \rightarrow \text{products} \tag{3.180}$$

$$A^= \rightarrow \text{products} \tag{3.181}$$

$$A^= + OH^- \rightarrow \text{products} \tag{3.182}$$

so the rate equation takes the form

$$\text{Rate} = k_1[H_2A^{\pm}][H^+] + k_2[H_2A^{\pm}] + k_3[HA^-] + k_4[A^-][OH^-] + k_5[A^-] \tag{3.183}$$

Hydrochlorothiazide (Mollica et al., 1971) also exhibits a profile of this type.

9.7. Subtype ABCDENOP

Claudius and Neau (1998) have shown that vancomycin decomposes in aqueous solution according to apparent first-order kinetics and exhibits a subtype ABCDENOP kinetic pH profile, with minima at $B = \text{pH } 4.7$ and $O = \text{pH } 8.0$ and local maximum $E = \text{pH } 6.9$.

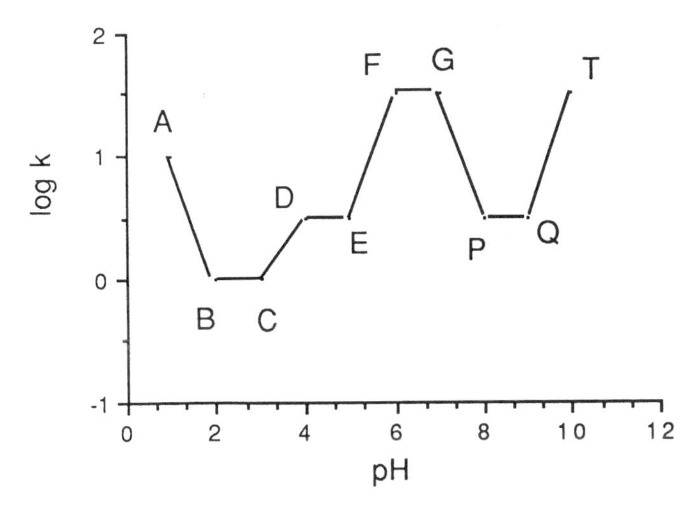

Fig. 18 Subtype ABCDEFGPQT. Units on both axes are arbitrary, and the graph is intended to be an *example* only.

9.8. Subtype ABCDEFGPQT (see Fig. 18)

This type has been reported for rolitetracycline (Vej-Hansen and Bundgaard, 1979). They are accounted for by

$$H_2B^{++} + H^+ \rightarrow \text{products (line segment AB)} \tag{3.184}$$

$$H_2B^{++} \rightarrow \text{products (line segment CD)} \tag{3.185}$$

$$HB^+ \rightarrow \text{products (line segment CD)} \tag{3.186}$$

$$B \rightarrow \text{products (line segment BC)} \tag{3.187}$$

$$B^{=} + \rightarrow \text{products (line segment DE)} \tag{3.188}$$

$$B^{=} + OH^- \rightarrow \text{products (line segment EF).} \tag{3.189}$$

9.9. Subtype AH

This subtype (Fig. 19) is hydroxyl ion catalyzed hydrolysis or decomposition. Unlike type AB, this is often typical over a long (but typical alkaline) pH range.

Apomorphine (Lundgreen and Landersjoe (1970) is of type AH over a pH range from 2.5 to 8.

Benzocaine hydrolysis (Hamid and Parrott, 1971; Higuchi and Lachman, 1955; Meakin et al., 1971; Winterborn et al., 1972; Smith et al., 1974) is type AH over a pH range of 8 to 13.

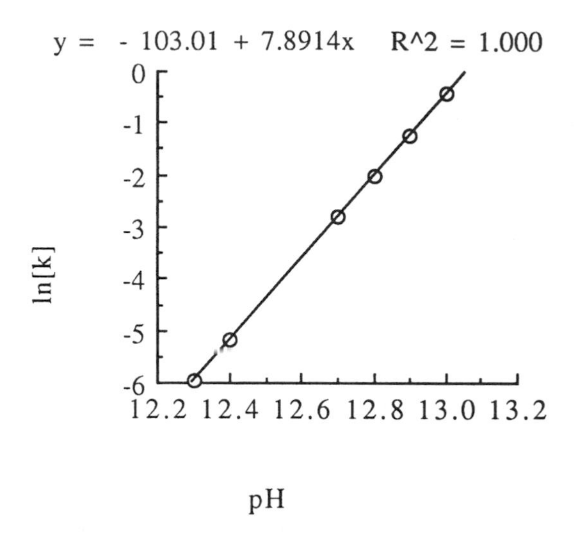

Fig. 19 Subtype AH. (Graph constructed from data published by Yang and Yang, 1994, on hydrolysis of 2-oxoquazepam.)

Amoxicillin (Bundgaard, 1977) decomposition is also of this type over a pH range of 8 to 12.5.

Epinephrine interaction with bisulfite (Higuchi and Schroeter, 1960) is of this type, but the slope fails to be +1, and the reaction is more complicated than a typical hydroxyl ion catalysis.

The compound tricyclo[4.2.2.02.5]dec-9-one-3,4,7,8-tetra-carboxylic acid diimide (Vishunuvajjala and Cradock, 1986) is of AH type with nonunity slope, and the interaction between isoniazid and reducing sugars (Devani et al., 1985) is of AH type with unity slope. Visconti et al. (1984) have studied the degradation profile of cadralazine in aqueous solution. The log k pH profile (investigated from pH 0 to pH 10) is a steadily increasing rate constant with nonunity slope. Arrhenius plots are straight lines at pH values from 5 to 10.

9.10. Subtype AHJ

Reactions of the type

$$\mathrm{HA} + \mathrm{H}^+ \rightarrow \text{products} \tag{3.190}$$

are the simplest example of this type. This gives rise to an observed rate constant of

$$k_{\mathrm{obs}} = k_+[\mathrm{H}^+][\mathrm{HA}] \tag{3.191}$$

i.e., is of the type where $k_- \approx 0$.

Chlorambrucil (Ehrson et al., 1980), promethazine (Underberg, 1978a,b; Stavchansky et al., 1983), mechlorethamine (Cohen et al., 1948), alpha-methyldopa (Sassetti and Fudenberg, 1971), and acylglucoronide (Tanaka and Suzuki, 1994) have profiles of this type also.

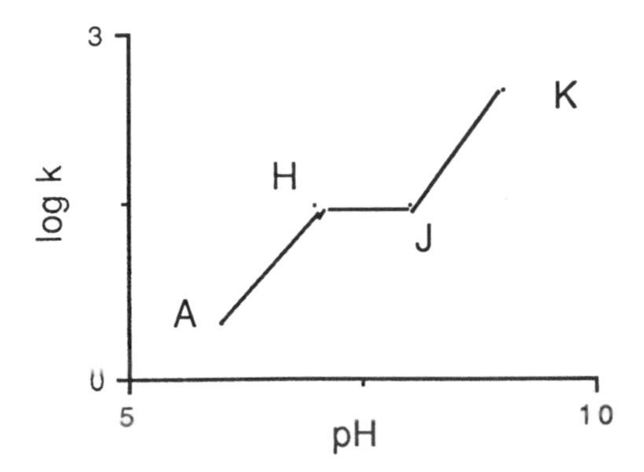

Fig. 20 Subtype AHJK. Units on both axes are arbitrary, and the graph is intended to be an *example* only.

9.11. Subtype AHJK

Methabarbital, barbital, and amobarbital (Garrett et al., 1971; Asada et al., 1980) exhibit this type (Fig. 20) of profile, and attribute it to the reactions

$$A^- \rightarrow \text{products} \tag{3.192}$$

$$HA \leftrightarrow \text{product} \tag{3.193}$$

all being OH^--catalyzed. The rate expression for this is

$$k = k_- K_w + \frac{k'_- K_a[OH^-]}{K_a + [H^+]} \tag{3.194}$$

Cyclophosphamide (Hirata et al., 1967) shows a loss of potency, methylprednisolone sodium succinate shows an acyl migration, and diethylpropion shows (a photolytically induced) decomposition of a profile of type AHJK. Hashimoto and Tanaka (1985) report moxalactam to be of this type.

9.12. Subtype AHJKL

Phenobarbital (Garrett et al., 1971) is of this type (Fig. 21), and is attributed to the rate equations

$$HA + OH^- \rightarrow \text{products} \tag{3.195}$$

$$A^= + OH^- \rightarrow \text{products} \tag{3.196}$$

i.e.,

$$k_{obs} = k_-[OH^-] f_{HA} + k_=[OH^-] f_{A=} \tag{3.197}$$

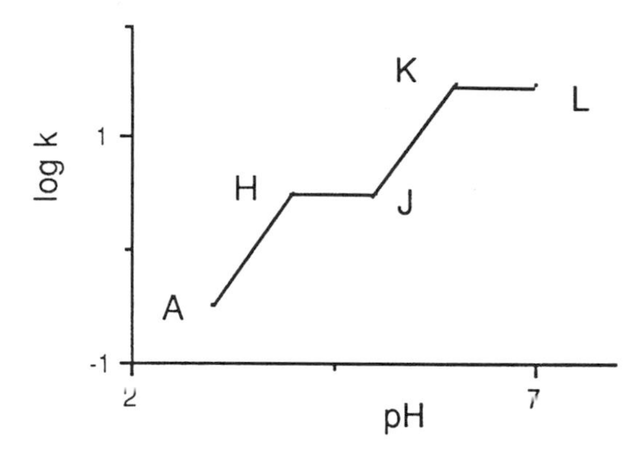

Fig. 21 Subtype AHJKL. Units on both axes are arbitrary, and the graph is intended to be an *example* only.

9.13. Subtype AHJD

An example of this type of profile is shown in Fig. 22 and has been reported on by Huang et al. (1993).

9.14. Subtype DENOPQ

Wang and Notari (1993) have shown that cefuroxime hydrolysis kinetics adhere to a DENOP pH profile. Cytarabine (Notari et al., 1972) exhibits (Fig. 22 less the last leg) this type of profile as well with unity slope, and follows the rate expression

$$k_{\text{obs}} = k_{-}[\text{OH}^{-}][\text{B}] + k_1[\text{BH}^{+}] + k_2[\text{B}]$$

Phenylbutazone (Stella and Pipkin, 1976) also exhibits this type of profile but with nonunity slope, partly due to the process being an oxidation, and partly due to enol–keto equilibria.

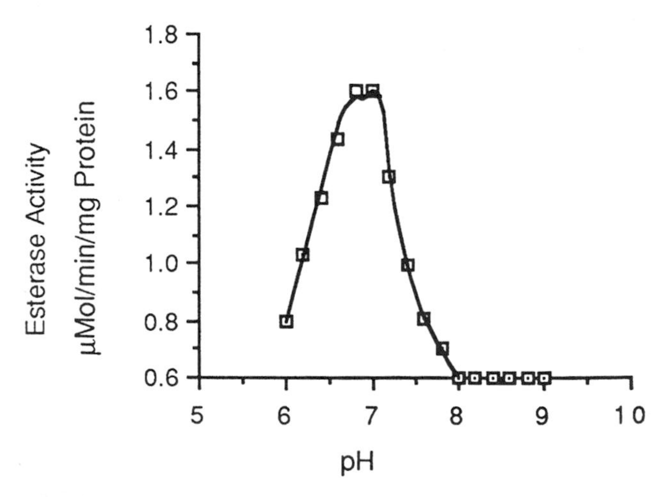

Fig. 22 Example of subtype AHJD. (Figure constructed from data by Huang et al., 1993.)

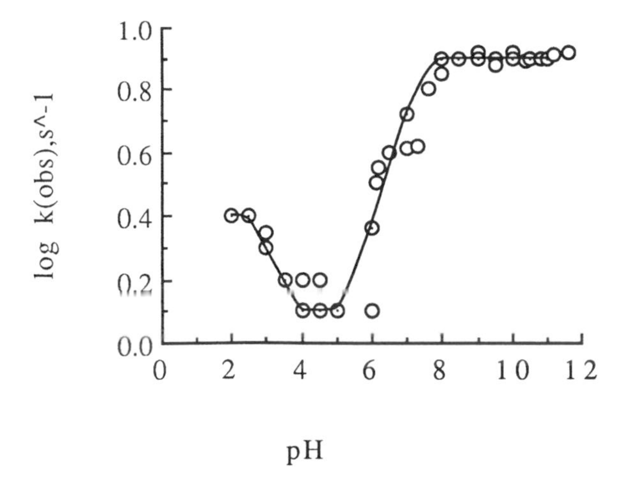

Fig. 23 Subtype DENOPQ. (Graph constructed from data published by Kurono et al., 1994.)

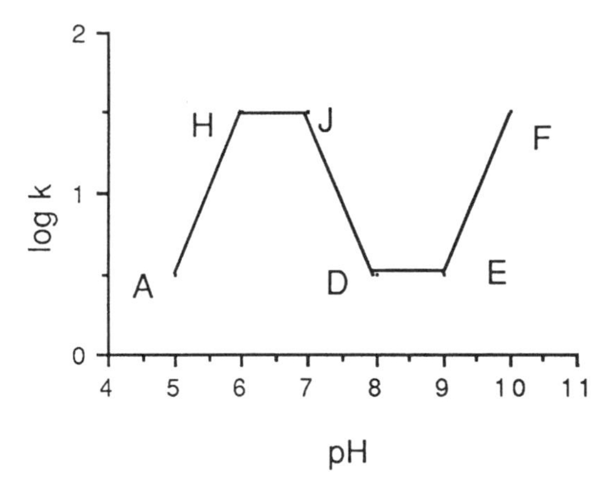

Fig. 24 Subtype AHJDEF. Units on both axes are arbitrary, and the graph is intended to be an *example* only.

Kurono et al. (1994) (Fig. 23) have published on the tautomerism kinetics of *N*-[2-{2-hydroxyethylimino(methyl)}-phenol]1-2-chlor-propamide in solution.

9.15. Subtype AHJDEF

This type of curve is illustrated in Fig. 24, and *p*-aminosalicylic acid is of this type (Liquori and Ripamonti, 1955; Tanaka and Nakagaki, 1961). Jivani and Stella (1985) reported the AHJD portion of this.

9.16. Subtype BCD

Wang and Yeh (1993) have reported that metrondiazole in solutions of pH from 2 to 10 is of the subtype BCD (Fig. 25).

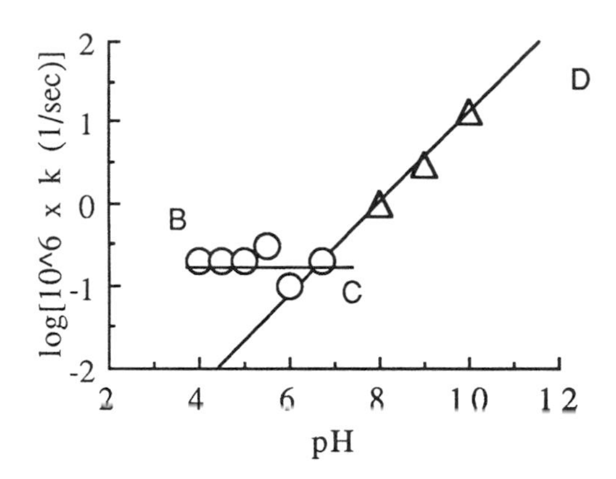

Fig. 25 Subtype BCD. Metranidazole in aqueous solution. (Graph constructed from data published by Wang and Yeh, 1993.)

Oxidation of captopril (Timmins et al., 1982; Kadin, 1982), pH 2–5, is of this type, but the slope is not −1 as in the decomposition of zenarestat 1-*O*-acylglucoronide in water at 37°C at pH 5, 6, 7.4, and 7.8, reported by Tanaka and Suzuki (1994).

Morphine oxidation (which may be better described by BCDE) is of this type and is described by

$$M + O_2 \rightarrow \text{products (rate constant } k_1) \tag{3.198}$$

$$MH^+ + O_2 \rightarrow \text{products (rate constant } k_2) \tag{3.199}$$

Echothiophate iodide (Hussain et al., 1968) pH 2–12, is of this type with slope 1 and is described by

$$k_{obs} = k_-[OH^-] + k_1 \tag{3.200}$$

Hydrocortisone-21-lysinate (Johnson et al., 1985) is of this type.

9.17. Subtype HJDE

Lee and Querijero (1985) report the rotational isomerism equilibration of *N*-thionaphthoyl-*N*-methyl glycinate to be of this type (schematized in Fig. 26).

This could typically be two noncatalytic reactions of two species of drug and one hydroxyl ion catalyzed reaction. Simpler systems exist, as will be seen in the following.

Methenamine (Tada, 1960) exhibits this type of profile, which may be ascribed to

$$BH^+ \rightarrow \text{products} \tag{3.201}$$

$$BH^+ + H^+ \rightarrow \text{products} \tag{3.202}$$

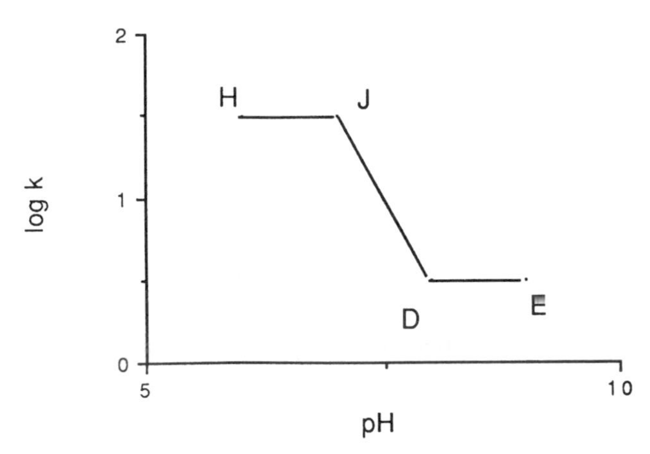

Fig. 26 Subtype HJDE. Units on both axes are arbitrary, and the graph is intended to be an *example* only.

i.e., with an observed rate constant of

$$k_{obs} = k + f_{HB} + [H^+] + k_1 f_{HB+} \tag{3.203}$$

Cycloserine (Kondrat'eva et al., 1971) follows this type of profile and is attributed to:

$$AH^{2+} \rightarrow \text{products} \tag{3.204}$$

$$AH \rightarrow \text{products} \tag{3.205}$$

$$A^{=} \rightarrow \text{products} \tag{3.206}$$

with an observed rate constant of

$$k_{obs} = \frac{k_{++}[H^+]^2 + k_+ K_{a1}[H^+] + k_1 K_{a1} K_{a2}}{[H^+]^2 + K_{a1}[H^+] + K_{a1} K_{a2}} \tag{3.207}$$

Mitomycin C (Underberg and Lingeman, 1983 a,b) decomposition involves

$$AH^{2+} + H^+ \rightarrow \text{products} \tag{3.208}$$

$$AH + H^+ \rightarrow \text{products} \tag{3.209}$$

$$A \rightarrow \text{products} \tag{3.210}$$

$$A^- + OH^- \rightarrow \text{products} \tag{3.211}$$

9.18. Subtype HJDEF

Hou and Poole (1969a,b) have shown that ampicillin, in solution, exhibits this type of pH profile.

9.19. Subtype HJKLM

Van der Houwen et al. (1997) reanalyzed data published by Quigley et al. (1994) and showed propanolol acetate to be of this type or of type HJKLM. Van der Houwen et al. (1997) reanalyzed data published by Tu et al. (1989, 1990) and showed metronidazole to be of this type or of type HJKLM.

9.20. Subtype JDEN

Oliiyai and Borchardt (1994) have shown the pH profile for Val-Tyr-Pro-Asp-gly-Ala hexapeptide to be of this type.

9.21. Subtype JDENOP

Carbenicillin (Zia et al., 1974; Hou and Poole, 1969a,b, 1971) exhibits this type of profile as does ampicillin (Hou and Poole, 1969a,b). The latter is attributed to noncatalyzed and to general acid base catalyses of the two charged species of the drug, i.e., the set of reactions

$$AH_2 + H^+ \rightarrow \text{products} \tag{3.212}$$

$$AH_2 \rightarrow \text{products} \tag{3.213}$$

$$AH^- + H^+ \rightarrow \text{products} \tag{3.214}$$

$$A^- \rightarrow \text{products} \tag{3.215}$$

$$A^- + OH^- \rightarrow \text{products} \tag{3.216}$$

and the rate becomes

$$\begin{aligned} k_{obs} &= k_{1+}[H_2A+][H^+] + k_{2+}[HA^\pm][H^+] \\ &\quad + k_3[HA^\pm] + k_4[A^-] + k_-[A][OH^-] \end{aligned} \tag{3.217}$$

Methotrexate (Chatterji and Galleli, 1978) has this type of profile and the rate equation is

$$\text{rate} = k_+[H^+] + k_1[HA] + k_-[OH^-][A^-] + k_2[A^-] \tag{3.218}$$

Carney (1988) reports closidomine to have a pH profile of this type. Kenley and Warn (1994) have reported on the acid catalyzed peptide bond hydrolysis (at 50°C) of recombinant human interleukin 11, as a profile to type JDENOP as shown in Fig. 27. (It should be pointed out that the authors consider this simply as ABCD profile, which indeed it may be.)

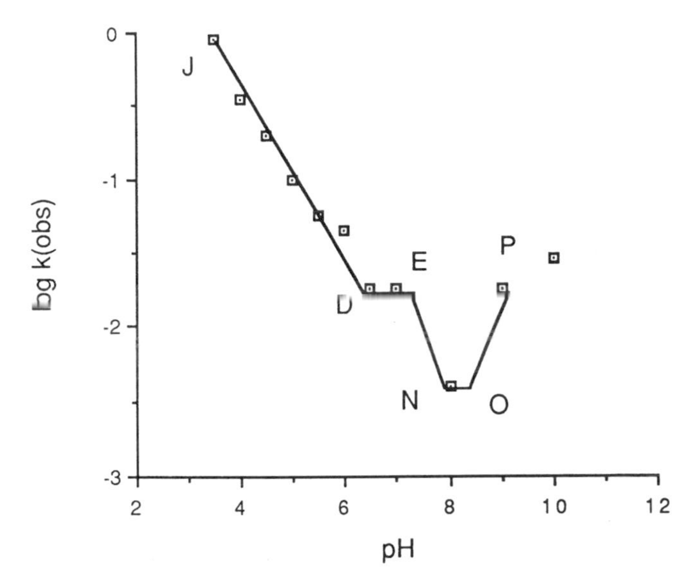

Fig. 27 JDENOP profile. (Graph constructed from data reported by Kenley and Warne, 1994.)

Ito et al. (1997) have shown that panipenem is first-order degraded in aqueous solution and that the rate constants follows this profile. They further demonstrated a lack of ionic strength effect.

10. EFFECT OF CONCENTRATION

Kinetic studies are usually carried out in quite dilute solutions. When solution kinetics are carried out as support for solid state stability work, then saturated solutions are employed. Here the pH profile may be quite different. Pralidoxime for instance at pH 1–3 exhibits a type AH profile with unity slope (Fyhr et al., 1986) as opposed to what was reported earlier, and aspirin (Carstensen and Attarchi, 1988a, 1988b; Carstensen et al., 1985) has a second-order hydrolysis rate constant virtually independent of acetic acid concentration from quite dilute to concentrated. *p*-aminosalicylic acid (Carstensen and Pothisiri, 1975) also exhibits a different kinetic profile in saturated solution.

As will be seen in the chapter dealing with disperse systems, drugs that form micelles will show, in general a shift in mechanism and rate above the critical micelle concentration.

11. EFFECT OF SOLVENT

It stands to reason that solvent will have an effect on the hydrolysis rate. Carstensen et al. (1971) demonstrated the effect of dielectric constant on the hydrolysis rate of benzodiazepines, and Hussain and Truelove (1979) showed the effect of dielectric constant on the hydrolysis rate of 2-tetrahydropyranyl benzoate (Fig. 28).

As mentioned, Brandl and Magill (1997) have reported on the pK values of ketorolac in isopropanol. They also showed the decomposition to be first order

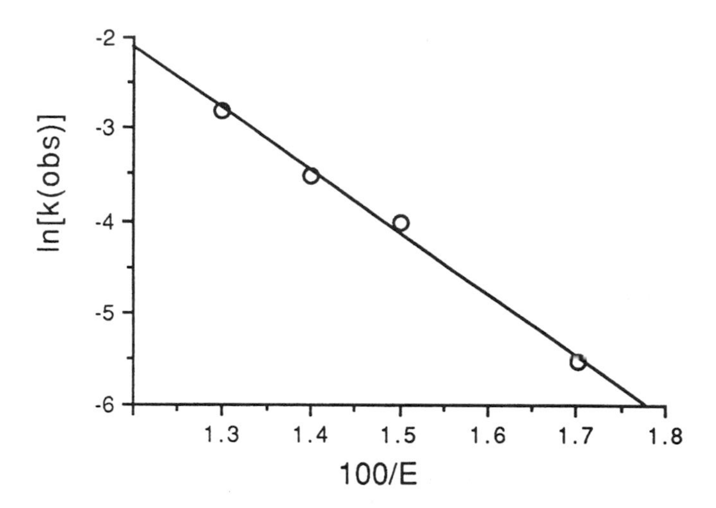

Fig. 28 Effect of dielectric constant on the hydrolysis rate of 2-tetrahydropyranyl benzoate. Least-squares fit $y = 5.98 - 6.74 \times (R = 1.00)$. (Graph constructed from data by Hussain and Truelove, 1979.)

in 35% isopropanol, although the acid form exists in equilibrium with its isopropyl ester and with its sodium salt.

Hou Poole (1969a,b) showed a linear effect of the reverse of the dielectric constant on the log of the rate constant for decomposition of ampicillin in solution.

12. KINETIC SALT EFFECT

Kinetic studies in solutions, as have been shown, should be carried out for at least two buffer concentrations, so that the pH profile can be constructed without confounding it with the catalytic effect of the buffer components. At times, however, the mere presence of an ionized substance may affect the kinetics of decomposition.

To investigate this, the ionic strength is varied by carrying out the decomposition in solutions of different concentrations of inert electrolyte. As mentioned, the definition of ionic strength is

$$\mu = 0.5 \cdot [\Sigma m_i z_i^2] \tag{3.219}$$

where m_i is the molarity of the ith species and z_i is its charge. One molar $CaCl_2$, for instance would be one molar in calcium ion and two molar in chloride ion, so that the ionic strength would be $[(1 \cdot 2^2) + (2 \cdot 129; 1^2)]/2 = 3$. For a 1 : 1 electrolyte the ionic strength will of course be the same as its molarity.

The decomposition reaction for a drug substance (A) interacting with an ionic solute (B) can be described as

$$A + B \rightarrow [AB^f] \rightarrow \text{products} \tag{3.220}$$

where $[AB^f]$ denotes a transition complex. In this case, the rate constant k, will be dependent on the ionic strength in concentrations below 0.01 M by the relation

$$\log k = \phi + 2.Q.z_A \cdot z_B \cdot (\mu^{1/2}) \tag{3.221}$$

where z denotes charge, and the vale of Q is given by

$$Q = 1.825 \cdot 10^6 \cdot \left[\frac{\rho}{(T\varepsilon)^3}\right]^{1/2} \tag{3.222}$$

where T is absolute temperature, ϵ is the dielectric constant of the solution, and ρ is its density. It is noted that at 25°C in aqueous solutions, the value of $2Q = 1.018$. This equation can only be expected to hold in the very dilute concentration range, and at higher concentrations of solute (up to ionic strengths of 0.1 molar), the Güntelberg equation rather than the Debye–Hückel equation is employed:

$$\log f_i = \frac{Q \cdot z_i \cdot (\mu^{1/2})}{1 + \beta(\mu^{1/2})} \tag{3.223}$$

where β is often 1, where i denotes a central ion and z_i its charge, and where Q is a constant. Figure 29 shows data by Wang and Yeh (1993) obtained by using Eq. (3.223). Brooke and Guttmann (1968) have found such linearity (but with a negative slope) for the decomposition of 3-carbomethoxypyridinium cation. The slope depends on whether the interacting ions are of the same sign or of opposite sign, as shown in Eq. (3.224).

Eq. (3.223) (Carstensen, 1970) when applied to rate constants, becomes

$$\log k = \phi + \frac{2 \cdot Q \cdot z_A \cdot z_B.(\mu^{1/2})}{1 + \beta(\mu^{1/2})} \tag{3.224}$$

where β is a constant related to the ionic diameters of the solutions. Curve fits are relatively insensitive to the size of β, which is most often close to 1, so that most authors (e.g., Czapski and Schwarz, 1962) make this assumption—in fact the original Güntelberg equation made this assumption.

At even higher ionic strength the relationship becomes

$$\log k = \log k_0 + q \cdot \mu \tag{3.225}$$

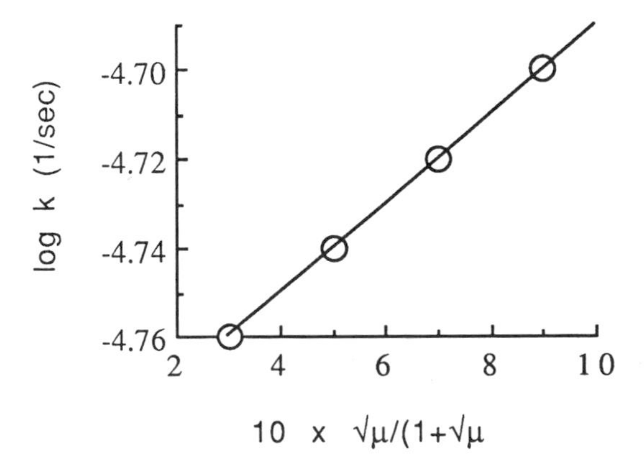

Fig. 29 Degradation metronidazole at pH 7.4. (Graph constructed from data published by Wang and Yeh, 1993.)

It is noted that in Eqs. (3.221) and (3.224) there will be a nonzero kinetic salt effect only if both A and B are charged. If either is uncharged, then the effect is zero. This is a useful tool, mechanistically, since it allows assessment as to the reaction species. If, for instance, a drug that can be either HA or A^- interacts with a species B^-, then a zero kinetic salt effect rules out $A^- + B^-$ as a possibility.

It is not to be expected that the slope of the kinetic salt effect lines will have the predicted slope. Since both species interacting must be charged, it follows that, for an acid for instance, the effect for HA is zero. The effect for A^- would be -1 if it were $[H^+]$ catalyzed and plus one if it were $[OH^-]$ catalyzed. However, depending on the pH, only part of the drug substance would be present as A^-. If this fraction is calculated, then the slope should equal either plus or minus that fraction.

There have been a small number of reports on kinetic salt effects in the pharmaceutical literature. Brooke and Guttmann (1968) reported on 3-carbomethoxy-1-methylpyridinium cation decomposition, Hussain et al. (1968) reported on echothiophate, Felmeister et al. (1965) on chlorpromazine, Finholt et al. (1962, 1963, 1965) on penicillin, Koshy and Mitchner (1964) on 2-(4-phenyl-1-piperazinylmethyl) cyclohexanone, Szulczewski et al. (1964) on 4,6-diamino-1-(3,5)-dichlorophenyl-1,2-2-2-dimethyl-1,3-5-triazine, Windheuser and Higuchi (1962) on thiamine, Finholt and Higuchi (1962) on niacinamide, and McRae and Tadros (1978) have shown the kinetic salt effect on triclofos hydrolysis rates. Hashimoto et al. (1984) have reported a kinetic salt effect for the empimerization of moxalactam. The concentration range is, however, too high for Eq. (3.209) to hold, and it is possible that it is actually Eq. (3.210) that holds.

Kirsch and Notari (1984) have discussed the effect of the ionic atmosphere on the relative amount of general base attack and how and to what extent this can influence the kinetic salt effect.

The best presentation mode for kinetic salt data are in a form exponential to the forms stated in Eq. (3.224). This is repeated here for convenience:

$$\frac{\ln[k]}{2.303} = \phi + \frac{2 \cdot Q \cdot z_A \cdot z_B.(\mu^{1/2})}{1 + \beta(\mu^{1/2})} \tag{3.226}$$

The traditional conversion from non-Briggsian to natural logarithms has been carried out and the exponential form of this is

$$k = \exp[2.3\phi] \exp \frac{2 \cdot Q \cdot z_A \cdot z_B.(\mu^{1/2})}{1 + \beta(\mu^{1/2})} \tag{3.227}$$

An example of this is shown in (Fig. 30).

The advantage in this presentation mode is that it is not biased. On the contrary, lest-squares fitting of a log-square root plot would be biased because neither x and y values would be centered about their means.

As mentioned, the kinetic salt effect depends on the charges of the interacting ions, and if one of these is zero, then there is no such effect. For instance, Ito et al. (1997) showed that degradation of panipenem in aqueous solution exhibited first-order rate constants that showed a lack of ionic strength effect.

As mentioned, Brandl and Magill (1997) have reported on the pH values of ketorolac in isopropanol. They also showed the decomposition to be first order

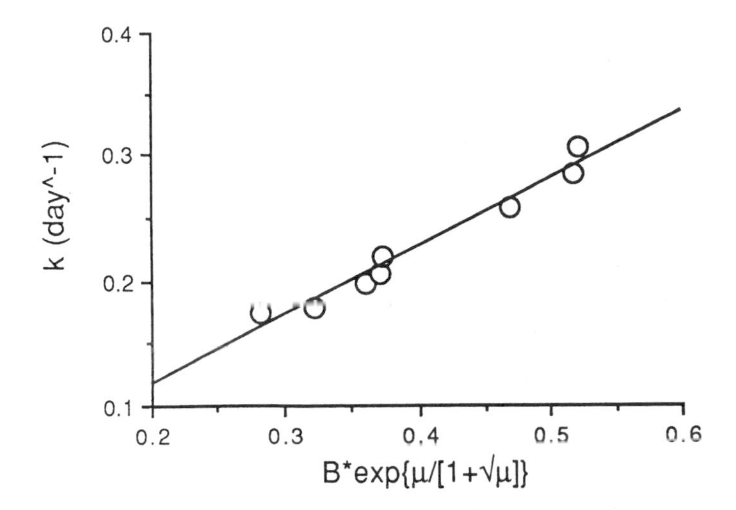

Fig. 30 Best presentation mode for kinetic salt effect data in the intermediate range. Least-squares fit is (with $R^2 = 0.967$) $k = 0.0085 + 0.55B\exp[\mu^{1/2}/(1+\mu^{1/2})]$. (Plot constructed from data published by Franchini, 1992.)

in 35% isopropanol, although the acid form exists in equilibrium with its isopropyl ester and with its sodium salt.

Hou and Poole (1969a,b) showed a linear effect of the reverse of the dielectric constant on the log of the rate constant for decomposition of ampicillin in solution.

There are other, both recent and older, references to the kinetic salt effect. Hou and Poole (1969a,b) have shown that ampicillin is subject to a kinetic salt effect, which differs, based on the pH of the solution. Since a positive effect exemplifies reaction between charged species, they were able to establish a reaction mechanism. van Maanen et al. (1999) have demonstrated a kinetic salt effect in the aquous solution decomposition of thiotepa.

REFERENCES

Amer, M. M., Takla, K. F. (1968). Bull. Fac. Pharm. Cairo Univ. 15:325.

Anderson, B. D., Fung, M.-C., Kumar, S. D., Baker, D. C. (1985). Asada, S., Yamamoto, M., Nishijo (1980). J. Bull. Chem. Soc. Japan 53:3017.

Avdeef, A., Comer, J. E. A., Thomson, J. S. (1993).Anal. Chem. 65:42.

Baeschlin, P. K., Etter, J. C. (1969). Pharm. Acta Helv. 44:348.

Barbosa, J., Bergés, R., Toro, I., Sanz-Nebot, V. (1997). Int. J. Pharm. 149:213.

Bates, R. G. (1964). Determination of pH. New York: John Wiley, pp. 20, 23.

Berge, S. M., Henderson, N. L., Frank, M. J. (1983). J. Pharm. Sci. 72:59.

Blaug, S. M., Hagratwata, B. (1972). J. Pharm. Sci. 61:556.

Brandl, K., Kennedy, C. (1964). J. Pharm. Sci. 53:345.

Brandl, M., Magill, A. (1997). Prod. Dev., Ind. Pharm. 23:1079.

Brodersen, A. C. (1947). Trans. Faraday. Soc. 43:351.

Brønsted, J. N. (1943). Fysisk Kemi, Munksgaard, København, pp. 262, 293.

Brooke, D., Guttmann, D. (1968). J. Pharm. Sci. 57:1677.

Bundgaard, H. (1977). Arch. Pharm. Suec. 14:47.

Bundgaard, H. (1980). Arch. Pharm. Chemi. Sci. Ed. 8:83.

Carney, C. F. (1988). J. Pharm. Sci. 77:393.
Carstensen, J. T. (1970). J. Pharm. Sci. 59:670.
Carstensen, J. T., Attarchi, F., (1988a). J. Pharm. Sci. 77:314.
Carstensen, J. T., Attarchi, F., (1988b). J. Pharm. Sci. 77:318.
Carstensen, J. T., Pothisiri, P. (1775). J. Pharm. Sci. 64:37.
Carstensen, J. T., Su, K. S. E., Maddrell, P., Johnson, J. B., Newmark, H. N. (1971). Bull. Parenteral. Drug Assoc. 25:193.
Carstensen, J. T., Attarchi, F., Hou, X.-P. (1985). J. Pharm. Sci. 74:741.
Carstensen, J. T., Franchini, M., Ertel, K. (1992). J. Pharm. Sci. 81:303.
Chatterji, D. C., Galleli, J. F. (1978). J. Pharm. Sci. 67:26.
Cho, M. J., Scahill, T. A., Hester, J. B., Jr. (1983). J. Pharm. Sci. 77:1365.
Chung, P. H., Chin, T. F., Lach, J. L. (1970). J. Pharm. Sci. 59:1300.
Claudius, J. S., Neau, S. H. (1998). Int. J. Pharm. 168:41.
Cohen, B., VanArtsdalen, E. R., Harris, J. (1948). J. Am. Chem. Soc. 70:281.
Cruz, J. E., Maness, D. D., Yakatan, G. J. (1979). Int. J. Pharmaceutics 2:275.
Czapski, K., Schwartz, M. (1962). J. Phys. Chem. 66:471.
Devani, M. B., Shishoo, C. J., Doshi, K. J., and Patel, H. B. (1985). J. Pharm. Sci. 74:427.
Edwards, L. J. (1952). Trans. Faraday Soc. 48:696.
Ehrson, H., Ehsborg, S., Wallin, I., and Nillson, S. O. (1980). J. Pharm. Sci. 69:1091.
Ellin, R. I. (1958). J. Am. Chem. Soc. 80:6688.
Ellin, R. I. Wills, J. H. (1964). J. Pharm. Sci. 53:995.
Ellin, R. I., Carlese, J. C., and Kondritzer, J. C. (1964). J. Pharm. Sci. 53:141.
Fallavena, P. R. B., Schapoval, E. E. S. (1997). Intl. J. Pharm. 158:109.
Fassberg, J., Stella, V. J. (1992). J. Pharm. Sci. 81:676.
Felmeister, A., Schaubman, R., and Howe, H. (1965). J. Pharm. Sci. 54:1589.
Finholt, P., Higuchi, T. (1962). J. Pharm. Sci. 51:655.
Finholt, P., Paulssen, R. B., Higuchi, T. (1963). J. Pharm. Sci. 52:948.
Finholt, P., Jurgensen, G., Kristiansen, H. (1965). J. Pharm. Sci. 54:387.
Franchini, M. (1992). M. S. thesis, University of Wisconsin, School of Pharmacy.
Fubara, J. O., Notari, R. E. (1998). J. Pharm. Sci. 87:1572.
Fyhr, P., Brodin, A., Ernerot, L. and Lindquist, J. (1986). J. Pharm. Sci. 75:608.
Garrett, E. R. (1957). J. Am. Chem. Soc. 79:3401.
Garrett, E. R. (1962). J. Pharm. Sci. 51:451.
Garrett, E. R., Seyda, K. (1983). J. Pharm. Sci. 72:258.
Garrett, E. R., Nester, H. J., and Somodi, A. (1968). J. Org. Chem. 33:3460.
Garrett, E. R., Bojarski, J. T., Yakatan, G. J. (1971). J. Pharm. Sci. 60:1145.
Gupta, V. (1983). J. Pharm. Sci. 72:205, 1453.
Guven, K. C., Aras, A. (1970). Eczacilik. Bull, 12:72.
Hamid, I. A., Parrott, E. L. (1971). J. Pharm. Sci. 60:901.
Hanilton-Miller, J. M. T. (1973). J. Pharm. Pharmacol. 25:401.
Han, W. W., Yakatan, G. J., and Maness, D. D. (1977). J. Pharm. Sci. 66:573.
Hansen, J., Bundgård, H. (1979). Arch. Pharm. Chemi. Sci. Ed. 7:135.
Hansen, J., Bundgård, H. (1980a). Arch. Pharm. Chemi. Sci. ed. 8:5.
Hansen, J., Bundgård, H. (1980b). Arch. Pharm. Chemi. Sci. ed. 8:91.
Hashimoto, N., Tanaka, H. (1985). J. Pharm. Sci. 74:69.
Hashimoto, N., Tasaki, T., Tanaka, H. (1984). J. Pharm. Sci. 73:369.
Heller, H. (1939). J. Physiol, 96:337.
Higuchi, T., Lachman, L. (1955). J. Am. Pharm. Assoc. Sci. Ed. 44:521.
Higuchi, T., Schroeter, L. C. (1960). J. Am. Chem. Soc. 82:1904.
Higuchi, T., Marcus, A., Bins, C. D. (1954). J. Am. Pharm. Assoc. Sci. Ed. 43:530.
Hirata, M., Kagowa, H. Baba, M. (1967). Ann Rept. Shionogi Res. Lab. 17:107.
Hou, J. P., Poole, J. W. (1969a). J. Pharm. Sci. 58:447.

Hou, J. P., Poole, J. W. (1969b).J. Pharm. Sci. 60:503.
Huang, T. L., Székács, T., Uematsu, T., Kuwano, E., Parkinson, A., Hammock, B. D. (1993). Pharm. Res. 10:639.
Hull, L. A., Rosenblat, D. H., Epstein, J. (1979). J. Pharm. Sci. 68:856.
Hussain, A., Truelove, J. (1979).J. Pharm. Sci. 63:235.
Hussain, A., Schurman, P., Peter, V., Milosovich, G. (1968). J. Pharm. Sci. 58:447.
Ismail, S., Simonelli, A. P. (1986). Bull. Pharm. Sci. (Assiut U.) 9:119.
Ito, N., Suzuki, M., Ikeda, M. (1997). Drug Stability 1:196.
Jivani, S., Stella, V. J. (1985). J. Pharm. Sci. 74:1274.
Johnson, K., Amidon, G. L., Pogany, S. (1985). J. Pharm. Sci. 74:87.
Kadin, H. (1982). In: Analytical Profiles of Drug Substances K. Florey, ed. New York: Academic Press, Captopril.
Kearney, A. S., Crawford, L. F., Mehta, S. C., Radebaugh, G. W. (1993). Pharm. Res. 10:1461.
Kenley, R. A., Warne, N. W. (1994). Pharm. Res. 11:72.
Khalil, S. A. H., El-Masry, S. (1978). J. Pharm. Sci. 57:1358.
King, S.-Y. P., Sigvardson, K. W., Dudzinski, J., Torosian, G. (1992). J. Pharm. Sci. 81:586.
Kirsch, L. E., Notari, R. L. (1984). J. Pharm. Sci. 73:896.
Kondrat'eva, A. P., Bruns, B. P., Libison, G. S. (1971). Khim' Farm. Zh. 5:38.
Koshy, K., Lach, J. (1961). J. Pharm. Sci. 50:113.
Koshy, K., Mitchner, H. (1964). J. Pharm. Sci. 53:1381.
Kreldgaard, B., Kisbye, J. (1974). Arch. Pharm. Chim. Sci. Ed. 2:38.
Kurono, Y., Taqmaki, H., Yokota, Y. U., Ida, M., Sugimoto, C., Kuayama, T., Yashiro, T. (1994). Chem Pharm. Bull, 42:344.
Lee, H.-K., Querijero, G. (1985). J. Pharm. Sci. 74:87.
Liquori, A. M., Ripamonti, A. (1955). Gazz. Chim. Ital. 85:589.
Longhi, M. R., deBertorello, M., and Granero, G. E. (1994). J. Pharm. Sci. 83:336.
Loy, H. W. (1952). J. Assoc. Off. Agr. Chemists 35:169.
Lund, W., Waaler, T. (1968). Acta Chem. Scand. 2:3085.
Lundgreen, P, Landersjoe, L. (1970). Acta Pharm. Sueccica 7:133.
Marcus, A (1960). J. Am. Pharm. Assoc. Sci. Ed. 49:383.
Marcus, A., Baron, B, (1959). J. Am. Pharm. Assoc. Sci. Ed. 48:85.
Maron, S. H., Prutton, S. H. (1965). Principles of Physical Chemistry, 4th ed. London; MacMillan, p. 445.
Martin, A., Swarbrick, J., Cammarata, A. (1993). Physical Pharmacy, 3rd ed. Philadelphia; Lea and Febiger, pp. 191, 202.
Maulding, H. V., Nazareno, J. P., Pearson, J. E., and Michaelis, A. F. (1975). J. Pharm. Sci. 64:278.
McRae, J. D., and Tadros, L. M. (1978). J. Pharm. Sci. 67:631.
Meakin, B. J., Transey, I. P., Davies, D. J. G. (1971). J. Pharm. Pharmacol. 23:252.
Mollica, J. A., Connors, K. A. (1967). J. Am. Chem. Soc. 89:308.
Mollica, J. A., Rehm, C. R., Smith, J. R., Goran, H. K. (1971). J. Pharm. Sci. 60: 1380.
Monkhouse, D. C., VanCampen, L., Aguiar, A. J. (1973). J. Pharm. Sci. 62:576.
Moore, W. J. (1963). Physical Chemistry. Englewood Cliffs, N. J.: Prentice-Hall, pp. 355, 399.
Muhammed, N., Adams, G., Lace, H. K. (1988). J. Pharm. Sci. 77:126.
Nakami, Y., Tanabe, T., Kobayashi, T., Tanabe, J., Okimura, Y., Koda, S., Moromoto, Y. (1987). J. Pharm. Sci. 76:208.
Nguyen, T. H., Shih, C., Himmelstein, K. J., Higuchi, T. (1984). J. Pharm. Sci. 73:1563.
Notari, R. E., DeYoung, J. L. (1975). J. Pharm. Sci. 64:1148.
Notari, R. E., Chin, M. L., Wittebort, R. (1972). J. Pharm. Sci. 61:1189.
Oliyai, C., Borchard, R. T. (1994). Pharm. Res. 11:75.
Ong, K. C., Robinson, R. A., Bates, R. G. (1964). Anal Chem. 36:1971.

Osterling, T. O. (1970). J. Pharm. Sci. 59:63.
Patel, R. M., Chin, T., Lach, J. L. (1968). Am, J. Hosp. Pharm. 25:256.
Pawelczyk, E., Kniatter, B., Alejska, W. (1979). Acta Polon. Pharm. 36:181.
Pipkin, J. D., Davidovich, M. (1982). Program Abstracts, Academy of Pharm. Sci., 35th National Meeting, Miami Beach, Florida: Abstract No. 60, Basics Section.
Pipkin, J. D., Barry, E. P. (1982). Program Abstracts, Academy of Pharm. Sci., 35th National Meeting, Miami Beach, Florida: Abstract No. 28, IPT.
Powell, M. F. (1986). J. Pharm. Sci. 75:901.
Powell, M. F. (1988). Pharm. Res. 4:42.
Quigley, J. M., Jordan, C. G. M., Timoney, R. F. (1994). Int. J. Pharm. 101:145.
Ravin, L. J., Simpson, C. A., Zappala, A. F., Gluesich, J. (1964). J. Pharm. Sci. 53:1064.
Sassetti, R. J., Fudenberg, H. H. (1971). Biochem. Pharmacol. 20:57.
Schulz, J. (1967). Methods Enzymol. 11:255.
Schwartz, M. A., Grannatek, A. P., Buckwalter, F. H. (1962). J. Pharm. Sci. 51:523.
Schwartz, M. A., Bara, E., Rabyez, I., Granatek, A. P. (1965). J. Pharm. Sci. 54:149.
Seydel, J. K. (1970). Antibiot. Chemotherapy 16:380.
Shedlovski, T., Kay, L. R. (1956). J. Phys. Chem. 60:151.
Siegel, S., Lachman, L., Malspeis, L. (1959). J. Am. Pharm. Assoc. Sci. Ed. 48:431.
Sing, S., Sharda, N., Mahajan, L. (1999). Int. J. Pharm. 176:261.
Sørensen, S. P. (1909a). Biochem. Z. 21:131.
Sørensen, S. P. (1909b). Compt. Rend. Trav. Lab. Carlsberg 8:1.
Smith, G. G., Kennedy, D. R., Nairn, J. G. (1974). J. Pharm. Sci. 63:712.
Smithuis, L. O. M. (1969). Pharm. Weekbl. 104:1097.
Stavchansky, S., Wallace, J. E., Wu, P. (1983). J. Pharm. Sci. 72:546.
Stella, V. J., Pipkin, J. D. (1976). J. Pharm. Sci. 65:1161.
Sunderland, V. B., Watts, D. W. (1984). Int. J. Pharmaceutics. 19:1.
Szulczewski, C., Shearer, B., Aguiar, A. (1964). J. Pharm. Sci. 53:1157.
Tada, H. (1960). J. Am. Chem. Soc. 82:255.
Tanaka, N. Nakagaki, M. (1961). Yakagaku Zasshi 81:591.
Tanaka, Y., Suzuki, A. (1994). J. Pharm. Pharmacol. 46:235.
Timmins, P., Jackson, I. M., Wang, Y. J. (1982). Int. J. Pharmaceut 11:329.
Tu, Y. H., Wang, D. P., Allen, L. V., Jr. (1989). J. Pharm. Sci. 78:3030.
Tu, Y. H., Wang, D. P., Allen, I. V. (1990). J. Pharm. Sci. 79:48.
Ugwu, S. O., Lan, E. L., Sharma, S., Hruby, V., Blanchard J. (1994). Int. J. Pharm. 102:193.
Underberg, W. J. M. (1978a). J. Pharm. Sci. 67:1133.
Underberg, W. J. M. (1978b). J. Pharm. Sci. 67:635.
Underberg, W. J. M., Lingeman, H. (1983a). J. Pharm. Sci. 72:549.
Underberg, W. J. M., Lingeman, H. (1983b). J. Pharm. Sci. 72:553.
van der Houwen, O. A., Beijnen, J. H., Bult, A. (1988). Int. J. Pharm. 45:181.
van der Houwen, O. A., Bekers, G. J., Beijnen, J. H., Bult, A., Underberg, W. J. M. (1991). Int. J. Pharm. 67:155.
van der Houwen, O. A., Bekers, G. J., Bult, A., Beijnen, J. H., Underberg, W. J. M. (1993). Int. J. Pharm. 89:R5.
van der Houwen, O. A., Bekers, G. J., Beijnen, J. H., Bult, A., Underberg, W. J. M. (1994). Int. J. Pharm. 109:191.
van der Houwen, O. A., de Loos, M. R., Beijnen, J. H., Bult, A., Underberg, W. J. M. (1997). Int. J. Pharm. 155:137.
van Maanen, J. M., Brandt, A. C., Damen, J. M. A., Beijenen, J. H. (1999). Int. J. Pharm. 179:55.
Vej-Hansen, B., Bundgaard, H., Hreilgaard, B. (1978). Arch. Pharm. Chemie Scie. Ed. 6:151.
Vej-Hansen, B., Bundgaard, H. (1979). Arch. Pharm. Chemie Scie. Ed. 7:65.
Visconti, M., Citerio, L., Borsa, M., Pifferi, G. (1984). J. Pharm. Sci. 73:1812.

Vishnuvajjala, B. R., Cradock, J. C. (1986). J. Pharm. Sci. 75:308.
Wang, J. C. T., Patel, B. (1986). J. Pharm. Sci. 75:204.
Wang, D.-P., Yeh, M.-K. (1993). J. Pharm. Sci. 82:95.
Windheuser, J. J., Higuchi, T. (1962). J. Pharm. Sci. 51:354.
Winterborn, I. L., Meakin, B. J., Davies, D. J. G. (1972). J. Pharm. Pharmacol. 24:133P.
Won, C. M. (1994). Pharm. Res. 11:165.
Yamana, T., Tsuji, A. (1976). J. Pharm. Sci. 65:1563.
Yamana, T., Mizukami, Y., Tsuji, A. (1968). Chem. Pharm. Bull. 16:396.
Yang, S. K., Yang, M. S. (1994). J. Pharm. Sci. 83:6290.
Zia, H., Tehrani, M., Zargarbachi, R. (1974). Can. J. Pharm. Sci. 9.112.
Zvirblis, P., Socholitsky, I., Kondritzer, A. (1956). J. Am. Pharm. Assoc. Assoc. Sci. Ed. 47:450.

4

Oxidation in Solution

JENS T. CARSTENSEN

Madison, Wisconsin

1. OXIDATIVE REACTIONS

The 1987 Stability Guidelines state that "It is also suggested that the following conditions be evaluated in stability studies on solutions or suspensions of the bulk drug substances. . . . High oxygen atmosphere."

Oxidation reactions are relatively rare in pharmaceutical dosage forms as a main reaction. Some oxidation probably takes place in many cases and results in small amounts of unidentified degradation products. When an oxidation reaction is one of the main reactions, then it is a serious matter, and formulating around such a problem can be difficult. Notable examples are liquid vitamin A products (although, strictly speaking, they are not solutions). Other examples will be cited shortly.

1.1. Oxidation Mechanisms

Oxidation, as the word implies, is an interaction between drug substance, A, and oxygen, and the net reaction would be

$$A + O_2 \rightarrow \text{products} \tag{4.1}$$

However, oxidation reactions are usually the sum of a series of reactions (at times chain reactions), and these start with one particular reaction (the initiation reaction), which usually does not involve molecular oxygen. Frequently, oxidations are catalyzed by metal ions M^{++}; an example of this is the oxidation mechanisms reported for captopril (Timmins et al., 1982). Captopril contains a thiol group and will be symbolized as ASH below:

$$2ASH + 2M^{++} \rightarrow 2\{AS\cdot\} + 2H^+ + 2M^+ \tag{4.2}$$

$$2\{AS\cdot\} \rightarrow AS \cdot SA \tag{4.3}$$

$$2M^+ + O_2 \rightarrow 2M^{++} + O_2^{2-} \tag{4.4}$$

$$O_2^{2-} + 2H^+ \rightarrow H_2O + 0.5O_2 \tag{4.5}$$

where $\{\cdot\}$ denotes free radical. Hence, the overall reaction is

$$2ASH + 0.5O_2 \rightarrow AS \cdot AS + H_2O \tag{4.6}$$

It is noted that there is no net consumption of metal ion. (This latter has been chosen as being divalent in the above but could have other valences depending on metal in question). There is, however, a consumption of oxygen (otherwise there would be no oxidation). Upon exhaustion of oxygen, the oxidation would cease.

If oxygen is abundant, then of course all the drug can decompose, and in this case, as will seen below, the kinetics can be simply first order, since $[O_2]$ becomes (virtually) constant.

When captopril decomposes without presence of metals, it undergoes an autooxidation:

$$2ASH \leftrightarrow 2AS^- + 2H^+ \tag{4.7}$$

$$2As^- + 2O_2 \rightarrow 2\{AS\cdot\} + 2\{O_2\cdot\}^- \tag{4.8}$$

$$2As^- + 2\{O_2\cdot\} \rightarrow 2\{AS\cdot\} + 2O_2^{2-} \tag{4.9}$$

$$2\{AS\cdot\} \rightarrow ASSA \tag{4.10}$$

$$O_2^{2-} + 2H^+ \rightarrow H_2O + 0.5O_2 \tag{4.11}$$

So the total reaction is

$$2ASH + 0.5O_2 \rightarrow ASSA + H_2O \tag{4.12}$$

Most oxidations in pharmaceutics occur in aqueous solution and are a function of the oxygen dissolved in the aqueous phase. In practice it is desired to minimize the decomposition, and it is a practice to remove oxygen from such systems by nitrogen flooding. Depending on the method used for the removal there can still be quite a bit of oxygen left in solution. Simple flooding is not effective, and to remove oxygen completely by boiling is rather difficult, and since oxygen has a low molecular weight, it can still be left present in sufficient (molar) excess to allow considerable oxidation. Hence, to free solutions of oxygen, nitrogen should be bubbled through them (the very best way of driving out dissolved oxygen). The use of antioxidants (bisulfite, ascorbic acid) is advocated where possible, in the case of oxygen-sensitive drug substances in solution.

In practice, complexing the heavy metals (e.g., by use of ethylene diamine tetraacetic acid) is often employed, since, as seen in the reactions sequences shown, metal ions are deleterious to drug stability in oxidative situations.

1.2. Quantitative Considerations in Oxidations in Two Phases

Most oxidation problems in the pharmaceutical sciences occur in systems (liquid or solid) that are in contact with a gas phase (usually air) that contains oxygen. As far as the methodology for following a decay, the monitoring of parent compound and decomposition products is what is usually carried out. If it is necessary to monitor oxygen concentrations, then Winkler's method (Winkler, 1888; Novaczyk et al., 1993) may be used, although it is cumbersome. Oxygen electrodes can also be used, but these are only good below 45°C. Finally, potentiometric titrations are possible. For work in the gas phase, Raman spectroscopy can be used. Most of what follows will concentrate on the disappearance of the parent drug.

Henry's law states, in the case of oxygen/water systems, that

$$P_{O_2} = K' X_{O_2} \approx K^* C \tag{4.13}$$

where C is the molar concentration and X the mole fraction of dissolved oxygen; K and K^* are Henry's law constants. For ordinary atmospheric conditions, $P_{O_2} = 0.22$ atm. At 25°C the value of K' for oxygen is $3.314\ 10^7$ atm, i.e., the mole fraction of oxygen dissolved water would be $0.22/(3.3 \times 10^7) = 6.6 \times 10^{-9}$. The molar concentration corresponding to this is given by

$$X_{O_2} = \frac{C_{O_2}}{55.5 + C_{O_2}} \approx \frac{C_{O_2}}{55.5}$$

i.e., C_{O_2} is about 3.7×10^{-7} molar. Equation (4.13) is often expressed in inverse form, i.e.,

$$C = KP \tag{4.14}$$

where $K = 1/K^*$; both presentation forms will be used in the following.

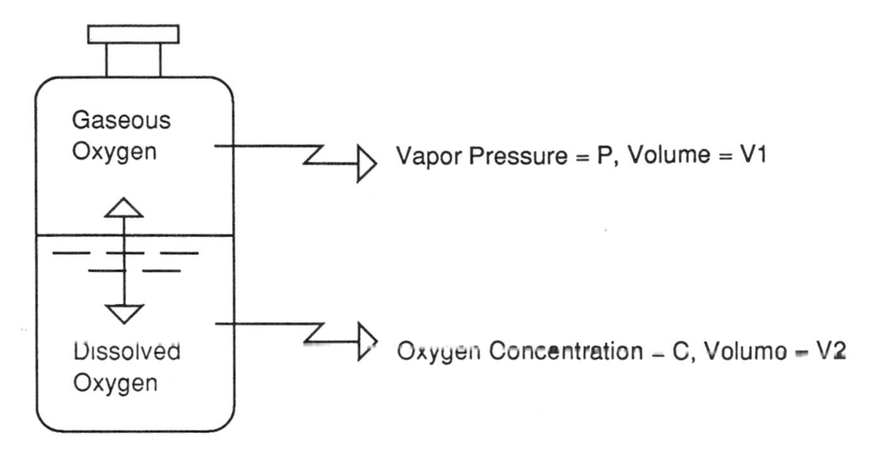

Fig. 1 Nomenclature used for kinetic schemes.

If a solution of a drug substance (A) is oxygen sensitive, then

$$A + O_2 \rightarrow B \tag{4.15}$$

where B represents oxidation products. In the first case considered, it is assumed that the system is closed (Fig. 1), that the head space is V_1 cm^3, and that the liquid volume is V_2 cm^3. The total number of moles of oxygen in the ensemble is the number of moles, n_1, in the gas phase plus the number of moles, n_2, dissolved.

The gas law gives the value for n_1 as

$$n_1 = \frac{PV_1}{RT} \tag{4.16}$$

where P in the following is understood to be the oxygen pressure in the atmosphere.

$$n_2 = V_2 C \tag{4.17}$$

so that the total number of moles, N, is given by

$$N = n_1 + n_2 = \frac{PV_1}{RT} + V_2 C \tag{4.18}$$

Introducing Eqs. (4.13) and (4.16) gives

$$N = \frac{PV_1}{RT} + V_2 K^* P = P\left(\frac{V_1}{RT} + V_2 K^*\right) \tag{4.19}$$

The oxidation is (a) simple or (b) catalytic or (c) autocatalytic.

1.3. Simple Oxidation in Solution in a Closed System

The simple case will be treated first (Franchini et al., 1993; Franchini and Carstensen, 1994). It follows from the general second-order reaction equation that

$$\frac{d[\mathrm{A}]}{dt} = -k_2[\mathrm{A}]C \tag{4.20}$$

where [A] is concentration of drug, k_2 is the second-order rate constant, and C is the oxygen concentration in solution at time t. If the initial concentrations are denoted by subscript zero, and the concentration of decomposition product is Y molar at time t, then

$$[\mathrm{A}] = A_0 - Y \tag{4.21}$$

and

$$d[\mathrm{A}] = -dY \tag{4.22}$$

so that now

$$\frac{dY}{dt} = k_2[A_0 - Y]C = k_2 A_0[1 - g \cdot Y]C \tag{4.23}$$

where

$$g = \frac{1}{A_0} \tag{4.24}$$

The number of moles decomposed is $V_2 Y$, so that the total number of oxygen molecules, $N(t)$, at time t is given by

$$N(t) = N_0 - V_2 Y \tag{4.25}$$

It is assumed that the distribution of oxygen between head space and liquid is rapid, so that at time t,

$$C = K\,P(t) \tag{4.26}$$

where $P(t)$ is the oxygen pressure remaining at time t. Mass-balancing of total oxygen gives

$$V_2 C + \frac{V_1 P(t)}{RT} = N_0 - V_2 Y \tag{4.27}$$

Introducing Eq. (4.26) into this gives

$$V_2 C + \frac{V_1 C}{RTK} = N_0 - V_2 Y \tag{4.28}$$

Solving for C gives the expression

$$C = q[1 - aY] \tag{4.29}$$

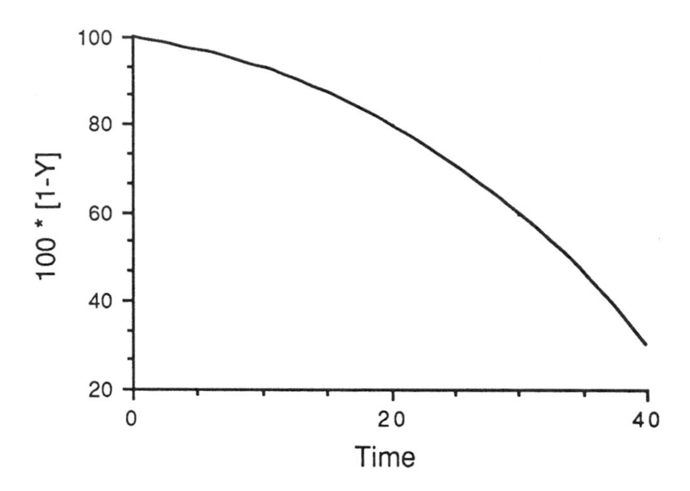

Fig. 2 Profile of $[1 - Y]$ versus t according to Eq. (4.32).

where

$$q = \frac{N_0}{V_2 + V_1/RTK} \tag{4.30}$$

and

$$a = \frac{V_2}{V_2 + V_1/RTK} \tag{4.31}$$

Introducing this into Eq. (4.23) gives

$$\frac{dY}{dt} = q \cdot k_2 A_0 [1 - gY][1 - aY] \tag{4.32}$$

This equation can be solved (in closed form as well as by graphical integration) and gives curves of the type shown in Fig. 2. Examples of this type of curve are the oxidation of promethazine at pH 3.2 (Underberg, 1978) and the oxidation of propildazine (Ventura et al., 1981). This latter is shown in Fig. 3.

If A_0 is large and fairly constant during the reaction, then

$$\frac{dY}{dt} \approx k_2 A_0 C \tag{4.33}$$

Introducing Eq. (4.29) then gives

$$\frac{dY}{dt} = k_2 A_0 q [1 - aY] \tag{4.34}$$

This can be rewritten as

$$d\left(\frac{1}{a} - Y\right) = a \cdot k_2 A_0 q \cdot dt \tag{4.35}$$

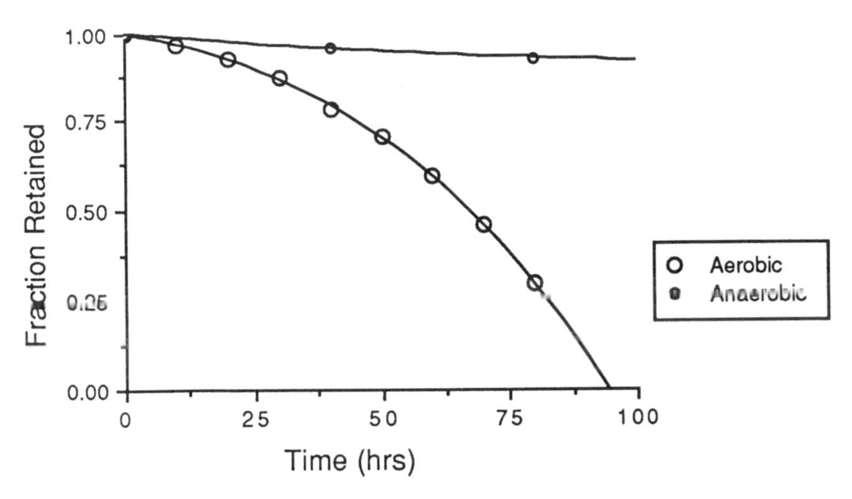

Fig. 3 Oxidation of propildazine. (Graph constructed from data by Ventura et al., 1981.)

which integrates to

$$\ln\left(\frac{1}{a} - Y\right) = -k_2 A_0 qat + \ln\frac{1}{a} \tag{4.36}$$

If a is close to unity, then this predicts first-order kinetics, as is for instance the case in work reported by Kassem et al. (1972) for ascorbic acid, and by Brown and Leeson (1969). The results of the latter are shown in Fig. 4.

It should finally be added that these are cases (which will be discussed in Chapter 7), where oxidation leads to zero-order profiles (Franchini and Carstensen,

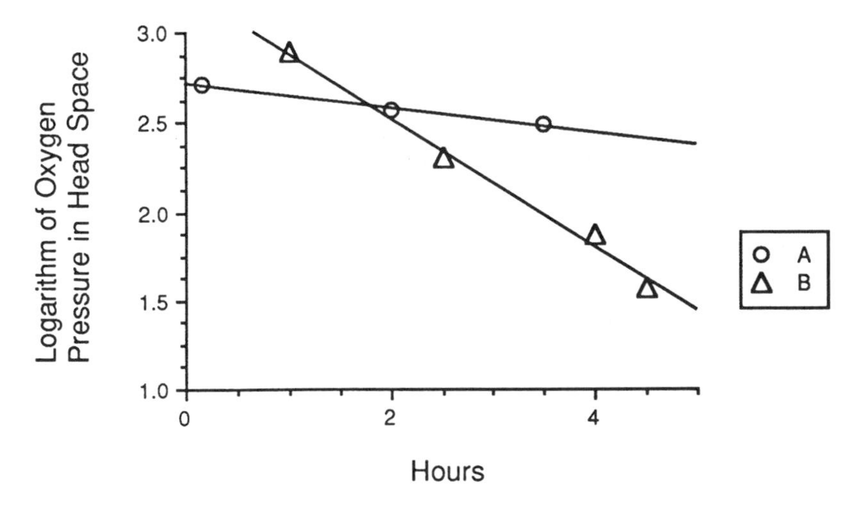

Fig. 4 Head space oxygen content versus time in (A) a system layered with nitrogen and (B) an unprotected system (which extrapolates to zero time at 21% oxygen as it should). (Graphs constructed from data published by Brown and Leeson, 1969.) The least squares fit lines are (A) $y = 2.71 - 0.067x$ and (B) $y = 3.24 - 0.36x$.

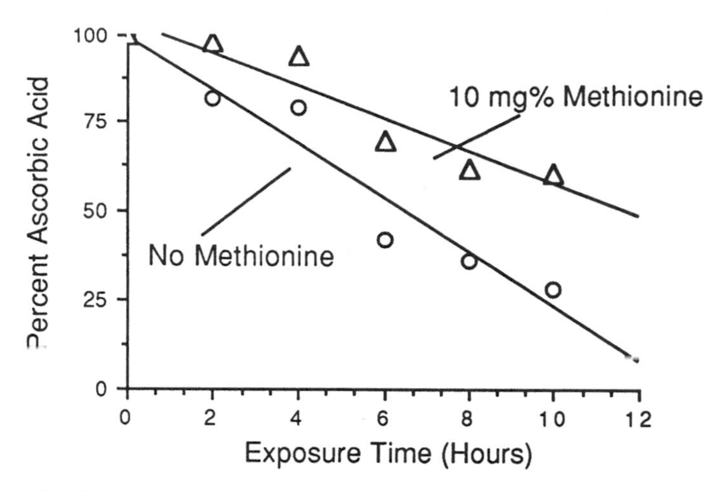

Fig. 5 Photolytic oxidation of ascorbic acid solutions. Least squares fit: 10 mg% methionine: $y = 99.38 - 7.64x$ ($R = 0.97$) and no methionine: $y = 104.19 - 4.67$ ($R = 0.95$). (Graph constructed from data by Asker et al., 1985.)

1995). These are cases where the drug is amphiphilic and where one of the non-micellar n-mers (A_n) oxidizes. In this case there is an excess of oxygen, and the concentration of the reaction species, A_n, is constant (since it will be replenished by the micellar species).

Cases where the oxidation is photochemically driven will often become zero order. An example of this is the work by Asker et al. (1985), whose data are shown in Fig. 5. It should be noted that the data are probably more suitably considered S-shaped curves, such as will be discussed shortly.

1.4. Oxidation in an Open System

It is always worthwhile determining k_2 in a separate experiment in an open system. This is done by bubbling oxygen through an aqueous solution of drug, so that C is the saturation concentration of oxygen (C_{sat}) at all times:

$$\frac{dA}{dt} = -k_2[\mathrm{A}]C_{\text{sat}} = -k_2[\mathrm{A}]KP \tag{4.37}$$

Since the amount of A present is the original amount of drug, $A_r m : o^*$ less the amount decomposed, Y^*, it follows that

$$\frac{dY^*}{dt} = k2[A_0^* - Y^*]KP \tag{4.38}$$

It is noted here that the A^* and Y^* values are V_2 times concentrations. Eq. (4.40) integrates to

$$\ln[A^* - Y^*] = -(k_2KP)t + \ln[A_0^*] \tag{4.39}$$

or

$$\ln\left[1 - \frac{Y^*}{A_0^*}\right] = \ln[\text{fraction retained}] = -(k_2KP)t \tag{4.40}$$

where the initial condition that $X=0$ at $t=0$ has been invoked. Equation (4.39) predicts (a) that the reaction is first order and (b) that the observed rate constant is

$$k_{obs} = k_2KP = k_2C \tag{4.41}$$

where C is the (now virtually constant) oxygen concentration in that aqueous phase. The situation described holds for any situation where the oxygen concentration changes so little that it becomes almost constant and the reaction is pseudo first order.

It has tacitly been assumed that the equilibrium between gaseous and dissolved oxygen is instantaneous, but at low agitation it may be diffusion controlled. A good example of this is the oxidation of sulfites reported by Schroeter (1963).

It can be seen from the data in Fig. 6 that at the lowest bubbling speed, the oxygen supply does not keep up with the decomposition. In general the observed rate constant should be a function of oxygen concentration, which in this model is constant. It can of course be varied by having different mixtures of oxygen and inert gas bubbling through the mixture, and in this case Eq. (4.41) becomes

$$\ln[k_{obs}] = \ln[C] + \ln[k_2] \tag{4.42}$$

Plotting either $\ln[k_{obs}]$ versus $\ln[C]$ or $[k_{obs}]$ versus C should give straight line plots. In the former case the slope would have to be unity. An example of this is (Fig. 7)

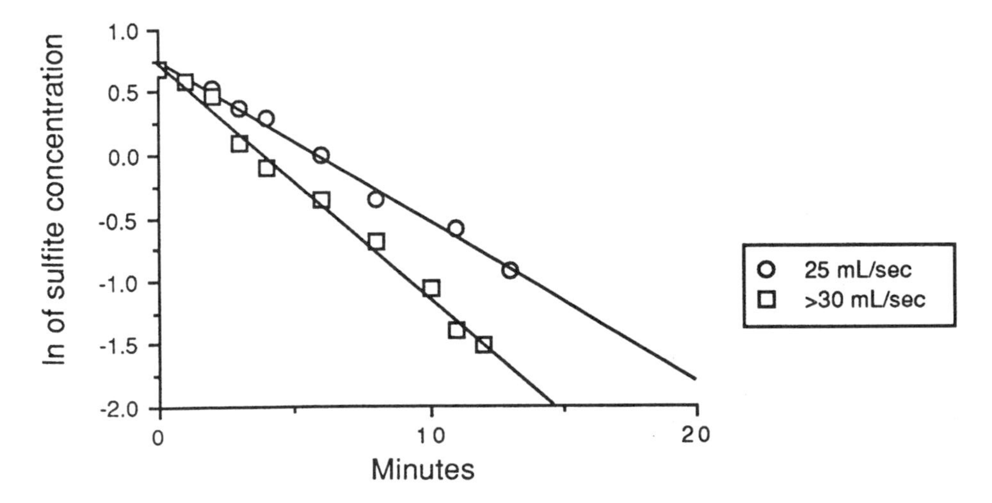

Fig. 6 Sulfite oxidation as a function of rate of bubbling. The least squares fit lines are 25 mg/mL: $y=0.75-0.13x$ and >30 mg/mL: $y=0.734-0.19x$. (Graph constructed from data by Schroeter, 1963.)

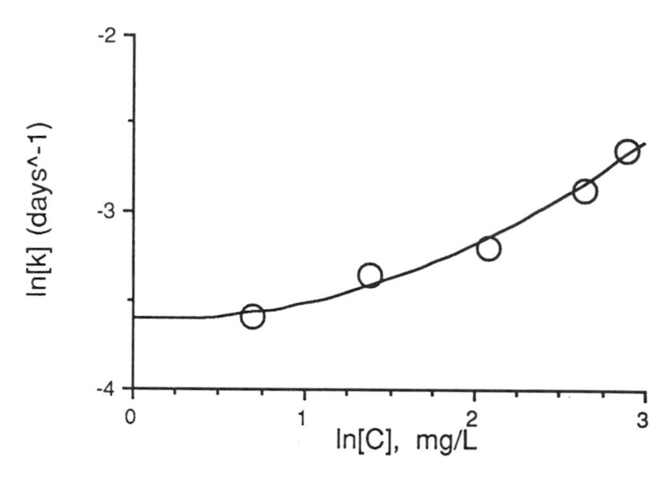

Fig. 7 Effect of oxygen concentration on oxidation rate. If plotted linearly the equation for the trace would be $y = -3.93 + 0.41x$. It is shown parabolically: $y = -3.6 - 0.04x + 0.125x^2$. (Figure constructed from data by Kassem, 1972.)

oxidation of ascorbic acid (Kassem et al., 1972) (where, however, the slope is not unity).

1.5. Oxidative Autocatalysis

In autocatalysis, the rate constant, k_2, is a function of Y. The most all-encompassing equation of this type is the Ng equation, Eq. (4.43) (Ng, 1975) which will be dealt with at another point. Eq. (4.23) assumed that the reaction rate was proportional to drug concentration and reactant concentration. Where this is in deviance from experimental results it may be assumed that it is dependent on powers of these, i.e.,

$$\frac{dY}{dt} = k_2 Y^n(1 - Y)^p \tag{4.43}$$

In many cases, the exponents are simply unity, and hence Eq. (4.43) becomes

$$\frac{dY}{dt} = k_2 \cdot Y[1 - Y] \tag{4.44}$$

where, here, Y denotes fraction decomposed. Profiles of this type are shown in Fig. 8.

There is an initial apparent lag time, and then a precipitous drop. Often, in real systems, it is difficult to duplicate the induction time point, t^*, at which the drop commences. The FDA has often quoted this situation as one that makes the Agency leery about extrapolations, because it is possible to get a good linear fit to the data at times $t < t^*$, and these would extrapolate to high retention values, which in the case cited would be incorrect. Eq. (4.44) is rearranged to

$$\frac{dY}{Y(1 - Y)} = k_2 t \tag{4.45}$$

To integrate Eq. (4.45) it is recalled that

$$\frac{1}{Y(1-Y)} = \frac{1}{Y} + \frac{1}{1-Y} \tag{4.46}$$

so that Eq. (4.45) becomes

$$d\ln[Y] - d\ln[1-Y] = k_2 t \tag{4.47}$$

which integrates to

$$\ln\left[\frac{Y}{1-Y}\right] = k_2(t - t_i) \tag{4.48}$$

where t_i is the point in time where the inflection point occurs. Eq. (4.48) may be expressed as

$$Y = \frac{\exp[k_2 t]}{1 + \exp[k_2 t]} \tag{4.49}$$

This type profile is comparable to a Henderson–Hasselbach curve, i.e., is an inverse S, and it would have the appearance shown in Fig. 8. If power functional dependence of k on Y is larger than 1, then the typical lag time curve will generate.

In the preamble to this section on oxidation, it was stated that oxidation was not the order of the day in pharmaceutical systems. This is exemplified by the number of publications. In J. Pharm. Sci., eg., for the period 1965 to 1990 there are only some 30 publications on the subject, and several of these do not deal with the kinetics but rather with the (very important) problem of proper assignment of degradation products. Some of the articles in the literature (beyond those quoted already) are listed below.

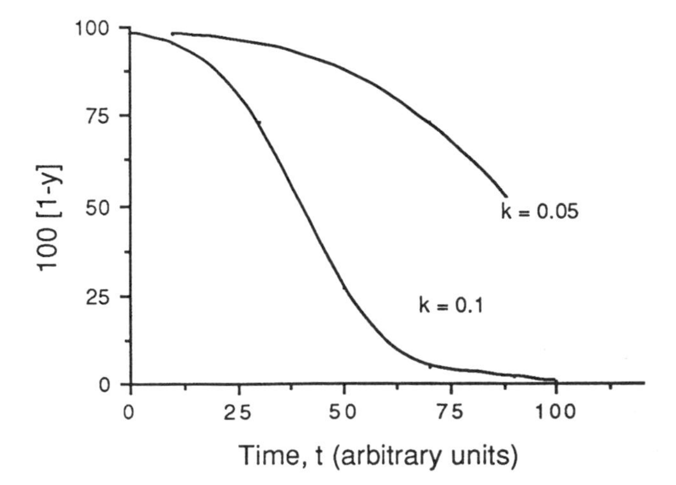

Fig. 8 Trace of Eq. (4.49), the function $\exp(-kt)/(1+\exp\{-kt\}]$ for various values of k_2. The profiles of the function have been generated for $k = 0.01$ and 0.005, and a starting value of 100.

Repta and Beltagy (1981) reported on the oxidation of 6-seleno guanosine in aqueous solution. Underberg (1978) reported on the oxidative degradation of promethazine. Duchene et al. (1986) have reviewed the effect of cyclodextrin complexes on drug oxidative stability. Szejtli et al. (1980) and Szejtli and Bollan (1980) demonstrated that for vitamin D_3 the cyclodextrine complex reacted less rapidly with oxygen. Swarbrick and Rhodes (1965) showed that the maximum oxidation rate of linoleic acid in aqueous systems containing the surfactant Brij 35 was a slightly decreasing linear curve.

Sadhale and Shah (1998) reported on the hydrolysis and oxidation of cefazolin at low (50 μg/g) and high (200 μg/g) concentrations. Under some of the conditions the plots are S-shaped or have downward curvature; in other cases the plots appear semilogarithmic. EDTA slows down the reaction, and if gels are present the degradation constants are 3–18 times lower. In the latter case, the reduced oxygen diffusion may be the reason. Similar results were obtained, although to a lesser degree, with cefuroxime.

2. MORE COMPLEX MODELS

A more general, semi-empirical equation for the treatment of autoxidative reactions (invoking some power dependence of Y and $(1 - Y)$ is

$$\ln\left\{\frac{Y^n}{(1-Y)^p}\right\} = -k'(t - t_i) \tag{4.50}$$

A more manageable form of this equation is

$$\ln\left\{\frac{Y^q}{1-Y}\right\} = -k(t - t_i) \tag{4.51}$$

where

$$k = \frac{k'}{p} \tag{4.52}$$

and

$$q = \frac{n}{p} \tag{4.53}$$

The problem exists as to how to treat Eq. (4.51) and obtain parameters from the untransformed Y versus t equation, since n and p and k are not known. The reason for recasting Eq. (4.50) is to reduce the number of iterants to one. Large number of iterants in a model are suspect, because even good but not quite rigorous first estimates may lead to secondary minima, and it is always a good practice to reduce the iterants to as few as possible. Equation (4.51) only requires one iterant, so that data treatment is not all that questionable.*

* Iteration with more than one iterant always leaves the question of whether the minimum obtained by iteration is a primary or a secondary minimum. An example of this is shown in Chapter 3.

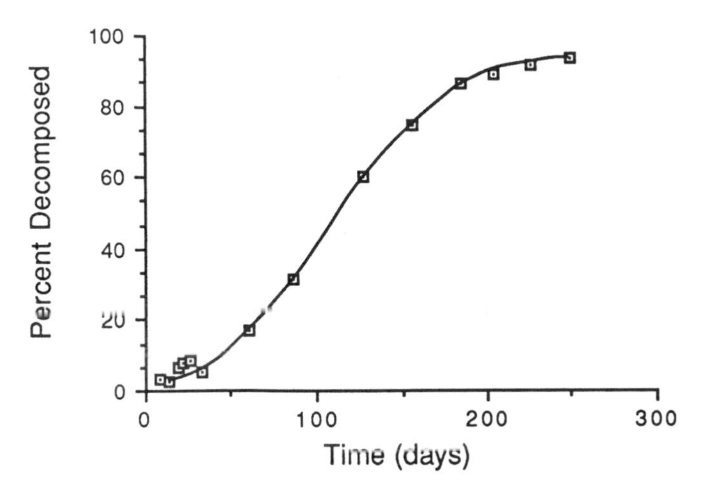

Fig. 9 Decomposition of retinoic acid at 37°C. (Graph constructed from data reported by Tan, Melzer and Lindenbaum, 1993.)

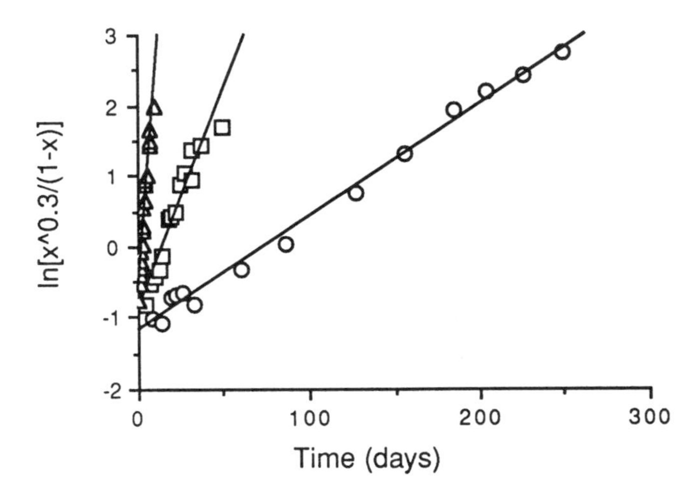

Fig. 10 Treatment of Equation (2.33) using $q = 0.3$. (Data reported by Tan et al., 1963).

But even then is it necessary to have a reasonable estimate of the iterant, q, before starting the iteration procedure. Such an estimate can be obtained (and in obtaining the estimate, often the iteration is unnecessary) by the considerations to follow. Consider the data (Fig. 9) reported by Tan et al. (1993).

Manual trial and error will often give reasonable estimates of iterants (in this case q), and the value $q = 0.3$, arrived at in a few tries, gives good results, as shown in Fig. 10. This value could then be used in computer iteration, as a first estimate.

The least squares fit lines are

$$37°\text{C}: \quad \ln\left[\frac{x^{0.3}}{1-x}\right] = -1.192 + 0.0160t \tag{4.54}$$

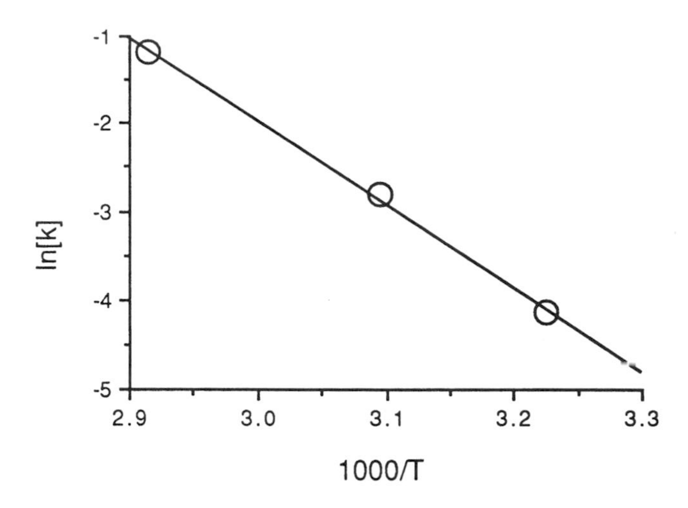

Fig. 11 Arrhenius plot of data in Fig. 4.10.

$$50°\text{C:} \quad \ln\left[\frac{x^{0.3}}{1-x}\right] = -0.881 + 0.0613t \tag{4.55}$$

$$70°\text{C:} \quad \ln\left[\frac{x^{0.3}}{1-x}\right] = -0.654 + 0.307t \tag{4.56}$$

When the slopes (the rate constants, in units of day^{-1}) are plotted versus absolute inverse temperature, a good Arrhenius fit is obtained as shown in Fig. 11. It is observed that the activation energy is about 18 kCal/mole, which is a reasonable figure.

2.1. Stability Prediction by Fractional Lives

It has been shown in Chapter 2 that for zero- and first-order reactions it is possible to assess and extrapolate stability from high-temperature data by use of fractional lives.

Tan et al. (1993) have extended this principle to oxidations, and Yoshioka et al. (1994) have extended it to complex reactions following neither of the more describable orders. They consider reactions where the fraction decomposed is a second-degree polynomial in time, t, and (more frequently) a triple exponential formation, i.e.,

$$x = A\exp(at) + B\exp(bt) + C\exp(ct) \tag{4.57}$$

or

$$x = a + bt + ct^2 \tag{4.58}$$

When treating decomposition of proteins by either of these equations, they obtain best parameters and hence can calculate the point in time where $x = 0.1$

(which is usually denoted t_{90}). They then show by this empirical mode that in the cases studied,

$$\ln[t_{90}] = \frac{Q}{T+q} \tag{4.59}$$

It is noted that the two presentation modes for the appearance of x with time do not contain what is usually thought of as a rate constant. Hence, unlike the treatment of zero- and first-order equations, there is no way of "invoking" a temperature dependence on k or t_{90}. The method has great practical value.

To the contrary, Tan et al. (1993) attempted to obtain an explanation for the fact that fractional lives (from $\beta = 0.1$ to 0.9) give linear plots, but they found that the lines change "activation energy". The argument put forth by Tan et al. (1993) is the following: the decomposition reaction may be characterized by the differential equation

$$\frac{dx}{\phi(x)} = k\,dt \tag{4.60}$$

where $\phi(x)$ is a function that is not "simple." If this is integrated (if it can be integrated), then the integral is denoted $F(x)$, i.e.,

$$F(x) = kt \tag{4.61}$$

or

$$F(\beta) = kt_\beta \tag{4.62}$$

where β is the fractional life. By now invoking that

$$k = Z\exp\left(\frac{-E_a}{RT}\right) \tag{4.63}$$

it follows that

$$F(\beta) = Z\left\{\exp\left(-\frac{E_a}{RT}\right)\right\}t_\beta \tag{4.64}$$

or

$$\ln[t_\beta] = \ln\left(\frac{F(\beta)}{Z}\right) + \frac{E_a}{RT} \tag{4.65}$$

i.e., pro forma, the linearity of $\ln[t_\beta]$ versus $(1/T)$ would appear to be proven. It should be noted that integrals of the type $dx/\phi(x)$ can rarely be solved in a manner such as to have only one "k-value."

If one takes the example by Yoshioka et al. (1994), where x is a polynomial in t, then one would have to express dx/dt as a function of x, not of t. Here

$$x = a + bt + ct^2 \tag{4.66}$$

If this is differentiated so that dx/dt is expressed as a function of x, not of t, then

$$\frac{dx}{dt} = b + ct \tag{4.67}$$

To eliminate t, Eq. (4.66) is solved with respect to t, and

$$t = -\frac{b}{2c} \pm \left[\frac{b^2}{4c^2} - \frac{a-x}{c}\right]^{1/2} \tag{4.68}$$

If

$$\left[\frac{a-x}{b}\right]^2 \ll \frac{b^2}{4c^2} - \frac{a-x}{c} \tag{4.69}$$

then

$$t \approx \frac{a-x}{b} \tag{4.70}$$

so

$$\frac{dx}{dt} = b + \left\{\frac{a-x}{b}\right\}c = A + Bx \tag{4.71}$$

where

$$A = b\left[1 + \frac{a}{c}\right] \tag{4.72}$$

and

$$B = -\frac{c}{b} \tag{4.73}$$

It is noted that even with a fairly simple function (a second-order polynomial) it is not possible to obtain an equation of the type of Eq. (4.60), where there is a definitive k. At best one might expect B to follow an Arrhenius equation.

Hence the method only has theoretical basis if $dx/\phi(x)$ can be presented in a form, and in such a way, that the rate constant, k, is a meaningful quantity.

The method, nevertheless, has great practical importance, because it allows extrapolations in a simple manner.

The data by Yoshioka are shown in Fig. 12. Again, the "activation energies" are high, but no theoretical importance should be placed on them.

3. ANTIOXIDANTS

Antioxidants work by consuming oxygen at a faster rate than the rate at which the drug substance reacts with oxygen; and in such cases they will protect the drug substance until they are completely used up.

This means that the use of antioxidants imposes a lag time upon the decomposition profile of the drug.

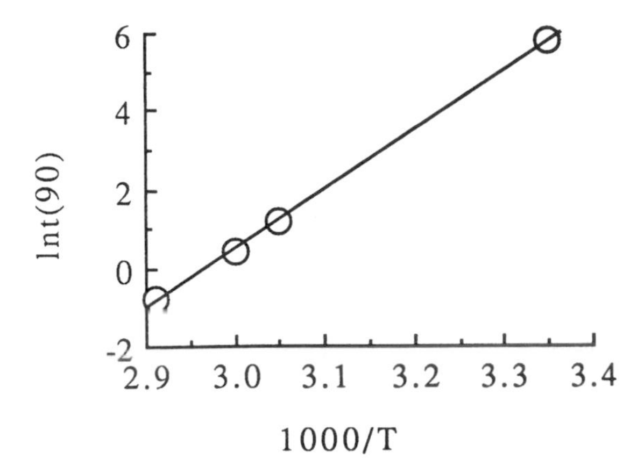

Fig. 12 Decomposition of a-chymotrypsin tablet. ln[t_{90}] plotted in Arrhenius fashion. The neat decomposition is not an integer order. (Graph constructed from data published by Yoshioka et al., 1994.)

An example is the work by Pudapeddi et al. (1992), who showed that sulfites act in a quantitative manner. Once the sulfites are consumed, the oxidation starts. Gonçalves et al. (1998) have reported on the antioxidant activity of 5-aminosalicylic acid in the presence of vitamins C and E of lipid peroxidation. They show typical S-shaped decomposition curves. Chakrabarti et al. (1993) have shown that hydroquinone, butylated deoxycholate, and ascorbyl palmitate stabilizes hyamycin (a polyene antifungal antibiotic).

Antioxidants can also act by interfering in e.g. the reaction schemes shown in Eqs. (4.2–4.11) so that the oxidative pathway is interrupted. In such a case they will themselves be regenerated, and will function in a manner independent of elapsed time. The most common antioxidants used are ascorbic acid, BHA, BHT, and sodium sulfite. The use of ethylene diamine tetraacetic acid as a chelator has already been mentioned.

4. OTHER WORK

The first step in oxidative (and any other) kinetic investigation is decomposition product identification. It is then possible later to study the kinetics of the system. For instance, Hooijmaaijer et al. (1999) have studied the peroxide catalyzed degradation of mocophenolate mofetyl in aqueous solution and have characterized the decomposition products.

As mentioned, until recently, the number of reports on oxidation in the pharmaceutical literature were scarce, but the gap is being filled. For instance, Bosca et al. (1992a) have described the oxidative decarboxylation of naproxen, and Bosca et al. (1992b) have described the photochemical byproducts. Vargas et al. (1992) have described the photochemical oxidation in light of nifedipine. Tereoka et al. (1993) have described the oxidation of ranitidine in acetate buffer.

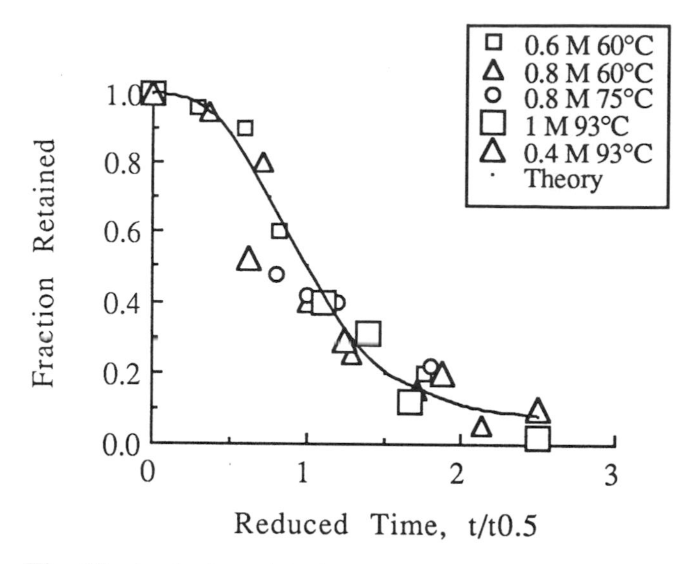

Fig. 13 In the legend *M* denotes molarity of phosphate buffer. The pH employed was 8.0 and the concentration of drug was 0.2 mg/mL. (Graph constructed from data published by Franchini and Carstensen, 1994.)

Franchini and Carstensen (1994) have studied the auto-oxidation (in the presence of trace quantities of heavy metals) of the oxygen-sensitive compound β-[2-(2-carboxyethyl)-thio-2]-[-8-2-phenylacetyl]-benzenepropanoic acid. This is denoted A in the following.

The degradation is an oxidation leading to a sulfoxide and a cinnamic acid. The mechanism that applies to the decomposition is

$$A + A \rightarrow A_2 \qquad \text{with equilibrium constant } K$$

and

$$A_2 + O_2 \rightarrow B \qquad \text{with rate constant } k_2$$

The ensuing rate equation is fourth order in A_{obs}. The authors studied the decomposition at a series of buffer concentrations and temperatures, and some of their data are shown in Fig. 4.13. They showed that the reaction was subject to kinetic salt effect, but that buffer catalysis was absent.

REFERENCES

Asker, A. F., Larose, M. (1987). Drug. Dev. Ind. Pharm. 13:2239.

Asker, A. F., Canady, D., Cobb, C. (1985). Drug. Dev. Ind. Pharm. 11:2109.

Bosca, F., Martinez-Manez, R., Miranda, M., Primo, J., Soto, J., Vano, L. (1992a). J. Pharm. Sci. 81:479.

Bosca, F., Miranda, M., Vargas, F. (1992b). J. Pharm. Sci. 81:181.

Brown, J. C., Leeson, L. (1969). J. Pharm. Sci. 58:513.

Chakrabarti, P. K., Harindran, J., Saraf, P. G., Wamburkar, M. N. (1993). Drug Dev. Ind. Pharm. 19:2595.

Duchene, D., Vaution, C., Glomot, F. (1986). Drug. Dev. Ind. Pharm. 12:2193.
Franchini, M., Carstensen, J. T. (1994). Int. J. Pharmaceutics 111:153.
Franchini, M., Unvala, H., Carstensen, J. T. (1993). J. Pharm. Sci. 82:550.
Gonçalves, E., Almeida, L. J., Dinis, T. C. P. (1998). Int. J. Pharm. 172:219.
Hooijmaaijer, E., Brandl, M., Nelson, J., Lustig, D. (1999). Drug Dev. Ind. Pharm. 25:361.
Kassem, M. A., Kassem, A. A., Ammar, H. O. (1969). Pharm. Acta Helv. 44:667.
Kassem, M. A., Kassem, A. A., Ammar, H. O. (1972). Pharm. Acta Helv. 47:97.
Ng, W. L. (1975). Aust. J. Chem. 28:1169.
Nowaczyk, F. J., Jr., Schnaare, R. L., Ofner, C. M., III, Wigent, R. J. (1993). Pharm. Res. 10.305.
Pudipeddi, M., Alexander, K., Parker, G.A., Carstensen, J. T. (1992). Drug Dev. Ind. Pharm. 18:2135.
Repta, A. J., Beltagy, Y. A. (1981). J. Pharm. Sci. 70:635.
Sadhale, Y., Shah, J. C. (1998). Pharm. Dev. Tech. 3:549.
Schroeter, L. (1963). J. Pharm. Sci. 52:559.
Swarbrick, J., Rhodes, C. T. (1965). J. Pharm. Sci. 54:903.
Szejtli, J., Bollan, E. (1980). Starke 32:386.
Szejtli, J., Bolla-Pusztai, E., Szabo, P., Ferenczy, T. (1980). Pharmazie 35:779.
Tan, X., Meltzer, N., Lindenbaum, S. (1993). J. Pharm. Biomed. Anal. 11(9): 817.
Tereoka, R., Otsuka, M. and Matsuda, Y. (1993). J. Pharm. Sci. 82:601.
Timmins, P., Jackson, I. M., Wang, Y. J. (1982). Int. J. Pharmaceut. 11:329.
Underberg, W. J. M. (1978). J. Pharm. Sci. 67:1133.
Vargas, F., Rivas, C., Machado, R. (1992). J. Pharm. Sci. 81:399.
Ventura, P., Parravicine, F., Simonotti, L., Colombo, R., Pifferi, G. (1981). J. Pharm. Sci. 75:308.
Winkler, L. W. (1888). Berichte der Deut. Chem. Ges. 21:2843.
Yoshioka, S., Aso, Y., Izutsu, K.-I., Terao, T. (1994). J. Pharm. Sci. 83:454.

5

Catalysis, Complexation, and Photolysis

JENS T. CARSTENSEN

Madison, Wisconsin

1. CATALYSIS

General and specific acid and base catalyses have been discussed in Chapter 3, and are one example of catalysis. When discussing catalysis, however, the metal induced decomposition is what comes to the pharmaceutical investigator's mind. In parenterals especially, great care is taken to exclude metals, because only slight decomposition caused by trace metals may cause sufficient discoloration to render the product unsatisfactory. Examples of this are thiamine hydrochloride injectables and ascorbic acid injectables.

Metals are most detrimental in oxidations, as shown in the previous chapter. Examples of metal catalyzed oxidation in pharmaceutical systems are cyanocobalamine (which is stabilized at very low concentrations, but destabilized at higher concentrations of ferrous ion), erythromycin (which is stabilized by such ions as mercuric, magnesium calcium, ferric, and aluminum, and destabilized by cobaltous, plumbic, zinc, and nickel) and (Kassem et al., 1969) ascorbic acid (which, in general, is destabilized by metal ions).

Figure 1 shows data by Kassem et al. (1969). Barcza and Lenner (1988) have shown that chloral hydrate forms hydrogen bonded complexes with halide ions

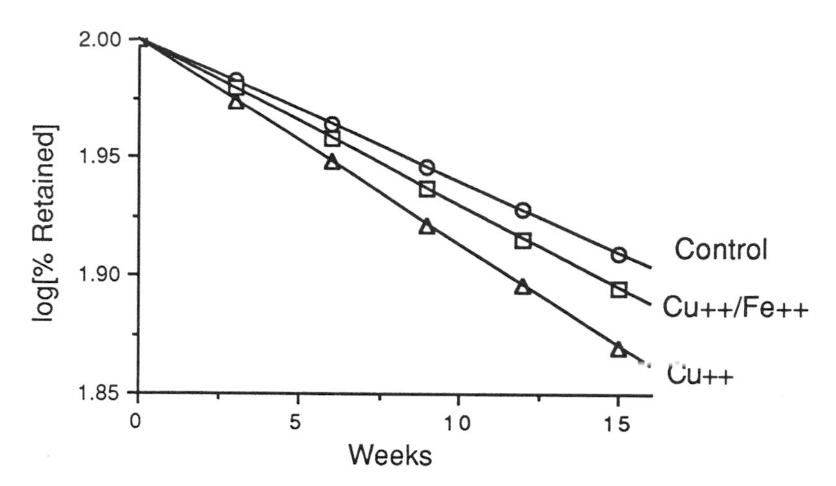

Fig. 1 Effect of metal ions on the decomposition of ascorbic acid. Least squares fit lines are (Control) $y = 2 - 0.006x$, (with copper and iron) $y = 2 - 0.007x$, (with copper alone) $y = 2 - 0.0087x$. (Figure constructed from data published by Kassem, 1969.)

in aqueous solutions. They are 1 : 1 complexes with relatively high stability constants. Upadrashta and Wurster (1988) used ethylene diamine tetraacetic acid to protect anthralin solutions from metal catalyzed oxidation. Tomida et al. (1987) showed that zinc ion increased the degradation of cephalosporins in tromethamine solution. The second-order rate constant, divided by $[Zn_o]$ plotted versus pH has unity slope from pH 7.5 to 8.5. They suggest the formation of a ternary complex (penicillin-Zn^{2+}-tromethamine).

Substances such as vitamin A and D are also prone to metal ion catalyzed decomposition.

Of other works in this area, McCrossin et al. (1998) reported on the effect of guanidine HCl on degradation of recombinant porcine growth hormone at alkaline pH and different concentrations and found it to be first order.

Fredholt et al. (1999) studied the catalytic effect of α-chymotrypsin on desmopressin decomposition and reported on the influence of concentration, pH, and cyclodextrin. The reaction is presumably A-B-C and the disappearance rate of the compound is first order. The pH profile is type AHJD with a maximum at pH 7.7.

2. COMPLEXATION

It is obvious that in many cases, drugs may complex with one or more of the ingredients in a solution dosage form. Sometimes this is intentional, e.g. bioavailability in certain instances may be improved (Levy and Reuning, 1964; Newmark et al., 1970). In other instances the stability of a drug is favorably affected by complexation, although in many cases the opposite is the case.

The basic principles of complex formation have been reviewed by Connors and Mollica (1964) and demonstrated by them as well. Only the formation and stability of 1 : 1 complexes will be covered here. For coverage of 1 : 2 and 2 : 1 complexes, the reader is referred to the work by Connors and coworkers (Rosanske and Connors, 1980; Connors and Rosanske, 1980; Pendergast and Connors, 1984).

In the following, A will denote drug (substrate) and B will denote complexing agent (ligand). (It should be noted that either could be called either, and that there is no generally accepted nomenclature).

If A complexes with B by the scheme

$$A + B \leftrightarrow AC \tag{5.1}$$

then the complex is denoted 1 : 1. The terminology 1 : 2 or 2 : 1 is then obvious. The equilibrium (stability) constant of the complex AB is given by

$$K = \frac{[AB]}{[A][B]} \tag{5.2}$$

In a 1 : 1 complex (and other types as well), it would be fortuitous if the stability of both the drug and the complex were identical. In other words, in degrading, there will be two decomposition reactions:

$$A \rightarrow \text{products (rate constant } k) \tag{5.3}$$

and

$$AB \rightarrow \text{products (rate constant } k^*) \tag{5.4}$$

The analytically measured quantity is

$$C = [A] + [AB] \tag{5.5}$$

except if the study is carried out by other than chemical means, e.g., if one species is charged, then conductimetry might elucidate the concentration of one of the species.

The analytically measured rate is

$$-\frac{dC}{dt} = k[A][B] + k^*[AB][B] = \{k + k^*K[B]\}[A][B] \tag{5.6}$$

where use has been made of Eq. (5.2) for the last step.

The apparent rate with which the reaction proceeds is given by

$$\begin{aligned} -\frac{dC}{dt} &= k_{obs}[B]C = k_{obs}[B]\{[A] + [AB]\} \\ &= k_{obs}[B][A]\{1 + K[B]\} \end{aligned} \tag{5.7}$$

where Eq. (5.2) has been used for the last step. Equating Eqs. (5.6) and (5.7) then gives

$$\{k + k^*K[B]\}[A][B] = k_{obs}[B][A]\{1 + K[B]\} \tag{5.8}$$

Dividing through by [A][B] gives

$$k + k^*K[B] = k_{obs}\{1 + K[B]\} \tag{5.9}$$

Table 5.1 Effect of β-Cyclodextrin on Benzocaine Stability

Concentration of cyclodextrin (%)	k_{obs} (h^{-1} M^{-1})
0	0.666
0.25	0.358
0.5	0.299
1	0.129

Source: Data from Lach and Chin (1964a).

from which we obtain

$$k_{obs} = \frac{k + k^* K[B]}{\{1 + K[B]\}} = k + \frac{(k^* - k)K[B]}{1 + K[B]} \tag{5.10}$$

which may be expressed reciprocally as

$$\frac{1}{k_{obs} - k} = \frac{1}{k^* - k} + \frac{1}{K(k^* - k)} \cdot \frac{1}{[B]} \tag{5.11}$$

If the amount of [B] complexed in small, then [B] is synonymous with the amount of [B] added. Hence if kinetic studies were carried out in a series of systems with different concentrations of [B], then a reciprocal plot should be linear.

Example 5.1.

Lach and Chin (1964a,b) reported kinetic data for benzocaine, complexed with betacyclodextrin (Table 1). Calculate the equilibrium (stability) constant between benzocaine and betacyclodextrin, assuming a 1 : 1 complex.

Answer.

First of all the concentration units have to be consistent. Beta-cyclodextrine is taken to have a molecular weight of 1700, so that e.g. 0.25% = 2.5 g/L = 2.5/1700 = 1.47×10^{-3} molar. This is shown in column 2 of Table 2. The sixth column lists the reciprocal of these figures, e.g., $1/0.00147 \times 680$ M^{-1}.

The values for $(k_{obs} - k)$ are listed in the fourth column, e.g., for the second entry, $0.229 - 0.666 = -0.43$.

The fifth column then lists $1/(k_{obs} - k)$, e.g. $1/(-0.537) = -1.862$.

The fifth column, $(k - k_{obs})$ is then plotted versus the sixth column, $1/[B]$. This is shown in Fig. 2.

It is seen that the slope $= -0.0027$ (M^2h) and that the intercept is -1.3822 (Mh). Equation (5.11) predicts that the slope-to-intercept ratio is K, i.e.,

$$K = \frac{\text{slope}}{\text{intecept}} = \frac{-0.0027}{-1.3822} = 2 \times 10^{-3} M^{-1} \tag{5.12}$$

Table 5.2. Treatment of Data in Table 1. Effect of β-Cyclodextrin on the Decomposition of Benzocaine in Solution

Concentration of cyclodextrin					
%	$10^3 \times$ Molar	k_{obs} $h^{-1}M^{-1}$	$(k_{obs} - k)$	$1/(k - k_{obs})$	$1/[B]$
0	0	0.666	0		
0.25	1.47	0.368	0.298	−3.247	680
0.5	2.94	0.299	−0.467	−2.288	340
1	5.88	0.129	−0.537	−1.862	170

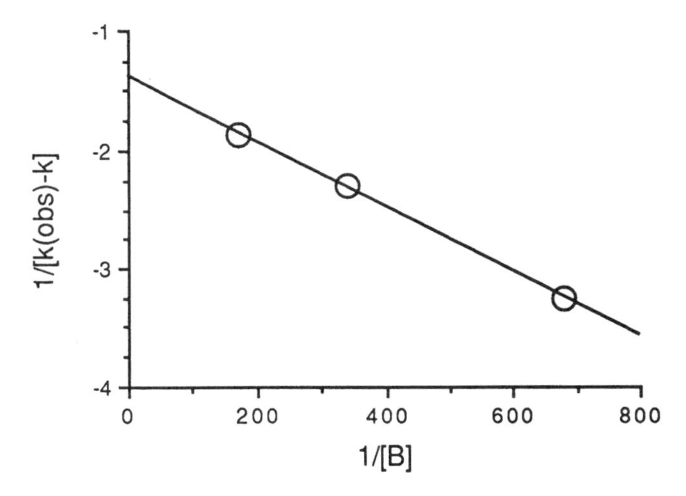

Fig. 2 Data from Table 2. (Graph constructed from data published by Lach and Chin, 1964.)

Complexation constants can also be deduced by determining the solubility of A in aqueous (or other) solutions of different concentrations of B. An example of this is the work by Chen et al. (1994).

Chen et al. (1994) have studied the complexation of adenine with a series of ligands. Their data from using caffeine as a ligand are shown in Fig. 3. The complexation constant is obtained by Eq. (5.13) by which

$$K = \frac{\text{slope}}{\text{intercept} \times (1 - \text{slope})} = \frac{0.26}{(0.0076 \times 0.74)} = 46.2\ (M^{-1}) \tag{5.13}$$

The pharmaceutical literature dealing with complexes is not abundant. One interesting example is urea including compounds, which were in vogue in the 1960s (and probably still are, although they are not as frequently published on), an example being the thiourea and urea inclusion compounds of alpha-lipoic acid methyl ester reported by Mima and Nishiwasha (1964). Other, presently quite researched, com-

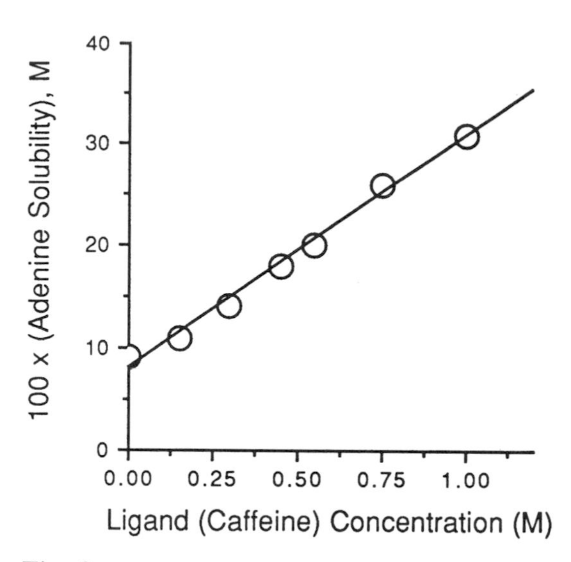

Fig. 3 Least squares fit: $y = 0.26x + 0.0076$, where x is solubility (rather than $100 \times$ solubility, as used in the figure). (Graph constructed from data published by Chen et al., 1994.)

plexes are those with cyclodextrins: for instance Connors and coworkers have published substantially in this area (Pendergast and Connors, 1984; Connors et al., 1982). Duchene et al. (1986) have reviewed the effect of cyclodextrin complexes on drug stability. Lach and Chin (1964a) have studied the complexes of cyclodextrins with benzocaine and found them to be 1 : 1 complexes. Higuchi and Lachman showed that benzocaine complexed with caffeine (1955). Lach and Chin (1964b) showed that a series of substituted benzoic acids complexed with cyclodextrin. Cyclodextrins are unique (alpha, beta, and gamma) in that they are doughnut shaped molecules that allow inclusion of the drug molecule. Other (non-macrocyclic) carbohydrate molecules also complex, for instance Gupta (1983) has shown that procainamide complexes with glucose, lactose, and maltose. All of these have a hemiacetal group, as opposed to sucrose and fructose. The complex formation was pH dependent.

Complexation constants may be obtained by spectrophotometric means as well. The complex can exhibit a specific maximum in the ultraviolet spectral range, so its concentration can be distinguished from that of the parent species such as in the case of the alendronate/Cu^{++} complex and the case of metal complexes of anhydrotetracycline (Siqueira et al., 1994).

2.1. Complexing Agents

Caffeine and polyvinylpyrrolidone were the most common complexing agents used in pharmaceutics for a series of years. In recent years the cyclodextrins have become of importance.

An example of this is the work by Van Der Houwen et al. (1994) dealing with the kinetics of 7-*N*-(*p*-hydroxyphenyl)mitomycin C (M-83) in the presence of γ-cyclodextrin. The pH profile of this is V shaped with a couple of extensions of lesser slope and with a minimum at pH 7. Cyclodextrins have also been used in stabilization of monoclonal antibodies (Ressing et al., 1992).

Cyclodextrins (CDs) are powerful complexing agents, and much work has centered about their use in drug solubilization and stabilization. They are the subject of a fair amount of pharmaceutical research. For instance, Scalia et al. (1998) have described the complexation of butyl-methoxydibenzoilmethane with hydroxypropyl-β-cyclodextrin. Veiga and Ahsan have described the interaction between tolbutamide and β-cyclodextrins. The tolbutamide phase solubility diagram, in this case, shows a maximum in solubility at a concentration of β-CD of 0.008 M. Such maxima are usually ascribed to the limit of solubility of the complex. Ventura et al. (1998) have described the interaction between papaverine and modified and natural β-cyclodextrins. The phase-solubility diagram is such that the papaverin solubility increases monotonically (in a straight line or curved) with β-CD concentration. Ventura et al. (1997) have shown the increase in aqueous solubility of ursodesoxycholic and chenodesoxycholic acids by complexation with β-CDs.

Miyake et al. (1999) have demonstrated the inclusion compounds of itraconazolone with 2-hydroxypropyl-β-cyclodextrin in propylene glycol solution. Antoniadou-Vyza et al. (1997) have reported that the hydrolysis rate constant of methocarbamol is reduced almost 50% when complexed with hydroxypropyl-β-cyclodextrin. Vianna (1998) et al. showed that complexes of dexamethasone acetate with cyclodextrin showed marked improvement in aqueous solution stability and gave rise to first-order decomposition. Másson et al. (1998) have shown that chlorambucil and indomethacin have greatly improved (first-order) aqueous stability when complexed with a variety of cyclodextrins. Diazepam behaved in different manners depending on the particular cyclodextrin used.

Less common ligands such as dextran stabilize porcine pancreatic elastase (Chang et al., 1993).

3. PHOTOLYSIS

Attention is given to light stability in the 1993 ICH guidelines, although at the time of this writing, testing methods are not finalized. This is exemplified in lines 471–473:

> Light testing should be an integral part of stress testing. [The standard conditions for light testing are still under discussion and will be considered in a further ICH document.] (471–473).

In photolysis, light ($h\nu$) is absorbed by the solution and activates a species in it. ($h\nu$) here represents a quantum of light, where h is Planck's constant ($h = 6.626 \times 10^{-27}$ erg s) and ν (in units of s^{-1}) is the frequency of the light. The activated species (denoted by [*]) then returns to ground state, it either

1. Emits light (of a different frequency ν'); this is referred to as fluorescence or phosphorescence
2. Causes the activated species to decompose, in which case one deals with photolysis. The simplest sequences in photolysis would, therefore, be

$$A + (h\nu) \rightarrow [A^*] \tag{5.14}$$

$$[A^*] \rightarrow \text{products} \tag{5.15}$$

$$[A^*] + A \rightarrow 2A \tag{5.16}$$

At times a second component of the system, B, may preferentially absorb light, in which case photosensitization may occur:

$$B + (h\nu) \rightarrow [B^*] \tag{5.17}$$

$$[B^*] + A \rightarrow B + [A^*] \tag{5.18}$$

If only (5.17) predominates, and is followed by

$$[B^*] \rightarrow \text{products} \tag{5.19}$$

or

$$[B^*] \rightarrow B + (h\nu'') \tag{5.20}$$

then B is referred to as a screening agent. In the last case, [B] will not decrease with time and hence will protect the other photosensitive compounds in the preparation during the shelf life of the product.

When a photolysis experiment is carried out, one should know the quantum yield, i.e., the number of molecules decomposed as a function of the number of quanta absorbed. To measure the latter, use is made of an actinometer (which is a photosensitive system with known quantum yield). A simple and reliable actinometer is the chloroacetic acid actinometer, where the reaction is

$$ClCH_2COOH + H_2O + (h\nu) \rightarrow HOCH_2COOH + HCl \tag{5.21}$$

This has a quantum yield of 1.0 at concentrations of 0.3–0.5 molar. The use is simple. A product is placed in a container, e.g., a clear ampul, or a specially designed apparatus; then, directly before or after, a chloroacetic acid solution is studied under the same circumstances, for a given length of time, t, (e.g., 5 minutes). The hydrochloric acid formed is then titrated and the number of moles calculated, and this is then converted to number of molecules, N. If the drug substance subsequently is irradiated for t^* minutes, then the number of quanta absorbed (provided the absorbency of preparation and actinometer is the same) is given by

$$(h\nu) = \frac{t^* N}{t} \tag{5.22}$$

If the UV spectrum of the substance is known, and if the spectral distribution of the light source is known, then the absorbency (fraction f) of the drug substance solution in relation to the chloroacetic acid solution can be calculated, and the number of quanta are in this case

$$(h\nu) = \frac{f t^* N}{t} \tag{5.23}$$

The 1993 ICH Stability Guidelines promote the general philosophy of attempting to establish a reaction order.

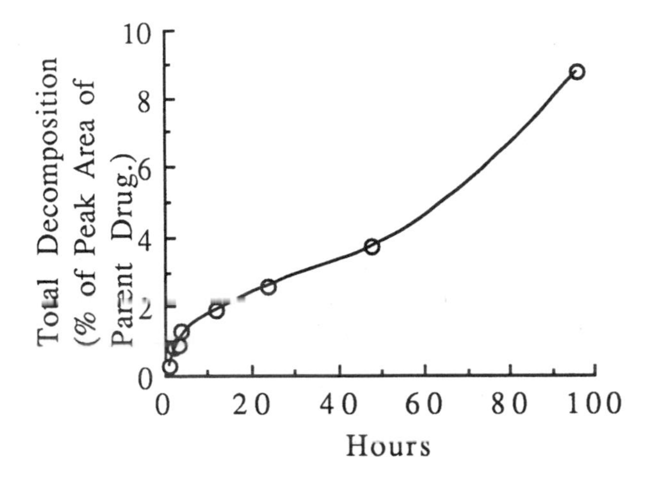

Fig. 4 Photolysis of ciprofloxacin. Irradiation with natural/artificial light. (Figure constructed by averaging of data from Table 2 of the publication by Tievenbacher et al. 1994.)

> The nature of the degradation relationship will determine the need for transformation of the data for linear regression analysis. Usually the relationship can be represented by a linear, quadratic or cubic function on an arithmetic or logarithmic scale. Statistical methods should be employed to test the goodness of fit on all batches and combined batches (where appropriate) to the assumed degradation line or curve (303–308).

In pharmaceutical systems, most reported photolysis has been first order. Examples of this are cefatoxime photolysis (Lerner et al., 1988) and the work by Mizuno et al. (1994). They point out that the wavelength of the irradiating light plays an important part in photodecomposition. They showed the wavelength influence on the photodegradation of ethyl 2-[4,5-bis(4-methoxyphenyl)thiazole-2-yl]pyrrol-1-ylacetate in solution.

Another example of photodegradation is the work by Tievenbacher et al. (1994) dealing with a series of antimicrobial quinolones. Their data are shown in Fig. 4.

It is noted that here the reaction is not first order. More rarely is the reaction zero order except when it is an oxidation, in which case it often is zero order. The work by Asker et al. (1985) and Asker and Larose (1987), and the oxidation of chlorpromazine (Ravin et al., 1978; Felmeister et al., 1965) are examples of this.

In general the time frame in photolysis is different from that in usual kinetics, and this is also addressed in the 1993 ICH Stability Guidelines:

> Frequency of testing should be sufficient to establish the stability characteristic of the drug product. Testing will normally be every three months over the first year, every six months over the second year and then annually (271–273).

It is obvious that photolysis is a stress situation and that trimonthly time protocols would be unrealistic. For solution products, the testing should be carried out both in bulk, unprotected solution (and that is what elucidates the kinetic schemes and rates) and in the final package. This latter is more truly a package test. The 1993 ICH Stability Guidelines addresses this as follows:

> The testing should be carried out in the final packaging proposed for marketing. Additional testing of unprotected drug product can form a useful part of the stress testing and pack evaluation, as can studies carried out in other related packaging materials in supporting the definitive pack(s) (276–279).

Some additional pharmaceutical examples of photolyses are the following: Fabre et al. (1993) have studied the photoisomerization kinetics of befurroxime axetil and found them to follow an A-B-C reaction. Matsuda and Masahara (1983) have shown that ubidecarenone is photochemically decomposed by a first-order process, and that the activation energy of a solution is different from that in solid state. Heat, light, and metal ions (e.g. copper) accelerate the oxidative decomposition of vitamin D3. This can be retarded and actually inhibited by the use of β-cyclodextrin (Szeijtli et al., 1980; Szejtli and Bollan, 1980). Vitamin A stability is also increased when complexed with cyclodextrins (Kyoshin, 1982). Asker and Larose (1987) have shown that uric acid increases the photostability of sulfathiazole sodium solutions, and Asker et al. (1985) have shown that dl-methionine increases the photostability of ascorbic acid in solution. Vandenbossche et al. (1993) have reported on the photostability of molsidomine in infusion fluids. Akimoto et al. (1985) have reported on the photostability of cianidanol in aqueous solution. In this case there is a leveling off, and the approach to the plateau is first order. Tønnesen et al. (1997) have reported on the photoreactivity of mefloquine hydrochloride in the solid state. Baertschi (1997) has discussed the quinine actinometry system embodied in the ICH guideline.

REFERENCES

Akimoto, K., Nakagawa, H., Sugimoto, I. (1985). Drug. Dev. Ind. Pharm. 11:865.

Antoniadou-Vyza, E., Buckton, G., Michaleas, S. G., Loukas, Y. L., Efentakis, M. (1997). Int. J. Pharm. 158:233.

Asker, A. F., Larose, M. (1987). Drug. Dev. Ind. Pharm. 13:2239.

Asker, A. F., Canady, D., Cobb, C. (1985). Drug. Dev. Ind. Pharm. 11:2109.

Baertschi, S. W. (1997). Drug Stability 1:194.

Barcze, L., Lenner, L. (1988). J. Pharm. Sci. 77:622.

Chang, B. S., Randall, C. S., Lee, Y. S. (1993). Pharm. Res. 10:1478.

Chen, A. X., Zito, S. X., Nash, R. A. (1994). Pharm. Res. 11:398.

Connors, K. A., Mollica, J. A. (1964). J. Pharm. Sci. 52:772.

Connors, K. A., Rosanske, T. W. (1980). J. Pharm. Sci. 69:564.

Connors, K. A., Lin, S-F, Wong, A. B. (1982). J. Pharm. Sci. 71:217.

Duchene, D., Vuation, C., Glomot, F. (1986). Drug Dev. Ind. Pharm. 12:2193.

Fabre, H., Isork, H. J., Lerner, D. A. (1993). J. Pharm. Sci. 82:553.

Felmeister, A., Schaubman, R., Howe, H. (1965). J. Pharm. Sci. 54:1589.

Fredholt, K., Østergaard, J., Savolainen, J., Friis, G. J. (1999). Int. J. Pharm. 178:223.

Gupta, V. (1983). J. Pharm. Sci. 72, 205, and 1453.

Higuchi, T., Lachman, L. (1955). J. Am. Pharm. Assoc., Sci. Ed. 44:521.

Kassem, M. A., Kassem, A. A., Ammar, H. O. (1969). Pharm. Acta Helv. 44:667.

Kassem, M. A., Kassem, A. A., Ammar, H. O. (1972). Pharm. Acta Helv. 47:97.

Kyoshin Co. (1982). Japan Kokai, JP 57:117 671, 1 November.

Lach, J. L., Chin, T. R. (1964a). J. Pharm. Sci. 53:1471.

Lach, J. L., Chin, T. R. (1964b). J. Pharm. Sci. 53:69.

Lerner, D. A., Bonneford, G., Fabre, H., Mandrou, B., and deBouchberg, M. S., (1988). J. Pharm. Sci. 77:699.
Levy, G., Reuning, R. H. (1964). J. Pharm. Sci. 53:924.
Másson, M., Loftson, T., Jónsdóttir, S., Fridriksdóttir, H., Petersen, D. S. (1998). Int. J. Pharm. 164:45.
Matsuda, Y., Masahara, T. (1983). J. Pharm. Sci. 72:1198.
McCrossin, L. E., Charman, W. N., Charman, S. A. (1998). Int. J. Pharm. 173:157.
Mima, H., Nishiwasha, M. (1964). J. Pharm. Sci. 53:931.
Mizuno, T., Hanamori, M., Akimoto, K., Nakagawa, H., Arakawa, K. (1994). Chem. Pharm. Bull. 42:160.
Mollica, J. A., Rehm, C. R., Smith, J. R., Goran, H. K. (1971). J. Pharm. Sci. 60:1380.
Myake, K., Irie, T., Arima, H., Hirayama, F., Uekama, K., Hirano, M., Okamoto, Y. (1999). Int. J. Pharm. 179:237.
Newmark, H. L, Berger, J., Carstensen, J. T. (1970). J. Pharm. Sci. 59:1249.
Pendergast, D. D., Connors, K. A. (1984). J. Pharm. Sci. 73:1779.
Ravin, L. J., Rattie, E. S., Peterson, A., Guttman, D. E. (1978). J. Pharm. Sci. 67:1523.
Ressing, M. E., Jiskoot, W., Talsma, H., VanIngen, C. W., Beuvery, E. D., Crommelin, D. J. (1992). Pharm. Res. 9:266.
Rosanske, T. W., Connors, K. A. (1980). J. Pharm. Sci. 69:564.
Scalia, S., Villani, S., Scatturin, A., Vandelli, M. A., Forni, F. (1998). Int. J. Pharm. 175:205.
Siequeira, J. M., Carvalho, S., Paniago, E. B., Tosi, L., Beraldo, H. (1994). J. Pharm. Sci. 83:291.
Szejtli, J., Bollan, E. (1980). Starke 32:386.
Szejtli, J., Bolla-Pusztai, E., Szabo, P., and Ferenczy, T. (1980). Pharmazie 35:779.
Tievenbacher, E.-M., Haen, E., Przybilla, B., Kurz, H. (1994). J. Pharm. Sci. 83:1471.
Tønnesen, H. H., Skrede, G., Martinsen, B. K. (1997). Drug Stability 1:249.
Tomida, H., Kohashi, K., Yasuto, T., Setsuo, K., Schwartz, M. A. (1987). J. Pharm. Sci. 76:147.
Underberg, W. J. (1994). Int. J. Pharmaceutics 105:249.
Upadrashta, S. M., Wurster, D. E. (1988). Drug. Dev. Ind. Pharm. 14:749.
Vandenbossche, G. M., deMuynk, C., Colardyn, F., Remon, J. P. (1993). J. Pharm. Pharmacol. 45:486.
Veiga, M. D., Ahsan, F. (1998). Int. J. Pharm. 160:43.
Ventura, C. A., Tirendi, S., Puglisi, G., Bousquet, E., Panza, L. (1997). Int. J. Pharm. 149:1.
Ventura, C. A., Guglisi, G., Zappalà, M., Maxxone, G. (1998). Int. J. Pharm. 160:163.

6

Solid State Stability

JENS T. CARSTENSEN

Madison, Wisconsin

The stability of drugs in solid dosage forms is the most important, since solid dosage forms are more common than the other types, and because the first clinical trials are usually carried out in this type of dosage form.

The author was, for several years, in charge of the investigational stability program at Hoffmann-La Roche, Nutley, N.J., and in these years accumulated a feel for the manner in which such dosage forms behaved. Some of these behavior profiles were, at the time, explainable, others not. The ones not explainable formed the basis for a great deal of research in the university setting he later enjoyed.

The general pattern that emerges is that solid dosage forms decompose by either first- or zero-order profiles, after adjustment has been made for initial events. Before attempting to elucidate why, it is necessary to know what happens to a solid compound itself, when it is exposed to adverse storage conditions.

The 1993 ICH Stability Guidelines pay particular attention to the intrinsic stability of the drug substance, the bulk drug, as witnessed by the following lines of the Guidelines:

> DRUG SUBSTANCE
> General
> Information on the stability of the drug substance is an integral part of the systematic approach to stability evaluation (36–39).
> Stress testing helps to determine the intrinsic stability of the molecule by establishing degradation pathways in order to identify the likely degradation products (41–43).

They formally introduce the concept of a re-test period for a bulk drug substance:

> Primary stability studies are intended to show that the drug substance will remain within specifications during the re-test period if stored under recommended storage conditions (46–48).
> The degree of variability of individual batches affects the confidence that a future production batch will remain within specification until the retest date (124–126).
> A re-test period should be derived from the stability information (165).

This requires that this be done on at least three batches of raw material manufactured at a minimum of pilot scale level:

> Stability information from accelerated and long term testing is to be provided on at least three batches. The long term testing should cover a minimum of 12 months duration on at least three batches at the time of submission. The batches manufactured to a minimum of pilot plant scale should be by the same synthetic route and use a method of manufacture and procedure that simulates the final process to be used on a manufacturing scale.
> The overall quality of the batches of drug substance placed on stability should be representative of both the quality of the material used in pre-clinical and clinical studies and the quality of material to be made on a manufacturing scale. Supporting information may be provided using stability data on batches of drug substance made on laboratory scale.
> The first three production batches of drug substance manufactured post approval, if not submitted in the original Registration Application, should be placed on long term stability studies using the same stability protocol as in the approved drug application (50–64).

It is not only the chemical but also pertinent physical parameters that must be monitored, such as possible polymorphic transformations:

> The testing should cover those features susceptible to change during storage and likely to influence quality, safety and/or efficacy. Stability information should cover as necessary the physical, chemical ... characteristics (66–69).

It also addresses the formation of decomposition products and their limits, a point frequently elucidated through kinetics:

> Limits of acceptability should be derived from the profile of the material as used in the pre-clinical and clinical batches. It will need to include individual and total upper limits for impurities and degradation products, the justification for which should be influenced by the levels observed in material used in pre-clinical studies and clinical trials (73–77).

The length of the studies are tied in with the anticipated use of the bulk drug:

> The length of the studies and the storage conditions should be sufficient to cover storage, shipment and subsequent use. Application of the same storage conditions as applied to the drug product will facilitate comparative review (79–81).

The ICH 1993 Stability Guidelines recommend a minimum set of testing conditions. It emphasizes that one of the virtues of accelerated testing is to ascertain that the effects of "excursions outside the label storage condition" such as might occur during shipping, can be assessed.

	Conditions	*Minimum Time Period at Submission*
Long term testing	25°C±2°C/60% RH ±5%	12 Months
Accelerated Testing	40°C±2°C/75% RH ±5%	6 Months

> Where 'significant change' occurs during six months storage under conditions of accelerated testing at 40°C±2°C/75 percent RH ±5 percent, additional testing at an intermediate conditions (such as 30°C±2°C/60% RH ±5%) should be conducted for drug substances to be used in the manufacture of dosage forms tested long term at 25°C/60 percent RH and this information included in the Registration Application. The initial Registration Application should include a minimum of 6 months data from a 12 month study. 'Significant change' at 40°/75 percent RH or 30°C/60 percent RH is defined as failure to meet specification.
>
> The long term testing will be continued for a sufficient period of time beyond 12 months to cover all appropriate re-test periods, and the further accumulated data can be submitted to the Authorities during the assessment period of the Registration application.
>
> The data (from accelerated testing or from testing at an intermediate condition) may be used to evaluate the impact of short term excursions outside the label storage conditions such as might occur during shipping (90–109).

The frequency of testing is also specified:

> Frequency of testing should be sufficient to establish the stability characteristics of the drug substance. Testing under the defined long term conditions will normally be every three months, over the first year, every six months over the second year and then annually (111–114).

The containers are assessed as well, but in this respect the effect of oxygen and moisture, as covered in the next chapter, are probably more à propos.

> The containers to be used in the long term, real time stability evaluation should be the same as or simulate the actual packaging used for storage and distribution (116–118).

The 1993 ICH Guidelines address the evaluation of data as well. It is noted here that it is necessary to attempt to establish the "degradation relationship", i.e., attempt to assess the mechanism of degradation:

> The nature of any degradation relationship will determine the need for transformation of the data for linear regression analysis. Usually the relationship can be represented by a linear, quadratic or cubic function on an arithmetic or logarithmic scale. Statistical methods should be employed to test the goodness of fit of the data on all batches and combined batches (where appropriate) to the assumed degradation line or curve (138–143).

There is the admission of limited extrapolation of data, but here again, it is necessary to know the degradation pattern:

> Limited extrapolation of the real time data beyond the observed range to extend expiration dating at approval time, particularly where the accelerated data supports this, may be undertaken. However, this assumes that the same degradation relationship will continue to apply beyond the observed data and hence the use of extrapolation must be justified in each application in terms of what is known about the mechanism of degradation, the goodness of fit of any mathematical model, batch size, existence of supportive data etc. (149—155).

The guidelines again emphasize the need for assessment of degradation products and "appropriate attributes":

> Any evaluation should cover not only the assay, but the levels of degradation products and other appropriate attributes (156–157).

Such "appropriate attributes" are most commonly, for a bulk drug substance, morphology, particle size, shape, and fractal dimension.

1. SIMPLEST DECOMPOSITION MODES OF PURE SOLIDS

If a solid is placed in a vacuum and exposed to temperatures at which it decomposes at a measurable rate, one of the following situations may arise:

I Solid → solid + solid
II Solid → solid + liquid
III Solid → liquid + liquid
IV Solid → solid + gas
V Solid → liquid + gas
VI Solid → gas + gas

Other schemes are theoretically possible, but not likely. Of the above, it is schemes IV and V which will be treated in some detail below, because they are the ones most investigated in the pharmaceutical sciences. It will later be shown that most pharmaceutical systems will not be of such a purist nature, but the experiences gathered from examining them will throw light on several important real-life situations.

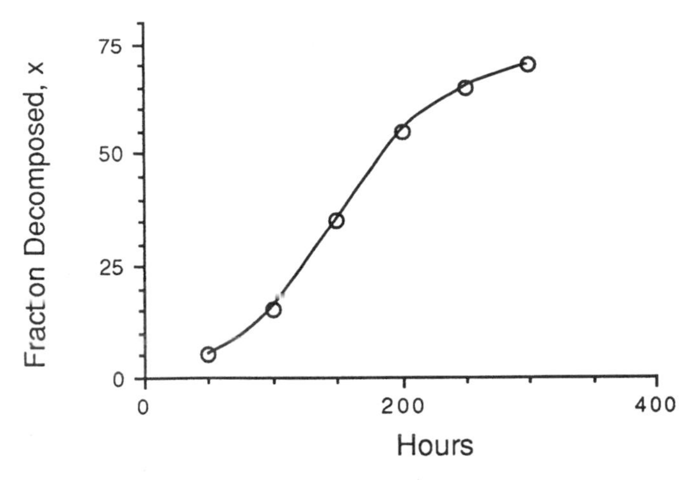

Fig. 1 Decomposition of *p*-aminosalicylic acid at 65°C in vacuo. (Figure constructed from data by Carstensen and Pothisiri, 1975.)

2. THE SOLID TO SOLID + GAS REACTION

This reaction has been investigated by Prout and Tompkins (1944) and later by Kornblum and Sciarrone (1964) and Carstensen and Pothisiri (1975). A typical example of such a reaction is that of p-aminosalicylic acid, shown in Table 1 and Fig. 1. It is noticed that the profile is S-shaped. The general explanation for this type of reaction is the following:

No solid has a smooth surface, i.e., there are always surface imperfections. These could be "steps" in the surface or they could be crystal defects. These sites are more energetic than the remaining sites. They are most likely to occur at surfaces, which in any event are populated with molecules that are unlike the molecules in the bulk of the crystal. For instance they have at least one less neighbor than bulk molecules. It is assumed that decomposition is more likely to occur at such "activated" sites (Fig. 2).

Once a molecule decomposes at an activated site it changes its geometry, and hence the neighboring molecules are more likely to decompose. There will then be a chain or plane of activated molecules forming, with a probability of a (second figure in Fig. 2). The rate of formation of activated molecules, N in number at time t, is dN/dt, and this is proportional to N. Initially this is then given by

$$\left[\frac{dN}{dt}\right]_0 = a \cdot [N + N_0] \tag{6.1}$$

Table 1 Decomposition of *p*-Aminosalicylic Acid

Time (hours)	0	50	100	150	220	260	325
Percent decomposed	0	4	18	36	60	70	77

Source: After Carstensen and Pothisiri, 1975.

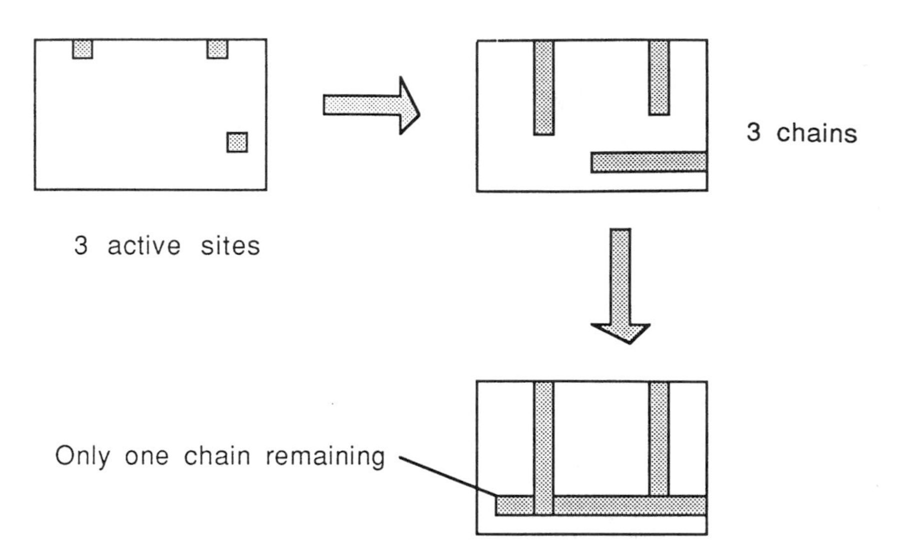

Fig. 2 The propagation of active site chains from three surface sites.

It is obvious that after even a short period of time N becomes much larger than N_0, so that this latter can be dropped at times even remotely larger than zero.

After a certain while (last figure in Fig. 2), planes will start to merge, and hence there will be a termination probability, b, so that at measurable times, Eq. (6.1) becomes

$$\frac{dN}{dt} = \{a - b\}N \tag{6.2}$$

Both a and b are functions of t (or what is equivalent, to the fraction decomposed x). It is reasonable to assume that

$$a = b \qquad \text{at} \qquad t = t_{1/2} \qquad (\text{or } x = 0.5) \tag{6.3}$$

i.e., at the point in time where one half of the substance has decomposed. Also,

$$b = 0 \qquad \text{at} \qquad t = 0 \qquad (\text{or } x = 0) \tag{6.4}$$

since there can be no termination probability at time zero. One (not necessarily the correct) function which satisfies this condition is

$$b = 2xa \tag{6.5}$$

When this is inserted into Eq. (6.2) one obtains

$$\frac{dN}{dt} = a[1 - 2x] \cdot N \tag{6.6}$$

The decomposition rate, dx/dt, is proportional to N, i.e., $dx/dt = k \cdot N$ or

$$N = \frac{1}{k}\frac{dx}{dt} \tag{6.7}$$

Equation (6.6) can now be written

$$\frac{dN}{dt} = \frac{a}{k}[1 - 2x]\frac{dx}{dt} \tag{6.8}$$

Chain differentiation of dN/dt gives

$$\frac{dN}{dt} = \frac{dN}{dx}\frac{dx}{dt} \tag{6.9}$$

Introducing Eq. (6.8) into Eq. (6.9) gives

$$\frac{dN}{dt} = \frac{dN}{dx}\frac{dx}{dt} = \frac{a}{k}[1 - 2x]\frac{dx}{dt}$$

dx/dt is cancelled out of the last part of this equation to give

$$\frac{dN}{dx} = \frac{a[1 - 2x]}{k} \tag{6.10}$$

which integrates to

$$N = \frac{a}{k}(x - x^2) \tag{6.11}$$

Equation (6.7) is now introduced to give

$$\frac{1}{k}\frac{dx}{dt} = \frac{a}{k}x(1 - x) \tag{6.12}$$

which integrates to

$$\ln\left[\frac{x}{1 - x}\right] = a(t - t_{1/2}) \tag{6.13}$$

The equations have a zero time problem, since the equation is not defined for $x = 0$. This is a consequence of neglecting N_0. Similar paradoxes exist in the scientific literature. The Gibbs adsorption isotherm, for instance, is not defined for $C = 0$, i.e., liquid without surfactant. In the case of solid state stability, it might be thought of in the vein, that as the material is being produced, i.e., at time zero (e.g., through recrystallization), it is already decomposing (however little).

Example 6.1.

A set of decomposition data for a sample of a 5.52 mmoles of a solid are as shown in Table 2. Obtain the rate constant and the value of $t_{1/2}$.

Answer.

The transformation of data into the form of Eq. (6.13) is shown in Table 3. These data are plotted in Fig. 3. The least squares fit of the line according to Eq. (6.13) is

$$\ln\left[\frac{x}{1 - x}\right] = 0.868(t - 4) \tag{6.14}$$

Table 2 Prout–Tompkins Data at 55.6°C

Time	0	1	2	3	4	4.5	5
Gas (mmoles)	0	0.48	0.86	1.56	2.77	3.71	4.96

Table 3 Data in Table 4.2 Treated According to Eq. (6.13)

Time	0	1	2	3	4	4.5	5
Gas (mmoles)	0	0.48	0.86	1.56	2.77	3.71	4.96
x	0	0.087	0.155	0.28	0.50	0.668	0.893
$\ln[x/(1-x)]$		−1.75	−1.17	−0.58	0	0.291	0.581

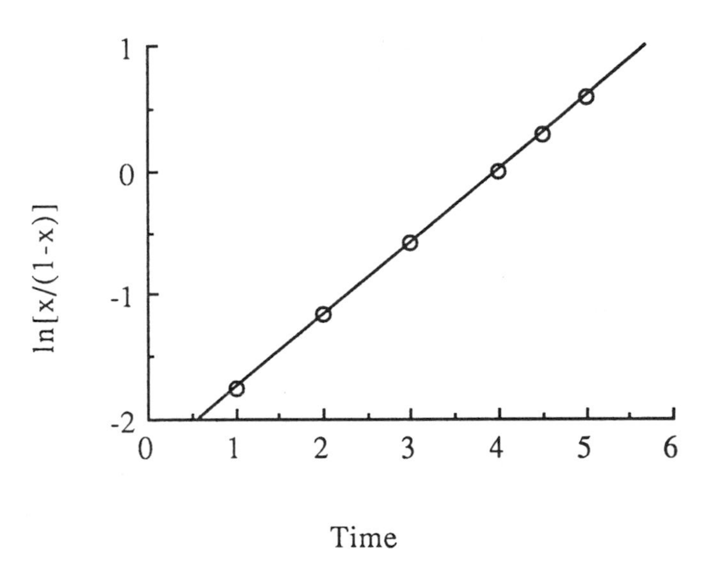

Fig. 3 Data from Table 2 plotted according to Eq. (6.13). The least squares fit is $\ln[x/(1-x)] = -2.334 + 0.5832t$ with an R value of 1.00.

This type of reaction embodies the dehydration of solid hydrates. Leung et al. (1998a,b) have shown that aspartame 2.5 hydrate cyclizes by Prout–Tompkins kinetics and that the rate constants follow an Arrhenius equation.

3. TEMPERATURE DEPENDENCE OF THE SOLID TO SOLID + GAS REACTION

The rate determining parameter in Eqs. (6.12) and (6.13) is a. In general, activation energies encountered in pharmaceutical systems are between 15 and 30 kCal/mole. However, the parameter a is a stoichiastic parameter and is not necessarily of this order of magnitude. Figure 4 shows the data from Table 3 extended to several temperatures. The least squares equation for the Arrehenius plot is

$$\ln[a] = 83.553 - 28.42\frac{1000}{T} \tag{6.15}$$

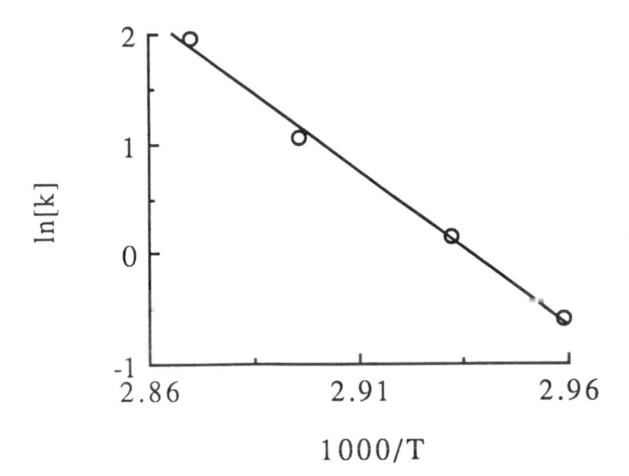

Fig. 4 Arrhenius plot of a values from data of the type shown in Table 2 carried out at 65 (Table 2), 68, 72, 75°C. Least squares fit: $\ln[k] = 83.553 - 28.421\{1000/T\}$ with $R = 0.997$.

and it is seen that $E_a/R = 28.42$ kCal, i.e., E_a is about 57 kCal. In most solid to solid + gas reactions the activation energy will be excessively high (up to 80 kCal/mole). This means that the decomposition will occur at a measurable rate over a narrow temperature range, T_1 to T_2. Below T_1 it is too slow to allow detection of the entire curve in a reasonable length of time, and above T_2 it is too fast to measure with precision. In the data in Fig. 4 the rate constants increase 10-fold in a 10° span. $T_1 - T_2$ is frequently denoted a *decomposition range*, under melting point, in chemical tables.

It should be pointed out that the solid to solid + gas reaction may be so only over a certain temperature range, or to a certain degree of decomposition. Fig. 5 shows the eutectic diagram of a compound with its solid decomposition product.

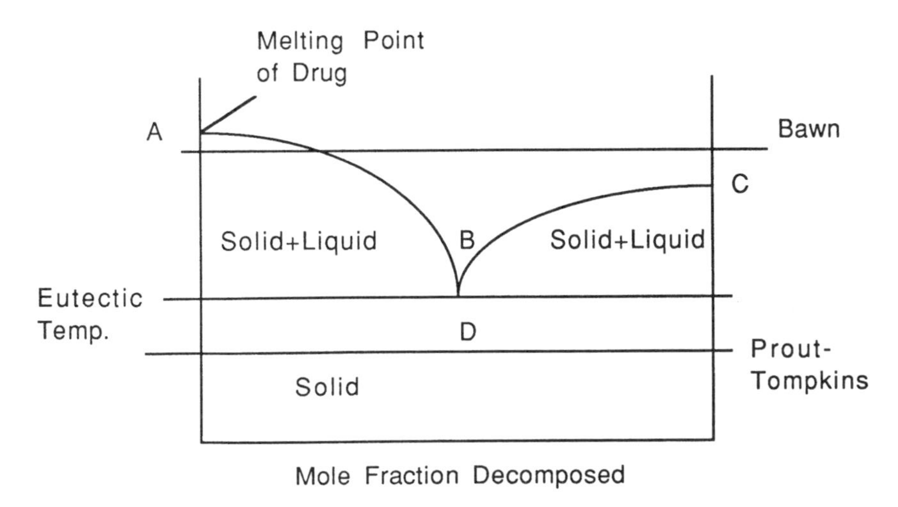

Fig. 5 Binary melting point diagram showing areas where Bawn kinetics apply and where Prout–Tompkins kinetics apply.

If the study is carried out at temperatures below the eutectic temperature, T^*, then the reaction will be solid to solid + gas. If above the eutectic temperature, then the reaction will be solid to solid + liquid + gas. The compounds reported in literature to be of the solid to solid + gas type are most often inorganic salts, e.g., potassium permanganate (Prout and Tompkins, 1944), silver permanganate (Goldstein and Flanagan, 1964), and some organic compounds, such as oxalic acid and *p*-aminosalicylic acid (Kornblum and Sciarrone, 1964; Pothisiri and Carstensen, 1975; Carstensen and Pothisiri, 1975).

Olsen et al. (1997) showed cefaclor monohydrate to decompose (as judged by related substances) by first-order kinetics. The rate constants could be plotted by Arrhenius plotting and were consistent with ambient rate constants. The reaction scheme, when amorphous material was present, was such that the rates were faster at early time points and then became equal to those of the crystalline modification. The conclusion was that the initial phase was decomposition of amorphous content parallel to conversion of amorphous to crystalline drug.

At times the solid state reaction cannot be completely specified yet can be descried in analytical terms. Tzannis and Prestrelski (1999) described the effect of sucrose on the stability of trypsinogen during spray-drying by plotting denaturation temperatures as a function of melting temperature and found a linear increase between residual activity after spray-drying and melting temperature. Adler and Lee (1999) have reported on the stability of lactate dehydrogenase in spray-dried trehalose.

4. THE SOLID TO LIQUID + GAS REACTION

Many more compounds seem to decompose by this reaction scheme than by the solid to solid + gas. The reaction kinetics are usually referred to as Bawn kinetics (Bawn, 1955). This situation at time t is as shown in Fig. 6, and as seen there will be a certain amount of liquid decomposition product. This amount corresponds to the amount of drug decomposed. However, the liquid decomposition product will dissolve parent compound to the extent, S (mole drug/mole decomposition product), to which it is soluble, so that the amount present in the solid state at time t is the original number of moles, A_0, minus the amount decomposed, A_0x, minus the amount dissolved, A_0Sx.

The rate of decomposition would be the sum of the rates of decomposition in the solid state (assumed first order with rate constant k_s, time^{-1} and in the dissolved state (assumed first order with rate constant k_1 time^{-1}. The rate equation, hence, is

$$\frac{dA}{dt} = -k_s[A_0(1-x) - A_0xS] - k_1[A_0xS] \tag{6.16}$$

Noting that

$$\frac{A}{A_0} = (1-x) \tag{6.17}$$

it follows, by division through by A_0, that

$$\frac{d(1-x)}{dt} = -k_s[1-x-xS] - k_1xS \tag{6.18}$$

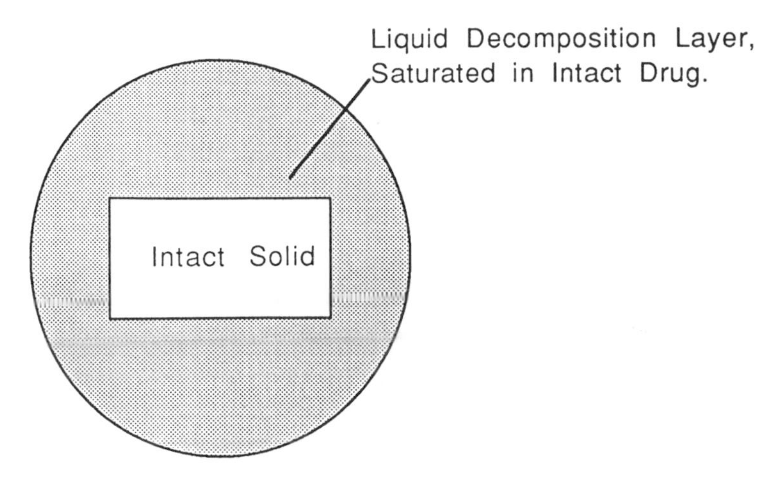

Fig. 6 Situation leading to Bawn kinetics.

or, noting that $d(1-x) = -dx$,

$$\frac{dx}{dt} = k_s[1 - x - xS] - k_1 Sx = k_s[1 + Bx] \tag{6.19}$$

where

$$B = \frac{k_1}{k_s} - 1 - S \tag{6.20}$$

Eq. (6.19) can be integrated, and it then yields

$$\ln[1 + Bx] = Bk_s t \tag{6.21}$$

Using B as an adjustable parameter, it is possible to find the value that makes the data profile through the origin, as dictated by Eq. (6.21), and also gives the best fit.

Figure 7 and Table 4 show an example of data from decomposition of *p*-methylaminobenzoic acid.

To plot this according to Eq. (6.21) it is necessary to assume values of B, plot the data, and assess the goodness of fit by some criterion. A different value of B is then chosen, and this process is repeated until a "best" value of B is arrived at. It can be shown that in general the sums of the squares of the deviations $[s_{yx}^2 = \Sigma(y - \hat{y}\wedge)^2/(n-2)]$ of the points from the ensuing line can be used as a criterion. A different criterion is the correlation coefficient. In many cases this is also *not* a good criterion, and criteria for linearity (e.g., Durbin–Watson statistics) are the best. With data fitting to Eq. (6.21), the line must pass through the origin. Fitting the data in this fashion is shown in the table for three values of B (0.1, 0.85, and 2.0). It is, of course, best to do this by computer, and a simple program in BASIC is shown in Table 5.

The number of data points are inserted, the assumed value of B is inserted, and the program is run. One can then in three or four tries arrive at a "best" value for B.

In the case of Eq. (6.21), using the correlation coefficient is not a good parameter, because it simply increases with increasing values of B up to a very high

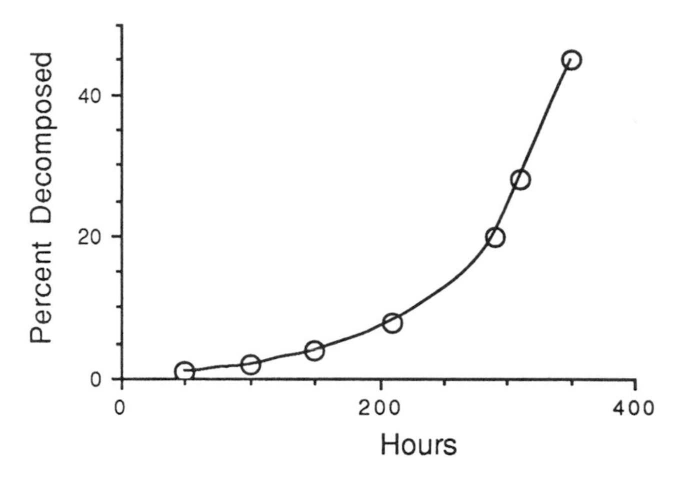

Fig. 7 Data from Table 4. Decomposition of *p*-methylaminobenzoic acid. (Graph constructed from data published by Carstensen and Musa, 1972.)

Table 4 Decomposition Data for *p*-Methylaminobenzoic Acid

Time (h)	0	50	110	150	210	290	310	350
% Decomposed	0	1	2	4	8	20.5	27.9	45

(unrealistic) value and also results in a very high intercept (see Table 6). All the correlation coefficients are good. The best criterion would be a criterion that dealt with curvature, but a simpler one, as stated, is simply to note the intercept, which should be zero (see Fig. 8).

Studies of this type are usually done on a vacuum rack. In this, the pressure is monitored as a function of time, and the sample can be observed. At a given point in time (which is quite reproducible), the last trace of solid will disappear. At this point in time, t^*, the amount not decomposed, $A_0(1 - x)$, is just sufficient to dissolve the amount of liquid, A_0x, present, i.e., at time t^*:

$$S = \frac{1 - x^*}{x^*} \tag{6.22}$$

where x^* is the mole fraction decomposed at time t^*. Therefore Eq. (6.21) is valid from time 0 to time t^*. If $t^* = 350$ (as in the example used here), and $x^* = 0.45$ at this point, it follows that

$$S = \frac{0.55}{0.45} = 1.22 \text{ moles/mole} \tag{6.23}$$

The slope in the above case is 0.01 h^{-1}. Since the slope is $[B \cdot k_s]$ it follows that

$$k_s = \frac{\text{slope}}{B} = \frac{0.01}{0.85} = 0.012 \text{ h}^{-1} \tag{6.24}$$

Table 5 Program for Obtaining Best Values by Manual Iteration

```
100   PRINT "Type in data as x,y, in 400 block"
110   INPUT "Number of Data Points=";N1
120   INPUT "Iteration Parameter, B=";B
130   PRINT "T";SPC(6);"X";SPC(6);"LN(1+BX)
140   PRINT "----------------------"
200   READ A, C
210   X=A
220   Y=LOG(1+B*C)
230   X1=X1+X
240   X2=X2+(X^2)
250   Y1=Y1+Y
260   Y2=Y2+(Y^2)
270   Z1=Z1+(X*Y)
280   N2=N2+1
300   PRINT X;SPC(6);C;SPC(6);Y
310   IF N2=N1 goto 700
400   DATA 50,1
410   DATA 100,2
420   DATA 150,4
430   DATA 210,8
440   DATA 290,20
450   DATA 310,28
460   DATA 350,45
700   Z2=X2-((X1^2)/N2)
710   Z3=Y2-((Y1^2)/N2)
720   Z4=Z1-(X1*Y1/N2)
730   Z5=Z4/Z2
740   Z6=(Y1-(Z5*X1))/N2
750   PRINT
760   PRINT "Slope="; Z5
770   PRINT "Intercept="; Z6
780   Z7=(Z4^2)/(Z3*Z2)
790   Z8=(Z7)^(0.5)
800   PRINT "Correlation Coefficient=";Z8
810   Z9=(Z3-((Z5^2)*Z2))/(N2-2)
820   PRINT "syx^2=";Z9
```

k_1 is now calculated from Eq. (6.20):

$$0.86 = \frac{k_1}{0.012} - 1 - 1.22 \tag{6.25}$$

i.e.

$$k_1 = 3.08 \times 0.012 = 0.037 \text{ h}^{-1} \tag{6.26}$$

Beyond t^* the system is a solution system and should decompose by first-order kinetics. The density of the liquid will actually change with time, but it is assumed that both parent drug and decomposition product have approximately the same

Table 6 Data in Table 5 Treated by Eq. (6.21)

	$\ln[1+Bx]$		
Time (h)	$B=0.1$	$B=0.85$	$B=2$
50	0.095	0.615	1.099
100	0.182	0.993	1.610
150	0.334	1.481	2.200
210	0.588	2.054	2.830
290	1.099	2.890	3.710
310	1.335	3.210	4.040
350	1.705	3.677	4.510

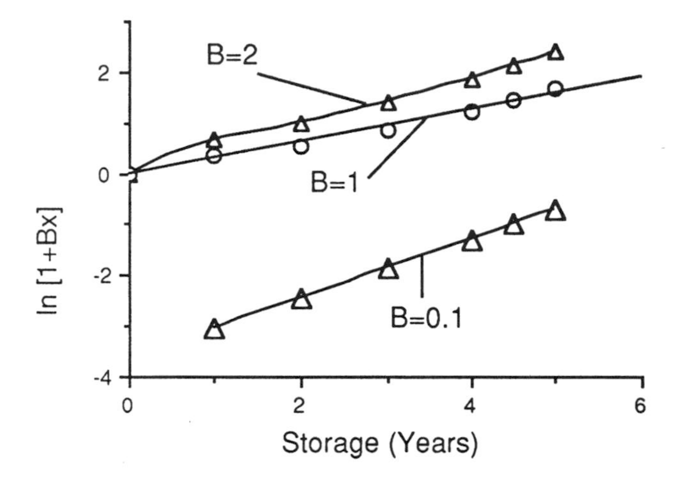

Fig. 8 Data from Table 4 treated by Eq. (6.21).

density. The common density is denoted ρ, and since there is a total number of A_0 moles, the volume of liquid is $A_0\rho$. The initial molar concentration (at time t^*) is, therefore, $A_0(1-x*)/[A_0/\rho] = (1-x*)/\rho$. The time is counted from $t=t^*$ and the concentration at time $t-t^*$ is $(1-x)/\rho$, so that

$$\ln\left[\frac{1-x}{\rho}\right] = -k_1 t + \ln\left[\frac{1-x^*}{\rho}\right] \tag{6.27}$$

or

$$\ln\left[\frac{1-x}{1-x^*}\right] = -k_1(t-t^*) \tag{6.28}$$

or

$$x = 1-(1-x^*)e^{-k_1 t} \tag{6.29}$$

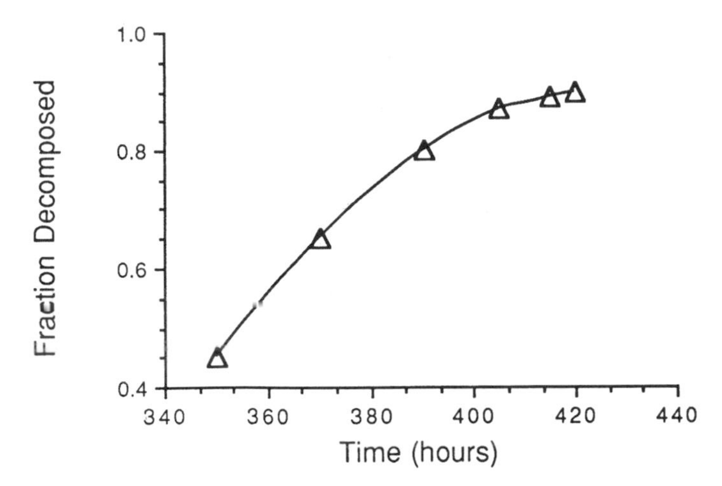

Fig. 9 Decomposition of *p*-methylaminobenzoic acid after t^* (350 hours) at which point $x = 0.45$ (i.e., $1 - x = 0.55$, as used in Fig. 10). (Graph constructed from data published by Carstensen and Musa, 1972.)

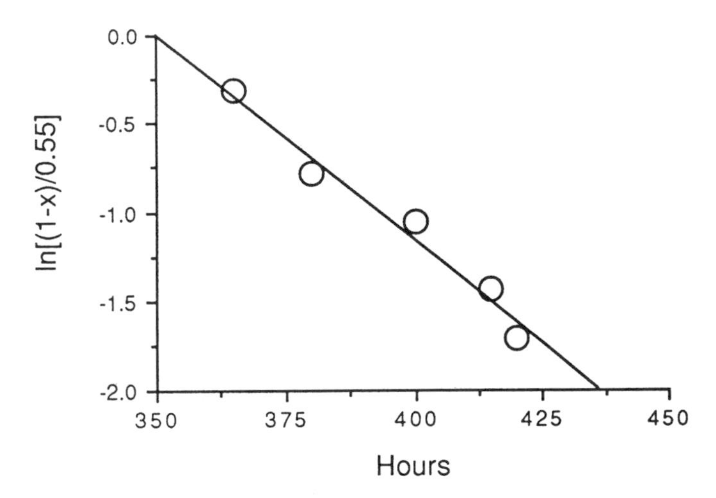

Fig. 10 Data in Fig. 9 treated according to Eq. (6.28). (Graph constructed from data published by Carstensen and Musa, 1972.)

Data of this type, for *p*-methylaminobenzoic acid, are presented in Figs. 9 and 10. It is seen that the data are quite first order. The first-order rate constant obtained from this plot is $k_1 = 0.040\ \text{h}^{-1}$ in quite good agreement with the value of 0.037 found from the first part of the curve.

When the total curve is plotted (i.e., when Fig. 7 and Fig. 9 are combined), then an S-shaped curve results. Unlike the Prout–Tompkins curve, the Bawn curve is a two-phase curve, one part relating to the phase where there is solid present, the other to the part where all solid has dissolved.

The values of x^* obtained at t^* will differ from temperature to temperature, since the solubility is a function of temperature. This is actually the value of the

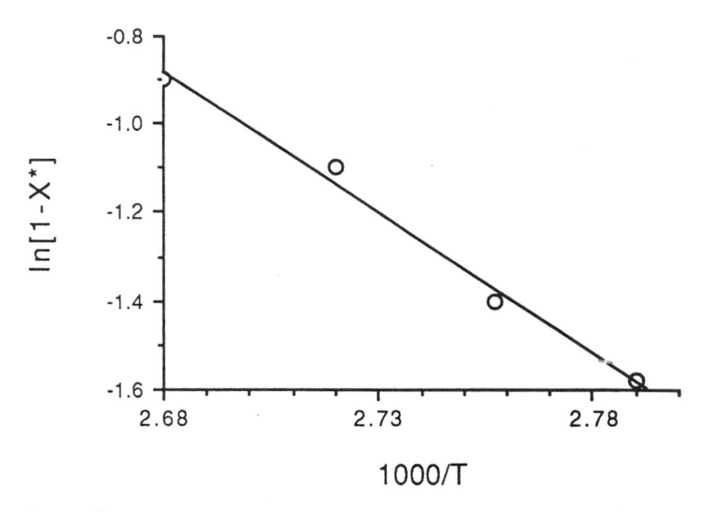

Fig. 11 $\ln[1-x^*]$ as a function of $1000/T$. Least squares equation: $y=16.19-6.37x$ ($R=1.00$). (Graph constructed from data by Carstensen and Kothari, 1983.)

liquidous line on a eutectic diagram. The melting point depression curve (Maron and Prutton, 1965) is given by

$$\ln(1-X^*)=\frac{\Delta H}{R}\left[\frac{1}{T_f}-\frac{1}{T}\right] \tag{6.30}$$

Such plots are quite linear, as shown in Fig. 11.

5. THE NG EQUATION

Ng (1975) suggested the following global equation for solid state decomposition:

$$\frac{dx}{dt}=x^n(1-x)^p \tag{6.31}$$

As pointed out earlier, a modification of this equation is

$$\ln\left\{\frac{x^n}{(1-x)^p}\right\}=-k'(t-t_i) \tag{6.32}$$

which may be written as

$$\ln\left\{\frac{x^q}{1-x}\right\}=-k(t-t_i) \tag{6.33}$$

where

$$k=\frac{k'}{p} \tag{6.34}$$

and

$$q = \frac{n}{p} \tag{6.35}$$

6. TOPOCHEMICAL REACTIONS

There are theories akin to the above, which simply, empirically state that (a) decomposition starts at the surface of the solid and works inwards. If, for instance, the solid were a cube originally with side a_0 cm, then, after a given time the side length, a, would be

$$a = a_0 - kt \tag{6.36}$$

i.e., it is assumed that the decomposition "front" progresses in a linear fashion. This is akin to physical phenomena such as crystal growth (the so-called McCabe law). At time t there will, therefore, be an amount undecomposed given by

$$N\rho a^3 = N\rho[a_0 - kt] \tag{6.37}$$

where N is the number of particles in the sample and ρ is the density of the solid. The original volume of the solid was Na_0^3, so that the fraction not decomposed, x, would be given by

$$x = \frac{N\rho a^3}{[N\rho a_0]^3} = \left[\frac{a}{a_0}\right]^3 = \left[1 - \left(\frac{k}{a_0}\right)t\right]^3 \tag{6.38}$$

It is noted from Eq. 38 that the rate constant (k/a_0) is particle size dependent. This property will be touched on frequently in the following.

An example of this type of decomposition pattern is aspirin in an alkaline environment (Nelson et al., 1974). This is shown in Fig. 12.

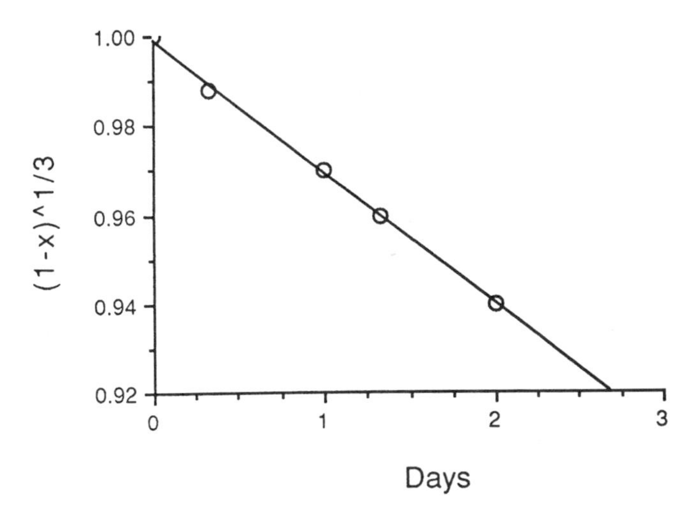

Fig. 12 Decomposition of aspirin in alkaline environment. Least squares fit: $y = 1.0 - 0.0295x$ ($R = 1.00$). (Graph constructed from data by Nelson et al., 1974.)

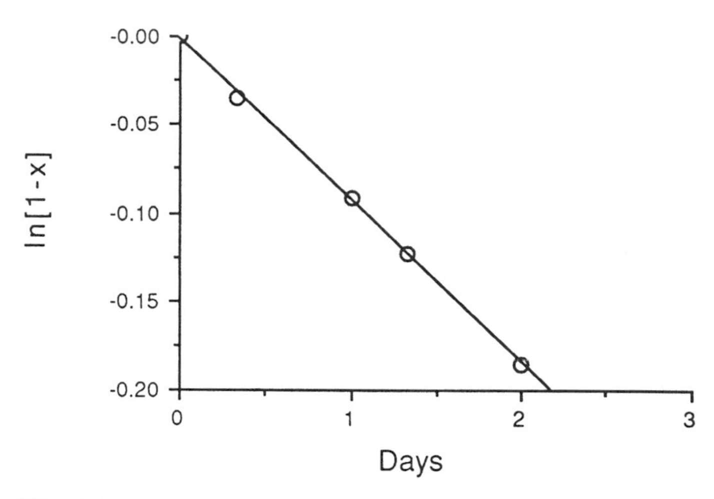

Fig. 13 Data in Fig. 12 treated as first order.

In general it is not possible to distinguish between a reaction of the type described by Eq. 38 and a first-order reaction. Only with excellent precision, with a fairly large number of assays, and with a sufficiently large decomposition will it be possible to distinguish between the two. The data in Fig. 12 are shown treated as first order in Fig. 13.

7. DIFFUSION CONTROLLED INTERACTIONS

Figure 14a shows a situation where a solid, A, is in contact with another solid, B. The contact area is assumed to be 1 cm^2. It is assumed that A can react with B in this situation, i.e.,

$$A + B \rightarrow C \tag{6.39}$$

As the reaction proceeds (Fig. 14b), decomposition product, C, will accumulate between A and B, and at a given time t, compound A must diffuse to the surface of compound B through a layer of compound C, h cm thick, in order for the reaction to take place. The density of B is denoted ρ. A layer of B h cm thick would contain $h\rho$ g of B, and hence

$$\frac{dB}{dt} = D\frac{dh}{dt} \tag{6.40}$$

By Fick's first law, dB/dt is inversely proportional to h, so that we may write

$$\rho\frac{dh}{dt} = \frac{q}{h} \tag{6.41}$$

or

$$h \cdot dh = \left[\frac{q}{\rho}\right]dt \tag{6.42}$$

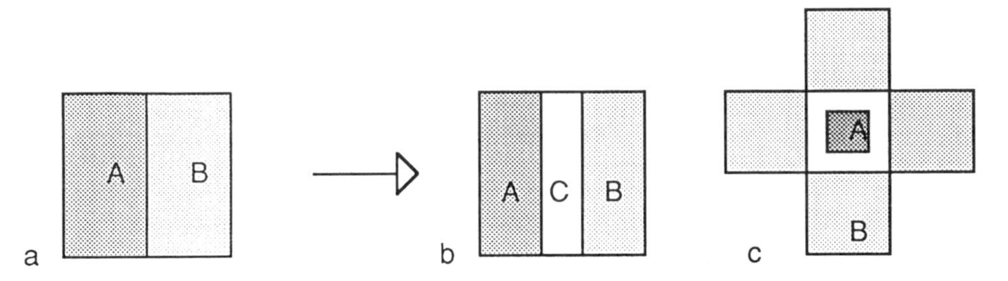

Fig. 14 Stages in Jander kinetics.

This may be integrated to

$$h^2 = \left[\frac{2q}{\rho}\right] t = k' t \tag{6.43}$$

or

$$h = [k' t]^{1/2} \tag{6.44}$$

where $k' = 2q/\rho$. If, as shown in Fig. 14, A and B are cubical, of side length a_0 initially, and a at time t, and if B is surrounded by A as shown, then

$$h = a_0 - a \tag{6.45}$$

The amount retained at time t is

$$\begin{aligned}(1 - x) &= \left[\frac{a}{a_0}\right]^3 = \left[\frac{a_0 - a_0 + a}{a_0}\right]^3 \\ &= \left[1 - \frac{h}{a_0}\right]^3 = \left[1 - \frac{\{kt\}^{1/2}}{a_0}\right]^3\end{aligned} \tag{6.46}$$

or

$$\{1 - (1 - x)^{1/3}\}^2 = \frac{kt}{a_0^2} \tag{6.47}$$

where x is fraction decomposed. It is seen that the rate constant is related to the particle size, i.e., the finer the particles the larger the rate constant. A system of this type is, again, the aspirin–sodium bicarbonate system, but at lower temperatures. At higher temperatures, the autodecomposition of aspirin is higher than the diffusion coefficient (related to q), and the reaction at higher temperatures then follows Eq. (6.40) (Nelson et al., 1974).

Recently it has become customary to compare polymorphic and pseudo-polymorphic transformation data with prevailing solid state equations, e.g. forms of the Ng-equation. Several such equations, some of them already alluded to, are listed in Table 7.

Table 7 Equations Relating to Decomposition in the Solid State

$\ln(x/(1-x)) = kt$	surface nucleation, Prout–Tompkins equation
$\{-\ln(1-x)\}^{1/n} = kt$	n-dimensional nuclear growth (Avrami)
	n-dimensional nucleus growth
$1-(1-x)^{1/n} = kt$	n-dimensional boundary reaction
$x^2 = kt$	diffusion in one dimension
$(1-x)\ln(1-x) + x = kt$	diffusion in two dimensions
$(1-(1-x)^{1/3})^2 = kt$	diffusion in three dimensions (Jander equation)

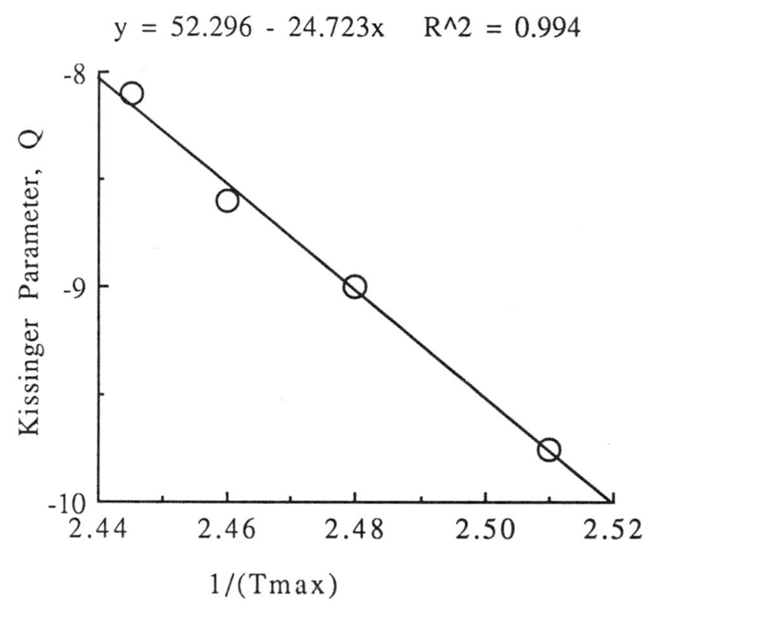

Fig. 15 Kissinger plot of polymorph II of glybuzole. (Plot constructed from data published by Otsuka et al., 1999.)

There has been a tendency in recent literature to simply fit data to several (or all) of these equation, and the equation that gives the "best fit" is then assumed to be the mechanism. Figure 15, for instance, shows a literature example of such data. It is claimed that this data best fits a Jander equation (and such treatment is shown in Fig. 16), but first of all the fit is not good, and secondly, it is obvious that the phase, C, in the Jander model (Fig. 14) cannot possibly apply to a polymorphic transformation where the reaction is simply $A \rightarrow B$, not $A + B \rightarrow C$.

It is emphasized here that sorting out mechanisms by statistical analysis can be dangerous.

Several recent investigations in this field have appeared in recent years. Fini et al. (1999) have studied the dehydration and rehydration of diclofenac salt hydrate at ambient temperature. Otsuka et al. (1999) investigated three forms of glybuzole

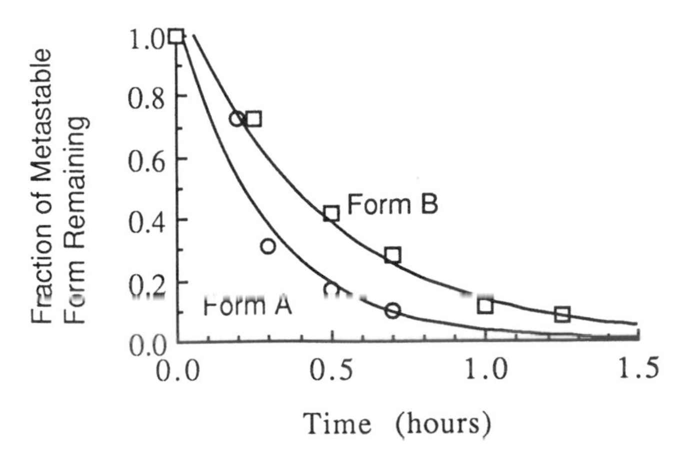

Fig. 16 Literature data dealing with two polymorphic transformations allegedly diffusional because it adheres (somewhat) to a Jander model.

(I, II, and amorphate) and found all to have fairly much the same solubility. Neither form I nor form II changed after storage at 40°C/75% and 0% RH for 2 months. DSC for form I showed no peak other than a sharp melting endotherm at 167.4°; form II showed a slight endotherm at 116.8°C and a sharp endotherm at 166.6°C. The amorphate showed a (slight) exotherm peaking at 81.5°C, presumably due to crystallization, and a sharp endotherm at 167.3°C. From this it would be reasonable to conclude that form II is stable at room temperature and transform to I at 116.8°C, this latter form being stable at the higher temperatures.

The authors estimated the polymorphic stability of form II by way of the Kissinger equation (Kissinger, 1956):

$$\partial\left\{\frac{\ln(\phi/T_{\max}^2)}{\partial(1/T_{\max})}\right\} = \frac{-E}{R} \tag{6.48}$$

where ϕ is the rate of heating, $T_{\max}$ is the temperature at the peak maximum in the DSC, E is the activation energy, and R is the gas constant. If the experiment is conducted at different heating rates, different $T_{\max}$ values result, and in the case of glybuzole there were four such values.

It can be seen from the graph that the activation energy is $24.723 \times 1.99 = 49.2$ kCal/mol. Otsuka et al. (1991, 1993, 1999) employed the Jander equation to explain crystallization rates of compounds, e.g., amorphous glybuzole. As mentioned above, however, the Jander equation is based on an assumption of a layer of "reaction product," and such a layer (i.e., such a model) cannot be conceived of in a polymorphic transformation. What would be the "reaction product"?

8. GENERAL INTERACTIONS IN DOSAGE FORMS

It is tempting to think of a tablet as an agglomeration of individual particles, independent of one another, but this cannot be the case. By their mere nature, particles

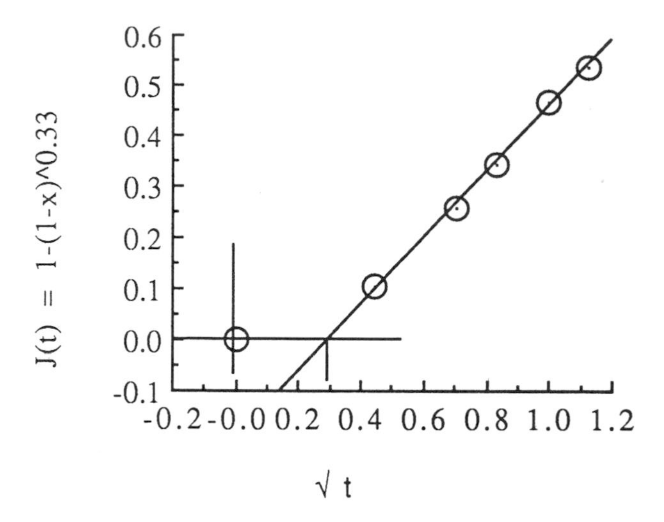

Fig. 17 Data from form B in Fig. 15 treated according to a Jander model. The curve follows the least squares fit equation: $J(t) = -0.194 + 0.652\sqrt{t}$.

are fused together (either by brittle fracture or by plastic deformation in tablets or tamping in capsules), and if the created contact area is between two different components of the tablet (one being the drug), then there is the possibility of interaction. It is highly likely that moisture plays a part in all of these. In fact, in one of the cases to be discussed below (tartaric acid + sodium bicarbonate) this is the case (although the tablet can, for all practical purposes, be anhydrous at the onset).

The most common type of interaction in solid dosage forms is actually between water and drug. This is a large topic in itself, and the following chapter is devoted to it. The topic discussed here will be of special cases where water is not the interactant (or the main interactant).

The following cases are illustrative examples:

1. Tartaric acid and sodium bicarbonate
2. Aspirin and phenylephrine
3. Aspirin and lubricants

In addition to the points made, note that in the curve in Fig. 17 a lag time has to be invoked for the data to linearize.

8.1. Tartaric Acid and Sodium Bicarbonate

This is a common combination in effervescent tablets. When the tablet is added to water, the acid and the base will react, forming carbon dioxide, which produces the desired bubble effect:

$$R_2(COOH)_2 + 2NaHCO_3 \rightarrow R2(COO^-)_2 + 2Na^+ + 2H_2O + 2CO_2 \qquad (6.49)$$

To be strictly correct, the left hand side should be written in ionic form as well.

Of course, it is necessary that this reaction not take place prior to the time it reaches the consumer, because if the reaction does occur in the solid state, then

(a) carbon dioxide will form in the container, (b) the tablet will become softer, and (c) upon "reconstitution" the bubble effect will be reduced to the extent that carbon dioxide was lost in storage.

The evolution of carbon dioxide would normally build up pressure in a glass bottle, but the tubes in which effervescent products used to be sold were not tight, and the carbon dioxide could escape. The same is true to a great extent in plastic bottles and in plastic blister packs, but the problem that the reaction (as will be demonstrated below) is catalyzed by moisture, in other words the fact that the container is not hermetic in this aspect, is a disadvantage. This is so sensitive that during manufacture extra precautions are taken to keep the relative humidity of the processing areas low. Hence one must also pack the products in hermetic containers, and aluminum foil has become a popular means of doing this. If, however, the initial moisture is not low enough, then the reaction will proceed, and in this case the internal pressure will cause the aluminum foil to balloon.

The solid-state reaction has been investigated by Usui and Carstensen (1985) and by Wright and Carstensen (1986). When the reaction occurs in the solid state, there are two questions that present themselves:

1. Is moisture important and if so in what sense?
2. What is the stoichiometry? Is it that of Eq. (6.48) or is it

$$R_2(COOH)_2 + NaHCO_3 \rightarrow R_2(COOH)COONa + H_2O + CO_2 \quad (6.50)$$

Usui checked the weight loss of heated samples in hermetic containers, utilizing different ratios of acid and base, and established that the stoichiometry is that of Eq. (6.50), i.e., the mole-to-mole interaction of tartaric acid and sodium bicarbonate.

He next studied the weight loss in open containers and demonstrated that the tartaric acid did not lose weight, and that the sodium bicarbonate, and the mixture of sodium bicarbonate and tartaric acid, lost weight at a low rate, corresponding to that of the sodium bicarbonate itself. In other words, in an open container there was no interaction, simply decarboxylation of the bicarbonate itself.

He next studied the effect of compression on the decomposition of sodium bicarbonate. Characteristic curves are shown in Fig. 18. It is noted that the decomposition rate are a function of applied pressure. In the following it is assumed that the particles are isometric and that the reaction rate is proportional to the surface area of unreacted sodium bicarbonate. The following nomenclature is used: there are M g of unreacted sodium bicarbonate at time t, and M_0 initially. There are N particles each of area a, volume v, and density ρ. The surface area is proportional to the two-thirds power of the volume by the isometry factor Q, i.e.,

$$a = Qv^{2/3} = Q\rho^{-2/3}m^{2/3} \quad (6.51)$$

The total area, A, hence is given by

$$A = NQ\rho^{-2/3}m^{2/3} = N^{1/3}Q\rho^{-2/3}M^{2/3} \quad (6.52)$$

It follows that

$$A_0 = N^{1/3}Q\rho^{-2/3}M_0^{2/3} \quad (6.53)$$

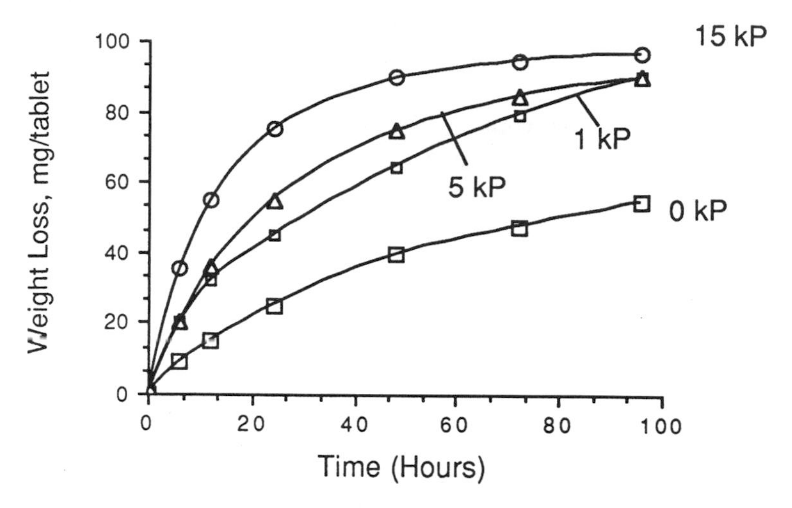

Fig. 18 Effect of tableting pressure on sodium bicarbonate decomposition at 70°C. (Graph constructed from data by Usui and Carstensen, 1985.)

The decomposition rate is proportional to the surface area at time t, i.e.,

$$\frac{dM}{dt} = -k''A = -k'M^{2/3} \tag{6.54}$$

where

$$k' = k''N^{1/3}Q\rho^{-2/3} \tag{6.55}$$

Rearrangement of Eq. (6.55) gives

$$\frac{dM}{M^{2/3}} = -k't \tag{6.56}$$

This can be integrated, and when initial conditions are imposed the following expression results:

$$\left(\frac{M}{M_0}\right)^{1/3} = (1 - X)^{1/3} = 1 - kt \tag{6.57}$$

where X is mole fraction decomposed, and where

$$k = \frac{k''N^{1/3}Q\rho^{-2/3}}{3M_0^{1/3}} \tag{6.58}$$

Eliminating N by inserting Eq. (6.55) into Eq. (6.58) gives

$$k = \frac{k''A_0}{3M_0} \tag{6.59}$$

The data should, therefore, plot by a cube root equation and Fig. 19 shows this, indeed, to be the case.

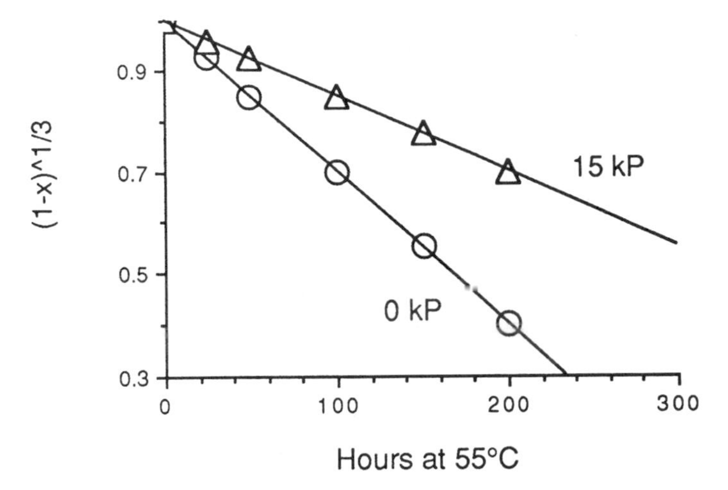

Fig. 19 Cube root plot of sodium bicarbonate decomposition at 55°C. Least squares fit equations: 0 kP: $y=1-0.0015x$ ($R=1.00$); 15 kP: $y=1-0.003x$ ($R=1.00$). (Graph constructed from data by Usui and Carstensen, 1985.)

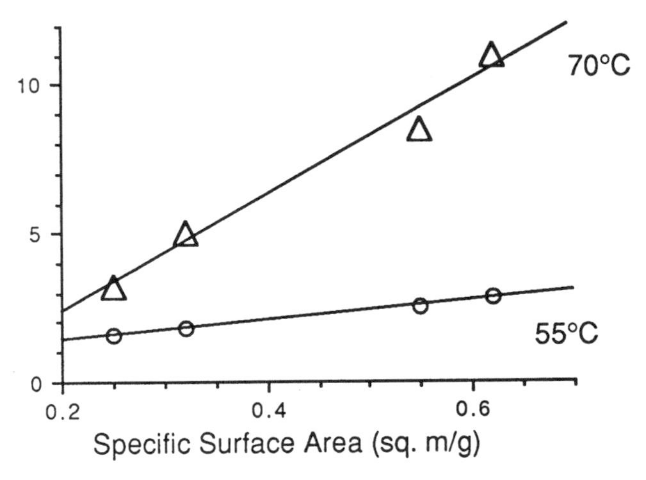

Fig. 20 Cube root constants from Fig. 19 versus specific surface areas. Least squares fits: 70°C: $y=-1.534+19.447x$ ($R=0.99$); 55°C: $y=0.788+3.188x$ ($R=1.00$) (Graph constructed from data by Usui and Carstensen, 1985.)

The rate constants according to Eq. (6.59) should be proportional to the surface area at time zero (A_0/M_0). That this is the case is shown in Fig. 20. The rate constants follow an Arrhenius plot (Fig. 21) and are in line with the data reported by Schefter et al. (1974).

In a closed system there is a rapid interaction between the sodium bicarbonate and the tartaric acid in compressed tablets. Even though the system is supposedly dry, it is assumed that there is a very slight amount (z moles) of water present in the table initially and that the reaction starts in a dissolved stage. If this is the case, then, since water is produced in the reaction, there will be an acceleration.

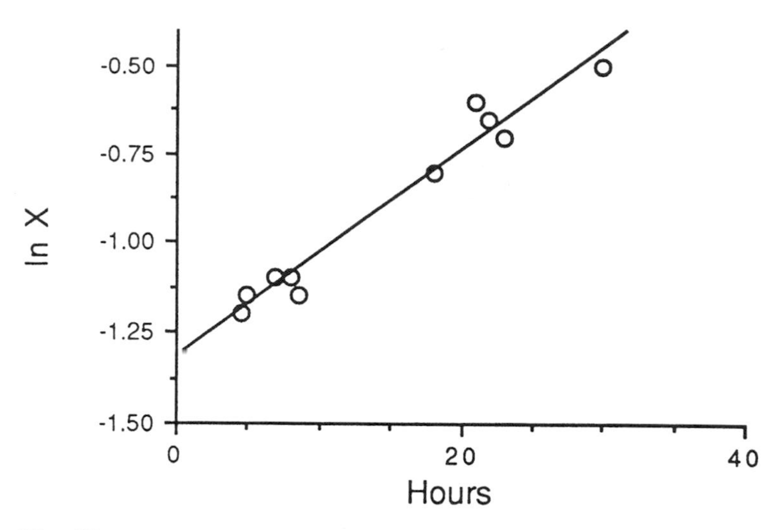

Fig. 21 Decomposition of tartaric acid plus sodium bicarbonate tablets at 55°C (5 kP force). Least squares fit: $\ln\{X\} = -1.3225 + 0.0291 \cdot t$ ($R = 0.98$). (Graph constructed from data by Usui and Carstensen, 1985.)

The data can be modeled in the fashion shown in the following. M' is the number of moles of sodium bicarbonate left at time t, M'_0 is the initial number of moles, S is its solubility in water, and C is the concentration in the water present at time t. S_1 is the solubility of the tartaric acid in water.

According to the reaction scheme the number of moles of water present at time t then is

$$(M'_0 - M' + z)\text{moles} = (M'_0 - M' + z)\ 0.018\ \text{liters} \tag{6.60}$$

The disappearance rate of sodium bicarbonate in solution is given by

$$\frac{-dC}{dt} = k_2 S_1 S \tag{6.61}$$

where k_2 is the second-order rate constant. To express this as number of moles decomposed, this figure is multiplied by the volume of water present, i.e., the expression in Eq. (6.60):

$$\frac{dM'}{dt} = -k^*(M'_0 - M' + z) \tag{6.62}$$

where

$$k^* = 0.018 k_2 S_1 S \tag{6.63}$$

Equation (6.62) can now be recast in the form

$$\ln(M'_0 - M' + z) = k^* t + \ln[z] \tag{6.64}$$

or employing X, the mole fraction decomposed:

$$\ln\left[X + \frac{z}{M_0}\right] = k^*t + \ln\left[\frac{z}{M'_0}\right] \tag{6.65}$$

Recalling that z is a small number, the term z/M_0 is small, and the Eq. (6.65) then simplifies to

$$\ln[X] = k^*t + \ln\left(\frac{z}{M'_0}\right) \tag{6.66}$$

Data are plotted in this fashion in Fig. 21. It is seen that the linearity is quite good. The value of z may be estimated from the intercept and comes to about 0.1 mg per tablet, which is a reasonable figure. This, in essence, shows that the theories suggested by Wright (1983) are correct.

It is obviously of pharmaceutical importance in most situations to slow down the reaction in the solid state and yet maintain the reactivity in the solid state. (An exception to this is when a reaction is purposely carried out during a granulation, for instance.) One way of retarding the reaction rate is to preheat the bicarbonate to 95°C for a certain length of time (White, 1963, Mohrle, 1980). This will react by the scheme

$$2NaHCO_3 \rightarrow Na_2CO_3 + H_2O \tag{6.67}$$

The water formed granulates the mixture and makes it easier to compress. But more importantly, the sodium carbonate formed can form double salts with the bicarbonate. These are dodecahydrates and act as moisture scavengers. They hence stabilize the acid/base mixture in the solid state (if reasonable moisture barriers are provided): any *small* amount of moisture created by a beginning reaction of the type of Eqs. (6.48) or (6.49) will react with a mixture of the carbonate and bicarbonate to form a double salt hydrate.

9. INCOMPATIBILITY PREVENTION TECHNIQUES

Frequently, interactions are particle size dependent. This stands to reason, because the finer a powder is, the more contact points there will be in the tablet mass, hence the larger the potential for interaction. Means of overcoming this are as follows.

In double granulation or pocketing techniques, one component is placed in one granulation, the other in another; keeping the granulations coarse will give fewer contact points, hence less interaction. This is a technique often used in vitamin granulations. Here the more famous incompatibilities are usually those involving (a) cyanacobalamine, iron, and ascorbic acid (b) vitamin A, (c) calcium pantothenate, and (d) tocopherol. The first of these cases is one where *pocketing* is used. This can be accomplished by actually coating (rather than just granulating) the iron salt (often ferrous fumarate) in order to separate it from the remaining ingredients. The other cases will be dealt with separately below.

Other means of separating incompatible ingredients is to make a compression-coated tablet. This consists of an inner tablet compressed in a coating granulation. This principle can be extended to a tablet within a tablet within a tablet

(Bicota^{TM}, Manisty). The machines used to make them are, however, slow. If a layer separation is necessary (and effective), then it is most often accomplished by triple-layer tablets. It should noted that these techniques are ineffective in the case of reactions that occur via the gas phase. (These types of reactions will be discussed in the following.)

As mentioned, coating is a special case of pocketing. Ferrous fumarate is sometimes coated, but the most famous case of coating is undoubtedly that of vitamin A esters. Prior to this technique, in the early 1950s vitamin A was added, with an antioxidant, to powder blends that were then encapsulated or tableted. The loss of the vitamin was excessive (frequently 50% in 6 months, plus a processing loss). In the early 1950s Hoffmann-La Roche and Pfizer (almost simultaneously) marketed a so-called beadlet. The coating of vitamin A was a bit different from that of other compounds, since the most common ester (acetate) is a liquid. The coating was therefore done by making an emulsion of the vitamin A in a solution of gelatin, spraying this onto an insoluble starch derivative (which rapidly absorbed moisture), and then further drying the beadlet. After drying, the starch derivative could be separate from the vitamin A by sieving. Later, the coated palmitate bead was introduced, and, with normal precautions, oxidation of the vitamin A (except for the droplets on or rather in the surface) was prevented. It follows from this that the finer the beadlet, the less stable will it be (because there will be more surface droplet of vitamin A). 40 mesh is about the coarsest than can be handled in tableting or encapsulation, and this mesh cut offers a good stability. Obviously compression will cause fracture of the beads to some extent and this is the actual stability problem in a dry tablet.

If moisture is present in the tablet, then the gelatin will soften and become more oxygen-permeable, and the stability will suffer. It is therefore always best to perform moisture stress tests in stability programs. At a point in development where enough tablets are available the following is done: four times the regular sample is taken, and this sample is subdivided into four equal portions, A, B, C, and D. A is placed on stability as is. B is exposed to water vapor in a desiccator, and removed and placed on stability when it has gained 0.5% in weight. (The tablets can be placed on Petri dished and weighed periodically.) The procedure is repeated with C to 1% and D to 2%. The information gained is valuable, because it aids in decisions of the following kind: (a) Should a desiccator bag be used? (b) What should the moisture specification on the product be? (c) If there is no effect of moisture, there would be less of a problem selecting plastic bottles for the product.

10. pH OF THE MICROENVIRONMENT

In the strictest sense, the term pH is not defined in a solid system. For it to have meaning, there must be some water mediation; tocopheryl acetate and calcium pantothenate are cases in point. The former is sensitive to high pH, the former to low pH. Calcium pantothenate is frequently admixed with magnesium oxide and granulated separately from the remaining ingredients. In this manner an alkaline microenvironment is created, which ascertains the stability of the vitamin.

In the case of tocopheryl acetate, the hydrolysis is accelerated by hydroxyl ions. Again it is noted that the reaction must be associated with some dissolution step in small amounts of water. The produced tocopherol is much less stable, and hence the hydrolysis and the presence of water are contraindicated. This is a particular case where the use of alkaline excipients (e.g. hydroxyapatite) can be deleterious at higher temperatures. In the absence of (or at low levels of) moisture the reaction may not proceed. It is also characteristic that often, higher temperatures are not indicative of what will happen at room temperature.

If it is desired to control the pH of the microenvironment then citric, tartaric, and fumaric acids are the acids of choice. They are, however, all corrosive, and their pharmaceutical handling is far from ideal. In the case of alkali, sodium bicarbonate, sodium carbonate, and magnesium and calcium oxides are common. They are not as corrosive as the acids mentioned, but they are abrasive, and they, too, are not the most ideal substances to handle in a tablet or a capsule.

For certain compounds it is necessary to control the microenvironment in even more drastic fashion. Gu et al. (1990) report on drug excipient incompatibility studies of moexipril hydrochloride and demonstrate that (even "wet") adjustment of the microenvironmental pH (i.e., adding small amounts of water to a mixture of the drug with sodium bicarbonate or sodium carbonate) did not sufficiently stabilize the mix. But when the mixture *was wet granulated*, and when *stoichiometric amounts of alkali were used*, then stabilization resulted. This essentially means that in the solid state *the sodium salt is stable* as opposed to the acid. It might be argued that in such a situation the sodium salt should be manufactured and used as such. It might be argued that it should be claimed as the active ingredient (equivalent to a certain amount of free acid, or in the case of amphoteric substances, the acid addition salt), but often the salt is very soluble and hygroscopic (e.g., potassium clavulanate) and hence difficult to produce. The situation is referred to in the Federal Register as a *derivative drug*.

11. INTERACTIONS INVOLVING A LIQUID PHASE

At times an active ingredient or a decomposition product in a solid dosage form is a liquid, and this may interact with other ingredients in the dosage form. A typical example is the work by Troup and Mitchner (1964) dealing with aspirin and phenylephrine. The authors showed that the decomposition of phenylephrine was linearly related to the formation of salicylic acid. They showed that the decomposition of phenylephrine was an acetylation. This can be thought of in many ways. There has to be some moisture present to allow for the hydrolysis of aspirin. If the salicylic acid is formed by interaction of aspirin with traces of water, then the acetic acid formed may react with the phenylephrine ($R(OH)_3$), again liberating water, so that the moisture does not play a part quantitatively in the overall reaction; in other words,

$$C_6H_4(OCOCH_3) + H_2O \rightarrow C_6H_4(OH)COOH + CH_3COOH \tag{6.68}$$

$$CH_3COOH + 1/3\,R(OH)_3 \rightarrow 1/3\,R(OCOCH_3)_3 + H_2O \tag{6.69}$$

$$C_6H_4(OCOCH_3) + 1/3\,R(OH)_3 \rightarrow 1/3\,R(OCOCH_3)_3 \tag{6.70}$$

An alternate explanation would be that phenylephrine interacted directly with aspirin in an anhydrous solid state to transacetylate, which is not probable. The question is whether the acetic acid (which has a sizable vapor pressure) interacts with the phenylephrine as a gas with a solid reaction (to be covered shortly) or as a liquid with a solid reaction.

There are other examples of the interaction of acetic acid with active ingredients, e.g., the work by Jacobs et al. (1966), in which acetylation of codeine in aspirin/codeine combinations was demonstrated. Again, whether the acetylation is achieved by acetic acid in the vapor phase or in the liquid state or (more unlikely) whether it is a direct solid to solid interaction, is not resolved at the present time. If it were the latter, then Jander kinetics should actually apply. But it is difficult to distinguish this and pseudo-first-order reactions. If it is an interaction in the liquid state, then it probably occurs by the phenylephrine dissolving in the acetic acid formed.

In more general terms, it is assumed that there are two drugs, A and D. A decomposes (e.g., by hydrolysis) to form a liquid decomposition product C. The reactions then are

$$\mathrm{A} + \mathrm{H_2O} \rightarrow \mathrm{B} + C \qquad \text{(rate constant } k) \tag{6.71a}$$

$$\mathrm{D} + C \rightarrow \text{decomposition} \qquad \text{(rate constant } k') \tag{6.71b}$$

C is the species that is liquid. In this case a saturated solution (S moles/mole) of D in C is formed, and it is assumed that dissolution is fast. Let A be the number of moles of drug #1 present at time t, let C be the number of moles of acetic acid, and let M denote the molarity of the liquid decomposition product (e.g., for acetic acid at 25°C the density is 1.05 g/mL, so that, since its molecular weight is 60, M would be $1005/60 = 16.75$). The rate at which D disappears is the question. It is assumed that the disappearance rate of A is pseudo–first order, i.e.,

$$A = A_0 \exp(-kt) \tag{6.72}$$

The disappearance rate of D depends on how much C is present, so the equation for C must first be established and solved. C is created at a rate of kA, but it is consumed by D. The rate of the latter step is given by a second-order reaction term. The concentration of D is S, and the concentration of C is M. The amount of C at time t is C, so that (in terms of moles)

$$\frac{dC}{dt} = kA - k'SCM \tag{6.73}$$

Inserting Eq. (6.73), and using and denoting

$$k'SM = a \tag{6.74}$$

where a is constant, the following equation is arrived at:

$$\frac{dC}{dt} = kA_0 \exp(-kt) - aC \tag{6.75}$$

Laplace transformation, using **L** notation, gives

$$s\mathbf{L} - 0 = \frac{kA_0}{\mathrm{s} + \mathrm{k}} - a\mathbf{L} \tag{6.76}$$

or

$$\mathbf{L} = \frac{kA_0}{a - \mathrm{k}}\left[\frac{1}{\mathrm{s} + \mathrm{k}} - \frac{1}{\mathrm{s} + a}\right] \tag{6.77}$$

so by taking anti-Laplacians,

$$C = \frac{kA_0}{k'SM - k}\{\exp(-kt) - \exp(-k'SMt)\} \tag{6.78}$$

It follows that the decomposition rate of D is given by

$$\frac{dD}{dt} = k'SCM = aC \tag{6.79}$$

i.e., by integrating Eq. (6.79) and multiplying by a, we obtain

$$D = \frac{kaA_0}{k'SM - k}\left[\frac{\exp(-k'SMt)}{k'SM} - \frac{\exp(-kt)}{k}\right] \tag{6.80}$$

An example of this is shown in Fig. 22 using $A = 50$, $k = 0.2$, and $k'SM = 0.1$. A different situation arises when an insoluble component interacts with a drug in solution (or vice versa). An example of this is the interaction between microcrystalline cellulose (R′CHOHR″) and substituted furoic acids (RCOOH) (Carstensen and Kothari, 1983). The furoic acids decompose when heated by themselves, into a liquid decomposition product and carbon dioxide. In the presence of microcrystalline cellulose, however, the mixture forms carbon monoxide:

$$\mathrm{RCOOH} + \mathrm{R'CHOHR''} \rightarrow \mathrm{Q} + \mathrm{Q'} + \mathrm{CO} \tag{6.81}$$

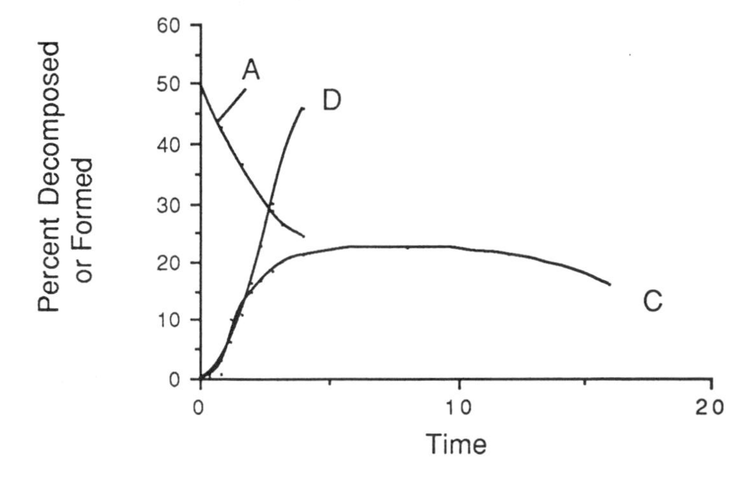

Fig. 22 Stability profile using $A = 50$, $k = 0.2$, and $a = 1$.

Q is a liquid that will dissolve furoic acid to the extent of its solubility and will spread over the microcrystalline cellulose. There will be a number of contact points, N, at which interaction can take place (essentially the "wetted" part of the microcrystalline cellulose). There will be a reaction probability, a, associated with each contact point. The reaction accelerates because the larger the extent it has reacted, the more liquid there will be to dissolve the furoic acid hence the more contact points. At a given point in time there will be overcrowding, since dissolved molecules will be next to contact points that have already reacted. Hence there is also a termination probability, b. But unlike the Prout–Tompkins model, this is finite at time zero.

It might be argued that the external surface of the microcrystalline cellulose would be insufficient to account for the total decomposition. There are however, two types of surface present in microcrystalline cellulose: nitrogen adsorption gives low surface areas (the external area), whereas for instance water isotherms give surface areas 100 times as large (Hollenbeck, 1978, Marshall et al., 1972, Zografi and Kontny, 1986).

By the decomposition at a contact point, it is assumed that the decomposition, creating one liquid decomposition molecule, will dislodge (dissolve) S molecules of furoic acid at the contact point. If the initial number of contact points is N_0 then

$$\frac{dN}{dt} = [-b + a(S - 1)]N = qN \tag{6.82}$$

where $q = -b + a(S - 1)$. The factor arises from the fact that when molecules react, aS new contact points are created and one (the one at which the reaction took place) is lost.

It follows then from integrating Eq. (6.82) (which can be done, since a and b are assumed constant), that

$$N = N_0 \exp(qt) \tag{6.83}$$

Since, at a given time t, the rate of decomposition is proportional to the number of contact points, then, L being the number of intact alkoxyfuroic acid molecules,

$$\frac{dL}{dt} = -gN \tag{6.84}$$

where g is a constant. From the definition of L it follows that the mole fraction, x, decomposed is given by

$$x = \frac{L_0 - L}{L_0} \tag{6.85}$$

or

$$\frac{dx}{dt} = -\frac{g}{L_0}\frac{dL}{dt} \tag{6.86}$$

Equation (6.84) inserted in this gives

$$\frac{dx}{dt} = \frac{1}{L_0} gN \tag{6.87}$$

Substituting Eq. (6.83) into this gives

$$\frac{dx}{dt} = \frac{gN_0}{L_0} \exp(qt) \tag{6.88}$$

This integrates to

$$x = \frac{gN_0}{L_0 q}[e^{qt} - 1] = A[e^{qt} - 1] \tag{6.89}$$

where the term $A = gN_0/L_0q$ has been introduced for convenience. Equation (6.89) is equivalent to

$$\ln[1 + Ax] = qt \tag{6.90}$$

Fig. 23 shows data treated in this fashion.

12. CASES OF INTERACTION OF A LIQUID WITH A POORLY SOLUBLE DRUG

There are cases where there are liquids in a solid dosage form. An example is panthenol in a multivitamin tablet. Here it is customary to adsorb the liquid onto a solid carrier, and in the case of panthenol, magnesium trisilicate is used. At elevated temperatures (and at room temperature under compression as well) the panthenol will ooze out of the carrier and come into intimate contact with other solids. If interaction potentials exist, then separation techniques such as triple-layer

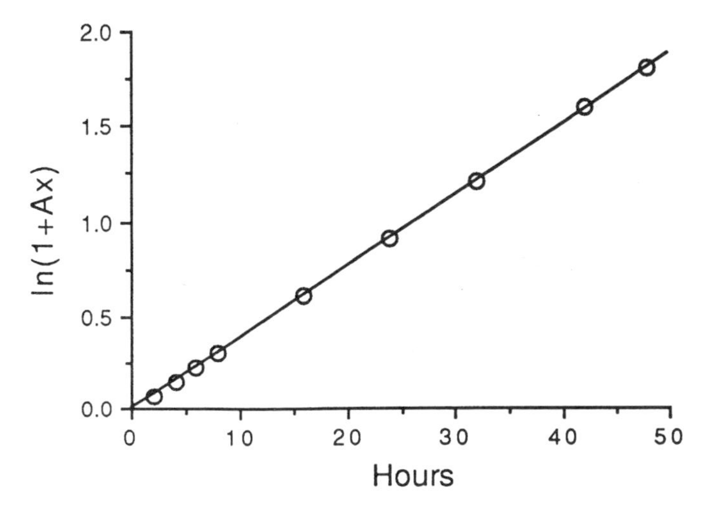

Fig. 23 Furoic acid data treated according to Eq. (6.90). (Graph constructed from data by Carstensen and Kothari, 1983.)

tablets (or compression coated tablets) are resorted to. In this case, the liquid will still ooze into the layer containing its interactant, but the process will be diffusion controlled. It can be shown (Jost, 1962) that the average concentration, C, of the liquid in the neighboring layer, with which it is in contact, is given by

$$\frac{C - C_f}{C_0 - C_f} = Qe^{-kt} \tag{6.91}$$

where C_f is the concentration at infinite time. The term on the right-hand side is actually the leading term of an infinite series.

13. REACTIONS VIA THE GAS PHASE

Sometimes the vapor pressure of a drug is sufficiently high that it may interact with other substances via the vapor phase. An example is ibuprofen (B). This is a Lewis acid, and it can interact with Lewis bases. Usual measures, such as e.g. triple-layer tablets, do not work in this case, since the interactant will be present in the gas phase.

If the reaction with another drug (D) is

$$\mathrm{D} + \mathrm{B} \rightarrow \text{decomposition} \tag{6.92}$$

then the initial reaction rate is given by

$$\frac{d\{D\}}{dt} = -kP_\mathrm{B}[\mathrm{D}]A \tag{6.93}$$

where $\{D\}$ is the surface density of D molecules (number of molecules per cm^2) at time t, and A is the surface area. As long as there is no penetration into the crystals, the reaction will therefore be a first-order reaction, since Eq. (6.93) integrates to

$$\ln[C] = -kAP_\mathrm{B}t + \ln[C_0] \tag{6.94}$$

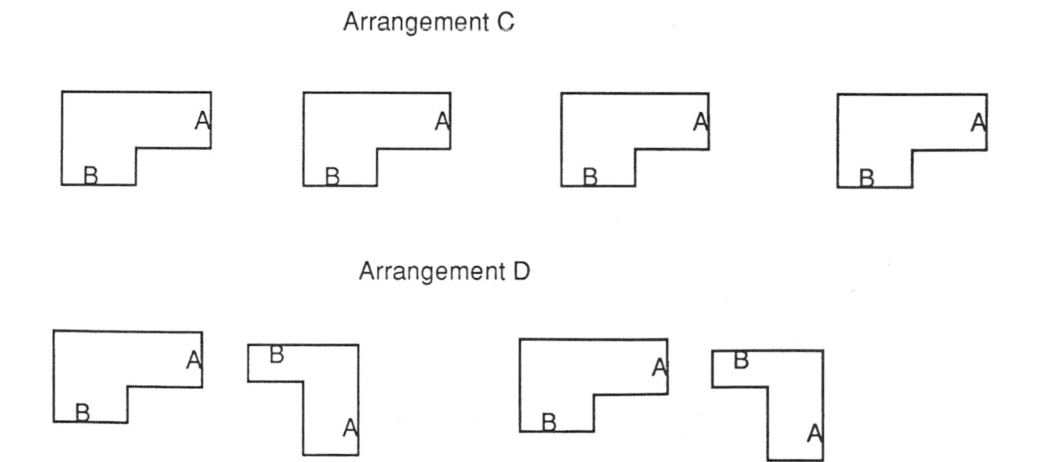

Fig. 24 Schematic of an example of molecular arrangement possibilities in a crystalline solid. If groups A and B can interact, then the situation in the upper arrangement is less prone to reaction, since A and B are at a greater distance.

where C_0 is the initial concentration. This will be true if only the surface of the solid interactant is affected. The extent of decomposition will be slight, because (unless the drug is extremely finely subdivided) only a small fraction of the molecules are on the surface. If, however, the ibuprofen penetrates the crystal, then Jander kinetics should prevail. A similar situation may be at work in the aspirin incompatibilities mentioned earlier.

14. AMORPHATES

As mentioned earlier, solids can occur either in crystalline form or as particulate amorphates. The chemical stability of the solid in crystalline form will differ from the same entity in amorphous form. In most cases the crystalline form, under the same conditions, will be more stable than the comparable amorphate.

The most interest and the largest body of work of amorphates is in the field of macromolecules. These usually possess a glass transition temperature*, T_g, and the states are referred to as "glassy" below† and "rubbery" above T_g.

Only a few articles have appeared on the subject of chemical stability of amorphates. In general, a compound is more stable in the crystalline state than in an amorphous state, but exceptions exist (Sukenik et al., 1975; O'Donnel and Whittaker, 1992; Stacey et al., 1959). There *are* cases that have been reported (Lemmon et al., 1958) where the crystalline state is less soluble than the molecule in solution, but they are rare.

In general, in a crystalline state, molecules are to a great extent fixed in position. In cases where the situation exists where a group from one molecule reacts with another group in a neighbor, the situation as shown in Fig. 24 arises.

Pothisiri and Carstensen (1975) have shown that in a situation such as the case of substituted benzoic acids the decomposition is between two groups in the same molecule.

Suppose parts A and B of the molecule depicted in Fig. 24 react. In such a case arrangement C would give better stability than arrangement D, because A would be further away from B in the former arrangement. Arrangement D can be more adverse than a random orientation as well, and if that is the case, then the amorphous form would be more stable than the crystalline (arrangement D). This is the exception rather than the rule.

In the presence of moisture, conversions from amorphous to crystalline modifications are promoted (Carstensen and Van Scoik, 1990; Van Scoik and Carstensen, 1990), and the material developed in the following all refers to anhydrous conditions.

In the work by Carstensen and Morris (1993), amorphous indomethacin was produced by melting a crystalline form of it to above melting (162°C) and recooling it to below 162°C. Amorphates made in this manner are morphologically stable down to 120°C‡ so that their chemical stability can be monitored. At a range of temperatures below this temperature, crystallization occurs too rapidly to allow for assessment of chemical stability. Amorphous samples were placed at several con-

---

* More than one glass transition temperature may exist.
† The highest T_g in the case of multiple glass transition temperatures.
‡ If the temperatures are lowered rapidly, then stable amorphates can be formed at room temperature, but kinetics cannot be followed easily because of the slow reaction rate at room temperature.

stant temperature stations (145, 150, 155, 165, 175, and 185°C) and assayed from time to time. The content of intact indomethacin was assessed by using the USP method of analysis.

The decomposition curves of amorphous indomethacin and a melt of indomethacin at different temperatures is shown in Figs. 25 and 26. It is noted that the pattern is strictly first-order. Of the few reports in the literature dealing with the chemical stability of compounds in the amorphous state, amorphous cephalosporins (Pfeiffer et al., 1976; Oberholtzer and Brenner, 1979; Pikal et al., 1977) also adhere to a first-order pattern. One purpose of the following writing

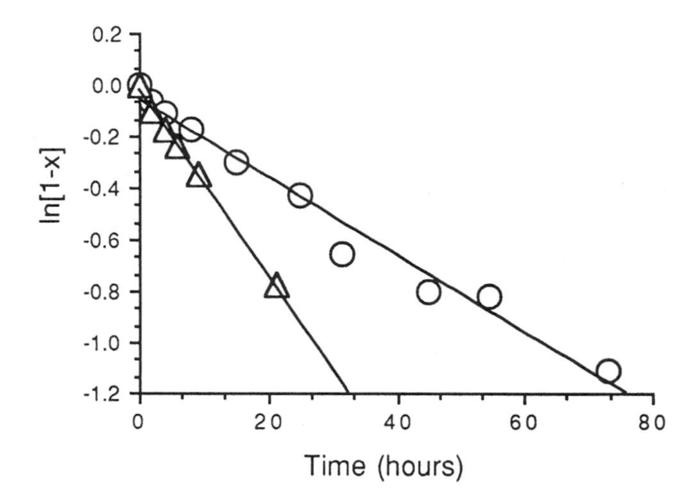

Fig. 25 Decomposition curves of solid, amorphous indomethacin at two of the three temperatures tested. ○: 145°C ($k = 0.015\ h^{-1}$); △: 155°C ($k = 0.036\ h^{-1}$).

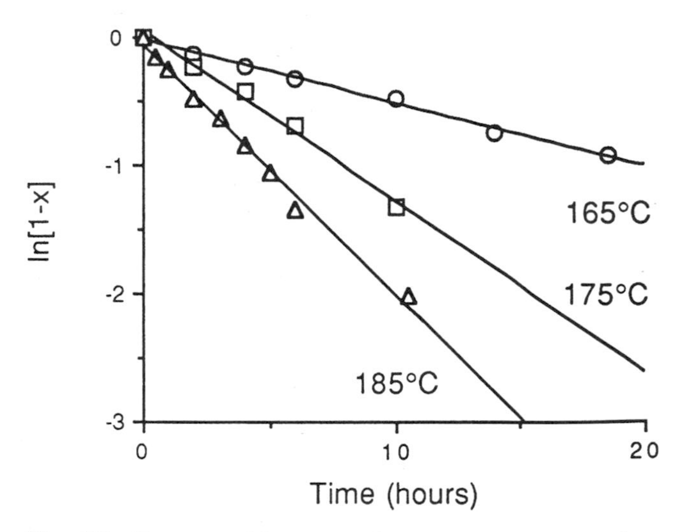

Fig. 26 Decomposition data of indomethacin in a molten state. Circles 165°C (rate constant $0.05\ h^{-1}$); squares: 175°C (rate constant $0.13\ h^{-1}$); triangles: 185°C (rate constant $0.19\ h^{-1}$).

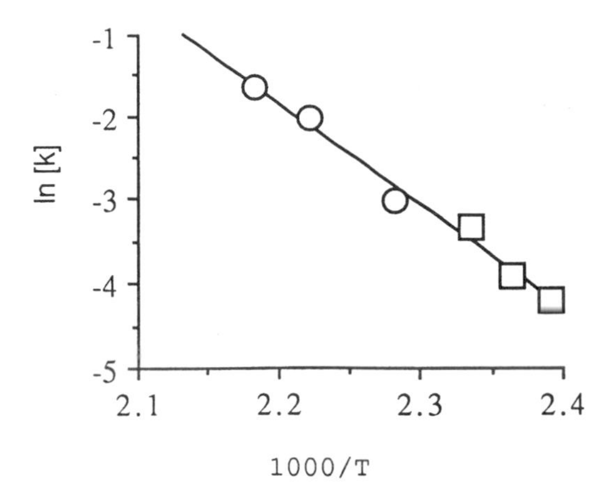

Fig. 27 Arrhenius plot of data from Fig. 24 and Fig. 25. Squares are melt and circles are amorphate. The least squares fit is $\ln[k] = 25.218 - (12{,}300/T)$.

is to seek an explanation for this pseudo-first-order (or indeed, truly first-order) pattern. The explanation must lie, in some manner, with the fact that in the rubbery state, the molecules can arrange in a random fashion, in a somewhat frozen (or much slowed) manner of that of the melt above the traditional melting point.

The results obtained from the melt are shown in Fig. 26, and against a first-order plot results. If an Arrhenius plot is drawn of the data from both Fig. 25 and 26, then Fig. 27 results.

It is seen that the Arrhenius plot of the amorphate continues into the Arrhenius plot of the melt. An attempt to explain this is made in the following.

If the substance in Fig. 24 were a crystalline solid, then the potential energy between molecules would be inversely proportional to a power function of their distance (the lattice constant) (Carstensen and Morris, 1993), i.e., would be akin to a Lennard-Jones potential (Lennard-Jones, 1931). However, in the amorphous state, if the decomposition is an intermolecular (rather than an intramolecular) reaction, then a group A in molecule I interacts with group B in the neighboring molecule II. The energy of the molecular pair will be dependent on the distance between the group A in one of the pair and group B in the other. These distances would be assumed to be randomly distributed, and a certain fraction $N_{>i}/N_0$ of the molecular pairs would be at or above a critical energy, E_i, necessary for reaction between A and B. The fraction of molecules that have this energy, E_i, is given by the Boltzmann distribution (Moelwyn-Hughes, 1961):

$$\frac{N_i}{N} = \frac{\exp(-E_i/RT)}{\sum_{k=0}^{k=\infty} \exp(-E_k/RT)} \tag{6.95}$$

where N is the total number of molecules and where the summation is over all energy levels. The fraction of molecules having energies in excess of $E_{>i}$ is then $N_{>i}$, given by

$$\frac{N_{>i}}{N} = \frac{\sum_{k=i}^{k=\infty} \exp(-E_k/RT)}{\sum_{k=0}^{k=\infty} \exp(-E_k/RT)} \tag{6.96}$$

There are several ways of approaching these summations, e.g., by considering the energy differences small and integrating. A discrete approach is to assume that the energy difference, ΔE, between adjoining energy states is constant. In this case Eq. (6.96) can be written:

$$\begin{aligned}\frac{N_{>i}}{N} &= \frac{e^{-E_i/RT} + e^{-(E_i+\Delta E)/RT} + \ldots}{e^{-E_0/RT} + e^{-(E_0+\Delta E)/RT} + \ldots} \\ &= \frac{e^{-E_i/RT}[1 + e^{-\Delta E/RT} + e^{-2\Delta E/RT} + \ldots]}{e^{-E_0/RT}[1 + e^{-\Delta E/RT} + e^{-2\Delta E/RT} + \ldots]}\end{aligned} \tag{6.97}$$

i.e.,

$$\frac{N_{>i}}{N} = \frac{e^{-E_i/RT}}{e^{-E_0/RT}} = e^{-(E_i-E_0)/RT} \tag{6.98}$$

Alternatively, if the difference between energy levels is large compared to the ground state energy, one can simply approximate the series in the numerator and denominator of these equations with their leading terms. This leads to the same result:

$$\frac{N_{>i}}{N} = \frac{\sum \exp(-E_i/RT)}{\sum \exp(-E_0/RT)} = \exp\left[\frac{-(E_i - E_0)}{RT}\right] \tag{6.99}$$

If, in a time element dt, a fraction of the molecules (dN/N) reaching E_i (or higher) react, then, denoting this fraction q,

$$\left(\frac{1}{N}\right)\frac{dN}{dt} = q\left[\frac{N_{>i}}{N}\right] = q\exp\left[\frac{-(E_i - E_0)}{RT}\right] = -k_1 \tag{6.100}$$

where k_1 (by definition in differential form) is a first-order rate constant, i.e., by integrating Eq. (6.100) and imposing $N = N_0$ at time $t = 0$,

$$\ln\left[\frac{N}{N_0}\right] = -k_1 t \tag{6.101}$$

i.e., first-order, where the rate constant is given by

$$k_1 = q\exp\left[\frac{-(E_i - E_0)}{RT}\right] \tag{6.102}$$

or its logarithmic equivalent,

$$\ln[k_1] = \ln[q] - \frac{E_a}{RT} \tag{6.103}$$

i.e., an Arrhenius equation where the activation energy is given by

$$E_a = E_i - E_0 \tag{6.104}$$

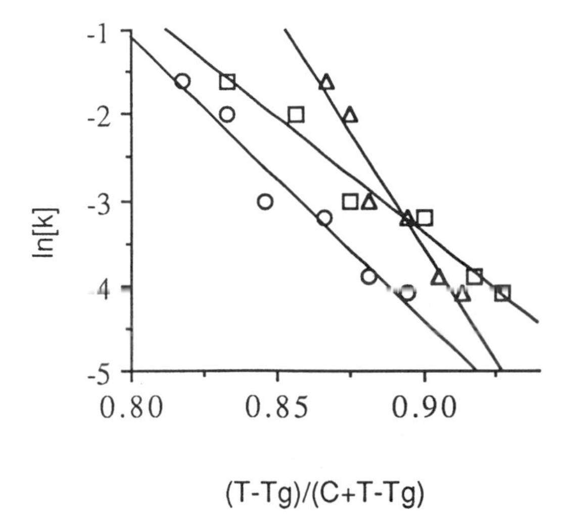

Fig. 28 Data from Fig. 26 plotted by the inverse function of the WLF equation. *Triangles:* $T_g = 80°$, $C_2 = 10$: $\ln[k] = 25.40 - 33.117\{T - T_g\}/\{C + (T - T_g)\}$; $R = 0.977$. *Circles:* $T_g = 100°$, $C_2 = 6$: $\ln[k] = 45.48 - 54.47\{T - T_g\}/\{C + (T - T_g)\}$; $R = 0.97$. *Squares:* $T_g = 120°$, $C_2 = 5$: $\{T - T_g\}/\{C + (T - T_g)\} = -0.771 \ln[k] + 0.0289$; $R = 0.982$.

The data in Figs. 24 and 25 demonstrate the correctness of Eq. (6.100), i.e., the expectancy of a first-order decomposition, and Fig. 27 demonstrates the correctness of Eq. (6.102).

There have been proposals (Moelwyn-Huges, 1961, Franks, 1989) that the stability of a compound near its T_g is best described in terms of the Williams-Landel-Ferry equation (Williams et al, 1955):

$$\ln[R] = \ln[R_g] + \left[\frac{c_1\{T - T_g\}}{\{C_2 + (T - T_g)\}}\right] \tag{6.105}$$

where C_2 and c_1 are constants. It is far from certain that this equation would apply to chemical reactions, but Fig. 28 shows its application to the data in Fig. 27. Several different values of C and T_g will give reasonable fits, as seen. It would seem intuitive that if the Arrhenius equation fits, then there would be values of C_2 that would make the WLF equation fit as well.

Schmitt et al. (1999) described the crystallization of amorphous lactose above the glass transition temperature to follow the Johnson–Mehl–Avrami (Johnson and Mehl, 1939; Avrami, 1939) equations:

$$[-\ln(1 - \alpha)]^{1/n} = k(t - t_i) \tag{6.106}$$

Pikal et al. (1977) employed solution calorimetry to determine the amorphous content of cephalothin sodium, cefazolin sodium, cefamandole naphtate, and cefamandole sodium. Since the amorphous forms are more energetic, they have a higher heat of solution, and the percent amorphate can be obtained if the heat of solution of amorphate and crystalline forms separately is known.

Lo (1976) showed that ampicillin trihydrate dehydrated to amorphous ampicillin, which had much poorer stability than the trihydrate. Upon storage the decomposition appears biphasic.

15. PSEUDO-POLYMORPHIC TRANSFORMATIONS

Dehydration, as mentioned above, may result in amorphous anhydrates, but it may also result in another crystalline phase (e.g., a lower hydrate or a crystalline anhydrate). These are, properly speaking, *pseudo-polymorphic transformations*. There are several steps in dehydration of a hydrate, and they can be summarized in the following manner, where S denotes solid, D denotes drug, V denotes vapor, and L denotes liquid (Han and Suryanarayanan, 1997).

$$D \cdot xH_2O \rightarrow D_S + xH_2O_L \qquad \text{(enthalpy of dehydration} = \Delta H_D)$$

$$D_S + xH_2O_L \rightarrow D_S + xH_2O_V \qquad \text{(enthalpy of vaporization} = \Delta H_V)$$

$$D \cdot xH_2O \rightarrow D_S + xH_2O_V \qquad \text{(enthalpy of transition} = \Delta H_T)$$

i.e.,

$$\Delta H_T = \Delta H_V + \Delta H_S \qquad (6.107)$$

so that different results can be obtained in DSC experiments depending on whether a crimped or open pan is used.

Bray et al. (1999) have shown such a diagram for [2(S)-[p-toluenesulfonyl amino]-3-[[[5,6.7.8-tetrahydro-4-oxo-5-{5-[2-(piperidin-4-yl)ethyl]-4-H-pyrazolo[1, 5-a][1,4]diazepin-2-yl]carbonyl]amino]-propionic acid.

Suihko et al. (1997) have employed DSC to show that dehydration of theophylline monohydrate is a two-step process.

16. POLYMORPHIC TRANSFORMATIONS

Polymorphic transformation rates have become of importance of late, and an example is a recent article by Agbada and York (1994) dealing with the dehydration kinetics of theophylline. The article by Ng (1975) is similarly instructive in the sense that it reviews all the equations that have been developed for polymorphic transformation kinetics.

In most cases the transformation kinetics are S-shaped curves, and before any model is imposed on the data, the following model should be considered. (This is comparable to the model proposed by Carstensen and Van Scoik, 1990). If the phenomenon that governs the transformation is essentially the nucleation lag time, then the curves may be considered as representing either a normal or a log-normal error curve, and the mean would be the mean (or geometric mean) nucleation time. What this states is that each particle, in a sense, acts as its own entity, that there is a nucleation time (with an error or a variance attached to it) and that the particle will endure the nucleation time and then decompose, individually, very rapidly.

Table 8 Data (Fig. 16) from Which Fig. 29 Was Generated

	Fraction remaining		Standard normal deviate	
Time (h)	A	B	Z_A	Z_B
0	1.00	1.00		
0.2	0.698		0.50	
0.25		0.72		0.58
0.3	0.33		−0.43	
0.5	0.17	0.42	−0.92	−0.2
0.7	0.08	0.28	−1.5	−0.58
1.0		0.11		−1.18
1.25		0.08		−1.4

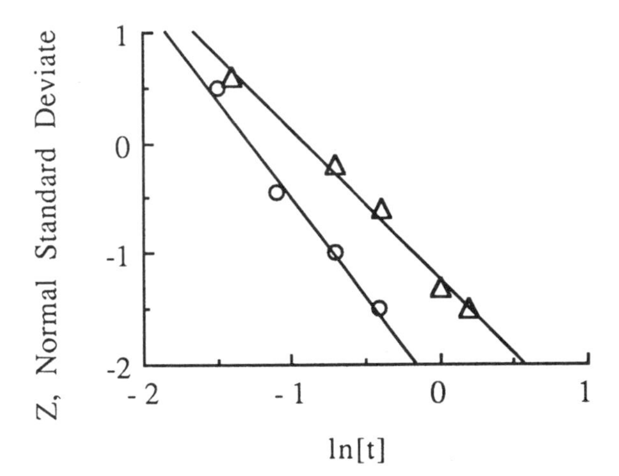

Fig. 29 Data from Fig. 16 treated as log-normally distributed in time.

The data from Fig. 16 are shown in tabular form in Table 8. These data are plotted log-normally in Fig. 29, and it is seen that there is excellent linearity. The model is much simpler and much more reasonable in the case of polymorphic transformations than other models relying on farfetched mechanistic assumptions.

The reason for the log-normal relationship is not difficult to understand. Solids are usually log-normally distributed. If the nucleation time is inversely proportional to size, then it too would be log-normally distributed.

Dehydration, at times, results in a morphic transformation. For instance, Lo (1976) showed that the transformation of crystalline ampicillin trihydrate to amorphous penicillin was primarily first order, it either was first order or followed a contracting cylinder model ($(1-x)^{1/2}$ being proportional to time).

16.1. Pseudo-Polymorphic Transformations. Dehydration Kinetics of Hydrates

A special case of polymorphism is pseudo-polymorphism which deals with hydrates. Anydrates and hydrates often crystallize in different crystal systems, but the molecule in solution is the same. In the solid state there is a difference in that water molecules form part of the matrix.

Dehydration kinetics of hydrates has had the attention of the pharmaceutical scientist for a while. As far back as 1967, Shefter and Kmack (1967) studied the dehydration of theophylline hydrate and found it to be first order. However, this is not the normal order of dehydration.

Byrn (1982) has developed a generalized kinetic theory for isothermal reaction in solids, and theophylline has been used as a model for several studies of this kind (Lin and Byrn, 1979; Suzuki et al., 1989; Agbada and York, 1994). In the recent work by Dudu et al. (1995), microcalorimetric methods were used and show a two-step process to take place. The predominant model used for this type of kinetics is the Avrami–Erofeev treatment leading to the equation

$$[-\ln(1-x)]^{0.25} = kt \tag{6.108}$$

and good adherence of the data for each step was found. This points to microcalorimetry as being a potent tool in the investigation of such projects.

There are vacuum/electrobalances on the market now, which have the capability of increasing the relative humidity and keeping the "new" relative humidity constant until the weight gain of a sample in the balance assembly has gained less than a preset quantity of water. When an anhydrate is placed in a vacuum in such a balance assembly and the relative humidity gradually raised, then the "up" and "down" portions of the curve will often have the profile shown in Fig. 30.

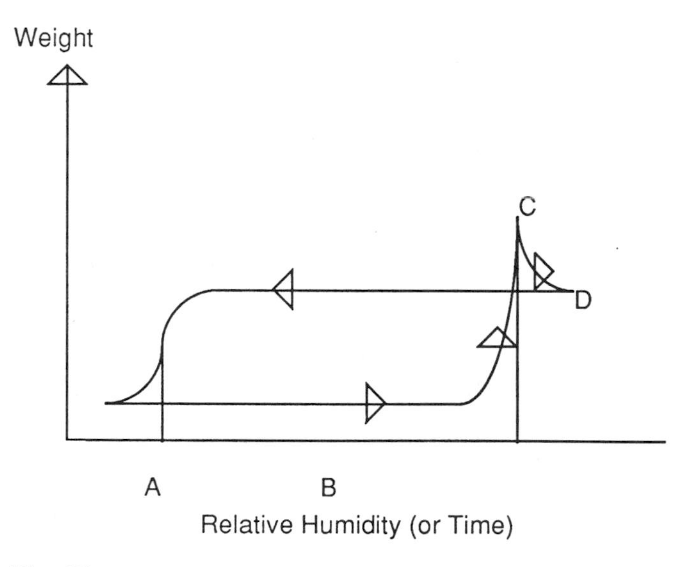

Fig. 30 A represents the relative humidity over the crystalline hydrate, and at this point hydration should actually occur. But a metastable situation usually occurs, so if the amorphate is exposed to higher humidities, then it is not until at B that it changes, and in this case it forms a (supersaturated) solution, which at C precipitates out as the crystalline hydrate D. The desorption isotherm then results in a constant composition until a lower humidity, where the hydrate starts dehydrating into the amorphous anhydrate.

It is to be expected that the weight of the anhydrate will not increase until the relative humidity of the salt pair is reached. If one only considers the part ABC, then this is logical.

It is noted that the abscissa is time as well as relative humidity, since this latter is changed as a function of time. The curve often goes beyond the weight of the anhydrate, and it is hypothesized that what actually happens is that the relative humidity at which the increase occurs is that of the saturated solution of the anhydrate. This is a metastable solution and will start precipitating out in time (point C) until the weight levels off at the theoretical weight of the hydrate. On the "down" curve, this hydrate then remains until the relative humidity of the salt pair is reached, and then it drops off.

So, if an experiment is carried out as shown in Fig. 30, it is not certain that the relative humidity at A is the equilibrium relative humidity of the hydrate/anhydrate salt pair, because as in a conventional isotherm, the "water channel" could act in the same fashion as a pore, and the "breakthrough" vacuum might be an indication of the effect of the Kelvin equation.

Only a few unreported studies have been carried out of this kind (Pudipeddi, 1995; Dali, 1995; Shlyankevich, 1995), but the method, in the sense of the preceding paragraph, could be of importance in assessing diameters of water channels and of interfacial tension between water and the organic and inorganic matrix molecule (via the Kelvin equation).

17. PHOTOLYSIS IN THE SOLID STATE

Not much systematic work has been reported on photolysis of solids. Lachman et al. (1961) pointed out that, most often, a solid tablet will decompose by photolytic decomposition only in the surface area, so that if one broke a "discolored," exposed tablet the color would be unaffected in the interior.

However, Kaminski et al. (1979) reported on a case where a combination of moisture and light caused an interaction between a dye and a drug that permeated the entire tablet.

Tønnesen et al. (1997) have reported on the photoreactivity of mefloquine hydrochloride in the solid state.

REFERENCES

Adler, M., Lee, G. (1999). J. Pharm. Sci. 88:199.
Agbada, C. O., York, P. (1994). Int. J. Pharm. 106:33.
Anderson, N. R., Banker, G. S., Peck, G. E. (1982). J. Pharm. Soc. 71:7.
Avrami, M. (1939). J. Chem. Phys. 7:1103.
Bawn, C. (1955). "Chemistry of the Solid State," p. 254, Academic Press, New York.
Bray, M. L., Jahansouz, H., Kaufman, M. J. (1999). Pharm. Devl. Techn. 4:81.
Byrn, S. R. (1982). Solid State Chemistry of Drugs. New York: Academic Press, pp. 59–70.
Carstensen, J. T., Kothari, R. (1983). J. Pharm. Sci. 72:1149.
Carstensen, J. T., Morris, T. (1993). J. Pharm. Sci. 82:657.
Carstensen, J. T., Musa M. N. (1972). J. Pharm. Sci. 61:273, 1112.
Carstensen, J. T., Pothisiri, P. (1975). J. Pharm. Sci. 64:37.
Carstensen, J. T., VanScoik, K. (1990). Pharm. Res. 7:1278.
Carstensen, J. T., Aron, E., Spera, D., Vance, J. J. (1966). J. Pharm. Sci. 55:561.

Dali, M.V. (1995). Personal communication.
Dudu, S. P., Das, N. G., Kelly, T. P., Sokoloski, T. D. (1995). Int. J. Pharmaceutics 114:247.
Fini, A., Fazio, G., Alvarez-Fuentes, J., Fernández-Hervás, J. T., Holgado, M. A. (1999). Int. J. Pharm. 181:11.
Franks, F. (1989). Process Biochem 24:3–8.
Goldstein, M., Flanagan, T. (1964). J. Chem. Ed. 41:276.
Gu, L., Strickley, R. G., Chi, L.-H., Chowhan, Z. T. (1990). Pharm. Res. 7:379.
Han, J., Suryanarayanan, R. (1997). Int. J. Pharm. 157:209.
Hollenbeck, R. G., Peck, G. E., Kildsig, D. O. (1978). J. Pharm Sci. 67:1599.
Jacobs, A., Dilatusch, A., Weinstein, S., Windheuser, J. (1966). J. Pharm. Sci. 53:893.
Jander, W. (1927). Z. Anorg. Chem. 163:1
Johnson, W. A., Mehl, R. F. (1939). Trans. Am. Inst. Min. Eng. 135:416.
Jost, H. (1962). Diffusion. New York: Academic Press, p. 45.
Kaminski, E. E., Cohn, R. M., McGuire, J. L., Carstensen, J. T. (1979). J. Pharm. Sci. 68:368.
Kissinger, H. E. (1956). J. Res. Nat. Bur. Std. 57:217.
Kittel, C. (1956). Introduction to Solid State Physics. 2d ed. New York: John Wiley.
Kornblum, S., Sciarrone, B. (1964). J. Pharm. Sci. 53:935.
Lachman, L., Weinstein, S., Swartz, C., Urbanyi, T., Cooper, J. (1961). J. Pharm. Sci. 50:141.
Lemmon, R. M., Gordon, P. K., Parsons, M. A., Mazetti, F. (1958). JACS 80:2730.
Lennard-Jones, J. E. (1931). Proc. Phys. Soc. (London) 43:461.
Leung, S. S., Padden B. E., Munson, E. J., Grant, D. J. W. (1998a). J. Pharm Sci. 87:501.
Leung, S. S., Padden B. E. Munson, E. J., Grant, D. J. W. (1998b). J. Pharm. Sci. 87:509.
Lin, C. T., Byrn, S. R. (1979). Mol. Cryst. Liq. Cryst. 50:99.
Lo, P. K. A. (1976). A study of the solid state stability of ampicillin. Ph.D. thesis, University of New York at Buffalo.
Maron, S. M., Prutton, C. F. (1965). Principles of Physical Chemistry, MacMillan, London, p. 322
Marshall, K., Sixsmith, D., Stanley-Wood, N. G. (1972). J. Pharm. Pharmacol. 24:138.
Moelwyn-Hughes, E. A. (1961). Physical Chemistry 2d rev. ed. New York: Pergamon Press, p. 31.
Mohrle, R. (1980). In: Lieberman, H. A., Lachman, L., eds. Pharmaceutical Dosage Forms: Tablets. Vol. 1, New York: Marcel Dekker. p. 24.
Nelson, E., Eppich, D., Carstensen, J. T. (1974). J. Pharm. Sci. 63:755.
Ng, W. L. (1975). Aust. J. Chcm. 28:1169.
Oberholtzer, E. R., Brenner, G. S. (1979). J. Pharm. Sci. 68:863.
O'Donnel, J. H., Whittaker, A. K. (1992). J. M. S.-Pure Appl. Chem. A29:1–10.
Oksanen, C. A., Zografi, G. (1993). Pharm. Res. 10:791.
Olsen, B. A., Perry, F. M., Snorek, S. V., Lewellen, P. L. (1997). Pharm. Dev. Tech. 2:303.
Otsuka, M., Teraoka, R., Matsuda, Y. (1991). Pharm. Res. 8:1066.
Otsuka, M., Onoe, M., Matsuda, Y. (1993). Pharm. Res. 10:577.
Otsuka, M., Ofsua, T., Yoshihisa, M. (1999). Drug Dev. Ind. Pharm. 25:197.
Pfeiffer, R. R., Engel, G. L., Coleman, D. (1976). Antimicrobial Agents Chemotherapy 9:848.
Pikal, M. J., Lukes, A. L., Lang, J. E., Gaines, K. (1976). J Pharm. Sci. 67:767.
Pikal, M. J., Lukes, A. L., Jang, J. E. (1977). J. Pharm. Sci. 66:1312.
Pothisiri, P., Carstensen J. T. (1975). J. Pharm. Sci. 64:1931.
Prout, E., Tompkins, F. (1944). Trans. Faraday Soc. 40:448.
Pudipeddi, M. (1995). Personal communication.
Roy. M. L., Pikal, M. J., Rickard, E. C., Maloney, A. M. (1990). International Symposium on Product Biological Freeze-Drying and Formulation, Bethesda, USA. In Develop. Biol. Standard. 74:323–340 (Karger, Basel, 1991).
Shefter, E, Lo, A., Ramalingam, S. (1974). Drug Dev. Comm. 1(1):29.
Schmitt, E. A., Law, D., Zhang, G. G. Z. (1999). J. Pharm. Sci. 88:291.

Shefter, E., Kmack, G. (1967). J. Pharm. Sci. 56:1028.
Shlyankevich, A. (1995). Personnel communication.
Stacey, F. W., Sauer J. C., McKusick, B. C. (1959). JACS 81:987.
Suihko, E., Ketolainen, J., Poso, A., Ahlgren, M., Gynther, J., Paronen, P. (1997). Int. J. Pharm. 158:47.
Sukenik, C. N., Bonopace, J. A., Mandel, N. S., Bergman, R. C., Lau, P.-Y., Wood, G. (1975). JACS 97:5290.
Suzuki, E., Shimomura, K., Sekiguchiki, I. (1989). Chem. Pharm. Bull. 37:493.
Tønnesen, H. H., Skrede, G., Martinsen, B. K. (1997). Drug Stability 1:249.
Troup, A., Mitchner, H. (1964). J. Pharm. Sci. 53:375.
Tzannis, S. T., Prestrelski S. J. (1999). J. Pharm. Sci. 88:351.
Usui, F. (1984). Master's thesis, University of Wisconsin, School of Pharmacy, Madison, WI 53706.
Usui, F., Carstensen, J. T. (1985). J. Pharm. Sci. 74:1293.
VanScoik, K., Carstensen, J. T. (1990). Int J. Pharmaceutics 58:185.
White, B. (1963). U. S. (Patent) 3 105 1792.
Williams, M. L., Landel, R. F., Ferry, J. D. (1955). JACS 77:3701.
Wright, J. L., Carstensen, J. T. (1986). J. Pharm. Sci. 75:546.
Zografi, G., Kontny, M. (1986). Pharm. Res. 3:187.

7

Interactions of Moisture with Solids

JENS T. CARSTENSEN

Madison, Wisconsin

1. PREDOMINANT REACTION ORDERS

The ICH 1993 Stability Guidelines "do not accept the term 'room temperature'," yet it is so common that it will be used in this text in its intuitive sense. At room temperature, for a product to be marketable, decompositions are less than 15% (in fact a good deal less than 10%). Most products exhibit good content uniformity, and usually decompositions will appear zero order, i.e., be pseudo-zero-order. The mathematical approximation

$$\ln[1 - x] \approx -x \tag{7.1}$$

explains this. A reaction that is truly first order, but where x is small (less than 0.1 or at most 0.15), will, by way of Eq. (7.1), appear zero order. If M is drug present at time t and M_0 the initial amount, then it follows that

$$x = \frac{M_0 - M}{M_0} \tag{7.2}$$

hence

$$\frac{M}{M_0} = 1 - x \tag{7.3}$$

This means that the equation takes the form

$$\ln\left[\frac{M}{M_0}\right] = \ln[1 - x] \approx -x = -\frac{M_0 - M}{M_0} = -k_1 t \tag{7.4}$$

where k_1 is the first-order rate constant. One may write this

$$M = M_0 - \{M_0 k\}t = M_0 - k_0 t \tag{7.5}$$

i.e., the data will appear pseudo-zero-order with a rate constant of

$$k_0 = M_0 k_1 \tag{7.6}$$

There are other causes for pseudo-zero-order behavior, chemical and physical reasons. The effect moisture has on solid state stability is, in its simplest form, visualized by moisture being sorbed on the particles. In this sense the water would behave like a solution, a bulk layer, and it is assumed that this is saturated in drug. This is known as the Leeson–Mattocks model (Leeson and Mattocks, 1958; Li Wan Po and Mroso, 1984; Carstensen and Li Wan Po, 1993). The decomposition follows the equation

$$M = M_0 - k_0 t \tag{7.7}$$

where k_0 is the pseudo-zero-order rate constant given by

$$k_0 = k_1 S V \tag{7.8}$$

In this equation S denotes the drug solubility in the bulk aqueous phase and V denotes the volume of the layer. Most often one deals with hydrolysis and pseudo-first-order is assumed.

An assumption is that the moisture is in such excess that the term V does not change while the stability is being studied. Investigators (e.g. Kornblum and Sciarrone, 1964) have assumed that k_1 equals that from solution-kinetic studies. Quite often this is not so, because the model dictates that it is the kinetics in concentrated solutions that matters.

As seen in Chapter 9, anhydrous solids usually exhibit sigmoid profiles, i.e., Prout–Tompkins kinetics (Prout and Tompkins, 1944) or Bawn kinetics (Bawn, 1955). This is mostly not the case with decompositions when moisture is present in "large" amounts.

2. KINETICS IN THE DRY VERSUS MOIST STATE

Pothisiri (1975) and Carstensen and Pothisiri (1975) studied the decomposition of substituted *p*-aminosalicylic acids as a function of moisture content and attempted to extrapolate the pseudo-zero-order rate constants down to 0% (anhydrous). The extrapolated value (as well as the mechanism) differs from that observed in experiments where moisture is excluded (pseudo-zero-order versus sigmoid behavior).

It is therefore of interest to define the nature of the water in the transition between very low moisture contents (the anhydrous state) and the other extreme (the very moist state). The models just mentioned are extreme cases of very "dry" and very "wet." Various stages of "wetness" are depicted in Fig. 1.

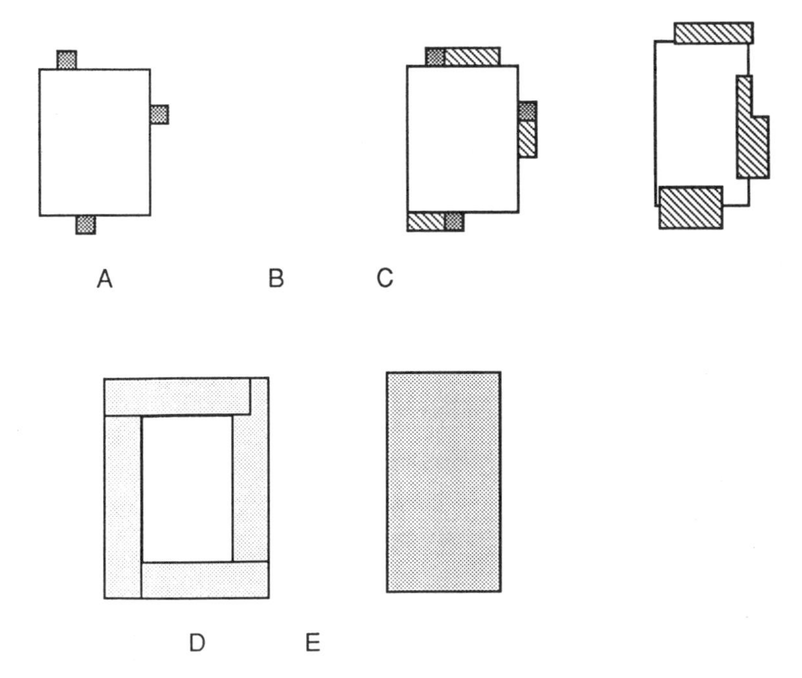

Fig. 1 (A) Anhydrous solid with active sites (cross-hatched); (B) solid with less than monolayer coverage of moisture; (C) solid with more than monolayer coverage, where the active sites have disappeared; (D) bulk sorbed moisture; and (E) moisture in an amount sufficient to dissolve the solid completely.

3. TYPES OF SURFACE MOISTURE

There are, broadly speaking, three types of situations: *Limited water*, where all the water is used up in the decomposition of the drug, but the amount is not enough to decompose all of the drug.

Adequate water, when there is enough moisture to decompose all of the drug substance.

Excess water, when the amount of water present is more than needed to dissolve the drug completely.

3.1. Excess Water

This is shown as E in Fig. 1, where the amount of water suffices to bring all of the drug into solution. This may not be applicable initially, but it occurs as the amount of parent drug decreases in time.

Examples of this are the work by Morris (1990), where the indomethacin/water system was studied in a closed system at 130°C. After a short period of time a eutectic consisting of indomethacin, decomposition products, and water is formed, and from this point in time the decomposition is first order as expected for solution kinetics (Fig. 2). The amount of time (t') required for the eutectic to form (for the mass to form a homogeneous liquid) is linear in water activity ($a = \mathrm{RH}/100$), i.e.,

$$t' = \beta - q'a \tag{7.9}$$

where β and q' are constants (Fig. 3).

Yoshioka and Uchiyama (1986a,b), Carstensen et al. (1987), and Yoshioka and Carstensen (1990a,b) have reported similarly in relationship to propantheline bromide. Yoshioka and Uchiyama (1986a) introduced *critical relative humidity* (CRH) as the point where the water activity just equals that of a solution saturated

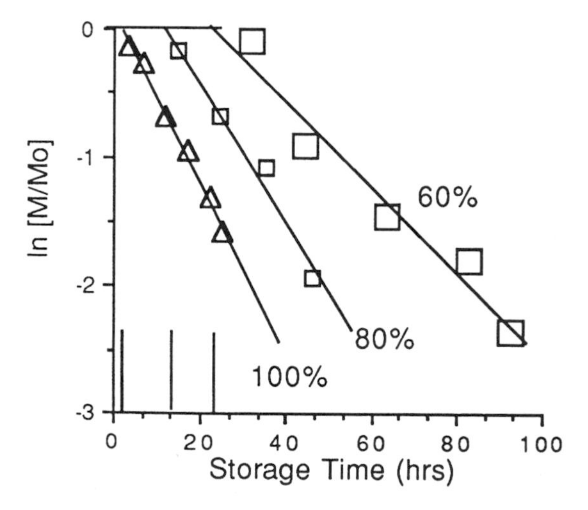

Fig. 2 Decomposition of indomethacin in the presence of moisture at 130°C. (Graphs constructed from data published by Morris, 1990.)

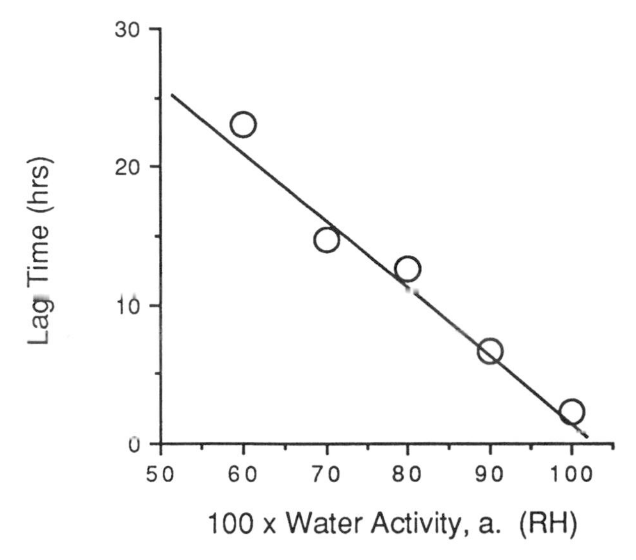

Fig. 3 Lag times from Fig. 2 plotted versus relative humidity. (Graph constructed from data published by Morris, 1990.)

in the drug (Carstensen, 1977) and they also showed that the mechanism changed at this point. At values higher than the CRH the degradation consists of (a) dissolution up to where dissolution is complete, after which (b) moisture condensation will continue until a concentration of the totally dissolved drug equals that of the RH of the atmosphere.

Koizumi et al. (1997) showed that the dependence of water concentration on the rate constant of decomposition of Lornoxicam tablets is log-log related to the log of the moisture content.

Carstensen et al. (1965) had shown this to be correct for vitamin A beadlets as well.

$$\frac{d[A]}{dt} = -k[A][H_2O]^n \tag{7.10}$$

4. THE LEESON–MATTOCKS MODEL

This is the most frequently applicable model and it assumes that sorbed moisture forms a layer about the particles. It corresponds to situation D in Fig. 1. One might argue that such a layer (a so-called bulk sorbed moisture layer) could not be created until the moisture content is high enough, so that the RH of the atmosphere surrounding solid equals or is in excess of the RH of a saturated solution of the drug. One might then conclude that the Leeson–Mattocks model only holds at RH values in excess of the critical relative humidity (CRH). However *rather than that being true it holds below the CRH*. For a certain range of RH values less than the CRH, the Leeson–Mattocks model applies, and degradations are pseudo zero order. Phenobarbital when it decomposes at 80°C in the presence of phosphate buffer at pH 6.7 is an example of a case where, in the initial stages of decomposition, this holds (Gerhardt 1990). Another case is that reported by Morris (1990) and Morris and Carstensen (1990a,b).

Equation (7.9) applies to the decomposition, hence one must know k_1, V, and S, which should allow for elucidation of the mechanism. This often holds true (Pothisiri, 1975; Pothisiri and Carstensen, 1975), but it has also been known to fail (e.g. Janahsouz et al., 1990).

Carstensen and Attarchi (1988) elucidated the discrepancy between the rate with which aspirin decomposes as a solid with water present and its behavior in saturated solution. If they presumed that the solubility in the moisture layer were three times that of the bulk solubility, then their calculated data corresponded to the experimental data. Whether it is possible that the condensed layers of water are so energetic that they would allow for such an increase is doubtful.

The speculation that the solubility might be increased in the sorbed moisture layer (Fig. 1D) might lead one to consider it akin to an amorphous state. An amorphous state would exhibit increased higher solubility over that of crystalline states, and would also possess a higher vapor pressure than would a crystalline form, the form that it would rest upon, posing the question of why and where the critical moisture content would exist.

A further extension of this, though, is that water dissolves into the solid. As mentioned, this happens for a wholly amorphous compound, but for a crystalline compound the crystallinity would have to be lost, an assumption that has no basis in fact. If the moisture molecules really created a "hot spot" of amorphous solid on the crystal surface, then at a certain given RH value the mass of moisture ad/absorbed should be equivalent to the composition of the amorphate/water in equilibrium at the RH in question.

Carstensen and VanScoik (1990) have demonstrated for small molecules (sucrose) that the water activity over this type of supersaturation of water in solid is simply an extrapolation of the RH values of saturated and unsaturated solutions at the other end of the diagram. One might consider this as water that is dissolved in the solid or as solid that is dissolved in the water, but in either view the important aspect is that (ideality assumed) the mass of water sorbed is linearly dependent on the RH. (For polymers of high molecular weight the isotherms are S-shaped, an example being microcrystalline cellulose reported by Hollenbeck et al., 1978, and by Marshall et al., 1972).

If an amorphate "hot spot" hypothesis were correct then the pseudo rate constants should be linearly related to the water activity (the relative humidity) used in the study. Figure 4 shows the profiles with which crystalline indomethacin decomposes at different water activities (RH values). Pseudo zero order obviously prevails, but the plot of rate constants versus water activities (RH values) is not a straight line (Fig. 5). As mentioned in the last chapter, one cannot "prove" a model by statistically comparing curve fittings, but one can eliminate models (Li Wan Po, 1984; Mroso et al., 1982) when a fit is lacking. An assumption made in Fig. 8.18 is, it is conceded, that ideality prevails, but for the "hot spot" model to hold, nonideality in the case would have to be drastic.

A different interpretation of the indomethacin data was forwarded by Morris (1990) and Morris and Carstensen (1990a,b), by demonstrating that the rate constants are related to a BET (or possibly to some other nonlinear) water adsorption

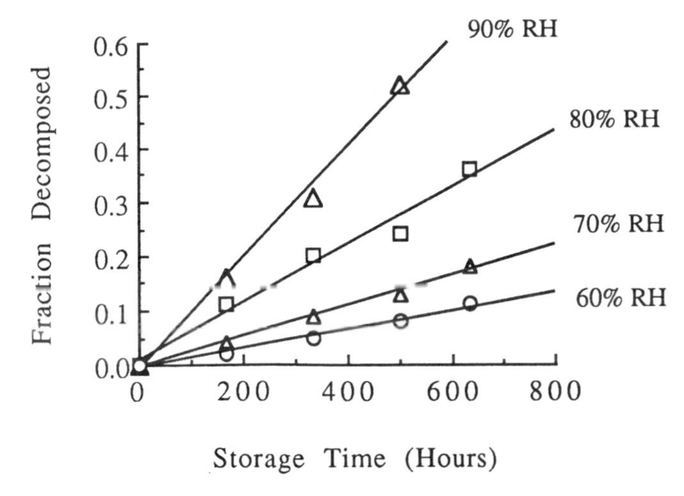

Fig. 4 Indomethacin decomposition at 115°C. This decomposition follows zero-order kinetics at the onset. (Graph constructed from data published by Morris, 1990, and Morris and Carstensen, 1990a,b.)

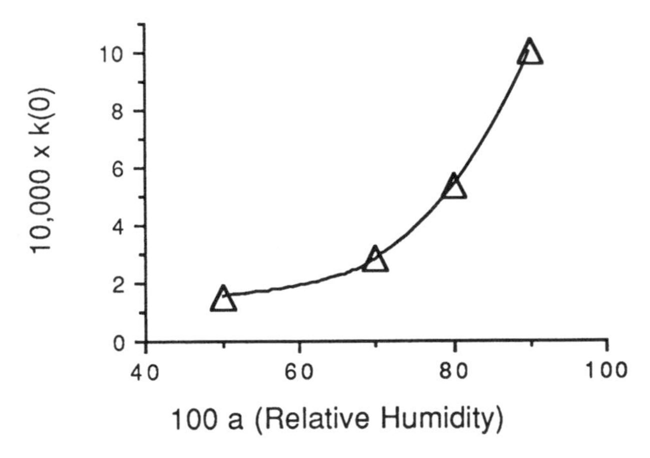

Fig. 5 Data from Fig. 4 of Chapter 10. Rate constants as a function of RH. (Graph constructed from data published by Morris, 1990, and Morris and Carstensen, 1990a,b.)

isotherm. The volume of water, V, adsorbed when a BET isotherm with high c value applies would be of the type

$$V = \frac{v_m}{[1 - a]} \tag{7.11}$$

v_m is here a monolayer volume, and the symbol a is used to denote the water activity (RH/100). The k_0 value should, therefore, be linearly related to $[1 - a]$. That this is the case is shown in Fig. 6.

It would therefore seem (at least in the case of indomethacin) that the amorphate "hot spot" model does not apply. In addition to this, Morris (1990) tested

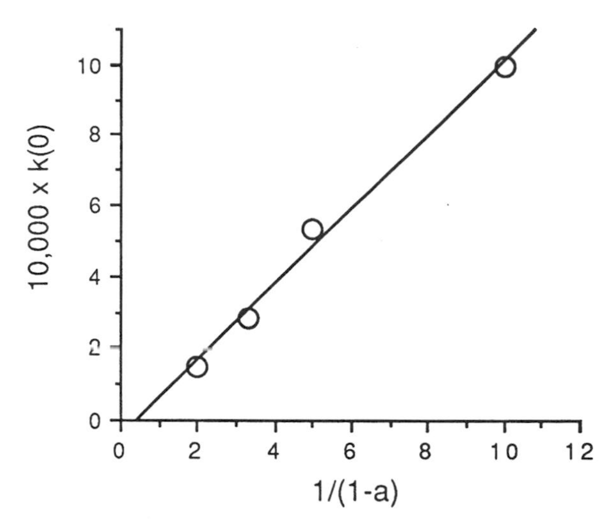

Fig. 6 Rate constants from graphs of the type shown in Fig. 2, plotted versus a BET-isotherm parameter $1/\{1-(1/a)\}$.

amorphous indomethacin with moisture present, but at high temperatures conversion to crystallinity was rapid to such an extent that kinetic data could not be obtained in reasonable time periods.

The "hot spot" theory is not new. In fact, work by Gluzman (1954, 1956, 1958) and Gluzman and Arlozorov (1957) postulated that "part of a surface of a solid was actually in a liquid like state"—in other words, in appearance being a solid, but with random molecular arrangement, and usually referred to as an amorphate.

Guillory and Higuchi (1962) hypothesized that if such a theory were correct, then the logarithm of the rate constant at a given temperature, T_d, of a series of analogous compounds in solid form should be inversely proportional to the inverse of the melting point, i.e.,

$$\ln[k] = -Q\left\{\frac{1}{T_d} - \frac{1}{T_m}\right\} \tag{7.12}$$

This has been found to be true in certain cases, e.g., for vitamin A esters at 55°C (Guillory and Higuchi, 1962) and substituted *p*-aminobenzoic acids (Carstensen and Musa, 1972), but in other cases, e.g., *p*-aminosalicylic acids, it does not hold well (Pothisiri and Carstensen, 1975).

More plausible than the "hot spot" amorphate theory is the hypothesis that the sorbed moisture layer acts as a solution layer and that degradation compounds (a) increase or decrease the drug solubility, (b) increase or decrease the kinetic parameter values of the drug, and (c) (noting that the degradants are solutes) cause a decrease in the water vapor pressure with which the moisture layer is in contact, so that in this manner the vapor pressure relationship is not violated. Gerhardt (1990) and Gerhardt and Carstensen (1989) have demonstrated that kinetic salt effects and salting-in of the drug into the moisture layer can explain the decomposition profiles exhibited by phenobarbital when moisture and buffers

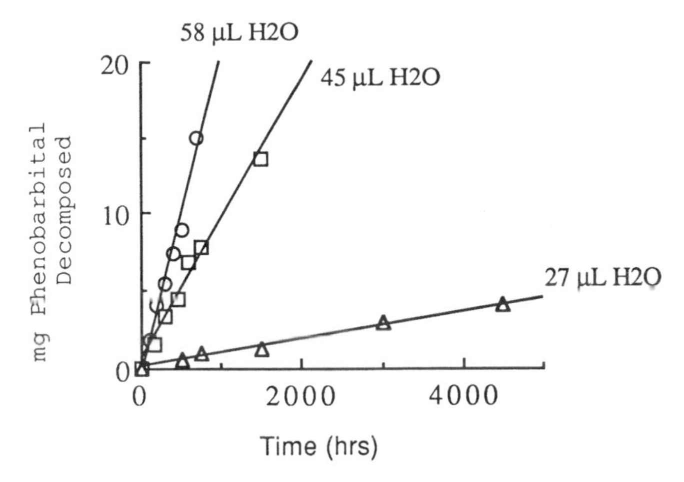

Fig. 7 Phenobarbital decomposition in the solid state at 80°C, with phosphate buffer present corresponding to a "pH" of 6.7.

are present. Carstensen and Pothisiri (1975) and Wright and Carstensen (1986) have done likewise.

In the case of very soluble drugs, e.g., ranitidine (Franchini and Carstensen, 1995; Carstensen and Franchini, 1995) the amount of moisture necessary to reach the CRH is small (i.e., the water activity (RH/100) over a saturated solution is of low magnitude). On the other hand, it is high for poorly soluble drugs.

5. KINETICALLY UNAVAILABLE (BOUND) WATER

Solid state rate constants often follow Eq. (7.2), in that they appear directly in proportion to the mass or volume of water the dosage form contains. Figure 7 presents data from the work of Gerhardt (1990) and Gerhardt and Carstensen (1989). The rate constants are pseudo zero order and are plotted versus moisture levels (Fig. 8). It is noted that the intercepts are nonzero, i.e.,

$$k_0 = k_1 S[V - w^*] \tag{7.13}$$

w^* is often called kinetically unavailable moisture or *bound water*. This is the case in many solid state reactions. The bound moisture, at times, is water of crystallization. For D,L-calcium leucovorin (Nikfar et al., 1990a,b), there are intermittent plateaus that correspond to a constant water activity (RH/100) for a series of water contents, i.e., akin to a salt pair. $[V - w^*]$ is denoted kinetically available, or more simply, *available moisture*.

Aso et al. (1997) have determined the decomposition rates of cephalotin in mixtures with pharmaceutical excipients and the effect of moisture. They found a linear relation between mobile water percentage and decomposition rate constants.

6. MICROENVIRONMENTAL pH

If a formulator is aware that a compound is more stable in an acid than in a neutral or basic environment one may often formulate it with solid acids (e.g., citric acid);

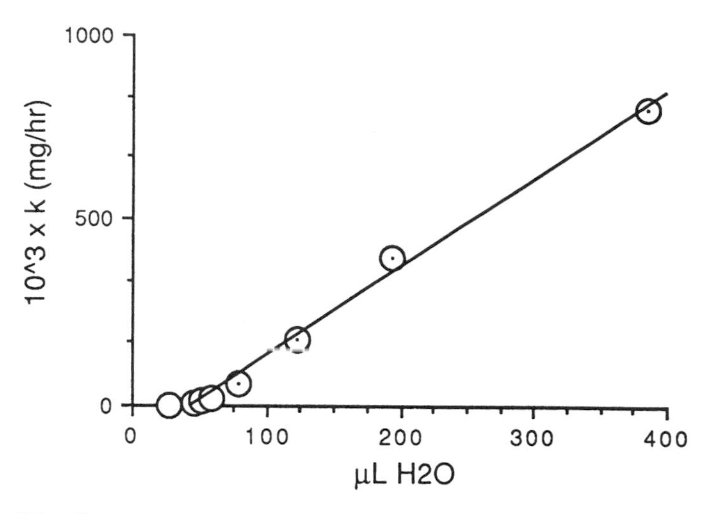

Fig. 8 Rate constants (pseudo zero order) from plots such as shown in Fig. 7 of Chapter 10, graphed versus added moisture. (Figure constructed from data published by Gerhardt, 1990, and Gerhardt and Carstensen, 1989.)

conversely, if it is acid-sensitive, one may employ bases (e.g., sodium carbonates) in an attempt to make an adjustment of "the microenvironmental pH." In the area shown as Fig. 1C, D, and E, if one may "buffer" a solid dosage form, Nikfar 1990, Nikfar et al. 1990a,b, Gerhardt 1990, Gerhardt and Carstensen 1989 have demonstrated the existence of a "solid pH-profile" that parallels (but is not identical with) the traditional pH profiles of the drug in solution. This is another piece of evidence of the sorbed moisture layer having solvent properties.

But how to define the microenvironmental pH? This a question that is not fully resolved yet. The shift in position of the kinetic pH profile in solution from the values obtained from solid state decomposition may be attributed to the fact that one assumes that the pH value of a saturated buffer solution is the same pH used for graphing of data from the moist solid. But the sorbed solution could be of a pH value displaced from that observed in solution.

There is also the possibility of a kinetic salt effect. It is seen from Fig. 9 (Nikfar, 1990; Nikfar et al. 1990a,b) that a displacement of 1.4 pH units applies to the rate constants in the solid state. The displaced values are symbolized by squares in Fig. 9, and if such a shift is made, then the data in solution would coincide with those in the solid state. In the work published by Gerhardt (1990) it would be necessary to force a 6 pH unit shift to obtain coincidence, so that are still unexplained factors at work.

7. VERY LOW MOISTURE CONTENTS

Such a case is shown in Figs. 1B and 1C. Nikfar (1990) and Nikfar et al. (1990a,b) suggested the term *immobile water* for cases such as the ones depicted in Fig. 1c. They have demonstrated that the decomposition in such a case translates into a pseudo-first-order profile (Fig. 10). At these levels of moisture the active sites in a Prout–Tompkins sense disappear by dissolution somewhat like what happens

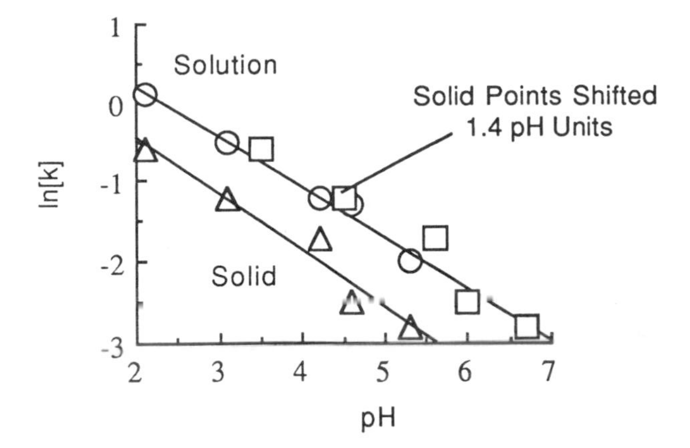

Fig. 9 pH-rate profile of first-order rate constants extracted from kinetics of decomposition of D,L-calcium leucovorin. The squares are points from solid-state decomposition shifted by 1.4 pH units.

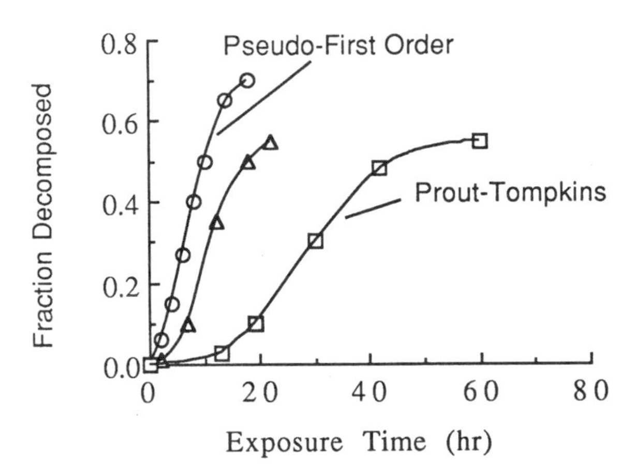

Fig. 10 D,L-calcium leucovorin with moisture and buffers added. ○: 5% water with a pH 2.2 buffer in the solid state (the buffer forms hydrates, and the water contents are percentages added and are not necessarily available moisture); △: intermediate moisture content; □: low moisture content.

in an etch-test of a metal. If one assumes that the aqueous solution is immobile, i.e., that only water molecules adjacent to intact drug molecules take part in the reaction, then first-order kinetics should prevail. One might also, at this level of moisture, consider the surface structure as amorphous, since amorphous substances in the presence of water degrade by first order (Pikal et al. 1977; Morris, 1990). Literature data are insufficient to distinguish if linearity or BET sigmoidness applies for rate constants when they are plotted as a function of relative humidity.

At even lower moisture contents (Fig. 10) the reaction profile takes on a sigmoid nature and can be explained by a surface-interaction model. The sigmoid profiles shown in Fig. 10 adhere well to Eq. (7.3). This can be explained by assuming the moisture to adsorb preferentially at the active sites (Fig. 1B and Fig. 11).

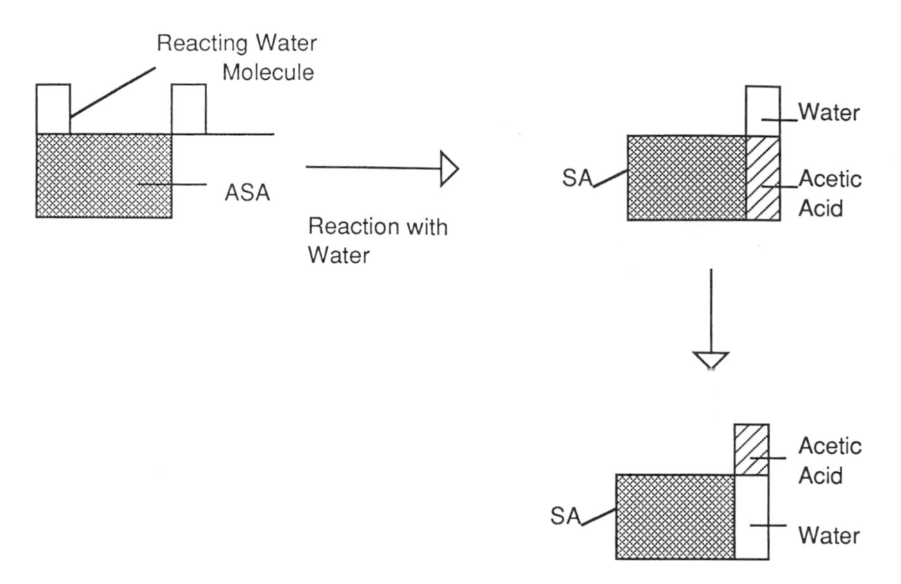

Fig. 11 Surface active site interaction using aspirin as an example. (Graph constructed from model proposed by Attarchi, 1984, and Attarchi and Carstensen, 1988.)

The amount of water does not suffice to "dissolve" the active sites, so the reaction is an interaction between moisture and drug at the activated site. The development of such a model has been published by Attarchi (1984) and Carstensen and Attarchi (1988). The applicable equation is Eq. (7.9):

$$\ln\left[\frac{x}{1-x}\right] = k(t - t_{1/2}) \tag{7.14}$$

The applicable model is presented in Fig. 11. Data plotted in this fashion is shown in Fig. 11, and the rate constants admirably follow an Arrhenius equation as shown in Fig. 12.

Obtaining the actual values of k and S in Eq. (7.2) is not as easy as might seem. As pointed out (and investigated) by Attarchi (1984) and by Carstensen and Attarchi (1988), both k and S are a function of amount of decomposition product. This was first pointed out by Pothisiri (1974), by Pothisiri and Carstensen (1975), and by Wright and Carstensen (1986). It is also a function of ionic strength, as pointed out by Gerhardt (1990) and by Gerhardt and Carstensen (1989), or simply a function of the composition of the sorbed moisture layer (Attarchi, 1984; Carstensen and Attarchi, 1988; Pothisiri, 1974).

8. DOSAGE LEVEL AND TOXICITY CONSIDERATIONS

In a great majority of cases the decomposition is zero order i.e., following Eq. (7.5). This means that the amount of decomposition product is linear in time.

If a product, for instance, is made in three dosage strengths, say 5 and 25 and 50 mg strengths, then after 3 years' storage at 25°C an amount of e.g. 0.075 mg has been decomposed, i.e., (assuming for simplicity equal molecular weights) 0.075 mg of decomposition product has formed (see Fig. 13). Since Eq. (7.5) is a

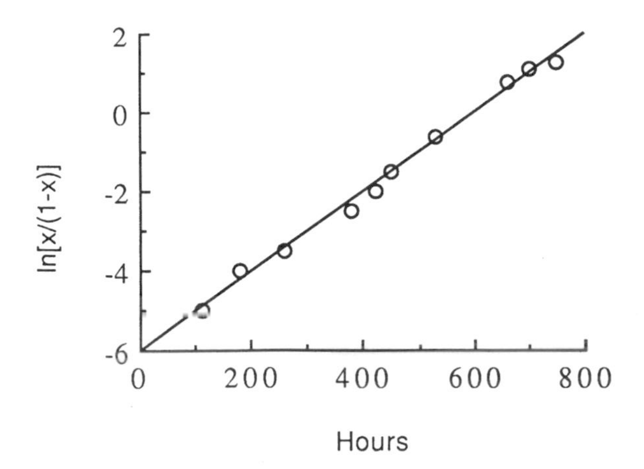

Fig. 12 Plot of aspirin decomposition data in the presence limited amounts (2.5%) of moisture. (Graph constructed from data by Carstensen and Attarchi, 1988.)

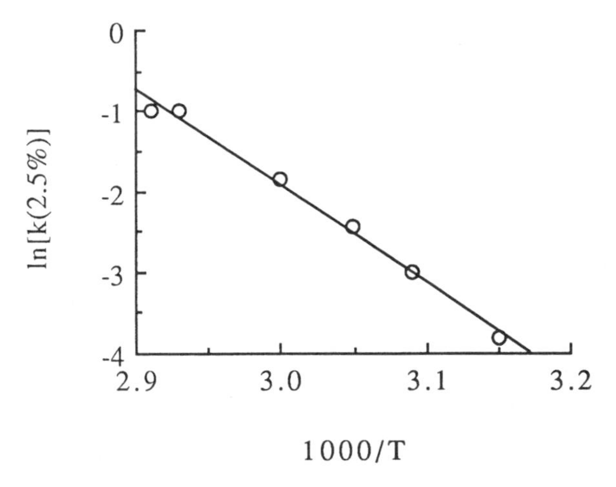

Fig. 13 Arrhenius plot of aspirin decomposition data in the presence of limited amounts of moisture. (Graph constructed from data by Carstensen and Attarchi, 1988.)

zero order reaction, this amount pertains to all the strengths, so that on a percentage basis, the 5 mg strength would have lost $100 \times 0.075/5 = 1.5\%$ of its original value, and hence (presuming an adequate precision and uniformity) would probably not meet the chemical requirement for a 3 year expiration period. The 25 mg and 50 mg strengths would have experienced 0.3% and 0.15% losses, respectively, and would be assumed to be satisfactory from a toxicity point of view.

However, often the decomposition product(s) is (are) unknown and as such are expressed as a percentage of the "main peak" if the method is HPLC (and this is usually the case). If, for instance, for toxicity reasons one assumes that the impurity peak may not exceed 1% of the area under the main peak, then in the above case, the 25 mg and 50 mg strengths would meet toxicity requirements, whereas the 5 mg would not. Nevertheless, the total amount of decomposition product is the same in all cases. One would assume that it would be the absolute (total) amount of decomposition product that would be of importance, so that expressing the toxicity limitation as a percentage is not appropriate.

In cases where the Leeson–Mattocks model holds (but not in other cases of zero-order reactions), the above dilemma may be prevented by making the smaller dosage forms proportionally smaller, since the term V in Eq. (7.13) then becomes smaller as well.

Regulations are never immutable, and it may well be that at some future date such regulations will be modified to meet the stated need.

9. NONSTOICHIOMETRIC INTERACTIONS WITH WATER

It is sometimes a pharmaceutical practice to "coat" labile pharmaceuticals. One such example is vitamin A beadlets, which are an emulsion of vitamin A ester in a gelatin solution, which has then been converted into drops and dried. The beadlets are therefore a matrix of gelatin with droplets of oil in the interior. The protection offered is one against oxidation.

10. PARENTERAL SOLID PRODUCTS

When an injectable product is insufficiently stable in solution to allow marketing of a ready-made solution, there are still ways to develop a marketed product.

In the far past, there were so-called powder-filled products. Here a solid drug substance was made under exceedingly "clean" conditions so that it emerged from its synthesis as "completely" free of foreign material. In such a case it could be filled into a vial, sterilized by suitable means (heat, ethylene oxide (not used of injectables anymore), or γ-ray sterilization). Excipients used (sodium chloride, for instance) would have to be equally clean, and the practice is not, to the author's knowledge, used much any more.

Aside from the sepsis issue, there was also the problem with rate of dissolution, and from both these aspects, lyophilization offers a better (but probably more expensive) alternative.

10.1. Lyophilized Products

The process is one where a solution of the drug (+ excipients) is made and aseptically filtered. The solution is then aseptically filled into vials, which are loaded into a sterile lyophilization oven. This has cooling coils in its shelves and can be evaluated to very high vacuum.

The vials containing the solution are transferred to the oven, and coolant at very low temperature ($< 30°C$) is flowed through the tray coils. The solution freezes, and then a vacuum is applied of such magnitude (P_v torr) that it is lower than the vapor pressure of ice at the given temperature (P_i torr). This causes the ice to sublime, and then there remains a cake that either is crystalline and has an exceedingly high surface or is amorphous and also possesses a high surface area.

When this, at time of use, is reconstituted with water or diluent, the dissolution is, in both cases, rapid and the "original" solution is regained.

There are two stability issues in this case: (a) how stable is the lyophilized cake and (b) how stable is the solution after reconstruction?

10.2. Stability of Crystalline and Amorphous Lyophilates

It has been seen in Chapters 2 and 3 that a drug product in solution will possess an optimum pH. It is noted that this is accomplished by studying stability of the substance at different pH values, and that these latter are arrived at by the use of different buffers. If, for instance, the drug substance is a weak acid, then approximately speaking one may write

$$\begin{aligned} k_{obs} = & k_0 + k_+[H^+] + k_-[OH^-] \\ & + k_{A^-}[A^-] + k_{HA}[HA] + k_{buffer}[HB] \end{aligned} \tag{7.15}$$

where HB refers to buffer concentration and k_{HB} is the part of the rate constant attributable to the buffer. It is simplified, because k_{HB} is a combination of two terms, k_B and k_{HB}, but for this purpose it suffices to employ one (or at most two) terms. As discussed in Chapters 2 and 3, drugs mostly are protolytic and partly exist in ionized (A^-) and unionized (HA) form giving rise to the terms involving $k_{A^-}[A^-] + k_{HA}[HA]$, and k_0 is the part of the rate constant, which is neither acid nor base dependent. At lower pH, the term $k_-[OH^-]$ the term falls out, and $[A^-]$ and [HA] are dependent on the pH of the buffer used and of the $pK_{(a)}$ of the acid at the concentrations given. It is recalled that the $pK_{(a)}$ is also a function of ionic strength, the pK_a value being the value of $pK_{(a)}$ from which the ionic effect has been eliminated.

If such a substance in solution is allowed to cool down, then first water will freeze out as ice. The solution, hence, becomes more and more concentrated in both buffer and drug substance, and the pH changes as well. At the eutectic point (or the collapse temperature) all freezes out.

The stability of the substance as the concentrations change of course changes as well, because the buffer concentration changes, because the pH changes, and because the pK of the species in solution changes as well. Hence the optimum manufacturing pH is not the same as that of the corresponding solution. The experimental procedure to use is to make solutions of the desired concentrations of buffer and other excipients at several, say four, different pH values straddling the optimum solution pH, and then produce the lyophilized cake. The stability of this cake is then determined, and the optimum lyophilization pH determined in this manner.

10.3. The Labelling Dilemma of Parenteral Products

The FDA usually takes the strong positional stand that a different "salt form" constitutes a different drug substance and hence a new NDA is required. The drug on the label is the form of the drug in the dosage form. If for instance a product is made with a tetracycline base, then the label must state that this is the source of the antibiotic (as opposed to for instance the use of the addition salt, e.g., the hydrochloride).

But what about a lyophilized product? If one used tetracycline hydrochloride (RHCl) and buffered it at its pK value (at the given ionic strength), then, first of all, the product would be present one half as positive ion (RH^+), one half as uncharged species (R). If the buffer is denoted HB, then, in concentrating a solution of this there would be two solubility products:

$$S_{RHCl} = [RH^+][Cl^-] \tag{7.16}$$

$$S_{RHB} = [RH^+][B^-] \tag{7.17}$$

aside from the solubilities S_{HB} and S_R. As the solution, hence, starts precipitating substances other than ice at the eutectic point, the species with the lowest S value or solubility product will at first precipitate out. This, for instance, could be RHB. As this precipitates out, both $[RH^+]$ and $[B^-]$ will decrease. At a given point R will start precipitating out. This will prevent further precipitation of RHB, because $[B^-]$ is now sufficiently low to be at the limit, had $[RH^+]$ not been affected. At a given point, because the amount of liquid water decreases as the process continues (freezing out of ice), the solubility limit of either HB or RHCl will be exceeded, and either species will then precipitate out until the remainder is left to freeze out as the last amount of water is solidified at the eutectic point.

The point is that the cake will contain four species: R, RHCl, HB, and RHB. And the question then is, under the present labelling policies, how does one properly label such a mixture?

11. OXIDATION

Oxidations are moisture mediated, as are hydrolyses. Often products that are oxidation sensitive are stored in glass rather than polymer bottles because, however good, these latter still allow permeation of oxygen.

In a glass bottle, if it is considered hermetic, and it often is, the oxygen in the head space will be consumed, and the amount of "initial" decomposition of the produce will tie in with the amount of oxygen available in the head space. It is a common phenomenon that solid dosage forms show an initial loss corresponding to the ratio between the amount of head space divided by the number of tablets in the bottle.

Often the oxygen is used up, and treatment of the data should be such that regression should be carried out on the data points after the *intial* drop.

Example 10.1.

A bottle contains 100 tablets and a head space of 25 mL of air. Each tablet contains 100 mg of drug substance of molecular weight 500. If the nonoxidative decomposition of the drug is 0.1% per month, how much would be expected, on the average, to remain after 3 years? Assume that one O_2 decomposes two drug molecules (i.e., $A + 1/2O_2 \rightarrow AO$).

Answer.

25 mL of air space at 25°C is $25/22.4 = 1.11$ moles of air, containing 22% of oxygen, so that the amount of available oxygen in the headspace is 0.22 millimoles.

This means that $0.22/100 = 2.44 \cdot 10^{-3}$ millimoles of oxygen will decompose an equal molar amount of drug *per tablet*. Each tablet contains $100/500 = 200 \cdot$

Over three years $36 \cdot 0.1 = 3.6\%$ of the drug will decompose by other means, so that a total of $2.4 + 3.6 = 6\%$ will decompose.

REFERENCES

Aso, Y., Sufang, T., Yoshka, S., Kojima, S. (1997). Drug Stability 1:237.

Attarchi, F. (1984). Decomposition of aspirin in the moist solid state. Ph.D. thesis. School of Pharmacy, University of Wisconsin, Madison, WI.

Bawn, C. (1955). Chemistry of the Solid State. W. Garner, ed. New York: Academic Press, p. 254.

Carstensen, J. T. (1977). Pharmaceutics of Solids and Solid Dosage Forms. New York: John Wiley, p. 12.

Carstensen, J. T., Attarchi, F. (1988). J. Pharm. Sci. 77:318.

Carstensen, J. T., Franchini, M. (1995). Drug Dev. Ind. Pharm. 21:523.

Carstensen, J. T., Johnson, J. B., Valentine, W., Vance, J. J. (1964). J. Pharm. Sci. 53:1050.

Carstensen, J. T., Li Wan Po, A. (1993). Int. J. Pharmaceutics 83:87.

Carstensen, J. T., Musa, M. N. (1972). J. Pharm. Sci. 61:273 and 1112.

Carstensen, J. T., Pothisiri, P. (1975). J. Pharm. Sci. 64:7.

Carstensen, J. T., VanScoik, K. (1990). Pharm. Res. 7:278.

Carstensen, J. T., Danjo, K., Yoshioka, S., Uchiyama, M. (1987). J. Pharm. Sci. 76:548.

Franchini, M., Carstensen, J. T. (1994). Pharm. Research 11:S238.

Gerhardt, A. (1990). Decomposition of Phenobarbital in the Solid State. Ph.D. thesis, School of Pharmacy, University of Wisconsin, Madison, WI, p. 61.

Gerhardt, A., Carstensen, J. T. (1989). Pharm. Research 6:S142.

Gluzman, M. (1954). Uch. Zap. Khar'kov Univ., 54, Tr. Khim. Fak. Nauch.-Issledovatel. Inst. Khim. 12:333.

Gluzman, M. (1956). Tr. Khim. Fak. Nauch.-Issledovatel. Inst. Khim. 14:197.

Gluzman, M. (1958). Z. Fiz. Khim. 32:388.

Gluzman, M., Arlozorov, D. (1957). Z. Fiz. Khim. 31:657.

Guillory, K., Higuchi, T. (1962). J. Pharm. Sci. 51:100.

Hollenbeck, R. G., Peck, G. E., Kildsig, D. O. (1978). J. Pharm. Sci. 67:599.

Janahsouz, H., Waugh, W., Stella, V. (1990). Pharm. Research 7:S195.

Koizumi, N., Adachi, T., Kouji, M., Itai, S. (1997). Drug Stability 1:202.

Kornblum, S., Sciarrone, B. (1964). J. Pharm. Sci. 53:935.

Leeson, L., Mattocks, A. (1958). J. Am. Pharm. Assoc. Sci. Ed. 47:329.

Li Wan Po, A., Mroso, P. V. (1984). Int. J. Pharmaceutics 18:287.

Marshall, K., Sixsmith, D., Stanley-Wood, N. G. (1972). J. Pharm. Pharmacol. 24:138.

Morris, T. (1990). Decomposition of indomethacin in the solid state. Ph.D. thesis; School of Pharmacy, University of Wisconsin, Madison, WI.

Morris, T., Carstensen, J. T. (1990a). Pharm. Research 7:S195.

Morris, T. Carstensen, J. T. (1990b). Pharm. Research 7:S196.

Mroso, P.V., Li Wan Po, A., Irwin, W.J. (1982). J. Pharm. Sci. 71: 1096.

Nikfar, F. (1990). Decomposition of D,L-calcium leucovorin in the solid state. Ph.D. thesis, School of Pharmacy, University of Wisconsin, Madison, WI.

Nikfar, F., Ku, S., Mooney, K.G., Carstensen, J.T. (1990a). Pharm. Research 7:S127.

Nikfar, F., Forbes, S.J., Mooney, K.G., Carstensemn, J.T. (1990b). Pharm. Research 7:S195.

Pikal, M.J., Lukes, A.L., Jang, J.E. (1977). J. Pharm. Sci. 66:1312.

Pothisiri, P. (1975). Decomposition of *p*-aminosalicylic acid in the solid state. Ph.D. thesis, School of Pharmacy, University of Wisconsin, Madison, WI.

Pothisiri, P., Carstensen, J.T. (1975). J. Pharm. Sci. 64:1931.

Prout, E.G., Tompkins, F.C. (1944). Trans. Faraday Soc. 40:489.
Wright, J.L., Carstensen, J.T. (1986). J. Pharm. Sci. 75:546.
Yoshioka, S., Carstensen, J.T. (1990a). J. Pharm. Sci. 79:799.
Yoshioka, S., Carstensen, J.T. (1990b). J. Pharm. Sci. 79:943.
Yoshioka, S., Uchiyama, M. (1986a). J. Pharm. Sci. 75:92.
Yoshioka, S., Uchiyama, M. (1986b). J. Pharm. Sci. 75:459.

8

Physical Characteristics of Solids

JENS T. CARSTENSEN

Madison, Wisconsin

1. STATES OF MATTER: CRYSTALLINITY AND AMORPHICITY

Prior to discussing the stability of drugs in the solid state, it is necessary to outline some characteristics of solids. A detailed discussion of the state of matter in regards to solids is outside the scope of this book. Suffice it here to say that solids may be characterized by being (a) crystalline or (b) amorphous. Crystalline solids are associated with a lattice, and amorphous solids are solids that are not crystalline. Some of the characteristics (those that apply to stability) of these two categories will be discussed in the immediate following.

There are seven crystal systems and two types of amorphates.

2. POLYMORPHISM

Inorganic (particularly ionic) solids usually are associated with one and only one crystal system. Well-known to all is that sodium chloride is cubic.

Organic solids, however, depending on how they are recrystallized, may occur in several different crystal modifications (polymorphs). There are two types of polymorphism, enantiotropes and monotropes. They are distinguished by their vapor pressure diagrams as shown in Figs. 1 and 2.

The situation referred to in Fig. 1 is one where there is a transition temperature, and DSC traces in such cases often have the appearance of either Fig. 2 or 3.

It is seen in Fig. 2 that two common situations may occur: first, the transformation may take place, so that there is an endotherm for the transformation followed by an endotherm for the melting. The melting point of form II (the room-temperature labile form) is recorded in this case as is the transition temperature.

The other possibility is that the transition is passed by, giving the melting point of the (now unstable) form I (lower trace). This forms an unstable melt, and often form II precipitates out, giving the exotherm shown in the lower graph followed by endotherm for the melting point of form II.

In some cases the exotherm is missing, and in such cases the melting endotherm of form II is also missing, i.e., the trace simply looks like the trace of melting of form

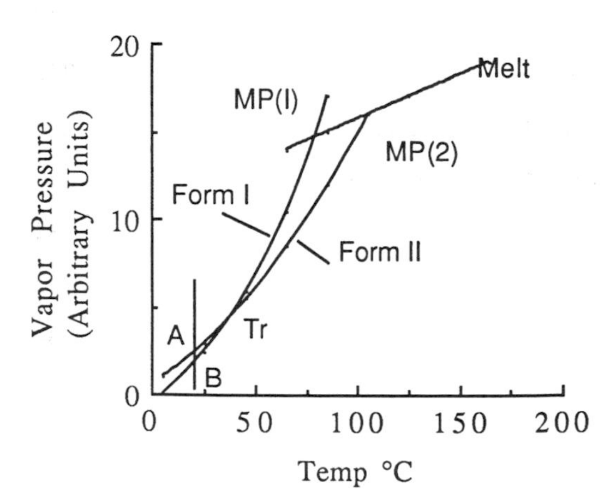

Fig. 1 Vapor pressure diagram of an enantiotropic pair.

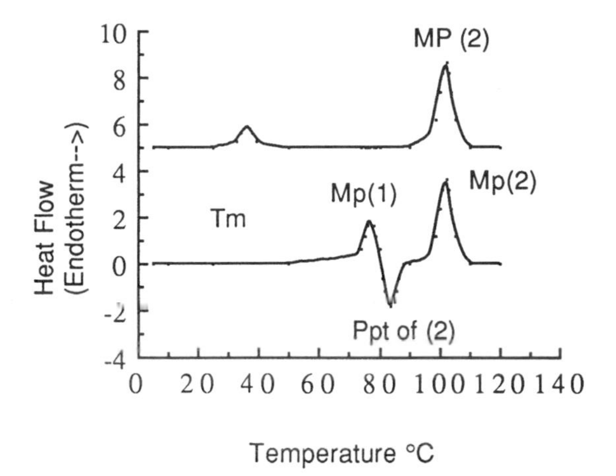

Fig. 2 Possible DSC traces resulting from heating of the room-temperature stable form of an enantiotropic pair.

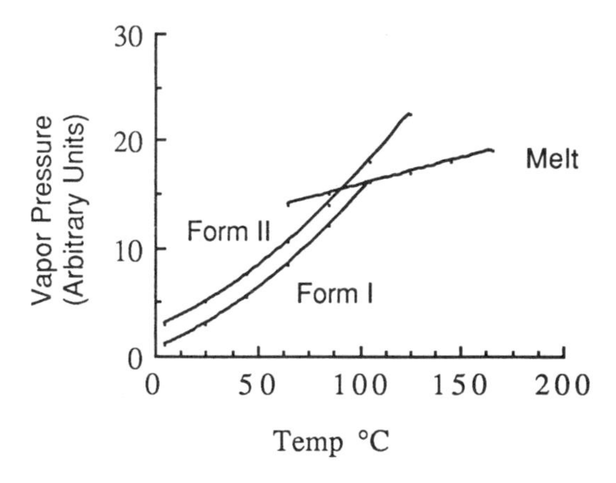

Fig. 3 Graph of vapor pressures for a monotropic pair.

I, as if no other form existed (similar to the bottom trace in Fig. 4). In such a case, if the compound is stable, it may be recooled, and the down direction melting point can be (and most often is) that of form II.

The other case is where one form (form II) is metastable throughout the melting range. This is exemplified in Fig. 3.

The DSC trace of such a pair may take one of several forms. The stable form will simply show up as a trace with one endotherm (the melting point of the stable form). Traces of the metastable form may either show up this way or as the lower trace in Fig. 4 .

As mentioned, if the compound is stable to melting, it is advisable to recool the mass and record the melting point on the down trace. Most often, however, decomposition of the solid and melt preclude conclusions from cooling curves.

The most powerful tool in polymorphic investigations, where it comes to determining whether two samples are of identical or different crystal systems, is

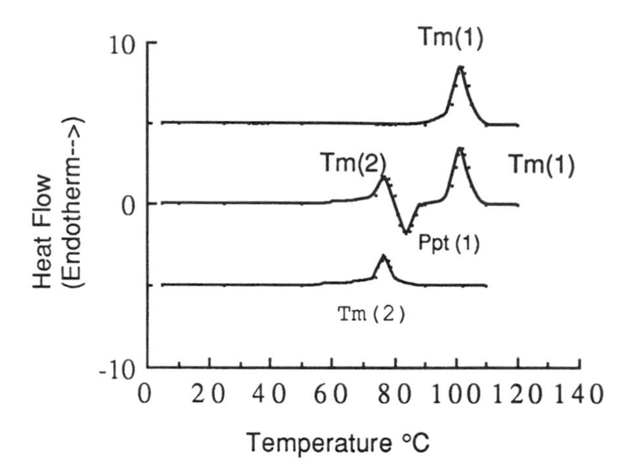

Fig. 4 Some possible DSC traces of the heating of polymorphs that are monotropic. The top trace is the heating of the stable polymorph in Fig. 3, and the two lower traces are the heating of the metastable polymorph, which may either simply melt (lower trace) or, as shown in the middle trace, melt and precipitate (exotherm) as the stable form I and then (second endotherm) remelt.

x-ray diffraction. Spacings in a crystal are related to the angle of the incoming beam (ϕ) by Bragg's law, which states that $2d\sin(\phi) = -k\lambda$, where λ is the wavelength of the x-rays used. The intensities are often used to monitor the amount of one form in another, or the amount present after a given time, t, when a conversion is taking place. An example of this is the work by Franchini and Carstensen on ranitidine (1994) where correlation was found between the content of form I in form II by the intensity at a 2ϕ value where form I did not "absorb" and where form II had a peak. Care should be taken in the interpretation of peak heights (or areas under the peak), since orientational factors can affect this. Orientation will, however, not affect the position of the peak.

It should be pointed out that in the strictest sense (Carstensen and Franchini, 1995, Martínez-Oharriz et al., 1994), there can only be true monotropism if the heats of solution are identical (and have the same temperature dependence). It is, therefore, advisable to perform heats of solution, calorimetrically, as was done for ranitidine by Franchini and Carstensen (1994) and for diffusinal by Martínez-Oharriz et al. (1993). If the heat of solution of the metastable polymorph in the pair is higher than that of the stable, the two curves may intersect at a temperature lower than the lowest temperature investigated. Of course, if this intercept is below absolute zero, then monotropism still prevails. If the heat of solution of the metastable polymoprh in the pair is lower than that of the stable one, then the two curves may intersect above the melting point, and in that case monotropism also prevails.

In the case of ranitidine, the two forms have identical solubilities (within experimental error), and what is denoted form I has a lower melting point. If the heats of solution are truly identical, this would then imply that form I is metastable over the entire temperature range and that it is a monotropic pair.

It follows from thermodynamics that the change in Gibb's energy by a path from metastable to stable form, ΔG, is given by

$$\Delta G = -RT \ln\left[\frac{P_{\text{metastable}}}{P_{\text{stable}}}\right] \tag{8.1}$$

It is negative, so the form with the highest vapor pressure at a given temperature is the least stable (metastable) compound. The term metastability (rather than instability) is used because under advantageous conditions the metastable compound may be "stable", i.e., not change for years or even decades.

3. SOLUBILITIES OF POLYMORPHS

It can be shown via Henry's law that solubilities are (approximately) linearly related to vapor pressures (actually activities as solubility are linearly related to fugacities). The graphs in Figs. 1 and 2 then become as shown in Figs. 5 and 6.

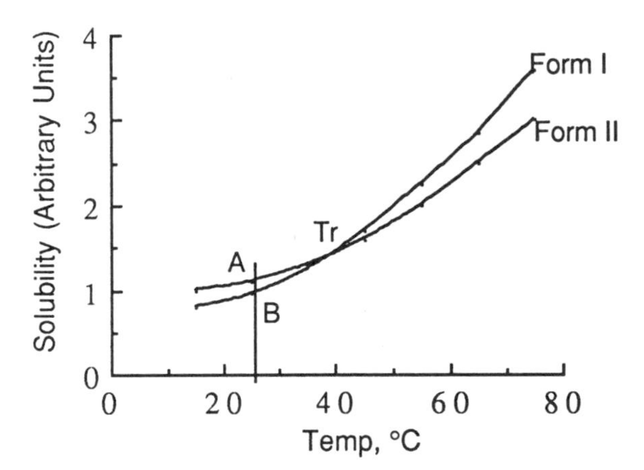

Fig. 5 Solubilities (in mass of solute per mass of solvent) of an enantiotropic pair.

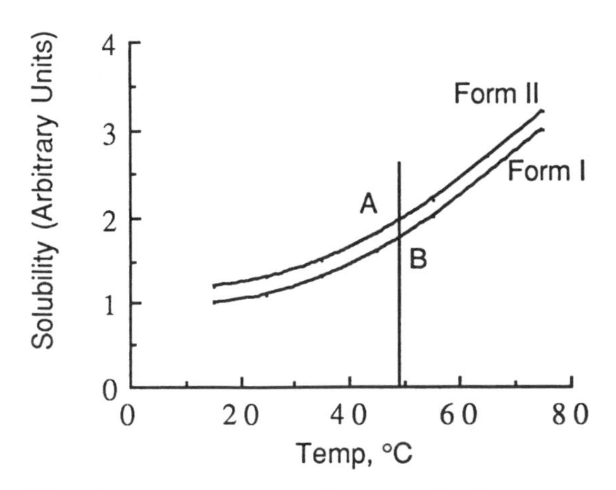

Fig. 6 Solubilities (e.g., in mass of solute per mass of solvent) of a monotropic pair.

If the Henry's law argument is applied to Eq. (8.1), then

$$\Delta G = -RT \ln\left[\frac{S_{\text{metastable}}}{S_{\text{stable}}}\right] \tag{8.2}$$

where S denotes solubility, R the gas constant, and T absolute temperature.

There are cases where the solubilities are close to one another over the entire temperature range, and in such cases it may be difficult to separate the two polymorphs in the final purification (recrystallization, reprecipitation), and there are cases where companies have been forced to suggest specifications that stipulate a minimum and a maximum of one polymorph in relation to another.

Increased solubility increases dissolution rates, and herein lay the initial interest in polymorphism in pharmacy. Shefter and Higuchi (1963) have shown the effect of solvates and hydrates on dissolution rates of several drug substances.

Pfeiffer et al. (1970) determined the solubility of cephaloglycin and cephalexin in binary mixtures and established that, depending on the composition of the medium, one or another polymorph would be stable (Fig. 7).

Poole and Bahal (1968) showed the differences in dissolution rates of anhydrous and dihydrate forms of ampicillin. The anhydrous form is amorphous, and hence would have a higher apparent solubility and hence a faster dissolution. Poole and Bahal (1970) have used Van't Hoff plots to show the conversion temperature between the anhydrous and dihydrate forms of an aminoalicyclic penicillin.

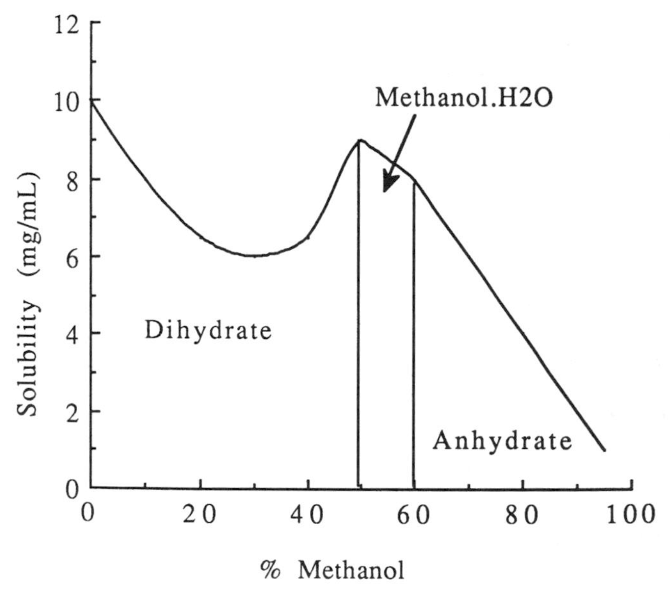

Fig. 7 The areas indicate the solid phase in which the solution is in equilibrium. (Graph constructed from data by Pfeiffer et al., 1970.)

4. RATES OF CONVERSION IN MOIST STORAGE

Good stability of a metastable compound can be achieved by (a) low temperature, (b) coarse crystals, and (c) dry storage. The moisture is the most significant contributor to conversion.

Moisture will condense onto the surface of the metastable form (II), which will then saturate the moisture layer to form a solution which is supersaturated in (I). This will eventually nucleate, and all of the II will convert to I.

The conversion rate is therefore a function of the nucleation rate in "solution," and it is well known (Mullin, 1961) that the nucleation rate, J, is inversely proportional to the viscosity of the solution and also to the supersaturation ratio, ΔS by the relation

$$J = A \exp\left[\frac{-q}{T^3 \ln \Delta S}\right] \tag{8.3}$$

For very soluble compounds, ΔS will be a very small number, and the tendency for one polymorph to change into another will be very small. An example of this is ranitidine.

5. EQUILIBRIUM MOISTURE CONTENT OF SOLIDS AND HYGROSCOPICITY

Hygroscopicity is the potential for moisture uptake that a solid will exert in combination with the rate with which this will happen. The condition of the atmosphere is an important factor as well, so a short, concise definition of hygroscopicity is not possible.

If a solid is placed in a room, moisture will condense onto it. If this moisture is simply a limited amount of adsorbed moisture, the substance is not hygroscopic under those conditions. These conditions exist if the water vapor pressure in the surrounding atmosphere is lower than the water vapor pressure over a saturated solution of the solid in question.

Often, however, the water vapor pressure in the atmosphere, P_a, is lower than that of the saturated solution, P_p. Then there will be a thermodynamic tendency for water to condense upon the solid. This is depicted in Fig. 8.

From a thermodynamic point of view, the situation shown dictates that moisture keeps on adsorbing until all solid has dissolved, and then continues until the solution is sufficiently dilute to have a vapor pressure of P_a. In this respect the moisture uptake curve differs from that of surface adsorption (polymers, and situations at atmospheric pressures below P_s), because these have asymptote at much lower levels.

The rate and extent of which moisture can condense on solids is usually collected under the term "hygroscopicity." In recent years a series of articles dealing with this phenomenon (e.g., Van Campen et al., 1980) have appeared in the pharmaceutical literature dealing with this subject. The purpose here is to derive a rational equation for the rate with which moisture is adsorbed onto a water-soluble solid.

As mentioned, if a solid is placed in an atmosphere that has a vapor pressure, P_a, higher than the vapor pressure, P_s, of the saturated solution of the compound,

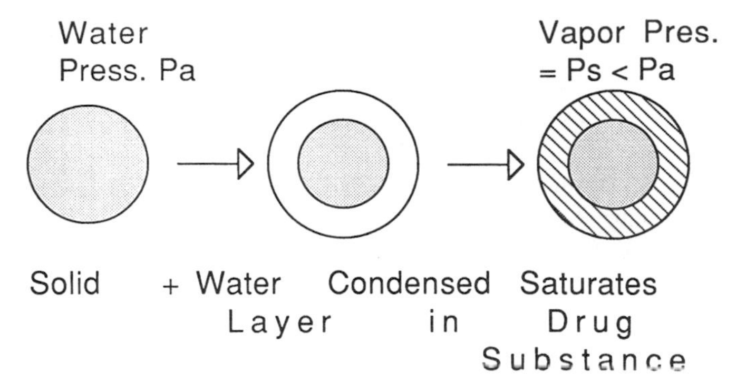

Fig. 8 Mechanism of moisture condensation.

the condensed water will dissolve solid. It will be assumed below that the sorbed solution is saturated at all times. The question is what sort of curve might be expected for the extent of moisture uptake with time (the moisture uptake rate curve, the MUR curve).

Assume that, at time t, a certain amount of moisture, w, has been adsorbed by a particular solid particle weighing m grams and of diameter, d_0, where the subscript denotes the condition prior to moisture adsorption. At time t, moisture will have adsorbed, some solid will have dissolved, and the diameter, d, of the solid itself will have decreased from its original value. The diameter of the ensemble, D, is the sum of the diameter of the remaining solid and the thickness, h, of the moisture layer.

It is assumed in the following that one gram of solid is studied and that the sample is monodisperse. Such a sample would consist of N particles, where

$$Nm = \frac{N\rho\pi d_0^3}{6} = 1 \tag{8.4}$$

The amount of solid present at time t is given by the original amount less the amount dissolved. If there are W grams of water adsorbed by one gram of solid (i.e. w gram dissolved per particle), then

$$N\left(\frac{\rho\pi}{6}\right)d^3 = N(m - wS) = 1 - WS \tag{8.5}$$

or

$$d^3 = \frac{1 - WS}{N(\rho\pi/6)} \tag{8.6}$$

The volume of liquid adsorbed by one solid particle has a volume of the total particle minus the solid particle, i.e.,

$$\begin{aligned} \frac{w}{\rho^*} &= \left(\frac{\pi}{6}\right)D^3 - \left(\frac{\pi}{6}\right)d^3 \\ &= \left(\frac{\pi}{6}\right)D^3 - \frac{\pi/6(1 - WS)}{N(\rho\pi/6)} \end{aligned} \tag{8.7}$$

where ρ^* is the density of the adsorbed liquid. Since it is assumed that it is always saturated, it is time-independent, and under ideal conditions it would be

$$\rho^* = (1 - x_s)\rho_0 + x_s\rho \tag{8.8}$$

where $1 - x_s$ and x_s are the volume fractions of liquid and solid, respectively, in the ensemble particle, and ρ_0 and ρ are the respective densities. It follows from Eq. (8.7) that the amount of moisture adsorbed per gram can be expressed in terms of diameters as

$$W = \rho^* N\left(\frac{\pi}{6}\right)D^3 - \frac{\rho^*(1 - WS)}{\rho} = QD^3 - F + FSW \tag{8.9}$$

where

$$F = \frac{\rho^*}{\rho} \tag{8.10}$$

$$Q = \rho^* N\frac{\pi}{6} \tag{8.11}$$

Equation (8.9) may be written

$$F + (1 - FS)W = QD^3 \tag{8.12}$$

or

$$D = \left\{\frac{F + (1 - FS)W}{Q}\right\}^{1/3} \tag{8.13}$$

The area, a, of the particle (solid plus liquid) is, hence,

$$a = \pi\left\{\frac{F + (1 - FS)W}{Q}\right\}^{2/3} = B[E + W]^{2/3} \tag{8.14}$$

where

$$B = \pi\left[\frac{1 - FS}{Q}\right]^{2/3} \tag{8.15}$$

$$E = \left\{\frac{FQ}{1 - FS}\right\}^{2/3} \tag{8.16}$$

The rate of condensation (dW/dt) is proportional to the pressure gradient, i.e., the difference between the water vapor pressure, P, in the atmosphere and the vapor pressure, P_s, over a saturated solution. At a given atmospheric milieu, this gradient is a constant.

It is also proportional to the surface area, a, by a mass transfer coefficient, k, so that we may write

$$\frac{dW}{dt} = ka(P_a - P_s) = k(P_a - P_s)B[E + W]^{2/3} \tag{8.17}$$

where Eq. (8.14) has been used for the last step. This may be written

$$\frac{dW}{[E + W]^{2/3}} = 3G\,dt \tag{8.18}$$

where

$$3G = k(P - P_s)B \tag{8.19}$$

Eq. (8.18) integrates to

$$[E + W]^{1/3} = Gt + [E]^{2/3} \tag{8.20}$$

where the initial conditions, $W=0$ at $t=0$, have been imposed. Equation (8.20) can be solved by iteration.

As an example of this, VanCampen et al. (1980) studied the moisture pickup in a vacuum system by using a Cahn balance and exposing the evacuated head space to relative humidities created by salt baths. They also reported moisture uptake rates of choline chloride at room temperature and different relative humidities using a desiccator method. An example of their results obtained by the latter method is shown in Fig. 9.

Jakobsen et al. (1997) have employed a highly sensitive microcalorimeter to evaluate the hygroscopicity of hydrophilic drug substances, such as flupentixol dihydrochloride (solubility>1 g/mL) as well as hydrophobic substances (such as

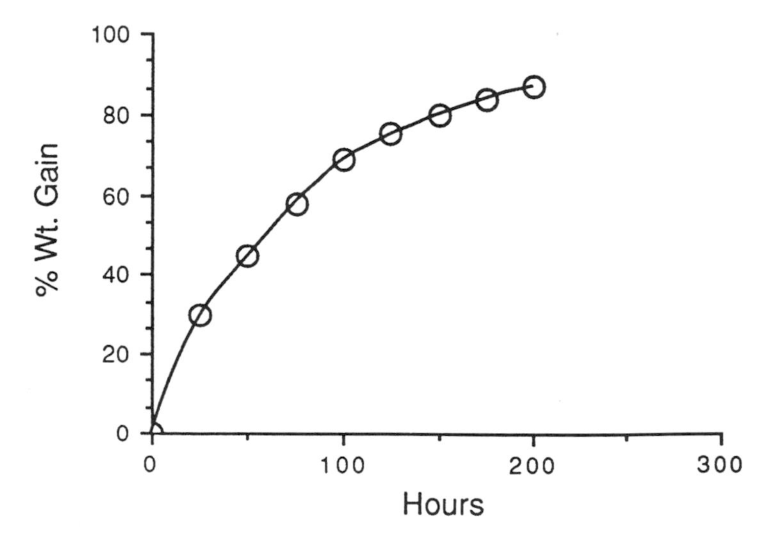

Fig. 9 Data for choline chloride moisture adsorption at 100% RH. (Drawn from data published by VanCampen et al., 1980.)

ertindole, solubility 10 μg/mL). Pinderre et al. (1997) have described coating powders with Eudragit to protect them against moisture uptake and have evaluated the coatings by way of moisture uptake rates.

6. CRITICAL MOISTURE CONTENT

There are humidities below which a solid will not adsorb (considerable amounts of) moisture, i.e., not form a "bulk-sorbed" layer. These are dictated by the solubility of the compound, as will be seen below.

Suppose a solid is placed in a room of a given RH, as shown in Fig. 10. If the RH were 30%, then it might pick up moisture at a given rate, at 50% RH at a higher rate, and at 80% RH at an even higher rate.

The rate with which it picks up moisture is determined by weighing the sample at given intervals, as demonstrated in Table 1. It is noted that there is a linear section of the curve (up to 6 days), as shown in Fig. 11 and 12. The slope of this linear segment is the moisture uptake rate (MUR). The actual uptake rates (determined from the linear portions) are shown in Table 2.

The uptake rates can simply be obtained by weighing the sample after a given time (6 days), but in such a case it is assumed that the moisture uptake is still in the linear phase. If, e.g., the weight gain is 5 mg per 10 g sample in 6 days, then the MUR is $5/10/6 = 0.083$ mg/g/day.

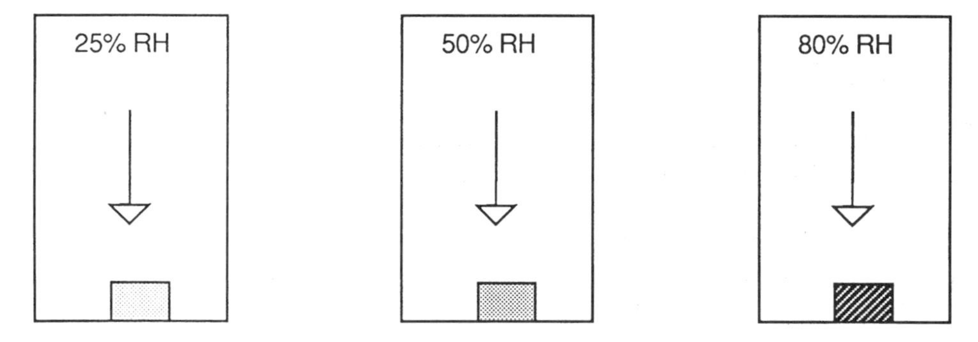

Fig. 10 Mechanism of moisture uptake.

Table 1 Moisture Uptake of a Water-Soluble Compound at 50% RH

Days stored at 50% RH	Moisture pickup (mg/g)
2	0.5
6	1.5
18	2.25
36	3.4
100	3.0
144	4.2
288	4.3

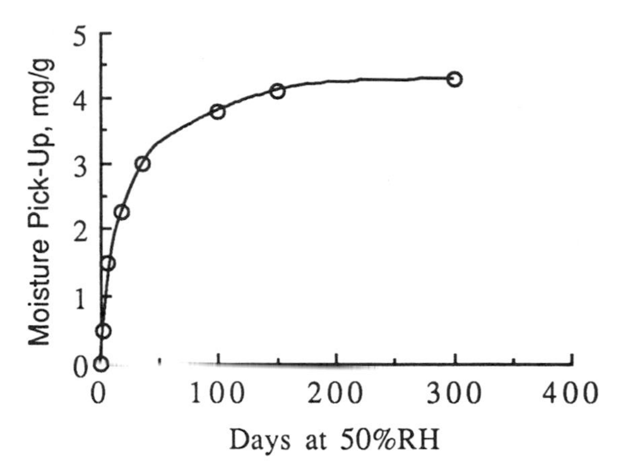

Fig. 11 Moisture uptake data from Table 1.

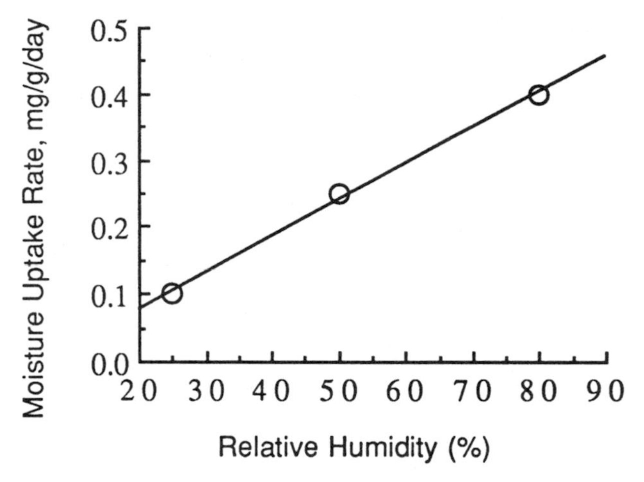

Fig. 12 Moisture uptake rate as a function of RH. Least squares fit is $y = -0.06264 + 0.006374x$, with $R^2 = 0.999$.

Table 2 Moisture Uptake Rate of Water-Soluble Compound

% RH	mg/g/day
25	0.1
50	0.25
80	0.45

If the MUR values are plotted versus RH, then a straight line results (Fig. 11). The curve intercepts the x-axis at 20% RH. This means that the compound can be stored without moisture pickup in atmospheres of less than 20% RH. In some cases the compound will dry out under such conditions (e.g., a hydrate), but in general

Table 3 Characteristics of Disodium Hydrogen Phosphate

Type	% Moisture in solid		$P(H_2O)$ (mm Hg)	Water activity (RH/100)
Anhydrous	0			
		Pair	9	0.38
Dihydrate	20			
		Pair	14	0.58
Heptahydrate	47			
		Pair	18	0.75
Dodecahydrate	60			
		Pair	22	0.92
Satd. solution (100 g water/4.5 g salt)				

the useful information reached from such a graph is the maximum RH that is satisfactory for storage of the products. 20% RH happens to be the relative humidity over a saturated solution of the compound (or over a salt pair, as will be discussed presently).

For inorganic compounds and hydrates, the curves are stepwise curves. For instance, for disodium hydrogen phosphate, the following situation exists: the compound can form three hydrates (2, 7, and 12) aside from being anhydrous. The percent of moisture in, e.g., the dihydrate, is calculated as follows: disodium hydrogen phosphate has a molecular weight of 142. The dihydrate hence has a molecular weight of $142 + 36 = 178$. Hence the moisture percentage is

$$100 \times \frac{36}{178} = 20\%$$

The moisture contents for the remaining hydrates are shown in Table 3.

7. EQUILIBRIUM MOISTURE CURVES FOR SALT HYDRATES

The previous section dealt with the *rate* with which moisture is taken up. As shown in Fig. 11, at longer time periods, the moisture level (the weight of the sample) will taper off and plateau at an equilibrium value. This equilibrium value is also a function of RH, and there are two types of curves that occur when equilibrium values are plotted against RH: salt pairs and continuous adsorption. The former will be discussed first.

It is seen in the table that the RH of the atmosphere above a mixture of anhydrous disodium hydrogen phosphate and the dihydrate is 9 mm Hg or $100(9/24) = 38\%$ RH. It is noted that any mixture of the anhydrous salt and the dihydrate will given this relative humidity. Hence disodium hydrogen phosphate containing from 0 to 20% moisture will have above it an atmosphere of 38% RH. Similarly, as shown in Table 3, the heptahydrate contains 47% moisture, and mixtures of di- and heptahydrate give rise to water vapor pressures of 14 mm Hg (58% RH). Proceeding in this fashion, a graph as shown in Fig. 13 results.

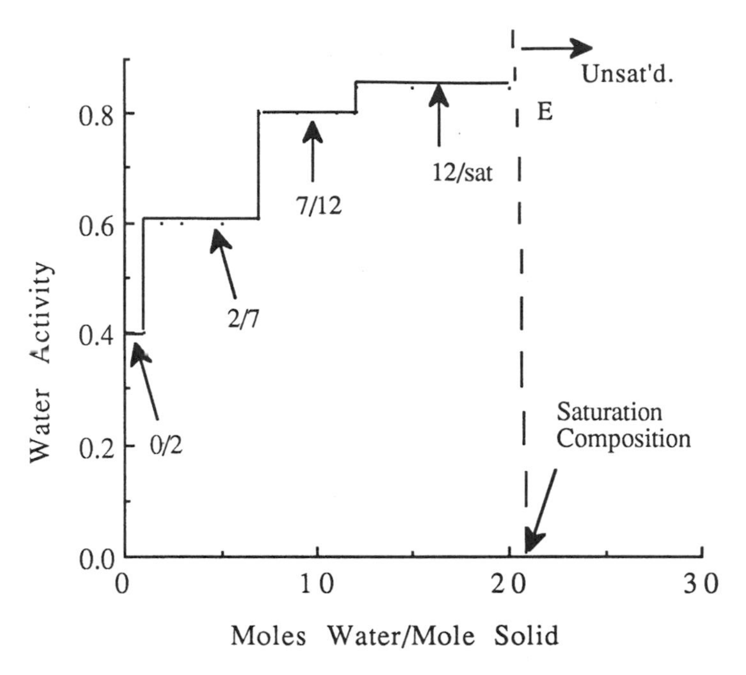

Fig. 13 Vapor pressure diagram of salt forming a dihydrate, a heptahydrate, and a dodecahydrate.

Two further points need to be mentioned. (a) If disodium hydrogen phosphate is stored at a relative humidity between 38 and 58% RH, it will not pick up moisture. Once the relative humidity is raised to (slightly above) 58%, then it will start picking up moisture until it has completely converted into the heptahydrate. (b) If the relative humidity is raised to (slightly above) 92% RH, then the dodecahydrate is converted to saturated solution. At higher RH values, the equilibrium will be dictated by the water vapor pressure over the now unsaturated solution.

At 100% RH the system in equilibrium is infinite dilution (pure water), and if a diagram such as this (and the following diagram for organic macromolecules) is carried out to 100% RH, then a sharply increasing curve should result at very high RH. The diagram in Fig. 12 is at a given temperature. Figure 13 shows a diagram of a dihydrate at different temperatures. At the temperature T_3, the line for the salt pair has caught up with that of the saturated solution. Essentially this means that the enthalpy of hydration for the solid is higher than the heat of vaporization of water from the saturated solution, since both have Clausius–Clapeyron type vapor pressures. Above T_3, therefore, the salt would have a higher vapor pressure than the saturated solution, but this is thermodynamically untenable, and T_3 is simply the highest temperature (and a triple point) where the dihydrate exists.

For a monohydrate as depicted in Fig. 14, the moisture content of the "salt hydrate" will increase drastically when the water vapor pressure is higher than that depicted by point H. Moisture keeps on condensing and converting the monohydrate to saturated solution, and this will continue until all is dissolved. After that the vapor pressure will increase so that it is always in equilibrium with the concentration in the (now) unsaturated solution.

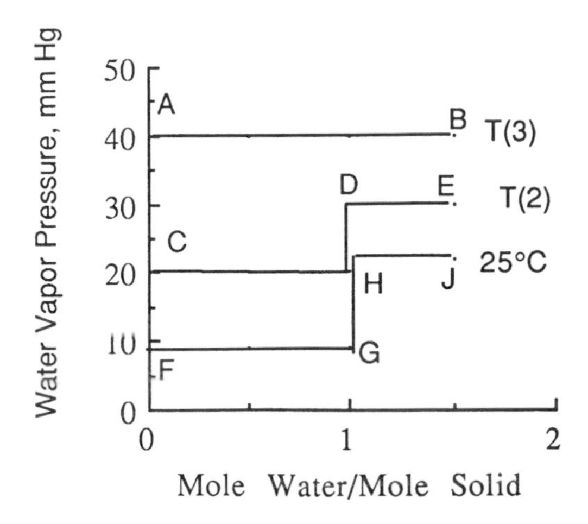

Fig. 14 Single salt pair (monohydrate) vapor pressures as a function of temperature. The line at point A has been drawn a mite to the left for graphical clarity. It occurs at 1 mole of water/mole of solid.

In general for an x-hydrate with a saturation concentration of $1/(x+y)$ moles of salt per mole of water, the reaction on the constant pressure isotherm is

$$B \cdot xH_2O + yH_2O \rightarrow B \text{ dissolved in } (x+y) \text{ moles of } H_2O \tag{8.21}$$

The solubility of B in water is $1/(x+y)$.

Beyond the solubility concentration, there will be a total of z moles of water and 1 mole of solid, so that the mole fraction of water will be $z/(1+z)$. The vapor pressure of the now unsaturated solution would be given by Raoult's law, i.e., $a = P/P_0 = z/(1+z)$, or, since a here is plotted versus z, it would be given by

$$z = \frac{a}{1-a} \tag{8.21a}$$

However, many authors plot a versus z.

Example 8.1.

The diagram in Fig. 15 is a vapor diagram of a drug substance that forms a pentahydrate. Comment on the following statements: (a) The pentahydrate is stable between 10 and 45%. (b) If the hydrate is exposed to a relative humidity of 81%, then it will lose water and become anhydrous. (c) It neither loses nor picks up moisture at 81%. (d) It gains moisture and forms a saturated solution.

Answer.

(a) and (d) are correct, but (d) is strictly correct only for 80% RH. (It will form a very slightly undersaturated solution.)

As an example of research on hydrate forms, it should be mentioned that Allen et al. (1978) have shown that erythromycin exists in crystalline form as an anhydrate, a

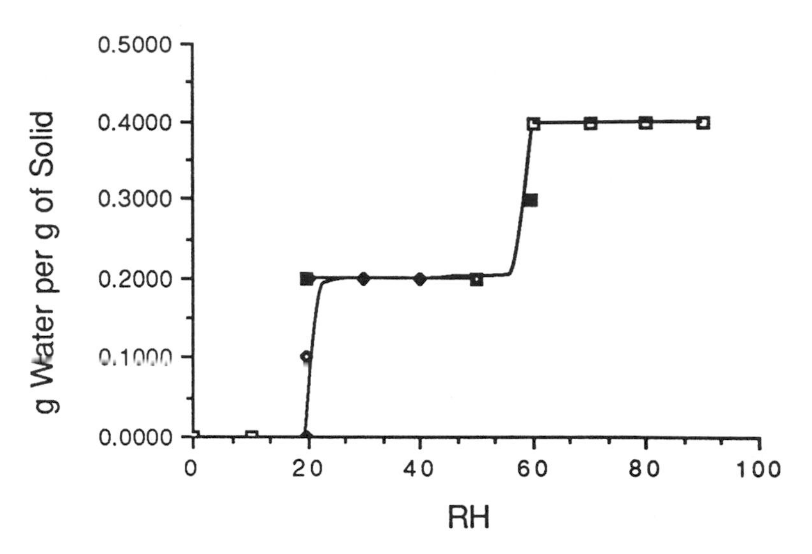

Fig. 15 Vapor phase diagram of an organic substance that forms a pentahydrate at 25°C (MW 360 + 90).

monohydrate, and a dihydrate. Shefter and Kmack (1967) showed that the dehydration kinetics of theophylline hydrate were first order.

Hemihydrates also exist. Wu et al. (1996) have reported on an anhydrous and a hemihydrate form of brequinar sodium. Both have fairly comparable solubilities. Loosely bound water is also present in the structure, and this is lost (in thermograms) at 90°C, and the water of hydration is released at about 175°C.

It should be mentioned that in some cases "bound" moisture is indeed held very tightly. Magnesium chloride tetrachloride is an example. Heating this substance to 80–100°C will remove two of the molecules of water. But further heating results in the removal of 2 moles of hydrochloric acid, leaving magnesium hydroxide behind.

8. MOISTURE EQUILIBRIUM CURVES OF A SMOOTH NATURE

There are substances such as gelatin and corn starch that give rise to moisture equilibrium curves of the type shown in Fig. 16. These are referred to as BET moisture isotherms.

As a dry sample is exposed to increasingly higher vapor pressures, P_u (u stands for "up"), moisture contents x_u will be in equilibrium with the sample. If the experiment is terminated at a pressure of P^*, and the vapor pressures in the atmospheres decrease, then, e.g. at P_d (d stands for "down") the moisture content will be x_d, i.e., higher than during the up curve. The hysteresis loops shown in Fig. 15 are exaggerated for graphical clarity. Such curves can be shown to be variants of the BET equation or the GAB equation (Guggenheim, Anderson, and deBoer) (Guggenheim et al., 1968; Zografi and Kontny, 1986; Grandolfi, 1986). It is noted that y_d is not an equilibrium condition. Obviously

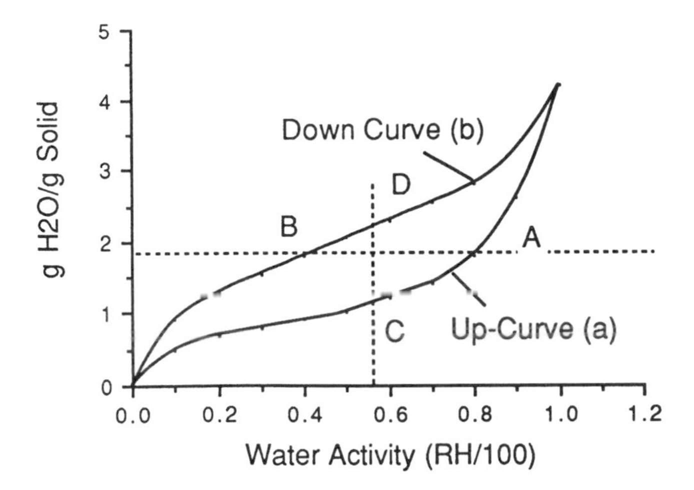

Fig. 16 BET up (adsorption) and down (desorption) moisture isotherm.

ΔG is negative in going from the down curve to the up curve, because

$$\Delta G = V \int_{P_u}^{P_d} V\,dP = V[P_u - P_d] < 0 \tag{8.22}$$

Several common tablet excipients give rise to Langmuir isotherms. An excipient study by Sangvekar (1974), when all the data are lumped together, gives an equation of the type

$$\frac{1}{y} = \frac{A}{P} + B \tag{8.23}$$

Usually, in pharmaceutical and engineering literature, the moisture equilibrium curves are shown in a sense opposite to that shown in Fig. 15, i.e.,

$$P = \phi(y) \tag{8.24}$$

The high RH tail of the curve is usually above 85% RH and therefore does not apply to most realistic pharmaceutical conditions, but it is applicable to one often-conducted test (40°C, 75%RH). Zografi and Kotny (1986) have described these types of moisture isotherms by either a BET equation or a GAB equation.

For routine isotherms, the high relative humidity tail is difficult to obtain with reasonable precision, and one approach (Carstensen, 1980) is to approximate them by Langmuir isotherms (i.e., not use the high-end portion).

9. AMORPHATES

Solids which are not crystalline are denoted amorphous. If one melts a (stable) solid and recools it, then it should crystallize when the melting point is arrived at.

This requires nucleation, and nucleation propensity is a function of the viscosity of the liquid in which it occurs. Materials that are viscous about their melting point are therefore prone to form supercooled solutions.

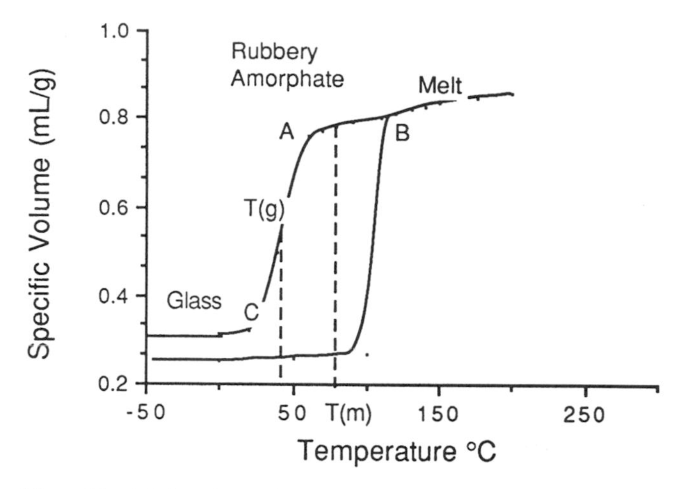

Fig. 17 Molecular volume as a function of temperature of a solid prone to forming an amorphate.

At a given high viscosity (attained at or below the melting point), the melt will have the appearance of a solid, and this is the type of material referred to as amorphous.

Right below the melting point, the molecules will have no specific orientation, and molecular movements will be random in direction and magnitude (within the limits of the system) as opposed to a crystalline material, where the molecules are arranged in lattices (ordered arrays), and where the orientation of each molecule is set.

At a temperature T_g, lower than the melting point, there will be a physical change in the amorphate. An example of this is shown in Fig. 17.

Between points A and B the properties of the amorphate are often like that of the melt. This is referred to as the "rubbery" state, and below C it is referred to as a glass.

10. WATER ABSORPTION "ISOTHERMS" INTO AMORPHATES

Amorphates are solids that are not crystalline. It is assumed at this point that the term "solid" is self-evident, although amorphates in the rubbery state (just below the melting point of the crystalline form of the compound) are actually highly viscous liquids. When exposed to humid atmospheres, they will pick up moisture in a fashion that is not like that of a BET isotherm (to be covered shortly). The moisture actually penetrates into the solid, which thus may be considered a "solution."

In an ideal situation, the water activity, a, will decrease linearly with $(1-x)$, where x is the mole fraction of solute. At a given point ($x=0.24$ in Fig. 17) the solution becomes saturated. (This concentration, of course, differs from compound to compound.) Beyond this concentration, the solution itself will be saturated, and the vapor pressure will not change with further addition of compound; rather, the composition will change, but the vapor pressure will stay constant. In this type of graph the coordinates are in the opposite direction of a usual isotherm.

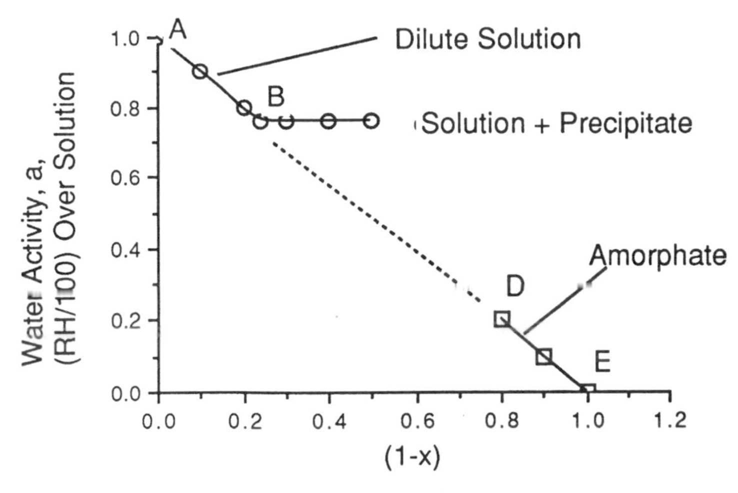

Fig. 18 Moisture isotherm for an amorphous solid. (Graph constructed from data published by Carstensen and VanScoik, (1990.)

If an amorphous form of the compound is produced and exposed to different relative humidities, then the "isotherm" is often quite linear, if the amount of water absorbed is expressed as a mole fraction (line DE in Fig. 18). As shown by Carstensen and VanScoik (1990) for amorphous sugar, this line is an extension of the solution vapor pressure line (AB in Fig. 18), and one may consider the moist amorphate as a highly concentrated, supersaturated "solution."

Due to the random arrangement and the mobility of the molecules in an amorphate as opposed to a crystalline modification, amorphates are usually less stable chemically than crystalline modifications (Carstensen et al., 1993).

Carstensen and VanScoik (1990) were the first to point out that for an amorphous substance, it is illogical to use the traditional moisture isotherms, because in this case it is probably not an adsorption, but rather an absorption, which is at play.

By exposing amorphous sucrose to various relative humidities, various moisture levels were reached. If these moisture levels were expressed as mole fractions of sucrose, then the vapor pressures fell in line with the vapor pressure curve of sucrose itself.

The fraction to the right of point B is the principle used for salt solutions to obtain constant relative humidity in desiccators. With electrolytes, the vapor pressure depression is larger (due to the 2- or 3-fold number of ionic particles, over that of the molarity of the salt), and the solubilities are often high, so that these are preferred for creating constant relative humidity in desiccators.

Zografi and Hancock (1993) have used this principle in their investigation of whether such an approach, i.e., solution theory, could be applied to macromolecules. To quote, "If one considers the absorption process to be completely analogous to the solution process, then it should be possible to use basic solution theories to model the data." Their data for PVP K30 are shown in Fig. 19.

First of all note that the "ideal solution" model advocated earlier is (probably) not applicable to macromolecules. (The concentrations, however, are not converted to mole fractions, but such a conversion would not make the plot linear.) The data

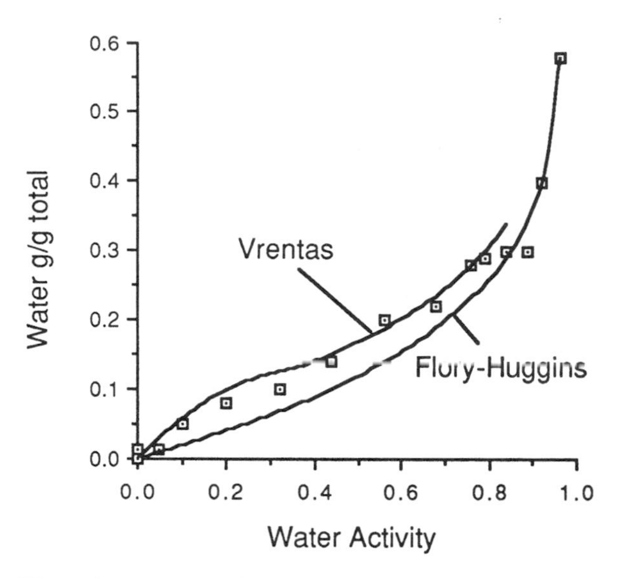

Fig. 19 Fit of vapor pressure data of aqueous solutions of PVP K30 at 30°C to the Flory–Huggins equation. The points are taken off Figs. 6 and 7 as accurately as possible, as is the trace of the Flory–Huggins equation. (Plot constructed from data published by Hancock and Zografi, 1993.)

obviously, fit the Vrentas equation better at low water activity, but the Flory–Huggins equation may be more applicable at high water activity. Data become slightly uncertain at such high humidities in any event.

It has been mentioned that one method of stabilizing a "solution" for marketing is to lyophilize it and thus increase its storage stability. Many lyophilizates are amorphous. The method for making a lyophilizate is first to make a solution, then to freeze it, and then to sublime off the moisture. In this process it is important that the solution stay sufficiently stable before and during freezing. Various lyoprotectants are used for such purposes, and Dekeyser et al. (1997), for instance, have shown that chymopapain is stabilized in the presence of different lyoprotectants such as maltodextrins.

Amorphates exhibit glass transition temperatures. These are a function of water content, as shown e.g. by Hancock and Dalton (1999) and in Table 4. These authors and others (e.g., Carstensen, 1995) compared moisture adsorption isotherms with the equations of Flory–Huggins, Vrentas, and Raoult.

Glass transition temperatures are always somewhat approximate. For instance, in contrast to the above, Hatley (1997) has reported the T_g of sucrose to be 64°C at 0.73% moisture.

11. MOISTURE EXCHANGE BETWEEN DOSAGE FORM INGREDIENTS

Gore and Ashwin (1967) were the first to report that for an excipient (in their case citric acid), "given a knowledge of the equilibrium moisture content for a particular moisture sensitive compound at the upper limit of its moisture specification, it would

Table 4 Glass Transition Temperatures of Water-Containing Sugar Amorphates

Water (%)	Lactose glass trans. temp. (°C)	Sucrose glass trans. temp. (°C)	Raffinose glass trans. temp. (°C)	Trehalose glass trans. temp. (°C)
0	112	74	103	115
1	102	60	92	101
2	94	50	83	90
3	83	32	75	80
4	80	<25	67	70
	71	<25	58	60

Source: Table constructed from data by Hancock and Dalton (1999).

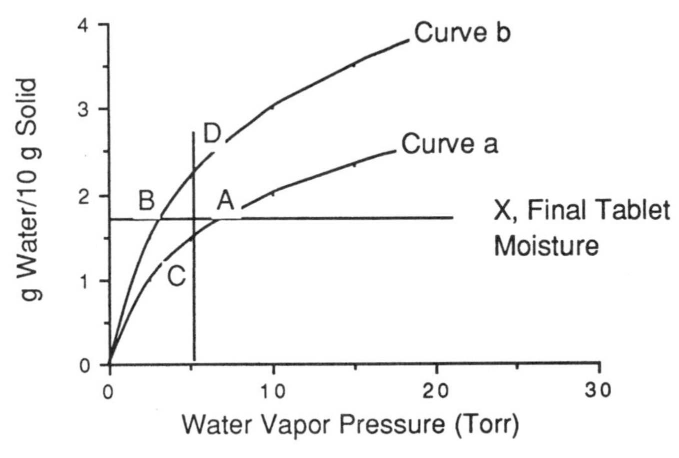

Fig. 20 Langmuir moisture isotherm presentation of initial part of a BET isotherm. From (Carstensen, 1980.)

be a simple matter to define moisture limits for citric acid in a particular formulation".

A solid dosage form (e.g., a tablet) is usually made to a given moisture content, e.g. (Fig. 20), 1.8 g per 100 g of solid. Since the drug and the excipients have different moisture isotherms, they will have different equilibrium RH values. There can, however, only be one RH condition in the pore space of the solid dosage form, so the result is that compound b will pick up moisture (move from B to D) and compound a will lose moisture (moving from A to C). The question is to estimate, quantitatively, where (at what RH) the line DC will be.

In Fig. 19, the two moisture equilibrium curves have (in an abbreviated fashion) been represented as Langmuir isotherms. This can be verified by inspection of Fig. 10, where lines OC and OB would both fairly well adhere to Eq. (8.25).

This may be used to estimate the moisture movement in a solid dosage form after it is manufactured. In consulting Fig. 20 and assuming that the up curve is that of drug (A) and the down curve that of excipient (B), there are m_A grams of A on an anhydrous basis, and A contains a fraction (on a dry basis) of q_A

moisture, i.e., a total of $m_A q_A$ grams of water. There are m_B grams of B on an anhydrous basis, and A contains a fraction (on a dry basis) of q_B moisture, i.e., a total of $m_B q_B$ grams of water.

The dry weight of the dosage form is therefore $mA + mB$, and as the dosage form (e.g., tablet) is made, it is made at a particular moisture content of a fraction (on a dry basis) of q moisture, i.e., a total of $mq = [m_B + m_A]q$ grams of water.

Since, as seen from the figure, the relative humidity (the vapor pressure, P) in the pore space must be one particular figure (P), it follows that A must give up moisture (from point A to point C) and B must take up moisture (from point B to point C).

The moisture isotherms are of the type

$$x(A) = \frac{q_A m_A}{m_A} = q_A = Q_A P_A^{1/n_A} \quad (8.25)$$

and

$$x(B) = \frac{q_B m_B}{m_B} = q_B = Q_B P_B^{1/n_B} \quad (8.26)$$

The values of n usually do not differ much (and the two isotherms can therefore be represented as only differing in the values of the Q's). It is noted that the areas have not been taken into account, and the isotherms apply to two samples of material. (To account for the area, plotting by BET would have to be done.)

Where a known amount of A, m_A, is mixed with a known amount of B, m_B, mass balance (assuming no loss of moisture) gives

$$x_C[m_A + m_B] = x_A m_A + x_B m_B \quad (8.27)$$

or

$$x_C = \frac{x_A m_A + x_B m_B}{m_A + m_B} \quad (8.28)$$

and the amount of moisture lost can then be gauged from

$$\text{moisture loss in A} = m_A(x_A + x_C) \quad (8.29)$$

and for B,

$$\text{moisture loss in B} = m_B(x_D - x_B) \quad (8.30)$$

Since x_C is known, P is then also known.

If for instance the two compounds are mixed together, moisture added (as in a granulation), and this then dried, then x_C is known. Mass balance about ACB in Fig. 20 then gives that the moisture loss experienced by A,

$$m_A(x_A - x_C) = m_A Q_A[P^{1/n} - P_C^{1/n}] \quad (8.31)$$

must equal the moisture gained by B, i.e.,

$$m_B(x_C - x_B) = m_B Q_B[P_C^{1/n} - P_B^{1/n}] \quad (8.32)$$

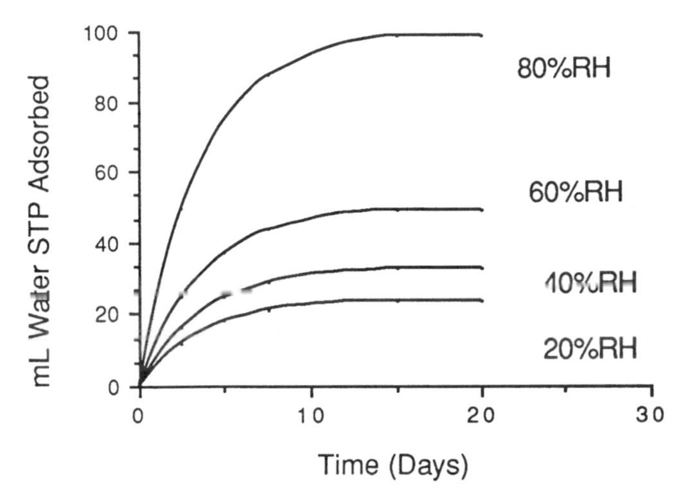

Fig. 21 Moisture uptake curves for a sample of silica gel at 20, 40, 60, and 80% RH.

All quantities are known, so that P [$= P_C = P_D$] can be calculated, i.e., both moisture losses and gains, and the final relative humidity can be calculated. In this latter case, the isotherms should be determined on samples that had been wetted and dried the same way the final mix had been wetted and dried (since the surface area changes).

12. EQUILIBRIUM MOISTURE CONTENTS FOR MACROMOLECULES

For an organic compound such as starch, a smooth equilibrium moisture curve will result. Here again there is a sharp upswing at very high relative humidities.

If experiments such as are exemplified in Table 1 and Fig. 11 are carried out on e.g. cornstarch, then curves of the *type* shown in Fig. 21 result. The figure shows moisture uptake rate curves at four different relative humidities: 20%, 40%, 60%, and 80%. When the moisture contents (x mg water/mg solid) of these levels are plotted as a function of relative vapor pressure, P/P^* (the relative humidity, divided by 100, the so-called water activity), then an isotherm results. This moisture isotherm has the shape shown in Fig. 16.

When $P/[x\{1-P\}]$ is plotted versus P, then a straight line results.

13. ADSORPTION ISOTHERMS OF SILICA

The curve in Fig. 21 eventually levels off. The equilibrium level is a function of the relative humidity at which the experiment is carried out. Table 5 shows an example of moisture uptake curves of a sample of silica, at various relative humidities. These levels are tabulated in the second column. It is customary in isotherm work to convert these adsorbed amounts to the volumes that would have been occupied at 0°C and 1 atm, and this can easily be done, e.g., for the first row. The number of moles is $n = (17.5 \times 10^{-3})/18 = 9.75 \times 10^{-4}$ moles. The volume of this at 25°C and 1 atm would be $V = nRT/P = 9.75 \times 10^{-4} \times 82 \times 298/1 = 23.8$ mL. These figures are shown in the third column and are denoted V.

Table 5 Data from Which Fig. 22 Was Constructed, and Conversion to BET Parameters

RH (%)	mg adsorbed	V (mL) (O°C, 1 atm)	RH/(V{100-RH})
20	17.5	23.8	0.01
40	23.9	32.5	0.021
60	36.1	49.2	0.030
80	72.6	98.9	0.040

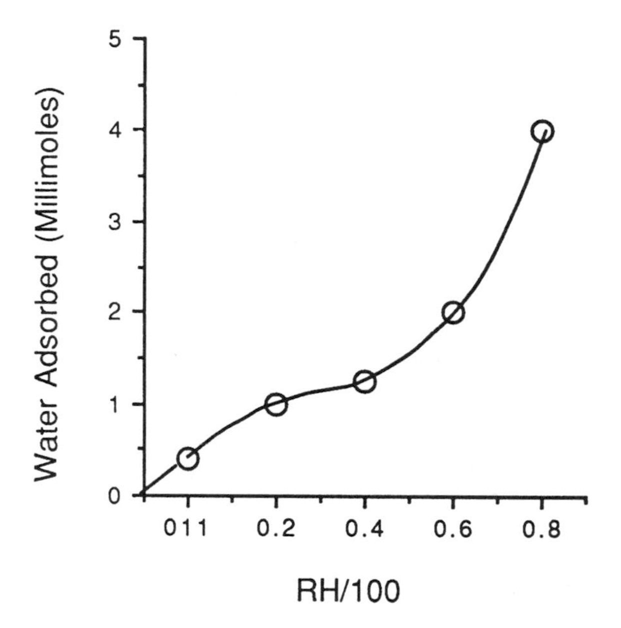

Fig. 22 The equilibrium levels in Fig. 20 plotted versus relative humidity.

The isotherms of this type are called BET isotherms. The data in the third column are shown in Fig. 22. It can be shown that such data follow the BET equation:

$$\frac{\mathrm{RH}}{V\{100 - \mathrm{RH}\}} = \phi + \frac{1}{V_\mathrm{m}}\left[\frac{\mathrm{RH}}{100}\right] \tag{8.33}$$

Treatment by this equation is shown in Fig. 23. V_m is here the volume (0°C, 1 atm) of water that just constitutes one layer on the entire surface of the solid sample. RH/[V{100-RH}] has been calculated in the table (last column) and is plotted in Fig. 23 versus RH/100.

The slope of the line is $1/V_\mathrm{m}$, so

$$\frac{1}{V_\mathrm{m}} = 0.05 \qquad \text{or} \qquad V_\mathrm{m} = 20\,\mathrm{mL} \tag{8.34}$$

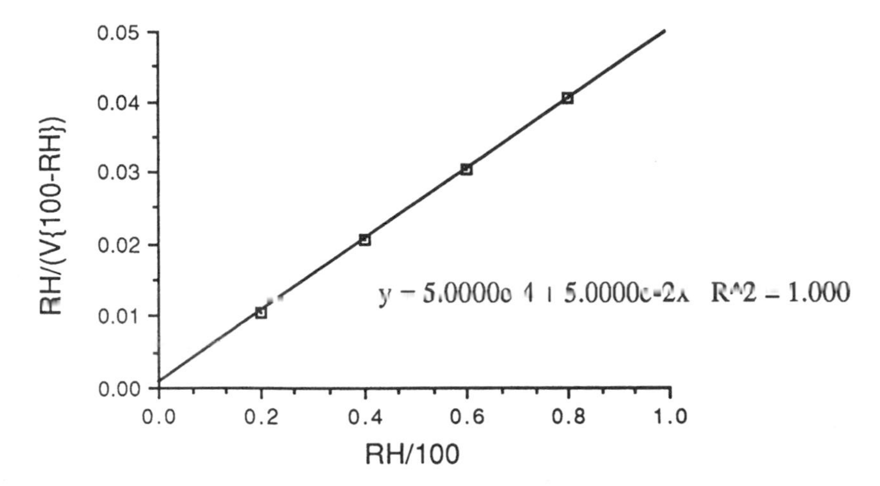

Fig. 23 Data from Table 5 treated by the BET equation.

This can be converted to moles (n) and then to molecules (N);

$$
\begin{aligned}
n_{\mathrm{m}} &= \frac{20PV}{RT} = 1 \times \frac{20}{82 \times 298} = 12.9 \cdot 10^{-4} \text{ moles} \\
&= 6 \times 10^{23} \times 12.9 \times 10^{-4} = 77 \times 10^{19} \text{ molecules}
\end{aligned}
\tag{8.35}
$$

Water molecules in a monolayer will position themselves so that their cross-sectional area is $10\,\text{Å}^2 = 10 \times 10^{-16}$ cm^2, so that in this case the entire surface area would be the number of molecules times the area of each molecule, i.e.,

$$77 \times 10^{19} \times (10 \times 10^{-16}) = 77 \times 10^4\,\text{cm}^2 = 77\,\text{m}^2$$

Most substances are not "hygroscopic" below 20% RH.

If a bag of silica is placed in a bottle with a dosage form, then, if there is a critical moisture content beyond which the dosage form becomes unstable, it is possible to calculate, from the isotherm of the dosage form, at which relative humidity this occurs. From the silica isotherm one can then calculate how much moisture is taken up by the silica bag at this point, and dividing this figure by the moisture penetration of the package, it is possible to calculate the length of time the product is good.

Moisture isotherms are of great significance in pharmaceutics. Cases in point are the moisture isotherms of PVP and of the complex of misoprostol and hydroxypropyl methylcellulose.

14. HYDROUS AMORPHATES

As mentioned, solids that are not crystalline are called amorphous. An important category of these are lyophilized cakes (for intravenous reconstitution). These are formed by freezing aqueous solutions. Upon such freezing (when part of the solid comes out as an atmosphere), ice will first freeze out, and then the remaining solution (which usually crystallizes as a eutectic) will supercool and will become "solid." But in this case the "solid" is simply a very viscous solution. Fig. 24 is an example of this and is constructed from data published by Her and Nail (1994).

The "solid" is (when dried) referred to as a lyophilized cake. The glass transition temperature can usually be arrived at from thermal analysis, as shown in Fig. 25. The collapse temperature is a temperature dictated by mechanical pro-

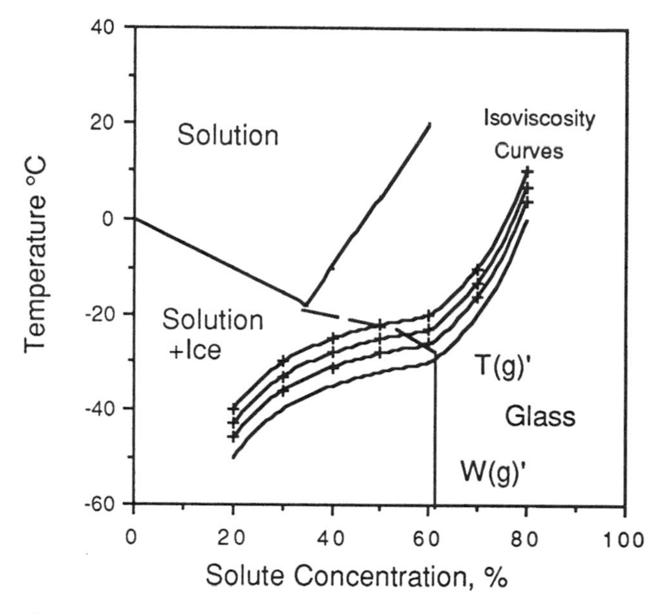

Fig. 24 Graph constructed from data published by Her and Nail (1994).

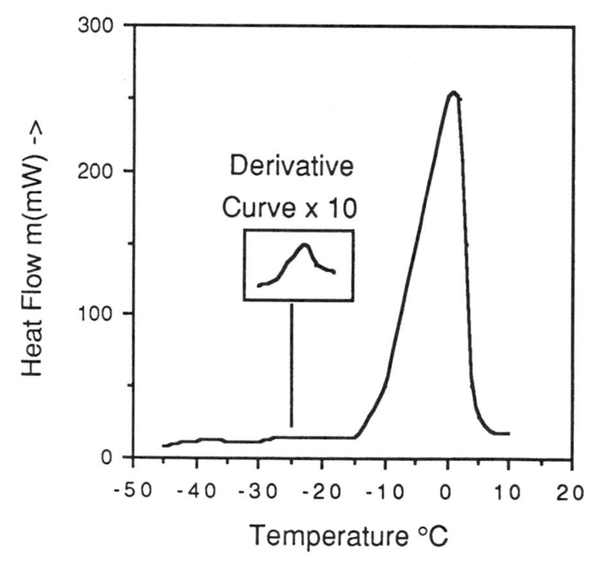

Fig. 25 Thermogram of aqueous solution of 10% PVP. The relative magnitudes of the endotherms for glass transition vis-à-vis melting is shown. (Graph constructed from data published by Her and Nail, 1994.)

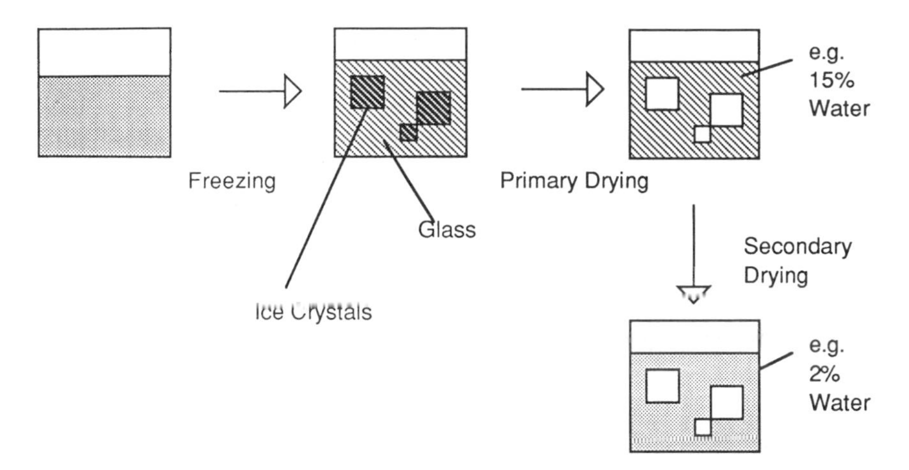

Fig. 26 Schematic of freeze-drying events.

perties. Just above the glass transition temperature, sucrose solutions, for instance, have viscosities of about 10^6 Pa/s, but below T_g this figure is 10^{12} Pa/s. The general sequence of events in freeze drying is shown in Fig. 26.

The primary drying (Fig. 26) consists of the evaporation of the crystalline ice, so that the cake is left with "holes" in it, and a glass of a water content in the range of 12–15% results. As mentioned, if the temperature is below the glass transition temperature, then this glass has a high viscosity and will dry slowly, since the diffusion coefficient, D, for evaporation of water, will be high.

If, after the primary drying, the initial freezing temperature were 240°K as shown in Fig. 26, and the solids content were 50%, then the composition would be at point C, Fig. 27, between the T_c and T_g curves. But if sublimation were continuously carried out at this temperature, then, at point B, the glass transition would be passed, and the viscosity would become very high, and sublimation would be very slow. The temperature is therefore continuously increased, so that the lyophilization temperature can stay within the bounds of the two curves.

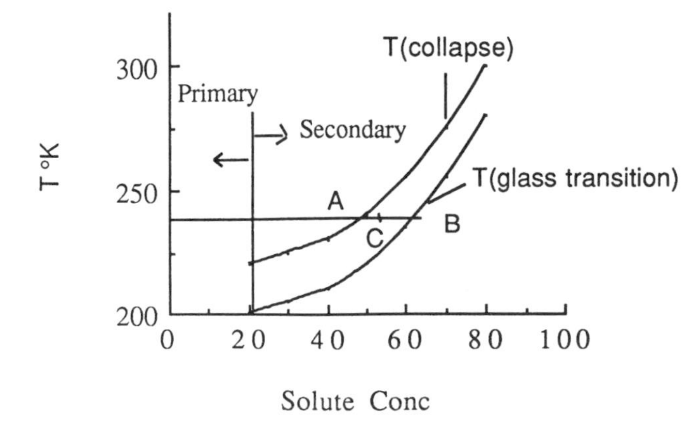

Fig. 27 Limiting phases in a lyophilization event.

Some proteins have stabilities that depend on the cooling rate, but this is primarily due to electrolytes (e.g., sodium chloride) and stabilizers (e.g., glycine) in the composition. These will crystallize out and give the cake structural strength so that T_s increases, but their presence, as well as the initial freezing rate, will modify the positions of the two curves, so that a slow cooling rate may provide a different (and sometimes worse) cake than when a fast cooling rate is employed.

These aspects have been discussed in detail by Franks (1990), Levine and Slade (1988), Mackenzie (1977) and Suzuki and Franks (1993).

REFERENCES

Allen, P. V., Rahn, P. D., Sarapu, A. C. Vanderwiele, A. J. (1978). J. Pharm. Sci 67:1087.
Carstensen, J. T. (1980). Drug Stability. 1st ed. New York: Marcel Dekker.
Carstensen, J. T. (1986). Pharmaceutical Technology 9 (September), 41.
Carstensen, J. T. (1995). Drug Stability. 2d ed. New York: Marcel Dekker, p. 218.
Carstensen, J. T., Kothari, R. (1981). J. Pharm. Sci. 70:1095.
Carstensen, J. T., Kothari, R. (1983). J. Pharm. Sci. 72:1149.
Carstensen, J. T., VanScoik, K. (1990). Pharm. Res. 7:1278.
Carstensen, J. T., Danjo, K., Yoshioka, S., Uchiyama, M. (1987). J. Pharm. Sci. 76:548.
Carstensen, J. T., Morris, T., Puddepeddi, M., Franchini, M (1993). Drug Dev. Ind. Pharm. In press.
Dekeyser, P. M., Corveleyn, S., Demeester, J., Remon, J.-P. (1997). Int. J. Pharm. 159:19.
Franchini, M., Carstensen, J. T. (1994). Drug Dev. Ind. Pharm. In press.
Franks, F. (1990). Cryo-Letters 11:93.
Gore, D. N., Ashwin, J. (1967) 27th International Congress of Pharmaceutical Sciences, Montpellier, France, Sept. 4–9, 1967.
Grandolfi, G. (1986). M. S. thesis, University of Wisconsin, School of Pharmacy.
Hancock, B. C., Dalton, C. R. (1999). Pharm. Dev. Tech. 4:125.
Hatley, R. H. M. (1997). Pharm. Dev. Tech. 2:257.
Her, L. M., Nail, S. L. (1994). Pharm. Res. 11:54.
Jakobsen, D. F., Frokjaer, S., Larsen, C., Niemann, H., Burr, A. (1997). Int. Pharm. 156:67.
Levine, H., Slade, L. (1988). Cryo-Letters 9:21.
MacKenzie, A. P. (1977). Dev. Biol Stand. 36:51.
Martînez-Oharriz, C., Martin, C., Goni, M. M., Rodrîguez-Espinosa, C., Troz de Olarduya-Apaolaza, M. C., Sanchez, M. (1993). J. Pharm. Sci. 81:83.
Mullin, J. W. (1961). Crystallization. London: Butterworths, p. 106.
Pfeiffer, R. R., Yang, K. S., Tucker, M. A. (1970). J. Pharm. Sci. 59:1809.
Pinderre, P., Cauture, E., Piccerelle, P., Kalantzis, G., Kaloustian, J., Joachim, J. (1997). Drug Dev. Ind. Pharm. 23:817.
Poole, J. W., Bahal, C. K. (1970). J. Pharm. Sci. 59:1265.
Shefter, E., Higuchi, T. (1963). J. Pharm. Sci. 52:781.
Shefter, E., Kmack, G. (1967). J. Pharm. Sci. 56:1028.
VanCampen, L., Zografi, G., Carstensen, J. T. (1980). Int. J. Pharm. 5:1.
Suzuki, T., Franks, F. (1993). J Chem Soc. Faraday Trans. 89:3283.
Wu, L.-S., Pang, J., Hussain, M. A. (1996). Pharm. Dev. Technl. 1:43.
Zografi, G., Hancock, P. (1993), Int. J. Pharm. 10:1263.
Zografi, G., Kontny, M. (1986), Pharm. Res. 3:187.

9

Preformulation

JENS T. CARSTENSEN

Madison, Wisconsin

Historically, preformulation evolved in the late 1950s and early 1960s as a result of a shift in emphasis in industrial pharmaceutical product development. Up until the mid-1950s, the general emphasis in product development was to development elegant dosage forms, and organoleptic considerations far outweighed such (as yet unheard of) considerations as whether a dye used in the preparation might interfere with stability or with bioavailability.

In fact, pharmacokinetics and biopharmaceutics were in their infancy, and although stability was a serious consideration, most analytical methodology was such that even gross decomposition often went undetected.

It was, in fact, improvement in analytical methods that spurred the first programs that might bear the name "preformulation." Stability-indicating methods would reveal instabilities not previously known, and reformulation of a product would be necessary. When faced with the problem of attempting to sort out the component of incompatibility in a 10-component product, one might use many labor hours. In developing new products, therefore, it would be logical to check, ahead of time, which incompatibilities the drug exhibited (testing it against common excipients). This way the disaster could be prevented in advance.

A further cause for the birth of preformulation was the synthetic organic programs started in many companies in the 1950s and 1960s. Pharmacological screens would show compounds to be promising, and pharmacists were faced with the task of rapid formulation. Hence they needed a fast screen (i.e., a preformulation program) to enable them to formulate intelligently. The latter adverb implies that some of the physical chemistry had to be known, and this necessitated determination of physicochemical properties, a fact that is also part of preformulation.

1. PREFORMULATION'S PLACE IN THE STABILITY FUNCTION

The approach of preformulation was so logical, indeed, that it eventually became part of the official requirements for INDs and NDAs (Schultz, 1984):

> New drug substances in Phase I submission. For the drug substance, the requirement includes a description of its physical, chemical or biological characteristics. We in the reviewing divisions regard stability as one of those characteristics. The requirement of NDA submissions ... of the rewrite stability information is required for both the drug substance and drug product. A good time to start to accumulate information about the appropriate methodology and storage stations for use in dosage form stations for use in dosage form stability studies, therefore, is with the unformulated drug substance.... Stress storage conditions of light, heat and humidity are usually used for these early studies, so that the labile structures in the molecule can be quickly

identified. . . . If degradation occurs, the chemical reaction kinetics of the degradation should be determined. . . . Physical changes such as changes from one polymorph to another polymorph should be examined. . . . With the drug substance stability profile thus completed, the information should be submitted in the IND submission.

2. TIMING AND GOALS OF PREFORMULATION

The goals of the program are therefore (1) to establish the necessary physicochemical parameters of a new drug substance, (2) to determine its kinetic rate profile, (3) to establish its physical characteristics, and (4) to establish its compatibility with common excipients.

To view these in their correct perspective, it is worthwhile to consider when, in an overall industrial program, preformulation takes place. The following events take place between the birth of a new drug substance and its eventual marketing (it is a fact, however, that most investigational drug substances never make it to the marketplace for one reason or another):

1. The drug is synthesized and tested in a pharmacological screen.
2. The drug is found sufficiently interesting to warrant further study.
3. Sufficient quantity is synthesized to (a) perform initial toxicity studies, (b) do initial analytical work, and (c) do initial preformulation.
4. Once past initial toxicity, phase I (clinical pharmacology) begins and there is a need for actual formulations (although the dose level may not yet be determined).
5. Phase II and III clinical testing then follows, and during this phase (preferably phase II) an order of magnitude formula is finalized.
6. After completion of the above, an NDA is submitted.
7. After approval of the NDA, production can start (product launch).

3. PHYSICOCHEMICAL PARAMETERS

Physicochemical studies are usually associated with great precision and accuracy, and in the case of a new drug substance would include studies of (a) $pK_{(a)}$ (if the drug substance is an acid or base), (b) solubility, (c) melting point and polymorphism, (d) vapor pressure (enthalpy of vaporization), (e) surface characteristics (surface area, particle shape, pore volume), and (f) hygroscopicity. Unlike in the usual physicochemical studies, an abundance of material is usually not at hand for the first preformulation studies: in fact, at the time this function starts, precious little material is supplied, and therefore the formulator will often settle for good estimates rather than attempt to generate results with four significant figures.

There is another good reason not to aim too high in the physicochemical studies of the first sample of drug substance. In most cases the synthesis is only a first scheme, and in later scale-up it will be refined; and in general the first small samples contain some small amount of impurities, which may influence the precision of the determined constants. But it is necessary to know, *grosso modo*, important properties such as solubility, pK, and stability. These are dealt with in order below.

3.1 pK_a and Ionizable Substances

The definition of $pK_{(a)}$ and pk_a have been discussed in chapter 2. For substances that are carboxylic acids (HA) it is advantageous to determine the pK_a, since this property is of importance in a series of considerations. For carboxylic acid the species A^- usually absorbs in the ultraviolet (UV) region, and its concentration can be determined spectrophotometrically (Underberg and Lingeman, 1983); HA on the other hand will absorb at a different wavelength.

The molar absorbances of the two species at a given wavelength are denoted ε_0 and ε_- (it is assumed that at the wavelength chosen $\varepsilon_0 < \varepsilon_-$), and it can be shown that if the solution is m_0 Molar in total A, then

$$\frac{A^-}{HA} = \frac{\varepsilon - \varepsilon_0 m_0}{m_0 \varepsilon_- - \varepsilon} \tag{9.1}$$

so that the ratio A^-/HA can be determined in a series of buffers of different pH. Hence the $pK_{(a)}$ can be found as the intercept by plotting pH as a function of $\log[(A^-)/(HA)]$ by Henderson–Hasselbach:

$$pH = pK_{(a)} + \log\left[\frac{A^-}{HA}\right] \tag{9.2}$$

If several buffer concentrations are used, extrapolation can be carried out to zero ionic strength, and the pK_a can be determined. For initial studies, however, a $pK_{(a)}$ in the correct range (i.e., $+0.2$ unit) will suffice, so that the determination above can be done at one buffer concentration only.

The conventional approach is to do titrations (Fig. 1), and this will yield graphs of fraction (neutralized (x) as a function of pH. Usually, the water is titrated as well (Parke and Davis, 1954), and what is presented in Fig. 1 is the "different." The $pK_{(a)}$ is then the pH at half neutralization (which is also the inflection point).

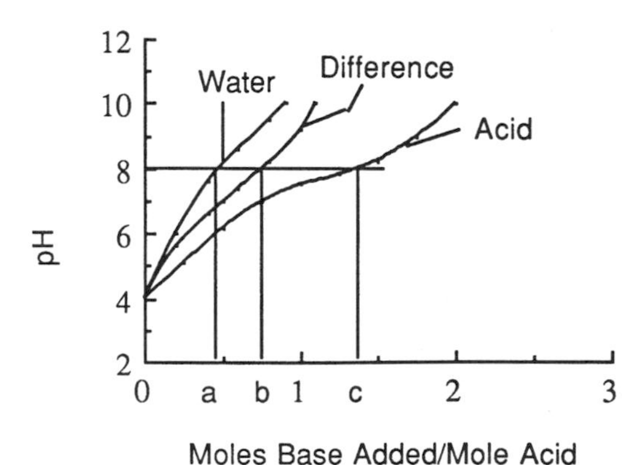

Fig. 1 Typical titration curves. The "water" curve indicates the amount of alkali needed to "titrate" the water, and the "acid" curve is a conventional titration curve. The difference curve is the horizontal difference between the "acid" and the "water" curve and is the adjusted titration curve. For example the point "b" is "c" minus "a". The $pK_{(a)}$ is the point of inflection, which is also the point where half of the acid is neutralized.

The pH solubility curve can now be constructed simply by determining the solubility of HA (at low pH) and A^- (e.g., of NaA) at high pH (e.g., at pH 10).

4. SOLUBILITY

One important goal of the preformulation effort is to devise a method for making solutions of the drug. Frequently, the drug is not sufficiently soluble in water itself to allow for the desired concentrations, for example for injection solutions. Solubilities are determined by exposing an excess of solid to the liquid in question, and assaying after equilibrium has been established. This usually is in the range 60 to 72 h, and to establish that equilibrium indeed has been established, sampling at earlier points is necessary. Unstable solutions pose a problem in this respect and will be dealt with in more detail later. Solubilities cannot be determined by precipitative methods (e.g., by solubilizing an acid in alkali and then lowering the pH to the desired pH) because of the so-called metastable (solubility) zone (Rodriguez-Hornedo, 1984). In the writing to follow, drug substances are subdivided into two categories: (1) ionizable substances, and (2) (virtually) nonionizable substances.

Solubility determinations are necessary both for stability reasons and for formulation reasons. It was noted, in the chapter dealing with stability of solids in the presence of water, that the solubility term becomes part of a rate constant. Since preformulation occurs in the early stage of development, the optimization of stability by way of compound selection (correct salt) is of importance, and often a drug product can be stabilized by keeping the solubility of the drug substance low. In the limit this, however, might affect bioavailability.

Probably among the most well-known examples of such stabilization are that of procaine penicillin and that of potassium clavulanate. In the latter case, the sodium salt, for instance, is unstable to such an extent that it cannot be utilized. The decrease in solubility of the potassium salt renders the product machinable (although low humidities must be observed in manufacturing).

4.1 Use of Salt Formation to Increase Solubility

It is noted that at a given pH the amount in solution in a solubility experiment is

$$S = S_{HA} + C_{A^-} \tag{9.3}$$

where S denotes solubility. The last term can be determined from knowledge of the pK_a, the pH, and the use of Eq. (2).

For drugs that are amines, the free base is frequently poorly soluble, and in this case the $pK_{(a)}$ is often estimated by performing the titration is a solvent containing some organic solvent (e.g., ethanol). By doing this at different organic solvent concentrations (e.g., 5%, 10%, 15%, 20%), extrapolation can be carried out to 0% solvent concentration to estimate the aqueous $pK_{(a)}$.

Usually, alkali metal salts of acids are more soluble than the free acids, and in the case of basic (e.g., amine type) drugs, the solubility of the acid addition salts are more soluble than the free bases. At times (e.g., in the case of enalipril) the compound is amphoteric. The acid addition salt is soluble, the free base is less soluble, and the sodium salt is, again, more soluble. In simple cases, the solubility curve

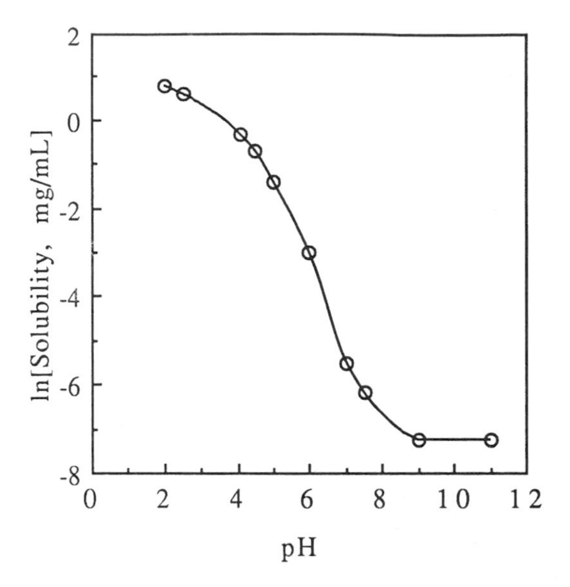

Fig. 2 pH solubility plot of imiquimod in water. (Graph constructed from data published by Chollet et al., 1999.)

will simply mimic the Henderson–Hasselbach equation. An example of this is imiquimod (Chollet et al., 1999), the pH solubility profile of which is shown in Fig. 2.

Streng and Yu (1998) have published a computer program for prediction of stability curves based on the pK_a value of the compound in question.

4.2. Nonionizable Substances

For hydrophobic, (virtually) nonionizable substances (i.e., those that show no ionic species of significance in the pH range 1 to 10, e.g., diazepam), solubility can usually be improved by addition of nonpolar solvents. Aside from solubility, stability is also affected by solvents either in a favorable or in an unfavorable direction (Bakar and Niazi, 1983). Theoretical equations for solubility in water (Yalkowsky and Valvani, 1983) and in binary solvents (Acree and Rytting, 1983) have been reported in the literature, but in general the approach in preformulation is pseudoempirical. Most often the solubility changes as the concentration of nonpolar solvent, C2, increases. For binary system, it may simply be a monotonical increasing function (Carstensen et al., 1971), as shown in Fig. 3.

The solubility is usually tied to the dielectric constant, and in a case such as that shown in curve A, the solubility is often log-linear when plotted as a function of inverse dielectric constant, ε, that is,

$$\ln S = -\frac{e_1}{\varepsilon + e_2} \tag{9.4}$$

where ε is the dielectric constant and the e terms are constants (Underberg and Lingeman,. 1983).

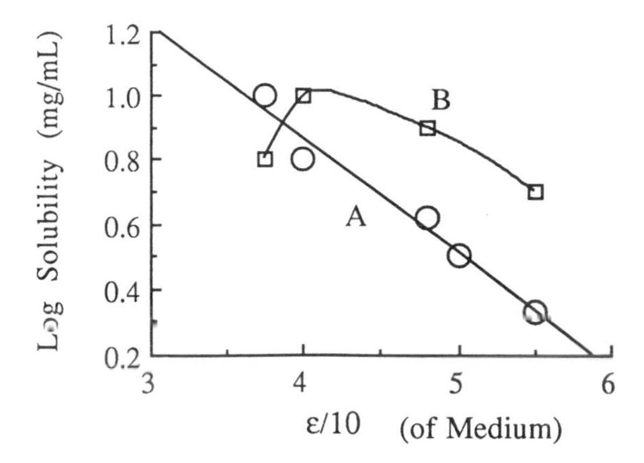

Fig. 3 (A) Solubility of 7-chloro-1,3-hydro-5-phenyl-2H-1,4-benzodiazepine-2-one-4-oxide in aqueous propylene glycol. (Data from Carstensen et al., 1971.) (B) Solubility of another benzodiazepine. (Unpublished data.)

Frequently, however, the solubility curve has a maximum (as shown in curve B in Fig. 3) when plotted as a function of C2 and ε (Paruta and Irani, 1964). In either case it is possible to optimize solubility by the selection of a solvent system with a given value of ε; that is, once the curve has been established, the optimum water/solvent ratio for another solvent can be calculated from known dielectric constant relationships (Cavé et al., 1979).

4.3. Ternary Systems and Optimization

Frequently, *ternary* solvent systems are resorted to. Examples are water–propylene glycol–benzyl alcohol or water–propylene glycol–ethanol. In such cases the solubility profile is usually presentable by a ternary diagram (Sorby et al., 1963). This type of diagram usually demands a fair amount of work; that is, the solubility of the drug substance in many solvent compositions must be determined. A priori, it would therefore seem that they would be out of place in a situation where only limited quantities of drug are available. However, their principle gives some validity to optimization procedures.

The diagram can be of one of two types, as shown in Figs. 4 and 5. In the first type, the solubility may be assumed to be of the type

$$S = a_{10} + a_{11}C_1 + a_{12}C_2 \tag{9.5}$$

where C denotes concentrations of nonaqueous solvents. An example of this is shown in Fig. 4. Here the subscripts to C denote the two nonaqueous solvents. Hence three solubility experiments would determine the relationship (with zero degrees of freedom). It is usual to do at least five, and determine possible curvature [i.e., inclusion of more terms in Eq. (9.6)].

In the second case, Fig. 5 each tie line will give a parabolic type curve as shown. Hence at a given concentration of C_2 the solubility can be approximated by

$$S = b_{10} + b_{11}C_1 + b_{12}C_1^2 \tag{9.6}$$

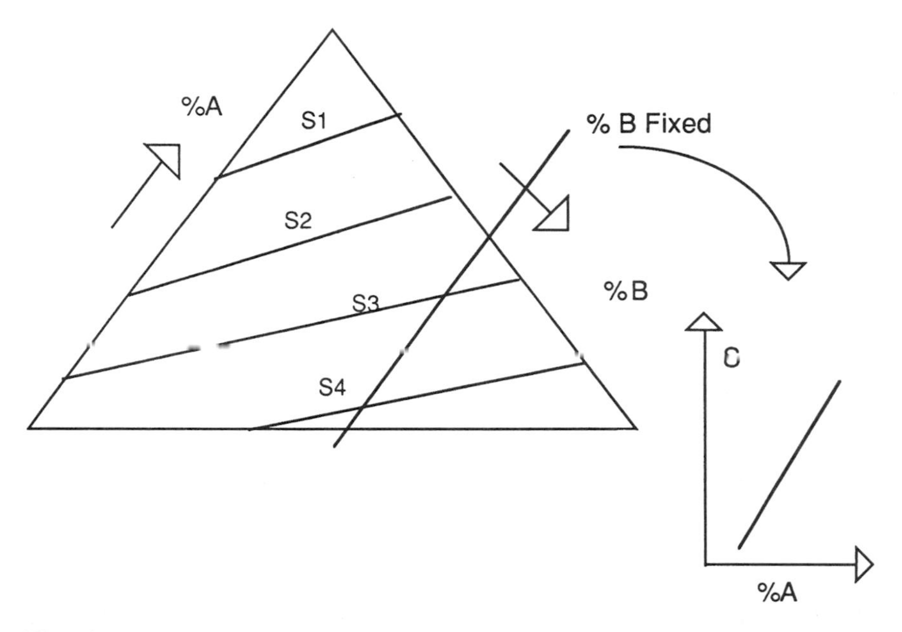

Fig. 4 Ternary diagram of solubility of a compound in a ternary mixture with linear solubility response. Inset: Concentration of drug in compositions with constant concentration of B. The composition of the solute is the constant concentration of B, the concentration of A in the abscissa, and the complement concentration of C (the third apex, not indicated in the figure). The drug solubility response is linear in the A concentration in this case.

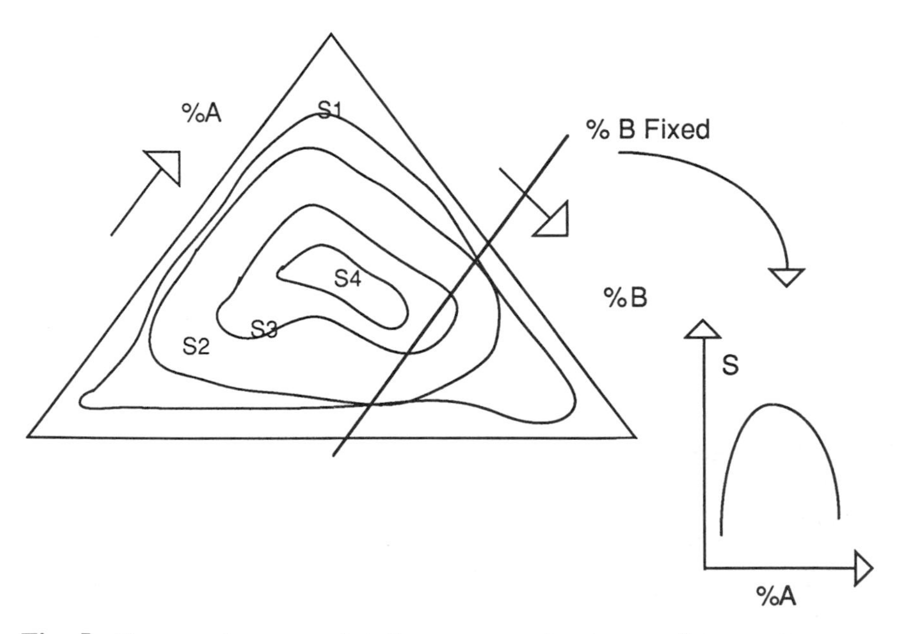

Fig. 5 Ternary diagram and tie line concentration in a nonlinear system.

where, in the simplest case,

$$b_{10} = C_{20} + C_{21}C_2 \tag{9.7}$$

Hence optimization can be achieved by five (or more experiments, with zero (or in general n-5) degrees of freedom.

4.4. Prediction of Solubility

It is advantageous, with a new drug substance, to be able to estimate what its solubility might be, prior to carrying out dissolution experiments. There are several systems of solubility prediction, e.g., the work by Amidon et al. (1974) and Yalkowsky et al. (1972, 1975). Their equation, for solubility of p-aminobenzoates in polar and mixed solvents, is a simplified two-dimensional analog of the Scatchard–Hildebrand equation and is based on the product of the interfacial tension and the molecular surface area of the hydrocarbon portion of a molecule.

More recently Bodor et al. (1989) have developed a semiempirical solubility predictor based on 14 variables (S = molecular surface in Å^2, I_a = indicator variable for alkanes, D = calculated dipole moment in Debyes, Q_n = square root of sum of squared charges on oxygen atoms, Q_o = square root of sum of squared charges on oxygen atoms, V = molecular volume in Å^2, S_2 = square of molecular surface, C = constant, MW = molecular weight, {O} = ovality of molecule, A_{bh} = sum of absolute values of atomic charges on hydrogen atoms, A_{bc} = sum of absolute values of atomic charges on carbon atoms, A_m = indicator variable for aliphatic amines, and N_h = number of N-H single bonds in the molecule).

The aqueous solubilities, W, of 331 compounds were found to follow the equation (with tolerances omitted)

$$\begin{aligned}\log W = &- 56.039 + 0.32235D - 0.59143I_a + 38.443Q_n^4 \\ &- 51.536Q_n^2 + 18.244Q_n + 34.569Q_o^4 - 31.835Q_o^2 + 15.061Q_o \\ &+ 1.9882A_m + 0.15689N_h + 0.00014102S^2 + 0.40308S - 0.59335A_{bc} \\ &+ -0.42352V + 1.3168A_{bh} + 108.80\{O\} - 61.272\{O\}^2\end{aligned} \tag{9.8}$$

Of the parameters listed only the ovality and the indicator value for the alkanes I_a are unfamiliar entities that are obtained from the literature (Bodor et al., 1989).

5. DISSOLUTION

The importance of dissolution is such (from biopharmaceutical considerations) that it is now used throughout the USP and is required in NDAs on solid dosage forms. According to Noyes and Whitney (1897),

$$\frac{dm}{dt} = \frac{VdC}{dt} = -kA(S - C) \tag{9.9}$$

where m is mass not dissolved, V is liquid volume, t is time, k is the so-called intrinsic dissolution rate constant (cm/s), and A is surface area of the dissolving solid.

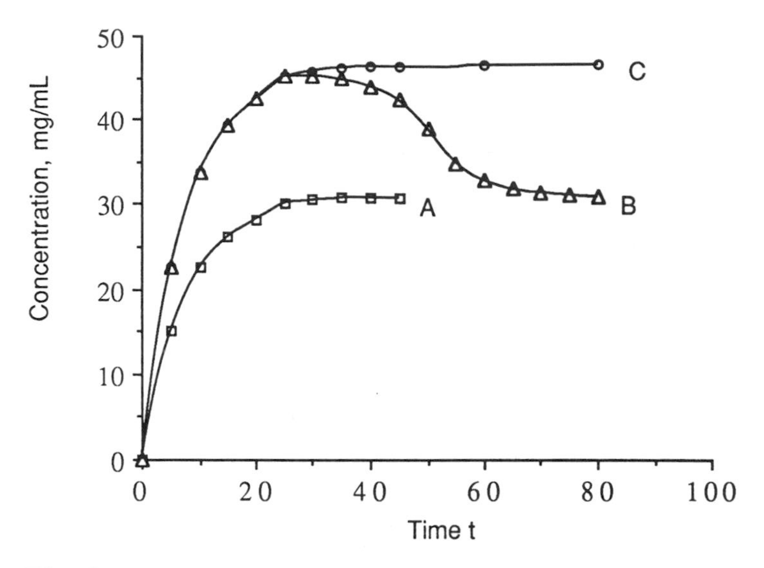

Fig. 6 Dissolution profiles obtained from the solubility determination of two polymorphic forms of the same drug substance. A is the stable form with solubility 31 mg/mL. B is the profile of the metastable form with solubility 46 mg/mL. This solubility, C, (circles) is not achieved in many instances, and precipitation of the stable form occurs at a point beyond the solubility of A, and the trace becomes B.

Many criticisms have been voiced against Eq. (9.9), but in general it is correct, and it will be assumed to be so in the following. Experimentation can be carried out with constant surface as when using a Wood's apparatus (Wood et al., 1963) or, with smaller amounts, making a small pellet and encasing it in wax and exposing only one face to a dissolution medium, or simply employing an excess of solid throughout the dissolution experiment. In such cases Eq. (9.9) may be integrated to give

$$\ln\left[1 - \frac{C}{S}\right] = -\frac{kA}{V}t \tag{9.10}$$

or

$$C = S\left[1 - \exp\left(-\frac{kA}{V}t\right)\right] \tag{9.11}$$

A typical curve following Eq. (9.11) is shown in Fig. 6.

In the critical time path for product development, solid dosage forms (tablets or capsules) must eventually be manufactured for the clinic (e.g., in clinical phase II). If possible, the drug substance per se is subjected to a dissolution test in a Wood's apparatus (Wood et al., 1963). This test is useful although it is quite dependent on hydrodynamic conditions. It consists of placing the powder in a special type of tablet die, compressing the tablet, and exposing the flat, exposed side of the tablet (with surface area A) to a dissolution liquid (usually water or N/10 HCl) in which it has a solubility *S*. Under these conditions (Carstensen, 1974), the intrinsic dissolution rate constant (cm/s) can be obtained by Eq. (9.10), which under sink conditions (i.e., when C is less than 15% of S) becomes

$$C = \frac{SkA}{V} t \tag{9.12}$$

It has been suggested (Riegelman, 1979) that if k is obtained under sink conditions over a pH range of 1 to 8 at 37°C in a USP vessel by way of Eq. (9.12) at 50 rpm, then if the dissolution rate constant (kA/V) is greater than 1 mg min^{-1} cm^{-2}, the drug is not prone to give dissolution-rate-limited absorption problems. On the other hand, if the value is less than 0.1, such problems can definitely be anticipated, and compounds with values of kA/V of from 0.1 to 1 mg min^{-1} cm^{-2} are in a gray area. For compound selectivity it is frequently useful to express dissolution findings in terms of k (i.e., in cm/s).

For a small amount of powder, dissolution of the particulate material can often be assessed (and compared with that of other compounds) by placing the powder in a calorimeter (Iba et al., 1991) and measuring the heat evolved as a function of time. The surface area must be assessed microscopically (or by image analyzer), and the data must be plotted by a cube root equation (Hixson and Crowell, 1931):

$$1 - \left[\frac{M}{M_o}\right]^{1/3} = -\frac{2kS}{\rho r} t \tag{9.13}$$

where M is mass not dissolved, M_0 the initial amount subjected to dissolution, ρ true density, S solubility, and r the mean "radius" of the particle. The method is simply comparative, not absolute, because the hydrodynamics are different in the calorimeter from what it would be in a dissolution apparatus.

It is obvious that the dissolution rate is a function of the exposed surface area, but how this changes during dissolution is not quite obvious. Sunada et al. (1989) measured the change in surface area during dissolution of *n*-propyl-*p*-hydroxybenzoate and found dissolution rates proportional to surface area.

5.1. Solubility of Unstable Compounds

Quite often a compound is rather unstable in aqueous solution. Hence the long exposure to liquid required for traditional solubility measurements will cause decomposition, and the resulting solubility results will be unreliable. In this particular case Nogami's method may be used. If a solution experiment is carried out as a dissolution experiment with samples taken at equal time intervals, δ, it can be shown (Nogami et al., 1966) that when the amount dissolved at time $t + \delta$ is plotted versus the amount dissolved at time t, a straight line will ensure. The following relationship holds:

$$C(t + \delta) = S[1 - \exp(-k\delta)] + \exp(-k\delta)Ct \tag{9.14}$$

hence such a plot as shown in Fig. 7 will give k from the slope; inserting this in the intercept expression will give S. The advantage of the method is that it can be carried out in a short period of time, and reduce the effect of decomposition; the disadvantage is that it is not as precise as ordinary solubility determinations.

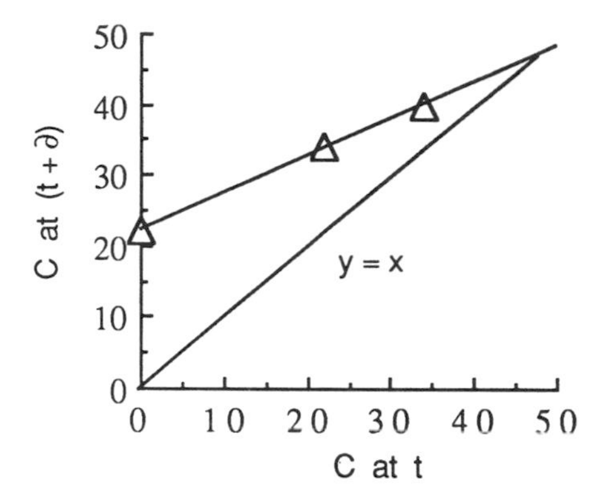

Fig. 7 The Nogami method applied to the upper curve in Fig. 5. The least-squares fit for the upper line is $y = 27.081 + 0.53139x$, so that $y = x$ when $x = y = 22.081/(1\text{-}0.53139) = 47.3$.

5.2. Solubility of Metastable Polymorphs

Polymorphism is an important aspect of the physical properties of drug substances. One of the characteristics of a metastable polymorph is that it is more soluble than its stable counterpart. The dissolution profile of it will be as the upper curve shown in Fig. 6; A is the stable form with solubility 31 mg/mL. B is the profile of the metastable form with solubility* 46 mg/mL. This solubility (circles) is often not achieved, and precipitation of the stable form occurs at a point beyond the solubility of A, and the trace becomes B.

In such cases, the Nogami method can be applied to the early points curve (Fig. 7), and the solubility, S', of the polymorph can be assessed. One of the important aspects of metastable polymorphs in pharmacy is exactly their higher solubility, since the dissolution rate will also be higher [Eq. (9.9)]. Hence the bioavailability will be increased where this is dissolution rate limited (Shibata et al., 1983).

5.3 Polymorphism

As mentioned at an earlier point, solids may exist either as amorphous compounds or as crystalline compounds. In the latter, the molecules are positioned in lattice sites. A lattice is a three-dimensional array, and there are eight systems known. Compounds often have the capability of existing in more than one crystal form, and this phenomenon is referred to as polymorphism.

If a compound exhibits polymorphism, one of the forms will be more stable (physically) than the other forms; that is, of n existing forms, n-1 forms will possess a thermodynamic tendency to convert to the nth, stable form (which then has the lowest Gibbs energy; it should be noted that in the preformulations stage it is not known whether the form on hand is the stable polymorph or not).

---

* It is noted that the "solubility" is not the equilibrium solubility. The solution is a supersaturated solution, but it is referred to as solubility, because conducting a solubility experiment on a metastable most often will give a reproducible figure. The supersaturated solution will eventually precipitate out as demonstrated in Fig. 5 of Chapter 12.

One manner in which different polymorphs are created is by way of recrystallizing them from different solvents, and at a point in time when sufficient quantities of material (and this need not be very much) are available, the preformulation scientist should undertake recrystallization from a series of solvents.

Knowledge of polymorphic forms is of importance in preformulation because suspension systems should never be made with a metastable form (i.e., a form other than the stable crystal form). Conversely, a metastable form is more stable than a stable modification, and this can be of advantage in dissolution [Eq. (9.11)].

6. VAPOR PRESSURE

In general, vapor pressures are not all that important in preformulation, but it should always be kept in mind that a substance may have sufficiently low vapor pressure to (a) become a lost to sufficient extent to cause apparent stability problems and content uniformity problems, and (b) exhibit a potential for interaction with other compounds and adsorption onto or sorption into package components (Pikal and Lukes, 1976).

Most drug substances are, substantially, not volatile. As an initial screen, it can be determined whether the drug is sufficiently volatile to cause concern, by placing a weighed amount of it in a vacuum desiccator and weighing it daily for a while. It is better to have a high-vacuum system for this, and the use of a vacuum electrobalance is best for this purpose. A good estimate of the vapor pressure can be obtained (Carstensen and Kothari, 1981) by using a pierced thermal analysis cell, placing it on a vacuum electrobalance, and monitoring the weight loss rate. A substance with known vapor pressure can then be used for calibration, the loss rates being proportional to the vapor pressures.

By using constant temperature TGA, graphs such as that shown in Fig. 8 will result. The weight rate (which should be established as due to evaporation of the compound) is given by

$$\frac{dW_a}{dt} = -kA'P_a \tag{9.15}$$

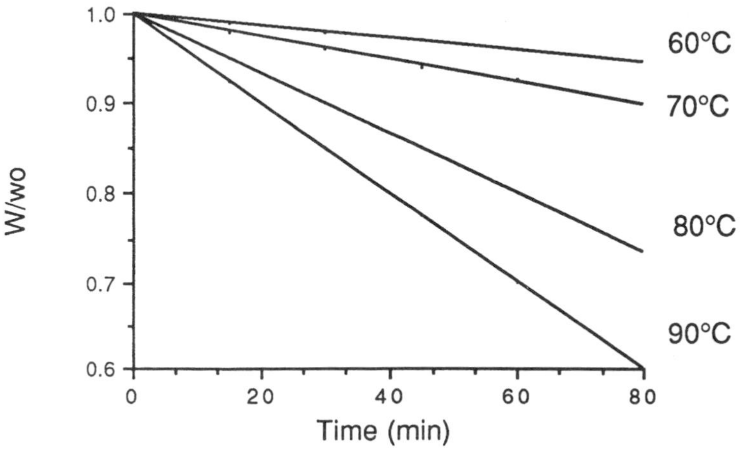

1.0
0.9
0.8
0.7
0.6
W/wo
0
20
40
60
80
Time (min)
60°C
70°C
80°C
90°C

Fig. 8 Weight loss curves from constant temperature TGA.

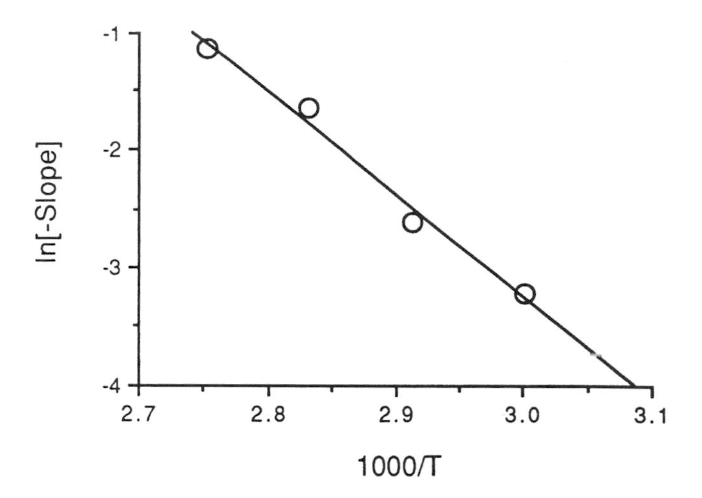

Fig. 9 Plot showing the Clausius–Clayperon treatment of the data in Fig. 7. The least-squares fit is 1n[-slope] = 22.861-8.7(1000/T.)

where W is weight, t is time, k is a heat/mass transfer coefficient, A' is surface area, and P_a is the vapor pressure of the compound at the given temperature.

If the specific surface area is

$$A_a = \frac{A'}{W_{ao}} \tag{9.16}$$

where W_{ao} is the original weight of the sample, we inserted this into Eq. (9.5) and obtain

$$\frac{W_a}{W_{ao}} = 1 - kA_aP_at \tag{9.17}$$

An example of such a plot is shown in Fig. 8.

If a compound with known vapor pressure (e.g., benzoic acid, subscript b) is subjected to the same conditions, then it will exhibit a weight loss curve given by a similar equation:

$$\frac{W_b}{W_{bo}} = 1 - kA_bP_bt \tag{9.18}$$

The ratio of the slopes is A_bP_b/A_aP_a, so if the specific surface area is known for each, then (since P_b of the reference is known), P_a can be calculated.

The heat of sublimation, ΔH, can be obtained by plotting the negative of the slopes by a Clausius–Clayperon equation, as shown in Fig. 9. In the cited case, the negative of the slope of this plot is $\Delta H/R = 8.7$ kCal/mol, so that the heat of sublimation is $\Delta H = 17.4$ kCal/mol. For this, no reference is necessary; it is only necessary to know and adjust for the weights of the samples studied.

7. PARTITION COEFFICIENT

Partition coefficients between water and an alkanol (e.g., octanol) should be determined in preformulation programs (Yalkowsky et al., 1983). The partition coefficient of a compound that exists as a monomer in two solvents is given by

$$K = \frac{C_1}{C_2} \quad (9.19)$$

If its exists as an n-mer in one of the phases, the equation becomes

$$K - \frac{(C_1)^n}{C_2} \quad (9.20)$$

or

$$\log k = n \log C_1 - \log C_2 \quad (9.21)$$

The easiest way to determine the partition coefficient is to extract V_1 cm^3 of saturated aqueous solution with V_2 cm^3 of solvent and determine the concentration C_2 in the latter. The amount left in the aqueous phase is $C_1 V_1 - C_2 V_2 = \mathrm{M}$, so that the partition coefficient is given by

$$K = \frac{M}{V_1 C_1} \quad (9.22)$$

If it is assumed that the species is monomeric in both phases, the partition coefficient becomes the ratio of the solubilities, and it is simply sufficient to determine the solubility of the drug substance in the solvent (since it is assumed that the solubility is already known in water):

$$K = \frac{S_1}{S_2} \quad (9.23)$$

8. HYGROSCOPICITY

Hygroscopicity is, of course, an important characteristic of a powder. It can be shown for a fairly soluble compound that the hygroscopicity is related to its solubility (Carstensen, 1977, VanCampen et al., 1980), although it has been shown that the heat of solution plays an important part in what is conceived as "hygroscopicity" (VanCampen et al., 1983a,b,c). As mentioned in Chapter 8, a hygroscopicity experiment is carried out most easily by exposing the drug substance to an atmosphere of a known relative humidity (e.g., storing it over saturated salt solutions in desiccators). Each solution will give a certain relative humidity (RH), and the test simply consists of weighing the powder from time to time and determining the amount of moisture adsorbed (weight gained). This does not work with drug substances that decompose as, for instance, effervescent mixtures, which start losing weight due to carbon dioxide evolution (Carstensen and Usui, 1984).

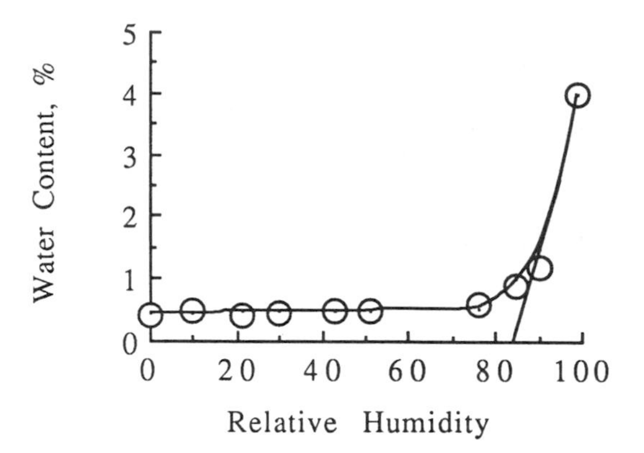

Fig. 10 Sorption isotherm of anhydrous droloxifene citrate. (Graph constructed from data published by Burger and Lettenbichler, 1993.)

It can be shown that if the air space is sufficiently agitated to prevent vapor pressure gradients, the initial uptake rate (g H_2O/g solid per hour) is related to the relative humidity by

$$L = a_{21}[\text{RH} - \text{RH}_0] \tag{9.24}$$

where RH_0 is the vapor pressure of a saturated solution of the drug substance in water. An example of this is shown in Fig. 10.

X_s can be estimated by an ideality assumption; that is, if the solubility is expressed as a mole fraction X_s, the vapor pressure over a saturated solution will be P' given by

$$P' = (1 - X_s)P_0 \tag{9.25}$$

where P_0 is water's vapor pressure at that temperature.

The experiments above are rather easy to carry out and should always be part of a preformulation program, since hygroscopicity can be so important that it will dictate whether a particular salt should be used. Dalmane, for instance, is a monosulfate, and is used as such since the disulfate, desirable in many other respects, is so hygroscopic that it will remove water from a hard-shell capsule and make it exceedingly brittle.

9. COMPATIBILITY TESTS

Prior to attempting the first formulation with a new drug, most research groups carry out compatibility testing (Carstensen et al., 1964). The principle is to make up reasonably ratioed mixtures of drug and excipient, to ascertain which excipients may be reasonably used with the drug. The original method used in the 1960s (Carstensen, 1964) consisted of visual observation of such mixtures, spectrophotometric assay, and TLC. The methods used nowadays have followed in step with analytical developments and are (a) chemical assay, (b) TLC, (c) HPLC, (d) DSC, and (e) microcalorimetric methods. The latter two have been of special interest in recent years and will be treated separately.

9.1. Use of DSC

Rustichelli et al. (1999) have employed DSC to obtain the phase equilibrium diagrams of the enantiomers of (a) verapamil HCl and (b) gallopamil HCl. In the former case the eutectic composition is at 90% (2S)-(-)-verapamil HCl and in the latter at 70% (2S)-(-)-gallopamil HCl.

Mura et al (1998) have used thermal analysis (DSC) to study compatibility of picotamid with common pharmaceutical excipients (palmitic acid, stearic acid, stearyl alcohol, PEG 20,000, and sorbitol) and showed that the interactions were primarily due to dissolution in the melted excipient.

9.2. Use of Microcalorimetry

Heat conduction microcalorimetry has been used as a method to evaluate stability and excipient stability by a series of researchers. Angerg et al. (1988, 1990, 1993), Hansen et al., (1989), and Wilson et al. (1995) have described the general method and results interpretation. For instance, Angberg et al. studied the oxidation of ascorbic acid in aqueous solution by microacalorimetry, and other researchers have used this method as well. Oliyai and Lindenbaum (1991) studied the decomposition of ampicillin in solution. Tan and Meltzer (1992) studied the solid state stability of 13-*cis*-retinoic acid by means of microcalorimetry and HPLC, and Pikal and Dellerman (1989) studied the kinetics of cephalosporin in the solid and solution states using the same method.

Seltzer et al. (1998) used the method to evaluate stability and excipient compatibility of (S)-(3-(2-4-(S)-(4-(amino-imino-methyl)-phenyl-4-methyl-2,5-dioxo-imidazolidin--1-y1)-acetylamino))-3-phenyl-propionic acid ethyl ester, acetate. The excipients used were potato starch, calcium hydrogen phosphate anhydrous, and colloidal silica.

They consider the reaction $A + B \rightarrow C + D$ and denote the concentration of C as x, the fraction decomposed at time t. The trace of heat evolved as a function of time is then characterized by

$$\frac{dx}{dt} = -k\{[A_0] - x\}^n \tag{9.26}$$

The amount decomposed is associated with (a usually exothermic) reaction enthalpy, ΔH, where the heat evolved is proportional to x, hence the heat flow is proportional to dx/dt and the proportionality constant is ΔH, so that

$$\phi = \frac{dQ}{dt} = \Delta H \frac{dx}{dt} \tag{9.27}$$

hence

$$\phi = \Delta H k \left\{ [\mathrm{A}_0] - \left(\frac{Q}{\Delta H} \right) \right\} \tag{9.28}$$

For a first-order reaction this becomes

$$\phi = \Delta H k [\mathrm{A}_0] - kQ \tag{9.29}$$

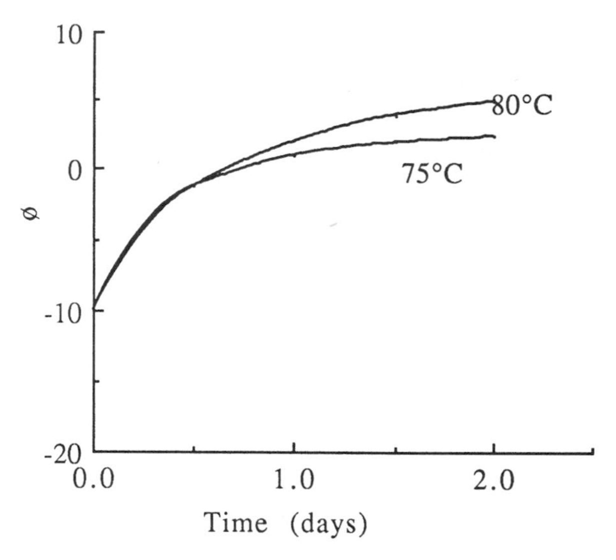

Fig. 11 Figure constructed from data published by Seltzer et al. (1998).

so that the terms k and ΔH may be deduced from nonlinear regression of a plot of heat flow versus time (Fig. 11).

If the reaction is zero order, then ϕ is simply a constant:

$$\phi = k\Delta H \tag{9.30}$$

This allows determination of kinetic profiles and Arrhenius plots of the studied reaction.

The value of Q at time t is obtained through integration (area under) of the curve from zero to t.

It should again be emphasized that at the onset of a new drug program, there are only small amounts of drug substance at hand. One of the first tasks for the preformulation scientist is to establish the framework within which the first clinical batches can be formulated. To this end it is important to know with which common excipients the drug is compatible. In the following, the distinction will be made between solid and liquid dosage forms.

The microcalorimetric methods give no direct information about the chemical nature of the reaction.

9.3. Compatibility Test for Solid Dosage Forms

It is customary to make a small mix of drug substance with an excipient, place it in a vial, place a rubber stopper in the vial, and dip the stopper in molten carnauba wax (to render it hermetically sealed). The wax will harden and form a moisture barrier up to 70°C. A list of common excipients characteristic of this type of test is shown in Table 1. At times it is possible to obtain quantitative relationships of excipient characteristics and interaction rates (Carstensen et al., 1964; Perrier and Kesselring, 1983). In addition to the test as described, a similar set of samples are set up where 5% moisture is added. A storage period of 2 weeks at 55°C is employed [except

Table 1 Categories for Two-Component Systems

			Worse		Total score	
		Ident-ical	17–27 mo at 25°C	10 days at 55°C	25°	55°C
Drug per se	Dry	15	4	1	38	31
	5% H_2O	9	8	3	49	38
+ Magnesium stearate	Dry	16	3	1	34	30
	5% H_2O	15	4	1	43	35
+ Calcium stearate	Dry	13	4	3	37	32
	5% H_2O	12	5	3	38	35
+ Stearic acid	Dry	15	5	0	42	31
	5% H_2O	7	11	2	60	38
+ Talc	Dry	14	5	1	38	30
	5% H_2O	10	8	2	45	34
+ Acid-washed talc	Dry	12	8	0	44	31
	5% H_2O	10	9	1	49	35
+ Lactose	Dry	12	5	3	38	32
	5% H_2O	9	7	4	65	56
+ $CaHPO_4$, anhydrous	Dry	12	6	2	46	36
	5% H_2O	9	8	3	66	53
+ Cornstarch	Dry	12	5	3	39	34
	5% H_2O	10	5	5	40	37
+ Mannitol	Dry	10	7	3	39	31
	5% H_2O	8	7	5	47	45
+ Terra alba	Dry	14	6	0	41	28
	5% H_2O	11	6	3	50	45
+ Sugar 4 ×	Dry	12	6	2	41	34
	5% H_2O	9	7	4	63	61

Source: Constructed from data published by Carstensen et al. (1964).

for stearic acid, where 45°C is used, and dicalcium phosphate, where 37°C is used (Toy, 1980)], after which time the sample is observed physically for (1) caking, (2) liquefaction, (3) discoloration, and (4) odor or gas formation. It is then assayed by thin-layer chromatography (or HPLC).

It is noted that one of the samples set up is the drug by itself. This is done for several reasons, one of which is that it is now required by the FDA for IND submissions (Schultz, 1984). One more reason is that at the onset of a program, the organic synthesis of the compound may lack the refinement it will later have, and it is not uncommon that there will be several weak TLC spots (impurities) on a TLC chromatogram of a compound obtained by initial laboratory synthesis. Hence, in selecting the excipients with which the drug substance is deemed to be compatible, it is customary to use as criteria that (after accelerated exposure of a drug/excipient mix) no new spots have developed and that the intensity of the spots in the drug stored under similar conditions (2 weeks at 55°C is the same as in the acceptable excipient. This type of program is used by many companies with good success (i.e., the formula developed based on the findings from the compatibility program is stable).

It should be noted that liquefaction at times occurs because of eutectic formation (e.g., often with caffeine combinations) and that this may not necessarily be associated with decomposition. On the other hand, discoloration (e.g., amines and sugars) usually is.

Finally, the reason for not forcing dicalcium phosphate (a very valuable formulation aid in direct compression) beyond 50°C is that at higher temperatures it converts to the anhydrate, a conversion that is, curiously enough, catalyzed by water. In other words, the dihydrate will be autocatalytic in this respect at elevated temperatures, and it should not be ruled out based on high-temperature findings.

Aside from magnesium stearate, dicalcium phosphate and lactose are the excipients that are the most often found incompatible with drugs. In the former case it is usually the pH effect, in the latter it is the formation of Schiff's bases with amines (and many drugs are amines), i.e.,

$$R_1CHO + H_2NR_2 \rightarrow R_1CH = NR_2 + H_2O$$

For instance, Eyjolfsson (1998) reported on the incompatibility of lisinopril with lactose.

Aso et al. (1997) have determined the decomposition rates of cephalotin in mixtures with pharmaceutical excipients and the effect of moisture. They found a linear relation between mobile water percentage and decomposition rate constants.

Several examples of this type of screening exist. Malan et al. (1997) have studied the compatibility of tablet excipients with albendazole and closantel. They prepared drug–excipient mixtures in a mixture and in 1 : 1 mixtures that were granulated with water and dried at 50°C. DSC and HPLC were used to evaluate the compatibilities. The excipients tested were colloidal silicon dioxide, microcrystalline cellulose, dibasic calcium phosphate monohydrate, starch, sodium starch glycolate, and magnesium stearate.

9.4. Compatibility with Containers

Compatibility studies may also include compatibility with container materials. Hourcade et al. (1997), for instance, reported that granisetron in concentrations of 1 mg/mL, when kept in polypropylene syringes, were quite stable, whereas dilutions with 0.9% NaCl or with 5% glucose resulted in unsatisfactory storage stability.

10. KINETIC pH PROFILES

pH profiles have been discussed in Chapter 3. Frequently, a broad screen of stability is performed on the initial small sample used for initial preformulation; this is frequently referred to as "forced decomposed studies" (Bodnar et al., 1983). In this the drug is exposed to "acid degradation," "base degradation," "aqueous degradation," "drug powder degradation," and "light degradation." More refined studies are eventually needed.

For any compound marketed by a pharmaceutical concern, at one time during its development, there should be a concerted project to establish a very exact pH profile. To do this correctly is a time-consuming undertaking. However, the information that can be gleaned from it is very important with regard to formulations, and it is therefore customary to carry out an approximate kinetic pH profile (Carstensen et al., 1992) early in the development stage. This will allow formulation of solutions for injections and for oral products as well, at a pH, and using buffers, that will give the best stability. Without it formulation is essentially guesswork.

11. LIQUID COMPATIBILITIES

The pH profile is the most important part of liquid compatibilities. However, two component systems are set up in aqueous (or other types of) solutions and treated as in Section 10 of Chapter 12. This is now required in the stability guidelines, which state that "it is suggested that the following conditions ... be evaluated in studies on solutions or suspensions of bulk drug substances: acidic and alkaline pH, high oxygen and nitrogen atmospheres, and the presence of added substances, such as chelating agents and stabilizers" and it is suggested "that stress testing conditions ... include variable temperature (e.g., 5, 50, 75°C)."

11.1. Aqueous Solution Compatibility

In general, such studies are carried out by placing the drug in a solution of the additive. These can be (and usually are) a heavy metal (with or without chelating agents present) or an antioxidant (in either oxygen or nitrogen atmosphere). Usually, both flint and amber vials are used, and in many cases an autoclaved condition is included. This will answer questions about susceptibility to oxidation, to light exposure, and to heavy metals. These are important questions as far as injectable compatibilities are concerned. Exposure to various plugs is frequently included at this point so that early injectable preparations can be formulated.

For preparations for oral use, knowledge of the desired dosage form is important, but compatibility studies with ethanol, glycerin, sucrose, corn syrup, preservatives, and buffers are usually carried out. This type of study also gives an idea of the activation energy, E, of the predominant reaction in solution. Arrhenius plots for compounds in solution are usually quite precise.

11.2. Nonaqueous Liquids

With transdermal dosage forms being of great importance of late, it is advisable to test for compatibilities with "ointment" excipients and with polymers (e.g., ethylvinyl polymer, if that is the desired barrier). In the case of transdermals, the dosage form is either directly placed in a stirred liquid or it is placed in a cell with an appropriate membrane (e.g., Cadaver skin) to estimate the release characteristics of the drug from the ointment (Chien et al., 1983).

It should be noted here that if the overall flux is J, then

$$\frac{1}{J} = \frac{1}{J_{\text{ointment}}} + \frac{1}{J_{\text{membrane}}} \tag{9.31}$$

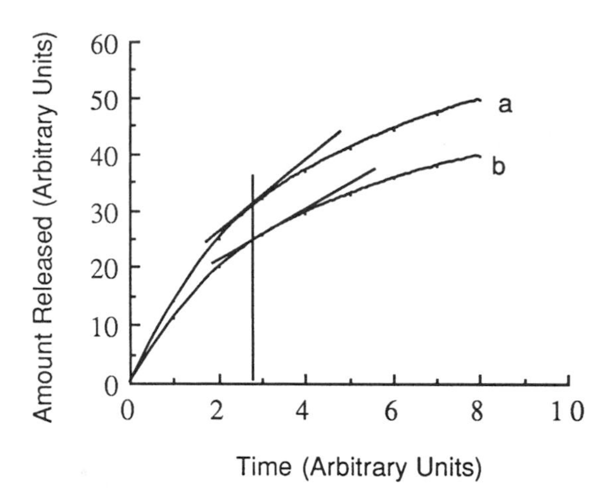

Fig. 12 Slope determination of flux of ointment release and release from ointment + membrane.

where subscripts refer to the respective phase J_{membrane} can be obtained from curves such as shown in Fig. 12 in the fashion that first the overall flux is obtained (with the membrane in place), giving the value of J, then the release is obtained without with membrane in place, giving J_{ointment}, that is

$$J = \frac{1}{A}\frac{dm_1}{dt} \tag{9.32}$$

and

$$J_{\mathrm{ointment}} = \frac{1}{A}\frac{dm_2}{dt} \tag{9.33}$$

J_{membrane} is then obtained as the reciprocal of the difference.

In vivo testing is usually carried out by applying the dosage form to hairless rats followed by subsequent sacrifice. Since the skin consists of a number of layers with differing hydrophilicity, the overall fate of the drug is of importance.

11.3. Emulsions

In the case of emulsions, the preformulation studies become very formulation oriented. Williams and Mahaguna (1998) have described preformulation studies of Freund's incomplete adjuvant (FIA), which is a water-in-oil emulsion. This included measuring the critical micelle concentration of the formulations to be investigated. Using ovalbumin (a model antigenic protein) in the interface, the surface activity of mannide monooleate, in the interface between water and oil phases, was determined.

11.4 Gels

Wong et al. (1997) studied the stability of cefazolin in Pluronic F-127 gels and found the decomposition to be first order, and for all Pluronic concentrations used (20, 25, and 30%), Arrhenius plotting was satisfactorily linear.

REFERENCES

Acree, W. E., Rytting, J. H. (1983). J. Pharm. Sci. 72:293.
Amidon, G. L., Yalkowsky, S. H., Leung, H. (1974). J. Pharm. Sci. 63:1858.
Angerg M., Nyström, C., Castensson, S. (1988). Acta Pharm. Suec. 25:307.
Angerg M., Nyström, C., Castensson, S. (1990). Int. J. Pharm. 61:66.
Angerg M., Nyström, C., Castensson, S. (1993). Int. J. Pharm. 90:19.
Aso, Y., Sufang, T., Yoshka, S., Kojima, S. (1997). Drug Stability 1:237.
Bakar S. K., Niazi, S. (1983), J. Pharm. Sci. 72:1024.
Bodnar, J. E., Chen, J. R., Johns, W. H., Mariani, E. P., and Shinal, E. C. (1983). J. Pharm. Sci. 72:535.
Bodor, N., Gabanyi, Z., Wong, C.-K. (1989). J. Am. Chem. Soc. 111:3783.
Burger, A., Lettenbichler, A. (1993). Eur. J. Pharm. Biopharm. 39:65.
Carstensen, J. T. (1974), in Dissolution Technology (Leeson, L., Carstensen, J. T., eds.), Academy of Pharmaceutical Sciences. American Pharmaceutical Association, Washington, DC, p. 5.
Carstensen, J. T. (1977). Pharmaceutics of Solids and Solid Dosage Forms, New York: Wiley-Interscience, pp. 11–15.
Carstensen, J. T., Kothari, R. (1981). J. Pharm. Sci. 70:1095.
Carstensen, J. T., Usui, F. (1984). J. Pharm. Sci. 74:1293.
Carstensen, J. T., Johnson, J. B., Valentine, W., Vance, J. (1964). J. Pharm. Sci. 53:1050.
Carstensen, J. T., Su, K. S., Maddrell, P., Newmark, H. (1971). Bull. Parenter. Drug Assoc. 25:193.
Carstensen, J. T., Franchini, M., Ertel, K. (1992). J. Pharm. Sci. 81:303.
Cavé. G., Puisieux, F., Carstensen, J. T., (1979). J. Pharm. Sci. 68:424.
Chien, Y. W., Keshary, P. R., Huang, Y. C., Sarpotdar, P. P. (1983). Drug. Dev. Ind. Pharm. 72:968.
Chollet, J. L., Jozwiakowski, M. J., Phares, K. R., Reiter, M. J., Roddy, P. J., Schultz, H. J., Ta, Q. V., Tomail, M. A. (1999). Pharm. Dev. Tech. 4:35.
Eyjolfsson, R. (1998). Drug Dev. Ind. Pharm. 24:797.
Hansen, L. D., Lewis, E. A., Eatough, D. J., Bergstrom, R. G., DeGraft-Johnson, D. (1989). Pharm. Res. 6:20.
Hixson, A., Crowell, J. (1931). Ind. Eng. Chem. 23:923.
Hourcade, F., Sautou-Miranda, V., Normand, B., Laugier, M., Picq, F., Chopineau, J. (1997). Int. J. Pharm. 154:95.
Iba, K., Arakawa, E., Morris, T., Carstensen, J. T. (1991). Drug Dev. Ind. Pharm. 17:77.
Malan, C. E. P, deVilliers, M. M., Lötter, A. P. (1997). Drug Dev. Ind. Pharm. 23:533.
Mura, P., Faucci, M. T., Manderioli, A., Furlanetto, S., Pinzauti, S. (1998). Drug. Dev. Ind. Pharm. 24:747.
Nogami, H., Nagai, T. Suzuki, A. (1966). Chem. Pharm. Bull. 14:329.
Noyes A., Whitney, W. (1897). J. Am. Chem. Soc. 23:689.
Oliyai, R., Lindenbaum, S. (1991). Int. J. Pharm. 73:33.
Parke, T., Davis, W. (1954). Anal. Chem. 25:642.
Paruta, A. N., Irani, S. A. (1964). J. Pharm. Sci. 54:1334.
Perrier, P. R., Kesselring., U. W. (1983). J. Pharm. Sci. 72:1072.

Pikal, M. J., Dellerman, K. M. (1989). Int. J. Pharm. 50:233.
Pikal, M., Lukes, A. L. (1976). J. Pharm. Sci. 65:1269.
Riegelman, S. (1979). Dissolution Testing in Drug Development and Quality Control, The Academy of Pharmaceutical Sciences, Task Force Committee, American Pharmaceutical Association, p. 31.
Rodriguez-Hornedo, N. (1984). Crystallization Kinetics and Particle Size Distribution. Ph.D. thesis, University of Wisconsin.
Rustichelli, C., Gamberini, M. C., Ferioli, V., Gamberini, G. (1999). Int. J. Pharm. 178:111.
Schultz, R. C. (1984). Stability of Dosage Forms, FDA-Industry Interface Meeting, Washington, D. C., Oct. 7, 1983.
Selzer, T., Radau, M., Kreuter, J. (1998). Int. J. Pharm., 171:227.
Shibata, M., Kokobu, H., Morimoto, K., Morisaka, K., Ishida, T., Inoue, M. (1983), J. Pharm. Sci. 72:1436.
Sorby, D., Bitter, R., Webb, J. (1963). J. Pharm. Sci. 52:1149.
Stability Guidelines, Congressional Record, May 7, 1984.
Streng, W. H., Yu, D. H.-S. (1998). Int. J. Pharm. 164:139.
Sunada, H., Shinohara, I., Otsuka, A., Yonezawa, Y. (1989). Chem Pharm. Bull. 37:467.
Tan, X., Meltzer, N. S. L. (1992). Pharm. Res. 9:1203.
Toy, A. D. F., (1980), Inorganic phosphorous chemistry. In Comprehensive Inorganic Chemistry (J. C. Bailar, Jr., H. J. Emelius, R. Nyholm, A. F. Trotman-Dickenson, eds.), A. Wheaton, Exeter, UK, pp. 389–543.
Underberg, W. J. W., Lingeman, H. (1983). J. Pharm. Sci. 72:553.
Van Campen, L., Zografi, G., Carstensen, J. T. (1980). Int. J. Pharm. 5:1.
Van Campen, L., Amidon, G. L., Zografi, G. (1983a). J. Pharm. Sci. 72:1381.
Van Campen, L., Amidon, G. L., Zografi, G. (1983b). J. Pharm. Sci. 72:1388.
Van Campen, L., Amidon, G. L., Zografi, G. (1983c). J. Pharm. Sci. 72:1394.
Williams, R. O. III, Mahaguna, V. (1998). Drug Dev. Ind. Pharm. 24:157.
Wilson, T. H., Wiseman, G. (1954). J. Physiol. 123:116.
Wilson, R. J., Beezer, A. E., Mitchell, J. C., Loh, W. (1995). J. Phys. Chem. 99:7108.
Wong, C.-Y., Wang, D.-P., Chang, L.-C. (1997). Drug Dev. Ind. Pharm. 23:603.
Wood, J. H., Catacalos, G., Lieberman, S. (1963). J. Pharm. Sci. 52:296.
Yalkowsky, S. H., Valvani, S. C. (1983). J. Pharm. Sci. 72:912.
Yalkowsky, S. H., Flynn, G. L. and Amidon, G. L. (1972). J. Pharm. Sci. 61:983.
Yalkowsky, S. H., Amidon, G. L., Zografi, G., Flynn, G. L. (1975). J. Pharm. Sci. 64:48.
Yalkowsky, S. H., Valvani, S. C., Roseman, T. J. (1983). J. Pharm. Sci. 72:866.

10

Physical Testing

JENS T. CARSTENSEN

Madison, Wisconsin

A great deal of space has been devoted to the subject of chemical testing. However, even if a product, chemically, is sufficiently stable to sustain e.g. a 3-year expiration date, physical changes may have occurred. In a solid dosage form, the dissolution may have slowed down to such an extent that the product is no longer as bioavailable as it was at the time of manufacture, and more importantly, it may not meet the minimum required for efficacy. For a solution, a precipitate may have occurred. This may not affect the chemical content, but for a parenteral product it would, obviously, be quite unacceptable, and for an oral solution it would also be unsatisfactory, because the dispensing pharmacist would rightfully question the integrity of the product. The caking of a suspension impairs the dispensing of a known amount of drug in a teaspoon, and a separated or broken emulsion or cream obviously will not have the same emollient properties as would a proper product.

Physical stability will be treated by product category in the same order as in the case of chemical stability.

1. PHYSICAL STABILITY OF SOLUTIONS

Solutions are broadly divided into two categories: oral and parenteral solutions. Appearance, in both cases, is an important factor. In the case of oral solutions, organoleptic properties are also of great importance. Organoleptic evaluation is usually done subjectively, i.e., a tester (operator, technician), will judge the product and score it, either numerically or descriptively or both. In the case of appearance of solutions, there should always be a subjective statement (quantitative or subjective description) even if more quantitative instrumental parameters are recorded. A few words are therefore in order regarding organoleptic and appearance testing.

1.1. Organoleptic Testing

For organoleptic testing it is important to establish a test panel early in the stability program. (Or if a stability program is in place, but no such testing is carried out, a test panel should be selected at the first opportunity when a product with important taste or odor properties is placed on stability.) Many companies utilize just one tester for the task of organoleptic testing, but this can be shortsighted, because the tester may leave, go on vacation, or become ill, and in that case the logical solution is to assign someone else to the task. There may be an evaluational bias between the two testers, and this should be established at the onset.

First of all, the depth of organoleptic capacity should be tested. This can be done by asking the tester to taste serial dilutions of a bitter substance (e.g., quinine). Hence a sensitivity level can be established. A control of e.g. water or high dilutions should always be part of the protocol.

It should be noted that the technicians are not taste testers in the ordinary sense. That is, it is not necessary to match their "likings" to that of the general public. Rather, it is important that they can (a) duplicate their results and (b) remember them, since they will be asked to taste a preparation that they originally tested 3 or 6 months earlier. In so doing they would have to score the degree of flavoring, e.g., is it less than originally present, i.e., is the flavor being lost? They would also have to be able to describe the flavor well originally. For example, if the chemical is slightly anesthetizing, the duration of the anesthesia would be important. If there is interaction with a plastic bottle, are off flavors appearing in the product? Finally it is important to screen several testers to ascertain that they give the "same result."

In describing the flavor, several categories can be used (degree of sourness, degree of saltiness, level of flavor, type of flavor). Each of these may be assigned to a level of e.g. 1–5. A flavor profile may hence be established, and this can then be reestablished at several time points in the room-temperature storage. It is not recommended to evaluate results from higher temperatures (although they may be carried out).

1.2 Subjective Appearance Testing

Solutions, particularly parenteral solutions, may have a tendency to discolor slightly. Often it is not possible, within analytical sensitivity, to establish either the source of the color or the level of the substance causing it. In this case it is a good practice to use a color standard to describe the "intensity" of the discoloration. Roche, for instance, uses the so-called Roche Color Standard (RCS), which uses a compound (the identity of which is a secret) that can be reliably reproduced and has exceptional color stability. Making up serial dilutions of this compound then gives solutions of different "slight" discolorations; they are denoted RSC#1, #2, etc., so that a solution can always be compared in this fashion. It is a bit like the old-fashioned Dubosque colorimeter (which can be used with advantage in this type of situation). The principle of the Dubosque colorimeter is to have a view of two test tubes from the top. One is the control, and the other is the solution being matched. It is possible to adjust the length of the light path in the second tube, and this is done until the intensity matches that of the standard. The length of the path is then an indication of the "concentration."

The RCS (and similar types of numbers) are difficult to analyze, but a Dubosque colorimeter gives numbers that follow Beer's law and are logarithmically proportional to concentration (although the proportionality factor cannot be known). In this fashion the "decomposition" could be represented simply as a first-order reaction, where the concentration, X, of the decomposition product would be given by

$$X = X_\infty[1 - \exp(-qt)] \tag{10.1}$$

or

$$\ln\left[1 - \frac{X}{X_\infty}\right] = -qt \tag{10.2}$$

where q is a constant, t is time, and X_∞ is found by iteration. This allows (from accelerated studies) a visual estimate of the worst appearance that a product could take on. The appearance of tablets can be treated differently and will be discussed later.

2. PARENTERAL SOLUTIONS

In parenteral solutions, physical stability includes interaction with a container and changes in chemical composition that give rise to physical changes. The latter will be discussed first.

One manifestation is slight discoloration. Thiamine hydrochloride solutions, for instance, may discolor slightly without showing detectable changes in content of parent compound. Such discolorations can be followed as described immediately

Table 1 Usual Concentrations of Antioxidants and Chelating Agents

Antioxidant	Usual concentration
Acetylcysteine	0.5%
Ascorbic acid	0.02–1%
BHT, BHA, and propyl gallate	0.005–0.02%
Citric acid (chelator)	Variable*
Sodium edetate (chelator)	0.01–0.075
Sulfites	0.1–0.15%
Thioglycerol	0.1–1.0%
Thiourea	0.5–1.0
Tochopherols	0.05–0.075

* Citric acid can be present in large amounts if it is present as a buffer (as well as present as a chelator).
Source: Table constructed from data published by Mendenhall (1984).

above, and at times they are detectable analytically. They are often oxidative in nature and metal ion catalyzed. Such a case in captopril (Lee and Notari, 1977).

Mendenhall (1984) has reviewed the stability aspects of parenteral products and has shown that discoloration is often either photochemical or oxidative. He has summarized the usually used antioxidants and chelating agents. These are shown in Table 1.

2.1. Swirly Precipitates

Often a parenteral solution will develop a swirly precipitate upon storage. This is most prevalent in vials and is usually an interaction with either the glass or the stopper. It may be difficult for the uninitiated to detect such slight changes, and the best person to use for this type of evaluation is a parenteral inspector. It is difficult to estimate the extent of the precipitate; it can be done by mechanical counting (e.g., with a Coulter counter), but the results are difficult to interpret. Often the count does not correspond to the "severity of the swirl." More to the point is how many swirls exist. If a box of e.g. 144 vials is placed on this type of stability, then the vials can be examined from time to time, and one may establish how many vials have become swirly. This number can then be treated in proper fashion to evaluate the severity of the problem, i.e., the stability parameter would be the number of swirly vials per box of 144.

Preferably there should be no swirls at all in the preparation, and if reformulation can be undertaken (which is wise), then an improved product would be the result. Otherwise, the stability program will establish the percentage probability of finding a vial with a swirl at the end of the expiration period. At times it is necessary to lyophilize products that are chemically stable, simply because the problem of swirls cannot be solved.

As mentioned, the occurrence of swirls is usually a container interaction, and a change in the stopper or the glass may often eliminate the problem. Vials should always be stored (a) upright, (b) on the side, and (c) upside down to check the interaction with the stopper. In this way primary evidence can be established as to the culpability of the closure.

2.2 Whiskers

McVean et al. (1972) reported on the case of a parenteral solution (morphine) where "whiskers" occurred at the tip of the ampul in a large percentage of ampuls upon room-temperature storage.

This is a defect that will occasionally occur in a product. It is due to pinholes in the glass. The solution wicks out, and the liquid evaporates on the outside. The solid that is formed serves to wick out more solution, and long crystals or "whiskers" may occur. One might ask why the pinholes have not been detected in the dye test used for autoclaved ampuls. There are two reasons. One is that the hole may be too small for detection (about 0.5 μm is the detection limit). The other is that the ampul was tight at the time of manufacture, but the heat sealing line was run too rapidly, or the flame temperature was incorrect, so that the glass did not have time to anneal properly, and the strain caused the crack during storage (not immediately after manufacture).

2.3 Cloud Times

Sometimes a cloud will appear in a product as the storage time progresses, and this is most often due to chemical changes in the system. If for instance an ester (e.g., polysorbate, which is a fatty acid ester) hydrolyzes, then the produced acid may be poorly soluble. If the solubility is denoted S, then the following holds: If the reaction in general is written

$$A + H_2O \rightarrow B \tag{10.3}$$

where A is a drug of initial concentration A_0 and B is the decomposition product with solubility S (which is assumed to be limited). Assuming first order, the concentration of B is then given by

$$[B] = A_0[1 - \exp(-kt)] \tag{10.4}$$

At time t^* the solubility will be exceeded, and t^* is what is denoted the cloud time. t^* is given by

$$S = A_0[1 - \exp(-kt^*)] \tag{10.5}$$

or

$$\ln\left[1 - \frac{S}{A_0}\right] = -kt^* \tag{10.6}$$

If $A_0 \gg S$ then this simplifies to

$$t^* = \frac{S}{kA_0} \tag{10.7}$$

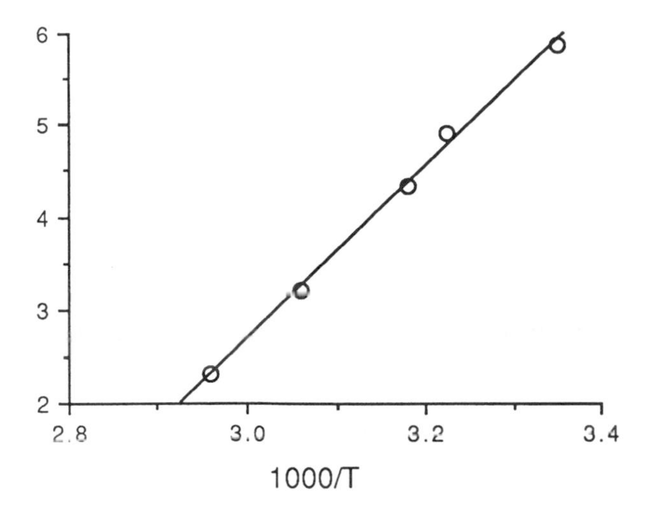

Fig. 1 Cloud times of a parenteral diluent containing polysorbate 80 and maleic acid. (Constructed from data published by Carstensen, 1972.)

Taking logarithms gives

$$\begin{aligned} \ln[t^*] &= -\ln[A_0] + \ln\left[\frac{S}{k}\right] \\ &= -\ln[A_0] + Q - \frac{\Delta H - E}{RT} = -\frac{\Delta H - E}{RT} + Q'' \end{aligned} \tag{10.8}$$

where Q and Q'' are constants. This shows that the cloud times can be plotted by Arrhenius plotting. Such plotting is quite predictive, as is shown in Fig. 1.

The precipitation may also occur by the solubility product being exceeded, or from any situation leading to a product with limited solubility.

There are other causes for precipitation on storage, one being the original use of a metastable form, so that the solutions in question, in fact, are supersaturated solutions. It was the author's experience, at his tenure at Hoffmann–la Roche in 1965, that a product to be introduced (Taractan Injectable) was in this category. Several pilot batches had been successfully made, but the first production batches precipitated, a more stable polymorph crystallizing out. This necessitated reformulation to a lower strength (corresponding to the lower solubility of the stabler polymorph) and subsequent resubmission of data to the FDA. This points out the importance of careful preformulation studies of the solubility of compounds. Errors of the above type are costly, both in terms of resubmission and in lost market time. Even official products fall into this category.

Calcium gluceptate is used to treat calcium deficiency and (USP XX, 1980) is highly water soluble (up to 85%). Solutions, however, show a tendency to precipitate on standing at room temperature (Muller et al., 1979). The storage time required for precipitation is a function of the commercial source, as is pointed out by Suryanarayanan and Mitchell (1981). It was shown that the precipitate was a sparingly soluble crystalline hydrate, and that the raw material was an amorphous (much more soluble) form of the drug. Seeds, and unfortuitous ratios of alpha and beta epimers of the calcium gluceptate, catalyzed the precipitation.

Precipitation is a nucleation and crystal growth phenomenon (Carstensen and Rodriguez, 1985, Rodriguez, Hornedo and Carstensen, 1985), and as such it can be impaired or prevented by inhibitors. These are often viscosity-impairing substances (carboxymethyl cellulose for instance), and hence the stability of the viscous component becomes important. The loss of this can be detected by following viscosity.

The viscosity of these agents is often Bingham bodies, i.e., they possess a yield value. The correct way of checking them is, therefore, with e.g. a cup-and-bob viscometer, so that a rheogram can be drawn. In this fashion it is possible to check both changes in yield value and slope of the rheogram (apparent viscosity). For very fluid solutions (dilute aqueous solutions) this is difficult, and most often it is best followed by the use of an Ostwald–Fenske pipette. Two pipettes (with different flow times) should be used in this case, because the difference in the measured viscosity is a measure of the yield value (although calculation of the yield value from the difference is a priori not possible). Both yield value and apparent viscosity are functions of concentration (Ben-Kerrour et al., 1980); in a multicomponent system there will usually be one main component responsible for viscosity, and it is the breakdown of this one compound that would be of importance. Often when drastic changes occur in viscosity, bacterial contamination can be suspected.

Precipitation is tied into solubility, as seen in the foregoing. Solubility can be augmented by various means. In the case of cloud times, the use of cosolvents (e.g., polyethylene glycol) will increase the value of S. Other methods are the use of a micellar approach and the use of complexation. A recent example of this latter is the work by Mehdizadeh and Grant (1984) on the complexation behavior of griseofulvin with fatty acids. Order of magnitude increases in solubility were reported.

2.4 Oral Solutions

The main types of changes in appearance of oral solutions (syrups, elixirs, etc.) are loss of dye, precipitation, and bacterial growth. Precipitation has already been dealt with to some degree, but some cases particular to oral solutions will be mentioned. Change in dye content will be treated below. Bacterial growth will be treated separately.

Scott et al. (1960) showed the loss of blue dye in a vitamin syrup, and showed that it could be treated exactly like a drug substance. Predictions by Arrhenius plotting are quite good in the case of degradation in solution, because the homogeneity is good. Figure 2 shows an example of this.

3. DISPERSE SYSTEMS

Disperse systems are suspensions and emulsions. The rationale for the physical tests carried out on these will be discussed below.

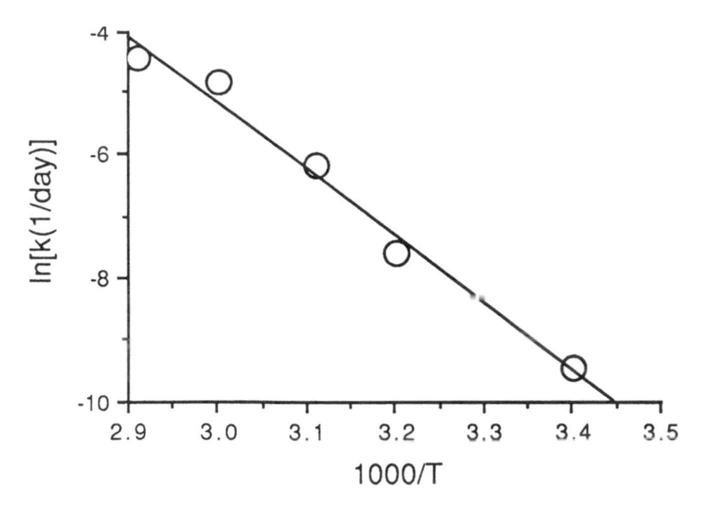

Fig. 2 Arrhenius plot of FDC Blue Dye #2 in a syrup. (Graph constructed from data by Scott et al., 1960.)

3.1 Suspensions

It would be desirable to have a suspension that did not settle (and there are such suspensions), but the general rule is that a suspension will settle, and therefore there are two parameters that are followed in this respect, namely sedimentation rate and sedimentation volume. When the sedimentation volumes are small, then there is a tendency for the suspension to cake, and hence various types of shaking tests are carried out.

Tests can be purely subjective, in that a tester notes that e.g. the suspension after three months' storage at 25°C was "difficult to resuspend, leaving some cake at the bottom." Such subjective tests should always be included in a program, but more quantitative means are desirable also. A typical quantitative test is to rotate the bottle under reproducible conditions. The type of setup used for solubility determinations is a good type apparatus for this purpose. The bottle is rotated x rotations, a sample of the supernatant is taken, and it is assayed. (This assay need not be stability indicating.) This is then repeated for twice the number of rotations, four times the number of rotations, and eight times the number of rotations. The time-relation of the assays is similar to that of a dissolution curve (although the phenomenon is redispersion), and it can often be represented by

$$Y = Y_\infty[1 - \exp(-kt)] \tag{10.9}$$

Y_∞, the asymptote value (found by iteration), should equal the dose, if caking has not occurred. The value of k is best found from the logarithmic presentation mode:

$$\ln\left[1 - \frac{Y}{Y_\infty}\right] = -kt \tag{10.10}$$

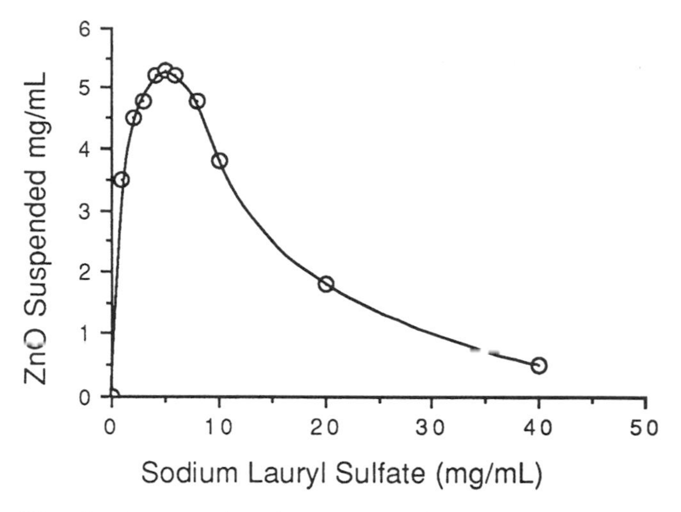

Fig. 3 Resuspension characteristics by controlled rotation. (Graph constructed from data published by Lemberger, 1967.)

k and Y_∞ can then be found by data treatment for extrapolated values or assessed in a room-temperature stability program to estimate the stability of the resuspendability parameter. Suspendability is also improved by the use of surfactants. Figure 3 shows a suspension isotherm (Moore and Lemberger, 1963) of the zinc oxide/sodium lauryl sulfate system. Such suspension isotherms should be carried out prior to the formulation of suspensions. They are in general not carried out in the preformulation effort, but rather by the formulator.

One way of accelerating the settling is to place the suspension product on a shaker at e.g. 37°C. This makes particle movement more rapid and allows the fine particles to slip into the interstices of the larger particles, hence promoting a close packing. This can then be used to judge qualitatively whether caking will take place.

It might be thought that centrifugation would be a good way in which to "accelerate" sedimentation, and the Stokes law indeed predicts this. However, it gives only an acceleration of the "initial settling rate," and the further settling, and the caking phenomena in which the formulator is interested, are not well predicted by this method.

Some caking is due to crystal growth, and this is accelerated by the use of freeze–thaw tests, i.e., alternating the temperature every 24 h from e.g. 25°C to −5°C (or some other low temperature above the freezing point of the product). The temperature cycle will promote crystal growth, and the effect of this on the product can be assessed. The freeze–thaw cycle has the advantage of emulating (and overstating) some real conditions to which the product could be exposed during shipping.

Zapata et al. (1984) have described the effect of freeze–thaw cycles on aluminium hydroxycarbonate and magnesium hydroxide gels. Coagulation after freeze–thaw cycles led to the formation of aggregates that were visible. These aggregates were particles in a primary minimum, and these were only reseparable by ultrasonic treatment. The freeze–thaw cycle affected content uniformity of both the gels, but the treatment did not alter the surface characteristics or the morphology (as judged by x-ray powder diffraction). It did cause a reduction in the acid neutralization rate, and the rate of sedimentation increased. The effect was pronounced after the first cycle (and indeed most of the effect occurred at this point). The duration of freezing was not important, but the aggregate size grew inversely with the rate of freezing. The use of polymers in the suspensions reduced the effects of the freeze–thaw cycle.

Freeze–thaw cycles (aside from being a stability monitoring tool) can be used to screen products as well, the best of a series of suspensions or emulsions being the one that stands up best to the test. This on the surface may be logical, but without a theoretical basis it is difficult to judge the generality of such a statement.

3.2 Sedimentation Volumes

If a suspension is particulate, then the particles will (approximately) settle by a Stokes law relation, i.e., the terminal velocity, v, is given by

$$v = d^2 \cdot g \frac{\Delta\rho}{18\eta} \tag{10.11}$$

where the constant g is gravitational acceleration, $\Delta\rho$ is the difference in density between solid and liquid, η is the viscosity of the liquid, and d is the diameter of the particle. The final apparent volume of the sediment, provided it is monodisperse, would be given by the fact that in cubical loose packing a sphere of diameter d will occupy the space of its confining cube, i.e., the sedimentation volume will be

$$V = n \cdot d^3 \tag{10.12}$$

where n is the number of particles per cm^3 of suspension. Since their density is ρ g/cm^3, then (denoting the dosage level Q g/cm^3) the following holds:

$$Q = \frac{\rho \cdot n \cdot \pi d^3}{6} \tag{10.13}$$

so that, solving for n,

$$n = \frac{Q \cdot 6}{\rho \pi d^3} \tag{10.14}$$

which inserted in Eq. (10.12) gives

$$V = \frac{6Q}{\rho\pi} \tag{10.15}$$

In this view, each particle touches its neighbors. The potential diagram from two particles is as shown in Fig. 4.

When the particles touch, the potential energy becomes exceedingly large ($x = 0$), and from an equilibrium point of view they will be trapped in the primary minimum, which is the deep minimum at short distance in Fig. 4. Hence it becomes difficult to separate them, and the precipitate becomes a cake. This would prevent redispersion by shaking and would make proper dispensing impossible. It is a formulation goal to prevent this from happening, and this is done by adjusting the

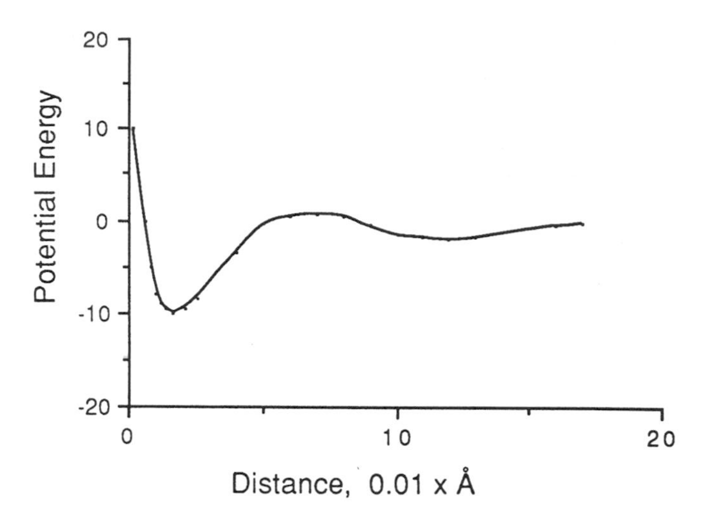

Fig. 4 Potential energy diagram for two particles.

zeta potential, as will be discussed shortly. From a formulation point of view, it is better to have the particles at larger distances, e.g., in the secondary minimum occurring at longer distances (Fig. 4).

A discussion of the connection between caking tendency and the so-called zeta potential is beyond the scope of this book. Suffice it to state the following: When particles are suspended in a liquid, they acquire a charge (and the liquid acquires a similar opposite charge, to maintain electroneutrality). The zeta potential is related to this charge, and caking is prone to happen if the charge potential is outside a range of −10 mV to +10 mV. If the zeta potential is high it can be lowered by the addition of negatively charged ions. Highly valent ions (e.g., citrate) are preferable. On the other hand, if the zeta potential is low, then it can be increased by the addition of positively charged ions (e.g., aluminium ions).

The zeta potential is measured with a zetameter. In this the particles are placed in an electrical field (between two electrodes, the voltage of which can be adjusted), they are tracked under a microscope, and their velocity is determined. The relation of velocity to voltage allows determination of the zeta potential.

It is worthwhile occasionally to check the zeta potential in a stability check of suspension (and emulsion) products. Counterions could be adsorbed and hence lose their capability of keeping the zeta potential close to zero, and this, in turn, could be the reason for subsequent caking.

When the zeta potential is close to zero, the suspension will be flocculated, i.e., the particles are positioned in the secondary minimum. The floccules are large and hence settle more slowly, but on the other hand the sedimentation volume is large. Since the particles are in the shallow minimum (small potential, i.e., easy to disrupt), they are easy to resuspend.

There are suspensions that do not settle. Here the yield value of the suspension is so large that the gravitational force does not exceed it. In this case it is very important to carry out complete rheological profiles at different time points in the stability program, to insure that the yield value is not changing. In such a system the yield value (Carstensen, 1973) is a function of the solids content and the viscosity of the medium. If the viscosity imparting substance deteriorates, or if the flocculation characteristic (the "diameter" of the particles) changes, then the yield value may change, and what originally was not prone to cake might at a later time have such a propensity.

It has been stated elsewhere that for Bingham bodies, a yield diameter of the bottle can be calculated and below this bottle diameter there will be no settling.

3.3. Sedimentation Rates

The rational treatment of sedimentation rates has been described by Carstensen and Su (1970). Since the suspension, when placed on stability, has just been well agitated, the floccule size is not the same as it will be at equilibrium (it will be smaller). The first part of a settling curve is, therefore, governed by the reforming of the equilibrium floccule, and the latter part is governed by settling towards the equilibrium sedimentation volume. A typical plot of the final settling phase of kaolin suspensions is shown in Fig. 5. The intercept does not correspond to full height, because the settling is the final phase. The first phase, as mentioned, consists of reflocculation of the equilibrium floccule (which does not exist at time zero, because the suspension has been thoroughly shaken at that point).

The sedimentation curve is, therefore, two-phasic, and the equation for the settling curve is

$$Y - Y_\infty = A_0 \exp(-k_0 t) + A_1 \exp(-k_1 t) \tag{10.16}$$

and the curve can be deconvoluted by feathering, or by programmed four-parameter techniques.

3.4. Preservation Stability

Methyl, ethyl, propyl, and butyl esters of 4-hydroxybenzoic acid are used in various combinations in antacid suspension (and other pharmaceutical) products. The

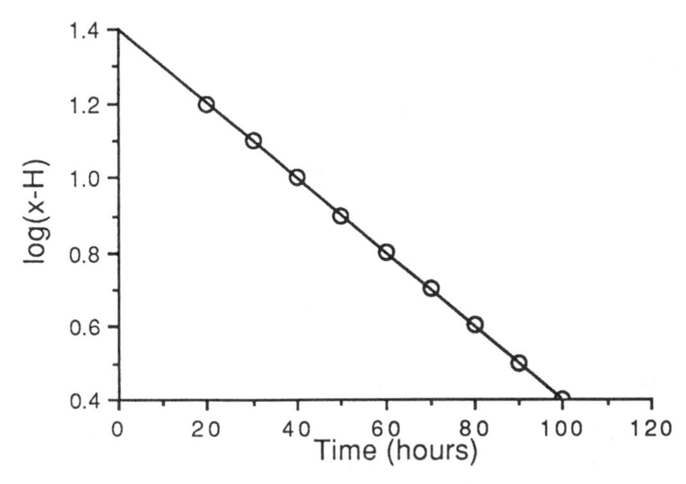

Fig. 5 Settling of kaolin suspensions. (Constructed from data published by Carstensen and Su, 1969.)

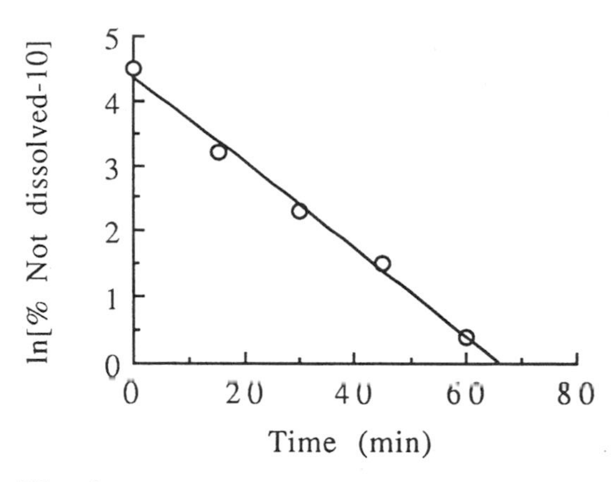

Fig. 6 The least squares fit equation is $y = 4.69 - 0.066x$ with $R^2 = 0.99$. (Graph constructed from data published by Cárdenas et al., 1994.)

antacids have high pH values, and hence hydrolysis of the esters occurs. The rationale for using several in combination is, exactly, to allow a certain amount to remain to retain preservative qualities of the suspension. An assay of the four esters and the parent acid (one of the decomposition products) in products where all occur has been described by Schieffer et al. (1984).

3.5. Dissolution of Suspensions

The 1987 Guidelines require testing of suspensions for dissolution. Cárdenas et al. (1994) have described the dissolution profiles from suspensions of benzoyl metronidazole, and a graph constructed from their published data is shown in Fig. 6. The curves should, by all rights, follow a cube root law without lag time, but they do not do so. If adjusted for amount not dissolved at the end (in the figure, 10%) they will adhere to a sigma minus plot.

3.6. Temperature Testing of Disperse Systems

A suspension is, as the name implies, a two-component system consisting of a solid and a liquid phase. (Gas phases are considered nonessential in this connection). Obviously, the solubility of the compound is a function of the temperature, and at a given temperature above 25°C this solubility will be reached. Testing about this temperature obviously has no meaning as far as suspension stability (neither physical nor chemical) is concerned. Prior to starting a program, this temperature should be established, so that unnecessary sampling stations can be avoided.

3.7. Semisolid Suspension Systems (Ointments, Suppositories)

Some semisolid systems (ointments and suppositories) are suspensions. Their testing is not different, in general philosophy, from what is described above, except that the rheology is checked differently. Davis (1987) has reviewed sophisticated means of checking the stability of such systems.

The factors checked for in stability programs of such products are the following:

1. Consistency, fell to the touch
2. Viscosity
3. Polymorphism

It is mentioned elsewhere that migration of a "disperse" phase within a semisolid product is quite possible when another phase is present. This situation may occur in the case of the use of benzocaine in, for instance, a suppository wrapped in aluminum foil coated with polyethylene. Polyethylene lining of aluminum wraps of suppositories is used to prevent contact between the metal and the suppository, and in most cases this has a positive effect.

However, a partitioning of drug or additive between the two phases may be possible if the drug or additive is suspended in the suppository. Denoting its solubility in the polyethylene S_p and the solubility in the suppository base S_s, the compound would disappear from solution in the suppository at a rate proportional to $S_p - S_s$, and "disappeared" compound would be replenished by dissolution from the solid phase.

The rate of disappearance would be governed in that the value of S_p would increase by a sigma minus relation (i.e., in the same manner as the appearance of decomposition product in a first-order reaction), and this then would be the overall "loss" of compound as a function of time. Since this is a first-order overall relationship, the "decomposition" would, initially, appear to be first order.

3.8. Ointments and Transdermals

Polymorphism can be followed by x-ray analysis and in some cases by thermal methods. There is, in fat systems, the possibility of trans esterification, and this can be tested for chemically.

The problem of morphology changes is often of particular importance and of particular frequency in the case of suppositories. In this type of product, it is also important to check for migration of suspended/dissolved substances. Often a substance is added to a suppository as a suspended particle, which is soluble in the suppository base to some extent. The phenomenon of dissolution will, of course, become evident by checking the particle size as a function of time. If a substance is soluble in the base, then it is preferable (if possible) to saturate the base with it at the onset. For this reason it is necessary to determine the solubility (S gm/gm) of the drug (or other) substance in the base. A Van't Hoff plot [solubility as a function of temperature (T°K), i.e., plotting $\ln[S]$ versus $1/T$] will allow extrapolation to room temperature. In manufacturing it is advisable to dissolve the drug (or other substance) to the extent of its solubility during the intermediate temperature phase of manufacturing (where the preparation is still quite fluid) and then suspend the rest at a lower temperature. An example is ascorbic acid, which is a good antioxidant in Carbowax bases. To exert its antioxidant action it must, however, be dissolved (and it is quite soluble in polyethylene glycols).

Dissolved drug (or other substance, e.g., benzocaine) will diffuse in the suppository base, and can, for instance, partition into polyethylene linings of the suppository wrap.

Release rates are important in many topical preparations, in particular in transdermal preparations. Here there are several investigational methods available. In-vitro methods involve placing the ointment on a membrane and measuring the appearance of drug in a receptor compartment on the sink side of the membrane. Hoelgaard and Møllgaard (1983) have, for instance, described the in-vitro release of linoleic acid through an in-vitro membrane. They mounted abdominal human skin in one case and skin from hairless rats in another to open diffusion cells. The dermal side was bathed with a receptor medium stirred at 37°C. The medium was 75 mL of 0.05 N phosphate buffer (pH = 7.4) which contained 0.05% Pluronic F68 and 0.01% butylhdroxytoluene, the latter two ingredients added in order to increase the lipid solubility. Linear, Ficksian diffusion curves were obtained. In a stability program, such tests are obviously useful and should be repeated periodically, but an "internal standard" or "calibrator" should be used, i.e., a stable test substance, the diffusion of which is known (e.g., salicylic acid). Other pseudo-in-vivo methods involve shaved or hairless rabbits, or cadaver skin. The interaction between ointment and container (patch) should also be part of the stability program.

Some of the testing applicable to semisolid emulsion systems is also applicable to ointment systems and will be discussed at a later point.

4. EMULSIONS

An emulsion should be thought of as a metastable system. In most cases the emulsion system (Fig. 7) is thermodynamically more energetic than the ground state system, which would simply be the totality of the two phases, separated. There will, therefore, always be the potential for oil droplets re-merging in an attempt to create the thermodynamically stable system.

Emulsion systems are taken orally (LipoGantracin™, Roche), parentally (as parenteral fat emulsions), and topically (creams).

4.1 The Emulsion Interface

The factors that stabilize the emulsion system are a layer of surfactant and protective colloid on the exterior of the droplet. The amount of these two must be such that they

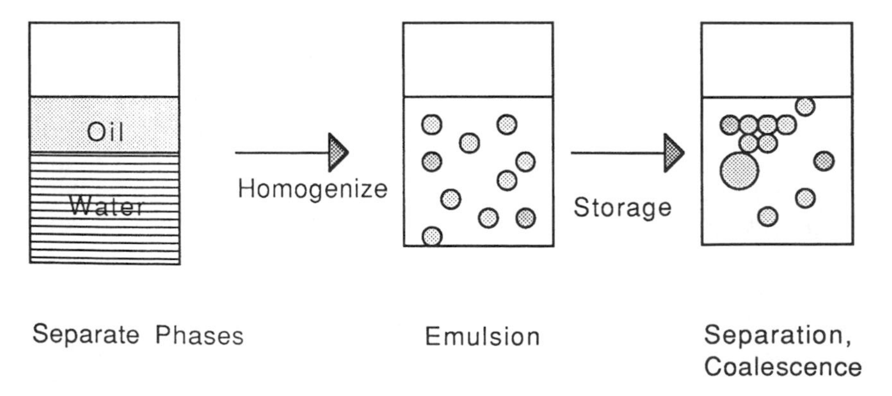

Fig. 7 Emulsion system.

cover the entire area of the droplets, otherwise coalescence will occur to the extent that the area, A, of the droplets will be reduced to such a point that it now will be completely covered by surfactant and protective colloid.

If, for instance, 1 g of emulsion contained W g of droplets of a size d μm and the oil had a density of ρ g/cm^3, then there would be n droplets per cm^3, where n is given by Eq. (10.14). Each particle has a surface area of πd^2, so that the total area is

$$A = n\pi d^2 = \frac{6}{\rho}\frac{Q}{d} \tag{10.17}$$

Example 10.1.

If the density of the oil is 0.9 g/cm^3, the amount of oil phase per cm^3, is 0.75 g, and the diameter of the oil globules is 10 μm (10^{-3} cm) what is the surface area of the oil phase?

Answer.

$$A = \frac{0.75}{10^{-3}}\frac{6}{0.9} = 5 \cdot 103\,\text{cm}^2 \tag{10.18}$$

Example 10.2.

If a surface active agent of molecular weight 800 and cross-sectional molecular area of 30 Å is present in a concentration of 0.2% will that suffice to cover the surface in Example 10.1?

Answer

$2\,\text{mg/cm}^3 = 2/800 = 2.5 \cdot 10^{-3}$ millimoles $= 2.5 \cdot 10^{-6}$ moles, which in turn equals $2.5 \cdot 10^{-6} \cdot 6 \cdot 10^{23} = 1.5 \cdot 10^{18}$ molecules $= 30 \cdot 1.5 \cdot 10^{18}\,\text{Å}^2 = 4500\,\text{cm}^2$, this is the surface the surfactant could cover. This is slightly less than the 5000 cm^2 surface area of the oil, so that the entire surface of the oil globules cannot be covered by the surfactant.

The above calculations are oversimplified. They assume, for instance, that all the surfactant is adsorbed onto the oil, which is not the case. It is important, however, to check, originally, whether enough surface coverage of the oil is provided for. If not, there will be an initial shrinkage of surface area (increase in droplet size) attributable to this. Hence, if the coverage of the droplets with surfactants and/or protective colloid is incomplete at the time of manufacture, then the droplets will grow in size as time progresses. Rowe (1965) for instance demonstrated that the globule size decreases with increasing surfactant concentration, as shown in Fig. 8.

4.2. Globule Size and Viscosity

The breakage of suspensions will be dealt with shortly, but (Fig. 7) it might be suspected that breakage would be a function of Stoksian motion [Eq. (10.11)], i.e., the globules move and collide and hence coalesce. This is true in a sense, but

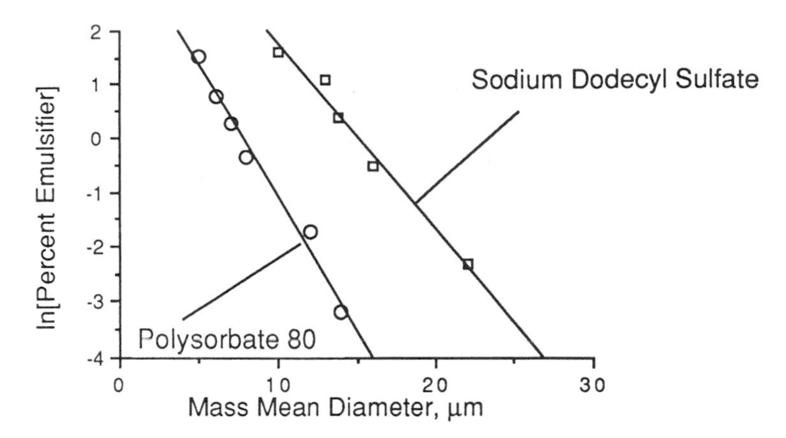

Fig. 8 Effect of emulsifier concentration of globule size. (Graph constructed from data by Rowe, 1965.)

the conclusion that might be drawn would then be that to increase viscosity [Eq. (10.11)] would reduce the severity of such impacts. However, Siragusa (1995) has demonstrated that although increased viscosity to some extent makes an emulsion more stable, the more important factor is the stability of the surfactant/protective colloid system at the interface. From a stability point of view, there is a correlation between the overall emulsion viscosity and the globule size. Figure 9 shows a typical example of viscosity as a function of droplet size and phase ratio (Sherman, 1964).

Hence, checking viscosity in the stability program, in a manner of speaking checks the globule size, which is the prime indicator of potential for progressing creaming and breaking. The viscosity is usually checked by a cup and bob method. The limitations of this will be discussed in the section dealing with semisolid emulsion systems.

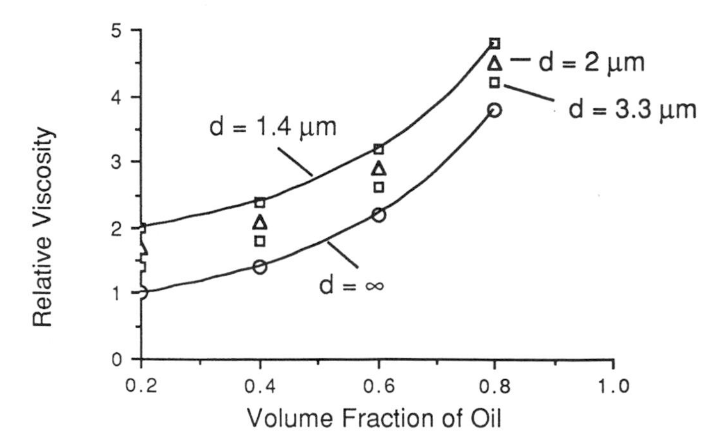

Fig. 9 Viscosity as a function of droplet size and phase ratio. (Constructed after data published by Davis, Sherman, 1964.)

Direct measurement of the droplet size can be accomplished in several ways: microscopy, electronic (Coulter) counters, photon correlation spectroscopy (for particles that are very small, Davis, 1967), diffuse reflectance spectroscopy (Akers and Lach, 1976) and the measurement of ultrasound (Rassing and Atwood, 1983). Davis (1987) points to the importance of choosing the proper techniques. He cites an example where a fat emulsion was tested for stability (as regards droplet size and distribution). The accelerated test used was a shaking test. The tests used were (a) microscopy (large globules), (b) electronic counting (medium size globule count) and (c) photon correlation spectroscopy (small particle count). Figure 10 shows the results.

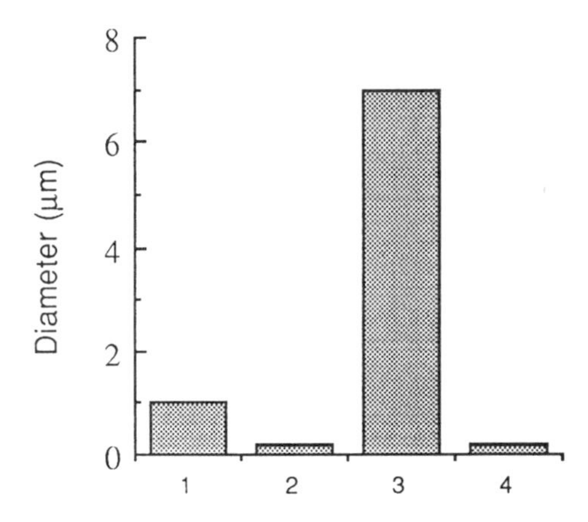

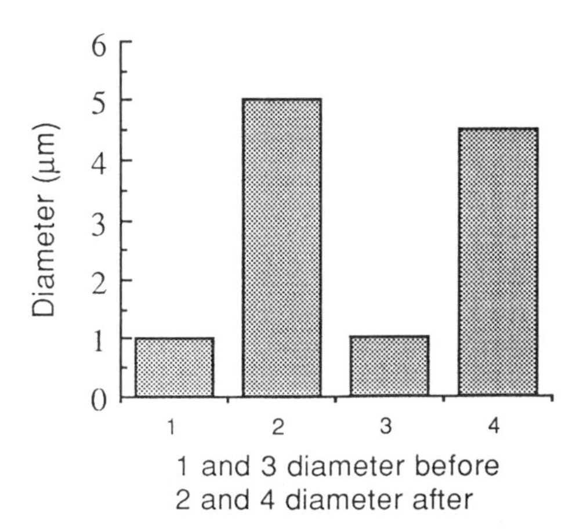

Fig. 10 Particle size analysis of accelerated test of emulsion system. (Constructed from data published by Davis, 1987.)

It is obvious that the small globule count does not change much, but that the intermediate count changes a lot. (The large globule count would then change in complement fashion to the intermediate count, and this was confirmed by the microscopy results.) What is important in this particular case (the system tested was a parenteral fat emulsion) is that there was formation of large oil droplets (not visible to the naked eye), and that these could have had a bearing on the toxicity of the product. This demonstrates that one method in itself is not enough, and that not one but several methods should be considered.

Reng (1984) has advocated electrical conductivity as an overall, common means of determining the state of dispersion of an emulsion system, and he shows that this parameter changes significantly over short periods of time, if the emulsion system is not satisfactory.

4.3. Stability of the Emulsifier/Protective Colloid System

The other phenomenon that may happen, which affects droplet size, is chemical breakdown in the surfactant. Nonionic surfactants are frequently used, and they are esters that may hydrolyze or interact with other components of the emulsion. Part of the formulator's job is, in independent experiments, to determine the pH profile and interaction potential of the surfactant (in a system simply consisting of the aqueous phase) with the other additives of the emulsion system. This can be done simply by cloud times (at accelerated temperatures) if the acid or alcohol from the hydrolysis or the interaction product is poorly soluble (as it is in the case of polysorbates and arlacels).

The problem with nonionic surfactant hydrolysis is exactly that it produces a fatty acid, which may become part of the oil phase and hence (aside from providing less coverage of the oil droplet) change the emulsion characteristics of the system.

In general the formulator also determines the HLB (hydrophilic/lyophilic balance) of the system he works with and matches it to the surfactant used [Atlas Chemical Company (now ICI Americas), 1963]. The HLB of the emulsifier can be adjusted by mixing two emulsifiers, e.g., arlacel 85 has an HLB value of 2.0 and polysorbate 80 one of 16.5. If an emulsion system required an HLB of 10 for instance, then the ratio of polysorbate (x) to arlacel ($1-x$) would be given by

$$2(1-x) + 16.5x = 10 \tag{10.19}$$

or

$$x = \frac{8}{14.5} = 0.55 \tag{10.20}$$

4.4. Emulsion Type

In emulsion formulation, the type of emulsion is of concern. If it is desired to make an oil-in-water emulsion (o/w, i.e., oil is the discontinuous phase), then it is important that phase inversion not occur. Investigating this possibility must be a task in the stability program (and is usually carried out by the formulator, not the preformulator). Most often phase inversion is associated with creaming and separ-

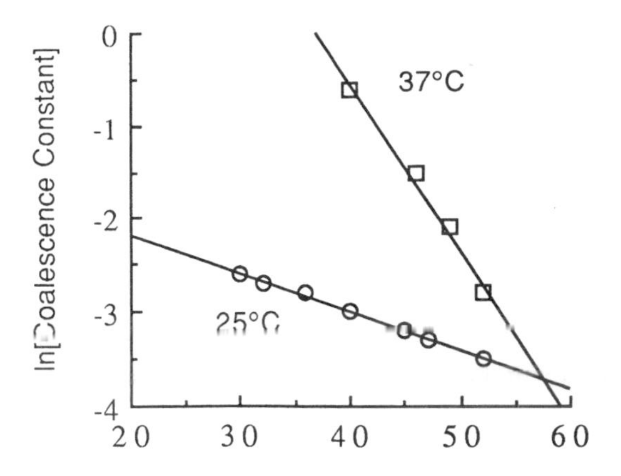

Fig. 11 Coalescence rate versus inversion temperature. (Graph constructed from data published by Enever, 1976.)

ation and will be noticed in the appearance testing of the emulsion. Such phenomena lead to graininess of feel. In some cases part of an emulsion will invert, another not, and then there is a distinct difference in appearances in various regions of the emulsion (creaming).

But the possibility for inversion should always be considered. It is the more likely the closer the system is to a close-packed system of spheres. In this connection, another of the formulator's tasks should be to determine the inversion temperature. (This is at times used to advantage in the manufacturing step, in that, in producing the emulsion, the inverse emulsion is produced at high temperature; this is then cooled, and at the inversion temperature, the "correct" type will result. Conversion in this manner gives rise to very small globules, and homogenization is then often not necessary.) If an inversion temperature exists, then accelerated testing above this temperature is meaningless. So preliminary testing is always advocated, if accelerated testing is contemplated, the philosophy being that there is no sense in testing a system above a temperature where it converts to a physical state that differs from that at room temperature (or recommended storage temperature). Enever (1976) has shown that there is a correlation between phase inversion temperature and the rate of coalescence (Fig. 11). It is possible to use a combination of sedimentation field flow fractionation and photon correlation spectroscopy to record droplet sizes in fat emulsions, and this would appear to be an excellent technique for studying the coalescence of finer spheres, and hence to obtain an extrapolatory tool early on in the storage of an emulsion system.

4.5. Rheological Properties

It has been mentioned that there is a gross correlation between viscosity and globule size. However, the rheological characteristics of an emulsion system in general depends on other factors as well (Sherman, 1955):

1. The viscosity of the internal phase

2. The viscosity of the external phase
3. The phase volume ratio
4. What emulsifiers are used and in what amount
5. The electroviscous effect
6. Distribution of particle sizes

4.6. Appearance of Emulsion Systems

The appearance of the emulsion will be a function of globule size, and Table 2 gives a gross correlation of these two factors. When an emulsion breaks, the hyponatant, rather than being a solution, will have one of the two first appearances in the table, i.e., will also be an emulsion, but with very fine droplets.

4.7. Breaking and Coalescence

It can be concluded from what has been mentioned that the reasons for breaking would include

1. Chemical incompatibility between the emulsifier and another ingredient in the emulsion system (Borax and gum acacia is a case in point)
2. Improper choice of surfactant pair (e.g., wrong HLB)
3. High electrolyte concentration
4. Instability of an emulsifier
5. Too low a viscosity
6. Temperature

As shown in the foregoing, breaking and creaming of emulsions are the typical defective criteria to be looked for in stability programs. Breaking implies that the emulsion separates into two distinct phases (Fig. 7). If this is a slow process, it often manifests itself in the appearance of small amounts of oil particles on the surface, and it then is referred to as *oiling*. When separation into two emulsions occurs (as described above), then the phenomenon is called *creaming*. A rapid test for this is to dip a finger into the preparation and notice if there are different "colors" present (Brown, 1953). Also, a creamed o/w emulsion will not drain off the skin with ease, and the converse holds for a creamed w/o emulsion.

A few words regarding the effect of ionic substances and the actual process of flocculation and coalescence are in order. Van den Tempel (1953) demonstrated that flocculation and coalescence are two different processes. Flocculation depends on electrostatic repulsion (and is akin to the zeta-potential considerations discussed previously). Coalescence depends on the properties of the interfacial film.

Table 2 Correlation Between Globule Size and Appearance of Emulsions

Globule size (μm)	Appearance
>0.005	Translucent (transparent)
0.005–0.1	Semitransparent, gray
0.1–1	Bluish-white emulsion
>1	Milky-white emulsion

Cations, as a whole, are less soluble in the oil phase than anions, and this gives rise to negatively charged droplets (akin to the creation of a zeta-potential in suspensions). The potential drop over the film depends on the nature of the electrolyte (and it should be noticed that there is a diffuse double layer in both liquids as opposed to the case of suspensions, where there is only one diffuse double layer).

Electrolytes may either improve or worsen the stability: If they eliminate the protection offered by the surfactant/protective colloid system then coalescence occurs. Most often electrolytes have the effect of reducing the emulsifying powers of surfactants and causing salting out or actually precipitating the surfactant. However, in some cases, electrolytes will favorably affect the potential drop over the two double layers, and in this case they may stabilize the suspension system.

4.8. Semisolid Dosage Forms

Semisolid emulsions (cold creams, vanishing creams) are not different, in general philosophy, from the above, except that the rheology is checked differently. Davis (1984) has reviewed sophisticated means of checking the stability of these types of systems. He lists the following properties as being important in stability programs for semisolid emulsions:

1. Particle size
2. Polymorphic/hydration/solvation states
3. Sedimentation/creaming
4. Caking/coalescence
5. Consistency
6. Drug release

Of these, particle size, sedimentation/creaming, caking-coalescence, and consistency have been discussed earlier.

Following viscosity as a function of time is here of particular interest. The problem is how to measure the viscosity, and what viscosity in essence means. Davis (1987) points out that changes in viscoelastic properties are much more sensitive than simple continuous shear measurements (Barry, 1974). He demonstrates this via data published by Eccleston (1976). Here (Fig. 12) the variation of the dynamic viscosity (η) and the storage modulus (ϕ) are shown and compared with the same type of graph for apparent viscosity (μ') from continuous shear experiments. It is obvious that the two former measurements are much more sensitive.

4.9. Transdermals

The most important concern about transdermals is the release of drug substance from them and the stability of this property. Other properties (stickiness, appearance, etc.) are of importance as well, but the release characteristic is paramount. Kokobo et al. (1991) have described a means of checking this in vivo by using a single diffusion cell. The volume could be, for instance, 2.5 mL, and the diffusion area could be of the order of 1 cm^2. The matrix is placed, e.g., in contact with a 40% polyethylene glycol solution, which can be, e.g., removed in 500 μL quantities.

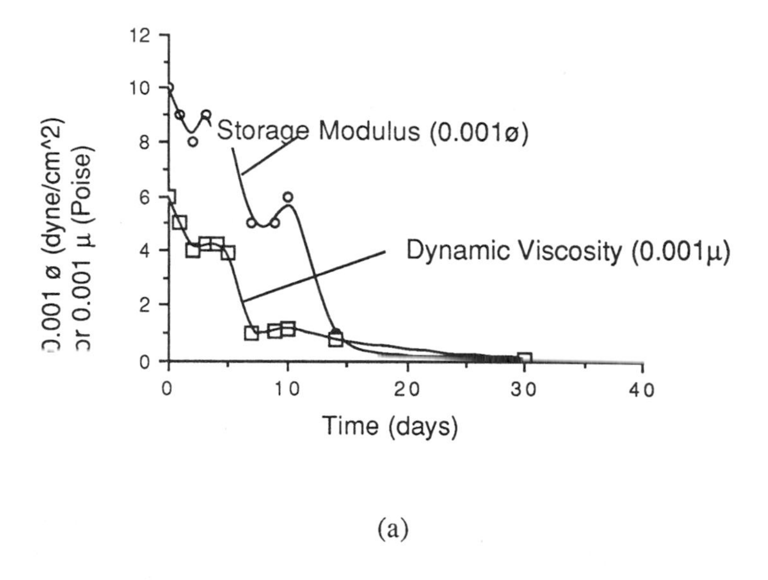

(a)

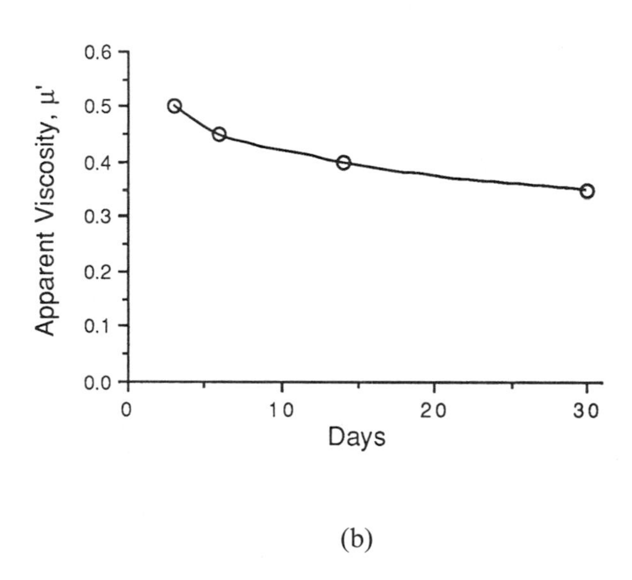

(b)

Fig. 12 (a) Dynamic viscosity (v) and storage modulus (ϕ) and (b) apparent viscosity (μ') as a function of storage time for cosmetic creams made from stearyl alcohol. (Graph constructed from data published by Eccleston, 1976.)

Kokobo et al. (1994) have reported on the interaction between pressure-sensitive adhesives and drug combinations used in transdermals. Their data are shown in Fig. 13. The data fit neither a diffusion equation (ln of retained versus time) nor a square root equation directly. It would appear that if one allows for either an initial dumping in the diffusion equation (or includes more than one term in the Barrer equation) or a lag time in a square root equation, then the data will

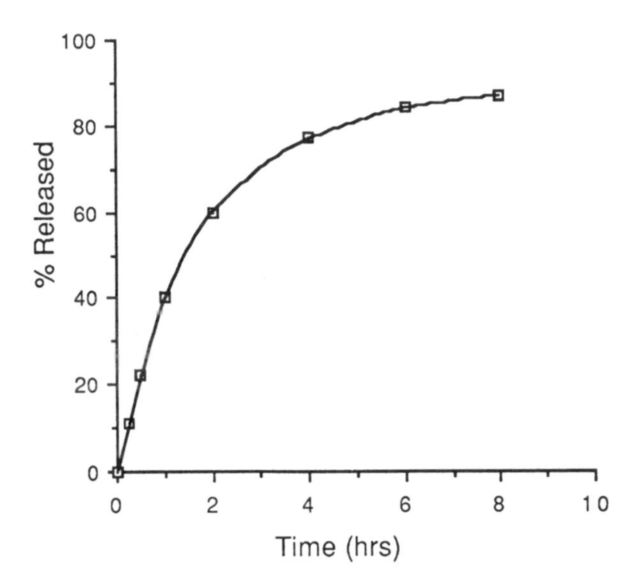

Fig. 13 Release of dipropylphthalate from 2-ethylhexylacrylate acrylic acid copolymer (2EHA/AA). (Graph constructed from data reported by Kokobo et al., 1994.)

fit either. A modified diffusion equation is probably the most likely. The authors suggest the use of the Williams–Landel–Ferry equation for fitting:

$$\{\log D\} - A = \frac{-896}{51.6 + (T - T_g)} \tag{10.21}$$

where D is the diffusion coefficient, A is a constant, and T_g is the glass transition temperature of the polymer.

5. ACCELERATED TESTING AND PREDICTION

Accelerated testing of physical properties of disperse systems is not as clear-cut as for instance chemical kinetics prediction. For instance, the stability of properties of semisolid materials is very difficult, for instance, for creams and ointments that give rise to bleeding there does not seem to be any reliable predictive test. Yet a series of stress tests are used for disperse systems. They include

Shaking tests
Centrifugal tests
Freeze–thaw tests
Elevated temperature tests

It should be cautioned that although these types of tests can be performed on a comparative basis (Is formula A better in one respect than formula B?), their interpretation, other than saying that A is better (or worse) than B, is uncertain, and predictive aspects are somewhat lacking, because the phenomena tested in the accelerated tests do not necessarily mimic what will happen in room temperature storage a/o shipping (Davis, 1987; Rhodes, 1979a and b).

For the freeze–thaw test, the question is what the minimum temperature should be, temperatures from −5° to +5°C being the most common. −5°C frequently gives rise to phase separation and irreversible changes that would not be seen in usual temperature ranges (Nakamura and Okada, 1976), but again, such tests may be used to select a "presumably best" formula from a series of preparations in product development. Results of a typical freeze–thaw cycle are shown in Fig. 10.

Centrifugation has been used by some investigators (Tingstad, 1964; Hahn and Mittal, 1979; and Ondracek et al., 1985). The general idea is that g can be increased in the terminal velocity predicted by Stokes's law (Eq. 11), but often the stresses caused by centrifugation may cause coalescence, which would not occur during normal collision stress.

Some investigators claim fair success in predictions by this means, but as Davis (1987) cautiously states, "as a general rule it can be stated that systems that withstand accelerated stress conditions should be stable under normal storage conditions. However the corollary is not necessarily true." That is, if the preparation fails the test it *may* still be all right, but if it passes the test it should be all right. Although this may be true overall, one can visualize that if a preparation is centrifuged right after manufacture, then the stress does not include the chemical changes (surfactant decomposition for instance) that occur on storage, and in this respect it may give too optimistic a prediction.

Buscall et al. (1979) have measured phase separation at several different centrifugal gs and have established from these data a so-called coalescence pressure. This (again recalling that the test does not account for chemical changes on storage) may be an appropriate parameter.

One predictive method in formulation is the correlation afforded by coalescence rates (Fig. 11), and this is rational in selecting the "best" of many formulations; in general the system with the highest phase inversion temperature is the best. The (nonchemical stability dictated) coalescence rate could theoretically be calculated prior to storage, and the difference between observed and calculated then attributed to chemical stability causes.

For emulsions, it should again be pointed out that rapid creaming and flocculation does not necessarily mean rapid coalescence. For emulsions there have been reports (Rhamblhau et al., 1977) that attempted to tie zeta-potentials to emulsion behavior on storage, but the generality of such an approach has been questioned (Davis, 1987).

The shaking test is usually carried out at 2–3 hertz (Davis, 1987), and the philosophy here is to intensify the collision frequency between globules (and to some degree also the intensity). This is therefore considered an accelerated test, but it actually is part of the product life (transportation). In any event, it should be included in protocols and simply reported.

6. AEROSOLS

Sciarra has reviewed pharmaceutical and cosmetic aerosols (1974). Aerosols are solutions, primary emulsions, or suspensions (i.e., suspensions in a suitable solvent such as ethanol) of active principle in chlorinated hydrocarbons, contained in a pressure can. Either a dip tube or a metering device connects the pressurized liquid contents to the valve. Upon activation of this, the internal pressure will force the liquid through the valve orifice and atomize the suspension. The chlorinated hydrocarbon and the primary emulsion or suspension vehicle will evaporate, and the drug, in finely divided form, will be administered to the location of treatment (lung, skin).

In general the physical instability of aerosols can lead to changes in (a) total drug delivered per dose or (b) total number of doses that may be obtained from the container. It is intuitively obvious that the particle size range must be fine (i.e., the particles will have to pass through the valve).

In general the primary disperse system is filled into a seamless aerosol can, the valve assembly is attached, and the halogenated hydrocarbon is filled by pressure through the valve. The under-the-cap filling method has been described by Boegli et al. (1969). The halogenated hydrocarbon can, alternatively, be "liquid filled" at low temperature. For products that are moisture sensitive, this presents the problem of condensed ice and water in the product.

As far as "cleanliness of operation," aerosol lines are usually kept separate from conventional filling lines (Sciarra, 1974) (or the product is contract filled). Some attempts have been made to use ethylene oxide sterilization of the can (Joyner, 1969a, 1969b), and aseptic fillings (Harris, 1968; Sciarra, 1967) can be carried out.

6.1. Aerosol Testing

Some testing methods are official in the USP (XXI). The Chemical Testing Manufacturers Association has developed a series of tests described in the ASCM Handbook (Aerosol Guide, 1981).

Several test methods are used to detect physical aerosol instability, viz., (1) unit spray content, (2) color and odor, (3) rate of leakage, (4) moisture and trace catalytical substances, (5) particle size distribution, (6) spray characteristics, (7) moisture and trace catalytical substances, (8) pH, (9) delivery rate, (10) microbial limit tests, and (11) container compatibility.

Of the above, leak testing is official in the USP (XXI). This consists of obtaining the weight loss after at least 3 days of storage and converting it to loss per year. If plastic-coated glass containers are used, the test should be done at constant humidity. A faster method is to use an eudiometer tube described in the CSMA aerosol guide. This has the advantage of speed and also is advantageous in that it distinguishes between leakage from crimp versus leakage from valve gaskets.

For spray characteristics a qualitative measurement is to spray onto paper that is treated with a mixture of dye and talc, as described in the CSMA Aerosol Guide. There are also radiotracer techniques (Smith et al., 1984) and TLC graphic techniques (Benjamin et al. 1983). The Aerosol Guide, p. 77 also describes a method whereby the spray is sprayed through a pie shaped wedge onto a rotator.

Particle size analysis is the most important characteristic and hence the most important aerosol stability test. Sciarra states that particle sizes are between 1 and 10 μm and mostly between 3 and 5 μm. Particle size affects stability of delivery rate, effective dose, mass of drug delivered and of course the stability of the suspension itself. The methods used are microscopy, sedimentation methods, light scattering, cascade impactors, and liquid impingers. If the particle size distributions are determined by electronic methods (e.g., Coulter counter, Malvern), then allowance for solubility should be made.

Polli et al. (1969) have shown that the spray particle size is reduced by decreased drug particle size, by concentration of drug, and by the valve orifice size. Higher propellant temperature, vapor pressure, and using a surfactant in the formula also made the spray particle size smaller.

Particle sizes are important for reasons other than physical stability. For inhalation aerosols, for example, it should be recalled that particles larger than 20 μm do not go past the terminal bronchioles, and particles at 6 μm do not reach the lower alveolar ducts. Particles 0.5 to 5 μm reach the alveolar walls and are intermixed with alveolar fluid (Idson, 1970). The chance for a 1 μm particle depositing is less than 50%. There is therefore, particularly for inhalation aerosols, a very narrow particle size range of effectiveness.

Moisture testing is of importance, and except for foams, pharmaceutical aerosols are nonaqueous suspensions. Devices exist that will allow the transfer of the content of the can directly into a Karl Fisher apparatus. This is preferable over transfer by cutting the can open, since this method would allow for condensation of water into the product (which is chilled at the time of the opening of the can). A description of the can device for piercing the can is to be found in the CSMA Aerosol Guide; it allows direct sampling from the content of the aerosol. The moisture is measured by Karl Fisher titration, and there are a number of commercially available instruments that can accomplish this.

Pressure testing is also an official USP XXI method. A prepressurized gauge is placed on the valve stem, and the valve is actuated so that it is all the way open. In the CSMA Aerosol Guide, pressure testing is described. One method employs piercing the can; the other tests directly through the valve.

Microbial limits are described in the USP, e.g., betamethasone valerate topical aerosol. The microbial limits must meet the requirements of the tests for the absence of *Staphylococcus aureus* and *Pseudomonas aeruginosa* under the Microbial Limit Tests.

Delivery rate is official in the USP XXI. The aerosol is allowed to temperature-equilibrate at 25°C. The weight is determined, the can is then actuated for 5 seconds, the weight is determined again, and the delivery rate is then calculated by difference. Delivery rates usually change on storage because of changes in elastomer hardness and gasket swelling. An apparatus is available from Peterson/Puritan Inc. that is accurate to four significant figures. In this assembly a solenoid can hold and actuate, and measure to 0.001 seconds by stop clock (Johnson, 1972).

Poiseuille's law applies to aerosol spray delivery rates: Fisher and Sheth (1973) have shown that delivery rate is linearly related to the container pressure and that it is inversely proportional to viscosity of the can content. Also for a satisfactory system, the delivery rate will not to any great extent be a function of how much of the can has been emptied out. Of course if a can is emptied in one fell swoop, then the cooling effect of expansion may slow down the rate. Also, fractionation of propellant mix occurs and may lead to increased variation of delivery rate.

Valve testing and evaluation should always be done on the final formula (i.e., not on selected solvent systems). A pure solvent will not fractionate, and hence the variation of spray rate may be smaller than with the final formula.

Finally there is the question whether there is an interaction between the can and the product. Can interaction and moisture content are closely related, since under adverse conditions, the halogenated hydrocarbon will react with water and form a halogen acid that may corrode the can. Coating of cans can slow down the rate of this corrosion but not necessarily eliminate it. The control of moisture is therefore important not only for this reason but often also for the reason of chemical stability of the drug.

6.2 Sprays

These are mentioned here in distinction from aerosols; they are mostly nasal sprays. In testing these, the droplet size is important in metered-dose sprays, since small droplets can reach bronchi and alveoli, which would be undesirable, e.g., for delivery of corticosteroid treatment of rhinal disease. Yu et al. (1984) have described a simple experimental setup used for determining the droplet size of flunisolid nasal spray. It is a glass chamber with an air inlet and a plastic stopper that has a hole matching that of the spray unit. This is connected via a conical cavity to a cascade impactor and an appropriate flow meter. This can be done (more expensively) by laser holography (Yu et al., 1983). Such instrumentation may be used to follow possible changes in droplet size distribution as a function of time.

VanOort et al. (1994) and Byron (1990) emphasize that the size of the particles is one of the most important factors in the efficiency of deposition of solids from inhalation aerosols. The FDA has called for a sampling chamber size of 500 mL (Adams, 1989). VanOort et al. (1994) have modified the Anderson Impactor as shown in Fig. 14 and have shown that the chamber volume greatly affects the percentage respirable dose.

VanOort et al. (1994) also tested the effect of the chamber volume, as shown in Fig. 15. In an Andersen Sampler (Andersen Sampler Inc., Mark II 1 ACFM Nonviable Ambient Sampler), the manufacturer recommends that, at a flow rate of 28.3 L/min, the effective cutoff diameters (ECD) are 9, 5.8, 4.7, 3.3, 2.1, 1.1, 0.7, and 0.4 μm for stages 0 to 7.

7. POWDERS

Pharmaceutical powders are for reconstitution into either suspensions or solutions. A prescription example of the former is chloramphenicol palmitate, where the reconstitution is carried out by the pharmacist prior to dispensing. An example of the latter is Metamucil, where the customer reconstitutes the product (e.g., in orange juice). Examples of solutions are Achromycin IM (which is a parenteral powder, i.e., not a lyophylizate). Over-the-counter examples of oral solutions of this type are older products such as Vi Magna Granules (Lederle^{TM}). Analogies in the food area are fruit drink powders, which are sold in packets and reconstituted by the consumer to a certain volume.

The main physical concerns in this type of product are appearance, organoleptic properties, and ease of reconstitution. Only the latter will be treated here.

There are several reasons a powder may change dissolution time as a function of storage time. The most common reasons are (a) cohesion, (b) crystal growth, and (c) moisture sorption, which causes a *lumping up* of powders. The latter is simply due to the dissolution and bridge-forming that occurs and is akin to what happens in wet granulation.

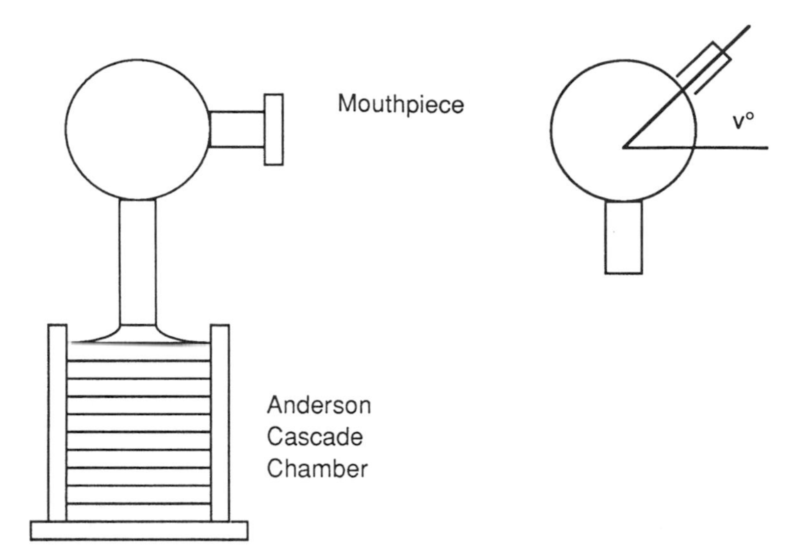

Fig. 14 Figure drawn from schematic published by VanOort et al. (1994).

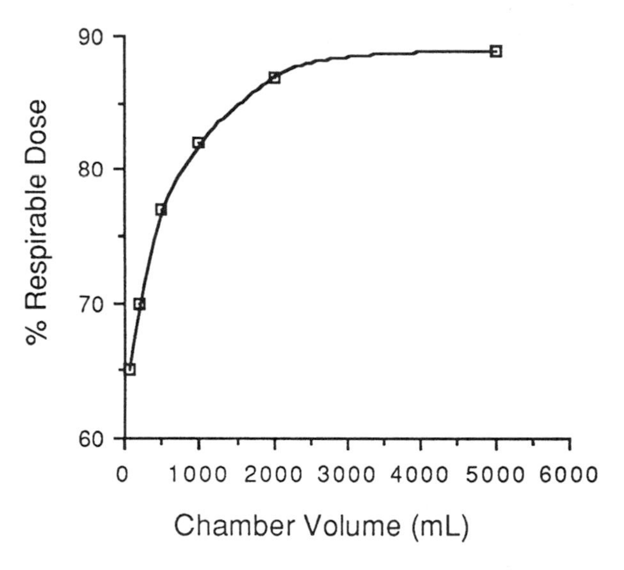

Fig. 15 Chamber volume vs. respirable dose. (Graph constructed from data published by VanOort et al., 1994.)

Cohesional force is the force between two particles, and *cohesion* in general is the stress (force per cm^2 of surface) that a particle experiences due to the surrounding particles. Problems due to cohesion are particularly predominant when a powder is fine, and great fineness of a powder is often required for dissolution reasons. Cohesional forces are inversely proportional to the square of the distance between the particles, so that in storage, where vibration, for instance, may consolidate the powder bed, these forces become large, and the powder "cakes up." This may give rise to problems in reconstitution.

There are two situations in *crystal growth*. One is due to the polymorphism. If the original product is either a metastable polymorph or amorphous, the conversion may occur in storage. For this to happen, some stress, e.g., the presence of moisture, must occur. The stress need not necessarily be moisture, conversion of a small amount of powder might occur in the filling head of the filling machine and then propagate in time.

If the content of the drug substance is such that there are no neighboring drug particles, then this conversion is limited. Particularly, contact points allow for propagation of conversions in situations where the spontaneous nucleation probability is low. The presence of moisture will accelerate conversions of this type, once a seed of the stable polymorph (or in the amorphate situation, once a crystal) has formed.

Crystal growth is, per se, not to be expected. It is true that, by the Ostwald–Freundlich equation, a larger crystal is thermodynamically favored over a smaller one; but the energy differences in the usual particle ranges is small and the activation energy high, so that the likelihood is rather low. If sufficient moisture is present so that the vapor pressure in the container exceeds that of a saturated solution, then some of the drug will dissolve in sorbed moisture. Fluctuations in temperature are never absent and would cause dissolution followed by precipitation, and this can lead to crystal growth. In cases where a drug substance is capable of forming a hydrate, and where an anhydrate is used, growth by way of hydrate formation is possible.

Ease of reconstitution is usually carried out subjectively, in that a tester carries out the reconstitution in the prescribed manner and records the length of time required to finish the operation. For this purpose it is important to have detailed directions on how the reconstitution is to be carried out, and to be sure that there is no operator-to-operator performance bias.

To insure the latter, a set of operators is usually selected for the operation at a point in the stability history. These operators will then be the test instruments for all testing of reconstitutability of oral powders.

The manner of screening operators could be as follows. A random sample is taken of a batch of a product. Random sets of four are taken from this random sample, and e.g. three operators tested. They are each given four samples to reconstitute on the first day, four on the second day, and four on the third day. It is a good policy to have two batches and mix them by day and operator, so as to carry out the test in a blind fashion. The results of such a screening could be as shown in Table 3.

Table 3 Screening of Operators for Reconstitution Testing. Reconstitution Time (min)

	Operator		
	1	2	3
Day 1	1.3 ± 0.3	1.5 ± 0.2	1.4 ± 0.5
Day 2	1.7 ± 0.5	1.4 ± 0.4	1.5 ± 0.3
Day 3	1.5 ± 0.6	1.3 ± 0.5	1.7 ± 0.4

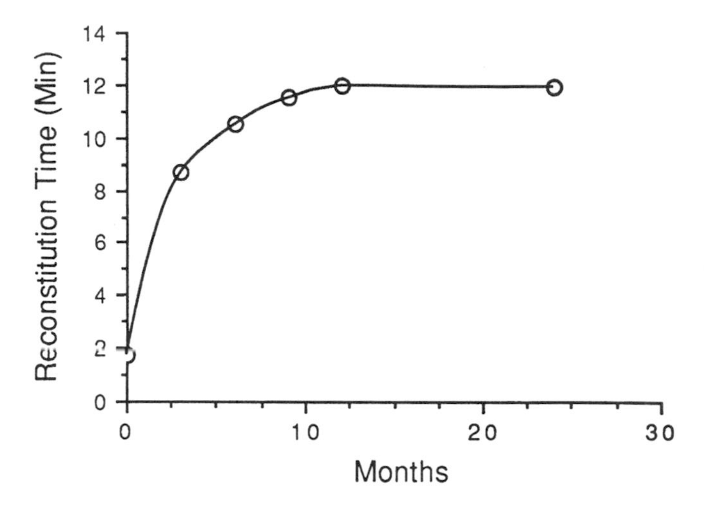

Fig. 16 Change of reconstitution time of a powder on storage.

The ranges shown denote standard errors of the mean. An F-test (Anova) will now fail to show a significant difference between operators. On storage, the reconstitution time could change as shown in Fig. 16.

As mentioned, the most common reason for increases in reconstitution time upon product storage is that the powder becomes more "lumpy" through cohesion developing over time or because it becomes coarser due to crystal growth. Both phenomena are associated with moisture content, and just as it is important to test the effect of the level of moisture content in the case of stability of a solid dosage form, so is it important to test it in the case of a powder.

If m is the number of mL of water adsorbed on one gram of powder, and if S is the solubility (in mg/mL) of all the soluble substances in the preparation, then, since the moisture layer is stagnant, the concentration of solubles at time t will be given by

$$C = S[1 - \exp(-qt)] \tag{10.22}$$

where q is the dissolution constant (kA/V).

The layer will have a higher viscosity (η), the more solid is dissolved, presumably by a power function:

$$\eta = \beta \cdot C^n \tag{10.23}$$

In analogy with the definition of viscosity, the force (F) needed to move two planes separated by a liquid is proportional to the viscosity. It would also be proportional to the amount of liquid, m, in the powder situation stated, so that combining this concept with Eqs. (10.22) and (10.23) gives

$$F = m \cdot \beta \cdot S^n \cdot [1 - \exp(-qt)]^n \tag{10.24}$$

so the reconstitution time would be proportional to this. The data in Fig. 16 follow this pattern.

8. TABLETS

The physical properties associated with tablets are disintegration, dissolution, hardness, appearance, and associated properties (including slurry pH). For special tablet products (e.g., chewable tablets) organoleptic properties become important. These have been described earlier, but in the case of tablets, the chewability and mouth feel also become of importance. The properties will be discussed individually below.

8.1. Tablet Hardness

The "hardness" properties of a tablet are usually assessed by subjecting the tablets to a diametral failure test. The tablet is placed (Fig. 17) between two anvils, one of which is stationary. The other anvil is moved at constant speed against the tablet, and the force (as a function of time) is recorded. The force, at which the tablet breaks is denoted the "hardness" and is usually measured in kp (kilopond = kilogram

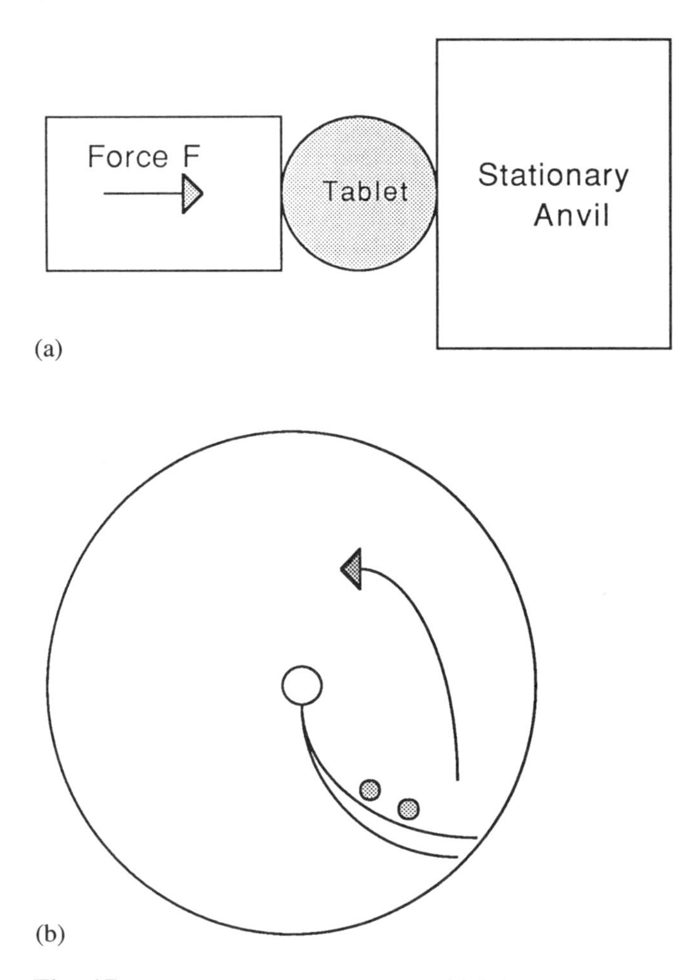

Fig. 17 (a) Hardness tester and (b) friabilator.

force). Other older units (Strong Cobb Units, SCU, or pound force) are used, usually when older instrumentation is used. Until recently, one limitation was that forces over 20 kp would simply register as $F > 20$ kp. Newer instrumentation allows for quantitation of higher forces. From a stability point of view this is important, since the better a parameter can be quantitated, the clearer the picture that emerges will be.

Tablets are made either by wet processing (wet granulation) or by dry processing (direct compression or slugging/roller compaction). In the former case a binder in solution is added to the powder mixture (or is contained in the powder mixture, and wetting then carried out). The binder forms soft bridges between particles, and when the granulation is dried then these bridges become hard. They form the bonds during the compression, and this is one of the reasons for the addition of the binder. The hardness of the tablet is tied in with the strength of the bond. The nature of the actual bond formation will be discussed presently.

In order for a bond to form, the particles or binder bridges must first be exposed to stresses (pressures) that exceed the elastic limit of the material. On failure, the material will either deform plastically or experience brittle fracture. A material that flows well and has a low elastic limit is, therefore, easy to transform into a tablet, and several such materials, known as direct compression ingredients, are used in the manufacture of pharmaceutical tablets. In these cases drug is simply mixed with the direct compression excipient (and other excipients), lubricated, and compressed. If the drug content is less than (approximately) 20% then the tablet will (generally) have the properties of the direct compression ingredient. At higher percentages, direct compression is usually only feasible if the drug substance itself is fairly compressible (i.e., has a low elastic limit).

The hardness of a tablet will be a function of the strength of the bond and the number of bonds. However, this is statistically oversimplified. If there are, for instance, many bonds in the bottom of a tablet and only a few in the top, then the tablet will break easily. Hence it is the average bond density and the standard deviation of the bonds that are really of importance. The same is true about the strength of the bond. Train (1957) has shown that the particle density in a tablet varies from spot to spot, and hence there is a variation in the density of the bonds (and probably in their strength as well).

If the hardness of a tablet is plotted versus the applied pressure, then a plot such as shown in Fig. 18 results. It is seen that the curve goes through a maximum. For good formulations, this maximum does not occur until very high pressures (outside the range of pressures used in pharmaceutical tableting). The maximum occurs because above the critical pressure, P^*, the tablet will laminate or cap, and a laminated tablet (Fig. 19) will contain strata of air and hence be thicker and weaker. Tablet thicknesses will respond in a manner opposite to the hardness, i.e., show a minimum (e.g., at 500 MPa in Fig. 18).

The reason for this phenomenon is the following: As applied pressure increases, the number of bonds, N, increases as well. But assuming that there is a maximum number of bonds, N^*, that can be formed, then the strength, H, of the tablet will asymptote as well. In a simplified manner the relations would be

$$N = N \cdot [1 - \exp(-qP)] \tag{10.25}$$

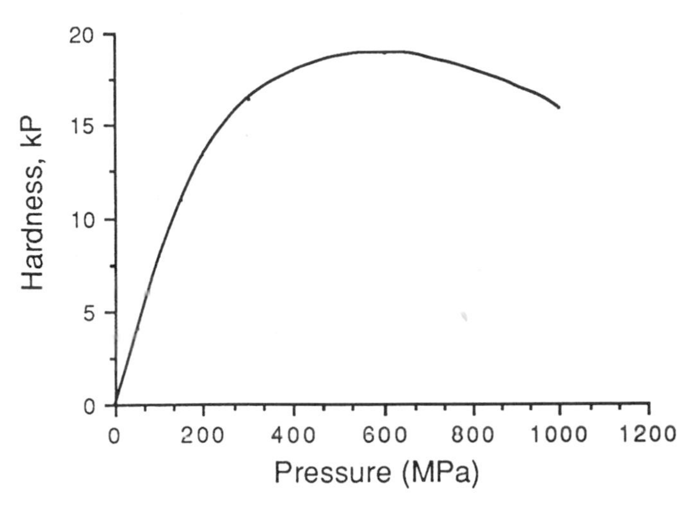

Fig. 18 Tablet hardness versus applied tableting pressure. (Graph constructed from data by Carstensen et al., 1986.)

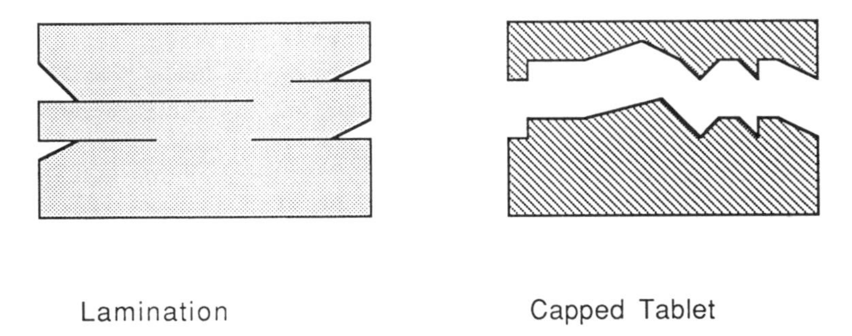

Fig. 19 Laminated and capped tablets.

If the hardness is assumed proportional to the number of bonds, then

$$H = \beta N = H \cdot [1 - \exp(-qP)] \tag{10.26}$$

where H is the capability of the tablet to withstand stress and β is a proportionality constant (Fig. 20).

During the tableting process, when the upper punch is released, a stress is exerted on the tablet, and this stress (S) is the larger, the larger the applied pressure, i.e.,

$$S = f(P) \tag{10.27}$$

At a given point, S becomes larger than H, and then fracture (lamination) occurs within the die, before the tablet is ejected.

There is a second type of stress that occurs during compression, and this happens upon ejection from the die. Here, many tablets expand, and this expansion is a stress that may also exceed H, i.e., laminated or capped tablets are formed.

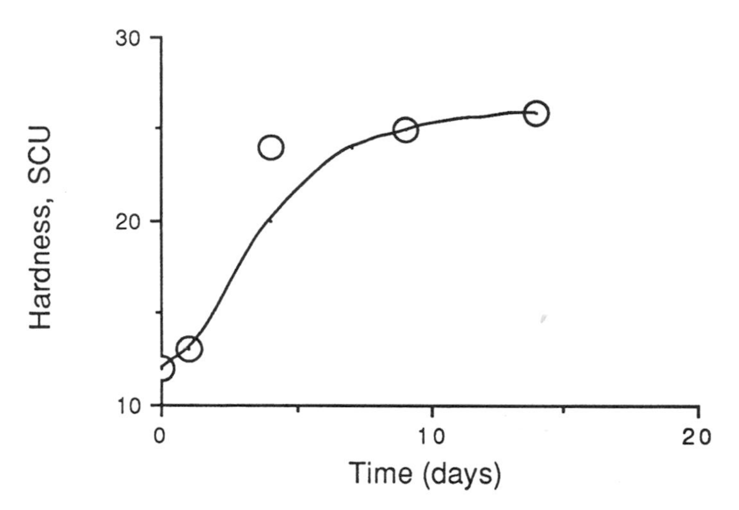

Fig. 20 Hardness as a function of time in pharmaceutical tablets. (Figure constructed from data published by Chowhan, 1979.)

On storage, this expansion can continue (Gucluyildiz et al., 1977), i.e., a tablet may become softer on standing for simple expansion reasons. Expansion is rarely checked as part of a stability program, and the cited article is one of the few published attempts to measure porosity as a function of time.

Frequently tablets will become either softer or harder within short periods of time after manufacture. Figure 20 shows hardness as a function of time for a series of tablets reported by Chowhan (1979).

Aside from the quoted instance of porosity changes and expansion, there are cases where crystallization of a soluble compound has occurred via the sorbed amounts of moisture in the tablet. This happens most often with very soluble compounds, and in such cases it is important to ascertain storage in a dry environment. A test that is now a requirement in the ICH Guidelines is storage in the final container at 40°C, 75% RH. During this test moisture is usually adsorbed by the tablets, and this can then cause softening of the binder bridge because of moisture uptake. At times, redrying will reinstitute the original hardness. Sometimes hardening occurs when the sorbed moisture causes recrystallization of a compound or excipient.

8.2. Softening

Softening can be associated with chemical interaction. Several furoic acids (Carstensen and Kothari, 1983), when tableted with microcrystalline cellulose, will cause a specific interaction leading to the formation of carbon monoxide (rather than decarboxylation of the acid). This interaction is not slow at 55°C, and it causes the tablets to crumble. At room temperature the effect is less pronounced yet significant.

Since a tablet, when produced, is not in equilibrium, there will be a redistribution of moisture. This could make the bonds of a lower or a higher moisture content, and there may for this reason be a change in hardness during a fairly short period of time after manufacture.

Table 4 Moisture Content of Selected Excipients

	Water content		
Excipient	TGA	Calcium carbide	Karl Fischer
Sta-RX 1500	10.8	9.7	10.4
Solca-Floc	5.8	4.6	6.4
CMC	9.2	3.7	14.9
Celutab	8.6	0	9.0
Microcryst. Cellulose	3.4	2.9	4.7
Polyvinyl-pyrollidone	5.4	2.9	6.4

Source. Table constructed from data published by Schepky (1974).

The moisture content of granules, when they are made initially, is a function of their particle size. Pitkin and Carstensen (1973) have shown that when granules are dried, each is associated with one given drying time, t^*. Since the drying (if it is countercurrent, or fluid bed) is a diffusional process, conventional diffusion theory predicts that the amount of moisture left in a granule, m, in relation to the initial amount, m_0, is given to a first approximation by

$$\frac{m}{m_0} = \exp\left[-\frac{D}{a^2}t\right] \tag{10.28}$$

where D is the diffusion coefficient of water in the granule and a is its diameter. The larger granules will hence, have a higher moisture content at the beginning, but the moisture will equilibrate, in most cases, on storage. However, Zoglio et al. (1975) have shown that in some cases (spray dried sucrose granules) there will be no redistribution of moisture between larger and smaller granules.

The moisture contents of various excipients have been reported by Schepky (1974) and are listed in Table 4.

In stability situations it is often the change in moisture as a function of storage time that is of importance. In such cases (Shepky, 1974), the thermogravimetric method may be of advantage.

It is of interest, in cases where moisture equilibrates and causes change in hardness on storage, to be able to assess the extent of moisture transfer within the tablet. As mentioned at an earlier point, the situation is that (in a simple case of a two-component tablet) the two components (I and II) have different moisture isotherms. These are approximated in linear fashion in Fig. 21. This is best illustrated by example.

Example 10.3.

A tablet is made of two components, I and II. It has a given moisture content, 10 mg of water per g of dry tablet weight. The moisture isotherms are as shown in Fig. 21. Calculate the final moisture contents of the two components after the moisture has equilibrated.

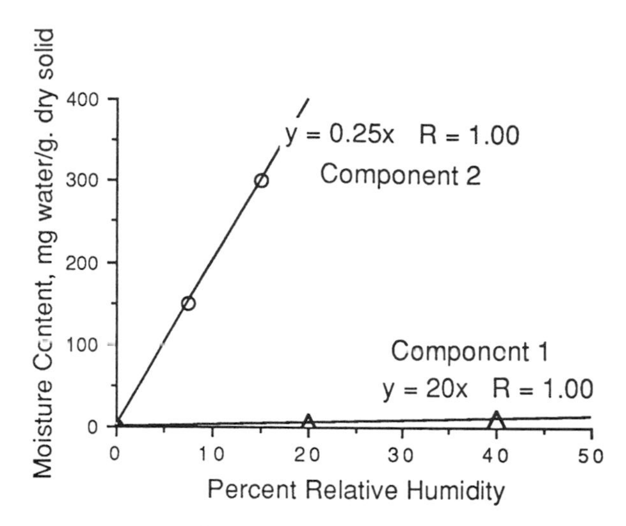

Fig. 21 Moisture exchange between ingredients.

Answer.

For component II, 10 mg of moisture per g of dry solid corresponds to a relative humidity of 40%, i.e., the equation for the isotherm for II is given by

$$y_2 = \frac{10}{40} x_2 = 0.25 x_2 \tag{10.29}$$

since it passes through the point (10, 40). The moisture isotherm for I has the equation

$$y_1 = \frac{300}{15} x_1 = 20 x_1 \tag{10.30}$$

since it passes through the point (15, 300). At the tablet moisture content (10 mg/g dry solid) it is in equilibrium with a gas phase of relative humidity

$$x_1 = \frac{10}{20} = 0.5\%\ \text{RH} \tag{10.31}$$

Hence the situation is not an equilibrium situation, because there is no common vapor pressure over the solids. Component I will, therefore, give up (q grams of) water, and II will pick up (q grams of) water until a common vapor pressure (X^*) in the porous and external vapor space has been achieved.

The equilibrium relative humidity is given by

$$X^* = 0.25(10 - q) \tag{10.32}$$

and

$$X^* = 20(0.5 + q) \tag{10.33}$$

and equating the two right hand sides then gives

$$2.5 - 0.25q = 10 + 20q \tag{10.34}$$

or

$$20.25q = 12.5 \tag{10.35}$$

or

$$q = \frac{12.5}{20.25} = 0.6\,\text{mg} \tag{10.36}$$

It simplifies the computation that the moisture content is given in mg of moisture per g of dry solid (i.e., not in percent, which would be related to mg of moisture per g of total weight). It should be noted that this situation is simplified by assuming the isotherm to be linear.

8.3. Disintegration

Tablets (whether coated or not) are usually subjected to a disintegration test. The disintegration was the first in-vitro test used by the U.S.P. It is now not obligatory compendially (but is recommended); in an obligatory sense it has been replaced by the dissolution test. This latter, hence, is the more important test, but it will be seen that there often is a correlation between the two, and since the disintegration test is much more easily carried out, a stability program will check disintegration frequently, and dissolution less frequently, primarily due to labor intensity.

The apparatus used (U.S.P. XX, p. 958) is shown schematically in Fig. 22. It is an apparatus where six tubes are placed in holders on a circular screen, which is then raised and lowered between 29 and 32 times per minute through a distance of 5.3–5.7 cm in a 1000 mL beaker containing the disintegration medium (either water or N/10 hydrochloric acid). The wire mesh oscillates so that it is 2.5 cm (or more) below the surface at the upstroke and 2.5 cm (or more) from the bottom of the

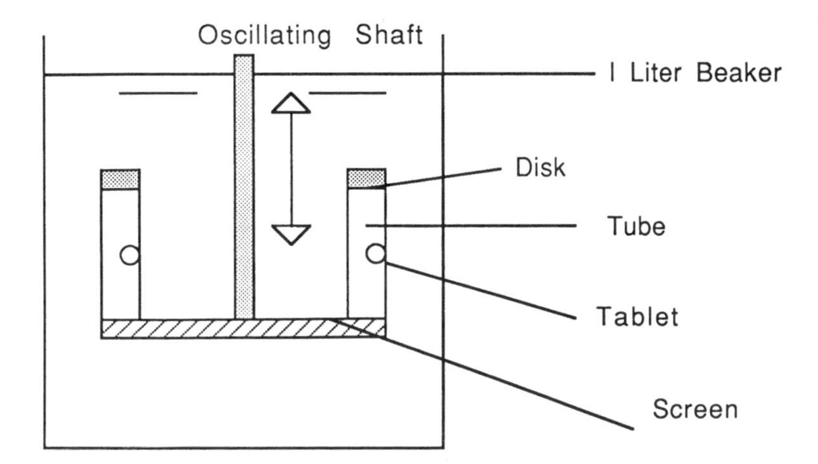

Fig. 22 Disintegration apparatus.

1000 mL beaker at the downstroke. The open-ended glass tubes are 17.75±0.25 cm long and have an inside diameter of 21.5 cm. The glass thickness is 0.2 cm.

Each tube is provided with a disk 95 mm thick and 20.7 mm in diameter, made of plastic of a specific gravity between 1.18 and 1.20 g/cm^3. There are five 2 mm holes in the cylinder (one of them in the axis). The disk also has notches in it and serves to keep the tablet within the tube and submerged during the stroke of the assembly.

To operate the apparatus, one tablet is placed in each of the six tubes, disks are added, and the apparatus is operated at 37°C in the immersion fluid. For quality control release purposes as well as for investigational purposes the time is noted when all tablets have disintegrated completely, and if not all tablets have disintegrated at the end of the specification limit, then the basket is removed and the tablets observed. If one or two tablets have failed, then 12 more tablets are tested, and these must all disintegrate within the limit. However, in stability testing it is important to note the time that each individual tablet disintegrates.

It should be pointed out that complete disintegration is defined as "that state in which any residue of the unit, except fragments of insoluble coating or capsule shell, remaining on the screen of the test apparatus is a soft mass having no palpably firm core." There are apparatuses on the market that have a sensor attached to the disk and can determine this state automatically and record the time at which it occurred. Such an attachment is strongly recommended for stability studies, since it provides an easy means of recording the time of disintegration of each tablet.

There are relatively few articles in the pharmaceutical literature that deal with the subject of the change in disintegration and dissolution upon storage, yet these qualities are as important as the retention of potency of the active compound. If a product falls short of specifications during its shelf life, it becomes unsatisfactory, regardless of the particular parameter that is shortfalling.

One fairly systematic study of this is the work by Chowhan (1979). Here disintegration and dissolution times of e.g. dicalcium phosphate based tablets were studied for prolonged times at 25 and 37°C. The pattern is a sigma minus type of pattern as shown in Fig. 23.

Carstensen et al. (1980a, 1980b) have shown that there often is a correlation between dissolution and disintegration, and Carstensen et al. (1978a, 1978b) have shown the theoretical basis for this. Figure 24 shows such a correlation of dissolution and disintegration times in a U.S.P. apparatus.

Couvreur (1975) has shown that the disintegration of a tablet is a function of several factors. If the tablet disintegrates by virtue of a disintegrant which expands, once it is wetted, then the most important attribute is the rate at which the disintegrating liquid penetrates the tablet, and hence the contact angle between the solid and the liquid is of importance.

8.4. Porosity of Tablets

Cruaud et al (1980) showed that there was a direct correlation between dissolution and porosity in a case where the correlation between disintegration and dissolution was not apparent.

There is one well documented case (Gucluyildiz et al., 1977) where the porosity was shown to change in a tablet as a function of time. What this indicates is that if

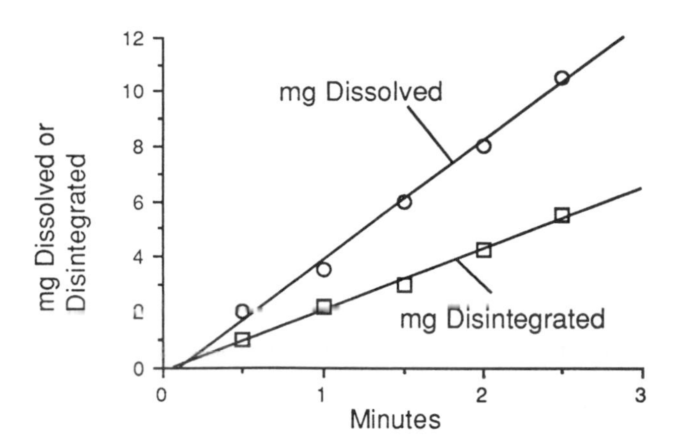

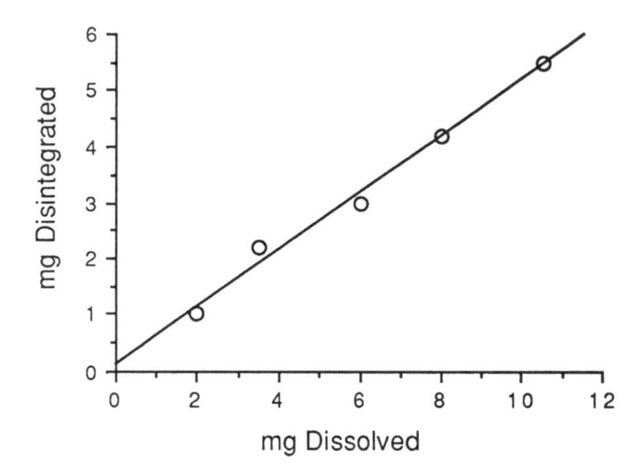

Fig. 23 Top figure: Correlation between dissolution and disintegration upon storage. (Graph constructed from data published by Carstensen et al., 1980a.) Bottom figure: Direct correlation between dissolution and disintegration.

dissolution and disintegration change on storage, then they may be functions of the change in porosity, if indeed porosity changes as a function of storage time.

A rational way of studying this would be to study mercury intrusion as a function of time in tablets of a drug, and to study simultaneously the disintegration and dissolution profiles.

There is a distinct effect of moisture uptake or equilibration on disintegration (and hence, indirectly, on dissolution). If the liquid penetrates an "average" pore, then it encounters, on its way, disintegrant particles. It is assumed that there are q disintegrant particles per linear length of pore. It is also assumed that N particles (per pore) must have been wetted before the tablet can break up (disintegrate).

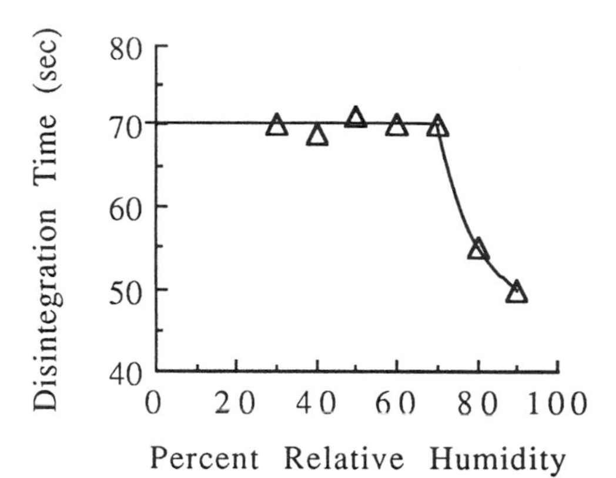

Fig. 24 Effect of storage at various relative humidities on disintegration time of an antidiabetic tablet. (Graph constructed from data published by Grimm and Schepky, 1980a.)

According to the Washburn equation (Washburn, 1921; Nogami et al., 1966; Carstensen, 1980), the length, L, of penetration of liquid at time t is given by

$$L^2 = \left(\frac{r \cdot f \cdot \cos\phi}{2\eta}\right)t = \beta \cdot dt \tag{10.37}$$

where

$$\beta = \frac{f \cdot \cos\phi \cdot r}{4\eta} \tag{10.38}$$

and where f is the interfacial tension, d is the average pore diameter, r is the pore radius, ϕ is the contact angle, and η is the viscosity.

The number of particles wetted n, is related to L by

$$q = \frac{n}{L} \tag{10.39}$$

or

$$L = \frac{n}{q} \tag{10.40}$$

where q is a proportionality constant. The disintegration time, t_N, occurs where $n = N$, so

$$LN = \frac{N}{q} \tag{10.41}$$

and

$$LN^2 = \beta \cdot t_N \cdot d \tag{10.42}$$

Combining these two equations gives

$$t_N = \frac{(N/q)^2}{\beta \cdot d} \tag{10.43}$$

Hence, the following hold for the disintegration time, t_N:

1. It is the larger the more disintegrating particles must swell to make the tablet disintegrate.
2. It is the longer the finer the pore (the smaller d is).
3. It is the smaller the larger the disintegrant concentration, q.
4. It is the smaller the larger the value of β (the smaller the contact angle and interfacial tension).

Of these, N may change, e.g., if the disintegrant becomes wetter, and partly expanded as a result of moisture uptake, this will affect the disintegration adversely. For instance, the Joel Davis test (40°C, 75% RH for three months) has an adverse effect on disintegration for this reason, although it is only true if the relative humidity of the testing station is above a certain critical moisture content (Grimm and Shepky, 1980a). This is demonstrated in Fig. 24.

8.5. Dissolution

The dissolution apparatuses used are usually USP Method I (basket apparatus) or USP Method II (paddle apparatus). Carstensen et al. (1976a,b) have pointed out that the hydrodynamics of the basket method is poor and results in highly different liquid velocities in different parts of the apparatus, and also causes a phenomenon known as coning: powder accumulates at the bottom of the dissolution vessel, where it is fairly stagnant and hence dissolves slowly. Most tests nowadays are therefore carried out with the paddle apparatus.

The assembly is described in USP XX p. 959 and is basically as shown in Fig. 25. The original apparatus could be operated at 50, 100, or 150 RPM, but the more up-to-date apparatus has a variable speed rheostat. In almost all instances the

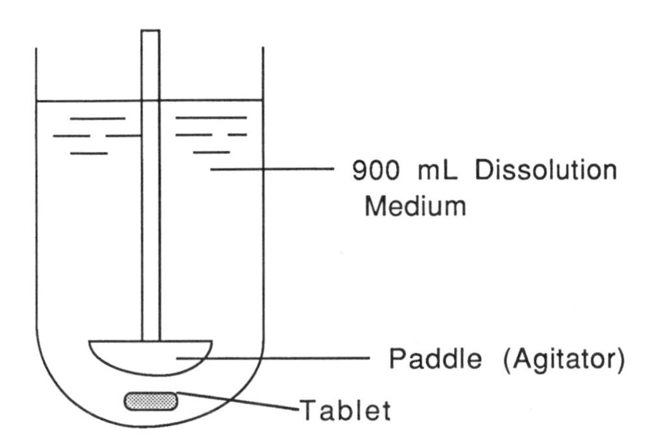

Fig. 25 USP dissolution apparatus.

FDA asks for 50 RPM (sometimes 100 RPM), but only rarely does it accept 150 RPM and insist on the test being "discriminating".

The apparatus should be subjected to a suitability test (USP XX, p. 959), using one tablet of the USP dissolution calibrator, disintegrating type, and one tablet of the USP dissolution calibrator, nondisintegrating type. The apparatus is satisfactory if the data are within the stated range of acceptability for each calibrator.

The dissolution medium is water, hydrochloric acid, or pH 7 buffer. These should be deaerated, since dissolved air may interfere with the dissolution rates.

The procedure used is to transfer usually 900 mL of the dissolution medium to the dissolution vessel and bring it to 37°C. After temperature is equilibrated, the thermometer is removed and one dosage unit is placed in the apparatus. Care is taken to exclude air bubbles from the surface of the tablet and to operate the apparatus right away. After given times, samples are removed from the supernatant and assayed, and the concentration is plotted as function of time. The results may be expressed as percentage of the highest possible concentration (D/V, where D is the dose and V is the volume of the liquid). Monographs specify, usually, a given time at which a certain percentage of label claim, Q, must be dissolved, and the term Q_{30} for instance is frequently used; it indicates the percent of label claim dissolved after 30 minutes. This is known as a one-point assay. For quality control purposes, this is acceptable, but for stability purposes, if rational graphing is contemplated, a dissolution curve rather than a one-point determination should be determined. This will allow determination of the dissolution constants, which will be discussed shortly.

It is often (particularly with slowly dissolving or sustained release products) of importance to have the value "at infinite time." This is usually imitated by increasing the rotational speed (e.g., to 150 RPM) and running the dissolution for an extra two hours. It is, in this scheme, assumed that all the drug will dissolve under such circumstances.

Shortcomings of the apparatus are still (a) that tablets made with excipients of high density will have a tendency to "cone," i.e., after disintegration accumulate in the dead spot just below the agitator; this gives false lows in dissolution rates, and (b) that capsules (and some tablet formula) may float. To avoid floating, a coil is usually placed about the capsule. It is interesting that some tablet formula with relatively small changes in composition (or compression pressure) will change their density, so that they float in one composition (or pressure) and sink in another (only slightly different) composition (or pressure). Expansion of tablets during storage may also change the density so that a table can change from a sinking to a floating composition which will give rise to an apparent slowing down of the dissolution rate.

It should be pointed out that dissolution testing of pharmaceutical products is carried out for several different reasons. In the early stages, the intent of dissolution testing is to get a feel for the comparative estimated bioavailability (on a rank order scale) of different formulations.

In preformulation, intrinsic dissolution rate constants are usually estimated. Although it is not possible, in a direct manner, to tie this in with an estimated bioavailability, it gives a feel for whether the drug substance will be exceedingly problematic, very problematic, problematic, or (in rare cases) not problematic. This feel is comparative with the intrinsic dissolution properties (obtained in a similar fashion) for other drug substances previously developed.

Table 5 Relative Rankings of Furosemide In-Vivo Versus In-Vitro (Random Cross-Over, 12 Patients)

	Rank dissolution	AUC (h-μg/mL)	C_{max} (mg/mL)	T_{max} (h)*
1	B	D	D	D
2	D	A	A=B=E	A=B=C=D
3	A=C	B=E	A=B=E	A=B=C=D
4	A=C	B=E	A=B=E	A=B=C=D
5	E	C	C	A=B=C=D

* Smallest *T* gets best rank, highest AUC and *C* gets best rank (i.e., lowest number in column 1).
Source: Table constructed from data published by McNamara et al. (1987).

In formulation, it is generally assumed that if e.g. three formulae, A, B, and C (Table 5) are developed, then the one that has the fastest dissolution rate should be the best. Whether this general statement is correct is debatable (Table 5), but lacking other criteria it is an accepted yardstick.

In postformulation, i.e. at the point where the new product is manufactured, and when the new product has become an established or old product, dissolution is in the domain of quality assurance. Here it is part of a specification, and the intent of conducting the test is to declare to the public (given the criterion that in-vitro dissolution within certain limits corresponds to in-vivo performance) that the product made on day X, year Y, is comparable to (and should perform in a manner similar to) the batches made year previously on day Z, year Q, when it was tested in the clinic.

If this premise were generally correct, then an in-vitro dissolution test would be universal for all formulae, and it will be seen below that that is an unwarranted extension. The question whether batches of the same formula fall under such a rank order rule is probably acceptable. In the history of a product, however, small changes are often made, and the question whether these small changes shift the in-vitro to in-vivo interrelation is, of course, not known a priori. What constitutes smallness is not clear (and actually is not determinable). Minor changes are defined, now, as changes that tighten specifications and do not involve change in procedure, equipment, or raw material.

If a "substantial" (major) formula change is made, then the bioequivalence between the clinical formula and the new formula must be established.

If subsequent formula changes are made (e.g., a bioequivalence study is carried out in year Y, then another in year Y + 1, etc.), then comparisons should be made with the original clinical formula, not with the previous formula. If the formulae are denoted A (clinical), B, C, etc., then if successive comparisons were made and performance were denoted P, then PB could be 0.8 times PA (and deemed equivalent), PC could be 0.8 times PB (and deemed equivalent), but PC would be $0.8^2 = 0.64$ times PA and hence no longer equivalent. Hence, equivalence testing should always test back to the formula used in the original clinical batches that were part of the medical scheme (and the results of which were approved by the Food and Drug Administration).

8.6. Percolation Thresholds

When a solid is compressed, then one might imagine that at "full" compression, the tablet would be similar to a perfect crystal, in that there would be no void space left in it. This is never achieved, however, and the fraction of void is called the porosity. This may be visualized as isolated pockets of void space or, as the porosity increases, strings of void, eventually terminating at the surface. The porosity at which this latter situation is achieved is denoted the threshold value.

Threshold values for a drug and its excipients in combination are important because they govern such properties as dissolution, hardness, and disintegration. For this purpose, percolation studies are often employed in pharmaceutical research.

Leuenberger and Leu (1992) and Leu and Leuenberger (1993) introduced the concept of drug percolation to the pharmaceutical sciences. By this, a pharmaceutical system is described as a bond/site system. In this concept, a cluster is defined as a group of nearest neighbor sites where all positions consist of the same component. There is a concentration where there is maximum probability that the clusters will start to percolate, and this is the percolation threshold. If the measured porosity of the tablet is denoted ε_m and (after dissolution) the porosity created by loss of dissolved matter is denoted ε^*, then the so-called β property is

$$\beta = -c\varepsilon_c + c\varepsilon \tag{10.44}$$

where $\varepsilon = \varepsilon_m + \varepsilon^*$ is the initial + developed (matrix) porosity, c is a constant, and ε_c is the critical porosity threshold for percolation. This ties in with the Higuchi type plot, the slope of which is b, and β is defined as

$$\beta = \frac{b}{[2A - \varepsilon S]^{1/2}} \tag{10.45}$$

where A is the drug load (g/cm^3 of total tablet) and S is solubility. When porosity is plotted versus β value, then a straight line ensues that cuts the x-axis at the percolation porosity.

The threshold for drug percolation may be obtained when more drug is available than that described in Chapter 9. Soriano et al. (1998) have described percolation methods that are done primarily by conducting dissolution studies with drug substance at various concentrations. They employ the method of Bonny and Leuenberger (1993) and Leuenberger and Leu (1992) for this purpose.

8.7. Multipoint Determinations

In post NDA testing, there is some reason for not carrying out dissolution at more than one time point, because of both human resources and equipment. In pre-NDA situations, however, as described e.g. by Prandit et al. (1994), the importance of carrying out multiple time points in dissolution cannot be stressed enough. Conclusions are difficult to reach if this is not done.

For instance Prandit et al. (1994) reported that aging affected the dissolution of nalidixic acid tablets and concluded that the effect was not attributable to an increase in disintegration time (as measured in a dissolution apparatus). Published data often suffer from being *one point data*, so dissolution/disintegration correlations cannot be deduced from the reported figures.

8.8. Dissolution Media

There is always the problem of what dissolution medium to use. For poorly soluble drugs there are several approaches: cosolvents, micellar systems, and/or large dissolution volumes. Naylor et al. (1993) studied the mechanism of dissolution of hydrocortisone in simple and mixed micelle systems by using a rotating disk and found the Levich equation to hold.

8.9. In-Vivo to In-Vitro Correlation

The problem of whether an in-vitro dissolution test generally measures in-vivo performance on a rank scale basis is still open to debate, when the problem is considered in general, i.e., if product A from manufacturer A has a dissolution rate curve "above" that of manufacturer B, will his product also have a better in-vivo performance as far as large magnitude (C_{max}) and short peak time (T_{max}) for the maximum of the blood level curve and high value for the area under the blood level curve (AUC)? The general premise is that the answer is yes, but as shall be seen below, this is not necessarily so. The correct general statement is that if two batches of the same product and formula are tested, then such a comparison is correct, i.e., that a "higher" dissolution curve implies at least one of the following: lower T_{max}, higher C_{max}, or higher AUC. An example of noncorrelation, when the formula is not the same, is the work by McNamara et al. (1987), in which furosemide from five manufacturers was tested against a solution. The relative rankings are shown in Table 5.

It is seen, then, that the best performer in vivo (D) is by no means the best performer in vitro, and that the worst performer in vitro (E) is not the worst performer in vivo.

The best and simplest method for correlation of in-vitro to in-vivo data would appear to be the mean residence time (MRT), and such comparisons have recently been described by Block and Banakar (1988). MRT is defined by many authors as shown in Fig. 26. The MRT factual definition is a measure of the "average" length of time a drug molecule is in the body (Fig. 26).

Mean residence time via statistical moment has also been described by Yamaoka et al. (1978). Podzeck (1993) has compared in-vitro dissolution profiles by calculating mean dissolution time and mean residence time.

Of late, deconvolution has been often reported and may form part of the 1995 USP. This method consists of comparing a blood level curve after solid dosage form administration with one after either solution or IV administration. The amount dissolved in the GI tract is then obtained by deconvolution. Sugawara et al. (1994) tested a series of controlled release preparations of prednisolone in alginate gel beads, all in a drug-to-alginate ratio of 1:4. As seen in Fig. 27, they were able to obtain in-vitro methods that "matched" the amount released in-vivo.

It is seen in the figure that for the fast releasing formulation (a), the in-vitro test, whether at pH 1.2 or at pH 6.8, follows the deconvoluted in-vivo results fairly well, but for the slow formula, it is only the pH 1.2 in-vitro test that correlates with the deconvoluted in-vivo dissolution test.

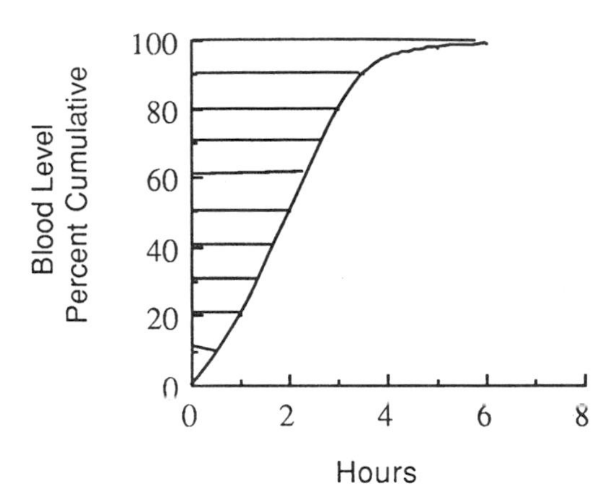

Fig. 26 Cumulative blood level curve (or urinary excretion curve, or dissolution curve).

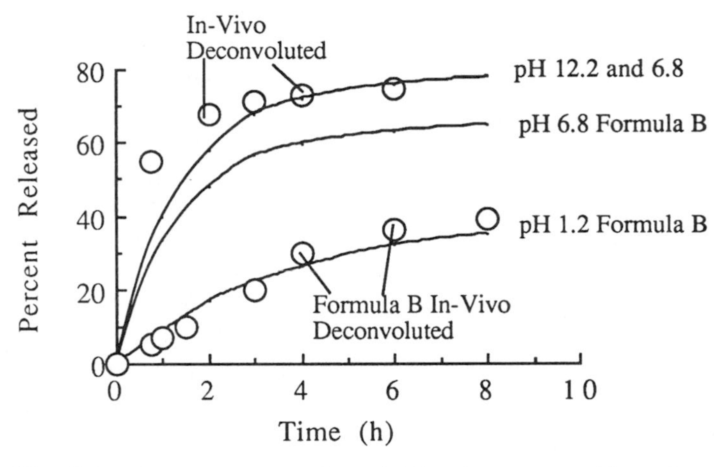

Fig. 27 Time vs. percent released. (Graph constructed from data published by Sugawara et al., 1994.)

The deconvolution method used was the one that has been described by Katori et al. (1991). Other, more recently developed methods are those of Gillespie and Cheng (1993). They first created a hypothetical clean curve with convolution. Then the absorption rates and cumulative amounts absorbed of the drug and metabolite were estimated by the proposed deconvolution method. For this purpose, polyexponential functions were fitted to the simulated data. The resulting parameters were compared by a multidimensional deconvolution program NDCREV (user-friendly IBM compatible).

8.10. Stability of Dissolution Curves

The problem from a stability point of view is that at times the dissolution curve will change as a function of storage time [as e.g. shown by Chafetz (1984) for hard shell capsules], but the bioavailability "stays the same." In such a case the in-vitro

MRT (or dissolution curve) would change but the in-vivo (either MRT or deconvolution curve) would not, so how can there be a correlation between the two?

The accelerated test in the ICH Guideline (40°C, 75% RH) is too severe a test for hard- and soft-shell capsules. Upon dissolution, a skin (a pellicule, or as some authors call it, a pellicle) will form around the capsule in the dissolution apparatus, and this will prevent dissolution. For instance, Dey et al. (1993) exposed etodolac capsules to the accelerated test so that they formed pellicules and showed that the dissolution was not affected when tested with enzymes, but that pellicules formed and dissolution decreased drastically on storage when nonenzymatic fluids were used. They showed that there was no difference in blood level curves of fresh, stored, and failed batches.

In the case of hard- and soft-shell capsules, gelatin can interact with substances in the fill. Gautum and Schott (1994) demonstrated an interaction of anionic compounds (substituted benzoic and sulfonic acid dyes) with gelatin. Capsule fills that contain, or on storage produce, keto groups will always show this phenomenon (Carstensen and Rhodes, 1993).

It should be pointed out that when disintegration of a dosage form changes on storage, it usually happens quite rapidly (usually within 12 weeks) at room temperature. Often, however, the tablet is not checked until 6 months after manufacture. There are then instances where it would seem to be logical to attempt an accelerated test at higher temperature. There has, to date, not been any convincing correlation between disintegration (and dissolution) profiles at higher temperature, vis-à-vis those at lower temperatures. Judging from the factors that affect these two properties, this is not surprising. But what is more to the point is that changes can usually be determined rapidly at room temperature. It is therefore more rational to determine disintegration at 4, 8, and 12 weeks at room temperature in stability programs, and to dispense with testing at higher temperatures.

Gordon et al. (1993) have reported on the effect of aging on the dissolution of wet granulated tablets containing superdisintegrants. Often the decay in dissolution efficiency is due to the lengthening of the disintegration time.

There is, obviously, a correlation between particle size and dissolution, and if the particle size changes as a function of storage time, there may be a correlation between accelerated temperature storage and dissolution. But in such a case the correlation should be established on the neat drug, as was done e.g. by Grimm and Shepky (1980b). Their data for oxytetracycline are shown in Fig. 28.

Dukes (1984) and Murthy and Ghebre Sellassie, (1993) have discussed storage stability of dissolution profiles in general, and Rubino et al. (1985) have described the specific storage stability of the dissolution of phenytoin sodium capsules. Carstensen et al. (1992) have discussed the mathematical basis for change in dissolution curves of dosage forms as a function of storage time. They employ the sigma minus model for dissolution, i.e.,

$$\frac{M}{M_0} = 1 - \exp[-k(t - t_i)] \tag{10.46}$$

where t is dissolution time, t_i is dissolution lag time, M_0 is initial amount in the dosage form, M is the amount left undissolved at time t, and k is a dissolution constant (time^{-1}). t_i is primarily a function of disintegration time.

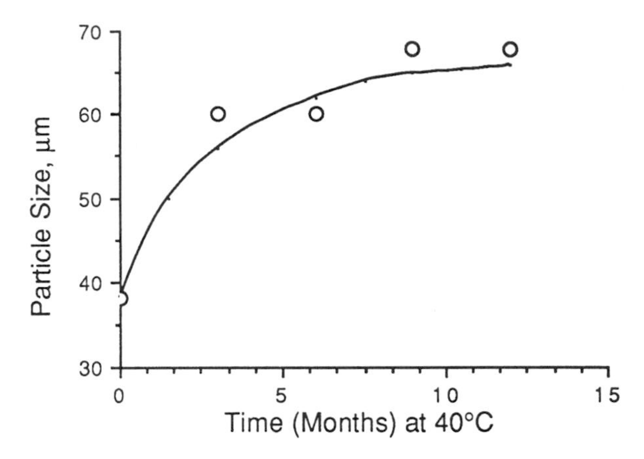

Fig. 28 Particle growth in accelerated storage of oxytetracycline. (Graph constructed from data published by Grimm and Schepky, 1980b.)

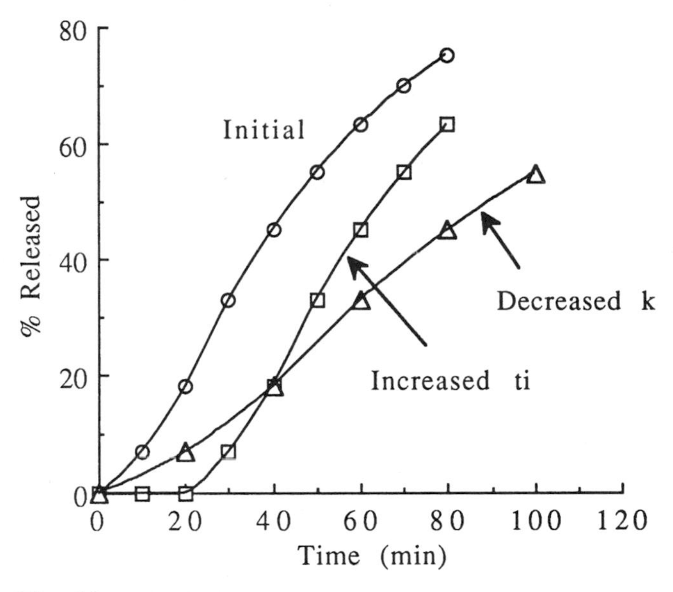

Fig. 29 Dissolution upon storage.

Frequently, upon storage, t_i may change, but k may not, in which case the dissolution curves simply move in parallel to higher and higher mean dissolution times. k, however, may change, and t_i may stay constant, in which case the curve becomes "flatter" (Fig. 29). Finally, both may change, giving rise to a flattening *and* a parallel displacement of the curves. If such parameters as t_{90} (the length of time for 90% to be dissolved) or Q_{45} (the amount dissolved at 45 minutes) are employed, then power function relationships result, and these are difficult to interpret. A better approach is to study and plot k and t_i as a function of storage time. If t_i on storage approaches 45, then the storage stability curve may have a shape as in Fig. 30.

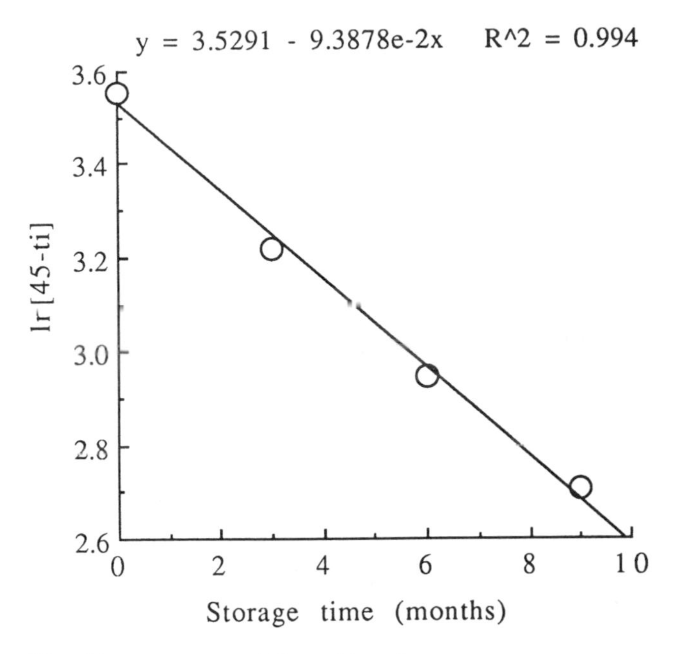

Fig. 30 Storage stability of lag time.

For instance, if a secondary parameter, such as the time needed for 50% dissolution, t_{50}, is followed, then

$$\ln 0.5 = -0.693 = -k(t_{50} - t_i) \tag{10.47}$$

or

$$t_{50} = \left(\frac{0.693}{k}\right) t_i \tag{10.48}$$

and if

$$t_i = t_i^{\infty}[1 - \exp(-q\phi)] + t_i^0 \tag{10.49}$$

where t_i^0 is initial lag time, t_i^{∞} is lag time at infinite storage time, ϕ is storage time, and q is the stability constant, then

$$t_{50} = \left(\frac{0.693}{k}\right) \{t_i^{\infty}[1 - \exp(-q\phi)] + t_i^0\} \tag{10.50}$$

If, now, the storage stability of k is of importance, the expression for t_{50} becomes even more complicated (but can obviously be deduced).

Jørgensen and Christensen (1996, 1997) have approached this problem by introducing a so-called Order Model. By this an order of reaction, n, is assigned to the dissolution curve, and the expression becomes

$$\frac{M}{M_0} = 1 - [1 - \{(1 - n)k(t - f(t_0))\}]^{1/(1-n)} \tag{10.51}$$

where $f(t_0)$ is the lag time function given by

$$f(t_0) = t_0\left[1 - \exp\left(\frac{-t}{t_0}\right)\right] \tag{10.52}$$

t is, again, the dissolution time.

8.11 Appearance of Tablets and Capsules

A stability program should record the appearance of tablets as a function of storage time. This is most often done by subjective description, or by a rating index (0 for unchanged, 5 for vastly changed). Quantitative methods exist and are the following:

Comparison with color chips or charts (Rothgang, 1974)
Dissolving the dosage form and measuring the solution spectrophotometrically (Hammouda and Salakawy, (1971)
Photography (Armstrong and Marsh, 1974)
Reflection measurements (Matthews et al. (1974/75), Carstensen et al. (1964), Carstensen (1964), Goodhart et al. (1967), Turi et al. (1972), Wortz, R. B., (1967)

In the case of the second and fourth methods, a qualitative appearance description is always necessary, because the instrument will "average" the product. Comparison with chips can be used but is somewhat subjective. Such color charts have triangularly arranged chips, and the operator matches the object with a chip, which has a coordinate number. In fact the degree of whiteness (L), redness, (a) and yellowness (b) can be calculated from this, and it will be seen later on that this will allow for quantitative treatment of the change of the color of a pharmaceutical tablet or capsule.

Photography, of course, is relying on stringent adherence to conditions (exposure, aperture, and development) to insure that it is actually the tablets that are being compared, not the procedure for making the photograph.

Reflection measurements are often carried out in tristimulus meters and have been used quite extensively with varying degrees of success. If a tablet (or other surface) is placed in the meter, then reflectance values at three spectral regions are registered and recorded as x, y and z values. Rowe (1985) has reviewed these and points out that the whiteness index is $4(100Z/Z_0) - 3Y$, and the yellowness index is $100[1 - (100Z/\{Y \cdot Z_0\})$, where $Z_0 = 118.1$. In actuality, the degree of whiteness, L, the degree of redness, a, and the degree of yellowness, b, are given by the formula (for a Hunter tristimulus meter):

$$L = 100\left(\frac{Y}{100}\right)^{1/2}$$

$$a = 175\frac{(X/98.041) + (Y/100)}{(Y/100)^{1/2}}$$

$$b = 70\frac{(Y/100) - (Z/118.103)}{(Y/100)^{1/2}}$$

Table 6 Tristimulus Parameters for Some Colors

Color	x	y	z	L	a	b
White	82.3	84.3	101.1	91.8	−0.7	−1.0
Light yellow	65.0	69.9	7.0	83.5	−7.4	53.5
Yellow ochre	32.9	28.7	7.5	53.6	16.1	29.3
Scarlet	36.2	20.6	4.3	45.4	62.2	26.2
Magenta	23.9	13.0	23.9	36.0	55.2	−13.9
Turquoise	13.8	21.0	44.7	45.8	−26.5	−25.7
Emerald green	17.7	30.1	12.3	54.8	−38.3	25.1

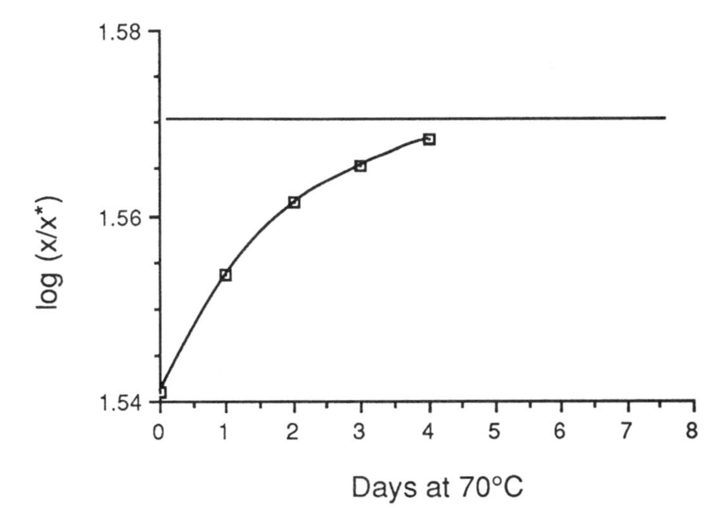

Fig. 31 Reflectance X values (with initially $X = X^*$) of vitamin C tablets as a function of storage time. (Graph constructed from data published by Carstensen et al., 1964.)

Komerup and Wanscher (1967) and Rowe (1983) give as an example the following values for some standard colors (Table 6).

The values calculated in the last three columns by Roe correspond quite well with those obtained from or listed in color charts (Komerup and Wanscher, 1967).

It should be pointed out first of all that reproducibility in reflectance meters is poor, and so results should always be obtained as averages of at least nine independent measurements. Since these are rapidly carried out, the labor is not all that intensive.

Changes in these values are difficult to interpret from a qualitative point of view, but the following procedure allows extrapolation, using x, y, or z (or composites).

Carstensen et al. (1964) have shown that the response values (Y) can be plotted as a function of storage time (t) to give graphs as shown in Fig. 31. This type of plot can be plotted as a sigma minus function:

$$Y = Y_\infty\{1 - \exp(-kt)\} \tag{10.53}$$

The k values can be plotted as an Arrhenius plot, i.e., one may, after short periods of time, at elevated temperature, calculate an extrapolated k value at room temperature. By sampling daily at 55°C, one can determine the Y value ($Y_{\text{lower limit}}$), which corresponds to the poorest appearance that is acceptable. Since k is known for room temperature (k_{25}), it is possible to calculate a "shelf life date" (t^*) based on appearance from inserting $Y_{\text{lower limit}}$ into Eq. (10.54):

$$\ln\left\{1 - \frac{Y_{\text{lower limit}}}{Y_\infty}\right\} = -kt^* \tag{10.54}$$

9. SUSTAINED RELEASE PRODUCTS

There are several types of sustained release principles used in pharmaceutical products, and a detailed description is beyond the scope of this book. What will be done here is simply to state the types of dissolution profiles that can be expected, and how the parameters could change with time.

9.1 Coated Beadlets and Granules

The coated nonpareil seed is the original sustained release form invented by SKF in the 1950s. Here a drug is applied (in the form of a sugar syrup) to monodisperse sugar crystals. Drying is carried out after each application step, so that the drug eventually is in a sugar matrix around the original seed. This beadlet is then coated with either a semipermeable film or an impermeable film with a soluble filler. The latter, upon exposure to dissolution medium, will allow the soluble filler to dissolve, so that pinholes are created in the film. Liquid then diffuses in through the film (or the holes in it), becomes saturated on the inside of the beadlet, and the dissolved drug then diffuses out. The diffusion takes place under an (approximately) constant concentration gradient (the solubility of the drug in the medium), as long as there is undissolved material inside the beadlet (and the concentration is low in the outside fluid creating sink conditions). Once the last drug has dissolved, the concentration inside the beadlet will decrease, and the diffusion slows down. It is, therefore, often, difficult to get the last 5–10% of material to release from this type (and other types of sustained release) dosage forms.

There are, obviously, three stages in the dissolution (Fig. 32):

$0 < t < t_i$: Penetration of liquid into the pellet. t_i is the time it takes for this to complete, and it is denoted the lag time.
$t_i < t < t_f$: t_f is the point in time where all the drug inside the pellet has dissolved.
$t > t_f$: This is the final period where dissolution is slower.

The general dissolution pattern in the period $t_i < t < t_f$ is

$$\ln\left[\frac{M}{M_0}\right] = -k(t - t_i) \tag{10.55}$$

M is the mass not dissolved (and M_0 is the dose) and is obtained by multiplying concentration with dissolution liquid volume and subtracting this (the amount dissolved) from M_0. k is the dissolution constant and will be the smaller (and t_i

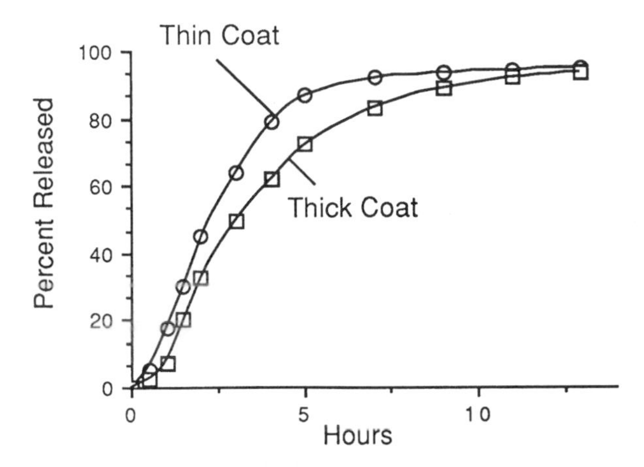

Fig. 32 Release patterns of thinly and thickly coated pellets.

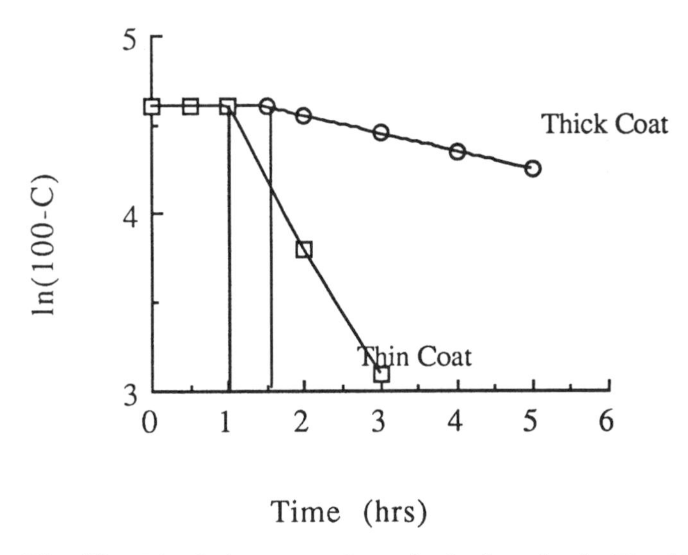

Fig. 33 Dissolution curve of sustained release beads. The thickness of the coat of II is 1.6 times that of I. The graph shows the decrease in slope and the increase in lag time with increased thickness. (Graph constructed from data published by Carstensen, 1973.)

the longer), the thicker the film and the lower the amount of soluble filler in the film. Figures 32 and 33 show this type curve (Carstensen, 1973). The first two points have been omitted, as have the last three points, i.e., $t_i =$ about 1.5 hours and $t_f =$ about 5 hours. The points omitted are such that the remaining points give the best linearity. It is seen from the graph that the least squares fit equations are

$$\text{I} \qquad \ln[M_1] = 5.09 - 5.04t \tag{10.56}$$

$$\text{II} \qquad \ln[M_2] = 4.8 - 0.29t \tag{10.57}$$

so that, as expected, the slopes are in a ratio of 5/3, i.e., the inverse of the ratio of thicknesses. The actual lag times are found by setting $\ln[M] = \ln[100]$, and they are $t_I = 0.67$ hours and $t_{II} = 0.96$ hours, i.e., again in the correct ratio. t_f is the point of inflection, i.e., occurs when all the drug inside the pellets will have dissolved (although not all will have diffused out).

In stability programs, t_i and k are the logical parameters to follow, i.e., complete dissolution curves should be determined. Again, it is wise to do this at room temperature storage at fairly short intervals at the onset (4, 8, and 12 weeks). Again, accelerated testing is not of much use.

9.2 Erosion Tablets

Tablets can be made of e.g. a waxy substance, which does not dissolve or disintegrate, but erodes away. The drug in the eroded portion will dissolve, and (in theory) the drug in the noneroded part will not have dissolved. There is, however, always some penetration of liquid into the waxy tablet, so that more than the eroded drug will often have dissolved. If pure erosion occurs, then the dissolution equation will be

$$M = M_0^{1/3} - K_e(t - t_i) \tag{10.58}$$

where K_e is an erosion constant (cube root dissolution rate constant) and t_i is the length of time of wetting. Both of these parameters can be calculated at different storage periods, and changes can be monitored in a logical fashion. Accelerated studies of this are not meaningful.

9.3 Insoluble Matrices

If a drug is enclosed in an insoluble matrix that is porous, then the release rate is given by the Higuchi square root law (Higuchi, 1963):

$$Q = K_i(t - t_i)^{1/2} \tag{10.59}$$

or

$$Q^2 = K_i^2 \cdot (t - t_i) \tag{10.60}$$

where

$$K_i^2 = a^2\left[2DS\varepsilon\left\{A - \frac{S\varepsilon}{2}\right\}\right] \tag{10.61}$$

a is here the surface area through which the diffusion takes place, ε is the porosity, and A is the loading, the amount of drug per cm^3 of dosage form. ε, the porosity, is the inherent porosity of the tablet plus the porosity created by the drug that has dissolved (i.e., A/ρ, where ρ is the density of the drug).

Eq. (10.61) applies to situations where the drug dosage, A, is larger than $S\varepsilon 2$. If this is not the case (Table 7), then the equation takes the form (Fessi et al., 1982)

$$Q^2 = a^2 Dt \tag{10.62}$$

Table 7 Dissolution According to Eq. (10.62)

Time (min)	Amount released (mg)	Square root of time ($min^{1/2}$)	Amount released
0	0	0	0
8	12.3	2.87	151
15	19.8	3.87	392
45	43.4	6.71	1884
78	59.4	8.83	3528
96	67	9.80	4489
128	75.5	11.31	5700
164	80	12.8	6400
216	84.9	14.7	7209
276	87.7	16.6	7691

Source: Data from Fessi et al. (1982).

Certain products are not porous but depend on the dissolution of the drug to create the porosity. In such cases there is a minimum drug content necessary for creating a porous network, and some of the drug will be occluded, i.e., will never release. A practical minimum is about 20% drug is such cases.

The derivation of Eq. (10.59) is based on the assumption that the penetration of liquid is faster than the dissolution of the drug. If, e.g., the contact angle (wettability) changes with storage (e.g., due to moisture redistribution), then this assumption could be rendered false.

Equation (10.59) applies only as long as there is undissolved material in the matrix (and until liquid has penetrated into the center of the tablet) The parameters K_i and t_i may be monitored at various periods of room temperature storage time. In the case of insoluble matrices, accelerated studies might be possible in certain instances (i.e., when neither matrix nor drug changes physically at the higher temperatures). Table 8 gives an example of Eq. (10.62).

These data are depicted graphically in Fig. 34. It is seen that the least squares fit is given by

$$Q^2 = -2.55 + 0.477t \tag{10.63}$$

when the linear points are used. [These are, again, obtained by successively omitting terminal points (beginning and end) until the best linear fit is obtained.] It is noted that the first two and the last three points have been omitted, i.e., $t_i = 2.55/0.477 = 5$ minutes and t_f (from the best, high point omitted) is 128 minutes. Again, 75.5% are released at this point, and this is quite characteristic, and it calculates out well for most such dosage forms as the point dissolved at the time the tablet has filled up with dissolution medium.

Curing of the product is at times necessary. The work of Omelczuk and McGinnity (1993) has, for instance, shown that matrix tablets containing poly(DL-lactic acid) change release pattern if thermally cured. Drug release from

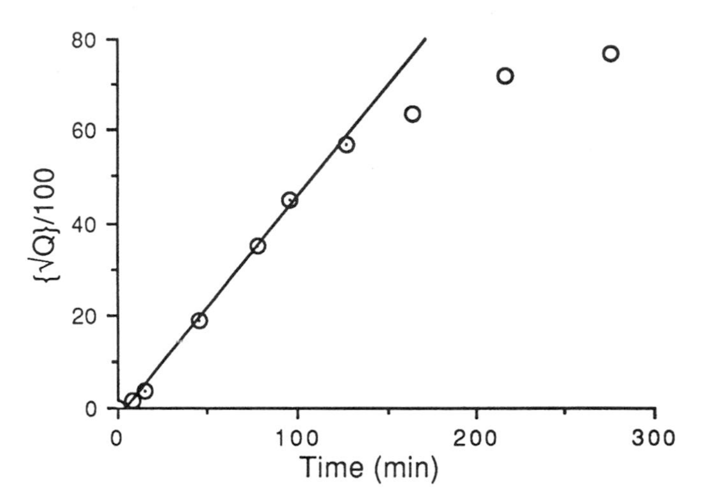

Fig. 34 Dissolution pattern in matrix dissolution. (Graph constructed from data published by Fessi et al., 1982.)

the tablets is neatly square root of time dependent as predicted by the Higuchi equation (Higuchi, 1962) shown in Eq. (10.58).

9.4. Osmotic Pump

The osmotic pump is a tablet coated with an impervious film, into which is (laser-)drilled a hole of exacting dimensions. Dissolution liquid will penetrate into the interior of the tablet, and a saturated solution will form. The excipients are chosen so that they have a given solubility and hence produce a given osmotic pressure. (The drug itself contributes to this as well.) This will be larger than the osmotic pressure in the outside liquid, and the difference between the osmotic pressure inside and outside will be the driving force by which liquid is being forced out through the hole. This gives rise to zero-order kinetics (which biopharmaceutically is an advantage), and the osmotic pump in many experimental situations, as well as in marketed situations, seems to be the dosage form that gives the most desirable release pattern, and also the one most likely to give in-vivo results that are predictable from in-vitro data. Dissolution data will therefore plot linearly, when amount released is plotted as a function of time. There may be a small nonzero (negative y-) intercept, i.e., a lag time. There will also be a point in time when there is no more solid drug inside the tablet, and deviations from linearity will occur from this point on.

There is no literature published on the stability pattern of this type of dosage form, but it is to be expected that it would be no more (and probably less) prone to change on storage than the other types mentioned.

9.5. Gel Forms

There are sustained release tablets that rely on gel-forming substances to accomplish the sustained release. In these cases the dissolution liquid will form a gel when it encounters the surface of the tablet. Drug must dissolve and diffuse out through this gel layer. As time goes on, the gel layer gets thicker, and the diffusion path becomes longer. The data can be represented by Eq. (14.49) (Bamba et al., 1979). There is also the possibility of some "sloughing off" (i.e., erosion) of gel, and in this case the release becomes a hybrid between erosion and diffusion through increasing thickness of gel.

10. COATED TABLETS

Tablets are often film coated and, less frequently, they are sugar coated. Sugar coating, when properly applied, provides an excellent moisture and quite an adequate oxygen barrier. Film coating does the same, but not quite as effectively. For instance, vitamin A beadlets are more stable in a (properly made) coated tablet, less so (but yet quite stable) in a film coated tablet, of course provided there are no drastic incompatibilities in the core.

Film coating offers many advantages and often is the coat of preference, because (a) its application is much less labor intensive (cycle times being in hours for film coated tablets, in days for sugar coated tablets); (b) they also provide the advantage of allowing an engraving to "show through," i.e., identification requires no extra operation. On the contrary, sugar coated tablets, for identification, require a separate printing step; and (c) there is an inherent advantage in film coating in that it allows the appearance of a deep color without the use of much dye. If an uncoated tablet is colored, the dye is present throughout the tablet, whereas in a film coated tablet it is only present in the outer layer (the film itself).

Enteric coated tablets belong in the category of coated tablets and will be treated below as well.

10.1. Film Coated Tablets

Film coated tablets are produced either in a coating pan or by column coating (Wurster coating). Most coatings, nowadays, are aqueous film coats (hydroxymethyl cellulose, hydroxypropyl methylcellulose). There are several types of defects that can occur originally (orange peel effect for instance). All coatings, essentially, are such that each applied coat is not complete, so that there are overlaps, and in essence there is always an orange peel effect, except in a "good" tablet this cannot be seen. It is simply assumed in this writing, that the tablets placed on stability are not defective.

From a stability point of view there may be changes in appearance, mostly due to dislodging or rupture of the film. Sometimes these changes are first seen in the engraving. To properly record changes in appearance of the film, descriptive means can be used, but it is often a good idea to take a photomicrograph originally of all coated tablets (be they sugar or film coated). If defects show up in the coating as a function of time, then the question arises whether this is due to the formula (film and uncoated tablet) or to the way in which it was made (initial defective procedure, possibly not noticeable). Most often, these problems result in efforts in the formulation area, and recording (visually or photographically) at many intervals (3, 6, 9, 12 months) is therefore advisable. In this manner reformulation can be carried out as soon as the problem is identified.

One property that should be monitored, both for film and for sugar coated tablets, is their gloss. This is usually done subjectively. Rowe (1987) has described a glossmeter that assesses the gloss, but points out that there is still a great deal of subjectivity in the use of it.

As a problem-solving tool, scanning electron microscopy is advised, because of the augmented detail it offers, a detail that often pinpoints the individual problem.

In some formulation setups, it is possible (e.g., with an Instron tester) to measure the force necessary to strip a film from a substrate. If this substrate is the tablet surface, it is possible to evaluate films, moisture contents, effect of additives, etc., to ascertain which is the proper way in which to reformulate the film.

The actual appearance (i.e., the color) of the film coated tablet can be checked by means of a reflectance meter (or by diffuse reflectance), as described in the previous section.

Dissolution and disintegration are, of course, sensitive parameters, because any change in the film will be reflected in these properties.

10.2. Sugar Coated Tablets

Detailed descriptions of sugar coating procedures are beyond the scope of this writing. In brief, in sugar coated tablets there is applied (usually in a coating pan) first a barrier coat (frequently shellac or other polymer), then a subcoat (frequently terra alba/gelatin, with talc used as a conspergent), then a dye coat (consisting usually of sucrose syrup and lake dye), then a finishing coat (usually sugar syrup), and finally a polish coat (usually beeswax either dry or in solvent solution). The latter is carried out in a canvas coated coating pan.

The typical defects on storage are chipped tablets and tablets that split in the periphery. The former can be tested for by using a friabilator test. In such cases, a correlation with an actual shipping test should be attempted. In such a shipping test, tablets are sent by various routes (rail, truck, air) from the plant to several destinations and then back again. In so doing, it is possible to observe whether the artificial stress test is comparable to the actual transportation test.

When tablets split in the periphery it is usually due to trapped moisture (i.e., the tablet may not have been quite dry at the time one of the coats was applied). Very often it is due to an improperly applied barrier coat.

Again, photomicrography (and in problem cases, scanning electron microscopy) is advocated as a reference for changes in appearance. The actual appearance (i.e., the color) of the coat can be checked by means of a reflectance meter (or by diffuse reflectance), as described in the previous section.

Dissolution and disintegration are, of course, sensitive parameters, because any change in the film will reflect in these properties.

10.3. Enteric Coated Tablets

An enteric coat is an attempt "to administer two doses in one tablet." This is done by placing an acid resistant film (e.g., a polymer containing a carboxyl group with a pK of 4–6) on an uncoated tablet and then sugar coating it. The first dose is contained in the core, and the second dose is applied in the sugar coat, which should release the material immediately.

Enteric coating is a delicate operation, and often there is, in the production write-up, a statement that in-vitro dissolution must be carried out after e.g., the seventh coat. The e.g. eighth coat may then be applied or not depending on the outcome of the in-vitro test. This latter is usually the USP test that calls for placing one tablet in each of the six tubes of the basket in water at room temperature for 5 minutes. The apparatus is then operated without discs in simulated gastric fluid at 37°C. After one hour the basket is removed and the tablets are observed, and they should show no sign of disintegration, softening, or cracking.

Next a disk is added to each tube, and the apparatus is filled with simulated intestinal fluid TS at 37°C for 2 hours (or whatever the monograph or the in-vitro to in-vivo relation calls for). If all of the tablets have disintegrated at the prescribed endpoint time, then the batch is acceptable, but if one or two tablets fail, then it is retested sequentially by testing an additional 12 tablets, all of which must pass.

Enteric coats (e.g., cellulose acetate phthalate) have tendencies to polymerize. (Shellac is particularly vulnerable in this respect.) Hence disintegration on storage should be monitored at all intervals (3, 6, 9, 12, 18, 24, and 36 months).

It is noted that the initial dose (in the coat) should be available immediately, and a check should be made (one or two points) to assure that disintegration of the coat also results in dissolution of the drug (which it usually does).

The behavior at accelerated temperatures is not necessarily indicative of (nor extrapolable to) room-temperature characteristics.

11. HARD AND SOFT SHELL CAPSULES

Grimm and Schepky (1980a) have demonstrated how, depending on the sorption isotherms of the capsule fill, a capsule shell can lose moisture to the capsule fill and become brittle, or conversely under opposite sorption isotherm conditions can draw moisture out of the fill and become soft.

As mentioned earlier, dissolution rate of the gelatin decreases in water. HCl, and aqueous buffer solution on storage, but gastric juice containing enzymes might well eliminate such a problem. A thorough review of the problem with cross-linking of gelatin and the occurrence of pellicule formation has been discussed by Digenis et al. (1994).

Ofner and Schott (1987) have studied the swelling of gelatin (Fig. 35) and have applied Eq. (10.64) to their considerations. If W grams of aqueous buffer solution is absorbed by 1 gram of gelatin at time t, then

$$\frac{t}{W} = A + Bt \tag{10.64}$$

where A and B are constants. The effect of additives can then be studied.

Vastly different behavior of gelatin was experienced with different drugs. This, obviously, is a powerful preformulation tool (when combined with data regarding the hygroscopicity of the drug, as demonstrated e.g. in Example 10.3).

York (1981) has reported on the moisture isotherms of gelatins. Knowing the moisture isotherm of the powder mixture in the gelatin, it is possible (as shown in the previous section dealing with tablets) to calculate the shift in moisture from shell to powder mixture (or vice-versa).

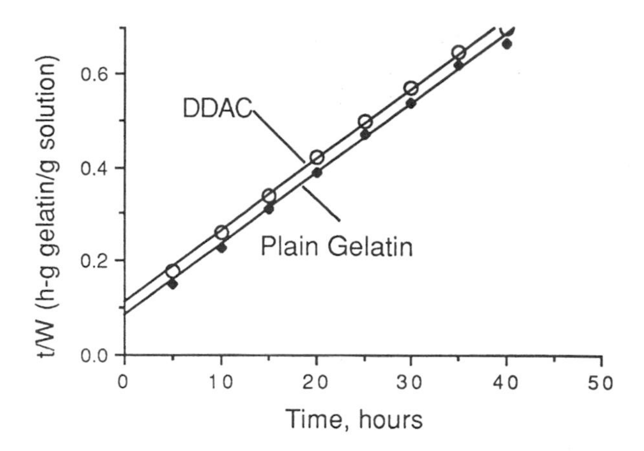

Fig. 35 Swelling isotherms of gelatin.

12. MICROCAPSULES

Microcapsules may decompose as a function of time. This has been reported by Makine et al. (1987) for the case of poly(L-lactide) microcapsules. Logical means of estimating the loss of intact polymer is (a) from the decrease in weight-averaged molecular weight, (b) by monitoring the loss in weight of polymer by gel permeation chromatography, and (c) by determining the amount of lactic acid formed. Figure 36 is an example of the decrease in weight-averaged molecular weight upon storage.

13. LIGHT SENSITIVITY TESTING

Both the ICH and the 1987 Guidelines advocate exposure of dosage forms to UV light, and although this might be instructive, it does not represent a test that simulates conditions in actual commerce (in general). There are exceptions: certain products are liable to be kept in handbags and kept out in the open, but these are the exception. In general products are considered to be kept in controlled plant environments, in warehouses or in controlled pharmacy conditions or in (short) transit.

To define a storage condition it is necessary to examine the actual conditions in the marketplace, and this has been done by Esselen and Barnby (1939), Lachman and Cooper (1959a, 1959b), and Lachman et al. (1960). They determined the spectral composition of light and light intensity in the typical American pharmacy, and in general it is assumed that the average foot-candles in a pharmacy is 5–15, and 10 is used as an average.

One could now proceed by checking a product for three years under such conditions, but rather than do that, it is desirable to accelerate the conditions so as to obtain an answer somewhat more rapidly. The guidelines' suggestion of using more energetic (UV) light is not good for such acceleratory attempts, because the more energetic light will (or may) give rise to reactions that would never take place in the light in a pharmacy (which is much more poor in ultraviolet light).

Lachman and Cooper determined that a #48 12 CWRS GE lamp 1.5″ in diameter and 48″ long produced a good average spectrum and produced 3250 lumens per

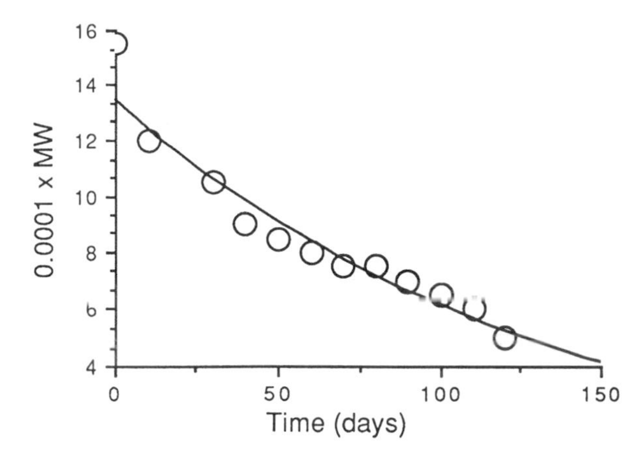

Fig. 36 Weight-averaged molecular weight of poly(L-lactide) microcapsules. (Figure constructed from data reported by Makine et al., 1987.)

Table 8 Appearance Change of a Tablet as a Function of Time in a Lachman–Cooper Light Cabinet

Storage (weeks)	Open dish	Amber bottle
0	0.302	0.302
4	0.212	0.315
8	0.188	0.312
12	0.190	0.309

60 watts. They suggested accelerating the test by increasing the lumen reached by the dosage form, and they increased this by placing it closer to the light source. The light quanta absorbed by the dosage form are inversely proportional to distance (d) of the dosage form from the light source squared, i.e., proportional to $1/d^2$. They describe a light cabinet, equipped with shelves close to the described light source, adequately ventilated so that the temperature does not rise substantially. Light meters on the shelves allow either movement of the shelf so that the light intensity is always the same, or simply serves as a device to control when the bulbs should be changed.

If used as described, 1 month in the light cabinet is equivalent to 24 months on the average pharmacist's average shelf. A typical set of data of the discoloration of a dyed tablet (as gauged by optical density) is shown in Table 8.

It is seen that the cabinet allows for evaluation of the means of preventing photoinduced changes or reactions to take place (use of the amber bottle). This could be accomplished by other means as well (coating of a tablet, using an opaque rather than a transparent capsule, using a capsule with a dye that screens out the part of the spectrum that causes the photoreaction).

In the case of uncoated tablets, most often the dye will discolor or fade in the upper layer only. If the tablet is broken, the color is still intact on the inside. If it fades all the way through, then this is indicative of photoinduced interaction. In the case of ethinyl estradiol and an FDC dye, Kaminski et al. (1979) showed interaction, but this interaction was first noticed in a light test and was photocatalyzed.

The regulations on light testing suggest the use of xenon lamps. In these a very high intensity light is applied in a very short period of time (minutes), and the effect on the dosage form (and the drug and the drug product) is recorded. The correlation of this with "real time and exposure" is unknown, and a database will have to be established before it can be rationally analyzed.

14. DIAGNOSTIC PAPERS

There is very little in literature regarding the stability of diagnostic product. In the case of diagnostic papers, the reagent is adsorbed on filter paper. The adsorption is usually governed by a Langmuir isotherm; there will, however, be active sites on the paper, and these may bind part (often a large percentage) of the reagent. Hence there will be an initial "loss" of reagent, and this will have to be compensated for by excesses, since only the reagent that is not chemisorbed will be available for the reaction in the diagnosis.

Stability data are gauged on the basis of initial assay (not theoretical content). Usually the stability is evaluated in a semiquantitative manner by an in-use test, i.e., an operator will carry out the diagnosis initially. For example, if it is a test that monitors sugar content (e.g., in urine), the filter paper will be judged against sucrose solutions of various concentrations. The instructions may state that the reaction is "positive" if a certain color is achieved (i.e., the concentration of sucrose is above a level, L) and negative if the response level is below another concentration, L^*, where $L^* < L$. The area in between is then "doubtful," and this would call for a retest. Initially the test should evoke the correct response (since the batch is, presumably, quality control released). As time goes by there will be a certain decomposition, which will vary from strip to strip, so at time t, a fraction of all the strips, q, may give an incorrect response. The best parameter to follow, stability-wise, is this parameter q, which should be such that it could be said at the expiration date with 95% confidence that q is below 5%.

15. EXPIRATION PERIODS

It was shown in the chapter dealing with statistics and expirations periods that there are mathematical means of calculating expiration periods from chemical stability data. This is not directly possible with physical testing. The reason for this is often the difficulty that exists in quantitating the physical property. Davis et al. (1977) state broadly that "the physicochemical changes that can occur ... upon storage or after processing or other external influence, should not be such that they can alter the therapeutic efficacy of the product." This is a good guideline, but the only way to test it is somehow to transform the experimental data into some quantity that can be extrapolated.

For instance, a suspension may start caking, but there are degrees of caking, and if it still can be shaken up in a reasonable length of time, then it should be all right. Here, a criterion must be set by the investigator, the quality control group, and the regulatory group within an organization. Such a criterion can be set up much like a test panel. Several containers of different degrees of caking can be evaluated, and the "worst acceptable" (akin to the lower acceptable quality limit in release philosophy) agreed to. The investigator now can do a rotation test on this in the fashion described by Moore and Lemberger (1963), and state e.g. that after 20 controlled rotations, an assay of the supernatant must not be less than L mg/mL. This type of test can then be carried out at various storage times at room temperature; L can be plotted versus time, and a usual statistical test performed on this number. In other words, for physical testing that has no number associated with it, it is important to attempt to find such a number.

In some cases this is not possible. A case in point is the particle size of an intravenous oil emulsion. Coalescence and formation of free oil will result in toxic manifestations, but it is next to impossible to determine an acceptable upper limit for droplet size (Davis, 1987).

REFERENCES

Adams, W. P. (1989). Division of Bioequivalence. Guidance for in-Vitro Portion of Bioequivalence Requirements for Mataproterenol Sulfate and Albuterol Inhalation Aerosols (Metered Dose Inhalers), FDA, Rockville, MD.
Aerosol Guide, 7th ed. (1981). New York: Chemical Specialties Manufactures Association.
Akers, M., Lach, J. (1976). J. Pharm. Sci. 65:216.
Armstrong, N. A., March, G. A. (1974). J. Pharm. Sci. 63:126.
ATLAS HLB System (1963). LD-91-R1-2M-1-69. 4th printing. Wilmington, DE: Atlas Chemical Industries (Now ICI Americas).
Bamba, M., Puisieux, F., Marty, J. P., Carstensen, J. T. (1979). Int. J. Pharmaceutics 3:87.
Barry, B. W. (1974). Adv. Pharm. Sci. 4:1.
Benjamin, E. J., Kroeten, J. J., Shek, E. (1983). J. Pharm. Sci. 72:381.
Ben-Kerrour, L., Duchene, D., Puisieux, F., Carstensen, J. T. (1980). Int. J. Pharmaceutics 5:59.
Block, L. H., Banakar, U. V. (1988). Drug Dev. Ind. Pharm. 14:2143.
Boegeli, R. T., Ward, J. B. Hutchins, H. H. (1969). J. Soc. Cosmet. Chem. 20:373.
Bonny, J. D., Leuenberger, H. (1993). Pharm. Acta Helv. 68:25.
Brown, T. (1953). Text. Color. 57:515.
Buscall, R., Davis, S. S., Potts, D. C. (1979). Colloid Polym. Sci. 257:1636.
Byron, P. R. (1990). Respiratory Drug Delivery. CRC Press, Boca Raton, FL.
Cárdenas, R. H., Cortés, A. S., Argotte, R. R., Luna, M. P., Dominguez, R. A. (1994). Drug Dev. Ind. Pharm. 20:1063.
Carstensen, J. T., Johnson, J. B., Valentine, W., Vance, J. S. (1964). J. Pharm. Sci. 53:1050.
Carstensen, J. T. (1972). Theory of Pharmaceutical Systems, I, Homogeneous Systems. New York: Academic Press. p. 39.
Carstensen, J. T. (1973a). J. Pharm. Sci. 62:1.
Carstensen, J. T. (1973b). Theory of Pharmaceutical Systems, II. Heterogeneous Systems. New York: Academic Press, p. 287.
Carstensen, J. T. (1973c). Theory of Pharmaceutical Systems, II, Heterogeneous Systems. New York: Academic Press, p. 153.
Carstensen, J. T., Kothari, R. (1983). J. Pharm. Sci. 72:1149.
Carstensen, J. T., Rhodes, C. R. (1993). Drug Dev. Ind. Pharm. 19:1811.
Carstensen, J. T., Rodriguez, N. (1985). J. Pharm. Sci. 74:1293.
Carstensen, J. T. Johnson, J. B, Valentine, W., Vance, J. J. (1964). J. Pharm. Sci. 53:1050.
Carstensen, J. T., Su, K.S. E. (1970). J. Pharm. Sci. 59:666, 671.
Carstensen, J. T., Lai, T. Y.-F., Prasad, V. K. (1976a). J. Pharm. Sci. 65:607.
Carstensen, J. T., Lai, T. Y.-F, Prasad, V. K. (1976b) J. Pharm. Sci. 65:1303.

Carstensen, J. T., Wright, J. L., Blessel, K., Sheridan, J. (1978a). J. Pharm. Sci. 67:48.
Carstensen, J. T., Wright, J. L., Blessel, K., Sheridan, J. (1978b). J. Pharm. Sci. 67:982.
Carstensen, J. T., Kothari, R., Chowhan, Z. T. (1980a). Prod. Dev. Ind. Pharm. 6:569.
Carstensen, J. T., Kothari, R., Prasad, V. K., Sheridan, J. (1980b). J. Pharm. Sci. 69:290.
Carstensen. J. T., Alcorn, G. J., Hussain, S. A., Zoglio, M. A. (1985). J. Pharm. Sci. 74:1239.
Carstenen, J. T., Franchini, M., Ertel, K. (1992). J. Pharm. Sci. 81:303.
Chafetz, L., Hong, W.-H., Tsilifonis, D. C. Taylor, A. K. Philip, J. (1984). J. Pharm. Sci. 73:1186.
Chowhan, Z. T. (1979). Drug Dev. Ind. Pharm. 5:41.
Couvreur, P. (1975). Thesis. Docteur en Sciences Pharmaceutiques. Louvain, Belgium: Univ. Catholique, p. 87.
Cruaud, O., Duchene, D., Puisieux, F., Carstensen, J. T. (1980). J. Pharm. Sci. 69:607.
Davis, S. S. (1984). In: Asche, H., Essig, D., Schmidt, P. C. eds. Technololgie von Salben, Suspensionen und Emulsionen. Stuttgart: Wissenschaftliche Verlagsgesellschaft, pp. 160–175.
Davis, S. S. (1987). In: Grimm, W., ed. Stability Testing of Drug Products. Stuttgart: Wissenschaftliche Verlagsgesellschaft, p. 40.
Davis, S. S., Khanderia, M. S., Adams, I., Colley, I. R., Cammack, J. Sanford, T. J. (1977) Texture Studies 8:61.
Dey, M., Enever, R., Kraml, M., Prue, D. G., Smith, D., Weierstall, R. (1993). Pharm. Res. 10:1295.
Digenis, G. A., Gold, T. B. Shah, V. P. (1994). J. Pharm. Sci. 83:915.
Dukes, G. R. (1984). Drug Dev. Ind. Pharm. 10:1413.
Eavens, T., Jones, T. M. (1970). J. Pharm. Pharmacol. 22:594.
Eccleston, G. M. (1976). J. Colloid Int. Sci. 57:66.
Enever, R. P. (1976). J. Pharm. Sci. 65:517.
Esselen, W. G., Barnby, H. A. (1939). Modern Packaging, September, p. 15.
Fessi, H., Marty, J. P. Puisieux, F., Carstensen, J. T. (1982). J. Pharm. Sci. 71:749.
Fisher, J. T., Sheth, B. B. (1973). Aerosol Age 18(2):28.
Gautum, J., Schott, H. (1994). J. Pharm. Sci. 83:316.
Gillespie, W. R., Cheng, H. (1993). J. Pharm. Sci. 82:1 1085.
Goodhart, F. W., Lieberman, H. A., Mody, D. S., Ninger, F. C. (1967). J. Pharm. Sci. 56:63.
Gordon, M. S., Rudraraju, V. S., Fhie, J. K., Chowhan, Z. T. (1993). Int. J. Pharmaceutics 97:119.
Griffiths, R. V. (1969). Manuf. Chem. Aerosol News. 40:29.
Grimm, W., Schepky, G. (1980a). Stabilitatsprufung in der Pharmazie. Aulendorf, Germany: Editio Cantor, p. 230.
Grimm, W., Schepky, G. (1980b) Stabilitatsprufung in der Pharmazie. Aulendorf, Germany: Editio Cantor, p. 229.
Grimm, W., Schepky, G. (1980c). Stabilitatsprufung in der Pharmazie, Aulendorf, Germany: Editio Cantor, p. 230.
Grimm, W., Schepky, G. (1980d). Stabilitatsprufung in der Pharmazie. Aulendorf, Germany: Editio Cantor, p. 216.
Gucluyildiz, H., Banker, G. D., Peck, G. E. (1977). J. Pharm. Sci. 66:407.
Hahn, A. U., Mittal, K. L. (1979). Colloid Polym. Sci. 257:959.
Hammouda, Y., Salakawy, S. A. (1971). Pharmazie 26:636.
Harris, R. P. (1968). Aerosol Age 18(1):36.
Higuchi, T. (1963). J. Pharm. Sci. 52:1145.
Hoelgaard, A., Møllgaard, B. (1983). J. Pharm. Pharmacol. 34:610.
Idson, B. (1970). Drug Cosmet. Ind. 107(1):46
Johnson, M. A. (1972). The Aerosol Handbook. Caldwell, NJ: Dorland.
Jørgensen, K., Christensen, F. N. (1996). Int. J. Pharm. 143:223.

Jørgensen, K., Christensen, F. N. Jacobsen, L. (1997). Int. J. Pharm. 153:1.
Joyner, B. C. (1969a). Mfg. Chem. 40(8):63.
Joyner, B. C. (1969b). Labo-Pharma. 17(9):71.
Kaminski, E. E., Cohn, R. M., McGuire, J. L., Carstensen, J. T. (1979). J. Pharm. Sci. 68:368.
Katori, N., Okudaira, K., Aoyagi, N., Takeda, Y., Uchiyama, M. (1991). J. Pharmacobio-Dyn. 14:567.
Kokobo, T., Sugibayashi, K., Morimoto, Y. (1991). J. Control Release 17:69.
Kokobo, T., Sugibayashi, K., Morimoto, Y. (1994). Pharm. Res. 11:104.
Komerup, A., Wanscher, J. H. (1967). The Methuen Book of Colour. London: Methuen.
Lachman, L., Cooper, J. (1959a). J. Am. Pharm. Assoc. Sci. Ed. 48:226.
Lachman, L., Cooper, J. (1959b). J. Am. Pharm. Assoc. Sci. Ed. 48:233.
Lachman, L., Schwartz, C. J., Cooper, J. (1960). J. Am. Pharm. Assoc. Sci. Ed. 49:226.
Lee, S., DeKay, H. G., Banker, G. S. (1965). J. Pharm. Sci. 54:1153.
Lee, T.-Y., Notari, R. E. (1987). Pharm. Res. 4:98.
Leu, R., Leuenberger, H. (1993). Int. J. Pharm. 90:213.
Leuenberger, H., Leu, R. (1992). J. Pharm. Sci. 81:976.
Lik, J., Cadwell, K. D., Anderson, B. D. (1993). Pharm. Res. 10:535.
Makine, K., Ohshima, H., Kondo, T. (1987). Pharm. Research 4:62.
Matthews, B. A., Matsumoto, S., Shibata, M. (1974/75). Drug Dev. Comm. 1:303.
McGinity, J. W. (1993). Pharm. Res. 10:542.
McNamara, P. J., Foster, T. S., Digenis, G. A., Patel, R. B., Craig, W. A., Welling, P. G., Rapaka, R. S., Prasad, V. K., Shah, V. P. (1987). Pharm. Research 4:150.
McVean, D., Tuerck, P., Christenson, G., Carstensen, J. T. (1972). J. Pharm. Sci. 61:1609.
Mehdizadeh, M., Grant, D. J. W. (1984). J. Pharm. Sci. 73:1195.
Mendenhall, D. W. (1984). Drug Dev. Ind. Pharm. 10:1297.
Moore, A., Lemberger, A. P. (1963). J. Pharm. Sci. 52:223.
Muller, R., Bardon, J., Arnaud, Y., Champy, J., Roland, P. (1979). Ann. Pharm. Fr. 37:301.
Murthy, K. S., Ghebre Sellassie, I. (1993). J. Pharm. Sci. 82:113.
Nakagawa, T., Uno, T. (1978). J. Pharmacobio-Dyn. 6:547.
Nakamura, A. Okada, R. (1976). Colloid Polym. Sci. 254:718.
Naylor, L. J., Bakatselou, V., Dressmann, J. B. (1993). Pharm. Res. 10:865.
Nogami, H., Nagai, T., Uchida, H. (1966). Chem. Pharm. Bull. 14:152.
Nurnberg, E. (1969). Pharm. Ztg. 4:128.
Ofner, C. M., III, Schott, H. (1987). J. Pharm. Sci. 76:715.
Omelczuk, M. O., McGinity, J. W. (1993). Pharm. Research. 10:542.
Ondracek, J., Boller, J., Zullinger, F. H., Niederer, R. R. (1985). Acta Pharm. Technol. 31:42.
Pengilly, R. W., Keiner, J. P. (1977). J. Soc. Cosm. Chem. 28:641.
Pietcsh, W. B. (1969). Eng. for Ind. 435.
Pitkin, C., Carstensen, J. T. (1973). J. Pharm. Sci. 62:1215.
Podzeck, F. (1993). Int. J. Pharmaceutics 97:93.
Polli, G. P. Grimm, W. M., Bacher, F. A., Yunker, M. H. (1969). J. Pharm. Sci. 58:484.
Prandit, J. K., Wahi, A. K., Wahi, S. P., Mishr, B., Tripathi, M. K. (1994). Drug Dev. Ind. Pharm. 20:889.
Rassing, J., Atwood, D. (1983). Int. J. Pharm. 13:47.
Reng, A. K. (1984). In: Asche, H., Essig, D., Schmidt, P. C., eds. Technololgie von Salben, Suspensionen und Emulsionen. Stuttgart: Wissenschaftliche Verlagsgesellschaft, p. 203.
Rhamblhau, D., Phadke, D. S., Doerle, A. K. (1977). J. Soc. Cosmet. Chem. 28:183.
Rhodes, C. T. (1979a). In: Banker, G. S., Rhodes, C. T., eds. Modern Pharmaceutics. New York: Marcel Dekker, p. 329.
Rhodes, C. T. (1979b). Drug Dev. Ind. Pharm. 5:573.
Richman, M. D., Shangraw, R. F. (1966). Aerosol Age 11(5):36.
Rodriguez-Hornedo, N., Carstensen, J. T. (1985). J. Pharm. Sci. 74:1322.

Rothgang, G. (1974). Dtsch. Apoth. Ztg. 114:1653.
Rowe, R. C. (1965). J. Pharm. Sci. 54:260.
Rowe, R. C. (1983). Pharmacy International 4:225, and 173.
Rowe, R. C. (1985). Pharmacy International 6:225.
Rubino, J. T., Halterlein, L. J., Blanchard, J. (1985). Int. J. Pharm. 26:165.
Schepky, W. (1974). Pharm. Ind. 36:327.
Schieffer, G. W., Palermo, P. J., Pollard-Walker, S. (1984). J. Pharm. Sci. 73:126.
Sciarra, J. J. (1974). J. Pharm. Sci. 63:260.
Sciarra, J. J. (1967). Aerosol Age 12(2):65; 12(3):45; 12(4):65.
Scott, M. W., Goudie, A. J., Huetteman, A. J. (1960). J. Am. Pharm. Assoc. Sci. Ed. 49:467.
Scott, M. W., Lieberman, H., Chow, F. S. (1963). J. Pharm. Sci. 52:994.
Seth, P. L., Munzel, K. (1959). Pharm. Ind. 21:417.
Shafer, E. G. E., Wollish, E. G., Engel, C. E. (1956). J. Am. Pharm. Assoc. Sci. Ed. 45:114.
Sherman, J. (1955). Research 8:396.
Sherman, J. (1964). J. Pharm. Pharmacol. 16:1
Shotton, E., Harb, N. (1966). J. Pharm. Pharmacol. 18:1175.
Siragusa, J. M. (1955). A Study of Some Emulsifiers for Pharmaceutical Emulsions. Ph.D. thesis, Univ. of Florida.
Smith, M. F., Bryant, S., Welch, S., Digenis, G. A. (1984). J. Pharm. Sci. 73:1091.
Soriano, M. C. Caraballo, I., Millán, M., Pinero, R. T., Melgazo, L. J., Rabasco, A. M. (1998). Int. J. Pharm. 174:63.
Sugawara, S., Imai, T., Otagiri, M. (1994). Pharm. Res. 11:272.
Suryanarayanan, R., Mitchell, A. G. (1981). J Pharm. Pharmacol. 33:112P.
Suryanarayanan, R., Mitchell, A. G. (1984). J. Pharm. Sci. 73:78.
Tingstad, J. E. (1964). J. Pharm. Sci. 53:995.
Train, D. (1957). Trans. Inst. Chem. Eng. 35:258.
Turi, P., Brusco, D., Maulding, H. V., Tausenfreund. R. A., Michaelis A. F. (1972). J. Pharm. Sci. 61:1811.
Van den Tempel, A. (1953). Stability of Oil-in-Water Emulsions. Rubber-Stichting, Oostingel 178, Delft (Netherlands), Communication No. 225.
VanOort, M., Gollmar, R. O., Bohinski, R. J. (1994). Pharm. Res. 11:604.
Walton C. A., Pilpel, N. (1972). J. Pharm. Pharmacol. 24:110P.
Washburn, E. H. (1921). Phys. Rev. 17:273.
Williams, M. L., Landel, R. F., Ferry, J. D. (1955). J. Am. Chem. Soc. 77:3701.
Wortz, R. B. (1967). J. Pharm. Sci. 56:1169.
York, P. (1981). J. Pharm. Pharmacol. 33:1269.
Yu, C. D., Jones, R. E., Wright, J., Henesian, M. (1983). Drug Dev. Ind. Pharm. 9:473.
Yu, C. D., Jones, R. E., Henesian, M. (1984). J. Pharm. Sci. 73:344.
Zapata, M. I., Feldkamp, J. R., Peck, G. E., White, J. I., Hem, S. L. (1984). J. Pharm. Sci. 73:1.
Zoglio, M. A., Streng, W. H., Carstensen, J. T. (1975). J. Pharm. Sci. 64:1869.

11

Development and Validation of HPLC Stability-Indicating Assays

DONALD D. HONG

Pharmaceutical Consultant, Raleigh, North Carolina

MUMTAZ SHAH

Trigen Laboratories, Salisbury, Maryland

Part I: Method Development

Part I: Method Development

1. WHAT IS A STABILITY-INDICATING METHOD?

According to the regulatory definition (1), a stability-indicating method is one of a number of

> Quantitative analytical methods that are based on the characteristic structural, chemical, or biological properties of each active ingredient of a drug product and that will distinguish each active ingredient from its degradation products so that the active ingredient content can be accurately measured.

Therefore a stability-indicating method is an analytical procedure that is capable of discriminating between the major active (intact) pharmaceutical ingredient (API) from any degradation (decomposition) product(s) formed under defined storage conditions during the stability evaluation period. In addition, it must also be sufficiently sensitive to detect and quantify one or more degradation products. A corollary may be added that the analytical method must be also capable of separating or resolving any other potential interfering peak such as an internal standard. With these criteria, then, the discriminating "nature" of the method indicates the method to be *stability-indicating* as well as *stability-specific*. Later in the discussion we will see that other methods may be stability-specific but not stability-indicating. Stressed testing may be used (1,2) to expedite the decomposition pathway(s) to generate decomposition product(s) for the API. However, stressed testing under forced conditions of oxidation, photolysis, hydrolysis, and varying pH values may form some decomposition products that are unlikely to be formed under accelerated or long-term stability storage conditions. The products generated nonetheless may be useful in developing and validating a suitable stability-indicating analytical method for the analysis of the drug substance and the drug product, expediting the availability of the completed analytical method.

It is paramount that the chosen analytical method used for stability evaluation be validated and discriminating to ensure efficacy of the subsequent stability evaluation. Confidence in the stability data is predicative on time invested up front to ensure a viable procedure as well as to conform to legal and regulatory requirements (2).

2. STRATEGY OF METHOD DEVELOPMENT

Development of a stability-indicating method should be predicated on the method's intended application as well as selecting a suitable technique designed to assess the API's stability requirements. Obviously the intended application of a stability-indicating method is for monitoring the stability of a given drug in a finished product and would require assessment of the method's stability-indicating properties. One specialty application of a stability-indicating method is cleaning validation testing, which would require assessment of its stability-indicating properties, as holding time (of the swaps) would be a critical factor. Other applications such as product release, performance testing (i.e., dissolution testing) and in-process testing do not require this assessment. Some pharmaceutical houses still (but now less commonly due to technological advances and overall industry practice) utilize a non-stability-indicating method such as UV, for product release, and an HPLC method for stability testing. However, whenever there is a hold time issue, common in dissolution or in-process testing, it would be prudent to assess the method for its stability-indicating properties before its intended application.

Other chromatographic separation methods, such as chiral chromatography (CC), thin-layer chromatography (TLC), gas chromatography (GC), and (increasingly) capillary electrophoresis (CE), are stability-indicating and stability-specific methods. Still the most prevalent technique is reversed-phase HPLC alone or coupled with ion-suppression, which accounts for 85% or more of the general pharmaceutical applications.

Nonchromatographic and spectroscopic techniques such as titrimetry, atomic absorption, UV spectrophotometry, and infrared spectroscopy, while precise, are not considered stability-indicating, and as such not suitable for stability assessment applications.

3. OVERVIEW OF THE METHOD DEVELOPMENT PROCESS

Before beginning with actual experimentation it would be advantageous to view method development from a broader perspective. The method development process can be visualized from a high-level process map perspective better to define the general steps encountered to achieving the end product, a stability-indicating method.

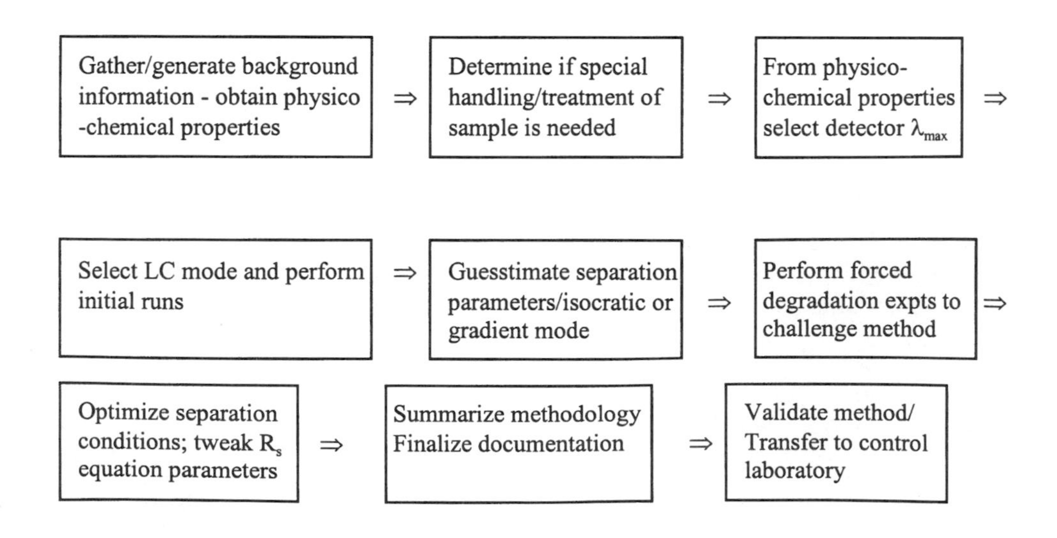

4. GETTING STARTED

It is probably best to approach method development with the intention of using the developed method for stability assessment as a final application, after the method has been validated. This approach entails determining the discriminating ability of the selected method up front before investing time and money in evaluating other analytical parameters prior to assessing the stability-indicating element of the method.

Reversed-phase HPLC is the method of choice for stability-indicating and stability specific methods, although thin-layer chromatography (TLC), gas chromatography (GC), and capillary electrophoresis (CE) are also acceptable choices. Reversed-phase HPLC coupled with ionic suppression account for probably over 85% of stability-indicating methodologies for small molecular weight pharmaceutical entities. This combination is well suited for applications in release testing, in-process as well as stability testing. Additional applications may be in cleaning validation and performance testing. Other techniques such as titration and UV spectroscopy, while commonly used for release testing, are generally considered nonspecific and thus are not considered for stability assessment.

Invariably when one is faced with finding or developing a method, one or two routes may be used depending on the nature of the chemical entity: modification or development. Modification is used when there is information or a method already exists for a similar entity. In this case, the existing method is modified or tweaked to accommodate the new entity. This may or may not be suitable; if not, development (starting from scratch) is the way to go. The goals of the separation should also be considered at this point.

4.1. Background Information

Knowledge of physicochemical properties of the API is invaluable to the method development process. Information on the various properties has been collected, either through a systematic program of generating the appropriate information in support of drug discovery (organic chemistry synthesis) on the one hand, or on the other, from a search of the literature or from company drug profiles, spectral libraries, or reports. Information such as dissociation constants, partition coefficients, fluorescent properties (if any), chromatographic behavior, spectrophotometric properties, oxidation-reduction potentials, formulation stability studies, and solubility studies are all very useful and can expedite the development process.

Dissociation constant and partition coefficients can be used to develop efficient liquid/liquid extraction procedures, and data on fluorescence, spectrophotometric, chromatographic, and oxidation-reduction properties can be used to determine the best means of measuring and quantifying the analyte of interest. Stability studies are performed on the drug substance, in solution and mixed with pharmaceutical excipients as part of compatibility studies. Labile functional groups are identified, and the susceptibility of the drug to hydrolysis, oxidation, thermal degradation, etc. is determined. Compatibility studies are performed to assess the stability of the API when mixed with common excipients and lubricants as well as to determine any interaction between the drug and the (inactive) raw materials. Solubilities should be determined in a number of solvents covering a range of polarities that are commonly used in method development.

Solubilities should be determined in aqueous and organic solvents, such as

Aqueous	*Organic*
Water	Ethanol/methanol
Buffers	Chloroform
0.1 N HCl	Cyclohexane
0.1 N NaOH	Acetonitrile
	Tetrahydrofuran

Spectral libraries are established, and information gleaned is useful for selection of initial conditions for an HPLC separation. On the other hand, however, sometimes this physicochemical information may not be known or available, so that an initial separation would have to be tried, based on prior experience, in order to determine a course of action for subsequent experimentation.

4.2. What Is Known About the Sample

Ideally, knowledge of the API's nature relative to composition and other properties would be beneficial. For example, information about the compound's synthetic route would shed light on any related product(s) and possible degradation product(s), as well as possible impurities; knowledge of the compound's chemical structure would reveal any possible stereoisomer which in turn would necessitate a different separation strategy, and so forth.

Table 1 shows typical information that would be helpful concerning the nature of the compound. The more information is available, the less empirical the approach to developing a separation method will be.

5. SEPARATION GOALS

To determine the separation goals, which should be clearly defined, a number of questions should be asked to help delineate the end purpose of the separation. Typical questions may include

What is the overall purpose of the method—quantitative, qualitative, or for isolation/purification of a compound (i.e., content assay, stability, impurities, cleaning assay, or for purification application)?

Table 1 Useful Physicochemical/Related Information Concerning the Compound

Wavelength of absorption (λ_{max})
Identity/number of compounds present (i.e., stereoisomers/chiral centers?)
Chemical structure (functionality); amphoteric
Molecular weight
pK_a values of compounds
Salt form of the drug
Solubility of compound
Purity of compound

What level of accuracy and precision would be needed?
The method is designed for what type of matrix? How many types of sample matrices are encountered?
Is the method developed using certain equipment transferable to the control laboratory, which may not have the same equipment?
Will the method be used for a few samples or many samples?
What chromatographic parameters are needed?
  How much resolution is needed?
  What is a suitable/acceptable separation time?
  What is a suitable column pressure?
  How much sensitivity is required?
  Is an internal standard needed?
  Are there any detection issues? Most analytes absorb in the UV region of the spectrum.
  Does integration use peak area or height?
  Is the mode isocratic or gradient?

6. SELECTION OF THE CHROMATOGRAPHIC MODE

6.1. The Different Modes of Liquid Chromatographic Methods (HPLC)

While there are a number of HPLC methods available to the development chemist, perhaps the most commonly applied method is reversed-phase. Reversed-phase and reversed-phase coupled with ion-pairing probably account for more than 85% of the applications for a typical pharmaceutical compound. The typical pharmaceutical compound is considered to be an API of less than 1,000 daltons, either soluble in water or in an organic solvent. The water-soluble API is further differentiated as ionic or nonionic which can be separated by reversed-phase. Similarly, the organic soluble API can be classed as polar and nonpolar and equally separated by reversed-phase. In some cases, the non-polar API may have to be separated using adsorption or normal phase HPLC, in which case the mobile phase would be a nonpolar organic solvent. For those "special" compounds that do not fall into this category (API>1000 daltons [biopharmaceuticals], isomers or enantiomers), other chromatographic modes may be necessary for separation. These include ion-exchange and chiral chromatography. In this discussion of developing a stability-indicating HPLC method, only reversed-phase will be discussed.

6.2. Reversed-Phase Chromatography

Thus given the limited number of methods with stability-indicating properties, it is probable that the method selected would be HPLC. Two very advantageous characteristics of HPLC, its discriminating power and its ability to operate at room temperature or at low elevated temperature, would not contribute to the degradation of the analyte. It is further assumed that the API is of low molecular weight (<1000 daltons), organic in nature (versus inorganic), and not a biopharmaceutical. These restrictions apply to a large percentage of the pharmaceuticals and enable them to be readily separated using reversed-phase HPLC, and sometimes with the aid of an ion-suppression agent, in roughly 85%

of the applications. The next question, then, is whether the chromatographic mode would be isocratic or gradient (see Fig. 3, Sec. 15).

6.3. Chiral Chromatography

Within this decade, since 1992, the FDA has published a position paper on the development of new stereoisomeric drugs (3). Prior to this time the majority of chiral synthetic compounds were marketed as racemic mixtures. This is because, until recently, it was not technically possible or economically feasible to separate racemic mixtures into their individual enantiomers. Experience has indicated that the individual enantiomers may exhibit different therapeutic effects. For example, the R-enantiomer of sotalol is antiarrhythmic while the S-enantiomer is a beta-blocker (4); and the dextro isomer *d*-propoxyphene (Darvon®, Lilly) is analgesic while the levo isomer *l*-propoxyphene is antitussive (but never developed into a marketed product) (5). However, with the FDA's position paper and current technological advances such as large-scale chiral separation techniques and asymmetric syntheses, new chemical entities (NCEs) containing a chiral center must be resolved into the different enantiomers and each enatiomer characterized and the drug product be composed of only one enantiomer instead of a racemate.

Thus, as contained in the International Conference on Harmonization (ICH) draft guideline on drug/drug product specifications (6), the tests in the table must be satisfied for new drug substances that are optically active:

Drug substance	*Test/specification requirement*
Impurities	Similar to other impurities
Assay	Enantioselective procedure or achiral method with appropriate means to control enantiomeric impurity
Identity	Test(s) should discriminate the enantiomers
Drug Product	
Degradation products	Control of other enantiomer if that enantiomer is a degradation product
Assay	If enantiomer is not a degradation product, an achiral method is acceptable, but chiral assay is preferred, or alternatively, achiral assay plus means to control the presence of the enantiomer
Identity	Test to verify the presence of the correct enantiomer

As such, in the development of a chiral method, the regulatory requirements must be considered. The reader is referred to decision tree #5 (page 62903) of the same reference for a schematic guide to development strategy and to the Wozniak (7) paper to determine what additional analytical information is needed for the development of chiral drug products.

6.4. Gas Chromatography

Gas chromatography, while *stability-indicating*, is not as versatile as HPLC, as the drug substance may not be volatile. On the other hand, increasing the temperature to effect volatility may cause degradation as well as effecting racemization. However, there may be a limited number of instances in which this technique would be useful, such as for small nonaromatic compounds that simple are not possible to separate by current HPLC and TLC techniques.

6.5. Thin-Layer Chromatography

Thin-layer chromatography (TLC) is a mature chromatographic technique and is still widely used throughout the pharmaceutical industry in research as well as in the control laboratory. It is used throughout the drug development process for determining the purity of the drug substance, reference standards, and intermediates. It possesses many advantages including simplicity, low cost, and a short run time. It is cost effective. Its main disadvantage is variability. Constanzo (8) has proposed a three-point window approach to optimize resolution, and thus to minimize the variability, by controlling the mobile phase composition.

TLC (limit test) is used to complement a non-stability-indicating procedure as indicated in the FDA Guideline for Submitting Samples and Analytical Data for Methods Validation (2).

6.6. Capillary Electrophoresis and Capillary Electrochromatography

As sciences, both capillary electrophoresis (CE) and capillary electrochromatography (CEC) today are probably where HPLC was 10 years ago. CE is a separations technique based on the mobility of ions through a buffer-filled capillary in an electrically charged environment. This would provide a separation of charged species. When CE is coupled with a stationary phase and high pressure, it is known as CEC, in which the separation is based on electrophoretic migration and chromatographic partitioning enabling the separation of neutral species. Both techniques are more applicable to biological systems, in biopharmaceutical and other R&D applications, than in quality assurance/product specification applications. The techniques are very sensitive and well suited for separations of small amounts of expensive biopharmaceuticals. On the other hand, they have less utility as a product release or stability test methodology, especially in product specification applications of small molecular entities where there is an abundance of samples and where sensitivity is not an issue.

The utility of the technique, however, lies in its ability to achieve high sensitivity and resolution through high efficiencies with minimal peak dispersion. Moffatt et al. (9) have reported unusually high efficiencies of up to 2.5 million plates per meter in the capillary electrochromatographic analysis of partially ionized anionic-neutral pyrimidine compounds using a standard C_{18} stationary phase.

The number of manufacturers of CE/CEC equipment are not nearly as many as for HPLC equipment. Major manufacturers include Unimicro Technologies, Thermo Bioanalysis, Beckman Coulter, and Micro-Tech Scientific. The last company's model Ultra-Plus II has an integrated, gradient capillary HPLC/CE/CEC system. This combination of gradient elution and electrophoretic migration provides a rapid analysis with high resolution (10).

7. ROLE OF FORCED DEGRADATION

7.1. Regulatory Basis

The 1987 edition of the FDA stability guidance document (1) stipulates that the API be subjected to a number of forced degradation conditions to include acidic, basic, and oxidative conditions. Workers in the field have also included temperature and light (photostability). The current draft stability guide (11), while not yet official, specifically includes photostability and temperature cycling requirements; no mention of acidic, basic, or oxidative conditions were made, however. The current ICH guidances (Q2A and Q2B) also do not specify how degradation studies are to be conducted; this was left to the discretion of the responsible companies.

7.2. Scientific Basis

Forced degradation should be one of the activities performed early in the development process to ensure that the method is discriminating before a lot of time, effort and money have been expended. The guidance documents do not indicate detailed conditions, so the conditions and interpretations are left up to the development scientist. Suggested forced degradative conditions are summarized in Table 2. Trial and error are needed to find the proper combination of stress agent concentration and time to effect a degradation, preferably in the 20–30% range. Depending on the API, not every stress agent may effect a degradation, but each agent has to be evaluated to determine whether degradation results.

Additional comments are warranted.

Adequate k'. The initially developed method should achieve a suitably retained peak, with a k' of about 4 to 10. This range allows a suitable time space in the chromatogram for degradants to elute before or after the active (major) peak. Since the polarity of the degradants relative to the major peak is not known, the k' of the major peak eluting in the middle of the chromatogram adds some assurance that the degradants would elute on either side of the main peak.

Degradation conditions. Unfortunately this is a trial and error process. Typical degradative conditions involve hydrolysis, photolysis, acid/base reactions, and temperature. The goal is to obtain about 20–30% degradation and not complete degradation of the active compound. Achieving 100% degradation would be too strenuous and could possibly cause secondary degradation, giving degradation products of the degradation product(s), which are not likely to be formed under normal storage conditions. Depending on the API, not all of the degradation conditions effect degradation, and after a reasonable effort (varying concentrations and time) to produce a degradation product with no success, one can move on to the next condition. For example, when chlorhexidine digluconate, an antimicrobial agent in mouthwash, was subjected to each of the above conditions, only degradants were isolated from heat, acid and light (12). While it was impervious to the other conditions, this was not known up front, so each of the conditions had to be tried.

Acid/base. Generally the concentration of the API is doubled to enable the reaction solution to be neutralized before injecting into the HPLC system to prevent damage to the silica-based chromatographic column.

Controls. Refer to Table 2. It is important that corresponding matrices and appropriate controls be treated in a similar fashion to identify possible interferences.

Table 2 Suggested Outline for Performing Forced Degradation Studies

Decide/select matrix for degradation

		Degradation conditions					
Product/matrix	Degradation	Acid	Base	Peroxide	Bisulfite	Photostability	Temperature
Product	Yes	√	√	√	√	√	√
Placebo/vehicle	Yes	√	√	√	√	√	√
API/raw material	Yes	√	√	√	√	√	√
Internal standard	No	—	—	—	—	—	—
Controls							
Product	No	—	—	—	—	—	—
API/Raw material	No	—	—	—	—	—	—
Blank solution	No	—	—	—	—	—	—

Decide/select degradation conditions/agents

Medium	Conditions*
1 N HCl, 10 mL	Reflux 30 minutes, neutralize with base
0.1 N NaOH, 10 mL	Reflux 30 minutes, neutralize with acid
3% Hydrogen peroxide, 10 mL	Reflux 30 minutes
10% Sodium bisulfite, 10 mL	Reflux 30 minutes
Light	Light chamber, 1 lumens (92.9 lux = 1000 ft-candles), 7 days
Temperature (dry heat)	80°C, 7 days

* Strive for 20–30% degradation.

Internal standard. Should the analytical method utilize an internal standard, it is not recommended to degrade the internal standard, but its k' should not interfere (elute) at any of the possible eluting k' peaks.

Evaluation of the degradation mixture is generally performed using a photodiode array detector. Assessing the purity of the major peak is very important and could be difficult in light of possible peak inhomogeneity after the degradation process. One must be assured that there is no degradation peak (hiding) under or unresolved from the major peak of interest. The utility of the diode-array detector is that the analyst can select a whole wavelength range, say from 200 to 350 mm, with a bandwidth of 80 nm. With just one single chromatographic run, all compounds absorbing within this range will be detected. With only one wavelength selected using a conventional UV detector, for instance at 280 nm, any compound not absorbing at this wavelength will not be detected. Figures 1A and 1B depict diode-array chromatograms for assessing peak purity.

Refer to Section 17.4 for further discussion.

8. PEAK PURITY

There is always that nagging question of whether the peak of interest (the major analyte peak) is pure or homogeneous. This is a difficult question, and many investigators have tried to prove the homogeneity of the major peak under stressed conditions during the method development and validation process. Various techniques have been used to characterize peak homogeneity, such as spectral suppression, absorbance ratio, spectral overlay (13), electrospray mass spectrometry (14,15) and dual detection (16).

9. SAMPLE PREPARATION

Sample preparation is a critical step in the overall chromatographic process, and can affect the chromatography if not developed or treated properly. This step encompasses sample filtration, sample extraction as well as sample derivatization, although the latter is not commonly used in the pharmaceutical quality laboratory. The purpose of this step is to prepare the sample so that the drug substance can be readily chromatographed, separated from other materials. Thus, it is a step to remove any interferences, to enhance the detection of the drug substance as well as to protect or enhance the life of the analytical column.

The following considerations are noted:

What is the matrix?
Ensure complete dissolution of the analyte in mobile phase or weaker solvent.
Miscibility and solubility.
Does the analyte precipitate in the buffer?
Some typical treatment modes are
- Direct injection
- Dilution
- Sonication
- Shaking

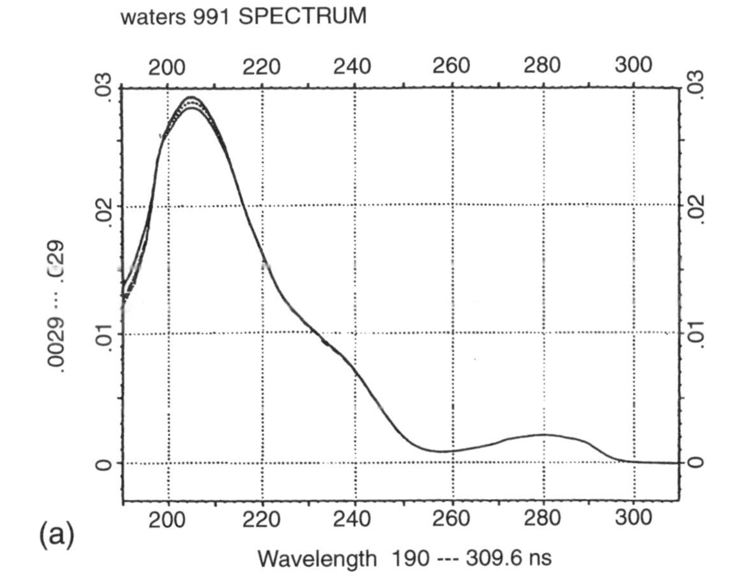

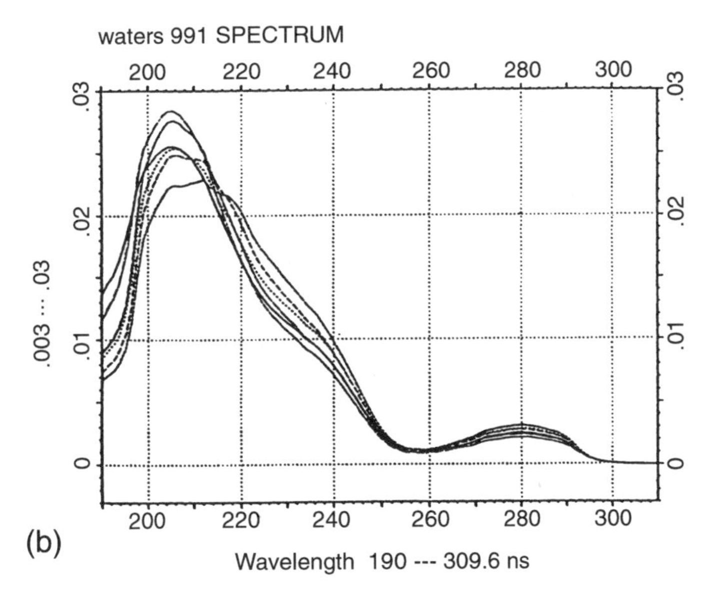

Figure 11.1 (a) No degradation (scans superimposable). (b) Presence of degradation (scans not superimposable).

Filtration/ultrafiltration
Extraction—Liquid or solid phase
Evaporation
Reconstitution
Derivatization
Heating/cooling

On rare occasions, the drug product, in a solution package form, may be injected directly or after an appropriate dilution. Typically, for a solid, such as a tablet or a capsule, the pretreatment would necessitate a comminution step, followed by extraction/sonication, filtration, and dilution. For example, an ointment may necessitate an extraction followed by evaporation, reconstitution, and dilution, or heating to dissolve the analyte/matrix, followed by cooling to precipitate the matrix, and then filtration.

While the typical dosage form—solid (tablet/capsule), semisolid (ointment/cream), or solution (cough syrup/ophthalmic solution)—utilizes a combination of the treatment modes mentioned earlier, solid-phase extraction (SPE) has become a recognized and viable technique for sample preparation methodologies, especially for biosamples and as an alternative to liquid-liquid extractions in many U.S. Environmental Protection Agency (EPA) methods. A recent supplement to *LC/GC* magazine was dedicated to advances in SPE (17).

It is very important that the sample preparation, prior to injection into the liquid chromatograph, be freed of particulate matter, through either filtration or centrifugation, and that the solvent be compatible with the HPLC system. If there is incomplete sample solubility or if the solvent is too polar, band distortions or tailing will result. Ensuring that the sample is completely dissolved in the proper solvent and then diluting the sample in mobile phase will eliminate these problems.

10. DEVELOPING THE SEPARATION—CHOOSING THE EXPERIMENTAL CONDITIONS

From Sec. 5, we assume that separation goals have been determined, such as resolution (at least baseline), reasonable run time (under 10 minutes), and ruggedness. These elements are further discussed below and developed in greater depth in Part II of this chapter under Validation. From Sec. 6.2 above, a case has been made that reversed-phase HPLC is suitable for our API of interest. The next step is to determine whether the API is typical. Referring to Secs. 6.1 and 6.2 above, let us further assume that the API is ionic and acidic. From a listing of generic separation conditions, see Table 3, conditions for an ionic and acidic compound are selected, and an initial exploratory run using gradient elution is made.

At this point, two options may be available to us before performing the exploratory run in the development of the desired stability-indicating procedure. First, there may be a method, either in-house or from the literature, already available for the same API or compound of interest. Useful information may be gleaned from here to modify to suit the specific compound on hand. On the other hand, sometimes established methods may not be optimal, so rather than modifying the method to suit our need, it may be better in the long run to develop a new method that is optimal and rugged.

Exploratory runs can be done manually or with computer software. Both are trial and error methods, but the latter is more systematic, quicker, and requiring fewer injections. When and after an initial exploratory run has been performed, the chromatogram is evaluated before proceeding with the next injection, and subsequent adjustments are made to the mobile phase composition. Each subsequent injection is thus based on the previous conditions, so that after a number of injections

Table 3 General Experimental Conditions for an Initial HPLC Run

	Initial Parameters		
Chromatographic variables	Neutral compounds	Ionic-acidic compounds (carboxylic acids)	Ionic-basic compounds (amines)
Column			
Dimension (length, ID)	25 cm × 0.46 cm	25 cm × 0.46 cm	25 cm × 0.46 cm
Stationary phase	C_{18} or C_8	C_{18} or C_8	C_{18} or C_8
Particle size	10 μm or 5 μm	10 μm or 5 μm	10 μm or 5 μm
Mobile phase			
Solvents A and B	Buffer-acetonitrile	Buffer-acetonitrile	Buffer-acetonitrile
% B (organic) isocratic	50%	50%	50%
% B (organic) gradient	20%–80%	20%–80%	20%–80%
Buffer			
Type	Phosphate	Phosphate	Phosphate
Concentration	50 Mm	50 mM	50 mM
pH	3.0	3.0 & 7.5 (gradient)	3.0 & 7 5 (gradient)
Modifier	10 mM triethylamine and 1% acetic acid, if needed	1% acetic acid	25 mM Triethylamine
Flow rate	1.5–2.0 mL/minute	1.5–2.0 mL/minute	1.5–2.0 mL/minute
Temperature	Ambient to 35°C	Ambient to 35°C	Ambient to 35°C
Sample size			
Volume	10 μL – 25 μL	10 μL – 25 μL	10 μL – 25 μL
Mass	<100 mcg	<100 mcg	<100 mcg

the proper conditions can be found (18). Refer to Sec. 12 below further discussion on software method development.

10.1. Key Variables—Resolution Equation Parameters

In reversed-phase/ion-pair chromatography, there are essentially 8–10 key variables that affect the separation, as depicted in the resolution equation, R:

$$R = \frac{1}{4} \cdot N^{1/2} \cdot (\alpha - 1) \cdot \frac{k'}{1 + k'}$$

where N, α and k' are referred to as the efficiency, selectivity, and retention (capacity) factors, respectively affecting the resolution of the analyte from other components in the separation. The efficiency is affected by the nature of the column, and both selectivity and retention are affected by the solvent. Column variables include length, particle size, and flow. Solvent variables are the nature of the sample, the mobile phase, and the column surface, i.e., bonded-phase (adsorbent type) such as C18, phenyl or cyano, etc.

These key variables include mobile phase strength, solvent type, column type/size, pH, temperature, ion-pair reagent (type and concentration), buffer, and mobile phase flow rate.

10.2. Isocratic or Gradient Mode

Either isocratic or gradient mode may be used to determine the initial conditions of the separation, following the suggested experimental conditions given in Table 3. Depending on the number of active components to be resolved or separated, the more complex the separation, the more gradient elution would be advantageous over isocratic mode, which is akin to a brute force application when trying to separate a complex mixture. When faced with developing a method to separate a complex mixture, the use of computer software is useful. This is further discussed in Sec. 12.

In deciding whether a gradient would be required or whether isocratic mode would be adequate, an initial gradient run is performed, and the ratio between the total gradient time and the difference in gradient time between the first and last component are calculated. When the calculated ratio is < 0.25, isocratic is adequate; when the ratio is > 0.25, gradient would be beneficial (19) as shown in Figure 2.

For complex mixtures (separations), when there are many degradation products, a long gradient run may be needed. In this case, a compromise may have to be made, using an isocratic method for product release and a gradient method for stability assessment. The isocratic method has generally a shorter run time, say under 15 minutes, and no degradation product would be monitored, assuming that none are formed initially. With time the degradation products are formed and must be monitored, which requires a gradient method to resolve completely the mixture (15 minutes and longer depending on the complexity of the degradation mix). The gradient method, then, would be the stability or regulatory method.

10.3. Role of pH

pH is another factor in the resolution equation that will affect the selectivity of the separation. In reversed-phase HPLC, sample retention increases when the analyte

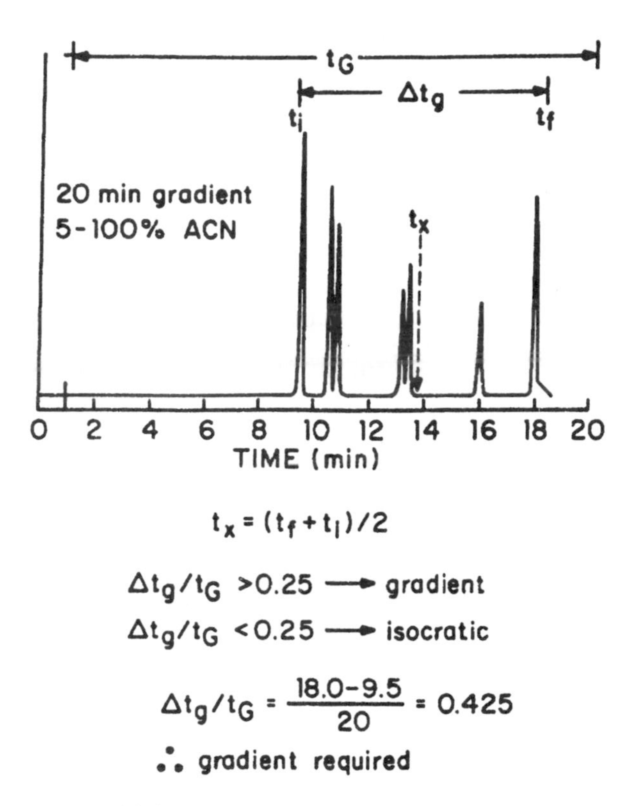

Figure 11.2 Isocratic or gradient? (From Ref. 19.)

is more hydrophobic. Thus when an acid (HA) or base (B) is ionized (converted from the unionized free acid or base) it becomes more hydrophilic (less hydrophobic, more soluble in the aqueous phase) and less interactive with the column's binding sites.

$$\begin{array}{lcl} \mathrm{HA} & \Leftrightarrow & \mathrm{H^+ + A^-} \\ \mathrm{B + H^+} & \Leftrightarrow & \mathrm{BH^+} \\ \text{Hydrophobic} & & \text{Hydrophilic} \\ \text{(more retained on column)} & & \text{(less retained on column)} \end{array}$$

As a result, the ionized analyte is less retained on the column, so that the k' is reduced, sometimes dramatically. When the $\mathrm{pH} = \mathrm{pK_a}$ for the analyte, it is half ionized, i.e., the concentrations of the ionized and unionized species are equal. As mostly all of the pH-caused changes in retention occur within ± 1.5 pH units of the $\mathrm{pK_a}$ value, it is best to adjust the mobile phase to pH values at least ± 1.5 pH units above or below the $\mathrm{pK_a}$ to ensure practically 100% unionization for retention purposes. The pH range most often encountered in reversed-phase HPLC is 1–8, normally considered as low pH (i.e., 1–4) and intermediate pH (i.e., 4–8). Generally, at low pH peak tailing is minimized and method ruggedness is maximized. On the other hand, operating in the intermediate range offers an advantage in increased analyte retention and selectivity. See Sec. 10.7 for further discussion. For a detailed treatment of retention as a function of pH, the reader is referred to the works of Lewis et al. (20) and Schoenmakers and Tijssen (21).

As stated in Table 1, it is important to know the salt form of the drug substance of interest or whether it is amphoteric. This information is invaluable to the development of the analytical methodology because it will aid in the optimization of the method to effect better separation, resolution, and chromatography. If the drug is amphoteric, the pH can be selected whereby the compound exists as a single species and not a mixture of species. Mixed species will lead to poor separations. On the other hand, if the drug has different salt forms, say the hydrochloride and the napsylate (e.g., *d*-propoxyphene hydrochloride, Darvon®, and *d*-propoxyphene napsylate, Darvon-N®, both drug products marketed by Eli Lilly), the problem is not as critical, for

The salts represent different products and are marketed separately for different pharmacokinetic effects, i.e., different absorption profiles, with the hydrochloride being more soluble, and thus showing a faster absorption and distribution.

In solution, both salts will be dissociated from the organic propoxyphene moiety so that the final analytical methodology is appropriate for the separation and detection (or titration) of the analyte free base. For example, in the USP monographs for the two propoxyphene (hydrochloride and napsylate salts) APIs and their several products (22), the final analytical methods, be they titrimetric or chromatographic, all detect the analyte free base propoxyphene, and the assay percentage is calculated using a molecular weight correction factor.

10.4. Role of Solvent Type

Solvent type (methanol, acetonitrile, and THF) will affect selectivity similarly for ionic and neutral analytes. Hence changing a solvent would be a useful variable in the separation. The choice between methanol and acetonitrile may be dependent on the solubility of the analyte as well as the buffer used. While THF may be the least polar of the three, it has the highest solvent strength. If that property is not essential, its odor and potential peroxide formation may be a deterrent.

10.5. Role of Mobile Phase

The mobile phase composition (percent aqueous to organic) as well as the solvent strength will affect both α (solvent selectivity) and k' (solvent strength). The sample solvent will have a similar effect as well and may lead to peak distortion if the polarity between the mobile phase and the sample solvent is great. Thus, if at all possible, it is best to dissolve the sample in the mobile phase, if not, at least to make the final dilution in the mobile phase.

Chromatographic separations thus vary with solvent properties and are related to sample solubility, polarity, and solvent strength. Solvents that interact strongly with the sample will increase the sample solubility and decrease the chromatographic retention as more sample ions exist in the solvent and are not able to be in equilibrium with the adsorbent surface. Thus changing the organic solvent will change the selectivity. Polarity is the summation of dipole and hydrogen bonding interactions, and in reversed-phase chromatography, less polar solvents exhibit greater solvent strength than polar solvents. The solvents water (most polar), methanol, acetonitrile, and tetrahydrofuran (THF) are placed in ascending order of polarity but reversed in their order of solvent strength.

These three organic solvents (methanol, acetonitrile, and THF) form the basis of the solvent selectivity triangle and exhibit differences in their relative interactions. They are also miscible with water and possess low viscosity and UV transparency. Collectively these three organic solvents along with water provide a four-solvent mobile phase optimization strategy. Each organic solvent in combination with water or water containing a buffer or additive(s) comprise the mobile phase. Sometimes the mobile phase may contain two organic solvents. The aqueous phase composition is commonly referred to % A and the organic phase as % B.

When the sample is eluted with a mobile phase of 100% B (organic), there is no separation, as thc sample is eluted in the void volume. This is because the sample is not retained; but retention is observed when the mobile phase solvent strength is decreased to allow equilibrium competition of the solute molecules between the bonded phase and the mobile phase. When the separation is complex, that is, many components are to be separated, and when the solvent strength is decreased and there is still no resolution between two close peaks, another organic solvent of a different polarity or even a mixture of two organics may need to be tried to effect separation. Additionally, mobile phase optimization can be enhanced in combination with bonded phase optimization (i.e., substituting C18/C8 with cyano or phenyl). A goal for the band spacing of a solute (k') should be in the range of 4 to 9 and a run time of about 15 minutes or 20 minutes at most for most routine product release or stability runs.

10.6. Role of Buffer

When the mobile phase contains only water and organic solvent, it is recommended that the pH of the mobile phase be controlled by using a buffer to provide capacity. Thus when selecting a buffer for a given application, the following considerations are important:

The buffer capacity is dependent on pH, buffer pK_a, and buffer concentration.
UV absorbance—UV transparent to below or at the wavelength of the organic solvent.
Other properties, such as solubility and stability of the buffer and its reactivity to the analyte and hardware components of the chromatographic system.

The buffer concentration, or ionic strength, will affect the selectivity. An increase in the buffer concentration can lead to a decreased retention as the ionic interaction between the analyte and silanols are swapped out by the increased buffer concentration. When selecting a given buffer, additive, or even the solvent, sufficient regard for their compatibility with the analyte or HPLC system must be considered (23).

10.7. Role of the Ion-Pair Reagent

Initially, when deciding whether to select reversed-phase HPLC or reversed-phase HPLC with ion-pairing, a good rule of thumb is to consider the nature of the analyte of interest. If the sample is neutral, begin with reversed-phase; and if the sample is ionic, use ion-pairing. Thus reversed-phase HPLC and reversed-phase HPLC with ion-pairing are similar except that the latter contains an ion-pair reagent in the mobile phase to improve the selectivity of ionic samples. The use of an ion-pair reagent is suggested only when separation is not adequate with reversed-phase HPLC. This is because using an ion-pair reagent introduces additional experimental parameters that need to be controlled, such as what ion-pair reagent to use and its concentration. Because of this added variable, reversed-phase HPLC should be utilized on any ionic analyte first before trying ion-pair reversed-phase HPLC.

The solubility of the ion-pair reagent may also be affected depending on the organic solvent used in the mobile phase. Methanol is generally preferred over acetonitrile or THF because it provides better solubility for the ion-pair reagent as well as for buffers and salts. In this case a suitable buffer is chosen, and at a concentration of about 25 mM, the pH and ion-pair reagent concentration are varied to provide optimal selectivity to the separation. These variables are not easily altered with commercial ion-pair "kits" as their pH and concentration have been standardized and ready to use.

Table 4 summarizes the types of ion-pair reagents and the conditions for their use.

The mechanism of retention imparted by the ion-pair reagent, such as an alkyl sulfonate, provides a change in the equilibrium between the ionized analyte and the ion-pair reagent that is attached to the silica adsorbent through the hydrophobic alkyl group and the negative charge of the sulfonate ion. The positively charged

Table 4 Listing of Types of Ion-Pair Reagents and Conditions of Use

Ion-pair reagent type and examples	Used for compound class	pH of mobile phase	Concentration of ion-pair reagent
Alkyl sulfonates (sulfonic acid alkyl salts)	Cationic samples (protonated bases)	3.5	0.005 M
Pentane sulfonate Hexane sulfonate Heptane sulfonate Octane sulfonate Examples PIC® B series (Waters) Q-Series (Regis)	Basic compounds		
Alkyl ammonium salts	Anionic samples (ionized acids)	7.5	0.005 M
Tetrabutylammonium phosphate Tetrabutylammonium hydrogen sulfate Examples: PIC® A series (Waters) Q-Series (Regis)	Acidic compounds		

(protonated) analyte ion competes for the negative site of the sulfonate ion. This altered equilibrium in effect imparts a change in the solubility of the analyte sample which in turn alters the retention as the analyte is now "attached" to the adsorbent so that it is eluted at a later time.

The pH of the mobile phase is closely associated with ion-pairing, whether the ion-pair reagent is positively charged (tetrabutylammonium, TBA^+) or negatively charged (C_5- or C_6- sulfonate/C_5- or C_6-SO_3^-), and dependent on whether the analyte is an acid or a base. Cationic samples (protonated base) or bases use the pentane, hexane, or a higher hydrocarbon sulfonate ion-pair reagent. Anionic samples (ionized acids) or acids commonly use tetraethylammonium or tetrabutylammonium hydroxide as the ion-pair reagent. Their optimization is pH dependent. For example, Waters® Chromatography ion-pair reagents operate in the low and intermediate pH ranges: PIC™ A (tetrabutylammonium phosphate) for acids operates at pH 7.5 and PIC™ B5 to B8 (pentane to octane sulfonic acid) for bases operate at pH 3.5.

For selection of the proper ion-pair reagent, alkyl chain lengths must be considered. The length of the alkyl chain enables selective separation of the analyte. The longer the chain, the more hydrophobic the counterion, and therefore, the greater the retention due to equilibrium between the counterion and the column adsorbent. Thus by selecting a reagent with a longer chain, selective solubility is obtained, enhancing the resolution.

10.8. Role of the Column

The HPLC column is the heart of the method, critical in performing the separation. The column must possess the selectivity, efficiency, and reproducibility to provide a good separation. All of these characteristics are dependent on the column manufacturer's production of good quality columns and packing materials. Properties of the silica (backbone) such as metal content and silanol activity produced in the manufacturing and bonding processes determine the properties of the finished bonded phase. A good silica and bonding process will provide the reproducible and symmetrical peaks necessary for accurate quantitation.

Commonly used reversed phases are C18 (octadecylsilane, USP L1), C8 (octylsilane, USP L7), phenyl (USP L11), and cyano (USP L18) (24). They are chemically different bonded phases and demonstrate significant changes in selectivity using the same mobile phase. Their properties vary from manufacturer to manufacturer, but given the state-of-the-art character of the vendor's manufacturing process, they show good quality control and provide batch-to-batch reproducibility. For example, no two L1 columns are the same, they vary from manufacturer to manufacturer relative to their pore volumes, pore sizes, surface areas, particle sizes (average range), carbon loads, whether end-capped or not, and the amount of bonded-phase coverage, as well as varying in their basicity and acidity characteristic. With state-of-the-art developments in column technology, most columns on the market exhibit good quality control and provide excellent column-to-column reproducibility and batch-to-batch reproducibility (25), and in some cases they give the chromatographer the option of using column selectivity as an alternative tool (besides mobile phase selectivity) to optimize the HPLC method development (26).

Column length also plays a role in the separation resolution. As column length changes, the efficiency (N) changes in direct proportion to the ratio of the column length (27). Resolution, as indicated in the resolution equation (vide supra), changes as a function of the square root of the change in N, and an estimate of the change in resolution as a function of column length can be approximated with the equation

$$R_{s2} = R_{s1} \cdot \left(\frac{L_2}{L_1}\right)^{1/2}$$

where R_{s1} is the resolution obtained from column 1 and R_{s2} is the estimated resolution with column 2.

Similarly the run time (RT) and column back pressure (P) will also change in direct proportion to a change in the column length by

$$RT_2 = RT_1 \cdot \frac{L_2}{L_1} \qquad P_2 = P_1 \cdot \frac{L_2}{L_1}$$

where RT_1 is the run time for column 1, RT_2 the run time for column 2, P_1 the pressure for column 1, and P_2 the pressure for column 2.

While most analytical columns are standardized to a 4.6 mm id, their lengths vary; they are available in lengths of 5 cm, 15 cm, and 25 cm, whereas the original Waters μBondapak® C18 column measures 30 cm × 3.9 mm id. A good selection of columns illustrating type and sizes can be found in most HPLC vendors' supply catalogs.

Claessens et al. (28) have reported on an extensive study on the effect of buffers on silica-based column stability in reversed-phase HPLC. As the analytical column has a silica-based backbone, it is not stable in alkaline pH. The authors reported that silica-based bonded phase packings variably degrade with buffers as a function of the type of anion, cation, pH, buffer type, and temperature.

10.9. Role of Temperature

While temperature is a variable that can affect selectivity, α, its effect is relatively small. Also, the k' generally decreases with an increase in temperature for neutral compounds but less dramatically for partially ionized analytes. Still, it may have some effect when there is a significant difference in shape and size between samples. Overall, it is better to use solvent strength to control selectivity than to use temperature; its effect is much more dramatic. Snyder et al. (29) reported that an increase of 1°C will decrease the k' by 1 to 2%, and both ionic and neutral samples are reported to show significant changes in α with temperature changes. Because of possible temperature fluctuations during method development and validation, it is recommended that the column be thermostated to control the temperature.

10.10. Role of Flow Rate

Flow rate, more for isocratic than gradient separation, can sometimes be useful and readily utilized to increase the resolution, although its effect is very modest. The slower flow rate will also decrease the column back pressure. The disadvantage is that when flow rate is decreased, to increase the resolution slightly, there is a corresponding increase in the run time.

11. OPTIMIZATION (OPTIMIZING THE SEPARATION)

Up to this point, efforts to develop a suitable stability-indicating HPLC method have revolved around the resolution equation (see Sec. 10.1). To optimize the method, the chromatographer must tweak the three variables in the equation. The capacity factor, k', can be affected with a change in the solvent. The efficiency factor, N, can be altered with a change in the column dimension, particle size, stationary phase, and flow rate. Lastly, the separation factor, α, can be modified with a change in the solvent, pH, ionic strength of the buffer, stationary phase, mobile phase additives, and temperature. These three factors need to be considered for optimizing the method, conveniently performed utilizing the Plackett–Burman design and computer software. The use of computer software for optimization is becoming more and more common. Refer to Sections 12 and 17.10 for further discussions on these topics.

11.1. Peak Area or Peak Height for Quantitation

The chromatographer can either select peak area or peak height for quantitation assuming that both modes have been properly calibrated and validated. It is suggested, however, that peak area be used for development and peak height for stability for the reasons stated in the table.

Development—Peak Area	*Stability Monitoring—Peak Height*
Suitable for simple, well-resolved mixtures	More accurate/sensitive than peak area
Less frequent standardization required	Requires less resolution of compounds
Generally more precise than peak height	More affected by instrumental variations
Generally best when simple equipment is used	Best suited for complex mixtures
Better suited for nonsymmetrical peaks	Use for trace analysis

11.2. Plackett–Burman Design

Often in method optimization it is necessary to consider various variables, such as environmental and experimental conditions, that affect the ruggedness of a given method. One such experimental design often used in ruggedness testing is the Plackett–Burman design named after the authors that first published their work more than a half a century ago. Refer to Sec. 17.10 in the *Validation* part of this chapter for further discussion on this subject.

12. COMPUTER SOFTWARE FOR METHOD DEVELOPMENT

The discussion in Sec. 10 presumes developing a method by manual trial and error, yet in a systematic manner. That is, the conditions for an initial run are noted, and, based on the outcome of the first run, modifications are made for the second run. Then based on the results of the second run, additional modifications are made for the third run, and so forth until a good separation is obtained. Thus a number of these trial and error runs may be needed to obtain the desired separation, which may conceivably be time consuming.

In the last fifteen years or so, the use of software for method development in reversed-phased HPLC has increased dramatically, with the intended purpose of separating complex mixtures by shortening the development time and optimizing the resolution based on a limited amount of experimental retention data. A number of these computer systems are commercially available. Many reviews on the subject have been published (30), and many references to using the DryLab™ have been reported (31). DryLab™ is a widely used computer simulation program that after a limited number of actual injections at different conditions can predict an optimal condition or separation at other conditions.

13. OTHER APPLICATIONS

13.1. Analytical Method for Cleaning Assessment

Normally, production equipment is shared to manufacture different pharmaceutical products. Thus cleaning processes following production of pharmaceutical products are critical to prevent cross-contamination. The analytical method used to assess the effectiveness of the cleaning process is usually the same stability-indicating method used for product release and stability monitoring, with some adjustments to increase its sensitivity. How sensitive and specific the method has to be is commonly determined from a joint effort between the pharmaceutical engineer and the analytical chemist to establish the necessary cleaning limit. The method developed must be capable of being validated and rugged enough to meet predetermined specifications consistently. In addition to HPLC, total organic carbon (TOC) analysis has become a widely used method for analyzing cleaning residues, and the Compendia have dedicated General Chapter <643>, Total Organic Carbon, to the subject (32). TOC, however, is not as specific as HPLC. Conductivity has also been used. Generally HPLC is the most accurate, reliable, and specific of all the analytical cleaning methods.

13.2. Physicochemical Characterization Method (Dissolution Method)

A liquid chromatographic method developed for product release or stability monitoring can be adapted for use with a dissolution assay. An HPLC method for dissolution assay testing is optimized for speed and is not intended for determination of degradation products or process impurities. Instead, the real utility of this combination (dissolution with HPLC determination) is that it eliminates interferences from formulation excipients. Assuming that the HPLC method has been developed and validated, the development process is bridged over to developing the dissolution methodology. A preliminary dissolution test is developed very early in the pharmaceutical development process to support formulation development. Primary dissolution parameters for development include selection of the filter, the apparatus type, the rotation speed, and the dissolution medium. Once these parameters have been established, they are to be validated as part of the total validation effort for the HPLC dissolution methodology. The reader is referred to the article by Skoug et al. (33) for an overview of the subject.

13.3. Nonchromatographic Methods

Approaches and guidelines used to develop and validate a chromatographic method can be applied to develop nonchromatographic methods (not stability-indicating) as well. It is equally appropriate to follow the guidelines of USP 23 General Chapter <1225>, Validation of Compendial Method (34), selecting and validating those analytical elements that are needed for a rugged method. These nonchromatographic methods include UV spectrophotometry, atomic absorption, infrared spectroscopy, and titrimetry.

Additional discussions on the validation of various nonchromatographic methods are found in Sec. 19.3.

Part II: Method Validation

14. REGULATORY AND COMPENDIAL BASIS OF METHOD VALIDATION—WHERE TO START

Analytical methods including chromatographic and nonchromatographic techniques are used to generate reliable and accurate data during drug development and post approval of the drug products. The testing, in general, includes the acceptance of raw materials and the release of drug substances and finished products, in process testing, and analysis of stability samples for establishing expiration dating. Therefore test methods that are used to assess the compliance of pharmaceutical products with established acceptance criteria must meet proper cGMP standards of accuracy and reliability as set forth by the regulatory agencies (35).

According to Section 501 of the Federal Food, Drug, and Cosmetic Act, assays and specifications in monographs of the USP and NF constitute legal standards. Under the Food Drug, and Cosmetic Act, the FDA can enforce the USP/NF standards of strength, quality, purity, packaging, and labeling. Therefore for compliance purposes, every analytical method should be validated according to pharmacopeial standard, because each method could be included in a drug monograph.

Method validation is a regulatory requirement. The Food and Drug Administration and the International Conference on Harmonization (ICH) of Technical Requirements for Registration of Pharmaceuticals for Human Use have published a series of guidelines on the validation of analytical procedures (36,37). In USP 23/NF 18, General Chapter <1225> has been allocated for validation of compendial methods (34). This chapter describes in detail as well as in summary how to evaluate particular performance parameters. In general, it is assumed that this chapter is applied to chromatographic methods of analysis, and that for nonchromatographic procedures some alternate guidelines should be used. However, in USP 23/NF 18 no such distinction has been made. Therefore performance parameters given in General Chapter <1225> can be used to evaluate the performance of any analytical method. However, one needs to be careful in selection of performance parameters. Also, methods described in the current USP are not stability-indicating in nature. Therefore for monitoring of stability studies, guidelines given in General Chapter ⟨1225⟩ can be used to validate these methods.

What constitutes validation? The validation of analytical method is the process in determining the suitability of a given methodology by laboratory studies that the method in question can meet the requirements for the method's intended use. Method validation is not simply a measure of procedure; method validation is a measure of performance of the total analytical system. Sections 211.165(c) and 211.194(a)(2) of the cGMP for method validation specify that any method adopted at the product development stage be verified under actual conditions of use, and that subsequent variations on existing methodology are subjected to validation.

General Chapter <1225> states, "Validation is the process of providing documented evidence that the method does what it is intended to do." In other words, the process of method validation ensures that the proposed analytical methodology is accurate, specific, reproducible, and rugged for its intended use.

The articles in the current revision of the Compendia are also recognized to be legal standards when determining compliance with the Federal Food, Drug, and Cosmetic Act. Regulated industries must perform method validation to comply with Compendial or other regulatory requirements, and the data generated becomes a part of the methods validation package submitted to the FDA.

Similarly, the general regulation, which is currently represented in 21 CFR 2.19, states, "Where the method of analysis is not prescribed in a regulation, it is the policy of FDA, in its enforcement programs to utilize the methods of analysis of AOAC as published in the current edition." Further, it is stated in the FDA's current Good Manufacturing Practices for Finished Pharmaceuticals regulations 21 CFR 211.165(e) and 21 CFR 211.194(a)(2) that if a firm is using AOAC-OMA or USP/NF methods of analysis, only minimal additional validation data is required (35).

15. VALIDATION PROTOCOL

While the text of Title 21 CFR Part 211, ICH Guidelines, and General Chapter <1225> all provide terms and definitions, there is no specific discussion of validation protocol and methodology. In ICH Guidelines (Q2B) on Method Validation Methodology, the applicant has been made responsible for the appropriate validation protocol and procedure suitable for their product. Therefore prior to initiating a validation study, a well-planned validation protocol is required. This protocol should consist of experimental design and elements required for validation of the proposed test method that have been reviewed for scientific soundness and completeness by qualified individuals and approved by appropriate company management authority. The validation protocol should include a detailed test procedure, basic experimental design, elements for validation, predefined acceptance criteria, reference of related methods, and management approval.

As mentioned earlier, description of the test method is very significant for successful validation. Therefore a test procedure is a description of the "analytical method" to be used as a guide in validating the method and serves as a basis for the preparation of the validation protocol. It should include

1. A listing of reagents, solvents, and other supplies
2. Instructions for the preparation of standards, samples, and solutions
3. A listing of equipment to be used or equivalents
4. Instrumental parameters and chromatographic conditions

5. System suitability requirements
6. Standard and sample analysis sequence
7. Calculation section to include results formatting

Prior to outlining the experimental design or protocol, however, it is necessary to make some basic assumptions as suggested by Swartz and Krull (38,39). These assumptions are that

Specificity or selectivity for the developed method has been demonstrated (i.e., forced degradation already performed).
The developed method has been optimized to the point where investing time and effort in validation is justified and feasible.
Evaluation of data generated by the developed method is performed by valid statistical approaches to remove some of the subjectivity of method validation.

Keeping in mind these assumptions, the current ICH methodology guidelines, and the requirements for validation depending on the type of analytical method, one can design a stepwise protocol. A typical protocol designed by Swartz and Krull is given in Fig. 3 (38,39).

As shown in Fig. 3, the first parameter to be evaluated is robustness. This parameter is usually evaluated during the method development stage, when the effect

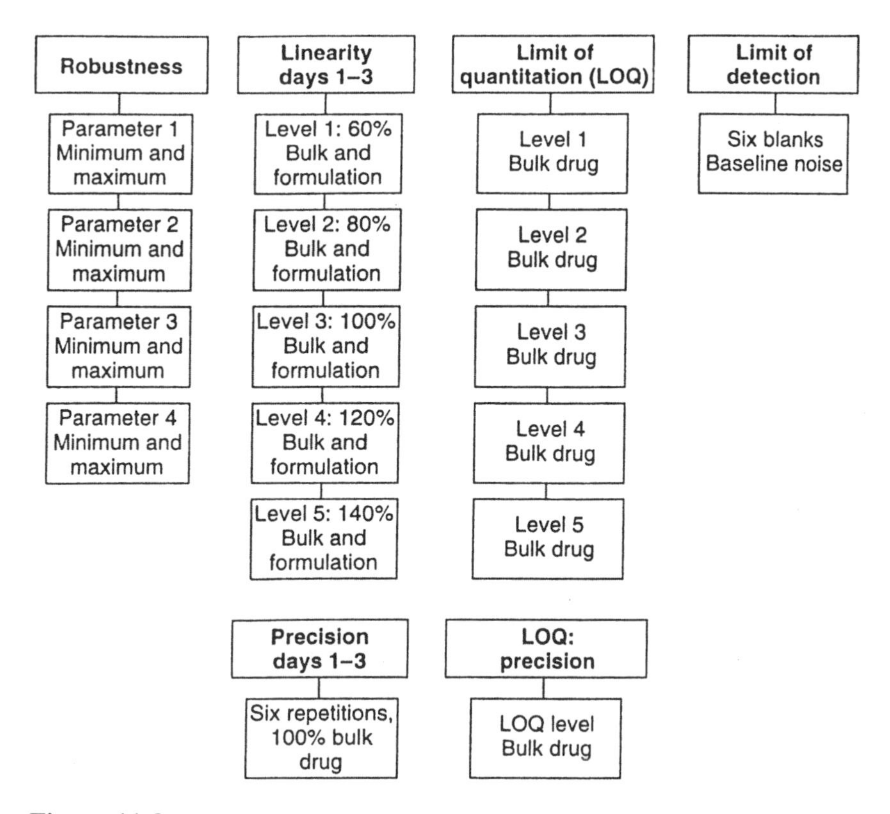

Figure 11.3 Sample method validation protocol. (From Refs. 38, 39.)

of different parameters on selectivity is studied. Method robustness can be evaluated in a stepwise univariate approach or as a part of an experimental design incorporating multivariate parameters.

Next, a linearity test over five levels for both the drug substance and the dosage form is performed. The range is determined according to the test method's intended use (34,36,38,39). Comparison of the results between the drug substance and the dosage form fulfills the accuracy requirements. A minimum of three measurements at each level should be made.

At the end of day 1, a minimum of six repetitions are performed at the 100% level of the drug substance for repeatability.

Steps 1 and 2 are repeated over additional days for intermediate precision. The detection limit (DL) and quantitation limit (QL) can then be determined if required. For calculation of these performance characteristics one can follow criteria given in the USP or ICH guidelines (Q2B). It is stated that this protocol is merely a generic example, and specific protocols or SOPs should be documented and followed for the particular method and its intended use.

16. VALIDATION PARAMETERS

Prior to conducting validation studies it is imperative to decide which parameters are required to be studied. These parameters are termed "analytical performance characteristics" or sometimes "analytical figures of merit." Most of these terms are familiar and are used daily in the laboratory. However, some may mean different things to different laboratory groups. Therefore a complete understanding of the terminology and definitions of these characteristics is important.

The selection of desired performance characteristics would depend on the type of analytical method and its intended use. For example, an assay method designed for finished product release should not be used for the determination of detection or quantitation limits of an active ingredient. However, if the method has been designed to monitor trace quantities of the active ingredient in cleaning validation samples, then knowledge of the detection and quantitation limits are appropriate and necessary.

Therefore, selection of validation parameters for each assay or test method should be made case by case, to ensure that parameters are appropriate for the intended use. This is even more important when validating stability-indicating methods, because such validations are complex, as these involve forced degradation studies, spiking of samples with known degradants and literature searches.

16.1. USP General Chapter <1225>, Validation of Compendial Methods

General Chapter <1225> (34) describes typical analytical performance characteristics, how they are determined, and which subset of data elements is required to demonstrate validity, based on the method's intended use. These performance characteristics can be referred to as the "Eight Steps of Method Validation." These analytical performance characteristics are

Accuracy
Precision
Specificity

Detection limit (DL)
Quantitation limit (QL)
Linearity and range
Ruggedness
Robustness

Compendial test and assay procedures vary significantly in type of analytical method used, and the type of information required for validation of a given analytical method will vary depending on the nature of the method. Consequently in General Chapter ⟨1225⟩ (34), the most common test and assay procedures have been divided into four categories for harmonization with the ICH guidelines:

Category I: Analytical methods for quantitation of major components of bulk drug substances or active ingredients (including preservatives) in finished dosage forms
Category II: Analytical methods for determination of impurities in bulk drug substances or degradation products in finished dosage forms, including quantitative assays and limit tests
Category III: Analytical methods for determination of performance characteristics (e.g., dissolution, drug release)
Category IV: Identification tests

Analytical variables that are normally required for method validation in each of these categories are listed in Table 5.

An evaluation of performance characteristics shown in Table 5 indicate that for assays in Category I, determination of DL and QL is not required because the major component or active ingredient to be quantitated is present at high levels. All other parameters are evaluated to obtain quantitative information needed. Assays in Category II are further divided into quantitative and limit tests subcategories. If quantitative information is required, measurement for DL is not necessary, but the remaining parameters are evaluated. For limit tests, on the contrary, no quantitation is required. Thus it is sufficient to measure only the DL and demonstrate specificity and ruggedness. The parameters to be determined under Category III are dependent upon the nature of the test. Dissolution testing, for example, falls into this category.

Table 5 USP Data Elements Required for Assay Validation

Analytical performance parameters	Assay Category I	Assay Category II		Assay Category III	Assay Category IV
		Quantitative	Limit tests		
Accuracy	Yes	Yes	*	*	No
Precision	Yes	Yes	No	Yes	No
Specificity	Yes	Yes	Yes	*	Yes
DL	No	No	Yes	*	No
QL	No	Yes	No	*	No
Linearity	Yes	Yes	No	*	No
Range	Yes	Yes	*	*	No

* May be required depending on the nature of the specific test (Ref. 33).

16.1.1. Stability-Indicating Nature of USP Assays

A word of caution. Assays appearing in USP monographs are not always stability-indicating. They may be for the innovator product, as submitted by the innovator company for inclusion as a USP monograph, which then becomes the benchmark. If another company wishes to market the same product, as a generic version, that company must validate the assay according to the validation parameters discussed in USP General Chapter <1225>, because that product is different from the innovator product relative to the source API and formulation.

16.2. ICH Guidelines

ICH Guidelines Q2A (Text on Validation of Analytical Procedures) and Q2B (Validation of Analytical Procedures: Methodology) were developed within the Expert Working Group (Quality) of the Requirements for Registration of Pharmaceuticals for Human Use. These documents present a discussion of the characteristics for consideration during validation of analytical procedures included as part of registration applications submitted within the European Union, Japan, and the United States.

ICH Guidelines Q2A also provides descriptions of typical validation parameters, how these are measured, and which subset of each parameter is suitable for validation of the analytical method, based on its intended use. The discussion of the validation of analytical procedures has been divided into three common categories of analytical procedures:

Identification tests
Quantitative tests for impurity content—Limit tests for the control of impurities
Quantitative tests of the active moiety in bulk drug substance or drug product or other selected component(s) in the drug product

As per ICH Guidelines Q2A, the objective of the analytical procedure needs to be clearly understood since this will govern the validation characteristics that need to be evaluated. Typical validation characteristics, which should be considered, are

Accuracy
Precision
Repeatability
Intermediate precision
Specificity
Detection limit
Quantitation limit
Linearity
Range
Robustness
System suitability

Analytical variables that are normally required for method validation is summarized in Table 6.

Table 6 ICH Validation Characteristics Versus Type of Analytical Procedures

Type of analytical procedure	Identification	Impurity testing		Assay
		Quantitative	Limit tests	
Accuracy	No	Yes	No	Yes
Precision				
Repeatability	No	Yes	No	Yes
Intermediate precision	No	Yes	No	Yes
Specificity	Yes	Yes	Yes	Yes
LOD	No	Yes	Yes	No
LOQ	No	Yes	No	No
Linearity	No	Yes	No	Yes
Range	No	Yes	No	Yes

Source: Refs. 35 and 37.

The difference in the USP and ICH terminology is for the most part one of semantics; however, there is one notable exception. In the ICH Guidelines, system suitability is part of validation, whereas the USP deals with system suitability under chromatography in the USP, called General Chapter <621> Chromatography (24). The FDA is already implementing the ICH Guidelines, and it is anticipated that the ICH definitions and terminology will become a part of the USP chapter on validation. It is probable that USP categories I and II will match the ICH categories of Assay and Impurity testing, respectively. The ICH has not yet chosen to address methods for performance characteristics (USP Category III) but has instead included analytical methods for compound identification. In this ICH category, it is only necessary to show that the method is specific for the compound being identified.

ICH Guidelines Q2B is complementary to ICH Guidance Q2A, which presents a discussion of characteristics that should be considered during the validation of analytical procedures. This guidance gives recommendations on how to consider the various validation characteristics for each analytical procedure. These recommendations will be discussed in detail under definition of validation parameters.

16.3. FDA Reviewer Guidance

The FDA Reviewer Guidance—Validation of Chromatographic Methods provides comprehensive description of typical validation parameters and how these are determined (40). This FDA guidance has similarities to the ICH Guidelines Q2A and Q2B, but has examples in form of tables or figures to demonstrate data representation for validation parameters. The purpose of this guidance is to present the issues to be considered when evaluating chromatographic test methods from a regulatory perspective. Examples of common problems, which can delay the validation process, have been included.

The validation characteristics to be evaluated according to this FDA guidance are

Table 7 Comparison of Analytical Parameters Required for Assay Validation

USP General Chapter <1225>	ICH Q2A Guidelines	FDA Reviewer Guidance
Accuracy	Accuracy	Accuracy
Precision	Precision	Precision
No	Repeatability	Repeatability
		Injection
		Analysis
No	Intermediate precision	Intermediate precision
No	No	Reproducibility
Specificity	Specificity	Specificity/selectivity
Detection limit	Detection limit	Detection limit
Quantitation limit	Quantitation limit	Quantitation limit
Linearity	Linearity	Linearity
Range	Range	Range
Ruggedness	No	No
Robustness	Robustness	Robustness
System suitability[a]	System suitability	System suitability
		Sample solution stability

[a] System suitability discussed separately in USP 23 General Chapter <621>.

Accuracy
Detection and quantitation limits
Linearity
Precision
Repeatability
Injection repeatability
Analysis repeatability
Intermediate precision
Reproducibility
Range
Robustness
Sample solution stability
Specificity/selectivity
System suitability specifications and tests

A comparative discussion of validation parameters given in the FDA and ICH guidelines will be made under Sec. 17, "Definition of Validation Parameters." Analytical parameters needed for method validation as described in the General Chapter <1225>, ICH Guidelines Q2A, and the FDA Reviewer Guidance are summarized in Table 7.

17. DEFINITION OF VALIDATION PARAMETERS

In the literature, there are many articles on definition and interpretation of validation parameters required for assay validation as published by Krull and Swartz (38,39,41,42). Persson et al. (43) have discussed the evaluation of method validation in an article titled "How good is your method?" In Sec. 17, definition of validation parameters is based on requirements stipulated in the ICH Guidelines Q2A, Q2B, the FDA Reviewer Guidance, and USP General Chapter <1225>.

Though many types of chromatographic techniques are available, the most commonly submitted method in NDAs and ANDAs is reversed-phase HPLC with UV detection. Therefore this method is selected here to illustrate parameters for validation. The criteria for the validation of this technique can be extrapolated to other detection methods and chromatographic techniques. For acceptance, release, or stability testing, accuracy should be optimized, since the need to show deviation from the actual value is of great concern.

17.1. Accuracy

Accuracy is the measure of how close the experimental value is to the true value. It is measured as the percent of analyte recovered by assay or by spiking samples in a blind study. For the drug product, this is performed by analyzing synthetic mixtures (placebos) spiked with known quantities of drug. Accuracy should be established across the specified range (that is, line of working range) of the analytical procedure. For the assay of the drug substance, accuracy measurements are made by comparison of the results with the analysis of a standard reference material or to compare the results obtained from a second well-characterized independent procedure, the accuracy of which is stated and/or defined. For quantitation of the impurity, accuracy is determined by spiking drug substance or drug product with known amounts of available impurities. In case it is impossible to obtain impurities or degradation products, comparison of results to a second well-characterized independent method is acceptable. The response factor of the drug substance can be used. Another approach is to perform specificity studies by forced degradation. This will be discussed under specificity. It should be decided up front how the individual or total impurities are to be reported, e.g., percent weight/weight or area percent, in all cases relative to the major analyte.

The FDA recommends that recovery be performed at the 80, 100, 120% of label claim as stated in the Guideline for Submitting Samples and Analytical Data for Method Validation (2). Recovery data, at least in triplicate at each level (80, 100, and 120% of label claim) is recommended. The data should be calculated as percent label claim, and the mean of the replicates along with % RSD for each level is reported to demonstrate accuracy and sample analysis precision.

ICH Guidelines Q2B recommend assessment of accuracy at three concentration levels covering the specified range (i.e., three concentration levels and three replicates at each level of the total analytical procedure). The data should be reported as the percent recovery of the known amount added or as the difference between the mean and true values with confidence intervals.

17.2. Precision

Precision is the measure of how close the data values are to each other for a number of measurements under the same analytical conditions. In USP 23/NF 18, General Chapter <1225>, precision is defined as "the degree of agreement among individual test results obtained by repeatedly applying the analytical method to multiple samplings of a homogeneous sample." Thus precision refers to the distribution of individual test results around their average. Precision is usually expressed as percent relative standard deviation (% RSD) for a statistically significant number of samples. Both the FDA and the ICH recommend that precision be measured at three different levels. No such recommendation is given in the USP.

17.2.1. Repeatability

Repeatability expresses the results of the method operating over a short time interval under the same conditions. Repeatability is also termed intra-assay precision. According to the FDA Reviewer Guidance, repeatability is evaluated for injector performance and analysis of samples. For injector repeatability, there must be a minimum of 10 injections with an RSD of not more than ±1%. Similarly, with the methods for release and stability studies, an RSD of not more than ±1% for at least five injections for the active drug is desirable. For low-level impurities, higher variations in RSD may be acceptable. For analysis repeatability, determinations are made on multiple measurements of a sample by the same analyst under the same analytical conditions. The FDA recommends that the study be combined with accuracy.

The ICH recommends that repeatability should be determined from a minimum of nine determinations covering the specified range for the procedure (e.g., three levels, three replicates each), or from a minimum of six determinations at 100% of the test or target concentration. The target concentration is defined as the concentration of the compound of interest given in the analytical method.

17.2.2. Intermediate Precision

Intermediate precision expresses within-laboratory variations. This was previously evaluated as part of ruggedness. This attribute evaluates the reliability of the method in an environment different from that used during the method development phase. Depending on time and resources, the method can be evaluated on different days, with different analysts and equipment, etc. The FDA recommends performing accuracy on two separate occasions to indicate the intermediate precision of the test method. The ICH recommends using an experimental design (matrix) so that the effects, if any, of the individual variables on the analytical procedure can be monitored.

17.2.3. Reproducibility

Reproducibility is assessed by performing collaborative studies between laboratories. Multiple laboratories are desirable, if possible. According to the FDA Reviewer Guidance, reproducibility is not required if intermediate precision is achieved. The ICH recommends that reproducibility studies be performed for standardization of an analytical procedure, for instance, for inclusion of procedures to pharmacopoeias. The ICH also recommends that documents in support of each type of precision should include the standard deviation (*S*), the % RSD, the coefficient of variation, and the confidence interval.

17.3. Specificity/Selectivity

The terms specificity and selectivity are often used interchangeably. The term selectivity has been used in General Chapter <1225> of the 1990 edition of the

USP (44), whereas in the 1995 edition the term specificity has replaced selectivity. Specificity is generally used to express a method's response for a single analyte, whereas the term selectivity of a method is a measure of the extent to which the method can determine a particular compound in the analyzed matrices without interference from matrix components. However, as both the USP and the ICH currently use the term specificity, it will also be used here to avoid any confusion.

The USP defines specificity as the ability to measure accurately and specifically the analyte of interest in the presence of other components in the sample matrix. These components may include other active ingredients, excipients, impurities, and degradation products. According to the ICH, the validation procedure should be able to demonstrate the ability of the method to assess unequivocally the analyte in the presence of impurities, matrix components, and degradation products. Lack of specificity of an individual procedure may be compensated by other supporting procedure(s) such as TLC.

Specificity has been divided into two separate categories by ICH:

A. IDENTIFICATION. Specificity is demonstrated by the ability to discriminate between compounds of closely related structures, which are likely to be present. The other approach is by comparison of results to a known reference material.

B. ASSAY AND IMPURITY TEST(S). For assay and impurity tests, specificity can be demonstrated by the resolution of the two components which elute closest to each other. Chromatograms obtained should be appropriately labeled to show individual components. For nonspecific assays, overall specificity may be demonstrated by use of other supporting analytical procedures. For example, where a titration is adopted to assay the drug substance for release, the combination of the assay and a suitable test for impurities can be used.

The ICH has also addressed issues of specificity for impurities. The approach is similar for both assay and impurities. If impurities are available, then it must be demonstrated that the assay is unaffected by the presence of spiked materials such as impurities and/or excipients. For the impurity test, the discrimination may be shown by spiking drug substance or drug product with appropriate levels of impurities and demonstrating the separation of these impurities individually and/or from other components in the sample matrix.

If the impurities or degradation product standards are not available, then specificity may be demonstrated by comparison of the test results to a second well-characterized pharmacopoeial or independent validated procedure. For the assay, the two results are compared. For the impurity tests, the impurity profiles are compared head to head.

For stability-indicating assays where potency and impurities are determined simultaneously mass balance must be taken into consideration. Any decrease in potency should be explained by mass balance. The following equation can be used to account for any loss of potency:

$$100\% = \text{Drug}\% + \text{Related substances}\% + \text{Water}\% + \text{ROI}\% + \cdots$$

In the FDA Reviewer Guidance, specificity/selectivity is established by showing that the analyte should have no interference from extraneous components and be well resolved from them. A representative chromatogram showing resolution of these extraneous peaks from the main analyte peak is required for submission. The origins of extraneous peaks in drug substance are process impurities (which include isomeric impurities) from the synthesis process, residual solvents, and other extraneous components from extracts of natural origins. For the drug product, sources of extraneous peaks include any impurities, degradation products, interaction of the active drug with excipients, residual solvents from both the active drug substance and the excipient, and so on.

17.4. Forced Degradation

In previous sections, we have defined specificity as discussed by USP Chapter <1225>, the ICH Guidance, and the FDA Reviewer Guidance. The discussion was limited to specificity studies in the presence or absence of impurities and excipients. A question that arises if nothing (i.e., no extraneous peaks) is observed is, What approach one might use to show the specificity and stability-indicating nature of the proposed method?

Both the FDA and the ICH recommend forced degradation/or stress testing of the drug substance and drug product. For these studies, acid and base hydrolysis, temperature, photolysis, and oxidation are recommended. Neither the ICH nor the FDA guidelines specify how to perform these forced degradation studies. Experimental conditions and the design of these studies have been left to the discretion of pharmaceutical companies. A generic protocol for these studies is shown in Table 2.

To demonstrate that the analyte chromatographic peak obtained after forced degradation or stress studies is a single entity, peak purity tests are recommended by the FDA and the ICH. Photodiode array detection can be used to demonstrate peak purity. The spectra collected across a peak are compared mathematically to establish peak homogeneity.

It is generally recommended that about 20–30% of analyte degradation, at least, in one medium be achieved. For some compounds, severe degradation conditions may be required.

17.5. Detection Limit (DL)

The detection limit (DL) is the lowest concentration of the analyte that can be detected, but not necessarily quantitated, under the stated experimental conditions. It is a parameter of limit test and specifies whether or not an analyte is above or below a certain value. In the current USP General Chapter <1225>, determination of limit of detection is described for instrumental and noninstrumental methods. For instrumental methods, one determines the signal-to-noise ratio by comparing test results from samples with known concentration of analyte with those of blank samples and establishes the lowest concentration at which analyte can be reliably detected. A signal-to-noise ratio of 2 : 1 or 3 : 1 is required. Another approach is to calculate the standard deviation for analysis of a number of blank samples. The standard deviation multiplied by a factor, usually 2 or 3, gives an estimate of limit of detection.

For noninstrumental methods, DL is determined by the analysis of samples with known concentrations of analyte. The minimum concentration at which the analyte can be reliably detected is the limit of detection. The ICH has recognized the signal-to-noise ratio convention but also lists several other approaches for determining DL, depending on whether the procedure is instrumental or noninstrumental. These approaches are as follows.

A. BASED ON VISUAL EVALUATION. Visual evaluation may be used both for instrumental and noninstrumental methods. It requires analysis of samples with concentrations of analyte and establishing the minimum level at which analyte can be reliably detected. Visual noninstrumental methods can include DL determined by techniques such as TLC or titration.

B. BASED ON THE STANDARD DEVIATION OF THE RESPONSE AND THE SLOPE. The detection limit may be calculated based on the standard deviation (SD) of the response and slope (S) of the calibration curve (a specific curve should be generated by using samples containing analyte in the range of detection limit), according to the formula

$$\text{Detection limit (DL)} = 3.3 \times \text{SD}$$

The SD of the response can be determined from the SD of the blank, the residual SD of the regression line, or the SD of the y-intercept of the regression line. The detection limit and method used to determine the detection limit must be documented and supported, and a suitable number of samples should be analyzed at the limit to validate it. The FDA is of the opinion that expression of the detection limit in terms of a signal-to-noise ratio of 2 or 3 is not very practical. The reason for this is attributed to differences in the noise level on a detector during the method development phase and when samples are analyzed on different detectors. Detector sensitivity can vary with the model number or manufacturer.

17.6. Quantitation Limit (QL)

The quantitation limit is the lowest concentration of analyte in a sample that can be determined with acceptable precision and accuracy under the stated experimental conditions of the method. This is a parameter of the quantitative assays for low concentrations of compounds in sample matrices such as impurities in bulk drug substances and degradation products in finished products.

In the current USP General Chapter <1225>, the quantitation limit, QL, which is similar to the detection limit, is expressed as the concentration of analyte in the sample, and precision and accuracy of the measurements are also reported. The QL is dependent on the type of procedure, i.e., instrumental or noninstrumental. For instrumental methods, sometimes a signal-to-noise ratio of 10 : 1 is used to determine the QL. However, it is pointed out that the determination of the QL based on signal-to-noise ratio criteria is a compromise between the concentration and the required accuracy and precision. In other words, as the QL concentration level decreases, the precision increases. For better precision, a higher concentration must be reported for the QL. This compromise is dependent on the analytical method and its intended use.

As with to the limit of detection, the ICH has recognized using a signal-to-noise ratio of 10 : 1 for quantitation. However, this approach can only be applied to analytical procedures that exhibit baseline noise. Again, as with the DL, the ICH lists the same two options that can be used to determine the QL. They are visual evaluation for both noninstrumental and instrumental methods; the latter method can

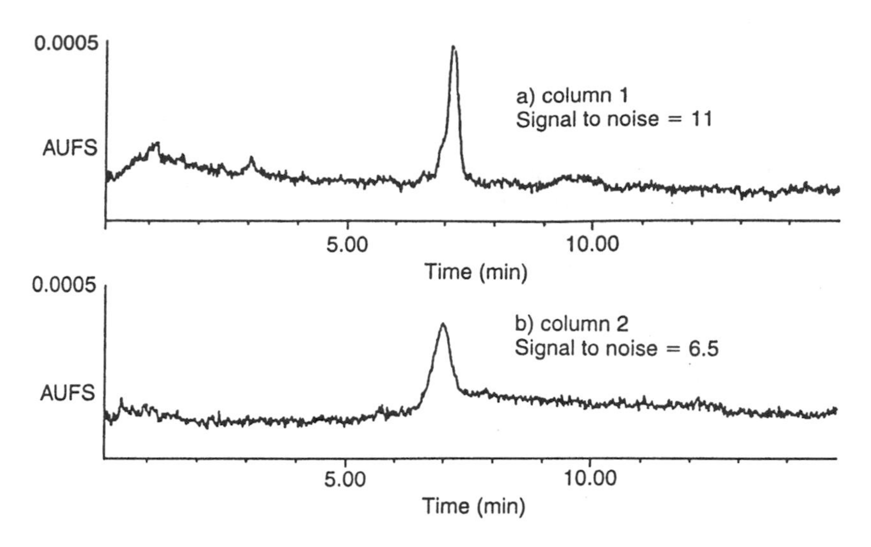

Figure 11.4 Effect of peak shape on LOD and LOQ. (From Ref. 38.)

be based on the standard deviation of the response and the slope. The formula is changed to

$$\mathrm{SD} = 10 \times \frac{\mathrm{SD}}{\mathrm{S}}$$

Determination criteria and requirements for documentation are the same as described under DL in Sec. 17.5, as well as comments by the FDA on the subject. In addition, the FDA Reviewer Guidance recommends that data for analysis repeatability and injection repeatability at the quantitation limit be generated. Further, the Guidance recommends that the use of an additional reference standard at the quantitation limit level be incorporated in the test method.

Additional points regarding the detection and quantitation limits are warranted. These parameters are affected by chromatography. Figure 4 shows the effects of peak shape and efficiency on the signal-to-noise ratio. Sharp peaks will yield a higher signal-to-noise ratio, thus lowering both the DL and the QL. Therefore for the chromatographic determination of these parameters, the age and type of the column and the age of the detector lamp need to be considered. Thus periodic maintenance of the chromatographic detector to maintain optimal results is required.

Finally, the DL and the QL should not be confused with sensitivity. Sensitivity is defined as the slope of the calibration curve, and as such does not usually reference the actual limit of detection or limit of quantitation.

17.7. Linearity

The linearity of an analytical procedure is its ability to obtain test results that are directly proportional to the concentration of analyte in the sample within a given range. Linearity is generally reported as the variance of the slope of the regression line calculated according to an established mathematical relationship from test results obtained by the analysis of samples with varying concentrations of analyte. The linear range of detection that obeys Beer's law is dependent on the compound analyzed and the type of detector used.

USP General Chapter <1225> gives general directions on the determination of linearity along with handling of the data. However, there are no concentration levels specified to monitor linearity. The ICH also has adopted an approach similar to that of the USP for the determination of linearity and data interpretation. The least squares method is recommended for evaluation of the regression line.

The correlation coefficient, y-intercept, slope of the regression line, and residual sum of squares should be reported. For linearity studies, a minimum of five concentrations is recommended. According to the FDA Reviewer Guidance, the linearity range depends on the intended use of the test method. For content assay, linearity should be performed between 80% and 120% of target concentration. The linearity range for the assay/impurities combination method based on area percent (for impurities) should be greater than 20% of the target concentration down to the limit of quantitation of the drug substance or impurity. A coefficient of correlation (r^2) value, an intercept, and a slope should be reported.

17.8. Range

The range of an analytical method is the interval between the upper and lower concentration levels of analyte (including these concentrations) for which the method as written has been shown to be precise, accurate, and linear. The range is usually expressed in the same units as test results obtained by the analytical method. According to USP General Chapter <1225>, the range of method is validated by verifying that acceptable precision and accuracy is obtained by the analytical method when actual analysis of samples containing analyte is performed throughout the intervals of the range.

The ICH recommends an approach similar to the USP for validation of range. It recommends specific ranges based on the intended use of the method, as follows.

1. For assay of a drug substance or drug product, the minimum specified range is 80% to 120% of the target concentration.
2. For content uniformity testing, the minimum range is 70% to 130%.
3. For the determination of impurity, the minimum range is from the reporting level of an impurity to 120% of the specification.
4. For a combination assay procedure for both active and impurity, where a 100% standard is used, linearity should cover the range from reporting level to 120% of the assay specification.
5. For dissolution testing, the recommended range is ±20% over the specified range of the test. That is, in the case of an extended release product dissolution test with a Q value of 20% after 1 hour, up to 90% in 24 hours, the range for validation will be 0 to 110% of the label claim.
6. For toxic or more potent impurities, the range should be commensurate with the controlled level. FDA recommendations for range are as discussed under the Linearity and Accuracy sections. These ranges can also be applied to other substances such as preservatives.

17.9. Robustness

The robustness of an analytical procedure is a measure of its capacity to remain unaffected by small but deliberate variations in some parameters and provide an assurance of its reliability during normal usage. The robustness of the method is investigated by varying some or all conditions, e.g., organic composition of the mobile phase, pH, ionic strength, column temperature, age of column, column type. ICH guidelines recommend that robustness studies be performed during the method development stage. Also, if measurements are affected by variations in analytical conditions, the analytical conditions should be suitably controlled or a precautionary statement should bc included in the test method.

Robustness can also be partly assured by good system suitability specification. Therefore, it is important to set tight but realistic system suitability specifications.

17.10. Application of Plackett–Burman Design to Ruggedness Testing

Ruggedness is normally defined as the lack of influence on test results by operational and environmental variables of the analytical method. Ruggedness is a measure of reproducibility of test results under normal operational conditions from laboratory to laboratory and from analyst to analyst. According to ASTM Guidance E 1169-89, "Standard Guide for Conducting Ruggedness Tests" (45), it is necessary to monitor the effects of environmental and experimental factors on the results obtained using the test method to assure method accuracy. Furthermore, the purpose of ruggedness testing is to determine which variables the method is susceptible to and how to control it. Ruggedness testing does not determine the optimum operational conditions for the test method. To determine the ruggedness of the method, the ASTM guidance recommends use of the experimental design as reported by Plackett and Burman. This guidance discusses effects of change on two levels per variable, as this design is easy to use and provide useful information needed for improvement of the test method. An example of ruggedness testing for an HPLC method is given in Tables 8 through 10.

Table 8 shows the various factors and their high and low limits to be considered in ruggedness testing. Table 9 shows the factors and their high and low limits in a +/− format. Lastly, Table 10 summarizes the results obtained when each of the eight combinations (rows across the spreadsheet) are experimentally performed.

Table 8 Ruggedness Testing—Typical HPLC Factors

Factor	Low value	High value
A. pH	3.0	4.0
B. Temperature	35°C	40°C
C. Mobile phase composition	45/55	55/45
D. Buffer concentration	0.05 M	0.1 M
E. Particle size	3 micron	5 micron
F. Column length	3 cm	5 cm
G. Flow rate	1.0 mL/min	1.5 mL/min

Table 9 Ruggedness Testing—Typical HPLC Conditions

Excel	A	B	C	D	E	F	G	H	I
1	Run/factor	A	B	C	D	E	F	G	Result
2	1	−1	+1	+1	+1	−1	−1	+1	99.8%
3	2	+1	−1	+1	+1	+1	−1	−1	101.1
4	3	−1	+1	−1	+1	+1	+1	−1	98.9
5	4	−1	−1	+1	−1	+1	+1	+1	99.5
6	5	+1	−1	−1	+1	−1	+1	+1	99.9
7	6	+1	+1	−1	−1	+1	−1	+1	98.5
8	7	+1	+1	+1	−1	−1	+1	−1	98.0
9	8	−1	−1	−1	−1	−1	−1	−1	97.0
10	Effect	0.575	−0.575	1.025	1.675	0.825	−0.025	0.675	

Table 10 Ruggedness Testing—Typical HPLC Conditions

Excel	A	B	C	D	E	F	G	H	I
1	Run/factor	A	B	C	D	E	F	G	Result
2	1	3	40	55/45	0.1	3	3	1.5	99.8%
3	2	4	35	55/45	0.1	5	3	1.0	101.1
4	3	3	40	45/55	0.1	5	5	1.0	98.9
5	4	3	35	55/45	0.05	5	5	1.5	99.5
6	5	4	35	45/55	0.1	3	5	1.5	99.9
7	6	4	40	45/55	0.05	5	3	1.5	98.5
8	7	4	40	55/45	0.05	3	5	1.0	98.0
9	8	3	35	45/55	0.05	3	3	1.0	97.0
10	Effect	0.575[a]	−0.575	1.025	1.675	0.825	−0.025	0.675	

[a] Content of cell = SUM PRODUCT(B2:B9,$I2:I9)/4. This takes the difference between the average test results for the "+" runs and the average test results for the " − " runs. Conclusion: Eight experiments performed compared to 56 individual experiments. The cell with the "highest" effect value indicates the most variable factor. In this example, it is Factor D, the buffer concentration, followed by Factor C, the mobile phase composition.

Results obtained are placed in a spreadsheet, such as Excel, and the effect calculated. The highest effect (i.e., largest value) in the column listed would indicate that factor to be the most critical, and special attention is needed to control its variability.

For a detailed discussion of Plackett–Burman design experimentation, readers should consult the ASTM guidance (45) and Torbeck (46).

17.11. Stability of Sample and Standard Solutions

The FDA recommends that solution stability of the drug substance (used as sample or in-house standard) or drug product after preparation according to the test method should be evaluated. This is considered critical as most HPLC analyses are automated. For the duration of an analytical run, the standard or sample will stay in solution for hours in the laboratory environment before all the samples are completely tested. Therefore monitoring of sample or standard stability will ensure that there is no degradation occurring due to hydrolysis, photolysis, or adhesion to glassware over the course of the run period. The FDA recommends that data to support the stability of sample or standard solution under normal laboratory conditions for a minimum period of 24 hours should be generated.

17.12. System Suitability Specifications and Tests

The accuracy and precision of HPLC data collected begin with a well-behaved chromatographic system. The system suitability specifications and tests are parameters that provide assistance in achieving this purpose. According to the ICH and the USP, system suitability testing is an integral part of chromatographic procedures. These tests are used to determine that the resolution and reproducibility of the system are adequate for the analysis to the performed. The basis for these tests is that the equipment, electronics, analytical operations, and samples to be analyzed constitute an integral system that can be evaluated as a whole. System suitability test parameters to be established for a particular procedure depend on the type of procedure being validated. In USP 23 General Chapter <621>, Chromatography, a section has been devoted to system suitability requirements. It is important to know what are regulatory requirements for system suitability tests and specifications for method validation. As stated earlier, system suitability involves checking a system to ensure it is performing adequately before or during the analysis of unknowns. To establish these required parameters [i.e., plate count, tailing factor, resolution (if by-products or impurity standards are available; otherwise a chromatogram from forced degradation studies may be used)], the reproducibility (% RSD) of five or six replicates is calculated and compared to predetermined specification limits. System suitability tests are performed prior to analysis of actual samples. These parameters are studied by analysis of a system suitability sample that is a mixture of main active drug and expected by-product or a known impurity. Table 11 summarizes the parameters to be measured and their recommended regulatory limits for the system suitability tests and specifications (38,40). Definition of terms for system suitability parameters is shown in Figure 5.

Table 11 System Suitability Parameters and Recommendations

Parameter	Recommendation
Capacity factor (k')	The peak should be well resolved from other peaks and the void volume, generally $k' > 2.0$.
Repeatability	RSD $\leq 1\%$ for $N \geq 5$ is desirable.
Relative retention	Not essential so long as the resolution is stated.
Resolution (R_s)	R_s of >2 between the peak of interest and the closest eluting potential interferent (impurity, excipient, degradation product, internal standard, etc.).
Tailing factor (T)	T of ≤ 2.
Theoretical plates (N)	In general should be >2000.

Source: Ref. 37 and 39.

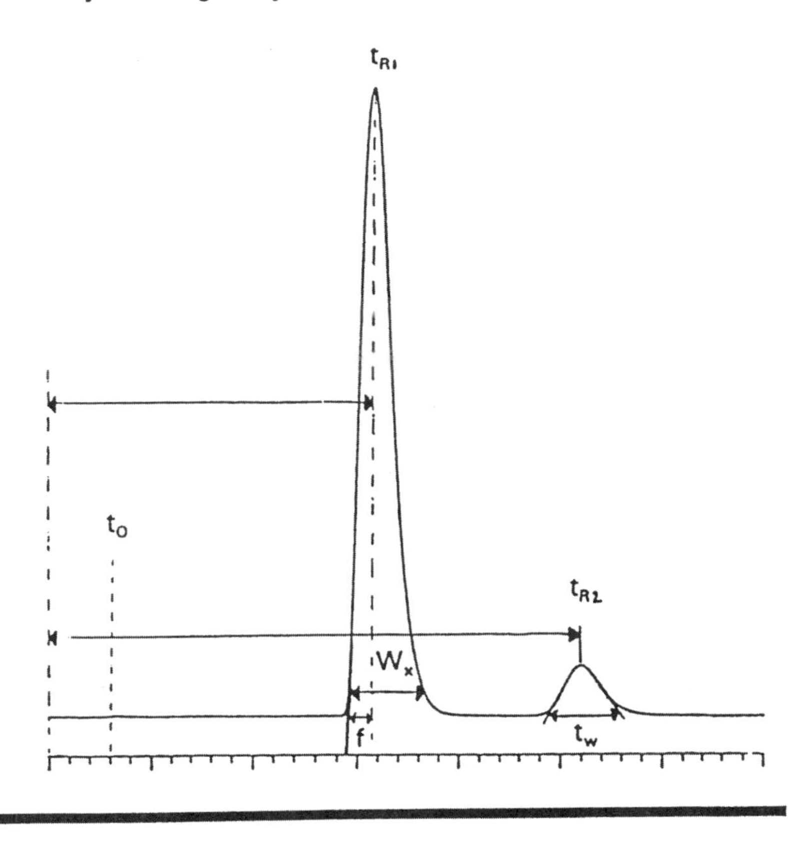

Where

W_x = width of the peak determined at either 5% (0.05) or 10% (0.10) from the baseline of the peak height

f = distance between peak maximum and peak front at W_x

t_o = elution time of the void volume or non-retained components

t_R = retention time of the analyte

t_W = peak width measured at baseline of the extrapolated straight sides to baseline

Figure 11.5 Definition of terms for system suitability parameters. (From Ref. 40.)

For accuracy and precision of analysis, all system suitability parameters play a significant role. Therefore, a critical evaluation of these parameters and their effect on a chromatographic separation are required. As an example, the effects of peak tailing and different resolution values on quantitation are depicted in Figs. 6, 7, and 8.

Resolution is a measure of how well peaks are separated from each other. For reliable quantitation, well-resolved peaks are essential. This parameter is very useful in determining if peaks can interfere in individual quantitation. As shown in Fig. 6, with a small resolution, accuracy of analysis will decrease.

Tailing peaks affect quantitation. With an increase in peak tailing the accuracy of quantitation decreases due to improper peak integration (the area under the peak will not be accurate). The effect of peak tailing is shown in Figs. 7 and 8.

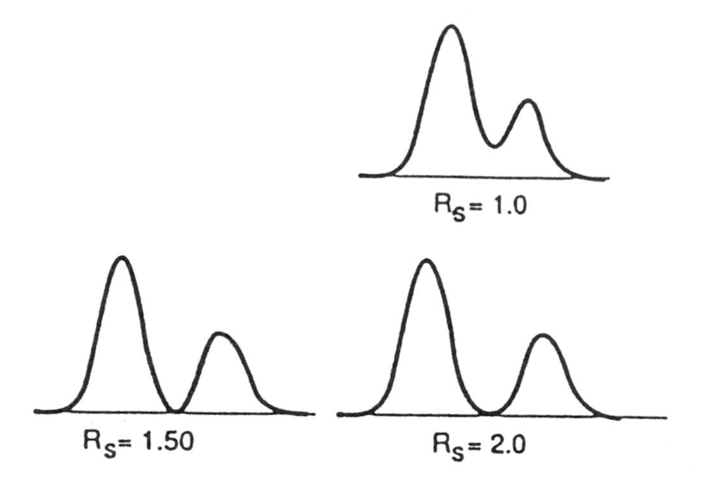

Figure 11.6 Separation of peaks as indicated by R. (From Ref. 40.)

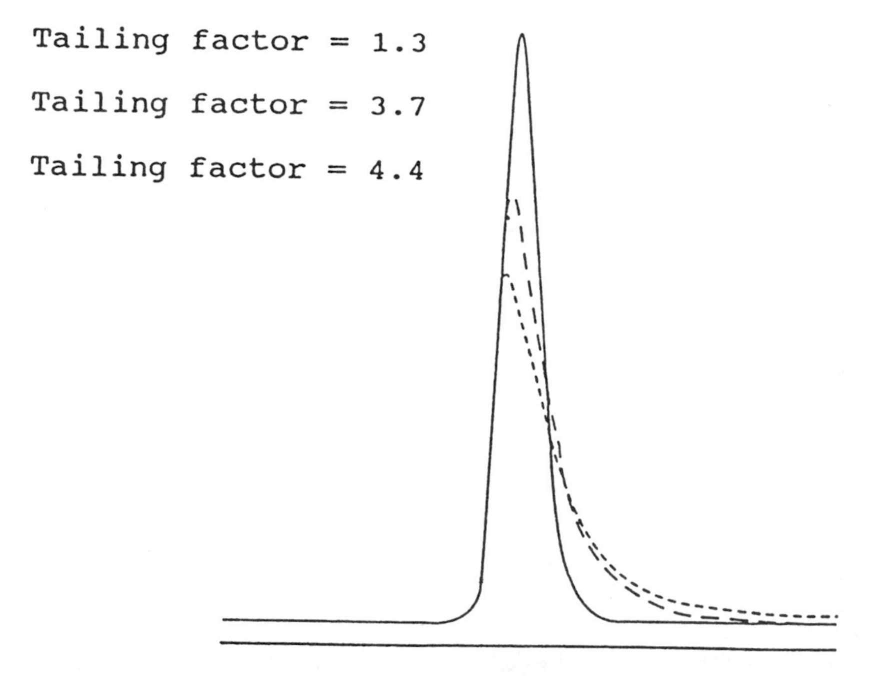

Figure 11.7 HPLC peak with various tailing factors. (From Ref. 40.)

18. POST VALIDATION ISSUES

The validation process does not end after experimental evaluation of the analytical parameters. Data must be evaluated to determine whether validation was successful or not. Does all the data generated meet the specified requirements or not? In the following subsections, steps required to finalize the validation process will be discussed.

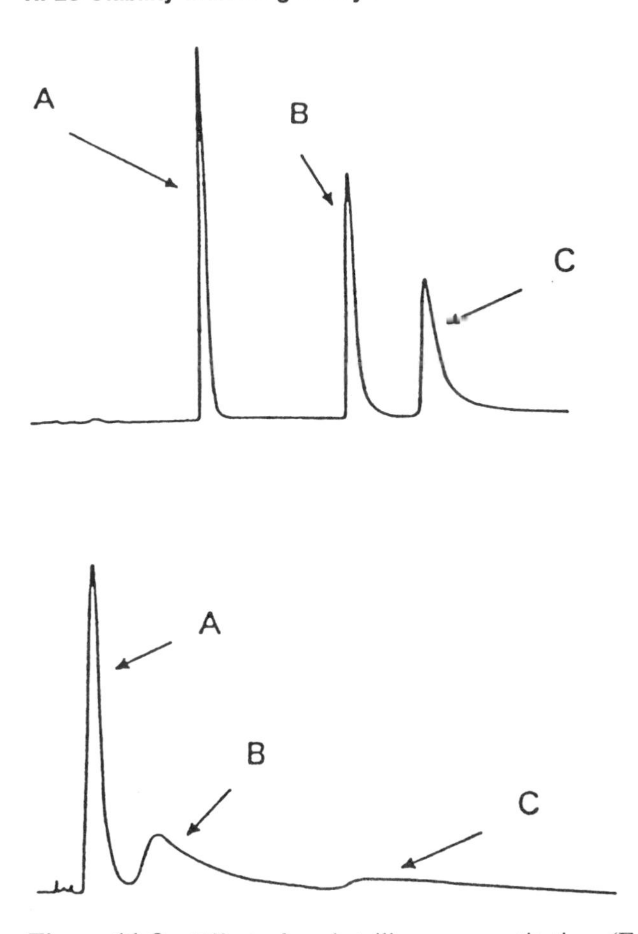

Figure 11.8 Effect of peak tailing on quantitation. (From Ref. 40.)

18.1. After the Laboratory Work

After completion of the laboratory work and documentation of the data in the analyst's notebook, it is very important that all data be carefully reviewed and audited by a qualified person. This process will ensure that data generated through the validation process is correct and meets all the requirements. Only after this step is completed can the next phase of the validation process be implemented.

18.1.1. Validation Report

The method validation report is a regulatory requirement and needs to be submitted to the FDA. The method validation report should be written by the method development group. The format for the report should be agreed upon at the onset of the validation process. This report must describe all the experimental procedures including equipment used, detector type, columns, information on reference standard, chemicals and composition of placebo for accuracy studies. All chromatograms and figures should be labeled properly. For forced degradation studies, conditions used and how this was performed must be explained.

18.1.2. Acceptance Criteria

For each step of the validation analytical parameters, acceptance criteria to determine success or failure of validation, are required. These acceptance criteria should be based on the intended use of the method. Also, regulatory implications should be taken into consideration. For acceptance criteria, it is imperative that responsible personnel with backgrounds in method development be involved. All validation steps should be evaluated against these acceptance criteria. Similarly, system suitability parameters should also accompany acceptance criteria. General acceptance criteria for each validation parameter have been discussed in Sec. 17.

18.1.3. Generation of the Test Method

Generation of the test method is the responsibility of the method's development group. The test method must include a detailed written procedure. For example, for a chromatographic procedure, preparation of mobile phase, column type, detector type, wave length, injection volume, flow rate, reference standard (USP/in-house), preparation of standard and sample solutions, reagent grade, and filters used for standard and sample solutions should be documented. If the method is designed to quantitate the main analyte and impurities simultaneously, then relative retention times for impurities should be given. If the main bulk active is light sensitive, then a precautionary note is required in the test method. Similarly, it should be reflected in the test method whether ambient or elevated temperature is required.

18.2. Revalidation

At some time during the lifetime of the method, for one reason or another, revalidation of the method may become necessary. For revalidation, reactive or proactive approaches may be used. Reactive validation will be required for changes in incoming bulk drug active, manufacturing batch changes, formulation changes, or other changes such as dilutions or sample preparation in the method. Recently, method change versus an adjustment has been the subject of discussion between regulatory agencies and industry (36,38,47).

This distinction is critical, as a process change requires method revalidation, whereas an adjustment does not. As a result of these discussions, limit changes for chromatographic changes do not require revalidation. Changes have been proposed under the following categories:

Aqueous buffer pH

1. Analytes without ionizable groups: ±1 unit
2. Analytes with basic or acidic groups and the buffer pH = pKa±2 units: ±0.2 unit
3. Analytes with basic or acidic groups and the buffer pH ⟨or⟩ pKa±2 units: ±1 unit

In each case a reference standard must be used to demonstrate that there is improvement in chromatography due to pH adjustment. No pH adjustment is allowed if standards are not available for all analytes of interest

Column dimensions

1. Length: ±70%
2. Inner diameter: ±25%, provided a constant linear flow velocity is maintained.

In addition, flow rate changes up to ±50% have been proposed. Work on limits for changes for mobile phase solvent ratios is underway. It is proposed that any method adjustment within these limits will not require additional validation. Proactive revalidation takes into consideration the availability of new technology or perhaps the automation of previously complex or time-consuming manual procedures. In such cases, revalidation may be more comprehensive, depending on the scope of the project.

18.3. Method Transfer

Method transfer is dependent on the intended use of a validated method. If other laboratories such as quality control or stability group are going to use this validated method, then a proper method transfer will be required. Under ideal conditions all laboratories involved should use an interactive approach to achieve method development, optimization, and validation goals in an efficient manner. If the end user has been involved in the development and validation process from the onset of this process as a participant or an observer, then it is convenient to place this method on line in a timely manner. Otherwise, reasonable time and effort will be required for the transfer process to be completed in a timely manner. Validation of a method demonstrates suitability of the method, whereas the method's evaluation and validity is approved by the end user.

The first step in a method transfer is to design a protocol, which is a document consisting of elements as outlined in the validation protocol, and other additional elements such as acceptance criteria, report format, and approval signatures of both the originating and the receiving laboratories. In addition, a detailed test procedure, design of experiments, sampling plan, analyst and equipment, interday and intraday ruggedness, and method transfer report form should also be included for method transfer studies.

Studies required for method transfer include system suitability, linearity, precision (day-to-day, within-day, analyst-to-analyst, analysis of multiple lots), collaboration of laboratories, developer user agreement on split sample results, and use of appropriate statistical standards, e.g., F-Test and t-Test, for evaluation of the method transfer process. The receiving laboratory should allocate enough time for the transfer, participate in interlaboratory studies, anticipate problems, and have a checklist ready of questions for the originating laboratory. For a successful method transfer, it is important to compare equipment or instrumentation in both laboratories. For example, for a chromatographic method, the age of the detector, the column, and the internal diameter of connecting tubing will play significant roles in the generation of comparable chromatograms.

Finally, where is method transfer required? In general, method transfer will be required for a new laboratory, a new method, new personnel, significant changes in a method, from company to a contract laboratory and from research and development group to quality control laboratory and stability group.

19. APPLICATION OF VALIDATION PRINCIPLES TO OTHER ANALYTICAL TECHNIQUES

19.1. Cleaning Method

To discuss cleaning validation in detail is beyond the scope of this work, as this has become an independent field. However, important steps required for cleaning validation will be briefly described. Readers are encouraged to research relevant publications on this subject. Recently, Kirsch has published an excellent review article on this subject (48). Cleaning validation is a regulatory requirement. The FDA has published a document titled "Guide to Inspections, Validation of Cleaning Processes" on this subject for field application. It is common industry practice to use the same equipment for production of a variety of products. The FDA has placed an increased emphasis on the cleanliness of the equipment to eliminate or minimize the risk of cross-contamination and adulteration of drug products.

Several analytical methods have been used by the pharmaceutical industry to demonstrate the cleanliness of process equipment surfaces. For low-level residues in rinse samples, the electronic conductivity technique are used. This technique is applicable to samples such as detergents and cleaning agents, which contain one or more ionic species. However, this technique is nonspecific and cannot be applied to neutral or highly polar compounds. Also, the FDA has specified a requirement that a correlation must be established between measurable conductance and concentration around the cleaning limit, which is a time-consuming process and not always possible for all analytes in a given formulation.

UV-visible spectrophotometry is another approach used for the detection of residue in rinse samples. This technique is sensitive but is dependent on the presence of a strong chromophore in the analyte for trace level determination. This is also a nonselective technique and not discriminating if more than one UV-active analytes are present in same sample.

Total organic carbon (TOC) has gained wide acceptance for cleaning applications (33,48). This technique is highly sensitive and specific for organic carbon-bearing analytes. TOC may be used in tandem along with conductivity, pH, and perhaps titrimetry to demonstrate the absence of both acid and alkaline detergents used for cleaning. TOC is only applicable to aqueous samples, and extra caution is required during sample acquisition and preparation to avoid bias in results due to carbon contamination.

High performance liquid chromatography (HPLC) has successfully been applied for cleaning residue samples. HPLC is sensitive for many pharmaceutical actives, and the necessary specificity can be obtained by this technique. For this technique, there are a variety of detection modes, such as spectrophotometric, electrochemical, fluorescence, and refractive index, to handle the diversity of pharmaceutical compounds.

Before the validation process begins, the appropriate predetermined level of cleanliness, i.e., at or below the limit at which equipment is considered clean, the final solvent used for the cleaning of equipment, and the type of swabs to be used should be chosen in consultation with the manufacturing group. It is critical that the limit agreed upon is practical and routinely achievable when an appropriate cleaning assay method is followed. Additionally, an acceptance limit that assures that the next product manufactured on the same piece of equipment is not adulterated or contaminated to the extent that its fitness for use is compromised, must be established. For determination of cleaning residues by HPLC, an appropriate laboratory procedure must be developed using available methodology and validated to meet certain acceptance criteria. For validation, a written protocol will be required.

The method should be sensitive enough to detect the analyte of interest at levels below or above the acceptance limit. The following studies are suggested for validation of the cleaning assay method:

1. Linearity of response for a wide range depending upon cleaning limit. Generally 50% of the cleaning limit to 10 times this concentration.
2. Specificity or selectivity to prevent false passing or failing results.
3. Precision and accuracy to assure correct results.
4. Limits of detection and quantitation.
5. Analyte stability before and after extraction from swab or in rinse samples.
6. The solvent used in the final rinse should be compatible with the assay mobile phase.

Since cleaning validation is considered a limit test, validation requirements may be less rigorous than an HPLC method used for bulk drug active, finished product, and stability samples.

For accuracy, swab and surface recovery approaches should be utilized and evaluated. In swab recovery, an appropriate number of swabs (minimum three) are spiked with amounts of analyte in rinse solution equivalent to the amount of analyte that should theoretically be removed from a known surface area. The spiked analyte is allowed to dried on the swab followed by extraction in a known volume of extracting solvent. Recovery studies should be performed at least for three levels, i.e., limit level, 50% of limit level, and 100% of limit level. The percentage recovery is calculated by using an external standard prepared at limit level concentration. The percent recovery obtained should not be less than 80%.

In surface recovery, 316 stainless steel or inert glass plates are used. A known surface area is spiked with a known amount of analyte by uniformly spreading analyte in rinse solution over the known surface area. The plate is dried under ambient conditions. The drying time will depend on the solvent used. However, overnight drying is preferable. The dried analyte residue is then swabbed using a premoistened swab across the spiked area. Swabbing can be done by horizontal, vertical, and zigzag motion of the swab. For better recoveries more than one swab may be required for removing analyte from the plate surface. The swabs are then placed in a 50 mL tube and extracted with the extracting solvent specified in the method. Recovery calculations are done against an external standard prepared at limit level. Recoveries can be performed at limit level, 50% of limit level, and 100% of limit level. It is pointed out that in the surface recovery approach, recovery values obtained are usually about 70 to 80%. Loss in recovery value is attributed to analyte solubility, analyte–metal binding strength, reactivity of surface with analyte, and swabbing technique. Also, streaking effects encountered in swabbing surfaces are detrimental and result in loss of the active.

19.2. Physicochemical Characterization Method (Dissolution)

USP 23 General Chapter <1225> designates dissolution testing under Category III. The validation parameters recommended in the Compendia are precision and ruggedness studies. Other studies are left to the discretion of the end user. It is common industry practice to verify a USP dissolution method by performing studies such as linearity of standard solutions, placebo interference, capsule shell interference (if a capsule) and reproducibility of response at specified times(s) for release. However, for an in-house developed dissolution method, proper validation studies are required. These studies would include specificity (interference from placebo), precision, linearity, system suitability, filter adsorption, and sample and standard stability. For automated dissolution systems, in addition to filter adsorption, there should be evidence of nonadsorption to active tubing used for delivery throughout the system and carryover effects.

It is difficult to perform recovery studies for dissolution, as spiking of placebo in vessels is not practical. Placebo excipients have a tendency to float on top of the dissolution medium. In addition, it is difficult to make single tablets unless a hand-held press is used. Hand filled capsules lack uniformity, and the procedure is tedious. In another approach, placebo along with label claim amount of active are placed in a 900 mL volumetric flask. The flask is filled to volume with dissolution medium and a magnetic stir bar is used to stir this mixture on a magnetic plate for the specified time period. Calculations of recovery are done against an external standard prepared in the dissolution medium. Acceptance criteria for precision, specificity, system suitability, and linearity are similar to assay validation.

19.3. Nonchromatographic Methods

There are a variety of guidelines available for the validation of the chromatographic procedures, but comparatively little information is available on validation of nonchromatographic methods. In general, it is assumed that USP General Chapter <1225> on analytical validation is only applicable to chromatographic methods. This assumption is incorrect, as USP General Chapter <1225> does not state that the validation parameters given in this chapter cannot be used for nonchromatographic techniques. By careful selection of parameters, a validation protocol can be designed for validation of nonchromatographic methods. Brittain has discussed validation issues and data elements required for validation of nonchromatographic methods (49).

19.3.1. UV Spectrophotometry

For UV spectrophotometric methods for assay, one needs to study parameters such as precision, accuracy, specificity, and linearity (49). For precision, a sufficient number of individual sample preparations should be assayed to permit the calculation of a statistically valid relative standard deviation. Accuracy can be determined by spiking a mixture of excipients (placebo) with known amounts of drug active at different concentration levels. Spike levels are, in general, similar to the linearity range. Spiked samples are prepared by following the "sample preparation" procedure and assayed against an external standard at the target level concentration. The accuracy is calculated from the test results as the percentage of analyte recovery by the assay. For specificity studies, intrinsic differences in chemical or physical properties are used to ensure accurate determination of analyte even in complex sample mixtures.

The purpose of a specificity study is to demonstrate that the method will yield reliable results even in the presence of interfering species. One approach to determine any possible bias in an assay is by comparison of results of assay value obtained in the presence of placebo excipients to assay value without placebo excipients. Assay bias is evaluated by calculating the percentage agreement between these two results by the formula (49)

$$\text{Percent agreement} = \frac{\text{TP}}{\text{TA}} \times 100$$

where TP = test results in the presence of placebo and TA = test results in the absence of placebo. A 100% agreement with show the absence of bias due to placebo or the potential interfering species. Agreement values >100% indicate positive bias, while agreement values <100% indicate negative bias in the assay procedure.

If standards for impurities or degradation products are not available, then the specificity can be determined by analyzing the samples containing the impurities or degradation products (from bulk drug) and comparing the results with those obtained by another independent and validated assay procedure. The independent assay is considered as the reference assay, and the degree of agreement between these two test results will dictate the specificity for the intended method. Calculations are similar to that described above. A percent agreement of 100% will be required for the absence of any bias in the intended method.

Linearity should be performed at least for five levels, including target level as 100%. Other concentrations should be 50%, 75%, 125%, and 150%. It is important that linearity responses obey Beer's law. Statistical evaluation of linearity is similar to that (as explained under linearity studies) for the chromatographic method validation.

19.3.2. Atomic Absorption Spectroscopy

Atomic absorption is used to determine heavy metals present in the drug substance. Heavy metals fall into the category of a limit test. Therefore rigorous validation may not be required. However, as these metals are present at trace levels, determination of limits for detection and quantitation is of significant importance for the validation of atomic absorption methods. Other validation parameters such as linearity, precision, specificity, and accuracy may be performed as described under Sec. 19.3.1.

The limits of detection and quantitation are determined by analyzing a number of samples prepared at low levels such as 2 ppm, 5 ppm, and 10 ppm. For each concentration level, multiple analysis is performed and standard deviation (SD) is calculated. All standard deviations are then averaged to calculate the mean standard deviation (MSD). To obtain an estimate of the noise level, the MSD is then divided by the slope of the calibration curve. For the detection limit the noise level obtained is multiplied by a factor of 3, whereas for the quantitation limit, a factor of 10 is used.

19.3.3. Infrared Spectroscopy (IR)

Infrared is used for identification of compounds. Currently in USP 23 there is a scarcity of monographs describing the use of IR for quantitation of analytes. For example, IR quantitation is used for the analysis of simethicon bulk drug active, tablets, oral suspensions, and capsules. As such, validation parameters required may be limited to interference studies. This interference may be due to the compound itself. For example, in an infrared spectrophotometric identification test, polymorphism may produce interference. Therefore, for compounds that exhibit polymorphism, it is critical that test samples and the reference standard have similar crystalline form. It then becomes obvious that for the infrared identification test, one should demonstrate that the method is insensitive to any polymorphic form of the material, or that the polymorphic effects have been taken into account. It is pointed out that unlike the chromatographic procedure, there are no official guidelines available on the validation of an infrared technique at present. An article by Ciurczak, "Validation of Spectroscopic Methods in Pharmaceutical Analyses," gives an overview of this subject (50).

19.3.4. Titration

USP 23 has several monographs that stipulate using titrimetry for release of bulk actives. These procedures are nonspecific and may not give accurate results in the presence of reactive impurities or degradation products. Therefore, for validation of these procedures an innovative approach will be required. The parameters to validate a titrimetric method include linearity, accuracy, blank determination, and insensitivity of the method to the amount of indicator used.

For linearity studies, different weights of the compound should be titrated, and the actual and theoretical results should agree. Alternatively, the titration could be done using a narrow range of compound weight, and then it should be stated in the method that the weight of the sample must be within this range. The accuracy should be studied by showing that the volumes of titrant for replicate titrations are very close to each other. In other words, small differences in volume of titrant required to reach an end or equivalence point does not introduce any significant error into the results.

As stated earlier, titrimetric procedures are nonspecific and cannot be used for simultaneous assay of active and impurities. In this case, impurities should be monitored by another independent validated procedure. For bulk active assay, comparison of results obtained by an alternate validated method and those obtained by the titrimetric procedure will demonstrate the validity of the titrimetric method.

19.4. General Considerations

The accuracies of chromatographic methods rely heavily on the purity of reference standards. Therefore a well characterized and highly pure standard is important. The FDA recognizes two categories of reference standards, i.e., compendial and noncompendial. The USP is the source of compendial standards. As these standards are well characterized, no further characterization is required. Noncompendial standards are also of high purity and can be obtained by reasonable effort and should be thoroughly characterized to ascertain their identity, strength, quality, and purity. Testing requirements for the reference standards are more rigorous as compared to bulk drug substance. The purity correction factors for non-USP standards should be included in any method calculations.

Quantitation of actives in chromatographic methods is based on either external or internal standards. An external standard method is used when the standard is analyzed on a separate chromatogram from the sample. Quantitation in this case will be by comparison of the sample and reference standard responses, i.e., peak area or peak heights for HPLC and GC or spot intensity in TLC for a given analyte of interest. External standard methods are generally used for samples with a single target concentration and narrow concentration ranges (acceptance and release tests). Simple sample preparation procedures or longer run times for detection of extraneous peaks, e.g., impurity test, HPLC methods for stability, and TLC methods also use external standards. For internal standard methods, a compound of known purity is added directly to the sample. However, it must be ensured that the compound being used as an internal standard does not interfere with any analyte of interest or degradation products in the sample. The response ratio between internal standard and analyte of interest in the sample is compared to the ratio of the internal standard and the analyte in the standard that is used for quantitation purposes. Internal standard methods are widely used for quantitation in biological samples and for low and wide sample concentration ranges, e.g., in pharmacokinetics studies.

There are some basic points that should be addressed in the test method.

1. The sample and the standard should be prepared in the mobile phase. If this is not possible, then the level of organic solvent used in the preparation of the sample and the standard must be lower than that present in the mobile phase.
2. The sample and standard concentrations should be close to each other.
3. Sample preparations often require filtration prior to injection onto the system. Filtration removes particulate matter that may clog the column. However, analyte adsorption on the filter can take place. This adsorption effect is important for low-level impurities. Therefore, data to validate this aspect will be required.

In conclusion, method validation is a dynamic process and should not be considered a one-time situation. The design and validation of the method should be such that they ensure its ruggedness or robustness throughout the life of the method. The accuracy of the data is affected by variations in the manufacturing process, the preparation of samples in the laboratory, and the instrument performance. With a well-designed validation and tight chromatographic system suitability criteria, the reliability of the data can be significantly improved. Variations, except from the drug product manufacturing process, can and should be minimized. Good, reliable validated methods will generate data that is trustworthy.

REFERENCES

1. FDA. Guideline for Submitting Documentation for the Stability of Human Drugs and Biologics, February 1987. Center for Drugs and Biologics, FDA, Department of Health and Human Services.

2. FDA. Guideline for Submitting Samples and Analytical Data for Methods Validation, February 1987. Center for Drugs and Biologics, FDA, Department of Health and Human Services.
3. FDA. FDA's Policy Statement for the Development of New Stereoisomeric Drugs, May 1, 1992. Center for Drug Evaluation and Research, FDA, Department of Health and Human Services (www.fda.gov/cder/guidance).
4. WE Heydorn. Developing racemic mixtures vs. single isomers in the U.S. Pharmaceutical News 2(2):19–21, 1995.
5. K Piezer. PV 3000 AMP. Eli Lilly and Company, Indianapolis, IN 46285.
6. Federal Register (November 25, 1997). International Conference on Harmonization; Draft Guidance on Specifications: Test Procedures and Acceptance Criteria for New Drug Substances and New Drug Products: Chemical Substances, Vol. 62, No. 227, 62890–62910.
7. TJ Wozniak, RJ Bopp, EC Jensen. Chiral drugs: an industrial analytical perspective. J Pharm Biomed Anal 9(5):363–382, 1991.
8. SJ Constanzo. Optimization of mobile phase conditions for TLC methods used in pharmaceutical analyses. J Chromatographic Sci 35(4):156–160, 1997.
9. F Moffatt, PA Cooper, KM Jessop. Capillary electrochromatography. Abnormally high efficiencies for neutral-anionic compounds under reversed-phase conditions. Anal Chem 71:1119–1124, 1999.
10. M Rouhi. Capillary electrophoresis. Chem Eng News 77(13):50, March 29, 1999.
11. FDA. Guidance for Industry: Stability testing of drug substances and drug products (DRAFT GUIDANCE), June 1998. Center for Drug Evaluation and Research (CDER) and Center for Biologics Evaluation and Research (CBER), FDA, Department of Health and Human Services (www.fda.gov/cder/guidance).
12. LK Revelle, WH Doub, RT Wilson, MH Harris, AM Rutter. Identification and isolation of chlorhexidine impurities. Pharm Res 10:1777–1784, 1993.
13. H Fabre, AF Fell. Comparison of techniques for peak purity testing of cephalosporins. J Liq Chrom 15(17):3031–3043, 1992.
14. WE Weiser. Developing analytical methods for stability testing. 1998 analytical validation in the pharmaceutical industry, suppl to Pharm Tech, pp. 20–29, 1998.
15. DK Bryant, MD Kingswood, A Belenguer. Determination of liquid chromatographic peak purity by electrospray ionization mass spectrometry. J Chrom 721(A):41–51, 1996.
16. A Gergely, P Horvath, B Noszal. Determination of peak homogeneity by dual detection. Anal Chem 71:1500–1503, 1999.
17. Supplement to LC/GC. Current trends and developments in sample preparation, May 1998.
18. LR Snyder, JJ Kirkland, JL Glajch. Practical HPLC Method Development. 2d ed. New York: John Wiley, 1997, pp. 402–438.
19. LR Snyder, JL Glajch, JJ Kirkland. Practical HPLC Method Development. New York: John Wiley, 1988, pp. 227–251.
20. JA Lewis, DC Lommen, WD Raddatz, JW Dolan, LR Snyder, I Molnar. Computer simulation for the prediction of separation as a function of pH for reversed-phase high performance liquid chromatography. J Chrom 592:183–195, 1992.
21. PJ Schoenmakers, R Tijssen. Modelling retention of ionogenic solutes in liquid chromatography as a function of pH for optimization purposes. J Chrom (A) 656:577–590, 1993.
22. USP 23/NF 18. Monographs for Propoxyphene Hydrochloride and Propoxyphene Napsylate and Products. Rockville, MD: United States Pharmacopeial Convention, 1995, pp. 1319–1327.

23. M El-Khateeb, TG Appleton, BG Charles, LR Gahan. Development of HPLC conditions for valid determination of hydrolysis products of cisplatin. J Pharm Sci 88(3):319–326, 1999.
24. USP 23/NF 18. General Chapter <621>, Chromatography. Rockville, MD: United States Pharmacopeial Convention, 1995, 9th suppl, pp. 4647–4654.
25. UD Neue, DJ Phillips, TH Walter, M Capparella, B Alden, RP Fisk. Reversed-phase column quality and its effect on the quality of a pharmaceutical analysis. LC/GC 12(6):468–480, 1994.
26. JJ DeStefano, JA Lewis, LR Snyder. Reversed-phase high performance liquid chromatography method development based on column selectivity. LC/GC 10(2):130–138, 1992.
27. Optimizing column conditions: the effect of column length on resolution. MAC MOD Forum 31(2):2, 1998.
28. HA Claessens, MA van Straten, JJ Kirkland. Effect of buffers on silica-based column stability in reversed-phase high-performance liquid chromatography. J Chrom (A) 728:259–270, 1996.
29. LR Snyder, JJ Kirkland, JL Glajch. Practical HPLC Method Development. 2d ed. New York: John Wiley, 1997, pp. 233–291.
30. LR Snyder, JJ Kirkland, JL Glajch. Practical HPLC Method Development. 2d ed. New York: John Wiley, 1997, pp. 439–478.
31. DryLab™ for Windows Tutorial Guide. Walnut Creek, CA: LCResources, 1994.
32. USP 23/NF 18, suppl 8. General Chapter <643>, Total organic carbon. Rockville, MD: United States Pharmacopeial Convention, 1995, p. 4320.
33. JW Skoug, GW Halstead, DL Theis, JE Freeman, DT Fagan, BR Rohrs. Strategy for the development of dissolution tests for solid oral dosage forms. Pharm Tech 20(1):58–72, 1996.
34. USP 23/NF 18. General Chapter <1225>, Validation of compendial methods, suppl 10. Rockville, MD: United States Pharmacopeial Convention, 1999, pp. 5059–5062.
35. Code of Federal Regulations. Title 21, Food and Drugs. Part 211—Current good manufacturing practices for finished pharmaceuticals. US Government Printing Office, Washington, 1998, pp. 85–104; Part 314—Applications for FDA approval to market a new drug or an antibiotic drug, pp. 99–179.
36. International Conference on Harmonization (ICH) Q2A. Text on validation of analytical procedures. March 1995.
37. International Conference on Harmonization (ICH) Q2B. Validation of analytical procedures: methodology. November 1996.
38. ME Swartz, IS Krull. Validation of chromatographic methods. Pharm Tech 22(3):104–119, 1998.
39. ME Swartz. Validation guidelines. Waters Website: www.waters.com
40. LL Ng. Reviewer Guidance: Validation of chromatographic methods. FDA Center for Drug Evaluation and Research (CDER), November 1994.
41. IS Krull, ME Swartz. Introduction: National and international guidelines in Validation Viewpoint. LC/GC 15(6):534–540, 1997.
42. IS Krull, ME Swartz. Introduction: National and international guidelines in Validation Viewpoint. LC/GC 16(5):464–467, 1998.
43. BA Persson, J Vessman, RD Mcdowall. How good is your method? in Question of Quality. LC/GC 15(10):944–946, 1997.
44. USP 22/NF 17. General Chapter <1225>, Validation of compendial methods. Rockville, MD. United States Pharmacopeial Convention, 1990, pp. 1710–1712.
45. ASTM E1169-89 Standard Guide for Conducting Ruggedness Tests (Plackett–Burman design). American Society for Testing and Materials (ASTM), 100 Barr Harbor Drive, West Conshohocken, PA 19428-2959, Tel. 610.832.9585.

46. LD Torbeck. Ruggedness and robustness with designed experiments. Pharm Tech 20(2): 168–172, 1996.
47. Drugs Directorate Guidelines—Acceptable Methods. Health Protection Branch, Health Canada 1994. (Contact: Drugs Directorate, Health Protection Branch, Health Canada, Health Protection Building, Tunney's Pasture, Ottawa, Ontario K1A0L2.)
48. RB Kirsch. Validation of Analytical Methods Used in Pharmaceutical Cleaning Assessment and Validation. 1988. Analytical Validation in the Pharmaceutical Industry, suppl to Pharmaceutical Technology, pp. 40–46.
49. HG Brittain. Validation of nonchromatographic analytical methodology. Pharm Tech 22(3):82–90, 1998.
50. EW Ciurczak. Validation of spectroscopic methods in pharmaceutical analyses. Pharm Tech 22(3):92–102, 1998.

APPENDICES

1. List of Guidance Documents, CDER. http://www.fda.gov/cder/guidance/index.htm
2. Useful websites:
   - www.pharmweb.net
   - www.waters.com
   - www.usp.org
   - www.ich.org
   - www.aoac.org

12

Stability Testing of Clinical Trial Materials

WOLFGANG GRIMM

Biberach, Germany

1. INTRODUCTION

1.1. General Requirements for Stability Testing

The aim of stability testing is to ensure the quality, safety, and efficacy of drug products up to their expiration date. This means that all

Organoleptic
Physicochemical
Chemical
Microbial

test results must be within the shelf life tolerance ranges up to the end of the shelf life. Extensive studies are needed for this purpose. Stability testing accompanies the development of a medicinal product from the first preliminary trials with the drug substance up to continuous production. If the stability program is scientifically well founded, systematically structured and logically coordinated stability information will be continuously augmented and become increasingly reliable.

The overall stability program can be divided into six steps (1):

Step 1: Stress and acceleration tests with the drug substance
Step 2: Preformulation and formulation finding for
  toxicological test samples
  clinical samples
  final dosage form
Step 3: Stress and acceleration tests with selected formulations
  toxicological test samples
  clinical samples
  final dosage form
  Selection of primary packaging materials
Step 4: Acceleration and long-term tests on drug substance and drug products up to marketing authorization
Step 5: Ongoing stability testing of drug substance, drug products
  marketing authorization batches
  production batches
Step 6: Follow-up stability tests on drug substance, drug products
  continuous production
  modifications during continuous production

Each stage covers eleven basic principles (2):

Selection of batches and samples
Test criteria
Analytical procedures
Specifications
Storage conditions
Testing frequency
Storage period
Number of batches
Packaging materials
Evaulation
Stability information

Some of these elements, such as selection of samples and batches, number of batches, storage conditions, testing frequency, and storage periods are firmly established, while others, such as test criteria, analytical procedures with validation, and specifications undergo further development. Combining the six stages and the eleven basic principles yields a systematically structured stability study schedule.

1.2. Specific Requirements for Stability Testing of Clinical Samples

Similarly to the requirements for proprietary medicinals, the aim of stability testing in this area of clinical samples is to maintain the quality and safety of these materials up to the end of phase I, II, and III clinical trials.

This means that all relevant test results must remain within the minimum shelf life specifications up to the end of the clinical trial.

In systematically structured stability programs, stability tests are carried out on clinical samples in steps 2 and 3.

Stability testing accompanies phase I to III clinical trials.

At first, neither the dosage formulation nor the dosage form is definitely established; they are gradually defined during the course of development.

The same applies to analytical procedures and specifications.

The effort and scope of stability testing must be tailored to the specific problem.

The stability program will always differ from that required to generate stability information (expiration date, etc.) for marketing authorization application documents for proprietary medicinals. This stability information is based on the results obtained with three representative pilot plant batches put into storage after the end of development. The stability information is then applicable to all production batches.

The stability information for clinical samples is required only for a small number of batches and for the duration of the respective clinical trial.

The shelf life determined is thus not a maximum shelf life ($\leq$5 years) at the end of which the acceptance criteria of the shelf life specifications for individual test parameters are reached, but a minimum shelf life at the end of which the tolerance limits usually have not been reached.

There are no defined official regulations stipulating effort, scope, and implementation, which are left to the manufacturer's discretion.

In the USA (6), in each phase of the investigation sufficient information should be submitted to ensure the proper identification, quality, purity, and strength of the investigational drug; the amount of information needed to achieve that assurance will vary with the phase of the investigation, the dosage form, and the amount of information otherwise available.

Therefore, although stability data are required in all phases of the IND to demonstrate that the new drug substance and drug product are within acceptable chemical and physical limits for the planned duration of the proposed clinical investigation, if very short term tests are proposed, the supporting stability data can be correspondingly very limited. It is recognized that modifications to the method of preparation of the new drug substance and dosage form, and even changes in the dosage form itself, are likely as the investigation progresses.

In the USA and Japan, clinical samples are not required to be marked with an open expiration date, as the period of use (use-by date, expiry date, or re-test date). Therefore in an initial phase I CMC submission, the emphasis should generally be placed on providing information that will allow evaluation of the safety of subjects in the proposed study. The identification of a safety concern or insufficient data to make an evaluation of safety are the only reasons for placing a trial on clinical hold on the CMC section.

Information to support the stability of the drug substance during the proposed clinical study(ies) should include the following: a brief description of the stability study and the test methods used to monitor the stability of the drug substance and preliminary tabular data on representative material. Neither detailed stability data nor the stability protocol need to be submitted. When significant decomposition during storage cannot be prevented, the clinical trial batch of drug product should be retested prior to the initiation of the trial, and information should be submitted to show that it will remain stable during the course of the trial.

This information should be based on the limited stability data available when the trial starts. Impurities that increase may be qualified by reference to prior human or animal data.

Development of drug product formulations during phase II should be based in part on the accumulating stability information gained from studies of the drug substance and its formulations.

The objectives of stability testing during phases I and II are to evaluate the stability of the investigational formulations used in the initial clinical trials, to obtain the additional information needed to develop a final formulation, and to select the most appropriate container and closure. This information should be summarized and submitted to the IND during phase II. Stability studies on these formulations should be well underway by the end of phase II.

At this point the stability protocol for study of both the drug substance and the drug product should be defined, so that stability data generated during phase III studies will be appropriate for submission in the drug application.

In stability testing during phase III IND studies, the emphasis should be on testing final formulations in their proposed market packaging and manufacturing site based on the recommendations and objectives of the ICH Stability Guideline.

It is recommended that the final stability protocol be well defined prior to the initiation of phase III IND studies.

In this regard, considerations should be given to establishing appropriate linkage between the preclinical and clinical batches of the drug substance and drug product and those of primary stability batches in support of the proposed expiration dating period. Factors to be considered may include for example source, quality, and purity of various components of the drug product, manufacturing process of and facility for the drug substance and the drug product, and use of the same containers and closures (6).

Clinical samples in the EU are required to be marked with an open expiration date (7).

The corresponding information for this open expiration date must be available before the packaging of the clinical trial batches.

The base for the stability information concerning clinical samples forms the Stability Profile of the drug substance derived from the stress investigations in step

1 of the systematically structured stability schedule, the stress and accelerated testing with the drug substance.

These studies are undertaken to elucidate the intrinsic stability characteristics of the drug substance with reference to physicochemical and chemical properties, to establish the degradation pathway in order to identify the likely degradation products, and to validate the stability-indicating power of the analytical procedures used.

The investigations are of general nature and not specific to any particulate dosage form.

The additional influence of the excipients is investigated in step 2 "Preformlation and formulation finding for clinical trial samples." The results are summarized in a research report, "Formulation evaluation."

From now on two different procedures can be pursued to provide the stability information for the open expiration date:

1. Stress and accelerated tests and long-term testing for confirmation.

Stress and accelerated testing are performed with selected batches to derive the minimum shelf lives for the phases I, II, and III. These minimum shelf lives are applicable to all batches of the relevant phase.

The predictions are then supported and confirmed by long-term testing under storage conditions presenting climatic zone II, 25°C/60% r.h.

This procedure has the following advantages.

- Instable formulations are recognized at an early stage and can be reformulated if required.
- The continuous development of the drug product is guaranteed without substantial time loss due to long-term testing.
- The stress and confirmation testing is performed in phases I, II, and III. Thereby an excellent linkage is given between the preclinical and clinical batches in support of the proposed shelf life for the registration batches.
- A series of possible influencing factors on the formulation can be assessed.
- The repeated confirmation of predicted minimum shelf lives by the results of long-term testing guarantees a high degree of reliability of the stability information.
- The number of investigated batches is limited; a prediction of the required capacity can be established.

If the results of the stress investigations indicate decomposition but no distinctive temperature dependence, then a prediction is not possible and a corresponding long-term testing has to be performed.

2. Accelerated and long-term testing with the clinical trial batches.

The clinical trial batches are put on accelerated and long-term testing.

- Only a limited prediction of the minimum shelf life, especially in the phases II and III, is possible, with the consequence that not the total period of the intended clinical investigation may be covered.
- In the USA no open expiration date is required. The clinical samples may be stored at 25°C/60%. r.h. and analyzed in parallel with the clinical trial investigation.
- If all data are within the specifications, nothing has to be done.

If the data, however, exceed the specification it may be necessary to replace the samples in the clinic by new ones. Therefore corresponding batches have to be kept in stock.

This is not easy to accomplish.

Therefore procedure 1 is preferred and consequently described in detail.

Stability information for clinical samples plays a very important role in the general assessment of the quality and safety of a medicinal product.

Continuous stability is assured for the transition from phase I to III, including pivotal and equivalence batches, to the finished drug product in the commercial form.

Stability programs for clinical samples now are based on the ICH Tripartite Guideline for Stability Testing of Drug Substances and Drug Products, although the guideline itself, as already mentioned, does not apply to clinical samples.

The basic principles of the ICH Tripartite Guideline correspond to the aforementioned eleven principles and now have to be adapted to deal with the specific problems encountered with clinical samples.

2. BASIC PRINCIPLES OF STABILITY TESTING APPLIED TO CLINICAL SAMPLES

2.1. Selection of Batches and Samples

The drug product is in the process of development. Several strengths are tested in clinical phase I and the formulation and dosage form are modified in the transition to II and III.

This developmental process has to be taken into account when selecting the batches. Especially in the initial phase of development, no representative batches are available. The following batches are put into storage with the aim of establishing the minimum shelf life for phases I to III:

Clinical phase I: experimental batches
Clinical phase II: clinical experimental batches
Clinical phase III: clinical or pilot plant batches

2.2. Test Criteria

The criteria of the product are investigated

That are potentially subject to change during the course of storage

That have a particular bearing on the quality, safety, or acceptance of the product.

The relevant test criteria will become apparent during the course of development from phase I to phase III.

2.3. Analytical Procedures

The analytical procedures themselves undergo a process of development from phase I to phase III. The same applies for the validation.

Three steps of validation are differentiated:

Orientational
Preliminary
Complete

Table 1 lists the extent of validation for the three steps.

Table 1 Extent of Validation During Development

	Extent of validation		
Validation characteristic	Orientational	Preliminary	Complete
Specificity	x	x	x
Linearity	x	x	x
Quantitation limit	x	x	x
Detection limit[a]	x	x	x
Accuracy		x	x
Range		x	x
Repeatability		x	
Intermediate precision			x
Robustness	x	x	x
Validation report			x

[a] Only for semi-quantitative procedures instead of quantitation limit

2.4. Specifications

Fixing specifications is an evolving process that accompanies the development of the new drug substances and drug products.

It can be described as a four step-procedure; see Table 2.

For all four steps one has to differentiate between

Release specification
Shelf life specification

Table 2 The Four Steps of the Specification

Step of development drug substance, drug product	Specifications	Characterization
Preclinical Clinical phase I	Orientational	Target values
Clinical phases II/III Pivotal batches	Preliminary	Broader acceptance criteria, ranges, numerical limits
Pilot plant batches Registration batches	Registration	Acceptance criteria focusing on safety and efficacy
Production batches after marketing authorization	Post approval	Experience gained with manufacture of a particular drug substance or drug product

This concept of different acceptance criteria for release versus shelf life specifications applies to products only, not for drug substances.

The ICH Guideline requires this distinction between release and shelf life specification.

Release specifications describe the quality after manufacture and include

Analytical variability
Manufacturing variability

The variability is described by

RSD of repeatability or intermediate precision of the analytical procedure. Accordingly, validation has to be performed, preliminary or complete.
Data of $\geq$3 batches to describe the manufacturing variability.

Shelf life specifications describe the quality at the end of the shelf life and include

Tolerable changes during storage and shipment.

Therefore corresponding stability data are required with organoleptical, physicochemical, chemical, and microbial tolerable changes.

2.5. Storage Conditions

A basic distinction is drawn between stress, accelerated, and long-term storage conditions; see Table 3.

Table 3 Stress, Accelerated, and Long-Term Storage Conditions

Type	Condition
Stress	Temperature: 10°C higher than accelerated temperature of 40°C, e.g., 50°C, 60°C, 70°C ≥ -10°C Temperature cycle 5–40°C Open storage at 25°C/60% r.h., 30°C/70% r.h. and 40°C/75% r.h. Xenon lamp 48 hours
Accelerated	40°C/75% r.h. (30°C/70% r.h.)
Long-term	25°C/60% r.h. 30°C/70% r.h.

The conditions used in stress and accelerated tests are above those of the relevant climatic zones, allowing

The discriminatory power of the analytical procedure to be verified.
Weaknesses of a formulation to be identified.
Stability information to be generated.

The last-named aspect is particularly important in designing a stability program for clinical samples. To ensure continuous development, specific stress and acceleration tests are carried out, which are then verified by long-term tests.

If stress and acceleration tests are to be successful, two aspects must be paid special attention:

Clear separation between the tests for
organoleptic and physicochemical stability
chemical/microbial stability
Use of packaging materials impermeable to water vapor for stress tests at elevated temperatures, allowing application of the laws of reaction kinetics

The reaction mechanism may change with higher temperature if the samples are dried.

It is therefore necessary to use packaging materials that are impermeable to water vapor to prevent solid formulations becoming dehydrated at higher temperatures or the active ingredient concentration of liquid formulations increasing due to loss of moisture.

The laws of reaction kinetics cannot be used to make stability predictions for organoleptic and physicochemical changes.

This is the reason that solid dosage forms, for example, are stored without packaging at 25°C/60% r.h. This induces the maximum possible changes due to absorption or loss of water. Semisolid and liquid dosage forms are stored at ≥ – 10°C, semiliquid at 5–40°C in order to detect irreversible changes.

Storage at 30°C/70% r.h. is not usually necessary because most clinical trials are performed in countries of climatic zones I and II.

2.6. Testing Frequency

The testing frequency is established to suit the problem being studied. Retest periods are different for stress, acceleration, and long-term tests.

2.7. Storage Period

The storage period depends on the required minimum shelf life. We can differentiate between stress, accelerated, and long-term storage period; see Table 4.

2.8. Number of Batches

With all the different strengths, dosage forms, and packaging materials examined during the development phase, it is not possible to provide three batches for each dosage form. Reliable information can be obtained nevertheless by applying

Table 4 Storage Periods

	Storage period (months)		
Clinical phase	Stress	Accelerated	Long-term
I		1.5	3
		3.0	6
II	3	6	12–18
III	3	6	24–36

scientifically based rationalization measures. This includes the expedients mentioned in the ICH Stability Guideline:

Bracketing
Matrixing

Both methods are based on the assumption that a reduced number of investigated samples is representative of the stability behavior of all samples.

In bracketing, only "limit samples" are tested, for example: the lowest and highest dosage, the smallest and largest container.

In matrixing, selection is performed according to a statistical procedure (random number).

A rational bracketing system for all dosage forms would be as in Table 5.

A rational matrixing system would be as in Table 6.

Table 5

Dosages	Samples tested
1–2	all
3–4	highest lowest
>4	highest middle lowest

Table 6

Test sample	Tests
Beginning, end	all
Intermediate values	1/3 or 2/3 design[a]

[a] At each testing point 1/3 or 2/3 of all samples are analyzed.

2.9. Packaging Materials

At higher temperatures, desorption and loss of moisture also occurs at higher relative humidities.

Unless packaging materials impermeable to water vapour are used for stress tests with solid dosage forms, the samples lose moisture at different rates in the temperature range 40–60°C, and the results are not suitable for a reaction kinetics calculation.

Packaging materials permeable to water vapor can however also result in a falsification of the results for semisolid and liquid dosage forms if varying degrees of weight loss occur that lead to differences in the active ingredient concentration or ion strength.

The use of inert standard packaging materials that are impermeable to water vapor is thus an important precondition for stress tests that are to be evaluated in terms of reaction kinetics, and on the results of which stability predictions are to be based.

An overview of the most important packaging materials for stability testing for clinical samples is given in Table 7.

2.10. Evaluation

A systematic approach should be adopted in the evaluation and presentation of the analytical results.

Tests for significant changes with the aid of statistics, reaction kinetics calculations, or linear regression analysis are valuable tools. The stress and accelerated test results are evaluated for each clinical phase taking into account the specific objective of the respective storage conditions.

On the one hand, there is the test for organoleptic and physicochemical stability, on the other hand for chemical or microbial stability. If, for example, discoloration, a decrease in hardness, an increase in dissolution, or phase separation is observed after 3 months's storage at 70°C, these changes will be recorded, but they are only of limited relevance for predicting stability. If there are no significant changes in the test for organoleptic and physicochemical stability, stability can be predicted by means of reaction kinetics calculations that are based mainly on the results obtained after storage at stress temperatures.

Considering these facts, all test criteria can be included in the stability prediction. A critical examination is conducted to determine whether relevant changes have occurred and whether the proposed minimum shelf life tolerance limits have been reached or exceeded.

Stability studies for clinical samples are based on stress and acceleration tests with the aim of speeding up, especially, chemical decomposition by storing samples at elevated temperatures. The results are then used to calculate the stability behaviour at 25°C/60% r.h. based on the laws of reaction kinetics.

The equations for a first-order reaction and the Arrhenius model are used. If decomposition levels are available for only one temperature (4,5), the expression ΔE: 83 kJ mol^{-1} is used for the activation energy.

Table 8 shows decomposition levels for 25°C/60% r.h. calculated from values obtained after storage at 40°C to 70°C. Reported in Table 6 are the decomposition determined after storage at accelerated and stress temperatures, and the decomposition for 25°C derived from these data.

After evaluating the data of one batch it is assessed whether the strengths or the dosage forms exhibit different stability behavior or whether the results of the batches can be combined to produce uniform stability information.

The packaging material and possible interactions have to be included in the evaluation.

The general stability information, the period of use, and if necessary storage instructions are based on

Primary data
: The results of the stress and accelerated investigations and later the results of the long-term testing for confirmation.

Supportive data
: Drug substance stability profile, which also includes orientational predictions regarding the chemical stability of the drug substance in solid, semisolid, and liquid dosage forms.

Table 7 The Most Important Packaging Materials for Stability Testing of Clinical Samples

	Dosage form, packaging material		
Clinical phase	Solid	Semisolid	Liquid
I Acceleration tests	Standard packaging material Glass container with twist-off closure	Standard packaging material Standard tube	Standard packaging material Ground-glass-stoppered flask or Glass ampoule or Injection vial with rubber stopper
Long-term test	Standard packaging material Glass container with twist-off closure PP tubes[a]	Standard packaging material Test packaging material	Standard packaging material Test packaging material
II Stress tests	Standard packaging material Glass container with twist-off closure	Standard packaging material	Standard packaging material
Acceleration tests	Standard packaging material Glass container with twist-off closure	Standard packaging material Testing packaging material	Standard packaging material Testing packaging material
Long-term test	Standard packaging material Glass container with twist-off closure PP tubes or Test packaging material: blister	Standard packaging material Test packaging material	Standard packaging material Test packaging material
III Stress tests	Standard packaging material Glass container with twist-off closure	Standard packaging material	Standard packaging material
Acceleration tests	Standard packaging material Glass container with twist-off closure	Standard packaging material Test packaging material	Standard packaging material Test packaging material
Long-term test	Standard packaging material Glass container with twist-off closure PP tubes or Test packaging material: blisters	Standard packaging material Test packaging material	Standard packaging material Test packaging material

[a] PP = polypropylene

Stability investigations during formulation findings for the clinical phases I–III, which provide specific information regarding the influence of excipients and the overall formulation on organoleptic, physicochemical, chemical, and microbial stability.

For phase II

Minimum shelf life for clinical phase I: prediction and confirmation.

For phase III

Minimum shelf life for clinical phases I and II: prediction and confirmation.

It is an important fact that the predicted minimum shelf lives will always be confirmed by the concurrently performed long-term tests.

Table 8 Reaction Kinetics Extrapolation

Clinical phase	Storage condition	Decomposition found/ extrapolated decomposition (%)					
I	40°C: 1.5 months	0.10	0.20	0.30	0.40	0.50	1.00
	25°C: 3 months	<0.10	<0.10	0.12	0.16	0.20	0.40
I	40°C: 3 months	0.10	0.20	0.30	0.40	0.50	1.00
	25°C: 6 months	<0.10	<0.10	0.12	0.16	0.20	0.40
II	60°C: 3 months	0.10	0.20	0.30	0.40	0.50	1.00
	25°C: 12 months	<0.10	0.18	0.27	0.35	0.44	0.89
	25°C: 18 months	0.13	0.27	0.40	0.53	0.66	1.30
II	40°C: 6 months	0.10	0.20	0.30	0.40	0.50	0.60
	25°C: 12 months	<0.10	<0.10	0.12	0.16	0.20	0.40
	25°C: 18 months	<0.10	0.12	0.18	0.24	0.30	0.60
III	70°C: 3 months	0.50	1.00	2.00	3.00	4.00	5.00
	25°C: 24 months	<0.10	0.10	0.20	0.30	0.40	0.50
	25°C: 36 months	<0.10	0.15	0.30	0.45	0.60	0.76

2.11. Stability Information

All the results and the stability information derived therefrom are complied in a stability report.

This contains

The batch information
The results
A critical assessment of the analytical procedures and of the results
The minimum shelf life
Storage instructions if necessary

Usually for each clinical phase a stability report is written.
Stability report:

Stress Testing and Long-Term Testing
Phase I
Phase II
Phase III

The results of stability tests on clinical samples are used to set minimum shelf lives and not expiration dates.

The duration of each clinical trial including logistics and provision of clinical supplies must be fully covered. In many cases, therefore, the minimum shelf life will not be determined by reaching a shelf life tolerance limit. In all these cases the expiration date can be extended, if necessary, on the basis of suitable studies.

The following minimum shelf lives are considered optimal:

Clinical phase I: 3–6 months
Clinical phase II: 12–18 months
Clinical phase III: 24–36 months

The established minimum shelf lives are valid for all the batches of the relevant clinical phase.

Since the product is in the process of development, the shelf lives are not based on three representative validation batches but on

batches comprising different dosages as tested by the bracketing procedure.
batches comprising different dosage forms obtained as development progresses.
batches of different origin and size. These are usually experimental and development batches.

If a critical examination reveals that the batches exhibit similar stability behavior, the result can be applied to the "same" batches of the relevant clinical phase with a high level of reliability.

It can be further stated that the factors batch size, technology, and equipment used in the manufacturing process affect primarily the quality of manufacture and not that of stability.

This means that they are identified in the analysis conducted immediately after manufacture that determines the quality in relation to the release specifications (Table 9).

Table 9 The Influencing Factors for the Individual Dosage Forms

Influencing factors	Possible influence on quality of the dosage forms		
	Solid	Semisolid	Liquid
Batch size	Appearance	Appearance	Appearance
Equipment	Content uniformity	Homogeneity	pH
Manufacturing process	Dissolution	Homogeneity within a container Preservation Chemical stability after manufacture	Preservation Chemical stability after manufacture

If all the analytical data obtained after manufacture are within the release specifications, the stability information obtained from the stress tests and acceleration tests can be considered generally applicable with a high degree of reliability.

It may be necessary to ensure compliance with the minimum shelf life by marking packs with storage instructions.

It is important to present this information unambiguously (Table 10).

Table 10 How Storage Instructions Should Be Worded

Storage instruction	Reason
Do not store above 30°C	Relevant changes were seen in the samples after storage at 40°C.
Do not store above 25°C	Relevant changes were seen in the samples after storage at 30°C/70%, but not after storage at 25°C/60%.
Store at ≤8°C in a refrigerator	Revelant changes were observed in the samples after storage at 25°C/60%.
2–8°C, store in a refrigerator, do not freeze	Relevant changes were observed in the samples after storage at 25°C/60% and −10°C.

2.12. Reliability of Minimum Shelf Lives

The shelf lives for the batches of clinical samples are established to cover the duration of the clinical trial plus a supplement to allow for logistics and the provision of clinical supplies. The shelf lives determined apply to all the batches of the relevant development stage, although only the batches in the final phase of development originate from a validated manufacturing phase and are therefore representative.

How reliable is shelf life and stability information?

This question can be answered as follows.

The shelf lives for clinical phases I and II (and in some cases III) represent a minimum shelf life, in other words, they still include a "reserve." A shelf life of 3 months for a first clinical trial does not mean that the batch may not be stable for longer periods. The shelf lives may be extended after appropriate storage and tests.

Minimum shelf lives are therefore associated with a lower risk than shelf lives at the end of which the sample may be unstable.

Furthermore, the principle of "semi-coverage" applies to clinical phase I, i,e., half the shelf life (3–6 months) is covered by storage at higher temperatures.

If there are several strengths, bracketing is performed, i.e., two to three strengths are tested simultaneously for stability.

If all two or three strengths exhibit the same stability behavior, a statement can be made regarding the reproducibility or the technological parameters.

If the stability information for two to three batches of different strengths or composition is identical, the information is naturally also applicable to identical batches.

As phases I through III progress, experience with analytical procedures and shelf life specifications increases. The results and stability information derived from them become steadily more reliable.

Finally, the stability program is designed in such a way that stability predictions are verified by the results of long-term tests.

There is thus still the possibility of replacing batches if necessary.

Like the overall system of stability testing, the stability program for clinical samples is systematically structured in such a way that the aggregate available information is continuously augmented (Table 11).

Table 11 Number of Dosages and Packaging Materials

Stage of development	Tests[a]	Number of dosages	Number of packaging materials	Total number	Derived information
Drug substance	Stress and acceleration tests	1	1	1	Retest date ≥2 years
Clinical phase I	40°C/1.5 months	≥2	2	≥4	3 months
	40°C/3 months	≥2	2	≥4	6 months
Clinical phase II	60°C/3 months	2[b]	2	4	12–18 months
Clinical phase III	70°C/3 months	1[b]	3	3	24–36 months
				≥16	

[a] A confirmatory long-term test is conducted concurrently with stress and acceleration tests.
[b] If more than 2 dosages are used in clinical phase II, ≥2 applies; if more than 1 dosage is used in clinical phase III, >2 also applies. This data in the table are therefore minimum limits.

2.13. Extension of the Derived Minimum Shelf Life

A minimum shelf life is determined for clinical trial samples and can be extended as necessary if the corresponding tests yield favorable results. Parallel to the stress and accelerated tests, samples are stored at 25°C/60% r.h. in order to confirm and support the predicted minimum shelf life.

If all the predicted data and the data confirmed in long-term tests are within the minimum shelf life limits, they can be extended if necessary. For this purpose, a new prediction is performed and samples are stored at 25°C/60% r.h. to confirm the new minimum shelf life (see Table 12).

3. PERFORMANCE

The performance of the stress and confirmation studies in phases I–III is described.

The base forms the Stability Profile of the NME, the corresponding drug substance.

Table 12 The Objective Is to Extend the Shelf Life by 30%

Derived minimum shelf life (months)	Extension by 30% to (months)
3	4
6	9
12	16
18	24
24	32
36	48

The Stability Report comprises the stability data of stress investigations with the active ingredient it represents the stability profile of the NME.

The following influencing factors were investigated: Moisture, temperature, moisture + temperature, moisture + temperature + drug substance concentration, pH, ionic strength, oxidation, and light.

The following stability information is derived.

For drug substance,

Test criteria for accelerated and long-term testing with the registration batches
Analytical procedures
Selection of packaging materials
Preliminary retest period
Storage instructions, if required

For drug product,

Solid, liquid and semiliquid dosage forms can be developed concerning chemical stability.

From the investigations of step 2, performulation and formulation finding for clinical trial samples, information is available on the additional influence of the excipients on the stability of the drug substance.

On the base of this comprehensive information, the stress and accelerated testing is planned and performed.

3.1. Solid Dosage Forms: Tablets, Capsules

3.1.1. Selection of Batches and Samples

Phase I: Experimental laboratory batches from the development laboratory
Phase II: Clinical batches, pilot scale from manufacturing clinical supplies
Phase III: Pilot plant batches, final formulation from pilot plant

3.1.2. Test Criteria

Organoleptic and physicochemical stability

Tablets: appearance, hardness, average mass, disintegration time, dissolution rate

Capsules: appearance, elasticity, average mass, average mass of content, average mass of filling, disintegration time, dissolution rate

Chemical stability

Tablets: appearance, hardness, disintegration time, dissolution rate, average mass, drug substance decomposition and assay

Capsules: appearance, elasticity, average mass, average mass of content, average mass of filling, disintegration time, dissolution rate, drug substance decomposition and assay

The organoleptic and physicochemical test criteria are included in chemical stability. The results may not be relevant to normal storage conditions.

3.1.3. Analytical Procedures

Stability specific information on degradation pathway is available from the stability profile. If possible the same analytical procedure is applied.

The validation is performed in three steps (Table 13).

If several strengths are investigated in phase I or II, the extent of validation can be reduced. If the final concentration of the analyte is the same after sample preparation, the validation is limited to one strength, usually the lowest.

Table 13 The Three Steps of Validation

Clinical phase	Step of validation	Validation criteria
At beginning phase I	Orientational	Specificity, linearity, quantitation limit ≙ reporting limit: 0.1%, robustness
At beginning phase II	Preliminary	In addition: accuracy, range, repeatability of assay and decomposition product, dissolution rate included
At beginning phase III	Complete	In addition: intermediate precision, complete robustness

3.1.4. Specifications, Table 14

Table 14

Clinical phase	Specifications
Clinical phase I	Orientational, target values
Clinical phases II/III	Preliminary

3.1.5. Storage Conditions, Storage Period, Testing Frequency

The same distinction is made for the storage conditions as for the test criteria.

For organoleptic and physicochemical stability testing, the samples are stored without packaging under the climatic conditions of long-term testing at 25°C/60% r.h. until equilibrium is reached. The maximum possible changes occur during this period.

The test for chemical stability is performed including also those samples that were stored without packaging at 25°C/60% r.h. until equilibrium was reached. With this approach, the influence not only of the temperature but also of humidity on the stability of the product can be examined.

Parallel to the stress and acceleration tests, samples are stored at 25°C/60% r.h. representing the conditions of climatic zone II for confirmation. The storage period corresponds to the planned expiration date. These measures allow the stability prediction to be checked (see Tables 15 and 16).

Table 15 Storage Conditions for Organoleptic and Physicochemical Stability

Clinical phase	Packaging material	Storage condition	Storage period
I, II	Without, open	25°C/60% r.h.	Until equilibrium (2 weeks)
III	Without, open	25°C/60% r.h. 30°C/70% r.h. 40°C/75% r.h.	Until equilibrium (2 weeks)

3.1.6. Packaging Material

For stress testing, tight containers are required. 50 mL glass containers with twist-off closure are suitable or tight equivalent.

For long-term testing and clinical trial samples, on the basis of the results of the stress tests for solid dosage forms, the sensitivity to moisture can be determined, and suitable packaging materials can be selected.

As a rule, no interactions are to be expected.

If the final packaging material has been selected, and samples packed in the final packaging material are available, the investigation on photostability should be performed.

The samples with and without container are irradiated with a xenon lamp (Suntest 250 W/m^2) for 22 hours (ICH Guideline on Photostability Testing).

The test criteria are appearance, drug substance decomposition and assay.

3.1.7. Number of Batches

There should be one batch per clinical phase.

If more than one strength is required in phases I or II, bracketing is applied.

3.1.8. Evaluation

If all the results of the test for organoleptic and physicochemical stability are within the shelf life specifications, the stability prediction depends exclusively on the chemical stability. If the water content of the pretreated samples (open storage, 25°C/60% r.h.) has no influence, or if the influence is acceptable, there is no restriction on the choice of packaging materials.

If the results of the test for organoleptic and physicochemical stability are outside the tolerance limits, or if the water content influences the chemical stability to an unacceptable degree, a packaging material impermeable to water vapor must be selected, e.g., glass bottle with screw closure, polypropylene or polyethylene tubes, aluminum blister, or aluminum/aluminum.

Table 16 Storage Conditions for Chemical Stability

Clinical phase	Minimum shelf life	Packaging material	Pretreatment	Storage conditions (°C)	(%)	Storage frequency, storage period									
I	3 months	Twist-off[a]	None	40	—	0	2	4	6						weeks
		Twist-off	25°C/60%	40	—	0	2	4	6						weeks
		Twist-off	None	25	60						12				weeks
		PP tubes	None	25	60						12				weeks
I	6 months	Twist-off	None	40	—	0	1	2	3						months
		Twist-off	25°C/60%[b]	40	—	0	1	2	3						months
		Twist-off	None	25	60					6					months
		PP tubes	None	25	60					6					months
II	12–18 months	Twist-off	None	60	—	0	1	2	3						months
		Twist-off	None	40	—		1	2	3	6					months
		Twist-off	25°C/60%[b]	60	—	0	1	2	3						months
		Twist-off	25°C/60%[b]	40	—		1	2	3	6					months
		Twist-off	None	25	60						12	18			months
		PP tubes or test packaging material	None	25	60						12	18			months
III	24–36 months	Twist-off	None	70	—	0	1	2	3						months
		Twist-off	None	60	—		1	2	3						months
		Twist-off	None	50	—		1	2	3						months
		Twist-off	None	40	—		1	2	3	6					months
		Twist-off	25°C/60%[b,c]	70	—	0	1	2	3						months
		Twist-off	25°C/60%[b]	60	—		1	2	3						months
		Twist-off	25°C/60%[b]	50	—		1	2	3						months
		Twist-off	25°C/60%[b]	40	—		1	2	3	6					months
		Twist-off	None	25	60						12	18	24	36	months
		PP tubes or test packaging material	None	25	60						12	18	24	36	months

[a] 50 mL glass bottle with twist-off closure or corresponding tight container.
[b] If stability data of phase I indicate that moisture does not influence the stability, this investigation can be deleted.
[c] In phase III, samples that have adsorbed the highest amount of water during open storage at 25°C/60%, 30°/70%, 40°C/75% r.h. will be included in the stress testing.

3.1.9. Stability Information

All the results and the stability information are compiled in a Stability Report.
Correspondingly three Stability Reports are available.

Stress Testing and Long-Term Testing phase I
Stress Testing and Long-Term Testing phase II
Stress Testing and Long-Term Testing phase III

If the data of the different stress investigations are comparable, it can be concluded that the quality of the clinical trial batches is comparable with the quality of the registration batches; the patient after marketing authorization will get the same quality as the patient during the clinical trials.

Furthermore, packaging materials can be recommended for the registration batches as in Table 17.

Storage instructions should be given if required.

Table 17

	Climatic zones		
Packaging material	I+II	III	IV
PVC/PVDC blister	x	x	—
Polypropylene tubes with polyethylene closure	x	x	x
Polyethylene bottle	x	x	x
Glass bottle with screw cap	x	x	x
Aluminum blister	x	x	x

3.2. Semisolid Dosage Forms: Creams, Ointments

3.2.1. Selection of Batches

This is as for solid dosage forms.

3.2.2. Test Criteria

Organoleptic and physicochemical stability; appearance, odor, homogeneity, consistency, pH, particle size (if active ingredient is in suspension), recrystallization.

chemical and microbial stability: Appearance, homogeneity, content uniformity within the container (tubes stored vertically are cut open, samples are taken from the beginning, middle and end of the tube and analyzed), drug substance decomposition and assay, preservative decomposition, assay

3.2.3. Storage Conditions, Storage Period, Testing Frequency, Packaging Material

Although the samples are stored differently for organoleptic/physicochemical and chemical/microbial stability, the tests overlap; see Tables 18 and 19.

Table 18 Storage for Organoleptic and Physicochemical Stability

Clinical phase	Packaging material	Storage condition	Storage period
I–III	Standard tube (aluminum tube with internal lacquering)	$\geq -10°C$	4 weeks
		5–40°C in a 24 hour cycle	2 weeks

Table 19 Storage Conditions for Chemical and Microbial Stability

Clinical phase	Minimum shelf life	Packaging material	Storage conditions temp. (°C)	Storage conditions rel. hum. (%)	Storage period, testing frequency									
I	3 months	Standard tube	40	—	0	2	4	6						weeks
			25	60					12					weeks
	6 months	Standard tube	40	—	0	1	2	3						months
			25	60					6					months
II	12–18	Standard tube	40	—	0	1	2	3	6					months
	months		25	60						12	18			months
														months
		Intended for	40	—	0			3	6					months
		application	25	60						12	18			months
														months
III	24–36	Standard tube	50	—	0	1	2	3						months
	months		40	—		1	2	3	6					months
			25	60						12	18	24	36	months
														months
		Intended for	40	—	0			3	6					months
		application	25	60						12	18	24	36	months

The standard tubes in phases II and III (aluminum tube with inert internal lacquering) are also stored vertically with the neck of the tube pointing upwards. After 1 and 3 months the samples stored at 40°C are subjected to a threefold analysis for homogeneous distribution of the active ingredient (content uniformity) by testing material taken from the beginning, middle, and end of the tube.

3.2.4. Number of Batches

These are as for solid dosage forms.

3.2.5. Selection of Packaging Material for Semisolid Dosage Forms

Suitable tests have to be carried out.

Packing: Aluminum tube internally lacquered, plastic tubes.

Problems: Corrosion of metal tube; interaction with internal lacquering; sorption; permeation of water vapor, oxygen, aromas, and essential oils.

Testing packaging material—dosage form: To test for corrosion, the filled metal tubes are stored horizontally, upright, and inverted at 40°C for 3 months and are then investigated.

To test for permeation and sorption, the filled plastic tubes are stored for 3 months at 40°C.

When selecting the packaging material, the climatic zone in which the product is to be introduced must also be taken into account.

Because of the problems arising with plastic tubes, aluminum tubes are preferred.

If the final packaging material has been selected, the investigations on photostability are performed.

The samples with and without container are irradiated with a Xenon lamp (Suntest 250 W/m^2) for 24 hours.

The test criteria are appearance, drug substance decomposition and assay.

3.2.6. Evaluation

If the results of the test for organoleptic and physicochemical stability are within the shelf life specifications, the stability prediction is only determined by the chemical and microbial stability. Content uniformity within a container and possible interactions with the packaging material must also be considered. Generally, stability predictions are most difficult for semisolid dosage forms, and this applies particularly to the use of reaction kinetics. It may also be necessary to ensure compliance with the minimum shelf life by marking packs with storage instructions.

3.2.7. Stability Information

This is as for solid dosage forms.

3.3. Solutions, Ampoules

3.3.1. Selection of Batches

This is as for solid dosage forms.

3.3.2. Test Criteria

Organoleptic and physicochemical stability: Appearance, clarity, pH.
Chemical and microbial stability: Drug substance decomposition and assay preservative decomposition and assay.

3.3.3. Storage conditions, storage period, testing frequency, packaging material (Tables 20 and 21)

3.3.4. Number of Batches

This is as for solid dosage forms.

3.3.5. Selection of Packaging Material for Liquid Dosage Forms

Packaging: ampoule, injection vial with rubber stopper, glass bottle or plastic bottle with screw closure or pilferproof closure and liner.

Problems: pH, leakage, desorption, sorption, permeation, interaction with rubber stopper, interaction with liner.

Tests packaging material—dosage form: To test for sorption, permeation, pH, and leakage, the final formulation solution is filled in the container, and for desorption placebo solution is used. The samples are stored vertically and inverted

Table 20 Storage Conditions for Chemical and Microbial Stability

Clinical phase	Minimum shelf life	Packaging material	Storage conditions temp. (°C)	Storage conditions rel. hum. (%)	Storage period, testing frequency									
I	3 months	25 ml ground-glass-stoppered glass bottle or glass ampoule or injection vial with rubber stopper, plastic bottle	40	—	0	2	4	6						weeks
			25	60					12					weeks
	6 months	25 ml ground-glass-stoppered glass bottle or glass ampoule or injection vial with rubber stopper, plastic bottle	40	—	0	1	2	3						months
			25	60					6					months
II	12–18 months	25 ml ground-glass-stoppered glass bottle or glass ampoule or injection vial with rubber stopper, plastic bottle.	60	—	0	1	2	3						months
			40	—		1	2	3	6					months
			25	60						12	18			months
			40	—	0			3	6					months
		Intended packaging material	25	60						12	18			months
III	24–36 months	25 ml ground-glass-stoppered glass bottle or glass ampoule or injection vial with rubber stopper, plastic bottle.	70	—	0	1	2	3						months
			60	—		1	2	3						months
			50	—		1	2	3						months
			40	—		1	2	3	6					months
			25	60						12	18	24	36	months
			40	—				3	6					months
			25	60						12	18	24	36	months
		Intended packaging material												

Table 21 Storage Conditions for Organoleptic and Physicochemical Stability

Clinical phase	Packaging material	Storage condition	Storage period
I	Ground-glass-stoppered bottle Glass ampoules	5°C	1 week
II–III	Injection vial with rubber stopper Plastic bottle	≥−10°C	4 weeks

under the following conditions: 30°C/70% r.h., 40°C, and 50°C for up to 12 weeks. Testing intervals: 0, 1, 2, 3 months.

If the final packaging material has been selected, the investigations on photostability are performed.

The samples in colorless glass and the original packaging material are indicated with a Xenon lamp (Suntest 250 W/m^2) for 24 hours.

The test criteria are appearance (colour of solution), clarity of solution, drug substance decomposition and assay.

3.3.6. Evaluation

If the results of the organoleptic and physicochemical tests are within the shelf life tolerance limits, reaction kinetics prediction presents few problems.

The influence of the packaging materials also has to be considered, especially when elastomers are used.

3.3.7. Stability Information

This is as for solid dosage forms.

4. STABILITY INFORMATION FOR COMPARATOR OR REFERENCE PRODUCTS

When an investigational medicinal product is compared with a marketed product, attention should be paid to ensure the integrity and quality of the comparator product (final dosage form, packaging materials, storage conditions, etc.). If significant changes are to be made to the product, data should be available (e.g., stability, comparative dissolution, bioavailability) to prove that these changes do not significantly alter the original quality characteristics of the product.

Because the expiry date stated on the original package has been determined for the medicinal product in that particular package and may not be applicable to the product where it has been repackaged in a different container, it is the responsibility of the sponsor, taking into account the nature of the product, the characteristics of the container, and the storage conditions to which the article may be subjected to determine a suitable use-by date to be placed on the label. Such a date is not later than the expiry date of the original package. In the absence of stability data or if stability is not followed during the clinical trial, such a date should not exceed 25% of the remaining time between the date of repackaging and the expiry date on the original manufacturer's bulk container or a six-month period from the date the drug is repacked, whichever is earlier (7).

According to the EU, GMP guideline stability data are necessary for comparator or reference drug products.

The following cases are differentiated:

The samples are repacked into packaging material that is as tight or tighter concerning moisture and light than the original packaging material. The original shelf life is used.

The samples are repacked into packaging material that is less tight than the original packaging material. Then the samples are tested for moisture sensitivity in the open at 25°C/60% r.h. and for photostability for 24 hours with the Xenon lamp (Suntest). Test criteria: average mass and appearance if no changes take place, the original shelf life is used; if changes take place, tighter or more protecting packaging material must be selected. Then the original shelf life is used.

Samples are reworked (tablets are ground and filled into capsules). There the stability protocol for phase I is applied with the difference that the samples are stored at 25°C/60% r.h. in the intended packaging material up to 18 months for phase II and 36 months for phase III (Table 22).

Table 22 Storage Conditions, Storage Period, and Testing Frequency for Reworked Comparators

Clinical phase	Minimum shelf life	Packaging material	Pretreat-ment	Storage conditions temp. (°C)	Storage conditions rel. hum. (%)	Storage frequency, storage period									
II	12–18 months	Twist-off	None	40	—	0	1	2	3	6					months
		Twist-off	25°C/60%	40	—	0	1	2	3	6					months
		Twist-off	None	25	60						12	18			months
		Intended packaging material	None	25	60						12	18			months
III	24–36 months	Twist-off	None	40	—	0	1	2	3	6					months
		Twist-off	25°C/60%	40	—	0	1	2	3	6					months
		Twist-off	None	25	60						12	18	24	36	months
		Intended packaging material	None	25	60						12	18	24	36	months

Testing specifications: Testing specification for release and stability testing of clinical samples.

5. STABILITY TESTING WITH PIVOTAL AND BIOEQUIVALENCE BATCHES

The stability information for a finished medicinal product is derived mainly from the primary data, i.e., the results obtained from the three registration batches. Usually these are representative pilot plant batches. After marketing authorization, three production batches are added.

Results from the development phase, supporting data, are also included in the application for marketing authorization to underpin the stability information.

The stability results obtained with clinical samples are a major factor for achieving a comprehensive assessment of the quality of a finished medicinal product. In this way it is possible to establish a link between the quality of clinical batches for phases I, II, and III and the quality of the finished drug product. If development is fully covered by stability data, then the quality, efficacy, and safety of the clinical batches will correspond to those of the finished medicinal product. Stability information gained by this broad-based approach thereby acquires a completely new dimension.

Pivotal and bioequivalence batches are also required for this comprehensive general strategy unless they are covered by batches from clinical phase III.

If this is not the case, pivotal and bioequivalence batches are included in the stability program.

Since the results are combined to produce a general statement, emphasis is placed not on the stress test but on the long-term test.

The stability program combines acceleration tests with long-term tests in accordance with the ICH Guideline. The storage period, however, is limited to 18 months (Table 23).

Table 23 Storage Conditions for Pivotal Batches

Storage conditions (°C)	(% rel. hum.)	Storage period and testing frequency (months)					
25	60	0	3	6	9	12	18
30[a]	70[a]		3	6			
40	(75)		3	6			

[a] These conditions are only used if significant change occurs after storage at 40°C(/75%).

The test criteria, specifications, and analytical procedures are the same as those used for batches of clinical phase III and the marketing authorization batches.

6. TIME TO AVAILABILITY OF STABILITY INFORMATION

The stability program for clinical samples is designed to produce stability information as rapidly as possible. The time required until minimum shelf lives and stability information are available is an important factor for planning clinical trials and establishing the data of manufacture.

In phases II and III it we can differentiate between preliminary and final prediction. The preliminary is based only on the data of the stress investigation, whereas the final includes also the data of the samples stored at accelerated condition.

Table 24 gives an overview.

7. REQUIRED CAPACITY

In the course of the strategic planning it is important to estimate the required capacity for the different stress investigations. Since each development is different it will always be a range.

Table 24 Time to Availability of Prediction of Minimum Shelf Life

Stage of development	Storage period		Analysis and evaluation	Stability report	Total period	
	Stress[a]	Accelerated			Preliminary*	Final
Phase I	—	6 weeks	1 week	2 weeks	—	9 weeks
Phase I	—	12 weeks	1 week	2 weeks	—	15 weeks
Phase II	12 weeks	24 weeks	1 week	2 weeks	15 weeks	27 weeks
Phase III	12 weeks	24 weeks	1 week	2 weeks	15 weeks	27 weeks
Registration batches	12 weeks	24 weeks	1 week	2 weeks	—	27 weeks

[a] In clinical phases II and III a preliminary prediction can be made based on the data of samples stored at stress conditions.

In Table 25 estimations for phases I, II, and III are given.

Table 25 Estimated Capacities for the Stress Investigations of Clinical Samples

Test sample batch	Dosage form	Analytical procedures, validation (weeks)	Analysis of test samples no. of dosages			Total required capacity no. of dosages		
			1 (week)	2–4 (weeks)	>4 (weeks)	1 (week)	2–4 (weeks)	>4 (weeks)
Phase I	Solid	6	5	10	15	11	16	21
	Semisolid	3	2	4	6	5	7	9
	Liquid	3	1	2	4	4	5	7
Phase II	Solid	6	9	17	25	15	23	31
	Semisolid	3	4	8	12	7	11	15
	Liquid	3	2	5	7	5	8	10
Phase III, final formulation	Solid	7	16	31	47	23	38	55
	Semisolid	3	6	11	17	9	14	20
	Liquid	3	5	10	15	8	13	18

8. SUMMARY

The stability program for clinical samples as presented in this chapter is based on the principles of the ICH Guideline "Stability Testing for New Drug Substances and Drug Products."

However, these principles have been adapted to suit the complex circumstances arising during ongoing development, as exemplified by the transition from clinical Phase I to III.

Storage conditions, storage periods, and the derived minimum shelf lives correspond to the duration of clinical trials in Phases I to III.

Shelf lives are established on the basis of stress and acceleration tests. Only with this approach can shelf lives be established rationally and all batches provided with an open expiration date. By consistently separating the storage conditions for organoleptic, physicochemical, and chemical-microbial test criteria, all stability-indicating test criteria can be integrated in the stability information.

The number of analyses can be reduced by bracketing if phases I and II are performed using several dosages.

Stability predictions based on stress and acceleration tests are supported by long-term tests conducted under the storage conditions representing climatic zone II, i.e., 25°C/60% r.h. The packaging material planned for commercial use is always included.

The analytical procedures and the specifications to be derived from the results also undergo a process of development.

For example, at the outset the validation has a preliminary character and includes specificity, linearity, recovery, and limit of quantitation, whereas on completion there is the completely validated specification for clinical samples and stability testing. In the same way, the specifications initially serve as a general guide that becomes increasingly specific. This flexible approach makes it possible to obtain reliable stability information while ensuring the rational use of resources.

Not carrying out stress and acceleration tests and replacing them by long-term tests would either cause major delays in clinical development or make it impossible to state an open expiration date.

An alternative would be to include all batches in stability testing, running the serious risk of having to replace batches during the clinical trial and, in addition to the great analytical effort involved, of always having to keep up-to-date stable batches available in order to safeguard the continuity of the clinical trial.

Summarizing, it can be stated that the systematic approach of proceeding in logically coordinated steps represents the best way of supporting the clinical trial by stability testing.

Furthermore, stability testing of clinical samples is a central factor for generating comprehensive stability information, i.e., an overall assessment of the quality of the finished medicinal product.

By applying the same principles to the stability testing of clinical samples and the finished medicinal product, the marketed drug, it is ensured that the results of the clinical trial can be considered applicable to the finished medicinal product; both products have similar stability and therefore quality.

REFERENCES

1. PMA's Joint-PDS Stability Committee "Stability Concepts." Pharmaceutical Technology, 1984.
2. Grimm W, Krummen K. Stability Testing in the EC, Japan and the USA. Wissenschaftliche Verlagsgesellschaft mbH, Stuttgart, 1993.
3. Futscher N, Schumacher P. Pharm. Ind. 34:479–483, 1972.
4. Dietz R, Feilner K, Gerst F, Grimm W. Drugs Made in Germany 36:99–103, 1993.
5. USP 23/NF 18 General Information 1151.
6. Guidance for Industry, Stability Testing of Drug Substances and Drug Products. Draft Guidance, June 1998. F.D.A.
7. Revised version of Annex 14, manufacture of investigational medicinal products of the EU-guide to GMP.
8. Grimm W. Drug Dev. Ind. Pharm. 24:313–325, 1996.

13

A Rational Approach to Stability Testing and Analytical Development for NCE, Drug Substance, and Drug Products: Marketed Product Stability Testing

WOLFGANG GRIMM

Biberach, Germany

1. INTRODUCTION: THE STRATEGIC PLANNING

Big efforts are necessary to reduce the period of time from the start of development for a new chemical entity, drug substance, or drug product to registration application and finally marketing authorization.

The results of the analytical development and the stability testing fro a NCE form an important part of a registration application.

To reduce the period of time and to provide information to assess variation and changes, a procedure was developed for analytical development and stability testing: the strategic planning.

By a strategic planning it is possible to secure a successful marketing authorization in the shortest period of time and in the most efficient way. The strategy is based on the ICH Harmonized Tripartite Guidelines (1).

Stability Testing of New Substances and Products	Q1A
Photostability Testing of New Substances and Products	Q1B
Text on Validation of Analytical Procedures	Q2A
Extension of the ICH Text "Validation of Analytical Procedures"	Q2B
Impurities in New Drug Substances	Q3A
Impurities in New Drug Products	Q3B
Rcsidual Solvents	Q3C
Specifications, Test Procedures and Acceptance Criteria for New Drug substances and New Drug Products: Chemical Substances	Q6A

It is also based on the Extension of the ICH Tripartite Guideline for worldwide marketing (2,6,7).

Thereby the strategy considers all aspects of analytical development and stability testing for a New Chemical Entity which are necessary for a registration application in the EU, Japan and the USA, and worldwide.

The overall development and stability program for the strategic planning has been divided into six decisive steps (3); see Table 1.

Furthermore, eleven basic principles have been established (4) that are decisive for stability testing and that are applicable to all stages of development, on all dosage forms.

The ICH tripartite stability guideline had taken over these principles and is likewise built upon them.

The 11 principles are as follows:

Selection of batches and samples

Table 1 Overall Development and Stability Program

Step	Analytical development and stability testing
1	Stress and accelerated testing with the drug substance
2	Preformulation and formulation finding for
	Toxicological samples
	Clinical trial samples
	Final dosage form
3	Stress and accelerated testing with selected formulations:
	Toxicological samples
	Clinical trial samples
	Final formulation
	Registration batches
	Selection of packaging material up-scaling, pilot-plant
4	Accelerated and long-term stability testing with registration batches up to registration application:
	Drug substance
	Drug product
	Transfer of analytical procedures to quality control
5	On-going stability testing with registration and production batches
	Drug substance
	Drug product
6	Follow-up stability testing
	Continuous production
	Variations and changes during continuous production

Test criteria
Analytical procedures
Specifications
Storage conditions
Storage period
Testing frequency
Number of batches
Packaging material
Evaluation
Stability information, statements

These principles apply to each of the six steps. At the same time they also pass through a process of development and have differing degrees of importance. Some of these principles, such as selection of samples and batches, number of batches, storage conditions, testing frequency, and storage periods are firmly established, while others, such as test criteria, analytical procedures with validation, and specifications undergo further development.

Combining the six steps and the eleven basic principles yields the systematically structured stability schedule as shown in Tables 2–4.

The big advantage of this systematic procedure is that the resulting data are in a logical relationship and can be put together like a puzzle. Therefore the information on the drug substance and drug product widens constantly.

Table 2 Systematically Structured Stability Schedule (Part 1)

	Step 1	Step 2	Step 3	Step 4	Step 5	Step 6
Basic principles	Stress and accelerated testing with the drug substance	Preformulation and formulation finding	Stress and accelerated testing with selected formulations, up-scaling	Accelerated and long-term stability testing up to registration application	On-going stability testing	Follow-up stability testing
Selection of batches and samples	Experimental batch	Experimental laboratory batches	Experimental and clinical batches, representative pilot-plant batches	Representative pilot plant batches	Representative batches	Representative and experimental batches
Test criteria	Corresponding to objective	Corresponding to objective	Corresponding to objective	Corresponding to results of step 1–3	As for step 4	As for step 4
Analytical procedures	Stability indicating, preliminary validation	Stability indicating, orientational validation	Stability indicating, orientational and preliminary validation	Stability indicating, completely validated	As for step 4	Stability indicating completely validated or revalidation
Specifications	Orientational specifications	Orientational specifications	Preliminary	Release and shelf life specifications proposed for registration	Post approval, release and shelf life specifications	Post approval, release and shelf life specifications
Storage conditions	25°C/60% 25°C/75% 30°C/70% 40°C/75% 40°C, 50°C, 60°C, 70°C	5°C 25°C/60% 30°C/70% 40°C/75% 40°C, 60°C	−10°C 5°C 25°C/60% 30°C/70% 40°C/75% 40°C, 50°C, 60°C 70°C	25°C/60% 30°C/70% 40°C/75%	25°C/60% 30°C/70% 40°C/75%	25°C/60% 30°C/70% 40°C/75%

Table 3 Systematically Structured Stability Schedule (Part 2)

	Step 1	Step 2	Step 3	Step 4	Step 5	Step 6
Basic principles	Stress and accelerated testing with the drug substance	Preformulation and formulation finding	Stress and accelerated testing with selected formulations, up-scaling	Accelerated and long-term stability testing up to registration application	On-going stability testing	Follow-up stability testing
Testing frequency	0, 1, 2, 3 months	Depending on problem	Depending on problem 0, 1, 2, 3, (6) months	0, 3, 6, 9, 12, (18) months	(18), 24, 36, 48 60 months 0, 3, 6, 9, 12, 18, 24, 36, 48, 60 months	0, 12, 24, 36, 48 60 months 0, 1, 2, 3, 6 months
Storage period	Up to 3 months	Depending on problem	Up to 3 months Up to 6 months	Up to 12 or 18 months	Up to 60 months	Up to 60 months Up to 6 months
Number of batches	1	Depending on problem	Depending on problem 1	3	3 3	1 per year 1–3
Packaging material	Open, standard packaging material	Depending on problem Standard packaging material	Depending on problem Standard packaging material	Commercial packaging material	Commercial packaging material	Commercial packaging material Standard packaging material

Table 4 Systematically Scheduled Stability Schedule (Part 3)

	Step 1	Step 2	Step 3	Step 4	Step 5	Step 6
Basic principles	Stress and accelerated testing with the drug substance	Preformulation and formulation finding	Stress and accelerated testing with selected formulations, up-scaling	Accelerated and Long-term stability testing up to registration application	On-going stability testing	Follow-up stability testing
Evaluation	Statistics reaction kinetics Stability report as drug substance stability profile Preliminary testing specification for stability testing of drug substance	Statistics reaction kinetic Research reports Orientational testing specifications	Statistics reaction kinetics Stability report and preliminary testing specification for release and stability testing of toxicological samples clinical trial samples final dosage form	Statistics reaction kinetics Stability report drug substance Testing specification for stability testing of drug substance Stability report drug product Testing specification for release and stability testing of drug product	Statistics Stability report	Statistics reaction kinetics Stability report
Stability information	Re-Test Period Storage instructions Selected test criteria for long-term testing Orientational shelf life predictions for drug products	Formulation selection	Period of use toxicological samples clinical samples phase I–III Shelf life prediction final dosage form Storage instructions Selection of packaging material Selection of test criteria for long-term testing	Shelf life prediction Storage instructions In-use stability Holding time for intermediates and bulk	Confirmation and extension of shelf life	Confirmation of shelf life Assessment of variations

The data are summarized in research reports and stability reports, testing specifications.

The stability reports of the different steps of development are built up and written in the same format. Thus all data can be cross-checked easily, and the final shelf life can be based on all these data.

By presenting all the available stability data in such a comprehensive way, considerable savings can be reached concerning different strengths, different packaging materials, and later with variations.

The derived stability information is based on a broad set of data and assures the quality of the drug product to the patient.

Table 5 gives an overview of the different documents that result during development.

The analytical procedures, the specification and the corresponding testing specifications are developed systematically in steps 1–4.

In step 4 they are transferred to quality control, which elaborates on this basis the testing specifications for quality control.

Therefore the registration application contains two types of testing specifications:

Those that have been applied during development for release of clinical trial samples and for stability testing and will be applied for on-going stability testing

Those that will be applied for quality control of running production and follow-up stability testing

The analytical procedures are usually not changed after the transfer into quality controls, but the format may be changed to consider the requirements of different countries.

After an overview of the required capacity (Table 6) and period for analytical statements (Table 7), each step is described in detail, practical examples are given, the required capacity calculated, and the period for analytical statements indicated.

2. Step 1: Stress and Accelerated Testing with the Drug Substance

2.1. Objective

Elucidation of the intrinsic characteristics of the drug substance with reference to chemical properties (physical properties are investigated separately)

Establishment of the degradation pathway, leading to identification of degradation products and hence supporting the suitability of the proposed analytical procedure

Investigation of the following influencing factors: moisture, temperature, moisture + temperature, moisture + temperature + drug substance concentration, pH, ionic strength, oxidation, light

The tests with the drug substance are of general nature and are not specific to any particular dosage form. Consequently the results are generally applicable. These investigations are required by the ICH Stability Guideline. The performance, however, is up to the applicant's discretion.

Table 5 Overview of the Documents that Result During Development

Step of development	Document	Needed for	ICH Guidelines
1 Stress and accelerated testing with the drug substance	Stability report as preliminary stability profile of drug substance	Predevelopment	
	Preliminary testing specification for stability testing of drug substance	CMC, Part II: IND, CTX, RA	Q2A, Q2B
	Stability report as stability profile of drug substance	CMC, Part II: IND, CTX, RA	Q1A, Q1B
2 Preformulation and formulation finding for toxicological samples clinical trial samples final dosage form	Orientational testing specification for toxicological samples clinical trial samples final dosage form		
	Research report	Development pharmaceutics,	(Q1A)
	Formulation evaluation	Development report	
	Research report	Development pharmaceutics,	Q6A
	Rationale for analytical procedure dissolution rate	CMC: NDA	
3 Stress and accelerated testing with selected formulation: toxicological samples clinical samples phase I–III final dosage form, representative batches	Stability report and preliminary testing specification for release and stability testing of		
	toxicological samples	GLP	
	clinical samples phase I	CMC, Part II: IND, CTX	(Q1A)
	clinical samples phase II	CMC, Part II: IND, CTX, RA	(Q1A)
	clinical samples phase III	CMC, Part II: IND, RA	(Q1A)
	final dosage form, representative batches	CMC, Part II: RA	(Q1A)
	Evaluation of packaging material	(Development pharmaceutics)	(Q1A)

RA: Registration application ≙ NDA.

Step of development	Document	Needed for	ICH Guidelines
Cleaning validation	Preliminary testing specification for cleaning validation	GMP	
Scaling-up, pilot plant optimization	Preliminary testing specification for release and stability testing	GMP	
validation	Testing specification for intermediate (IPC)	GMP (RA)	Q6A
	Research report scaling-up	GMP, Development pharmaceutics	
4 Accelerated and long-term Testing with registration batches up to registration application:	Testing specification for stability testing of drug substance	CMC, Part II: RA	Q1A, Q2A, Q2B, Q3A
drug substance	Rationale testing specification drug substance	CMC, Part II: RA	Q6A, Q3A
drug product		CMC, Part II: RA	Q1A, Q1B
	Stability report of drug substance	CMC, Part II: RA	Q1A, Q2A, Q2B, Q3B, Q6A
	Testing specification for release and stability testing of drug product	CMC, Part II: RA	Q2A, Q2B
	Validation report, drug substance, drug product	GMP: (RA)	Q2A, Q2B
	Rationale testing specification drug product	CMC, Part II: RA	Q6A, Q3B
	Stability report of drug product	CMC, Part II: RA	Q1A, Q1B
5 On-going stability testing			
drug substance registration batches production batches	Stability report drug substance	CMC, Part II: RA	Q1A
drug product registration batches production batches	Stability report drug product	CMC, Part II: RA	Q1A
6 Follow-up stability testing			
during continuous production	Stability report	GMP	
variation and changes	Stability report	Regulatory authorities	

RA: Registration application ≙ NDA.

Table 6 Summarized Required Capacity in Steps 1–5

Step of development	Stage of development	Required capacity (weeks)	Required capacity (weeks × 1.3)[a]	Capacity per step/total capacity (weeks)	Capacity per step/total capacity (weeks × 1.3)[a]	Capacity per step/total capacity (years ≙ 42 weeks)[b]
1	Stress and accelerated testing with drug substance					
	Preliminary stability profile	8	10			
	Stability profile	9	12	17	22	0.53
2	Preformulation and formulation finding	23	30	23	30	0.71
3	Stress and accelerated testing					
	clinical phase I	24	31			
	clinical phase II	24	31			
	clinical phase III	32	42			
	final formulation	18	23			
	cleaning validation	5	7			
	scaling-up	23	30	126	164	3.90
4	Accelerated and long-term testing					
	drug substance	9	12			
	drug product	17	22			
	PAI preparation	6	8	32	42	1.00
5	On-going stability testing					
	drug substance	20	26			
	drug product	42	55	62	81	1.93
Total 1–4				198	257	6.1
Total 1–5				260	338	8.1

[a] To calculate the actual time span, the remaining work that is indirectly related with the NCEs has to be considered, such as SOPs, qualifications, literature. Therefore 0.75% of a week is used corresponding to the factor of 1.3.
[b] One year is calculated with 210 days ≙ 42 weeks.

Table 7 Stability Testing on the Critical Path

Step of development	Stability information	Needed for	Time to availability from start
1	Preliminary stability profile of drug substance	Start of predevelopment	6 weeks
	Stability profile of drug substance	Base for minimum shelf life phase I	16 weeks
3	Minimum shelf life toxicological samples	Release of toxicological samples	8 weeks
	Minimum shelf life clinical samples phase I	Release of clinical trial batch phase I	9, 15 weeks
	Minimum shelf life clinical samples phase II	Release of clinical trial batch phase II	15, 27 weeks
	Minimum shelf life clinical samples phase III	Release of clinical trial batch phase III	15, 27 weeks
4	Stability report for registration application	Filing data for registration application	(Release of 3d batch for accelerated and long-term testing) 15 months

The drug substance to be investigated should have been selected as the most suitable salt form.

2.2. Application of the Basic Principles

Selection of batches and samples: experimental batch; it must comply with the acceptance limits of the preliminary testing specification, as far as they are available. The impurity profile, particle size distribution, and the surface area are especially important.

Test criteria: appearance, physical properties, assay, decomposition.

Analytical procedure:

Specific for stability testing

Orientational validation at the beginning specificity, linearity, quantitation limit $\geq 0.05\%$ of the drug substance.

Preliminary validation at the end in addition: accuracy, range, repeatability, robustness for drug substance and decomposition products

Specifications: at the end: preliminary release specifications.

Storage conditions:

In open containers: 25°C/60%, 25°C/75%, 30°C/70%, 40°C/75%.

In standard packaging material: 40°C, 50°C, 60°C, 70°C.

Storage period: Until equilibrium is reached at open storage, < 3 months.

Testing frequency: ≤ 4.

Number of batches per investigation: 1.

Packaging material: flask with ground-glass stopper, glass container with twist-off closure, glass container lined with polyethylene foil.
Evaluation: statistics and reaction kinetics are applied in evaluating the results. Assessment of observed decomposition products, whether they may be formed under accelerated and long-term testing. Establishment of degradation pathway of selected decomposition products; elucidation of their structure. Assessment of applied analytical procedures. The data and the derived stability information are summarized in a stability report, as the stability profile of the investigated drug substance. The stability report is part of CMC for CTX, IND and for registration application.
Stability information: the stability report contains the following stability information:
Stability prediction drug substance
Preliminary prediction of the retest period
Storage instructions if required
Test criteria, packaging material, assessed analytical procedure and stability test protocol for accelerated and long-term testing of registration batches
Stability predictions drug product
Orientational prediction of the chemical stability of the drug substance in solid, semisolid, liquid dosage forms

2.3. Practical Examples

The influencing factors to be investigated are now illustrated by means of a practical example.

To start the predevelopment as soon as possible, the investigations are carried out in two steps:

Preliminary investigation
Complete investigation

2.3.1. Preliminary Investigation

The experiments are organized so that the preliminary stability profile is available within 6 weeks.

Necessary amount: about 25 g.

2.3.1.1. Moisture

Sample: Drug substance stored in open container for 1 week at 25°C/75% r.h
Test criteria: Appearance, mass, DTA. If the drug substance has absorbed water it is investigated further in 50 mL glass container with twist-off closure under 2.3.1.2

2.3.1.2. Temperature, Moisture

Sample: Drug substance with and without adsorbed water in 50 mL glass container with twist-off closure
Test criteria: Appearance, decomposition, assay, DTA at the end
Analytical procedure: Orientational validation at the beginning
Storage temperature: 70°C

Storage period: 4 weeks
Testing frequency: 0, 2, 4 weeks

2.3.1.3. pH

Sample: 1% aqueous solution or slurry in 0.1, 0.01 M HCl, in McIlvaine's buffers (0.1 M citric acid, 0.2 M disodium phosphate), pH 3, 4, 5, 6, 7, 0.01 M NaOH in volumetric flask with ground glass stopper
Test criteria: Appearance, decomposition, assay
Storage temperature: 60°C
Storage period: 3 weeks
Testing frequency: 0, 1, 3 weeks

2.3.1.4. Oxidation

Sample: 1% aqueous solution or slurry in 0.3% H_2O_2 solution in 25 mL glass flask with ground glass stopper
Test criteria: Appearance, pH, decomposition, assay
Storage temperature: 50°C
Storage period: 3 weeks
Testing frequency: 0, 1, 3 weeks

2.3.1.5. Photostability (Xenon Lamp, Atlas Suntest, 250 W/m²)

Test sample: Drug substance spread in colorless and brown glass across the container to give a thickness of not more than 3 mm
1% aqueous solution (or inert organic solvent) with and without N_2 gasing in colorless glass flask with ground glass stopper
1% aqueous solution (organic inert solvent) in brown glass flask with ground glass stopper as control
Test criteria: Appearance, decomposition, assay
Storage period: 48 hrs, xenon lamp
Testing frequency: 24, 48 hrs

The investigations are summarized and listed in the stability test protocol, preliminary investigations including the applied analytical procedures (Table 8).

If the drug substance decomposes fast (>10% after the first test point), storage at lower temperature should be considered.

If the drug substance is not wettable or does not dissolve at all, the stability information will be limited. The drug substance may appear very stable because it did not dissolve at all.

In these cases the procedure has to be modified accordingly to reach wettability and increase the solubility.

If an enantiomer is present, testing for racemization is performed during the course of stability studies.

The results are summarized in the Stability Report as Preliminary Stability Profile of Drug Substance.

2.3.2. Complete Investigation

The stability protocol based on the results of the preliminary investigations has to be adjusted accordingly.

Necessary amount: about 50 g.

Table 8 Stability Test Protocol, Preliminary Investigations

Batch No.	Influencing factor	Test sample	Packaging material	Storage conditions	Storage times [weeks]	Anal. procedures
S991	Moisture	Pure drug substance	Open container	25°C/75% r.h	0, 1	No. X
	Temperature	Pure drug substance	50 mL glass container with twist-off closure	70°C	0, 2, 4	No. X
	Temperature + moisture	Pure drug substance with absorbed water at 25°C/75%	50 mL glass container with twist-off closure	70°C	0, 2, 4	No. X
	pH	1% aqueous solution pH 1, 2, 3, 4, 5, 6, 7, 8 0.1, 0.01 M HCl, McIlvaine's buffer (0.1 M citric acid, 0.2 M dibasic sodium-phosphate, 0.01 M NaOH	25 mL glass flask ground glass stopper	60°C	0, 1, 3	No. X
	Oxidation	1% aqueous solution in 0.3% H_2O_2 solution	25 mL glass flask ground glass stopper	50°C	0, 1, 3	No. X
	Light	Pure drug substance	Open petri dish	xenon lamp	24, 48 hours	No. X
			Brown glass flask	xenon lamp	24, 48 hours	No. X
		1% aqueous solution gassed with N_2	Colorless glass flask with ground glass stopper	xenon lamp	24, 48 hours	No. X
	Light	1% aqueous solution	Colourless glass flask with ground glass stopper	xenon lamp	24, 48 hours	No. X
			Brown glass flask with ground glass stopper	xenon lamp	24, 48 hours	No. X

Table 9 Required Capacity

Required capacity for a single analysis	Required capacity for an analysis in stress tests (serial factor 20%)
5 h ≙ 0.19 weeks	4 h ≙ 0.13 weeks

Analytical procedures, orientational validation	Analysis of test samples documentation	Total required capacity
2 weeks	4 weeks	6 weeks

Table 10 Time to Availability of Preliminary Stability Report

Storage period of sample	Evaluation of data Stability report	Total time period
4 weeks	2 weeks	6 weeks

2.3.2.1. Moisture

Sample: Drug substance stored in open containers until equilibrium is reached at 25°C/60%, 30°C/70%, 40°C/75%. The sample with the highest water adsorption is investigated further in 50 mL glass container with twist-off closure under 2.3.2.2

Test criteria: Appearance, mass, DTA

2.3.2.2. Temperature, Moisture

Sample: Drug substance with and without moisture in 50 mL glass container with twist-off closure

Drug substance in 50 mL glass container lined with polyethylene foil* with twist-off closure (70°C only)

Test criteria: Appearance, clarity (70°C only), decomposition, assay, DTA of 70°C 12 weeks samples

Storage temperatures: 40°C, 50°C, 60°C, 70°C

Storage period: 12 weeks

Testing frequency: 0, 4, 8, 12 weeks

* The polyethylene foil is selected which will be applied for storage of bulk drug substance in corresponding containers.

2.3.2.3. Temperature, Moisture, Drug Substance Concentration

Sample: 1% and 5% aqueous solution or aqueous suspension or slurry (only 1%) in 25 mL glass flask with ground glass stopper

Test criteria: Appearance, pH, decomposition, assay

Storage temperatures: 50°C, 70°C

Storage period: 12 weeks

Testing frequency: 0, 4, 8, 12 weeks

2.3.2.4. pH and Buffer Concentration

Sample: At the optimum pH obtained from preliminary investigation, 1% aqueous solution or slurry in McIlvaine's buffer, double buffer concentration (0.2 M citric acid, 0.4 M disodium phosphate) in glass flask with ground glass stopper
Test criteria: Appearance, decomposition, assay
Storage temperature: 60°C
Storage period: 4 weeks
Testing frequency: 0, 2, 4 weeks

The investigations are summarized and listed in the stability test protocol, complete investigations (Table 11).

Evaluation:

Degradation products are assessed whether they are formed during storage, shipment or accelerated and long-term testing. Thereby reaction kinetic calculation's and regression analysis are applied if scientifically justified.

The structures of the selected degradation products are elucidated and the degradation pathway established. The applied analytical procedures are assessed.

The test criteria, the packaging material, and the stability test protocol are established for accelerated and long-term testing with the registration batches.

All results, conclusions, and the derived stability information are summarized in the stability report "xy drug substance active ingredient stability profile."

Stability information: the stability report contains the following stability information:

Stability predictions drug substance:

Preliminary retest period
Storage instructions
Test criteria, assessed analytical procedures, packaging materials, stability test protocol for accelerated and long-term testing.

Stability prediction for drug products: Orientational stability prediction for drug substance in drug products.

The preliminary data are incorporated in the stability report as the stability profile of the drug substance.

2.3.3. Confirmation Testing for the Derived Preliminary Retest Period

The results of the stress investigations are finally confirmed by the data of the registration batches. To bridge the time in between, confirmation testing is performed under long-term storage condition of climatic zone II. These data support the retest period for the reference samples and batches of drug substance applied during development and clinical trial investigation with the drug product (Table 14).

Table 11 Stability Test Protocol, Complete Investigations

Batch No.	Influencing factor	Test sample	Packaging material	Storage conditions	Storage times [weeks]	Anal. procedures
S992	Moisture	Pure drug substance	Open petri dish	25°C/60% r.h 30°C/70% r.h 40°C/75% r.h	0, 2	No. X
	Temperature	Pure drug substance	50 mL glass container with twist-off closure	70°C 60°C 50°C	0, 4, 8, 12 0, 4, 8, 12 4, 8, 12	No. X
			50 mL glass container lined with polyethylene foil and twist-off closure	70°C	12	No. X
	Temperature + moisture	Pure drug substance with adsorbed water	50 mL glass container with twist-off closure	70°C 60°C 50°C	0, 4, 8, 12 4, 8, 12 4, 8, 12	No. X
	Temperature + moisture + drug substance concentration	1% and 5% aqueous solution	25 mL glass flask with ground glass stopper	70°C 50°C	0, 4, 8, 12 4, 8, 12	No. X
	Ionic strength	1% aqueous solution pH 5, (0.1 M citric acid, 0.2 M dibasic sodium phosphate)	25 mL glass flask with ground glass stopper	60°C	0, 4, 8, 12	No. X
		1% aqueous solution pH 5, double buffer concentration, (0.2 M citric acid, 0.4 M dibasic sodium phosphate)	25 mL glass flask with ground glass stopper	60°C	0, 4, 8, 12	No. X

Table 12 Required Capacity

Required capacity for a single analysis and documentation	Required capacity for an analysis in stress tests (serial factor 20%) and documentation
5 hrs ≙ 0.19 weeks	4 hrs ≙ 0.14 weeks

Analytical procedures evaluation, orientational and preliminary validation	Analysis of test samples	Total required capacity
3 weeks	5 weeks	8 weeks

Table 13 Time to Availability of Stability Information

Storage period of sample (months)	Evaluation Stability report (months)	Total time period (months)
3	1	4

Table 14 Stability Test protocol

Batch	Packaging material	Storage condition	Storage period testing frequency	Test criteria
Reference sample, laboratory or pilot plant batch	Selected from data of stress investigation	25°C/60%	0, 6, 12, 18, 24 months	Selected from data of stress investigation

The retest period of the reference sample is fixed to 6 months until the data of the confirmation testing are available.

3. Step 2: Preformulation and Formulation Finding for the Toxicological and Clinical Samples, Final Dosage Form

The tests are based on the drug substance stability profile. Therefore it is a prerequisite for step 2.

This stage covers the process of development from the first preformulation tests to the toxicological and clinical samples and the potential final formulation for the desired dosage form. Consequently, it includes the actual pharmaceutical–technological development.

Since a different dosage form is frequently used for the clinical samples than is planned for later market introduction, e.g., capsules for clinical trials, tablets for the market, a double-track approach is necessary in some cases. From the analytical viewpoint, this gives rise to the following objective:

3.1. Objective

Dosage form related tests with the drug substance
Specifications for individual excipients
Investigations of external factors that have an influence on individual excipients
Compatibility tests with the drug substance
Compatibility tests with preliminary formulations
Clarification of technological influencing factors
Selection of formulations for toxicological samples
Development of the analytical procedure for dissolution rate testing
Selection of formulations for clinical samples
Selection of formulation and optimization of the potentially final formulation

Before establishing the schedule of tests and before the practical implementation, all the results already available are evaluated. If this procedure is consistently followed, a broad spectrum of evaluable data, especially for frequently used excipients, will soon be available.

The requirements placed on the different dosage forms are highly variable. Furthermore, an individual approach is adopted for each specific case, which is shaped by the corporate philosophy as well as the skill and experience of the developer.

A schedule of tests can thus only be sketched in very general terms:

3.2. Application of the Basic Principles

Selection of batches and samples: drug substance, excipients, active ingredient–excipient mixtures, dosage forms the drug substance and excipient batches used must comply with the specifications of the quality control testing specifications.
Test criteria: appearance, relevant physicochemical parameters, drug substance decomposition and assay.
Analytical procedures: specific to stability testing orientationally validated. Methods for physicochemical determinations are optimized and preliminarily validated, e.g., dissolution rate.
Specifications: not relevant; tests are performed to identify changes.
Storage conditions: standard storage conditions for stress tests and long-term tests, relevant for the dosage form.
≥−10°C, 5°C
5–40°C temperature cycle
25°C/60%
30°C/70%, 40°C/75%, 40°C, 60°C

It is essential not to use any other storage conditions, so that these results can be built upon in the subsequent development stages.

Storage period: Problem-oriented.
e.g. until equilibrium is reached for open storage
Not more than 3 months.

Testing frequency: Individually determined, depending on the stability behavior.

Evaluation and stability information:

- Orientational testing specification.
- Dosage form related drug substance profile.
- Parameters and stability results of individual excipients.
- Research report of selected formulations for clinical samples and potential final formulations.
- Research report of rational for analytical procedure dissolution rate.
- These reports are part of the development pharmaceutics or a development report.

3.3. Practical Examples

The tests necessary at this stage can only be carried out in close cooperation between the pharmaceutical-technological and the analytical laboratory. The tests also depend on

- The active ingredient, its solubility and stability
- The dosage form to be developed
- The experience and skill of the developer. Only a few points can be given as examples.

3.3.1. Dosage Form Related Active Ingredient Profile

This concerns all the solubility problems of an active ingredient under consideration:

- Dissolution profile with respect to pH, particle size, surface area
- Tests to improve the apparent solubility or dissolution rate

3.3.2. Stability Behavior of Excipients

- Moisture sorption tests
- Sensitivity to oxygen
- Sensitivity to light
- Sensitivity to pH

3.3.3. Selection of Formulation

- Compatibility tests active ingredient—excipients, preferably preliminary formulations
- Planning of tests, factorial design
- Technological influencing factors
- Optimization of pH for liquid, semisolid and, if appropriate, solid dosage forms
- Evaluation of shelf life specifications for antimicrobial preservatives
- Oxygen and sensitivity to light-discoloration
- Content uniformity
- Phase separation with semisolid forms
- Solubility problems with semisolid and liquid dosage forms
- Optimized in vitro procedure for determination of dissolution rate

Table 15 Required Capacity

Analytical procedures evaluation, orientational validation	Analysis of test samples	Total required capacity
3 weeks	20 weeks	23 weeks

The capacity can be given only as a rough estimation because it depends on a high degree on the dosage form and the characteristics of the drug substance (see Table 15).

If different formulations have to be investigated, the 23 weeks have to be multiplied correspondingly.

A time to availability of stability information cannot be given at this stage of development.

4. Step 3: Stress and Accelerated Testing with Selected Formulations, Selection of Packaging Material, Up-scaling Pilot Plant, Registration Batches

These tests follow directly upon step 2. This is the main area of focus in development analysis.

It is necessary to establish whether the formulation selected in screening in step 2 is not only stable in relative terms but also absolutely stable enough for toxicological investigations, clinical trials, and possible commercialization.

4.1. Objective

Optimization and validation of analytical procedures
Selecting those degradation products that may be formed under accelerated and long-term testing within the anticipated shelf life
Establishment of the degradation pathway and elucidating the structure of the selected degradation products
Identification of the weaknesses of the formulation and parameters that may have a limiting effect on the anticipated shelf life
Identification of problems that could arise during storage and especially during transport
Establishment of tolerances for changes during storage
Evaluation of the robustness of the formulation
Selection of suitable packaging materials
Establishment of minimum shelf lives (period of use) and if necessary storage instructions for
  toxicological samples
  clinical trial samples for phases I–III.
Selection of test criteria for accelerated and long-term testing with the registration batches in step 4
Establishment of the stability test protocol for the registration batches in step 4

Establishment of registration release and shelf life specifications for the registration batches in step 4
Establishment of the anticipated shelf lives for the stability investigations in step 4 and 5

The stress tests must be thought out and planned very carefully. The individual stability protocol depends on the stage of development, the dosage form, the number of strengths, and the anticipated shelf life.

Two aspects however must be considered for all stress investigations:

Storage conditions. If a stability prediction of all test criteria is to be derived, a distinction must be drawn between storage for
- Organoleptic and physicochemical stability, where the laws of reaction kinetics do not apply.
- Chemical (drug substance, preservatives) stability, where the laws of reaction kinetics are applicable.

Tight containers to prevent loss of moisture.

4.2. Application of the Basic Principles

The drug product is in the process of development with different formulations, strengths, dosage forms, batch sizes, and equipment.

Therefore a variety of different batches is included in these stress investigations.

A prerequisite for all batches is that the used drug substances and excipients have been released by quality control.

Selection of batches and samples: the following batches may be included:
- Experimental laboratory batches
- Toxicological samples
- Experimental clinical batches
- Clinical or pilot plant batches
- Final formulation batches
- Representative registration batches

Test criteria: the relevant test criteria will become apparent during the course of development. Included are the four groups of test criteria:
- Organoleptic
- Physicochemical
- Chemical
- Microbial

Overview on general test criteria for solid, semisolid, and liquid dosage forms used in stability testing.

Solid dosage forms: tablets, capsules:
- Organoleptic and physicochemical stability
  - Tablets: Appearance, average mass, water content, disintegration time, dissolution rate, hardness (resistance to crushing strength)
  - Capsules: Appearance, elasticity, average mass, average mass of filing, water content of capsule shell and filling, disintegration time, dissolution rate

Chemical stability
Tablets: Drug substance: decomposition and assay
Capsules: Drug substance: decomposition and assay.
Semisolid dosage forms: Creams and ointments
Organoleptic and physicochemical stability
Appearance, odor, homogeneity, consistency, pH, particle size (if active ingredient is in suspension), recrystallization, content uniformity within container (tubes stored vertically are cut open, samples are taken from the beginning, middle and end of the tube and analyzed)
Chemical and microbial stability
Drug substance decomposition and assay, preservative assay
Liquid dosage forms: solutions, ampoules
Organoleptic and physicochemical stability
Appearance, colour of solution, clarity, pH.
Chemical and microbial stability
Drug substance decomposition and assay, preservative assay.
Analytical procedures: the analytical procedures have to be stability indicating. They undergo a process of development. The same applies for the validation.
Three steps of validation are differentiated (Table 16):
Orientational
Preliminary
Complete

Table 16 The Extent of Validation for the Three Steps

	Extent of validation		
Validation characteristic	orientational	preliminary	complete
Specificity	x	x	x
Linearity	x	x	x
Quantitation limit	x	x	x
Detection limit[a]	x	x	x
Accuracy		x	x
Range		x	x
Repeatability		x	
Intermediate precision			x
Robustness	x	x	x
Validation report			x

[a] Instead of quantitation limit for semiquantitative procedures.

Specifications: fixing specifications is an evolving process which accompanies the development of the new drug substances and drug products. Thereby it has always to be differentiated between
Release specifications
Shelf life specifications

Fixing specifications can be described as a four-step procedure as listed in Table 17.

Table 17 Procedure for Fixing Specifications

Step of development drug substance, drug product	Specifications	Characterization
Preclinical Clinical phase I	Orientational	Target values
Clinical phases II/III Pivotal batches	Preliminary	Broader acceptance criteria, ranges, numerical limits
Pilot plant batches Registration batches	Registration	Acceptance criteria focussing on safety and efficacy
Production batches after marketing authorization	Post approval	Experience gained with manufacture of a particular drug substance or drug product

Storage conditions, storage period: it has to be differentiated between stress, accelerated, and long-term storage conditions, Table 18.

Table 18 Various Storage Conditions

Type	Storage condition
Stress	Open storage at 25°C/60%, 30°C/70% r.h., 40°C/75% r.h. ≥ -10°C 5°C Temperature cycle 5–40°C Xenon lamp 48 hours 40, 50, 60, 70°C
Accelerated	40°C/75% r.h. (30°C/70% r.h.)
Long-term	25°C/60% r.h. 30°C/70% r.h.

If a stability prediction for all test criteria is to be derived, a distinction must be drawn between storage for

Organoleptic and physicochemical stability where the laws of reaction kinetics do not apply

Chemical stability (drug substance, preservatives) where the laws of reaction kinetics may be applicable

Table 19 Overview of Storage Conditions and Storage Period for Solid, Semisolid, and Liquid Dosage Forms

Stability investigation	Dosage form	Storage condition	Storage period
Organoleptic and physicochemical stability	Solid	Storage in open container until equilibrium is reached at 25°C/60%, 30°C/70%, 40°C/75%	1–2 weeks
	Semisolid	5°C	4 weeks
		≥ −10°C	4 weeks
		5°C–40°C temperature cycle within 24 hrs	2 weeks
		40°C (content uniformity)	3 months
	Liquid	5°C	4 weeks
		≥ −10°C	4 weeks
Photostability	All	Xenon lamp (Atlas Suntest, 250 W/m^2)	48 h
Chemical stability	Solid	40°C, 50°C, 60°C, 70°C	3 months
	Semisolid	30°C, 40°C, 50°C	3 months
	Liquid	40°C, 50°C, 60°C, 70°C	3 months

Testing frequency: Individually determined, depending on the problem and stability behavior. When testing for chemical and micobial stability ≤4 determinations including initial analysis.

Number of batches: basically one batch per formulation. If several strengths have to be investigated for clinical phase I or II, bracketing is applied, the expedient mentioned in the ICH Stability Guideline.

It is also used even if the composition of the individual strengths may differ.

A rational bracketing system for all dosage forms would be as in Table 20.

Table 20 Bracketing System for Dosage Forms

Dosages	Samples tested
1–2	All
3–4	Highest Lowest
>4	Highest Middle Lowest

Examples:

If minimum shelf lives are required for 10, 20, 40, 60, 80, 120 mg, then 10, 40, 120 mg are tested.

If 3–4 strengths have to be investigated, the two extremes, the highest and lowest, are fully tested. The most probable final strength, however, will be a middle strength. Therefore also a middle strength will be included with a reduced stability protocol, e.g., chemical stress testing only at 60°C.

Packaging material: in selecting packaging material, the following has to be considered:

At higher temperatures, desorption and loss of moisture also occurs at higher relative humidities.

Unless packaging materials impermeable to water vapor are used for stress tests with solid dosage forms, the samples lose moisture at different rates in the temperature range 40–60°C and the results are not suitable for a reaction kinetics calculation.

Packaging materials permeable to water vapor can however also result in a falsification of the results for semisolid and liquid dosage forms if varying degrees of weight loss occur that lead to differences in the active ingredient concentration or ion strength.

The use of inert standard packaging materials that are impermeable to water vapor is thus an important precondition for stress tests that are to be evaluated in terms of reaction kinetics, and on the results on which stability predictions are to be based.

Most of the stress tests are carried out in standard packaging material.

The following standard packaging materials are used:

Solid dosage forms: 50-mL glass container with twist-off closure
Polypropylene tube

Semisolid dosage forms: Standard tube
Small volumetric flask
Aluminum tube, inert internal lacquering

Liquid dosage forms: 25 mL volumetric flask with ground-glass stopper

However, further investigations for the selection of the final packaging are necessary.

Selection of packaging material for solid dosage forms.

On the basis of the results of the stress tests for solid dosage forms, the sensitivity to moisture can be determined and suitable packaging materials can be selected.

As a rule, no interactions are to be expected.

If the final packaging material has been selected and samples packed in the final packaging material are available, the investigation of photostability should be performed.

Photostability: The samples with and without container are irradiated with a Xenon lamp (Atlas Suntest, 250 W/m^2) for 24 hours.

Test criteria: appearance, drug substance decomposition and assay.

Selection of packaging material for semisolid dosage forms.

Suitable tests have to be carried out.

Packaging: Aluminum tube internally lacquered, plastic tubes.

Problems: Corrosion of metal tube; interaction with internal lacquering; sorption; permeation of water vapor, oxygen, aromas, essential oils.

Tests packaging material—dosage form: To test for corrosion, the filled metal tubes are stored horizontally, upright, and inverted at 40°C for 3 months and are then investigated.

To test for permeation and sorption, the filled plastic tubes are stored for 3 months at 50°C, 40°C, 30°C/70%.

When selecting the packaging material, the climatic zone in which the product is to be introduced must also be taken into account.

Because of the problems arising with plastic tubes, aluminum tubes are preferred.

If the final packaging material has been selected, the investigations on the photostability are performed.

Photostability: The samples with and without container are irradiated with a Xenon lamp (Atlas Suntest, 250 W/m^2) for 24 hours.

Test criteria: appearance, drug substance decomposition and assay.

Selection of packaging material for liquid dosage forms.

Packaging: ampoule, injection vial with rubber stopper, glass bottle or plastic bottle with screw closure or pilferproof closure and liner.

Problems: pH, leakage, desorption, sorption, permeation, interaction with rubber stopper, interaction with liner.

Tests packaging material—dosage form: To test for sorption, permeation, pH, and leakage, the final formulation solution is filled in the container, and for desorption placebo solution is used. The samples are stored vertically and inverted under the following conditions: 50°C, 40°C, 30°C/70% for up to 12 weeks. Testing intervals: 0, 1, 2, 3 months.

If the final packaging material has been selected the investigations on the photostability are performed.

Photostability: The samples in coloress glass and the original packaging material are irradiated with a Xenon lamp (Atlas Suntest, 250 W/m^2) for 24 hours.

Test criteria: appearance (colour of solution), clarity of solution, drug substance decomposition and assay.

Evaluation: a systematic approach is adopted in the evaluation and presentation of the analytical results.

Thereby all test criteria are included.

Investigations for organoleptic and physicochemical stability.

Where possible the data are evaluated for significant changes with the aid of statistics.

Investigations for chemical and microbial stability.

If decomposition and fall in assay has taken place, the equations for a first-order reaction and the Arrhenius equation are used. If decomposition levels are available for only one temperature, the activation energy ΔE: 83 KJ $\times$ mol^{-1} (8) is applied.

The decomposition levels for the required storage temperatures are calculated.

Organoleptic and physicochemical changes that have taken place at stress temperatures at 40–70°C are recorded, but they are only of limited relevance for predicting stability.

Stability information: all results and the stability information derived therefrom are compiled in a stability report. The different stability reports are structured and written in the same format.

They contain

Summary
Material and methods
Results and evaluation
Conclusions

The stability information corresponds to the stage of development (Table 21).

Table 21 Correspondence of Stage of Development with Stability Information

Stage of development	Stability information
General	Assessment of the formulation General prediction of shelf lives Storage instructions if required Proposal for packaging materials
Toxicological samples	Minimum shelf life (period of use) for toxicological samples in climatic zone I and II
Clinical phase I	Minimum shelf life (period of use) for phase I in climatic zone I and II Storage instructions if required Proposal of suitable packaging materials Robustness of the formulation
Clinical phase II	Minimum shelf life (period of use) for phase II in climatic zone I and II Storage instructions if required Proposal of suitable packaging materials Robustness of the formulation
Clinical phase III	Minimum shelf life (period of use) for phase III in climatic zone I and II Storage instructions if required Proposal of suitable packaging materials Robustness of the formulation Selection of those decomposition products that may occur under accelerated and long-term testing of the registration batches in step 4 Elucidated structure and degradation pathway of the selected decomposition products Establishment of test criteria for accelerated and long-term testing of the registration batches in step 4 Registration release and shelf life specifications for registration batches in step 4 Establishment of storage conditions for accelerated and long-term testing in steps 4 and 5 Suitable packaging materials for the registration batches in step 4 Expected shelf lives for the registration batches in steps 4 and 5 and storage instructions if necessary Establishment of stability test protocol for registration batches
Registration batch	Comparison and evaluation of the results of the stress investigation with laboratory, clinical trial, and registration batches Robustness of the formulation

4.3. Practical Examples

4.3.1. General

4.3.1.1. Solid Dosage Forms

Stability test protocols (Tables 22, 23, 24)

Table 22 Organoleptic and Physicochemical Stress Testing

Packaging material	Storage conditions	Storage period testing frequency	Test criteria
Open container	25°C/60%	0, 2 weeks	organoleptic, physicochemical
Open container	30°C/70%	0, 2 weeks	organoleptic, physicochemical
Open container	40°C/75%	0, 2 weeks	organoleptic, physicochemical

Table 23 Photostability Stress Testing

Packaging material	Storage conditions	Storage period testing frequency	Test criteria
Open container	Xenon lamp Atlas Suntest 250 W/m^2	24, 48 hours	Appearance, drug substance decomposition and assay

Table 24 Chemical Stress Testing

Packaging material	Pretreatment	Storage conditions	Storage period testing frequency	Test criteria
50 ml glass container with twist-off closure or equivalent tight container	none	70°C	1, 2, 3 months	chemical
		60°C	1, 2, 3 months	chemical
		50°C	0, 1, 2, 3 months	all
		40°C	1, 2, 3 months	all
	30°C/70%[a] or 25°C/60%[a] or 40°C/75%[a]	70°C	1, 2, 3 months	chemical
		60°C	1, 2, 3 months	chemical
		50°C	0, 1, 2, 3 months	all
		40°C	1, 2, 3 months	all

[a] Samples that have adsorbed the highest amount of water during open storage at 25°C/60%, 30°C/70%, 40°C/75% are used.

4.3.1.2. Semisolid Dosage Forms

Stability test protocols (Tables 25, 26, 27, 28)

4.3.1.3. Liquid Dosage Forms

Stability test protocols (Tables 29, 30, 31, 32)

Table 25 Organoleptic and Physicochemical Stress Testing

Packaging material	Storage conditions	Storage period testing frequency	Test criteria
Standard tube	≥ −10°C	0, 4 weeks	organoleptic, physicochemical
	5–40°C in 24 hrs cycle	0, 2 weeks	organoleptic, physicochemical
	40°C, tube stored vertically, with the neck of the tube pointing upwards	0, 3 months	content uniformity (homogeneous distribution of active ingredient, assay of material taken from the beginning, middle, end of the tube)

Table 26 Photostability Stress Testing

Packaging material	Storage condition	Storage period testing frequency	Test criteria
Open container	Xenon lamp Atlas Suntest 250 W/m^2	24, 48 h	Appearance, drug substance decomposition and assay

Table 27 Chemical and Microbial Stress Testing

Packaging material	Storage condition	Storage period testing frequency	Test criteria
Standard tube	50°C	1, 2, 3 months	Organoleptic, chemical, microbial
	40°C	0, 1, 2, 3 months	all
	30°C	3 months	all

Table 28 Selection of Packaging Material

Proposed packaging material	Composition	Storage condition	Storage period testing frequency	Test criteria
Metal tube	verum	40°C stored horizontally, upright, and inverted	0, 3 months	Appearance, drug substance decomposition, appearance of internal surface of metal tube for corrosion, interaction with internal lacquering
Plastic tube	placebo	50°C	0, 1, 2, 3 months	Loss in mass (water permeation), aromas, essential oils, desorption
		40°C	1, 2, 3 months	essential oils, desorption
		30°C/70%	3 months	essential oils, desorption
	verum	40°C	0, 3 months	Organoleptic, loss in mass, drug substance decomposition and assay

Table 29 Organoleptic and Physicochemical Stress Testing

Packaging material	Storage conditions	Storage period testing frequency	Test criteria
25 mL ground-glass stoppered glass bottle or equivalent	5°C	0, 4 weeks	organoleptic, physicochemical
	≥ −10°C	0, 4 weeks	organoleptic, physicochemical

Table 30 Photostability Stress Testing

Packaging material	Storage condition	Storage period testing frequency	Test criteria
25 mL ground-glass stoppered colorless glass bottle	Xenon lamp Atlas Suntest 250 W/m^2	24, 48 hrs	Organoleptic, physico-chemical, drug substance decomposition and assay

Table 31 Chemical Stress Testing

Packaging material	Storage condition	Storage period testing frequency	Test criteria
25 mL ground-glass stoppered glass bottle	70°C	0, 1, 2, 3 months	all
	60°C	1, 2, 3 months	all
	50°C	1, 2, 3 months	all
	40°C	1, 2, 3 months	all

4.3.2. Toxicological Samples

The anticipated minimum shelf life of 12 weeks is derived from the data of accelerated testing and confirmed by storage at 25°C/60% r.h. (climatic zone II).

Stability test protocol (Table 33)

Table 32 Accelerated and Confirmation Testing

Batch	Packaging material	Storage condition	Storage period testing frequency	Test criteria
Experimental laboratory batch	Standard	40°C/75% r.h.	0, 2, 4, 6 weeks	all
		25°C/60% r.h.	12 weeks	all

Table 33 Selection of Packaging Material

Intended packaging material	Composition	Storage condition	Storage period testing frequency	Test criteria
Glass bottle with screw closure or pilferproof with liner	placebo	50°C upright and inverted	0, 3 months	Desorption of liner components
	verum	50°C, 40°C upright and inverted	0, 3 months	all
Plastic bottle with closure	placebo	50°C 40°C 30°C/70% upright and inverted	0, 1, 2, 3 months 1, 2, 3 months 3 months	Appearance, loss in mass, desorption of plastic components
	verum	50°C 40°C	0, 2, 3 months 2, 3 months	all
Ampoule	verum	50°C 40°C	0, 3 months 3 months	all
Injection vial with rubber stopper	placebo	50°C 40°C stored inverted	0, 3 months 3 months	Organoleptic, physicochemical, desorption of rubber components
	verum	50°C 40°C stored inverted	0, 3 months 3 months	all

4.3.3. Clinical Trial Samples, Phase I–III (5)

4.3.3.1. Solid Dosage Forms

4.3.3.1.1. PHASE I. The anticipated minimum shelf life (period of use) is 3 to 6 months. The minimum shelf life is derived from data of stress and accelerated testing after 6 weeks or 3 months, respectively, and then confirmed by the data of samples stored at 25°C/60% r.h. (climatic zone II) up to 3 or 6 months.

Stability test protocols (Tables 34, 35)

Table 34 Organoleptic and Physicochemical Stress Testing

Batch	Packaging material	Storage condition	Storage period testing frequency	Test criteria
Experimental laboratory batch	open container	25°C/60% r.h.	0, 2 weeks	organoleptic, physicochemical

Table 35 Chemical Accelerated and Confirmation Testing

Batch	Packaging material	Minimum shelf life	Pre-treatment	Storage condition	Storage period testing frequency	Test criteria
Experimental laboratory batch	50 mL glass container with twist-off closure	3 months	none	40°C	0, 2, 4, 6 weeks	all
	,,		25°C/60% r.h.	40°C	0, 2, 4, 6 weeks	all
	,,		none	25°C/60%	3 months	all
	PP tubes		none	25°C/60%	3 months	all
	50 mL glass container with twist-off closure	6 months	none	40°C	0, 1, 2, 3 months	all
	,,		25°C/60% r.h.	40°C	0, 1, 2, 3 months	all
	,,		none	25°C/60% r.h.	6 months	all
	PP tubes		none	25°C/60% r.h.	6 months	all

If several strengths are to be investigated, bracketing is applied.

4.3.3.1.2. PHASE II. The anticipated minimum shelf life (period of use) is 12–18 months. The minimum shelf life is derived from the data of stress testing after 3 months, firstly confirmed by 6 months data of 40°C and finally by 12 or 18 months data of samples stored at 25°C/60% r.h. for climatic zone II.

Stability test protocols (Tables 36, 37)

Table 36 Organoleptic and Physicochemical Stress Testing

Batch	Packaging material	Storage condition	Storage period testing frequency	Test criteria
Experimental clinical batch	open container	25°C/60% r.h.	0, 2 weeks	organoleptic, physicochemical

Table 37 Chemical Stress and Confirmation Testing

Batch	Packaging material	Pre-treatment	Storage condition	Storage period testing frequency	Test criteria
Experimental clinical batches	50 mL glass container with twist-off closure	none	60°C	0, 1, 2, 3 months	all
	,,	none	40°C	1, 2, 3, 6 months	all
	,,	25°C/60%	60°C	0, 1, 2, 3 months	all
	,,	25°C/60%	40°C	1, 2, 3, 6 months	all
	,,	none	25°C/60%	12, 18 months	all
	PP tubes or test packaging material	none	25°C/60%	12, 18 months	all

If several strengths are to be investigated, bracketing is applied.

4.3.3.1.3. PHASE III. The anticipated minimum shelf life (period of use) is 24–36 months. The minimum shelf life is derived from the data of stress testing after 3 months, firstly confirmed by 6 months data of 40°C and finally by 24 or 36 months data of samples stored at 25°C/60% (climatic zone II). Usually in phase III the final formulation is applied. Therefore the stability information for the registration batches has also to be considered.

Stability test protocols (Tables 38, 39, 40)

Table 38 Organoleptic and Physicochemical Stress Testing

Batch	Packaging material	Storage condition	Storage period testing frequency	Test criteria
Clinical or pilot plant, final formulation	open container	25°C/60%	0, 2 weeks	organoleptic, physico-chemical
		30°C/70%	0, 2 weeks	,,
		40°C/70%	0, 2 weeks	,,

Table 39 Photostability Stress Testing

Batch	Packaging material	Storage condition	Storage period testing frequency	Test criteria
Clinical or pilot plant, final formulation	open container proposed test or final packaging material for marketing	Xenon lamp Atlas Suntest 250 W/m^2	24, 48 hrs	Appearance, drug substance decomposition and assay

Table 40 Chemical Stress and Confirmation Testing

Batch	Packaging material	Pre-treatment	Storage condition	Storage period testing frequency	Test criteria
Clinical or pilot plant batch	50 ml glass container with twist-off closure	none	70°C	1, 2, 3 months	chemical
			60°C	1, 2, 3 months	chemical
			50°C	0, 1, 2, 3 months	all
			40°C	1, 2, 3, 6 months	all
		30°C/70%[a] or 25°C/60% or 40°C/75%	70°C	1, 2, 3 months	chemical
			60°C	1, 2, 3 months	chemical
			50°C	0, 1, 2, 3 months	all
			40°C	1, 2, 3, 6 months	all
	50 ml glass container with twist-off closure	none	25°C/60%	12, 18, 24, 36 months	all
	PP tube or test packaging material	none	25°C/60%	12, 18, 24, 36 months	all

[a] The samples that have adsorbed the highest amount of water during open storage at 25°C/60%, 30°C/70%, 40°C/75% are tested.

If more than one strength is included in phase III, full testing is performed with all strengths. On the base of these data it may then be possible to apply bracketing or matrixing with the registration batches in step 4.

4.3.3.2. Semisolid Dosage Forms

4.3.3.2.1. PHASE I. The anticipated minimum shelf life (period of use) is 3–6 months. The minimum shelf life is derived from the data of stress and accelerated testing after 6 weeks or 3 months, respectively, and then confirmed by the data of samples stored at 25°C/60% (climatic zone II) up to 3 or 6 months.

Stability test protocols (Tables 41, 42)

Table 41 Organoleptic and Physicochemical Stress Testing

Batch	Packaging material	Storage condition	Storage period testing frequency	Test criteria
Experimental laboratory batch	Standard tube	5°C	0, 4 weeks	organoleptic, physicochemical

Table 42 Chemical and Microbial Accelerated and Confirmation Testing

Batch	Packaging material	Minimum shelf life	Storage condition	Storage period testing frequency	Test criteria
Experimental laboratory batch	Standard tube	3 months	40°C	0, 2, 4, 6 weeks	all
			25°C/60%	3 months	all
		6 months	40°C	0, 1, 2, 3 months	all
			25°C/60%	6 months	all

4.3.3.2.2. PHASE II. The anticipated minimum shelf life (period of use) is 12–18 months. The minimum shelf life is derived from data of stress testing after 3 months, firstly confirmed by 6 months data of 40°C and finally by 12 or 18 months data of samples stored at 25°C/60% (climatic zone II).

Stability test protocols (Tables 43, 44)

Table 43 Organoleptic and Physicochemical Stress Testing

Batch	Packaging material	Storage condition	Storage period testing frequency	Test criteria
Experimental clinical batch	Standard tube	5°C	0, 4 weeks	organoleptic, physicochemical

Table 44 Chemical and Microbial Stress and Confirmation Testing

Batch	Packaging material	Storage condition	Storage period testing frequency	Test criteria
Experimental clinical batch	Standard tube	50°C	1, 2, 3 months	Chemical and microbial
	Standard tube	40°C	0, 1, 2, 3, 6 months	all
	Standard tube	25°C/60%	12, 18 months	all
	Proposed packaging material	25°C/60%	12, 18 months	all

If several strengths are to be investigated in phase I or II, bracketing is applied.

4.3.3.2.3. PHASE III. The anticipated minimum shelf life (period of use) is 24–36 months. The minimum shelf life is derived from data of stress testing after 3 months, firstly confirmed by 6 months data of 40°C and finally by 24 or 36 months data of samples stored at 25°C/60% (climatic zone II). Usually in phase III the final formulation is applied. Therefore the stability information for the registration batches has to be considered.

Stability test protocols (Tables 45, 46, 47)

Table 45 Organoleptic and Physicochemical Stress Testing

Batch	Packaging material	Storage condition	Storage period testing frequency	Test criteria
Clinical or pilot plant batch, final formulation	Standard tube	5°C	0, 4 weeks	organoleptic, physicochemical
		≥ −10°C	0, 4 weeks	organoleptic, physicochemical
		5–40°C in 24 h cycle	0, 2 weeks	organoleptic, physicochemical
		40°C, tube stored vertically, with the neck of the tube pointing upwards	0, 1, 3 months	Content uniformity within the tube. Assay of material taken from the beginning, middle, end of the tube

Table 46 Photostability Stress Testing

Batch	Packaging material	Storage condition	Storage period testing frequency	Test criteria
Clinical or pilot plant batch, final formulation	Open container Proposed test or final packaging material for marketing	Xenon lamp Atlas Suntest 250 W/m^2	24, 48 h	Appearance drug substance decomposition and assay, preservative assay

Table 47 Chemical and Microbial Stress and Confirmation Testing

Batch	Packaging material	Storage condition	Storage period testing frequency	Test criteria
Clinical or pilot plant batch, final formulation	Standard tube	50°C	1, 2, 3 months	Appearance, chemical, microbial
		40°C	0, 1, 2, 3, 6 months	all
		30°C	6, 12 months	all
	Standard tube	25°C/60%	12, 18, 24, 36 months	all
	Proposed test or final packaging material for marketing	25°C/60%	12, 18, 24, 36 months	all

4.3.3.3. Liquid Dosage Forms

4.3.3.3.1. PHASE I. The anticipated minimum shelf life (period of use) is 3–6 months. The minimum shelf life is derived from the data of stress and accelerated testing after 6 weeks or 3 months, respectively, and then confirmed by the data of samples stored at 25°C/60% (climatic zone II) up to 3 or 6 months.

Stability test protocols (Tables 48, 49)

Table 48 Organoleptic and Physicochemical Stress Testing

Batch	Packaging material	Storage condition	Storage period testing frequency	Test criteria
Experimental laboratory batch	25 mL ground-glass stoppered glass bottle	5°C	0, 4 weeks	organoleptic, physicochemical

Table 49 Chemical, Microbial Accelerated and Confirmation Testing

Batch	Packaging material	Minimum shelf life	Storage condition	Storage period testing frequency	Test criteria
Experimental laboratory batch	25 mL ground-glass stoppered glass bottle	3 months	40°C	0, 2, 4, 6 weeks	all
	25 mL ground-glass stoppered glass bottle		25°C/60%	3 months	all
	Proposed test packaging material		25°C/60%	3 months	all
	25 mL ground-glass stoppered glass bottle	6 months	40°C	0, 1, 2, 3 months	all
	25 mL ground-glass stoppered glass bottle		25°C/60%	6 months	all
	Proposed test packaging material		25°C/60%	6 months	all

4.3.3.2. *Phase II*

The anticipated minimum shelf life (period of use) is 12–18 months. The minimum shelf life is derived from data of stress testing after 3 months, firstly confirmed by 6 months data of 40°C and finally by 12 or 18 months data of samples stored at 25°C/60% (climatic zone II).

Stability test protocols (Tables 50, 51)

Table 50 Organoleptic and Physicochemical Stress Testing

Batch	Packaging material	Storage condition	Storage period testing frequency	Test criteria
Experimental clinical batch	25 mL ground-glass stoppered glass bottle	5°C	0, 4 weeks	organoleptic, physicochemical

Table 51 Chemical, Microbial Stress and Confirmation Testing

Batch	Packaging material	Storage condition	Storage period testing frequency	Test criteria
Experimental clinical batch	25 mL ground-glass stoppered glass bottle	60°C	0, 1, 2, 3 months	all
		40°C	1, 2, 3, 6 months	all
	25 mL ground-glass stoppered glass bottle	25°C/60%	12, 18 months	all
	Proposed test packaging material	25°C/60%	12, 18 months	all

If several strengths are to be investigated in phase I or II bracketing is applied.

4.3.3.3. *Phase III*

The anticipated minimum shelf life (period of use) is 24–36 months. The minimum shelf life is derived from data of stress testing after 3 months, firstly confirmed by 6 months data of 40°C and finally by 24 or 36 months data of samples stored at 25°C/60% (climatic zone II). Usually in phase III the final formulation is applied. Therefore the stability information for the registration batches has to be considered.

Stability test protocols (Tables 52, 53, 54)

Table 52 Organoleptic and Physicochemical Stress Testing

Batch	Packaging material	Storage condition	Storage period testing frequency	Test criteria
Clinical or pilot plant batches, final formulation	25 mL ground-glass stoppered glass bottle	≥ -10°C	0, 4 weeks	organoleptic, physico-chemical

Table 53 Photostability Stress Testing

Batch	Packaging material	Storage condition	Storage period testing frequency	Test criteria
Clinical or pilot plant batch, final formulation	25 mL ground-glass stoppered colorless glass bottle Proposed test or final packaging material for marketing	Xenon lamp Atlas Suntest 250 W/m^2	24, 48 h	Organoleptic, physicochemical, drug substance decomposition and assay

Table 54 Chemical, Microbial Stress and Confirmation Testing

Batch	Packaging material	Storage condition	Storage period testing frequency	Test criteria
Clinical or pilot plant batch, final formulation	25 mL ground-glass stoppered glass bottle	70°C	0, 1, 2, 3 months	all
		60°C	1, 2, 3 months	all
		50°C	1, 2, 3 months	all
		40°C	1, 2, 3, 6 months	all
	25 mL ground-glass stoppered glass bottle	25°C/60%	12, 18, 24, 36 months	all
	Proposed test or final packaging material for marketing	25°C/60%	12, 18, 24, 36 months	all

4.3.4. Comparator or Reference Drug Products

According to the newest EU GMP Guideline (9) stability data are also necessary for comparator or reference drug products.

The following cases are differentiated:

When the samples are repacked into packaging material that is as tight or tighter concerning moisture and light than the original packaging material, the original shelf life is used.

The samples are repacked into packaging material that is less tight than the original packaging material. Then the samples are tested for moisture sensitivity in the open at 25°C/60% and for photostability for 24 hours with the Xenon lamp (Suntest). Test criteria: average mass and appearance.

If no changes take place the original shelf life is valid.

If changes take place, tighter or more protecting packaging material must be selected. Then the original shelf life is used.

If samples are reworked (tablets are ground and filled into capsules), stability protocol for phase I is applied with the difference that the samples are stored at 25°C/60% in the intended packaging material up to 18 months for phase II and 36 months for phase III; Table 55.

Testing specifications: Testing specification for release and stability testing of clinical samples.

Table 55 Stability Protocol for Reworked Comparator Drug Products

Clinical phase	Minimum shelf life	Packaging material	Pretreatment	Storage conditions Temp. (°C)	rel. hum. (%)	Storage frequency Storage period
II	12–18 months	Twist-off	none	40	—	0 1 2 3 6 months
		Twist-off	25°C/60%	40	—	0 1 2 3 6 months
		Twist-off	none	25	60	12 18 months
		Proposed packaging material	none	25	60	12 18 months
III	24–36 months	Twist-off	none	40	—	0 1 2 3 6 months
		Twist-off	25°C/60%	40	—	0 1 2 3 6 months
		Twist-off	none	25	60	12 18 24 36 months
		Proposed packaging material	none	25	60	12 18 24 36 months

If stability data are not available and it is not intended to consider or perform further stability investigations during the clinical trial, then the minimum shelf life cannot be longer than 25% of the remaining shelf life of the comparator or maximally 6 months whichever is shorter.

4.3.5. Registration Batches

The stability information derived for a registration application should be based on the results of preclinical development, the clinical batches and the registration batches. Thereby it can be assured that the patient after marketing authorization gets the same quality as the patient during the clinical investigation.

Therefore it is important not only to confirm the predicted quality by the data of the accelerated and long-term testing with the registration batches but to compare stress data directly. Consequently stress tests are performed also with registration batches.

Usually the first registration batch is tested.

The extent depends on the available stress data, but mostly a reduced stability protocol is sufficient.

4.3.5.1. Solid Dosage Forms

Stability test protocols (Tables 56, 57)

Table 56 Organoleptic and Physicochemical Stress Testing

Batch	Packaging material	Storage condition	Storage period testing frequency	Test criteria
Registration batch	open container	fixed from available data where highest amount was adsorbed	0, 2 weeks	organoleptic, physicochemical

Table 57 Chemical Stress Testing

Batch	Packaging material	Pretreatment	Storage condition	Storage period testing frequency	Test criteria
Registration batch	50 mL glass container with twist-off closure	none	70°C 50°C	1, 2, 3 months 0, 1, 2, 3 months	chemical all
		30°C/70% or 25°C/60% or 40°C/75%	70°C 50°C	1, 2, 3 months 0, 1, 2, 3 months	chemical all

4.3.5.2. *Semisolid Dosage Forms*

Stability test protocols (Tables 58, 59)

Table 58 Organoleptic and Physicochemical Stress Testing

Batch	Packaging material	Storage condition	Storage period testing frequency	Test criteria
Registration batch	Proposed final packaging material for marketing	≥ -10°C	0, 4 weeks	organoleptic, physicochemical
		5–40°C in 24 h cycle	0, 2 weeks	organoleptic, physicochemical

Table 59 Chemical and Microbial Stress Testing

Batch	Packaging material	Storage condition	Storage period testing frequency	Test criteria
Registration batch	Proposed packaging material for marketing	50°C	0, 1, 2, 3 months	appearance, chemical and microbial

4.3.5.3. *Liquid Dosage Forms*

Stability test protocols (Tables 60, 61)

Table 60 Organoleptic and Physicochemical Stress Testing

Batch	Packaging material	Storage condition	Storage period testing frequency	Test criteria
Registration batch	Proposed packaging material for marketing	≥ -10°C	0, 4 weeks	organoleptic, physicochemical

Table 61 Chemical Stress Testing

Batch	Packaging material	Storage condition	Storage period testing frequency	Test criteria
Registration batch	Proposed packaging material for marketing[a]	70°C	0, 1, 2, 3 months	all
		50°C	1, 2, 3 months	all

[a] If plastic bottle, a 25 mL ground-glass stoppered glass bottle is used for chemical stress testing.

Evaluation

The results of the stress tests are evaluated carefully. Where scientifically justified, statistics, reaction kinetics, and linear regression analysis are applied.

The analytical procedures are assessed, and then the data are evaluated as follows:

Organoleptic properties
Physicochemical properties
Chemical (and microbial) properties
Packaging material properties
Robustness of the formulation

The results of the stress investigations are also assessed relating to the robustness of the formulation.

During the development the formulation was challenged by the following factors:

Different batches of drug substance and excipients
Different strengths
Different compositions
Different manufacturing processes with different types and sizes of equipment
Different sites of manufacture
Scale of manufacture, laboratory, pilot plant
Batch size

The resulting batches manufactured with all these different influencing factors have been investigated in stress and confirmation testing. Therefore considerable information is available on the possible influence of these factors on the stability of the drug product, and the robustness of the formulation can be evaluated.

These data are an important base to assess later during running production the influence of variations and changes on the stability.

Finally it is decided whether it is necessary to ensure compliance with the minimum shelf life that has been established by providing storage instructions to be placed on the packaging material.

All the results, the information on the tested batches, the applied analytical procedures, and the derived stability information, are compiled in stability reports.

All the stability reports are structured in the same format.
The following stability reports are written:

Stress Testing and Long-Term Testing with the drug product phase I
Stress Testing and Long-Term Testing with the drug product phase II
Stress Testing and Long-Term Testing with the drug product phase III including registration batches.

The stability information of these three reports is then summarized in the

Stability profile of the drug product.

Stability information

The following information is available:

Assessment of the analytical procedures optimized and validated during development of the drug product
Establishment of the degradation pathway, under the influence of the excipients, of degradation products that may be formed under accelerated and long-term testing
Establishment of shelf lives. Minimum shelf lives that have been derived from the results of stress investigations and confirmed by the data of long-term testing: clinical phase I: 3–6 months; clinical phase II: 12–18 months; clinical phase III: 24–36 months. Expected preliminary shelf life for the registration batches in step 4: e.g., 24 months. Anticipated shelf life for the drug product in steps 4 and 5: e.g., 5 years
Storage instructions: clinical phase I–III; registration batches in step 4
Identification of problems that could arise during storage and especially during transport.
Selection of suitable packaging materials for clinical trial batches; for registration batches for all relevant climatic zones in step 4
Selection of test criteria for the accelerated and long-term testing with the registration batches in step 4
Establishment of specifications for tolerable changes during storage, registration shelf life specifications
Establishment of stability test protocol for the registration batches
Robustness of the formulation

Table 62 Required Capacity

Dosage form	Required capacity for a single analysis and documentation	Required capacity for an analysis in stress tests (−15% serial factor) and documentation
Solid	17 h ≙ 0.45 weeks	15 h ≙ 0.40 weeks
Semisolid	12 h ≙ 0.32 weeks	10 h ≙ 0.27 weeks
Liquid	8 h ≙ 0.21 weeks	7 h ≙ 0.19 weeks
Clinical trial for solid dosage forms	20 h ≙ 0.53 weeks	—

Table 63 Time to Availability of Prediction of Minimum Shelf Life (Weeks)

Stage of development	Storage period		Analysis and evaluation	Stability report	Total period	
	Stress[a]	Accelerated			Preliminary[a]	Final
Toxicological samples	—	6	1	1	—	8
Phase I	—	6	1	2	—	9
Phase I	—	12	1	2	—	15
Phase II	12	24	1	2	15	27
Phase III	12	24	1	2	15	27
Registration batches	12	24	1	2	—	27

[a]In the clinical phases II and III a preliminary prediction can be made based on the data of samples stored at stress conditions.

5. Step 4: Accelerated and Long-Term Testing with Registration Batches up to Registration Application for Drug Substance and Drug Products

Step 4 is the central part of the stability testing.

The samples are usually stored under conditions equivalent to the climatic conditions of the corresponding climatic zone, and thus binding stability statements can be derived from the results. These form an important part of licensing or registration documents.

5.1. Objective

Confirmation of the results of stress and accelerated testing.
Derivation of re-test-periods for the drug substance.
Determination of the influence of batch size on the stability.
Derivation of holding time for the bulk drug product or intermediate stages.
Derivation of use life, overage, if necessary.
Complete validation of the analytical procedures.
Testing specification for release and stability testing.
Derivation of the shelf lives for the final formulation of the drug product.

5.1.1. Drug Substance

5.1.1.1. Application of the Basic Principles

Selection of batches and samples: representative batch for registration application (registration batch).

The batches of a minimum of pilot plant scale should be by the same synthetic route and use a method of manufacture and procedure that simulates the final process to be used on a manufacturing scale.

The overall quality of the batches of drug substance placed on stability should be representative of both the quality of the material used in pre-clinical and clinical studies and the quality of material to be made on a manufacturing scale.

Most important are the impurity profile, the particle size and particle size distribution, possibly surface area.

Test criteria: appearance, physical criteria, decomposition and assay
Analytical procedures: specific for stability testing, completely validated: Specificity, linearity, quantitation limit, accuracy, range, intermediate precision, robustness.
Specifications: release specifications
Storage conditions: according to the four climatic zones, the following storage conditions were established for these zones:

Climatic zone	Storage conditions
I and II	25°C/60% r.h.
III and IV	30°C/70% r.h.

Which of these are used depends on where and how the drug substance will be stored, shipped and used. 30°C/70% storage only if the drug substance is applied in climatic zone III or IV.
Testing frequency: first year every 3 months, second year every 6 months, then annually.
Number of batches: 3.

If the information provided in the Registration Application does not contain data of production batches, information should be provided at a later stage on three early production batches manufactured for marketing.
The packaging material should be the same as or simulate the actual material used for storage and distribution.

Small fiber drum lined with polyethylene foil for drug substance not sensitive to moisture
Glass bottle lined with polyethylene foil with screw cap or twist-off for drug substance sensitive to moisture to simulate tight containers.

Stability test protocols: see Tables 64, 65.

Table 64 Climatic Zone II ≙ ICH

Batch	Packaging material	Storage condition	Storage period, up to registration	Testing frequency (months) on-going	Testing specifications
3 registration batches	Simulating proposed bulk storage container	25°C/60% r.h. 40°C/75% r.h.	0, 3, 6, 9, 12, (18) 1, 3, 6	(18), 24, 36, 48, 60	No:

Table 65 Climatic Zone III and IV

Batch	Packaging material	Storage condition	Storage period, up to registration	Testing frequency (months) on-going	Testing specifications
3 registration batches	Simulating proposed bulk storage container	30°C/70% r.h.	0, 3, 6, 9, 12, (18)	(18), 24, 36, 48, 60	No:

Evaluation: the data are evaluated carefully. If scientifically justified, statistics, reaction kinetics, and linear regression analysis are used.

If however the data show so little degradation and so little variability it is normally unnecessary to go through a formal statistical analysis but merely provide a full justification for the omission.

All the results obtained during the course of development are then compared with the primary accelerated and long-term stability test results. If they confirm fully the predicted data from the stress investigation the general stability information can be based on the following primary and supportive data and the derived stability information.

Primary data and derived stability information

The results of the three registration batches derived: the confirmed preliminary retest period

Supportive data and derived stability information

The results of stress testing, the stability profile derived: preliminary retest period

The results of confirmation investigation derived: preliminary confirmation of preliminary retest period

The primary and supportive data are also assessed relating to the robustness of the manufacturing process.

During the development the manufacturing process was likely challenged by the following factors:

Synthetic route that changed during development

method of manufacture and procedure with different types and sizes of equipment

site of manufacture, laboratory, pilot plant, chemical production with different types and sizes of equipment

scale of manufacture and batch size

The resulting batches manufactured with all these different factors have been investigated in stress, confirmation, accelerated, and long-term testing. Therefore extensive information is available on the possible influence of these factors on the stability of the drug substance.

Finally it is decided whether it is necessary to ensure compliance with the retest period which has been established by providing storage instructions to be displaced on the label.

All the results, the information on the tested batches, the applied analytical procedure, and the derived stability information are compiled in the Stability report of the investigated drug substance.

According to the two different stability protocols for climatic zone II and the climatic zones III and IV, two separate stability reports are written if required.

All the Stability reports are structured and written in the same format.

Stability information: in the dossier for registration application are included:

Primary and supportive stability reports with the derived stability information

Testing specifications for stability testing
Validation report for the analytical procedures

Table 66 Required Capacity

Dosage form	Required capacity for a single analysis	Required capacity for a time point accelerated and long-term testing and on-going	
		Climatic zone II	Worldwide
Drug substance	8 h ≙ 0.21 weeks	10 h ≙ 0.27 weeks	15 h ≙ 0.40 weeks

Testing specifications including full validation	No. of batches	Analysis of test samples (weeks)		Total required capacity (weeks)	
		Climatic zone II	Worldwide	Climatic zone II	Worldwide
3 weeks	1	1	2	4	5
	3	3	6	6	9

5.1.2. Holding Time for Intermediate Stages or Bulk Drug Product

5.1.2.1. Holding Time Period for Intermediate Stage

The expiration period of a production batch should be calculated from the date of release of the batch.

The date of such a release should, under normal circumstances, not exceed 30 days from the date of production of that batch.

If batches are released exceeding 30 days from the production date, the date of production, as defined below, should be taken as the start of the shelf life.

The date of production of a batch is defined as the date that the first step is performed involving combining the active ingredient with other ingredients. But there may be exceptions.

Intermediates are manufactured

In big quantities
At a different site of production
Over a longer period of time

Accordingly they may be stored over a longer period of time and can be regarded as starting materials. Under these circumstances, testing specifications are necessary.

A holding time has to be determined. These data may not be necessary for the registration application; it is mainly a GMP measure.

5.1.2.1.1. APPLICATION OF THE BASIC PRINCIPLES

Selection of samples: the samples should be part of representative batches (pilot plant batches).

Test criteria: the criteria are investigated that are potentially susceptible to change during the course of storage.

Table 67 Basic Principles and Conditions

Production in climatic zone	Storage condition	Packaging/container	Testing frequency, storage period (months)				
I–IV	40°C/75%	intended for storage	0	1	2	3	
I–II	25°C/60%	open		0.5			
	25°C/60%	intended for storage				3	6
III–IV	30°C/70%	open		0.5			
	30°C/70%	intended for storage				3	6

Analytical procedures: the analytical procedures for the dosage forms are applied or adapted accordingly.

Specifications: release specifications.

For storage conditions, testing frequency, and storage period, see Table 67.

Number of batches: 2.

Packaging: the packaging should simulate the actual packaging intended for storage. The steel container may be simulated by a tight glass container, lined with the applied polyethylene foil.

Evaluation: results should be within release specifications.

Stability information: holding time according to storage period at 25°C/60% or 30°C/70%.

The shelf life is then calculated from the date of release of the related batch of finished product, if this release date does not exceed 30 days from the date that the intermediate is introduced into the manufacture of the finished drug product. If this 30 day limit is exceeded, the shelf life is calculated from the date that the intermediate is introduced into the manufacture of the finished drug product.

One batch of the drug product should be produced from intermediate stored for the full holding period, and this batch should be monitored during follow-up stability testing under long-term testing conditions.

5.1.2.2. Holding Time for Bulk Drug Product

Usually the bulk drug product is not packed immediately after manufacturing or only partly or it is shipped for packaging to another site. Therefore corresponding stability investigations are necessary.

These data may not be necessary for the registration application it is mainly a GMP measure.

5.1.2.2.1. APPLICATION OF THE BASIC PRINCIPLES

Selection of samples: the samples should be part of representative batches (registration batches)

Test criteria: the criteria are investigated that are potentially susceptible to change during the course of storage.

Analytical procedures: the analytical procedures for the dosage forms are applied or adapted accordingly.

Specifications: shelf life specifications.

Table 68 Storage Conditions, Testing Frequency, and Storage Period

Production in climatic zone or shipment	Storage condition	Packaging container	Testing frequency storage period (months)				
I–IV	40°C/75%	intended for storage or shipment	0	1	2	3	
I–II	25°C/60%	open		0.5			
	25°C/60%	intended for storage or shipment				3	6
III–IV	30°C/70%	open		0.5			
	30°C/70%	intended for storage or shipment				3	6

For storage conditions, testing frequency, and storage period (according to requirement), see Table 68.

Number of batches 2.

Packaging: the packaging should simulate the actual packaging intended for storage or shipment. The steel container may be simulated by a tight glass container, lined with the applied polyethylene foil.

Evaluation: if results are out of release specifications the data have to be compared with those of the registration batches to investigate whether the stability is at least the same. Only under this precondition the same shelf life can be applied.

Stability information: shelf life according to registration batches. The expiration period is calculated from the date of release, and the storage period of the bulk has to be subtracted accordingly.

5.1.3. Drug Product

5.1.3.1. In-Use Stability

The derivation of in-use stability is necessary for

Drug products that are reconstituted into a usable form before administration
Drug products whose stability is jeopardized once the container is opened

The in-use stability of reconstituted drug products is usually limited by the chemical stability, whereas the corresponding drug products in multiple-use containers may be limited by the microbial stability. For both cases preliminary experiments are performed during the development stage.

5.1.3.1.1. APPLICATION OF THE BASIC PRINCIPLES

Selection of batches and samples: representative pilot-plant batch, usually the first one that is put on stability for registration application.

Test criteria: usually the test criteria are covered by the stress, the accelerated and long-term testing and only those should be followed that are potentially susceptible to change due to reconstitution or opening of the container.

Reconstituted dry product: appearance, clarity, pH, drug substance decomposition and assay, for suspension also dispensability and particle size distribution.

Multiple use container: appearance, drug substance decomposition and assay, microbial preservative challenge test.

Analytical procedure: the testing specification for the final formulation of the dosage form.

Specifications: shelf life specifications.

For storage conditions, testing frequency, and storage period, see Tables 69–72.

Table 69 Reconstituted Dry Product

Storage condition	Testing frequency, storage period (weeks)
5°C	0, 2, 4

Table 70 Reconstituted Powder for Injection with Antimicrobial Preservation

Storage period	Storage condition (days)
5°C	28

Multiple use container

Before storage, treatment is simulated for a duration of up to about 4 weeks, whereas 1 dosage is withdrawn daily.

Table 71 Multiple Use Container

Climatic zone	Storage condition	Testing frequency, storage period (months)
I–II	25°C/60%	1[a] + 5
III–IV	30°C/70%	1[a] + 5

[a] This 1 month represents the period where 1 dosage is withdrawn daily. No analyses after this period.

Table 72 Injection with Antimicrobial Preservation

Climatic zone	Storage condition	Storage period (days)
I–II	25°C/60%	28
III–IV	30°C/70%	28

Number of batches: 2.

Packaging material: commercial packaging material.

Evaluation: the results may be part of the stability report for the registration application. If not, such results should be available in case the regulatory authorities request them.

Stability information: use life,

at least 4 weeks after reconstitution store in the refrigerator 2–8°C

maximal 28 days

at least 6 months

5.1.3.2. Photostability

The investigations are undertaken as confirmatory investigations if data in the final packaging are not yet available.

5.1.3.2.1. SCHEDULE OF TESTS

Selection of batches and samples: representative registration batch in commercial packaging. Preferably after storage for 6 months at 40°C/75%.

Test samples:

drug product outside immediate pack in colorless glass container, tablets, or capsules spread in single layer

drug product in dark container as control sample.

If decomposition:

drug product in blister pack (if intended for marketing)

drug product in immediate packaging

If decomposition:

drug product in marketing packaging

Storage condition

Xenon test or corresponding

Overall illumination: 1.2 million lux hours ≙ 22 h Xenon test > 400 nm; integrated new ultraviolet energy not less than 200 watt hours/m^2; ≙ 9 h Xenon test 300–400 nm (Atlas Suntest, 250 W/m^2).

5.1.3.3. Accelerated and Long-Term Testing

5.1.3.3.1. APPLICATION OF THE BASIC PRINCIPLES

Selection of batches and samples: representative batches for registration application (registration batch). The batches should be put on stability within 4 weeks after release.

Manufacturing process: should meaningfully stimulate that which would be applied to large scale batches for marketing. The process should provide product of the same quality intended for marketing, and meeting the same quality specifications as to be applied for release of material.

Batch size: two of the three at least pilot scale: one tenth that of full production or for solid dosage forms 100,000 tablets or capsules. One of the three may be smaller, e.g., 25–50000 tablets or capsules for solid dosage forms.

Drug substance: where possible different batches should be applied for manufacturing.

The representative batches

Must be representative of the manufacture and packaging.

Should be produced by a validated manufacturing process.

Must meet the specifications required for the release of material.

Must be homogeneous and consist of random samples.
The container put into storage must be representative of the batch.
The samples investigated from a container must be representative of all the samples in the container at the time of analysis.

Besides the batches for registration application it may be necessary to select representative batches from laboratory or scaling up.

Test criteria: in stability testing, the criteria of a drug product are investigated that are potentially susceptible to change during the course of storage that are especially important for quality, safety, efficacy.

Analytical procedures: the analytical procedures must be stability specific and fully validated, e.g., the following validation characteristics must be taken into account:

specificity
linearity for drug substance and decomposition product
accuracy for drug substance and decomposition product
range for drug substance and decomposition product
quantitation limit 0.1% according to reporting limit
intermediate precision for drug substance and decomposition product
robustness

Specifications: it has to be differentiated between release specifications (quality after manufacture) and shelf life specifications (quality up to the expiration date).

To fix the shelf life specifications may be not easy. If possible the specifications are derived from the results of steps 1–3. But it may happen that the shelf life specifications cannot be fixed before 12 or 18 months' data are available, especially for decomposition products.

Storage conditions:

Climatic zones I and II: 25°C/60%, 40°C/75%.

If significant changes take place at 40°C/75%, storage at 30°C/70% is necessary.

Climatic zones III and IV: 30°C/70%, 40°C/75%.

It is differentiated between the stability protocol for the ICH countries (climatic zone II) and the extension to countries of the climatic zones III and IV with 30°C/70% r.h.

Testing frequency: up to registration application: 0, 3, 6, 9, 12, (18) months.

Storage period: the long-term testing should cover at least 12 months duration at time of submission. It is continued up to the intended shelf life after registration application (ongoing stability testing).

Stability test protocols: see Tables 73, 74.

Table 73 Climatic Zone II ≙ ICH

Batch	Packaging material	Storage condition	Storage period, up to registration	Testing frequency (months) on-going	Testing specifications
3 registration batches	Proposed packaging material for marketing	25°C/60% r.h. 40°C/75% r.h.	0, 3, 6, 9, 12 1, 3, 6	18, 24, 36, 48, 60	No:

Table 74 Climatic Zones III and IV

Batch	Packaging material	Storage condition	Storage period, up to registration	Testing frequency (months) on-going	Testing specifications
3 registration batches	Proposed packaging material for marketing	30°C/70% r.h.	0, 3, 6, 9, 12, 1, 3, 6	18, 24, 36, 48, 60	No:

Usually it can be predicted from the data of stress investigations whether a significant change will take place at the accelerated storage at 40°C/75% r.h. and correspondingly storage at 30°C/70% is required. If however there may be a risk samples are stored in parallel at 30°C/70% but only analyzed if significant change has taken place after 3 or 6 months.

Number of batches: 3.

If more than one dosage is intended to be put on the market, the number of batches can be reduced by applying scientifically based rationalization measures. This includes the expedients mentioned in the ICH Stability Guideline, bracketing and matrixing.

A stability protocol is elaborated by applying bracketing or matrixing to reduce the number of batches. This stability protocol may be discussed with the authorities before starting step 4.

If the information provided in the Registration Application does not contain data of production batches, information should be provided at a later stage on three early production batches manufactured for marketing.

Packaging material: the testing for registration application should be carried out in the final packaging proposed for marketing, or when justified in packaging that simulates the final one.

A drug product is frequently marketed in several packaging materials, which may differ in both type and size. However they do not all have to be included in a stability program. This applies particularly for solid dosage forms. If information about the sorption behavior is available from the stress tests, conclusions can be drawn through inference and by analogy. In these cases the stability program can be reduced by applying bracketing or matrixing, as mentioned in the ICH Tripartite Stability Guideline (the study protocol may be discussed in advance with authorities).

Solid dosage forms

The following packaging materials are used:

Blisters: PVC, polypropylene, composite film
Plastic tubes: polypropylene, polyethylene
Glass bottles
Aluminum foil, aluminum blister

Which of these are used depends on the stability of the drug product, the climatic zone for marketing, and the market strategy.

If the drug products are known not to be extremely sensitive to moisture on the basis of the results of stress tests, the packaging materials can be used as follows in the individual climatic zones, Table 75.

Table 75 Use of Packaging Materials in the Climate Zones

Packaging material	Climatic zones II	Climatic zones III	Climatic zones IV
Blisters	x	x	
Plastic tubes	x	x	x
Glass bottles	x	x	x
Aluminum foil, aluminum blister	x	x	x

Semisolid dosage forms
Frequently used packaging materials:

Aluminum tube, internally lacquered
Plastic tubes
Plastic containers

Table 76 General Suitability of Frequently Used Packaging Materials in the Climatic

Packaging material	Climatic zones II	Climatic zones III	Climatic zones IV
Plastic tubes	x		
Plastic containers	x		
Aluminum tubes, internally lacquered	x	x	x

For the packaging materials for semisolid dosage forms it may be necessary to include different sizes in the stability programme.

Liquid dosage forms
Frequently used packaging materials:

Glass bottles with closure
Ampoules
Ampoules, glass bottles with rubber stopper
Plastic containers

Table 77 General Suitability of Frequently Used Packaging Materials in the Climatic zones

Packaging material	Climatic zones II	Climatic zones III	Climatic zones IV
Plastic containers	x		
Glass bottles	x	x	x
Ampoules	x	x	x

Different sizes are of no importance for ampoules or glass bottles.

Evaluation: a systematic approach should be adopted in the presentation and evaluation of the stability information.

The long-term testing of the representative batches for the registration application should cover at least 12 months duration at the time of submission.

Extrapolation of the shelf life up to at least 2 years may be acceptable, where supported by stress, accelerated, and real-time data.

This may be especially necessary as long as the expiration date has not yet been covered by real time data for the full storage period.

An acceptable approach for quantitative characteristics that are expected to decrease with time is to determine the time at which the 95% one-sided confidence limit for the mean degradation curve intersects the acceptable lower specification limit.

Statistics and reaction kinetics are also valuable approaches depending on the results.

The assay, levels of degradation products, and other appropriate attributes must be included in the evaluation.

Where the data show so little degradation and so little variability that is apparent from looking at the data that the requested shelf life will be granted, it is normally unnecessary to go through the formal statistical analysis but only to provide a justification for the omission.

All the results obtained during the course of development are then compared with the primary accelerated and long-term stability test results. If they confirm fully the predicted data from the stress investigations during development, the general stability information for the drug product will be based on the following primary and supportive data and the derived stability information:

Primary data and derived stability information:

The results of the three registration batches derived: the confirmed preliminary shelf lives.

Supportive data and derived stability information:

The data of the drug substance stability profile derived: preliminary retest period

The data of the three drug substance registration batches derived: the confirmed preliminary retest period

The data of the stress and confirmation investigation during the clinical development derived: minimum shelf life (period of use) predicted and confirmed for the clinical trial phases I–III

The data of the stress investigations with the final formulation including one registration batch derived: preliminary shelf lives for registration batches

The primary data are also assessed relating to the robustness of the formulation.

It is challenged by different batches of drug substance and excipients.

Finally it is decided whether it is necessary to ensure compliance with the expiration date established by providing storage instructions to be displayed on the pack.

Storage instructions

Storage instructions to be displayed on packs must reflect a genuine necessity, must take into account the prevailing temperatures of the respective climatic zone, and must be derived directly from tests. A logical relationship to the storage conditions thereby results.

For Japan, drug products shown to be stable up to 3 years need not carry specific storage statements. In the EU, if there is evidence that batches of the stored product as packed for sale are stable at temperatures up to 30°C, the product need bear no special temperature storage instructions. Within the USA, all drug products usually require statements on storage conditions.

In Table 78 the storage instructions to be used in the EU are listed. lt means long-term testing, acc means accelerated testing.

In principle, medicinal products should be packaged in containers that ensure stability and protect from deterioration. A label statement should not be used to compensate for inadequate or inferior packaging. Nevertheless, the statements in Table 79 may be used to emphasise the need for storage precautions to the patient.

All the results, the information on the tested batches, the applied analytical procedures, and the derived stability information, are compiled in the Stability report of the investigated drug product.

Table 78 Storage Instructions to Be Used in the EU

Testing conditions where stability has been shown		Required label	Additional label,[a] where relevant
20°C ± 2°C/60% r.h. ± 5% (lt) 40°C ± 2°C/75% r.h. ± 5% (acc) ↓	→	No labelling to be used	Do not refrigerate or freeze
25°C ± 2°C/60% r.h. ± 5% (lt) 30°C ± 2°C/60% r.h. ± 5% (acc) ↓	→	Do not store above 30°C	Do not refrigerate or freeze
25°C ± 2°C/60% r.h. ± 5% (lt)	→	Do not store above 25°C	Do not refrigerate or freeze
5°C ± 3°C (lt)	→	Store at 2°C–8°C	Do not freeze
Below zero	→	Store in a freezer[b]	

[a] Depending on the dosage form and the properties of the product, there may be a risk of deterioration due to physical changes if subjected to low temperatures. Low temperatures may also have an effect on the packaging in certain cases. An additional labeling statement may be necessary to take account of this possibility.

[b] At a justified temperature.

Table 79 General Storage Statements in the EU

	Storage problem	Additional labelling statements[a] depending on the package	Comment
1	Sensitivity to moisture	Keep the container tightly closed	E.g., plastic bottles
2	Sensitivity to moisture	Store in the original package	E.g., blisters
3	Sensitivity to light	Store in the original container	
4	Sensitivity to light	Keep container in the outer carton	

[a] A rationale for the labelling statement should be given in the package leaflet.

According to the two different stability protocols for climatic zone II and the climatic zones III and IV, two separate stability reports are written.

One covers the ICH countries of climatic zone II with the EU, Japan, and the USA and a second stability report for climatic zones III and IV. This report contains the data for climatic zone II as supportive data.

For the climatic zones III and IV, different packaging materials and different specifications may be required.

To illustrate the amount of information that will be available by applying the strategic planning, an example is given.

The stability information derived for a registration application is based on the results of Table 80.

Table 80 Stability Information

Batches	Investigations	Derived shelf life
Three strengths of laboratory batches	Stress and long-term	6 months phase I
Three strengths of clinical batches phase II	Stress and long-term	18 months phase II
One batch phase III	Stress and long-term	24 months phase III
One registration batch	Stress	24 months
Three registration batches	Accelerated and long-term	24 months

Thereby it can be assured that the patient after marketing authorization gets the same quality as the patient during the clinical trial investigations.

The stability report for the drug product is structured and written in the same format as those during development.

Stability information

In the dossier for registration applications are included:

Primary and supportive stability reports with the derived stability information.

Testing specifications for release and stability testing.

Validation report for the analytical procedures.

Table 81 Required Capacity

Dosage form	Required capacity for a single analysis and documentation	Required capacity for a time point accelerated and long-term testing and on-going	
		climatic zone II	worldwide
Solid	17 h ≙ 0.45 weeks	20 h ≙ 0.53 weeks	32 h ≙ 0.85 weeks
Semisolid	12 h ≙ 0.32 weeks	18 h ≙ 0.48 weeks	27 h ≙ 0.72 weeks
Liquid	8 h ≙ 0.21 weeks	12 h ≙ 0.32 weeks	18 h ≙ 0.48 weeks

Dosage form	Testing specifications including full validation (weeks)	No. of batches	Analysis of test samples (weeks)		Total required capacity (weeks)	
			climatic zone II	worldwide	climatic zone II	worldwide
Solid	4	1	2.7	4.3	7	8
		3	8.1	13	12	17
Semisolid	4	1	2.4	3.6	7	8
		3	7.2	11	11	15
Liquid	3	1	1.6	2.4	5	6
		3	5	7.2	8	10

The required capacity is derived for 1 dosage and 1 packaging material.

6. STEP 5: ONGOING STABILITY TESTING

6.1. Objective

Confirmation and extension of the anticipated retest period, shelf life.

Monitoring of the stability characteristics for as long as the anticipated retest period/shelf life and the storage period are not yet identical, the data being generated by those studies already initiated or referred to in the registration application.

Stability testing of three production batches manufactured for marketing after marketing authorization, if registration batches were not full scale.

The ongoing stability testing is a continuation of accelerated and long-term testing; it concludes the development of a drug substance or drug product.

6.2. Application of the Basic Principles

Selection of batches and samples
- Representative batches of step 4.
- Representative production batches manufactured for marketing.

Test criteria, analytical procedure, specifications as for step 4.

Stability test protocols (Tables 82, 83).

Drug substance and drug product

Table 82 Climatic zone II

Batches	Packaging material	Storage conditions	Storage period, testing frequency (months)	Testing specification
3 registration batches	Simulating proposed bulk storage container (drug substance)	25°C/60% r.h.	18, 24, 36, 48, 60	corresponding to step 4
3 production batches	Final packaging material for marketing (drug product)	25°C/60% r.h. 40°C/75% r.h.*	0, 3, 6, 9, 12, 18, 24, 36, 48, 60 3, 6	corresponding to step 4

* Drug Product only

Table 83 Climatic Zones III and IV

Batches	Packaging material	Storage conditions	Storage period, testing frequency (months)	Testing specification
3 registration batches	Simulating proposed bulk storage container (drug substance)	30°C/70% r.h.	18, 24, 36, 48, 60	corresponding to step 4
3 production batches	Final packaging material for marketing (drug product)	30°C/70% r.h. 40°C/75% r.h.	0, 3, 6, 9, 12, 18, 24, 36, 48, 60 3, 6	corresponding to step 4

[a]Drug product only.

Number of batches: 3 pilot plant from step 4; 3 production batches.

Packaging material: drug substance: simulating proposed bulk storage container. Drug product: final packaging material for marketing.

Evaluation: the same methods and procedures are applied as for the registration batches. It may be necessary to derive post approval specifications from the data. The information on the robustness will be extended for the drug substance and the drug product by the following factors:

Drug substance:
- Batch size of starting materials
- Site of manufacture, chemical production
- Synthetic route, method of manufacture and procedure: final process
- Scale of manufacture: production scale

Drug product:
- Different batches drug substance from final process and production scale
- Site of manufacture: pharmaceutical production
- Manufacturing process: final manufacturing process
- Scale of manufacture: production scale

Stability reports with the extended stability information are written on demand, e.g.,

Yearly updating of the stability report
At the end of the anticipated shelf life/retest period
Stability information: confirmation and extension of the shelf life/retest period and storage instructions if necessary.

7. STEP 6: Follow-Up Stability Testing

7.1. During Continuous Production

7.1.1. Objective

Monitoring of the stability of the continuous production
Confirmation of the derived stability information

7.1.2. Application of the Basic Principles

Selection of batches and samples: representative production batches.
Test criteria: selection according to results of the steps 4 and 5.
Analytical procedures, specifications as for steps 4 and 5.
Storage conditions, testing frequency, storage period, as in Table 84.
Number of batches
Since the first year of production is covered by batches for the ongoing stability testing, the follow-up stability testing starts in the second year.
drug substance: 1 batch every second year
drug product: 1 batch per year.
If several strengths are marketed, only the most sensitive is investigated or matrixing design (1/3 design) is applied.
It is advisable to start with the storage of the batch for a drug product in the same months every year.
Packaging material: commercial. If several are marketed, only the most sensitive is investigated, or matrixing design (1/3 design) is applied.
Evaluation: the results are compared with those of the steps 4, 5 and summarized in a yearly quality report.
Stability information: confirmation of the stability information, shelf lives, retest periods. The follow-up stability testing is not regulated by a stability guideline. It is a GMP measure to guarantee the quality.

Table 84 Storage Conditions, Frequency, and Storage Period Testing

Climatic zone	Storage conditions		Testing frequency, storage period (months)						
	(°C)	(% r.h.)							
I–II	25	60	0		12	24	36	48	60
	40	75		6					
III–IV	30	70	0		12	24	36	48	60
	40	75		6					

7.2. Variations and Changes

Stability testing for variations and changes to a marketing authorization, post approval changes. The scope and design of the stability studies for variations and changes are based on the knowledge and experience acquired on active substances and drug products.

Drug substance

Variations and changes of the synthetic route or manufacturing process of the drug substance may influence

impurity profile, mainly by-products, intermediates, degradation products
particle size and particle size distribution
surface area
polymorphism

The following information is available:

Justification of specifications

Justification was given for each analytical procedure and each acceptance criterion.

The justifications referred to

relevant development data
test data of batches used in toxicology and clinical studies
results from stress testing
results from accelerated and long-term testing
results from production batches

The batches for justification have been manufactured according to

different starting materials
different synthetic route
different method of manufacture and procedure
different site of manufacture
different scale of manufacture

Stability information

The stability information is based on the following data and derived stability information:

The results of stress testing, the stability profile with derived preliminary retest period
The results of confirmation investigations with preliminary confirmed preliminary retest period
The primary data with accelerated and long term testing up to registration application with derived preliminary retest period
The ongoing stability data of the 3 registration batches, long-term testing with derived retest period of 60 months
The ongoing stability data of the 3 post approval production batches, accelerated and long-term testing with confirmed retest period up to 60 months.

Robustness of manufacturing process

The batches investigated in stability testing were manufactured according to the following variations and changes:

Different batches with different batches and batch size of starting materials

Different synthetic routes that changed during development
Different method of manufacture and procedure with different types and sizes of equipment
Different site of manufacture, development laboratory, pilot plant, chemical production with different types and sizes of equipment
Different scale of manufacture, laboratory, pilot plant, production scale

On the basis of this detailed information it is usually possible to evaluate the influence of the variations and changes on the stability of the drug product. This is valid all the more for stable drug substances (stay within initial specifications 6 months 40°C/75% r.h. and ≥2 years 25°C/60% r.h.).

There may be an influence on the following test criteria which have to be investigated carefully after production at release: particle size, particle size distribution, surface area, polymorphy, purity, dissolution rate.

To be on the safe side the general procedure is proposed as in Table 85.

Table 85 Drug Substance

Variation, change in	Possible influence on quality, release specifications, immediate investigation	Assessment of results	Possible influence on stability, retest period Stability investigation
Equipment	Particle size Surface area Polymorphy	Quality unchanged All data within specifications	none
Manufacturing site	Purity Dissolution rate		none
Process			none
Route of synthesis			2 batches 40°C/75% 3 months

Drug products

Basically the same amount of information is available as for the drug substance.

Justification of Specifications

Justification was given for the selection of the test criteria, each analytical procedure and each acceptance criterion. The justifications referred to

Relevant development data
Test data of batches used in toxicology and clinical studies
Results from stress and conformation studies
Results from accelerated and long-term testing
Results from production batches

Stability information

The stability information is based on the following data and derived stability information:

The results of stress and confirmation studies during the clinical development:

different strengths and composition phase I

derived: 3 or 6 months predicted and confirmed

different strengths phase II

derived 12 or 18 months predicted and confirmed

final formulation phase III including registration batch

derived: 24 or 36 months predicted and confirmed; 24 months predicted for registration batches

The primary data of the registration batches, accelerated and long-term testing derived: 24 months preliminary shelf life confirmed

The ongoing stability data of the 3 registration batches long-term testing derived: self life up to 60 months

The ongoing stability data of the 3 post-approval production batches accelerated and long-term testing derived: confirmed shelf life up to 60 months

Robustness of formulation

The batches investigated in stability testing had been manufactured according to the following "variations and changes":

Different batches of drug substance and excipients including drug substance from production

Different strengths

Different compositions

Different packaging material, type and size

Different manufacturing process with different types and sizes of equipment

Different sites of manufacture: development laboratory, manufacturing site of clinical supplies, pilot plant, pharmaceutical production with different types and sizes of equipment

Different scale and batch site of manufacture laboratory, pilot plant, production

On the basis of this detailed information it is very often possible to evaluate the influence of the variations and changes on the stability. In the individual case it depends on the dosage form and the type of formulation.

The SUPAC and CPMP guidelines and requirements do not refer to this broad base of information. By applying the strategic planning; scientifically based information can be provided in most cases. This information provides a higher degree of certainty than formal stability testing of 1–3 batches stored at 40°C/75% 3–6 months.

It is always necessary to investigate these batches carefully after production to see whether all data are well within release specifications. It may be also necessary to perform special investigations.

Nevertheless the following general procedure is proposed to be on the safe site:

Table 86 Solid Dosage Forms

Variation, change in	Possible influence on quality, release specifications, immediate investigation	Assessment of results	Possible influence on stability, shelf life specifications. Stability investigation
Equipment	Appearance Content uniformity Dissolution rate	Quality unchanged All data within release specifications	2 batches 40°C/75% or 30°C/70% up to 3 months
Manufacturing site			none
Process			2 batches 40°C/75% up to 6 months
Excipient Qualitativc Quantitative			

Table 87 Semisolid Dosage Forms

Variation, change in	Possible influence on quality, release specifications: immediate investigation	Assessment of results	Possible influence on stability, shelf life specifications. Stability investigation
Equipment Manufacturing site Process Excipient: Qualitative Quantitative	Appearance Homogeneity Content Uniformity within container Chemical Stability Preservation	Quality unchanged All data within release specifications	2 batches 40°C/75% (30°C/70%) up to 6 months

Table 88 Liquid Dosage Forms

Variation, change in	Possible influence on quality, release specifications: immediate investigation	Assessment of results	Possible influence on stability shelf life specifications. Stability investigation
Equipment	Appearance pH Chemical Stability	Quality unchanged All data within release	none
Manufacturing site	Preservation	specifications	2 batches 40°C/75% up to 6 months
Process Excipient: Qualitative Quantitative			

Table 89 Change in Packaging Material

Variation, change in	Possible influence on quality, release specifications, immediate investigation	Assessment of results	Possible influence on stability, shelf life specifications. Stability investigation
Immediate packaging Same material different size	*Solid dosage forms* Tightness of Container	Quality unchanged All data within release specifications	none
	Semisolid dosage forms Homogeneity Content Uniformity		none
	Liquid dosage forms Chemical Stability		none
Different material	*Solid dosage forms* Permeability O_2, H_2O, Light	Equal or less permeable	none
		Higher permeability	2 batches 40°C/75% up to 3 months
	Semisolid, liquid d.f. Permeability O_2, H_2O, Light Interaction	Equal or less permeable, no interaction	2 batches 40°C/75% or 30°C/70% up to 3 months
Test procedure	Specificity Sensitivity Validation	Corresponding validation data equal or better	none

After these general statements the application of the basic principles follows.

7.2.1. Application of the Basic Principles

Selection of batches and samples: drug substance: pilot plant; drug product: the manufacturing process to be used should meaningfully simulate that which should be applied to large scale batches for marketing; the quality must meet all release specifications.
Test criteria: corresponding to original formulation.
Analytical procedures: corresponding to original formulation.
Specification: corresponding to original formulation.
For storage conditions, testing frequency, and storage period, see Table 90.
Number of batches: 2.
Packaging material: commercial unless packaging material has been changed.
Evaluation: the results are compared with the corresponding data of the original formulation.
If no new data (≥ 3 years) are available, 1 batch of the original formulation is investigated together with the changed formulation.
Stability information: if the variation causes no influence on the quality, all data are within specifications; the stability information of the original formulation is still valid.
The stability is persued by the corresponding follow-up stability program, but in the first year of production two representative batches are put on stability. Furthermore the testing frequency at 40°C is 3 and 6 months.
If the changed formulation is less stable the shelf life has to be shortened accordingly.

Table 90 Conditions, Frequency, and Storage

Storage condition (°C)	(% r.h.)	Testing frequency, storage period (months)			
40	75	0	1	3	(6)
30[a]	70[a]			3	(6)
25	60				6 up to shelf life

[a] Only if at 40°C significant changes are expected.

REFERENCES

1. The Tripartite Guideline on
Stability testing of new drug substances and products endorsed, Q1A
Photostability testing, step 4, Q1B
Test on validation of analytical procedures, step 4, Q2A
Extension of the text on validation of analytical procedures, step 4, Q2B
Impurities in new drug substances, step 4, Q3A
Impurities in new drug products, step 4, Q3B

Residual solvents, step 4, Q3C
Specifications, step 4, Q6A

2. W Grimm. The extension of the ICH Tripartite Guideline, Second International Meeting of the Southern African Pharmaceutical Regulatory Affairs Association, 15–17 March 1995, Pretoria.
3. PMA's Joint-PDS Stability Committee "Stability Concepts," Pharmaceutical Technology (1984).
4. W Grimm, K Krummen. Stability testing in the EC, Japan and the USA. Wissenschaftliche Verlagsgesellschaft, 1993, p. 17.
5. W Grimm. Stability testing of clinical samples. Drug Development Ind. Pharm. 22:851–871, 1996.
6. WHO Expert Commitee on Specifications for Pharmaceutical Preparations 34. Report, WHO, Geneva 1996.
7. W Grimm. Extension of the ICH Tripartite Guideline for Stability Testing of New Drug Substances and Products to countries of climatic zones III and IV. Drug Development Ind. Pharm. 24(4):319–331, 1998.
8. W Grimm. Drug Develop. Ind. Pharmacy 19:2795–2830, 1993.
9. Revised version of Annex 14, Manufacture of investigational medicinal products, of the EU-guide to GMP.

14

Packaging, Package Evaluation, Stability, and Shelf-Life

D. A. DEAN

Consultant, Beeston, Nottingham, England

1. INTRODUCTION

Although no pharmaceutical product can be licensed without a tested and approved pack, the complex functions of packaging are frequently not fully recognized. Since most in industry are likely to have at least an indirect relationship with a packed product, all should be aware of what packaging basically sets out to achieve. Often this is initiated by some form of shock treatment, i.e., have you ever seen water which is not in a pack—yes—as a puddle on the ground—thereby indicating that all liquids need to be contained in an effective way. Having indicated this obvious need for a pack, let us now provide a clear basic definition of packaging. Packaging is the economical means by which a product is

Protected
Presented

Provided with identification, information, containment, convenience, and compliance during
STORAGE
DISTRIBUTION
SALE
USE

with due attention to ultimate DISPOSAL (i.e., the current environmental issues). This invariably equates with the product shelf life.

Since the final shelf life, which is declared dependent on the pack employed, all development activities must be checked against the ICH (International Conference of Harmonization) Guidelines (1). This chapter therefore highlights these, indicating where there are possible deficiencies in the guidelines, which have to be covered as part of the pack approval program. Topics are also stressed where the guidelines make "assumptions" that certain work, critical to general product and drug substance approval, is made.

Although all the factors identified in the definition of packaging contribute to the overall function and performance of the pack, the word protection tends to be most closely aligned with shelf-life. Protection is related to

A. Physical
B. Climatic

C. Biological
D. Chemical
hazards and is therefore a major function, which is described in detail first.

2. PACKAGING PROTECTION FUNCTIONS

2.1. Physical Hazards

Packs meet physical hazards both when they are static (stored, displayed, etc.) and in motion (during distribution and any stages of handling). The main physical hazards can be quantified under the headings of

1. Impact/shock (dynamic hazards)
2. Compression (static hazards)
3. Vibration
4. Puncture

2.1.1. Impact/Shock

Impact involves forces imposed on the product and pack by acceleration/deceleration. These forces may cause breakage, fractures, distortion, dents, etc. and may be transmitted via the pack to the product, i.e., damaged tablets, broken glass bottles, damage to closure systems, etc.

2.1.2. Compression

This arises from stacking loads both during warehousing and in combination with vibration during transportation. Whether this causes problems depends on the way the stack is constructed, the stacking period time, the nature of the bearing surface on which the stack is built (i.e., may relate to the type of pallet and its "footprint"), the environmental conditions around the stack, and the nature of the goods under the compressive force. The strength of the stack may also be modified by dynamic hazards such as vibration and impact.

2.1.3. Vibration

Stress caused by vibration depends on amplitude and frequency and may arise from vehicle engines, road surfaces, etc. It is usually measured in hertz—i.e., one cycle per second. Major damage usually arises between 3 and 30 hertz. Vibration can cause both pack problems—rub of surface, decorated or undecorated, possible loosening (rather than tightening) of screw caps, increased electrostatic on polymers, etc. and product problems—particle size separation, product powdering, etc.

2.1.4. Puncture, Tear and Snagging

These problems may occur externally or internally to the pack. The risks from the above can be quantified by actual travel and warehousing tests, simulated (laboratory) type tests, coupled to the use of instruments in packs during tests. The latest instruments, called data loggers or acquisition units, can quantify various actions as functions of time, i.e., temperature, humidity, impacts (drop heights), and vibration levels. Such information can then be utilized to assist simulated testing.

As indicated earlier, the above "hazards" can occur in isolation, but more often they cause problems in combination. Typical stability testing involves basically static climatic tests, which do not include any of the physical challenges just mentioned. It is, however, essential that these be evaluated, usually as part of a package investigational program. This may also include storage periods under a range of climatic conditions followed by analysis to check that no critical changes have occurred either to the product or to its pack. The author recalls one experience where a fine powder was adsorbed onto a larger crystalline carrier. This was packed into a sachet and immediately analyzed. 100% of active was identified. The sachets were then transported 80 miles to the stability testing area, where the product, when dissolved in water (as an oral drink), returned only 95% active present. The other 5% was found adsorbed to the walls of the sachet. Initially this was counteracted by a 5% overage followed by ultimate reformulation of the product. A non-travel tested product did not initially show this active loss. Parallel problems can arise with closures, which may change in seal efficiency according to the influences of top pressure (compression), vibration, impact, etc., coupled to expansion or contraction due to temperature/humidity variations.

2.2. Climatic Hazards

The previously mentioned physical hazards can be further influenced by the prevailing climatic conditions surrounding the pack. For example, an atmosphere of high humidity can reduce the stacking (and compression strength) of corrugated boxes by well over 50%. Dimensional changes, due to temperature and humidity increases, can also arise, especially with polymer based materials, where expansion is significantly greater than for metals. Material permeability changes also vary according to circumstances and hence may not follow the various "orders" of reaction or directly relate to an Arrhenius plot. The climatic hazards, which may influence product stability therefore, have to be discussed in terms of pack properties and characteristics.

2.2.1. Moisture—As Liquid Water or Relative Humidity (RH)

Although glass and metal (when free from perforations) are totally impermeable, all plastics are to some degree permeable. This may result in either moisture loss or moisture gain, depending on the nature of the product, the characteristics and the thickness of the polymer, and the "gradient" between the inside and outside of a pack. An aqueous product containing certain "salts," when packed in a plastic pack and stored at 40°C 75% RH, may show no moisture loss if the internal vapor pressure is at equilibrium with the external atmosphere. Storing at 40°C 15% RH could show a significant moisture loss, as there is a very positive "gradient" between the two atmospheres.

Where products take up moisture, the greatest moisture gain will arise when the external temperature and humidity is high, i.e., 40°C 75% RH, 38°C 90% RH, etc.

In the biscuit (U.S., cookie) industry, where a product such as Rich Tea picks up moisture, becomes soggy, and looses its "break" and crispness, an acceleration factor of 2 × is often used for each rise of 10°C and 10% RH, provided these factors do not exceed an acceleration factor of 8. If this was used to compare 40°C 75% RH with a typical U.K. biscuit storage condition of 20°C 45% RH, it would give a (false) acceleration factor of 2 × 2 (temperature) and 2 × 2 × 2 (RH), giving a factor of 32, which may give some indication as to the severeness of such a condition based on simple moisture gain and not involving a chemical reaction. Permeation also depends on the surface area exposed, the material thickness, and how the permeating material is removed from the far "environment" so that the gradient conditions are retained. Bottle-to-bottle contact (especially with rectangular, square, flat surfaces), or shelf-to-bottle contact (particularly if the shelf has no perforations or is enclosed) will influence the rate of permeation—since contact with the permeating layer is impeded or reduced. This thought can be extended further, as most packs have a series of enclosures (the secondary packaging) all of which slow down any loss or gain significantly, i.e.,

1. A blister pack
2. Blisters in a carton
3. Cartons in a display outer
4. Display outers in shipping outer
5. Shipping outers on a pallet stretch or shrink wrap on a pallet.

This results in a steady improvement in the total barrier for the product.

Some of the above can be additionally overwrapped, thus building up the barrier properties.

The author once did an experiment in which a blister (exposed "naked" at 37°C 90% RH) had a shelf life of three weeks, indicating that a blister was an unsuitable pack. The same pack, stored somewhat like the above but with a carton and display outer overwrap, was placed in the company's warehouse where the product was still in "specification" after 6 years (3 weeks to 6 years represents an "acceleration factor" of over 100).

40°C with extremes of humidity, 75% RH to 15% RH (hot moist to hot dry conditions), provides a high challenge to packaging materials, components, and finished packs. Cellulose based materials are particularly susceptible to moisture/temperature changes, as most properties are related to moisture content (and grain direction). It is therefore essential to "condition" such materials to 23°C 50% RH if comparative testing is to be meaningful. Conditioning may be advisable for other materials (e.g., plastics), if properties can be changed by external factors, 20°C 65% RH being used for polymers.

Like products, moisture is a critical challenge to many packaging materials—as even in the form of condensation (on glass, plastic, metal, foils, etc.) problems can arise (i.e., encourage bioburden, alter surfaces, set up erosion, corrosion, etc.). Some of these aspects will be discussed further under packaging materials and general combination effects.

Moisture (and gases) can be soluble (S) in and diffuse (D) through a polymer, hence permeation (P) is related to both these factors ($P = SD$). Moisture can also act as a carrier for other permeants, i.e., foreign or actual product flavors and aromas, organic and inorganic gases, depending on their solubility levels.

2.2.2. Gases

Gases, inorganic or organic, can pass through certain materials. The resulting interactions may relate to chemical changes, pH, flavor and aroma, color change, etc., any of which could be undesirable. Although oxygen is often the more critical, other gases or vapors that may be present, must not be excluded from consideration without some prior assessment. For example, the permeability of the common gases through plastic generally follows the ratio 1 : 4 : 20 for nitrogen, oxygen, and carbon dioxide. This usually means that an aqueous solution of a nonbuffered product will slowly move from say pH 7.0 to around pH 4.5 due to the presence of carbonic acid formed by CO_2 permeation, if it is stored in a plastic pack. In the case of LDPE (low density polyethylene), this may only take 3–6 months, whereas it occurs more slowly with a PET (polyester) bottle. If the pH then drops further, to pH 4.3 or lower, this is usually an indication of polymer degradation involving the shorter chains. Many polymers can suffer from oxidative changes and hence may contain antioxidants, stabilizers, which reduce such effects, especially when coupled to thermal challenges. The EP (European Pharmacopoeia) refers to grades of low-density polyethylenes which are free from antioxidants, but these need extra assessment related to potential degradation. Note certain polythene films if used out of doors may show rapid degradation due to combined effects of light, oxygen, and other gases, heat, moisture, etc., but if kept indoors, little deterioration occurs over a 5–10 year period. This also puts emphasis on the possible need for "surface" analysis, since gases can also alter the surface properties of any material, e.g., glass, metal, plastic. The author now argues that all materials are different at the surface when this is compared with the main "body" of the material. The surface is also the main interface between packaging material and product, where any interaction/exchange starts. Thus gases, which might not change the product, can have a greater influence on the properties of the pack. (Have you noticed how many natural polymers, when placed out of doors, can rapidly deteriorate, i.e., film, strapping and even rigid containers?)

2.2.3. Light

The pack and its packaging components may involve two questions with reference to light:

1. Do they exclude or reduce light from reaching and possibly influencing the product? For example, amber glass filters out a large proportion of UV light.
2. Does the pack absorb any particular range of light rays, and does this change the properties of the packaging materials involved?

For example, amber glass absorbs IR (infrared) rays, which can heat up the product and pack: colored surfaces may discolor, i.e., darken or fade, depending on the intensity of the light.

Again, light may operate in conjunction with other factors to cause various forms of degradation or change, which again start with surfaces, e.g., rubber products are influenced by heat, light, and oxygen and hence may react differently to needle penetration, fragmentation, coring, etc. after "aging." Retests for these are essential during prolonged periods of any product storage—and further changes may also arise from product contact.

Tests for light exposure may involve simulated tests, or actual conditions, i.e., behind glass, north or south facing windows, or total exposure to all the elements (wind, rain, sun, general atmosphere, temperature, etc.). Controls may consist of total darkness (a condition found in many climatic rooms and cabinets) or an area of the product-pack masked by a 100% light barrier material. It is important to understand whether this acts as a "white body" or a "black body": one reflects the heat/light rays, whilst the other is absorptive and hence may cause heat related differences. Materials themselves will also have heat reflective or heat absorptive properties, etc.

2.2.4. Pressure

Atmospheric pressure differentials can occur both internal and external to a pack. These can arise due to a number of reasons:

a. Natural fluctuations for any specific height above or below sea level, e.g., changes in altitude due to geographical location.

b. Transport in an unpressurized aircraft flying at 10,000 feet—creates a negative pressure of approximately $\frac{1}{2}$ atmosphere.

c. Transport in a pressurized aircraft. Such aircraft are pressurized to an equivalent of 8000 feet, which is 3.8 psi (pound per square inch) or roughly $\frac{1}{4}$ atmosphere. (One atmosphere is 14.7 psi.) This means that items are subjected to $\frac{1}{4}$ atmosphere on ascent and $+1/4$ atmosphere on descent to sea level.

d. Hot fill plus effective closuring and cold fill plus effective closuring can create negative (vacuum) and positive pressures within the pack, when the pack is subsequently stored under normal conditions.

e. Moist heat sterilization of filled packs, where the pack may vent and then reseal or simply expand and then not fully contract—as this may create a negative pressure within the pack. With certain flexible polymers this may give rise to pack "dimpling" or partial collapse due to negative pressure. (This can be partly controlled by an over pressure or a balanced pressure autoclave.)

Negative external pressures cause flexible and pliable packs to extend, and positive external pressures cause such packs to be under pressure or compression. Under these conditions packs that are not effectively sealed will "breathe," i.e., let "air"/moisture etc. in and out, according to the changing conditions.

2.2.5. Other Airborne Contamination

This can involve particulates and bioburden. Particulates can arise from various sources and also be a carrier for bioburden. Particulates can come from the packaging material itself (i.e., fibers from board and paper, wood based materials, pallets, glass particles from cutting glass, etc.) or a secondary source (hairs, fragments of skin from human beings, etc.).

The generation of particles from basic materials can occur in many ways, i.e., when they are subjected to vibration/abrasion, cutting/guillotining/punching out, breakage/impact/chipping, etc., hence the inference that the least handling and processing will probably lead to the lowest particulate levels. In converse, greater handling will encourage particulates, and this is partly counterbalanced by more use of "vacuum extraction" and environments using "filtered air." Study of air filters is then one way of providing better identification of the particles removed. On production lines, such operations as unscrambling, use of vibratory bowls, abrasion from moving conveyors, etc. can all be a source of particles, much of which "may" be removable by effective "vacuum extraction." The materials on which (wooden pallets) and in which packaging components are delivered (fiberboard) and stored tend to be a serious potential source of extraneous particles. These materials may require ongoing inspection, cleaning to remove surface dirt, or actual banning from certain types (of classified) production areas. Pallet control can be an important part of this operation, both for incoming supplies and outgoing finished stock—as too frequently this can be overlooked. Isolating filling and closuring from other activities can also improve cleanliness.

Particulate control, particle identification, and particle sourcing, together with the same for bioburden, are all part of cGMP (current good manufacturing practice). Finding an excess of particles (or bioburden) at any stage of development through clinical trial supplies, formal stability tests, or final production, is a clear indication of poor cGMP. However, finding a product with an acceptable level of particles does not necessarily mean that this "state" cannot change. Particles can reduce (possibly by adsorption) or increase during both storage and transportation, for a number of reasons. Particles can arise from packaging materials that have a surface growth, as crystals, as a bloom, simple migration, etc. This has been known to occur with rubber based materials, plastics generally, and even glass by a process of surface erosion. These will be further discussed under properties of packaging materials, e.g., surface active constituents in polymers that can be chemically or physically removed by surface abrasion.

All of the above relates to general "cleanliness," which often means producing materials "clean" and then handling them under controlled conditions, using specified procedures in order to optimize total cleanliness. The alternative is to produce effective cleaning processes on line, which may involve actual "washing" or some alternative means. For example, "air blowing" must use pressurized, dry, oil-free, particle-free air of a controlled temperature, coupled with vacuum extraction, using inverted containers (invariably inverted in the case of glass), if effective cleaning is to be achieved. Again, this air should be "filtered" and "captured" so that checks can be made for types of particulate and possible sources by a regular follow-up, i.e., it provides a continuous learning opportunity. "Cleanliness" has also to be balanced against general "hygiene," especially in terms of training, clothing for operators, working practices, etc.

2.2.6. Printing and Decoration

If a packaging material undergoes a decoration or printing stage, this is often seen as a relatively "dirty" operation, hence contamination can arise from these processes. Contamination may involve particulates (which may be ink particles), solvents (although there is now a tendency towards nonsolvent inks), and general migratory substances, which may be of polymeric origin. Abrasion of printed surfaces (rub) is a further source of particulates. Once rub starts, the particles produced can increase the possibility of abrasive effects. Checking hygiene and cleanliness is therefore part of all stages of product-pack evaluation and general good manufacturing practices.

2.2.7. Aging

Aging is traditionally accepted as change(s) against elapsed time, hence is based on more than one cause and effect. Like all of us, both products and packs, and the constituents from which they are manufactured, can age. This is currently being recognized with packaging materials; many companies are now restricting "life" by a "reexamination date" or "reinspection date" until a factual shelf-life can be firmly established. As indicated earlier, any change can be related to either the material surface or the main "body" or both. Knowledge of these aging effects is currently in its infancy, particularly as to whether they will have any influence on the product-pack shelf-life.

Surface evaluation can include chemical analysis, appearance, and in many cases microscopical examination for cracks, pores, reticulation, smoothness, etc. and physical examination, e.g., coefficient of friction, etc. In the case of metals, where surface area can be increased by abrasive substances, corrosion may be increased. Stress cracking of polymers is another factor that can be influenced by the nature of the surface.

2.3. Biological Hazards

These can include microbiological (bacteria, molds, yeasts, etc.) toxins and related substances produced by them, damage by animals, including vermin, birds, insects, reptiles, etc., and hazardous materials associated with them (e.g., excreta), and last but not least, human beings. Packaging aspects, related to the latter, may involve prevention or restriction of access (e.g., tamper resistance, tamper evidence, child resistance, i.e., security related) or the opposite where ease of access and/or reclosure is of importance (e.g., for the ever-growing elderly population which survive by taking regular medication). This must be supported by unambiguous, easy to understand, legible instructions, coupled to the avoidance of misuse or abuse. Since microbiological control may involve the use of preservatives, aseptic presterilized materials or terminal sterilization materials have to be selected to meet these demands. The possible presence of pyrogens, mainly associated with liposaccharides from dead gram-negative bacterial cell walls, also has to be considered.

The use and pack function in the administration or drug delivery aid is an important evaluation stage, as any changes in product and pack must be assessed both separately and together, i.e., does a new pack or an aged pack effectively deliver a product at the beginning and the end of its shelf-life. This also involves risks associated with any form of contamination. However, assessment may not be this simple, if the component delivering the drug can be influenced by intermittent or infrequent use (e.g., a pump based system or an aerosol) or is used in a separate device (powder inhaler or a nebulizer unit) that may be subject to "cleaning" by the user. Such delivery systems may differ as follows:

1. The pack directly acts as an administration aid (drops, sprays, prefilled syringes, aerosols, pump systems).
2. The pack is a "feeder unit" for a separate device—Glaxo-Wellcome Spindisk for Diskhaler.
3. There is a separate device into which product is transferred for administration purposes (nebulizers).
4. The device is manufactured containing a feeder unit (Glaxo-Wellcome Accuhaler).

Each of the above can lead to complex procedures to check that the total performance of pack, product, and administration system function satisfactorily.

The above form "gray areas" with true devices, and hence consideration of "device" guidelines may be essential in the evaluation of these systems.

2.4. Chemical Hazards

Chemical hazards initially relate to compatibility between the product and pack and any "exchange" that may occur from product to pack or pack to product. These may involve various chemical interactions or simple migration/leaching where no initial chemical change occurs. These types of situation may cause changes in appearance (e.g., corrosion of metals), aroma/flavor, microbiological integrity, preservative activity, pH, etc., as well as product potency, bioavailability, etc. Like products, packaging materials and components have levels of purity, impurity, and possible residues arising from the conversion process or method of manufacture (i.e., polymerization process in the case of plastics), plus any constituents added to modify or improve the material in some way (glass, metal, and plastics may all contain such constituents). Materials that are from sources of natural origin, i.e., ores, metals, earths, will tend to vary in the impurities (types and levels) according to their geographical source. Although these factors may not be critical in their general use, they do tend to be partly overlooked and could be important when the end use is for large- or small-volume parenteral products, where the risk of the "user" and "company" may be the greatest.

Product purity is an essential part of the ICH Guidelines. Normally any substance present at a 0.1% level as an impurity or degradation product needs identification and characterization with reference to its safety. Purity of packs has yet to receive similar attention, although work may be carried out on "extractives" from a chemical and biological point of view (5).

3. THE ICH GUIDELINES

Since some of the hazards that can affect pack do not form part of any formal stability program, it is often assumed that they are adequately covered in the research and product-pack development stages, by investigational/feasibility testing or evaluation. However, these assumptions cover many simple basic factors that may act in a variety of combinations. Hence it is relatively easy to overlook an important point or even points. It is therefore essential to have an "overviewer" who understands all the aspects of development and research, whereby there is a total data philosophy covering all stages from discovery of a new chemical entity (NCE) to ultimate product withdrawal from the market. This total data philosophy provides for accumulation of information from all stages, i.e., drug discovery, initial screening and safety-testing, human volunteer studies, early clinical evaluations, formulation development, formal clinical studies, investigational tests between product and pack, formal stability tests, etc. These cover the Phase I to IV stages from which the drug substance and the product dosage form are finally registered and licensed as a packed product for specific diseases, etc. This data should clearly indicate that the product is safe, effective, and of the correct quality, well before any formal stability test is carried out. This latter stage should primarily establish what already has been proved in terms of the shelf-life and stability of the packed product. A formal stability test should not therefore contain any surprises, assuming that adequate prior testing has been carried out. Reducing the cost of complex stability programs by matrixing and bracketing also makes similar assumptions.

However, the role of the pack will vary according to the product category, i.e., ethical, generic, OTC (over the counter) verterinary, etc., its area of use/application in the body, the potency of the drug substance, which in turn can be related to risks to the user/patient and the pharmaceutical company producing it (product liability). It is hence possible to produce a table that reflects this risk (especially for ethical products) and the level or intensity of testing advisable to minimize the risks. Such a table invariably is headed by large-volume parenterals with IV solutions at the top, followed by small-volume parenterals, with solid dosage forms at the foot of the table, where risks are minimal. To date most guidelines are related to ethical products, and their generic counterparts, because these carry higher risks than OTC type products, which are involved in self-medication. Since ethical products usually start with a research program involving drug discovery and development, leading to a new chemical entity, this has to be fully characterized and an approved pack established. Both drug substance and the product dosage form must therefore be effectively evaluated prior to entering a formal stability program. Although both programs involve a wide range of challenges, in the case of the product dosage form, this usually comes under the heading of preformulation studies. The important activity of selecting the correct pack extends beyond compatibility between product and pack and covers functional aspects such as efficient warehousing and distribution, good production line performance, high compliance and convenience in use, etc., plus general aesthetics. These may involve use of decorative techniques, color, size, shape, instructions, legal requirements, particularly where essential to sale, including name, brand image, and overall recognition features to encourage good presentation. Such factors may also be required to assist general compliance and help instill additional confidence in the product for both the professional persons involved (doctors, dentists, nurses, pharmacists, etc.) and the final user/patient.

4. MATERIAL AND PACK SELECTION

Material and pack selection requires a good basic knowledge of all materials, a QC plus (QC plus covers any range of tests to fully quantify a previously little-known material or component) system that quantifies exactly what is being used in any test, and an effective schedule by which product packs are evaluated.

An outline of the main packaging materials follows. In practice this knowledge should cover:

1. The raw basic materials
2. Any purification of these raw materials
3. Conversion processes used to create basic packaging materials, components, etc.
4. How materials or components should be packed
5. General storage and distribution
6. Delivery and unscrambling on production line
7. General storage and distribution
8. Delivery and unscrambling on production line
9. General assembly methods into a finished product/pack
10. Methods of storage and distribution and relative packaging methods covering handling and use options

11. Disposal options for each; including recovery, recycling, reuse, and awareness of energy and pollution levels
12. Supportive test methodology for all of the above

As basic knowledge builds up with experience, confidence in the pack properties and characteristics should increase. Some of this pack detail is therefore outlined below.

4.1. Glass Covering Blown and Tubular Containers

Glass exists as

Type I	neutral or borosilicate glass
Type II	treated—as Type I but surface treated by sulfating or sulfurizing
Type II	glass—soda glass—surface treated by sulfating or sulfarizing
Type III	or soda glass
Type IV	or NP (non parenteral in the USA)

Soda glass is mostly used for ointments, creams, and oral and topical products such as liquids, emulsions, and solid dosage forms generally. It has a typically alkaline surface and may react with alkaline substances and leach out alkaline earths, especially if subjected to multiple autoclaving.

Type I or neutral glass is widely used for small-volume parenterals, where low leaching is essential. Treated Type I glass is also found where an occasional container might have failed the Type I test for hydrolytic resistance.

Type II glass was widely used for IV solutions but is unlikely to withstand multiple autoclaving, as the "neutral" surface can be lost. (Type II is a soda glass treated by a sulfuring or sulfating process to give a thin layer of coating which reduces the alkalinity.)

Alkaline substances such as sodium or potassium tartrate, citrate, or salicyclate can cause greater extraction or create glass "flakes," especially where soda glass is used,and the pack is subjected to steam autoclaving.

Possible extractives from glass are now receiving more attention.

4.2. Metal

Metal containers can be made from tinplate, aluminum, or alloys of aluminum and stainless steel for large vessels. Surface attack can usually be eliminated by the use of external enamels or internal lacquers. Both of these consist of polymers of a thermosetting or thermoplastic origin and hence need to be checked for extractives or possible interaction with the product, both short and long term. Possible corrosion effects also need to be assessed. Although aluminum collapsible tubes are still used, multilayer laminates are being introduced for many ointments.

Aluminum foil is widely used. Most foils are surface coated with a thin polymer layer (wash or key coating) which is added to assist the "key" of the printing inks or other coatings (heat seal layers and laminations) and to reduce surface scuff. Foils of around 17 μm and above are usually commercially free of pinholes—which increase in number with the thinnest foils (down to 6 μM). Moisture permeation through these can be negligible, provided the foil is laminated to a plastic ply. However, thinner foils can stretch, perforate and suffer from sealing deficiencies that may allow moisture exchange (in or out), depending on the type of equipment employed. Metal-lisatin and other coatings are replacing some typical foil uses.

4.3. Plastics

Today, plastics tend to be preferred to metal or glass, provided an adequate shelf-life can be achieved. Basically all plastics are to some degree permeable to moisture and gases and need special requirements if they are to exclude light. They are also less inert than glass and may suffer from migration, leaching, absorption, and adsorption under certain circumstances. Few plastics consist of the pure polymer (the EP mentions LDPE as one such material), and hence they may contain residues from the process of polymerization, additives (substances deliberately added to change certain properties), processing aids (added to aid processing), or master batch constituents (e.g., where a concentrate is added and "let down" to the required content, e.g., titanium dioxide of 1–3%). Having "full" knowledge of these factors is relatively noncritical for solid dosage packs but can be critical for large- and small-volume parenterals, eye preparations, etc. Among the additives, there are constituents that are "active" by being present at the surface of the polymer, e.g., antistatic agents, slip agents and antislip agents, etc. These can be removed by surface abrasion. Many "poly" bags have a surface layer of a lubricant, typically "stearamide" or a metallic stearate. The loss of certain preservatives by absorption (loss into) or adsorption (loss onto the surface), particularly with LDPE, is generally well recorded, i.e., phenol, chlorbutol, 2 phenyl ethanol are absorbed and thiomersal can be adsorbed. The latter also depends on the nature of the surface—the larger the surface area the greater the loss. Such a phenomenon does occur with other polymers. Certain polymers, especially LDPE, may also suffer from ESC (environmental stress cracking), which may occur when a polymer is stressed in contact with a stress cracking agent. The stress may be in-built or applied, and typical stress cracking agents include wetting agents, detergents, and certain volatile oils.

Exchange of constituents between product and polymer can occur either way—product constituent(s) into the polymer, and polymer constituent(s) into the product. This always requires a thorough check, as some exchange, however minute, is almost inevitable. As a broad guide "like tends to absorb like," e.g., EVOH (ethylene vinyl alcohol) readily absorbs water and to some degree alcohol. Plastics may undergo changes during conversion and general processing. e.g., certain plastics need a pretreatment stage to improve print key. This is usually a surface oxidative process achieved by gas flaming (containers) or corona (high voltage) treatment (films). Both processes modify the surface properties, which may reflect on how any exchange occurs. The treated surface usually loses its properties with time.

Sterilization processes tend to need even greater investigation and control as all either tend to modify the polymer in some way or leave possible residues (i.e., as in ethylene oxide sterilization). Gamma and beta irradiation can cause changes in the polymer chain (i.e., cross-linking, chain scission) and the constituents that they contain (e.g., antioxidants). Intense UV light (as used for surface cleaning) can also cause problems. It should be stressed that plastics provide an excellent service to pharmaceutical packaging, provided proper attention is paid to the above factors and many others. A book, however, could be written on this topic. Possible "new" polymers include:

PEN (polyethylene naphthalate), LCP (liquid crystal polymers), which are probably used in conjunction with polyethylene or polyester, COCs (cyclic olefin copolymers), various polymers produced by metallacene catalysts, Resin CZ (a cyclopentane based polyolefin from Daikyo Seiki), and various coated versions of existing, widely used, polymers.

4.4. Multilayer Materials and Coatings

The last decade or so has seen a significant growth in multilayer materials, new coatings (silicon oxide, SiO_x, carbon coatings, etc.), and new coating techniques. The multilayer materials may involve layers of paper, foil, plastic film, and coatings. In most of these, the product contact layer is a polymer, hence previous comments apply. However, there are well over 200 polymers that might be employed, each involving a range of individual grades. The subject therefore can become very complex. Various combinations can be made by lamination, but only polymers can be employed where coextrusion is involved.

4.5. Rubber Based Materials and Elastomers

Rubber or elastomeric based materials are generally more difficult to clear than polymers. This is partly because they are involved in sterile products and have selected properties (resealability after needle penetration) that have not been achievable with plastics. They are widely used as plungers in syringes and for closures in injectibles. Each use requires specific properties, i.e., ease of movement, ease of insertion, freedom from particulates, low in coring and fragmentation, easy needle penetration, effective resealability, etc. Some of these requirements are in conflict with other factors, so a compromise has to be achieved. Synthetic based rubbers (now mainly butyl, chloro, and bromobutyl) offer good inertness, good barrier properties to gas and moisture, compatibility with preservatives, good age resistance, etc., but have poorer resealability, coring and fragmentation when compared with the earlier natural based rubber materials. Adding a facing of PTFE (polytetrafluoroethylene) can further improve inertness but puts extra demands on closure efficiency. Most rubber based materials (natural or synthetic) contain a variety of constituents all of which need screening in terms of possible extractives and safety. Rubber based materials find uses other than for sterile products. Sterile products remain most difficult to "clear" in terms of all the materials used for a pharmaceutical product, including changes that may arise due to aging. Many of the tests are now covered by new EU or ISO (International Standards Organization) standards.

4.6. Standards for Sterile Products

Standards for sterile products include limits on bioburden prior to sterilization, tests for sterility, and a range of tests on basic materials that include packaging components. In certain areas, ISO standards are taking over from National standards and Compendial standards. The more recent ISO standards include those in the table.

ISO8362	Injection containers for injectibles and accessories
ISO8536	Infusions
ISO8871	Elastomeric parts
ISO8362	Covers hardness, fragmentation, self sealability, needle penetration, seal integrity, etc.
ISO8871	Covers extractive tests, including chemical and biological assessment, UV, reducing substances, ammonia, nonvolatiles, pH, zinc, halides, turbidity, volatile sulphur, etc.

Validation is also of particular importance to sterile products and to all processes of sterilization, covering temperature recordings, bioindicators, dosimeters, pressure sensors, chemical methods of monitoring, and special methods of measuring filter integrity, etc. In the validation of sterilizers the now accepted stages of validation must be included, i.e., design, installation, operation, and performance qualifications, etc., i.e. DQ, IQ, OQ, PQ.

5. TESTING THE FINISHED PRODUCT IN ITS PACK

Filled packs can be subjected to a variety of tests depending on the pack and the product. Key tests include climatic challenges against a time scale involving accelerated testing. Inevitably these should include the ICH challenges both on the unpacked and the packed products as this can establish the true effectiveness of the pack. Intermediate climatic conditions (those outside the ICH Guidelines) may also be needed to achieve a full evaluation of how the packed product is likely to perform. Tests are essential for all the features not covered by the ICH Guidelines. Since most packs consist of a container and a closure (screw cap, child resistant closure, rubber stoppers with an overseal, heat seal, etc.), checking for closure integrity is a particularly important function of the total pack. A subsection on this aspect therefore follows.

5.1. Leakage Detection Methods

Leakage (egress and ingress) can be related to liquids, solids, or gases including bioburden and the methods associated with their detection. The methods for detection are becoming increasingly complex and more sophisticated with a trend, certainly for production materials, towards nondestructive testing. The list below covers a wide range of tests, which are not listed in any particular order of importance.

1. Observation of visual defects, e.g., pinholes, capillaries, etc., using human or automated inspection systems.

Vision systems could be included under this heading—see also 12 Microscopy should also be included.

2. Weight change—loss or gain against time under specifically defined conditions. A slow but reliable method.

3. Pressure—vacuum changes by the application of pressure or vacuum under defined conditions (including fluctuating conditions), e.g., pressure change detection; liquid ingress/egress—using various liquids including water, with and without wetting agents and dyes; Release of gases—e.g., air—bubble type tests.

4. Gaseous detection tests, i.e., helium (sniffing), e.g., using mass spectrometry leakage down 10^{-12} Pa-m^2/s can be detected. Oxygen, e.g., Mocon Ox-tran uses a dry nitrogen stream whereby oxygen is detected coulometrically. Carbon dioxide, e.g., Mocon Permatran C detects carbon dioxide in another dry gas by infrared (lightwave). Moisture vapor, e.g., Mocon Permatran W or dynamic water vapor tester measures moisture by a photoelectric sensor. Halogens (e.g. use of hydrofluorocarbons, HFCs), e.g., Krypton 85—high sensitivity is reported.

5. Microbial integrity. Various procedures have been developed (under normal pressure and vacuum) to check whether highly contaminated liquid, gel, or media based material will grow-back or penetrate closure systems. Although these tests may be useful, variable results suggest that alternative methods to detect leakage are preferable.

6. Crack, pinhole, capillary detection. Conventional dye/vacuum immersion tests were widely used to check the seal efficiency of ampoules. These have largely been replaced by electrical conductivity and capacitance type tests. Typical equipment includes the Nikka Densok ampoule inspection machine, which employs a high frequency at high voltage. This distinguishes between good glass (which is a nonconductor) and areas of cracks or pinholes where current will flow between the inner and outer glass surfaces. Other machines use the principle of capacitance and dielectric constants where a material with defects (penetrating cracks) will show a higher dielectric constant than a solid non-defective material.

7. Acoustical. Tests can be based on sonic or ultrasonic energy of a gas escaping from a defect. Mainly used for checking pipes, pressure lines, and ducts, but under further investigation.

8. Thermal conductivity. These use a thermister bridge balanced against air and is subsequently upset if another gas leaks into the air.

9. Chemical tracer tests. Used with materials that can be detected by interaction, i.e., strong ammonia on one side, concentrated hydrochloric acid gas on the other (formation of white cloud of ammonium chloride indicates transfer and leakage). This type of test has been used for pinhole detection.

10. Thermocouple gauges. Mainly used to detect a drop in temperature when solvent type systems escape under vacuum. Could also be used to detect the presence of a warmer gas.

11. Dry ice (carbon dioxide) tests. Detection of loss.

12. Physical or mechanical assessment procedures. These have been left at the end of the list, not because they are the least important, but because they involve the widest variety of factors, e.g., screw caps are controlled by application torque or by removal torque. However, these change according to material, design, environmental factors, etc., some of which may not have been fully investigated. (The assumption is often made that they have.)

This equally applies to any closure system that relies on compression, interlocking, interference forces, etc. in order to make and maintain a "seal." Thus certain forces can readily be measured, e.g., torque, force to push in a plug or pull out a plug type system, force to apply a press over closure, force to remove a press over closure.

The variables associated with most of these systems tend to be quite large and complex.

Whether the measured forces, torque, compression, etc., equate with an effective seal will vary according to the quality of the materials employed and the perfection or imperfection of the surfaces involved between the two interfacing materials. In a paper on screw caps the author identified over 150 variables that could influence closure efficiency.

13. Visual inspection systems. The ability to inspect for and inspect out visual defects is constantly improving. These defects cover such factors as stones, glass inclusions, cracks, scratches, glass filaments, etc. in glass containers, and splits, burn marks, and molding deficiencies in plastics. These visual inspection procedures are also capable of high speeds, i.e., 400 units per minute and over. Automated visual inspection methods are gradually replacing the previous human inspection in which each unit was individually viewed by humans (frequently under 2× magnification) against a white or black background. Although not necessarily bearing a direct relationship with assessment of closure integrity, these methods can eliminate likely suspect (imperfect) packs.

5.2. Evaluating Rubber Stopper Assembly

There are two basic test procedures (SCT and SFT)

1. The seal compression tester (SCT). The instrument measures the compression of the rubber-metal over cover combination, as a result of the sealing operation. This can be measured as a direct figure or as a percentage of rubber element compression (% REC). The latter is achieved by dividing the compression figure by the pre-seal thickness multiplied by 100.

2. The seal force tester (SFT). This instrument measures the static force exerted by the rubber element in the sealed closure. The very first downward movement of the metal cap is quantified as the value of F at that instant and equates with the residual static force in the rubber component. This test is generally preferred to the SCT test.

However, it should be remembered that a pack has other roles beyond that of basic protection, as shown in the definition given in the introduction to this chapter. These include aspects related to identification, information, convenience, compliance, meeting broad environmental needs, and fulfilling various performance functions. The latter include storage, distribution and production line efficiency, all of which have to be achieved in an economical manner. Changing any of these "performance" functions may bring about a need for a reevaluation of the shelf life achieved. Distinguishing between a minor change and a major change often creates problems.

Closures, however, are critical to the total performance of the pack; hence they not only need special attention but are still one of the major reasons for a product recall.

6. LEGISLATION AND GUIDELINES RELATING TO PACKAGING

There is an increasing demand for more guidelines to assist the selection and clearance of a suitable pack for any type of product. Such information has to be submitted under the chemistry and pharmacy submission to a licensing authority.

Recent European (EU) Directives (2) have introduced a document covering the terminology to be used in drug submissions. (Notes for guidance EU/III/3593/91, e.g., the name for a rubber injection closure is a stopper, not a bung.) Allowance is also made for the use of Drug master Files (3)—see (EU/III/3836/89. These may cover both product actives and excipients, and possibly packaging materials and components.

EU Guideline III/9090 (4) puts emphasis on packaging materials with particular reference to the use of plastics. It lists the information required in a drug submission under what could be considered as vague and sometimes difficult-to-define terms. Similar requirements are listed in a U.S.A. document entitled "Guidelines for Submitting Documentation for Human Drugs and Biologics" issued by the Center for Drugs and Biologics of the FDA in February 1987. A new document dated 1997 is currently under the review stage covered under CFR (Code of Federal Regulation) references. For example, a plastic parenteral container would typically require the following information:

Name of manufacturer
Type of plastic
Composition, method of manufacture of the resin and finished container, plus a full description of analytical controls
Physical description (size, shape)
Light transmission (USP)
USP tests
  Biological
Physicochemical
Permeation
Vapor transmission test, if appropriate
Toxicity studies, in addition to those included in the USP
Compatibility including leaching, migration, plus sampling plans, acceptance specifications, etc.

Although the above terminology is generally recognized in the trade, much is not clearly or absolutely defined and hence is subject to "judgements."

The European Pharmacopoeia provides help in that it lists "permitted" plastics and also lists certain "permitted" additives. However, it includes in the list polymethyl methacrylate (PMMA) or Perspex (trade name) which the author has never found in contact with a pharmaceutical product (long term storage). A "permitted" additive, in the author's opinion, does not mean that it is suitable for contact with a certain product until this has been established by contact tests. Guidelines also advise that polymers used must have basic "food grade approval," otherwise expensive additional toxicity studies would be required.

EU Food approval is covered by various directives, i.e., 90/128, amended by 92/39 and others.

These set specific migration limits as extracted under various conditions using a range of simulates, such as distilled water, 3% acetic acid, 15% ethanol, and olive or sunflower oil.

Section B of the directive identifies monomers and starting materials currently allowed, but which may be deleted if positive safety data is not supplied within an agreed period.

6.1. Typical Packaging Information Normally Required

Primary or immediate packaging

1. The nature of the packaging material, indicating the qualitative composition
2. Description of the closure (nature and method of sealing)
3. Description of the method of opening, and, if necessary, safety devices
4. Information on the container (single or multidose) and dosing devices
5. A description of any tamper evident closure and child resistant closure

6.2. Development Pharmaceutics

This data should justify the choice of pack and include

1. Tightness of closure
2. Protection of contents against external factors
3. Container/contents interaction
4. Influence of the manufacturing process on the container (e.g., sterilization conditions)

6.3. Packaging Materials (Primary or Immediate Packaging)

Specifications and routine tests, i.e., covering

1. Construction, listing components
2. Type of materials identifying nature of each
3. Specifications, which may vary in detail according to product nature and route of administration.

Routine tests are likely to include identification, appearance, dimensions, performance, bioburden, etc.

6.4. Scientific Data

Listed under general and technical information for plastics is general information such as

1. Name and grade as used by manufacturer
2. Name of plastic manufacturer (parentals, ophthalmics)
3. Chemical name of material
4. Qualitative composition
5. Chemical name(s) of monomer(s) used

Material should have Food grade approval, otherwise additional toxicological data will be required.

Technical information required will include

1. Characteristics, general description, solubility in various solvents
2. Identification usually by infrared (the material) of the main additives and any dyes

3. Tests such as general tests, mechanical tests, physical tests, and extractive procedures using suitable solvents
4. Name of manufacturer/converter

6.5. Migrations and Interactions

Again, certain phrases are subject to interpretation. These can raise some challenging questions, i.e.,

1. If plastics are identified in the EP, does it mean that they are suitable for all pharmaceutical products? No (not without compatibility tests).
2. If in the EP a plastic is identified with a list of approved antioxidants, are they approved with your product? No (not without further tests with product).
3. Stability and harmonization (ICH). Do the selected conditions of temperature and humidity provide an effective challenge for the product and the pack? Sometimes yes, but generally no.
4. Do all bought in supplies have an identified and proven shelf-life? If not, does your company put a reexamination date on the label? Often another assumption.
5. What importance does your company place on retention samples? Under what conditions are they stored and are they stored in suitable packs? Often not asked.
6. Are retention samples kept of all packaging materials used both for tests and ongoing production supplies? Especially important during investigational tests.
7. Are contact materials, such as piping, mixing vessels, intermediate storage containers, packaging equipment, etc., given the same approval checks as patient/user packs? Usually, no.
8. Have adequate controls been used in all testing procedures? Essential to consider controls for investigational tests.
9. Has attention to high science and high technology meant that simpler factors can be overlooked? Perhaps this happens too often.

This also questions the possible need of suitable "controls" at any stage during product pack development, i.e.,

6.6. Control Options

Different batches. Different starting materials/origins. Differing methods of assembly. Different equipment.

Changing scales of manufacture
Modifications in processes
Different locations of manufacturing
Different people
How information is taken and recorded
How long samples are held prior to analysis
Changing climatic conditions during processing

Beware "production manufacture" carried out over weekends, possibly with makeshift staff (i.e., manufacture made under production conditions ideally simulating the final production process to be used when the product is "sold").

Remember the philosophy of thinking simply.

Note: No list is all embracing.

GLASS as a CONTROL! When plastics are used, this answers the question whether it would have been more stable in glass.

Note: There is a case for using sequentially numbered samples during the production of stability batches, as any unexpected differences from normal can then possibly be equated with the sequence of production, (i.e., any changing factors).

7. GENERAL OBSERVATIONS ON PACKAGING SELECTION

Product-pack evaluation can extend outside the pack that contains the product to be sold or in which the bulk drug substance is to be held and distributed. All contact materials must have some form of investigation from discovery of a new chemical entity through all stages to product launch followed by outgoing regular stability evaluation.

Contact materials may involve reagents used in analytical tests, vessels, piping etc, used in processing and any intermediate or temporary storage containers (at any stage).

This observation and way of thinking will need to be extended in the future. It also means that all investigational or feasibility evaluations may (will) have to be improved. This work is unlikely to reduce but become more intensive—hence a need for better "pack" checks will be inevitable. The ICH guidelines often assume that this work is currently done and is adequate, but there are and will be occasions where this is not the case. There is undoubtedly a need for better packaging research.

There will be an increasing need to have a "competent overviewer" who surveys the total scene and discusses possible deficiencies with all concerned. This will require both "internal and external" partnerships to reduce risks and currently improve the total level of information—as part of a total data philosophy.

Packaging technologists will therefore need to

Recognize where "assumptions" were made in the past
Improve their knowledge and information gathering
Improve their powers of observation
Effectively record all activities in detail
Draw relevant conclusions

Packaging in therefore a critical part of product appraisal and shelf-life and must be treated as such.

The fact that it has been seen as the "Cinderella" of the industry must disappear and be replaced as a sound technological and scientific subject with a higher recognition as to its fundamental importance to pharmaceuticals, irrespective as to whether ethicals, generics, OTC, or veterinary products are involved.

This knowledge may already be less than adequate with certain apparently well established drug substances, but each must be considered against the risk to user/patient and the company.

A key question is whether you would be satisfied by the data presented. Remember there are many ways in which "improvements" (the philosophy of quality assurance) can be achieved, but note even the most apparently logical arguments can contain flaws.

A typical "assumption" is that if all components are within the dimensional tolerances on a drawing then it is simply "within specification." However, in certain cases, especially those related with closures and containers—where it is within specification (minimum, mean, and maximum) may influence the results. The author has seen few reports where dimensional detail has been included.

For example, a pack with a screw cap is removed from a cycling cabinet, allowed to equilibrate with the laboratory environment, and analyzed. The results do not fit the normal expected pattern. What information could have possibly helped to explain the result—think about it! Could it have been a lack of attention to packaging and environmental storage detail? Try creating a list of variables, as once this list exceeds 20 factors then you know that you are beginning to really think!

8. LIKELY IMPACT OF THE ICH GUIDELINES ON PACKAGING

The majority of pharmaceutical products show good stability over a wide range of climatic conditions. A minority, however, display instability that may be related to a single challenge or a multiplicity of challenges such as variations in temperature, exposure to oxygen, light, moisture, etc. These challenges may be combated by storage under defined conditions (e.g., below x°C, in a refrigerator, etc.) or in packs that act as effective barriers or restraints, to the critical deterioration factor(s). These cause and route of this is normally determined for the Drug Substance and the Product Dosage Form, early in the "Research" and "Development" stages, hence it should be relatively easy to find suitable storage conditions and a suitable form of "protective packaging." However, the full role of the pack extends beyond that of basic protection, as there are factors associated with overall quality, production performances, including assembly with particular emphasis on an effective closure systems, economics, etc., right through to ultimate use and final disposal. A packed product must therefore be seen as an effective marriage between the product and the pack, since the pack makes a major contribution to overall performance and to the declared shelf-life. The importance of the pack must therefore not be underrated. Having placed due emphasis on the role of the pack, it is now possible to consider the most likely impact of the ICH Guidelines on the future of the pack and packaging in general, and vice versa. The ICH Conditions are often equated with a "worst" situation or theory both in the chosen testing conditions and in the selection of the lower line in regression analysis as representative of the product shelf-life. Variations within and between packs can also contribute to this situation by lack of effective evaluation and analysis within a stability program. An example of this can arise where a pack on test is removed from a hot storage condition and is allowed to cool prior to chemical analysis being carried out. At the higher temperature of storage the closure fit could be loose or looser and then tighten up as it cools, hence giving a false impression of effectiveness, if only checked immediately prior to analysis. To cover this type of situation requires cap forces and physical appearance to be checked as follows:

a. On immediate removal from the storage condition
b. After conditioning to the laboratory environment
c. Plus a possible further examination after product removal

Test (a) and (b) have to be carried out on separate samples. This makes an assumption that such work has not been evaluated during development investigational stages. With such a background any resulting variations in analysis, then have a better chance of being coupled to possible pack variations or defects. It should be noted that there can be an assumption that there is significant data on plastic screw caps on plastic screw threaded bottles. This can often be untrue as the plastic used on both cap and bottle may vary, in addition to any dimensional detail. Experimentation is also complex as it needs to cover a range of torques (maximum, mean, minimum), immediate torques after removal from the storage condition, and torque after cooling or warming up (packs can also be stored at colder temperatures). This work needs to be supported by full dimensional details as the tolerances allowed within the specification may also contribute.

Since torque evaluation is a destructive type of test, i.e., a cap can only be removed once, a full statistically planned experiment can involve thousands of samples. To find out what happens if the cap is reapplied and further stored and removal is yet another experiment.

The storage condition chosen under the ICH guidelines, 40°C 75% RH, can be a very severe environment for a pack (try sitting in it for one hour and see how you like it!) An indication of its severity has previously been given by reference to a typical crisp biscuit like Rich Tea, where shelf-like is related to its direct pickup of moisture, when a typical storage condition of 15 to 20°C with an RH of 40% is compared with storage at 40°C 75% RH can be a near equilibrium condition, should a product have an aqueous base containing dissolved "salts" in a permeable pack, hence little moisture loss might occur. Reducing the storage condition to say 10 to 20% RH at 40°C, which increases the humidity/vapor pressure gradient between the external atmosphere and the aqueous product, will significantly increase the moisture loss, and the concentration of the active can increase. Again such a condition applies a severe challenge to both the product and the pack, thereby giving an acceleration factor of a high order. In each of these examples one must consider whether the product itself contains any free water, which can exert a vapor pressure and introduce a contra-effect as temperatures increase. The potential for moisture loss or gain primarily depends on the nature of the product, the gradient between the inside to outside of the pack, the efficiency of the closure, and the permeability of the pack. In general, materials that have a higher absorption to water have greater permeabilities. However, differences between real world climatic conditions and the ICH Guidelines must be constantly kept in mind. The guidelines normally only involve tests that employ the primary or immediate pack. The influence of the secondary or surrounding pack as required for display, warehousing, and distribution is normally not included in any tests. In the majority of circumstances these will serve to increase the shelf-life, particularly where effects due to moisture or gas permeation are involved. The surrounding packaging materials will also provide some "insulation" from the diurnal extremes in that the product is unlikely to reach the lowest and highest temperatures or be fully exposed to the humidity extremes. This combination of insulation and enclosure, whereby certain possible exchanges/interactions are slowed down, can improve the product shelf-life significantly. This however, can conflict with the fact that testing under the current ICH Conditions may suggest that certain packs need better barrier properties than were necessary previously. For instance, the author has recently been approached by three different companies whose blister packs have failed to provide an adequate shelf-life either under the standard 25°C 60% RH and/or 30°C 60% RH tests. This information is in conflict with some earlier products that are already being successfully marketed in blister packs (especially applies where hard gelatin capsules are the product form).

The ICH Guidelines also use one temperature and humidity at a time with no normal types of fluctuation other than those within the defined specification, i.e. ±2°C and ±5% RH. As mentioned previously, most storage conditions experience the normal diurnal changes that typically arise during the day–night period. How the pack and the associated basic materials react to both static and fluctuating conditions requires more intensive study, particularly in terms of possible acceleration effects. Certain critical differences between products and packs need highlighting, i.e., packs can dimensionally change according to the storage conditions, and certain properties can significantly change as moisture content reduces or increases. These are usually independent of Arrhenius type orders of reaction.

Materials can also change or "age" with the passage of time. The monitoring of possible surface changes can be especially critical and require the use of selected analytical techniques to examine the interface between product and pack. This trend towards more closely evaluating the product–pack relationship will inevitably vary according to the risk perceived to both the patient and the company and the use of "intensity of testing tables" will become more accepted and more critical. Packaging technology will therefore need more science and a more scientific approach. This will need to be supported by better validation and improved traceability.

The fact that the committees creating the guidelines had little knowledge of packaging, as no experts were directly employed, will therefore generate a new era in packaging where attention to detail must inevitably increase and aspects not evaluated previously will have to be considered. As previously mentioned above, this will extend to more surface analysis and also involve estimations that quantify the purity and impurities in packaging materials more effectively.

In conclusion, the advent of the ICH has undoubtedly more clearly defined formal stability testing, which with the use of matrixing and bracketing can lower potential costs. Although these costs have always been high, and any savings can be significant, the guidelines are likely to put more emphasis on the level of testing required prior to formal stability, both in research and in development. Since the accumulation of data from the discovery of a new chemical entity to the ultimate withdrawal of a product (following a successful period of "life") provides support both to product registration and its ongoing retention on the market, this work is also going to extend. The balance between saving on stability and spending more on general testing needs to be qualified and quantified as cost savings, long term, may be unlikely. Improving the data available at the earliest stage will, however, be commensurate with the earliest launch date. This will apply both to research and development and cover such factors as faster development and better analytical methods, improved chemical substance screening, clearer understanding of the product form required, including release characteristics, extension of the use of surface analysis techniques, etc.

Areas which require special attention (especially where not covered by the ICH Guidelines) include

A. Influences due to storage (stacking, drops, compression) and distribution (impacts, compression, vibration, puncture, etc.). Note top compression, and the influence of vibration, can be particularly critical.

B. Cycling conditions as found over various parts of the world (i.e., those in cold/dry conditions are significantly different from hot/dry and hot/humid areas of the world). Just testing under the static ICH Conditions is unlikely to be adequate.

C. Patient use, testing and final disposal. Again, this may need to cover temperature, humidity extremes, and cycling conditions as indicated in Ref. 2. However, possible patient misuse or abuse may be less easy to define, i.e., will patients blow hot moist breath through certain types of "lung" inhalation devices prior to use in order to "clean" off any possible adhering debris? Although checking for such possible activities may be uncovered through patient questionnaires, patients can be oblivious of what they do, hence patient observation may be an essential part of any evaluation.

Advice on disposal may also be required, especially where unique materials are used, or large quantities could create a disposal problem (e.g., large quantities of neutral Type I glass can cause severe problems if incorporated into other types of glassware, i.e., types II, III, and IV). Potent low-dosage drugs can also become dangerous substances when shipped in bulk.

D. Improve evaluation and quantification of pack details and any associated pack variations. This basically means that knowledge of raw materials, pack conversion processes and components, and assembled packs must be more effectively assessed and then held in databases. Unfortunately many experiences of the past tend to be anecdotal and are lost as people retire. Achieving effective control of such basic detail is particularly important for Third World countries, who frequently suffer from the sophisticated world's often long lost or forgotten earlier problems. Such information also tends to be edited out of books, as they are seen as outdated unnecessary information. Creating special books for the Third World also tends to be economically nonviable, as there is often no guarantee of sufficient sales.

However, it can be said that attention to packaging detail, including levels of "purity" and traceability, remains a weak link in the general clearance of a product-pack. Checking of the pack, when samples are withdrawn for stability assessment, is one further aspect, which requires special consideration.

It should be noted that packs do not necessarily follow any normal rules of degradation, e.g., polymers expand and contract according to conditions of temperature and, to a lesser extent, humidity.

E. How batches of product are prepared for formal stability also needs special attention with full coverage of all details. However, it should be noted that materials used in any type of test from initial research to the end of development should undergo a full QC clearance assessment (called QC plus by the author). Until a "final" specification (there is no such thing—all specifications should be subject to constant review) is made, there are many factors that may need quantifying and qualifying prior to them being deemed either a necessary or an unnecessary part of the "ultimate" specification.

Finally, the challenging conditions advised under the ICH Guidelines (no real conditions are at a constant 40°C 75%/15% RH or even 30°C 60% RH) may infer that more protective packs may be required in the future. Whilst the author suggests that general feedback from the market place does not support this observation, the means by which barrier properties could be improved needs, at the very least, to be identified and considered. The basic options may therefore involve choosing from the following:

F. A better barrier material, e.g., glass or metal to replace plastic, a plastic bottle to replace a blister pack, etc.

G. Use of the same or similar materials, with an additional barrier coating or a more effective barrier coating.

H. Placing one or more additional "overwraps" around the primary pack or improving the barrier properties of the secondary materials.

I. Recognizing and proving that the secondary packaging materials can add to the shelf-life of the product.

J. Accept a reduction in shelf life.

9. RISK VERSUS INTENSITY OF TESTING

It would appear reasonable that the risk (to patient/user and company) could and should bear some relationship to the intensity of the tests involved in the clearance of the materials from which the primary or immediate pack is made.

Intravenous (IV) solutions head this list as with the large volumes involved as even a small level of "extractive" will be rapidly in contact with the most sensitive body organs, i.e., heart and brain.

e.g., 1000 mL extractive 0.1% = 10 gm (IV solution)
20 mL extractive 0.1% = 20 mg (nasal spray)

Although this example has been oversimplified, it should confirm a need for different intensities of testing according to the product category/route of administration. This will also be influenced by

A. The product-to-pack contact area
B. The nature of the product (chemical–physical characteristics)
C. The contact time and temperatures (etc.) of storage and the nature of the extracted constituent

The list below suggests a product "intensity of testing" order, which is headed by sterile products. However, it must be stressed that the position in this order may change according to any special product characteristics, contact area, contact time, temperature, etc. The list is therefore given to create a greater awareness of how "risks" to "benefits" might be considered in package-product clearance. Some additional "other factors" are included as a separate list.

10. INTENSITY OF TESTING LIST

1. Sterile Products
   Large-volume parenterals (LVPs)
   IV solutions
   Irrigation solutions
   Dialysis solutions
   a. Small-volume parenterals (SVPs)

IV additives
Intravenous
Intramuscular
Subcutaneous
Intrathecal
Etc.
b. Implants
c. Ophthalmic products
Multi and unit dose
d. Nebulization solutions (may be sterile or preserved)
e. Wound care products (may be sterile or of low bioburden)

2. Nonsterile Products
   Powder products administered to the lungs
   Transdermal patches
   Vaginal and rectal products
   Buccal products
   Local and topical products
   Liquid oral products
   Solid oral products
   Note: Lists are not comprehensive and are only examples are given
3. Intensity of Testing Other Factors
   a. Patient type, i.e., elderly, adult, child, neonate
   b. Treatment point or period
   c. Volume of product to pack contact area
   d. Nature of products, e.g., oils tend to be more extractive than water based materials (rubber/plastics)
   e. Storage conditions—ICH conditions and "others"
   f. Storage period
   g. Processing factors, e.g., terminal vs. aseptic processing
   h. Type of pack, i.e., bulk pack for dry substance
      Bulk pack for sage form
      Original pack (OPD) for dosage form (unit of use pack)
4. Other Test Variants to Consider
   a. Type of product—ethical (NCE)
      Ethical (product extension)
      Ethical reformulated modified process
      Change in pack—major or minor
      OTC
      Veterinary
   b. Pack size variants—volume, weight, or number
   c. Storage period and storage status: unopened or opened—including variations in fill
   d. Storage conditions—controlled (simulated) ("others" and "real"); cycling (simulated); ambient (actual)
   e. Storage and transit conditions
   f. Methods of use—Instructions—use—misuse, abuse
   g. Process assembly variants
   h. Pack position—upright, on side, inverted

11. SOME POINTS REQUIRING SPECIAL CONSIDERATION

1. Beware of assumptions
2. The need to quality control all materials used in tests/evaluations by "QC plus"
3. The need to quantify all activities—keep good records
4. The importance of traceability—background information

The issuance of EU, WHO, and ICH Guidelines for formal stability programs makes many assumptions. For example, it is assumed that a company builds up in-depth knowledge on any new chemical entity or drug substance before it is put on a formal stability program, so that the chance that it will fail or not live up to expectations, when stability tested, should be remote. In the case of an established drug (as used for OTC products or generics), initial confidence is achieved by literature searches or cross-references to existing data. This assumes that the drug substance is synthesized by the same route, but if this route changes, then additional extensive testing may be required. Formulated products follow a familiar pattern. With ethical products the initial stability of product and pack(s) are achieved by accelerated and then longer term tests on the product, as associated with safety testing on animals, human volunteer studies, initial and ongoing clinical evaluations and general investigational studies, which are likely to include reformulation challenges, pack-product investigational studies, etc. Through all these studies the product-pack has to be supported by a shelf-life calculation, based on defined specifications. The various guidelines therefore make assumptions that this type of work, often covered under the heading of a "total data philosophy," has built up high confidence at all stages of research and development covering all the materials employed. Part of this confidence requires that all incoming materials related to product and pack undergo a full QC examination—what the author calls "QC plus," since it is on this information that a major part of subsequent specifications are developed. Knowing exactly what is being used in tests and evaluations generally, requires greater in-depth knowledge than what will be required at the final marketing stage, i.e., at early stages one tends to quantify all possible factors, as those that are most critical and need including in the final specifications will only emerge following experimentation and experience. This is particularly important where the product and/or pack are totally new with little or no prior historical experiences. This need to build up information must be applied at all stages, being based on effective

1. Observation/information
2. Recording in detail
3. Drawing any relevant conclusions

This should include how products are made, how packs are assembled, with what materials, how evaluated, etc., thereby being covered by the broad term "traceability," i.e., the ability to trace the history or background of any material or operation. The challenges that achieve this information invariably involve the use of such interrogative pronouns as how, where, what, by whom, why, which, when, etc., as part of the recorded information required. One also has to decide from this detail which answers are most relevant to a particular situation. The guidelines therefore make assumptions that such work has been effectively completed before formal stability testing is started on the drug substance and the product. The more critical aspects of this work should be written up as formal documents, which may form both part of a regulatory submission and part of the justification of selecting a product in a specific pack form. In the investigative work that supports the above, it may be necessary (and indeed advisable) to use both ICH and other challenging conditions, which should include adequate controls. These controls may involve selected conditions (many products, though not all, are more stable in a refrigerator) and other types of packs. If a product exhibits some deterioration in plastic, a glass control answers the question "Would it have been more stable in glass?", hence a glass control is advised in early investigational stages, etc., e.g., a plastic screw-capped bottle containing a product shows discoloration of bottle threads when the cap is removed: it raises the question whether this is product or pack material related. Controls could be samples of pack with no product, plus separate retention samples of bottle and cap, as this enables the source of the problem to be quantified. Packs may also have to be stored in various positions, i.e., upright, on the side, upside down, as these may also influence product stability, etc.

Product evaluation and stability can also relate to many other possible variables, i.e.

a. People and their responsibilities/functions
b. Materials employed
c. Equipment—instrumentation—machinery
d. Documentation
e. Environmental conditions and their control, as used for tests. Environmental conditions of any operation or process
f. Services, and their control
g. Facilities, and their general quality
h. General legislation, including packaging and waste directive, child resistance, etc.

Each of these factors could be in-depth discussed as part of the aspects of R and D. In any area we can be guilty of making assumptions, as too often all of us believe that we always do an excellent job and that there are no problems associated with our equipment, instrumentation, services, facilities, etc. Even with today's emphasis on validation, deficiencies can still arise. Moisture loss or gain in a climatic cabinet can vary according to the type of shelving used (solid or perforated) or whether samples are tightly packed or loosely packed (have air spaces around them), and this is one area which is rarely investigated. Such conditions are also difficult to validate fully, as conditions may change depending on whether the area is full, part full, or nearly empty, the nature of the material therein (absorptive or nonabsorptive of moisture, heat, etc.) and other factors. Such simple factors can be overlooked as the sophisticated analytical methodology now available usually has greater interest. The use of regression analysis, where the worst line of fit may be used for the shelf-life, can often be influenced by outliers, which could be packaging related, e.g., tightly packed samples will show slower moisture loss or gain when sampled at 3 or 6 monthly intervals at the beginning of a stability test, whereas the samc when sampled annually (year two onwards) from a less tightly packed stock may show a different pattern of moisture exchanges.

12. THE DEMAND FOR GUIDELINES

Experience has indicated that most people would like to ask the question: "Is there a set of rules or guidelines to which I can work with reference to product-pack stability?" Although the answer to this is now a broad "yes," there will be certain exceptions, i.e., factors, that are special to each product-pack-administration system that needs identification. This means that specifications and the relevant test procedures cannot be totally set by a fixed approach. What is relevant will therefore vary from product to product. This thinking also applies to what is generally termed "product quality," including "elegance" as this also will vary according to the product, its manufacture, and the situations in which it is stored, transported, displayed, and used.

Guidelines are also under constant review where areas have to be updated as new ideas of information, recognition of points of weakness, etc., become identified. In these circumstances, a product registered in territory A five years ago may have an area of deficiency that has to be investigated and cleared before it can be registered in another territory. The standards of yesterday are therefore unlikely to satisfy totally those of tomorrow.

13. CONCLUSIONS

There are many process procedures that can modify the properties of plastic. The more obvious ways in which plastic performance can be influenced include

1. Changes in the solubility of the absorbing materials in the polymer
2. Changes in the solubility of the absorbing materials in the surface layer
3. Changes in the surface conditions
4. Mobility of permeate within the plastic
5. Opening up a molecular structure by swelling (increases permeation)
6. Movement of internal constituents to the surface (or vice versa)
7. Changes in mass or density
8. Molecular changes due to degradation, including chain fission
9. Changes in particle size or dispersion of constituents, etc.

If any of the above or combinations of the above occur, changes may arise to physical, physicochemical, or chemical properties, e.g., in terms of diffusion, permeation, or migration/leaching.

Thorough retesting is, therefore, sometimes essential to detect changes between the "processed" and "unprocessed" (as supplied initially) plastic.

It has also to be recognized that stored plastics may change or "age" with the passage of time prior to their use to pack a product. This can mean that greater attention needs to be paid to how materials are packed, stored, etc.

REFERENCES

1. United States Pharmacopeia (USP23/NF18). Rockville, MD: United States Pharmacopeial Convention, 1959–1963.
2. European Union Guideline, Notes for Guidance EU/III/359/91.

3. European Union Guideline, Drug Master Files EU/III/3836/89.
4. European Union Guideline, Plastics III/9090/90.
5. United States Pharmacopeia (USP23/NF18). Rockville, MD: United States Pharmacopeial Convention, 1959–1963, pp. 1781–1790.

15

Industrial Stability Testing in the United States and Computerization of Stability Data

SHRI C. VALVANI

Pharmaceutical Consultant, Kalamazoo, Michigan

1. INTRODUCTION

Stability testing is an integral part of the pharmaceutical development process. It is routinely performed on drug substances and drug products. Drug substances can be derived from chemical synthesis, from plant or animal sources, or from biological, biotechnology, or recombinant DNA technology. Stability testing is carried out on drug products from chemical or biological sources, prescription drug products for human health and for veterinary use, over-the-counter drug products, and medical devices. Stability testing principles are applicable to all of the above types of drug substances and drug products. Specific unique requirements covering each of these areas exist. A thorough description of the stability system covering these areas is beyond the scope of this chapter. This contribution primarily deals with the stability of pharmaceutical new chemical entities and drug products for human health. The chapter relies on principles and concepts discussed in the recent draft of stability guidelines (FDA, 1998b) published by the Food and Drug Administration in the United States. It is recognized that the FDA guidelines are in draft format, which have not been finalized as of this writing. It is not known whether the guidelines will have been finalized by the time this monograph is published. It is this author's assumption that there will be changes in the final guideline from the current draft, but it is my expectation that the current draft guideline will retain most of the key contents without significant modifications. In addition, the guidelines covered in the Tripartite Report on the International Conference on Harmonization are applicable to the chapter contents (ICH, 1994).

The primary purpose of stability testing in an industrial setting is to provide supporting evidence on stability behavior of chemical or biological entities as well as to study the stability behavior of pharmaceutical drug products. Stability testing as a function of time against a variety of environmental factors such as temperature, humidity, light, and combinations of these parameters is critical for the establishment of recommended storage conditions, retest periods, expiry dates, and shelf lives of pharmaceutical products. Stability behavior in broad terms refers to overall quality of drug ingredient or drug product in terms of strength, purity, identity, safety, apparent degradation, physical or biological changes, and their effect on biological performance of drug products.

Many factors drive the process of industrial stability testing for pharmaceutical products. These include the generation of background information for the development of new products, the support of manufacture and distribution of pharmaceutical products in different regions of the world, the development of suitable packages to insure the quality, strength, purity, and integrity of the product, the satisfaction of the submission of stability information requirements for regulatory agencies around the globe, the determination of shelf life, and several other factors.

This chapter is divided into three sections. The first section, "Stability Testing and Product Development," covers general aspects of stability principles involved in the product development process. It provides the objectives of stability studies during various stages of drug substances and drug products as they evolve from the discovery phase to clinical formulations, to drug product for marketing, and revised product post approval. The next section, "Development of the Stability Testing Function," covers the development and establishment of the industrial stability testing function. It includes various activities and associated regulatory guidance information related to written testing programs, standard operating procedures, stability protocol generation, testing frequency, selection of batches, matrixing, bracketing, and other factors that must be addressed in establishing the industrial stability testing function. A discussion of the documentation required to minimize the chances of regulatory citations is included. This section provides information related to stability test conditions and stability rooms, cabinets, and chambers as well as a table with a list of suppliers and manufacturers of constant temperature/humidity rooms, cabinets, and chambers in the United States. In addition, principles for photostability testing and a list of the suppliers and/or manufacturers of the photostability testing chambers are included. The third section, "Management of Stability Information Using Computers" provides an account of the principles involved in the development of computerized stability information systems. Factors involved in the design of stability systems from data operations, performance factors, systems analysis, systems design, validation, maintenance, and other principles are included. In addition, a table with a list of developers and suppliers of stability system software and laboratory information management systems for pharmaceutical applications is included.

2. STABILITY TESTING AND PRODUCT DEVELOPMENT

Stability testing function is an evolutionary concept covering the life cycle of pharmaceutical product development. The objective of stability testing varies during various stages of product development. For example, during the early discovery phase, the primary focus is to generate stability characteristics of a chemical ingredient or biological entity, which will be helpful in the design and development of drug or biological products.

During later stages of product development, the goal of stability testing is to establish shelf life for formulations packaged in a final package intended for commercial introduction. Stability testing principles can be subdivided into various stages of development. The objective of stability testing differs in each stage of development. The following sequence describes several stages of drug development during which stability characterization should be carried out.

Stability testing of new chemical or new molecular entities (discovery phase)
Stability testing of formulations for preclinical safety testing or toxicology testing (preclinical stage)
Preformulation stability testing of new chemical entities (pre-IND stage)
Accelerated and normal storage testing of clinical formulations (IND stage)
Stability testing of early formulations, formulation mixtures, and packaging materials evaluation (product development stage)
Long-term primary stability testing of final marketable formulations in proposed packages for marketing for product registration and approval (NDA stage)
Stability testing of production batches post regulatory approvals (approved product stage)
Stability testing of revised products (revised product stage)
Stability testing of new chemical or new molecular entities (discovery phase)

During the discovery phase, stability testing of a new chemical entity or a new molecular entity is required to help select the most satisfactory chemical or molecular entity possessing the right pharmacological, toxicological, and pharmaceutical profiles. The pharmaceutical profile is mostly focused towards the optimum chemical and physical stability characteristics, a good preformulation profile, and satisfactory manufacturing potential. The emphasis initially is not only towards selecting the right chemical entity but also towards selecting the appropriate physical form (for example a satisfactory polymorphic form if applicable, or a physical form with the most desirable handling behavior), and, where applicable, the base, salt, ester, hydrates, solvates or other forms with optimum stability. It is not uncommon to find vast differences in stability characteristics of different salts of the same chemical entity. In addition, preliminary information regarding particle size, distribution, crystal shape, crystal habits, mechanical properties, and specific surface area can play key roles in affecting the stability, manufacturability, and biological performance of the chemical entity. These studies help establish the boundaries within which one must operate to design formulations for toxicology and for initial clinical testing. In the current competitive accelerated drug development environment, early physicochemical investigation has to be completed not only in the shortest time possible but also with limited quantities of drug substance.

2.1. Stability Testing of Formulations for Preclinical Safety Testing or Toxicology Testing (Preclinical Stage)

For a new drug to be evaluated for preclinical safety assessment, often in several animal species, it must be formulated in a dosage form or a delivery system that will deliver the drug in a way that maximizes the availability of drug at the site of action. The development of early dosage forms for preclinical testing and eventually for early testing in humans requires an extensive stability evaluation and interplay of physicochemical, biological, and dosage form considerations. Good laboratory practices (GLP's) dictate that all substances or formulations administered to animals during preclinical safety testing remain stable for the duration of toxicological testing. Preliminary stability testing on all formulations must be carried out using stability indicating assays in accordance with GLPs to insure that toxicology formulations are not adversely affected during toxicology testing. Since many toxicology formulations are suspensions, homogeneity and content uniformity are critical stability parameters, which must be evaluated. Suspensions must possess good homogeneity to deliver uniform dosing and should demonstrate acceptable physical stability, especially minimal settling tendency with good resuspendibility. Minimum stability testing for all toxicology formulations used in GLP studies requires that an entrance assay prior to the initiation of toxicological testing, and an exit assay, must be performed at the end of the studies.

2.2. Preformulation Stability Testing of New Chemical Entities (Pre-IND Stage)

The formal development process usually begins with preformulation characterization and initial stability testing of a drug substance. Preformulation testing is the study of physicochemical properties, which could affect the drug performance, processing, design, and development of efficacious dosage forms. Most of the preformulation characteristics are molecular properties of the drug, which are affected by environmental and physical–chemical variables. Preformulation characteristics are critical to successful formulation design and development.

Preformulation and stability evaluation of a chemical entity often includes systematic exposure in the solid state at various temperatures (normal storage temperature and accelerated condition), at high humidity (75% relative humidity or higher), as well as photodegradation exposure under standard conditions defined in the ICH Guidelines (ICH, 1994). It is important to carry out rigorously defined solution stability testing at various pH's and ionic strengths and to conduct an evaluation of selected additives (to elicit chemical instability due to oxidative, photolytic, hydrolysis, isomerism, racemization, decarboxylation, polymerization, deamidation, and various other pathways) and formulations likely to be considered in formulation development. In addition, an evaluation of microbiological contamination needs to be performed. This is particularly important for new chemical entities for sterile dosage form development.

During the initial stability evaluation phase, stability testing under a variety of heat, humidity, light, and solution conditions (pH, ionic strength, additives) provides a systematic evaluation of the stability characteristics of a new chemical entity. The main aim of stability evaluation during this phase is to help establish the rate and mode of degradation pathways and if possible the mechanism of degradation including the identification and quantitation of degradation products. These studies are critical for the development of suitable analytical methodology that serves as the basis of stability assay procedures.

In addition to normal preformulation evaluation, forced degradation studies under highly stressed stability conditions (at much higher temperature than the usual accelerated conditions) is undertaken. For example, stress testing at 50°C, 60°C, or at higher temperatures is carried out. Generally a single batch of drug substance is sufficient for the conduct of forced stress testing. While it is recognized that the degradation pathways of forced stress stability are unlikely to be encountered during normal storage or under accelerated temperature conditions of long-term stability testing, they do help establish the worst case scenario, and importantly they help in the development and validation of stability indicating analytical procedures.

2.3. Accelerated and Normal Storage Testing of Clinical Formulations (IND stage)

Regulatory guidelines in the United States as well as in other parts of the world require that formal stability testing of early formulations for human use must be initiated.

For example, the regulatory requirements for Investigational New drugs (IND's) state that information sufficient to support stability of the drug substance during the toxicological studies and the planned clinical trials should be available. The goal of these studies is to elucidate identity, quality, strength, and purity, and to help establish the retest period for drug substance. Similarly for clinical formulations, the agency requirements state that the information to assure the product's stability during the planned clinical studies should be available. The stability data generated during this early phase is used to establish storage conditions and specifications. Stability information for submission during this phase should include a brief description of the stability study, test methods for monitoring the stability of the clinical formulation in the proposed container/closure, and tabular stability data of the representative batches.

Accelerated and normal storage temperature testing of drug substance and for clinical formulation must be initiated prior to the initiation of clinical studies. The goal of these studies should be to generate information to insure that the clinical formulations are likely to remain stable during the planned clinical studies. Generally at the time of filling an IND, at least one-month stability data at accelerated conditions for a single batch of the clinical formulation should be available. This is consistent with the Food and Drug Administration's policy, which states that the regulation does not preclude a sponsor from conducting stability tests on an investigational drug product concurrently with clinical investigations of the product. However, the agency does expect that, by the time clinical studies have begun, the sponsor will have submitted to the FDA at least preliminary evidence (obtained from accelerated studies) to show that the product is likely to remain stable for the duration of the study.

> Further insight regarding the gradual nature of the investigation can be gained from the FDA's comments regarding INDs, which state "the regulations at 21 CFR 312.23(a)(7)(i) emphasize the graded nature of manufacturing and control information. Although in each phase of the investigation sufficient information should be submitted to assure the proper identification, quality, purity, and strength of the investigational drug, the amount of information needed to make that assurance will vary with the phase of the investigation, the proposed duration of the investigation, the dosage form, and the amount of information otherwise available. For example, although stability data are required in all phases of the IND to demonstrate that the new drug substance and drug product are within acceptable chemical and physical limits for the planned duration of the proposed clinical investigation. If very short tests are proposed, the supporting stability data can be correspondingly very limited (FDA, 1995a).

2.4. Stability Testing of Early Formulations, Formulation Mixtures, and Packaging Materials Evaluation (Product Development Stage)

The goal of the pharmaceutical scientist is to design and develop a dosage form that is stable (for the entire shelf life or until the expiry date), that contains the precise amount of drug that will be delivered in the most available form, and that can be manufactured on a large scale in an economic manner to meet regulatory approvals and market needs. Stability testing must be performed to ensure that the drug products not only remain chemically stable (maintaining their fully potency) but also retain satisfactory physical stability attributes in terms of elegance, appearance, lack of discoloration, as well as in-vitro and in-vivo performance characteristics (for example, dissolution). During product development stage, stability testing of several early clinical formulations leading up to marketable formulations should be performed. Also a systematic evaluation of container/closure systems and packaging systems is initiated to select an acceptable package. Packaging materials evaluation makes use of the preformulation information on the drug substance as well as stability evaluation of clinical formulations. The objective during this phase is to identify the stability limiting factor(s) and help establish shelf life for products.

Since many products go through several distinct steps during manufacturing, it is very common to have intermediate granulation, powder mixes, solutions that need to be stored prior to the next manufacturing step. Stability evaluation of intermediate mixtures that are likely to remain in the intermediate state for more than a few hours should be performed. For example, when a mixture of active and inactive ingredients or a granulated mixture waiting to be compressed into tablets or to be filled into capsules is produced, a short term stability evaluation in an appropriate packaging system needs to be performed. Similarly, if the finished dosage forms are stored in interim containers for an extended period of time prior to filling in marketed containers, real time stability data should be generated to justify the length and condition for acceptable storage. Often this type of short-term stability evaluation is undertaken during pilot plant scale-up or validation operations. Generally, interim storage of finished product exceeding six months is not appropriate. The primary driving force for this consideration is that the expiry date is computed from the date of quality control release of the batch, but no later than 30 days beyond the date of manufacture, and is independent of the packaging date.

2.5. Long-term Primary Stability Testing of Drug Substance and Final Marketable Formulations in Proposed Packages for Marketing for Product Registration and Approval (NDA Stage)

Early stability studies, preformulation studies, and ongoing development studies with a drug substance help establish the formal stability program for generation of stability data for registration applications. Critical stability characteristics and stability results should be used to determine the appropriate storage conditions, retest periods for internal use, and expiration dates for drug substances being sold commercially. The reader is referred to the detailed discussion in Section 3.5 on the selection and number of batches. Stability of these batches should be determined using stability indicating assay methodology. Test intervals should follow the minimum frequency as covered in Section 3.10. A stability program should include a commitment to follow the stability of a number of batches annually. The stability of drug substances should be evaluated in containers that simulate the storage container used for warehousing or the container used for marketing. For example, if the drug substance is stored in polyliners within fiber drums, stability samples should be stored in smaller polyliner bags of the same material and smaller size fiber drums or containers of similar or identical composition to the large-scale drums. Care should be exercised in selection of the size, surface-to-volume ratio, and air space, as they could affect the relative stability, possibly because of differences in relative humidity or because of specific interactions with polyliners. Changes in manufacturing site, materials, or manufacturing process require careful evaluation, as they could affect stability.

Experience gained from clinical formulations of drug product during earlier phases and the knowledge of stability behavior of drug substance help establish the stability program for the drug product for registration purposes. Critical stability characteristics and stability results should be used to determine the appropriate storage conditions, test intervals, and batches for the product registration stability program. It should be noted that strength in terms of chemical stability is not the only criterion of drug product stability. Drug products must maintain various chemical and physical properties to preserve the effectiveness, safety, elegance, and overall acceptance of the drug. Properties such as physical appearance, crystalline form, particle size, solubility, disintegration rate, dissolution rate, pH, sterility, viscosity, palatability (taste and odor), homogeneity, resuspendibility, and stability after reconstitution may be stability related and thus require testing and the setting of specific storage conditions and limits. In addition, tests may also be needed to determine the absence or presence of harmful degradation products. Safety testing may dictate an upper limit on the level for degradation product, which may be a stability limiting characteristic. Stability testing parameters should include those characteristics that are likely to influence quality, strength, safety, and/or efficacy. The reader should consult the detailed discussion concerning the selection and the number of batches as included in Section 3.5.

2.6. Stability Testing of Production Batches Post Regulatory Approvals (Approved Product Stage)

The goal of the stability evaluation program during this phase is to confirm and/or extend the expiration dating of the drug product that has been approved. In addition to the continuation of the monitoring of primary stability batches, a commitment should be made that long-term stability for the first three post-production batches will be generated using an approved stability protocol to justify the confirmation and extension of the expiration dating period. The commitment also includes monitoring of the stability of annual batches after the first three production batches. The commitment actually constitutes an agreement with the Food and Drug Administration indicating that the plan outlined in the product registration document will be strictly complied with. The commitment also requires submission of stability data at periodic intervals and/or annually as specified in the application. Finally, the commitment mandates that any batches that are found outside of the specifications approved with the product will be withdrawn from the market. If the deviation from the approved specifications is observed to be a single occurrence and is not expected to affect the safety and efficacy of the product, regulatory guidance allows notification to and discussion with the approving regulatory division with detailed justification for distribution of the product.

2.7. Stability Testing of Revised Products (Revised Product Stage)

Few products if any are marketed as approved for a long time before requiring some changes. Most products undergo some type of revision after the approval of the drug product application. Some of these changes may be internally driven, such as changes in package design, the addition of a new package, changing the shape and size of dosage forms, changes in manufacturing site, changes in manufacturing process, and so on. They may be externally driven, for example, deletion of dyes, regulatory restrictions on formulation ingredients, formulation changes. A key principle in the evaluation of stability for revised products is partially to revalidate the test method and confirm the applicability of stability indicating assay methodology or undertake modifications of the assay method.

Most post-approval changes for a revised product require some type of stability evaluation and comparative stability along with a commitment to monitor the stability of the first one or three production batches of drug substance or drug product and annual batches thereafter. The FDA has published a detailed account of the type of stability study required as well as the type of commitment through SUPAC-IR guidelines (FDA, 1995b). The type of stability data required for submission ranges from no additional stability data for a manufacturing site change within the same facility with the same equipment to 3 months of comparative accelerated stability data and available long-term on three batches of drug product with the proposed change. Likewise the stability commitment ranges from none to the first three production batches and annual batches thereafter on long-term studies. Overall, the goal is to apply good scientific principles and generate comparative stability data at accelerated conditions and long-term stability data with the proposed changes and a commitment to follow up long-term stability with the production batches.

3. DEVELOPMENT OF THE STABILITY TESTING FUNCTION

Development and establishment of a stability testing function in an industrial setting requires careful consideration of several areas of interest. Drug stability studies require a lot of resources; they add a lot of expenses to the development cost and require a long time for the generation of stability data. As discussed in the previous section, many batches of drug substance and drug products are put on long-term and accelerated stability testing from the initiation of stability studies during the discovery phase to clinical phases and finally the primary stability batches. Multiple batches are evaluated to ensure that a product will consistently remain within specifications for its entire expiration dating period. Stability studies are routinely conducted to investigate the effect of strength, package, batch, storage condition, and storage time on the stability of the dosage form.

For primary stability data generation a typical study will have three batches, several strengths, several packages, and several storage conditions. Thus if every batch by strength by package combination is tested for every storage condition, a substantial expense is involved. To reduce expenses, a bracketing design is commonly used; for example only the smallest and largest bottles are tested. In addition, the application of sound statistical principles by using fractional factorial-type matrixing designs for conducting the stability studies can substantially reduce the amount of testing required without compromising the outcome of results (Fairweather et al., 1995; Nordbrock, 1992). For complex stability studies involving many variables, a combination of bracketing and matrixing design can be used to gain some further economies in testing. Stability programs need to be well managed for efficient and effective operations. It is critical that a thorough review of ongoing stability studies be undertaken on a periodic basis to reduce and/or control costs.

Establishment of the stability testing function requires an extensive development of documentation to maximize the compliance and to minimize the chances of regulatory citations. It requires setting up multiple stability test conditions and stability rooms/cabinets/chambers. A table with a list of suppliers and manufacturers of constant temperature/humidity rooms/cabinets/chambers in the United States is included (Table 1). In addition, a list of the suppliers and or manufacturers of the photostability testing chambers is included (Table 2).

The following is not intended to be a comprehensive list. It covers several critical items that must be addressed. Specific guidelines for many of these exist in the guidance documents issued by the Food and Drug Administration and the ICH.

3.1. Written Stability Testing Program

The guidelines in Section 211.166 of 21 CFR (Code of Federal Regulations) state, "there shall be a written testing program designed to assess the stability characteristics of drug products. The results of such stability testing shall be used in determining appropriate storage conditions and expiration dates. The written program shall be followed and shall include ..." A written stability testing program is critical for the establishment of the stability testing function. The written program must address key issues such as the selection of batches, the development of stability indicating assay methodology, the testing frequency, the retest strategy, the selection of container/closures, and the stability of reconstituted products. The written procedure should also include the review and approval of stability data, the procedure for storage of raw stability data, and the periodic physical review of samples. In addition, the storage of samples, the procedure for handling out-of-specification results, review, action plan and sign-off procedure for deviations, equipment and storage malfunction, and many other issues specific to organization and product line should be part of the written documentation. Many other topics warrant attention in documentation. Every action performed in the management of the stability function must be documented in a SOP or some other form. Most regulatory agencies maintain that activities may not be claimed as having been performed without documentation.

3.2. Stability Protocol and Commitment

Generally a stability protocol should be generated for every batch that is being followed on long-term stability testing. The stability protocol should record the purpose for conducting the stability test, the method used, the testing frequency,

Table 1 Suppliers and Manufacturers of Stability Testing Rooms, Chambers, and Cabinets, Environmental Chambers, Humidity Chambers

Name	Address	Phone/fax	E-mail/web site
Absolute Control Systems	5168 Parfet Street, Unit G Wheat Ridge, CO 80033	Phone: 303-420-8600 Fax: 303-420-8692	E-mail: absolute@rminet.com Web site: http://www.absolutecontrolsys.com
ARS Enterprises	12900 Lakeland Road Santa Fe Springs, CA 90670	Phone: 562-946-3505 Fax: 562-946-4120	E-mail: arssubpci@aol.com
Atlas Electric Device Co.	414 N Ravenswood Avenue Chicago, IL, 60613	Phone: 773-327-4520 Fax: 773-327-5787	E-mail: mmacbaeth@atlas-mts.com
Barnstead/ Thermolyne	2555 Kerper Blvd Dubuque, IA 52001	Phone: 319-556-2241 Fax: 319-589-0516	
Benchmark Products Inc.	531 Bank Lane Highwood, IL 60040	Phone: 847-433-3500 Fax: 847-433-3545	E-mail: dmn@benchmarkprods.com
Biocold Environmental Inc	1724 Westpark Center Fenton, MO 63026	Phone: 314-349-0300 Fax: 314-349-0419	Web Site: http://members.aol.com/biocold
Brinkmann Instruments, Inc.	One Cantiague Road P.O. Box 1019 Westbury, NY 11590-0207	Phone: 516-334-7500 Fax: 516-334-7506	E-mail: info@brinkmann.com Web Site: http://www.brinkmann.com
Caron Products & Services, Inc.	PO Box 715, Products Lane Marietta, Ohio 45750	Phone: 740-373-6809 Fax: 740-374-3760	E-mail: caron@frognet.net Web Site: http://www.caronproducts.com
Clean Air Solutions	5011 Falling Leaf Trail North Syracuse, NY 13212	Phone: 315-452-4609 Fax: 315-458-0162	E-mail: dfeikert@ix.netcom.com Web Site: http://www.clean-air-solution.com
Cole-Parmer Instrument Co.	625 E Bunker Court Vernon Hills, IL 60061	Phone: 847-549-7600 Fax: 847-247-2929	E-mail: info@coleparmer.com Web Site: http://www.coleparmer.com
Controlled Environments, Inc.	612 W. Stutsman Street Pembina, ND 58271	Phone: 800-363-6451 Fax: 204-786-7736	
Constant Temperature Control Limited	220 Industrial Parkway South, Unit 22, Aurora, Ontario, L4G 3V6, Canada	Phone: 905-841-7749 Fax: 905-841-1669	
DJS Enterprises	110 West Beaver Creek, Unit 14 Richmond Hill, Ontario, Canada L4B 1J9	Phone: 905-764-7644 Fax: 905-764-7654	
Edge Tech Moisture + Humidity Instruments	455 Fortune Blvd. Milford, MA 01757	Phone: 800-276-9500 Fax: 508-634-3010	E-mail: h2o@edgetech.com Web Site: http://www.edgetech.com

Name	Address	Phone/fax	E-mail/web site
Electro-Tech Systems Inc (ETS)	3101 Mt Carmel Avenue Glenside, PA 19038	Phone: 215-887-2196 Fax: 215-887-0131	Web Site: http://www.electrotechsystems.com
Environmental Growth Chambers	510 E Washington Street Chagrin Falls, OH 44022-4448	Phone: 800-321-6854 Fax: 440-247-8710	E-mail: sales@egc.com Web Site: http://www.egc.com
Environmental Specialties Inc.	4412 Tryon Road Raleigh, NC 27606-4218	Phone: 919-829-9300 Fax: 919-833-9476	E-mail: esi@esionline.com Web Site: http://www.esionline.com
ESPEC Corp.	425 Gordon Industrial Court Grand Rapids, MI 49509-9506	Phone: 616-878-0270 Fax: 616-878-0280 Toll Free 800-537-7320	E-mail: espec-ad@espec.com Web Site: http://www.espec.com
Forma Scientific Inc.	P.O. Box 649 Marietta, OH 45750	Phone: 740-373-4763 Fax: 740-373-6770	E-mail: fmarketing@forma.com Web Site: http://www.forma.com
Gilson Co. Inc.	P.O. Box 677 Worthington, OH 43085	Phone: 800-444-1508 Fax: 740-548-5314	E-mail: gilson@coil.com Web Site: http://www.globalgilson.com
Hotpack	10940 Dutton Road Philadelphia, PA 19154	Phone: 215-824-1700 Fax: 215-637-0519	E-mail: Hotpack@hotpack.com Web Site: http://www.hotpack.com
Lunaire	2121 Reach Road Williamsport, PA 17701	Phone: 570-326-1770 Fax: 570-326-7304	E-mail: marketing@lunaire.com Web Site: http://www.lunaire.com
Lunaire Environmental	1719 Route 10 East, Suite 302, Parsippany, NJ 07054	Phone: 800-586-2473 Fax: 973-540-0367	
Luwa Lepco Inc.	1750 Stebbins Drive Houston, TX 77043-2807	Phone: 713-461-1131 Fax: 713-464-1148	E-mail: sschleisman@luwalepco.com Web Site: http://www.luwalepco.com
Powers Scientific Refrigeration & Incubators Inc.	P.O. Box 268 Pipersville, PA 18947	Phone: 215-230-7100 Fax: 215-230-7200	E-mail: psioff@voicenet.com Web Site: http://www.powersscientific.com
RKI/Clean Room Services	P.O. Box 16372 Rochester, NY 14616-0372	Phone: 800-447-4754 Fax: 776-621-2778	E-mail: clnrmsrvs@aol.com Web Site: http://www.cleanroomservices.com
Rotronic Instruments Inc.	160 E Main Street Huntington, NY 11743	Phone: 516-427-3898 Fax: 516-427-3902	E-mail: sales@rotronic-usa.com Web Site: http://www.rotronic-usa.com
Sanyo Gallenkamp	900 N Arlington Heights Road, Suite 320 Itasca, IL 60143	Phone: 630-875-3543 Fax: 630-775-0427	E-mail: tharding@compuserve.com

Name	Address	Phone/fax	E-mail/web site
Scientific Industries Inc.	70 Orville Drive Bohemia, NY 11716	Phone: 516-567-4700 Fax: 516-567-5896	E-mail: info@scind.com Web Site: http://www.scind.com
Servicor	830 Bransten Road San Carlos, CA 94070	Phone: 650-591-0900 Fax: 650-591-3121	E-mail: servicor@cleanroom.com Web Site: http://www.cleanroom.com
Spectrum Quality Products	14422 S San Pedro Street Gardena, CA 90248	Phone: 310-516-8000 Fax: 310-516-9843	E-mail: sales@quick.net Web Site: http://www.spectrumchemical.com
Sun Electronic Systems	1900 Shepherds Drive Titusville, FL 32780	Phone: 407-383-9400 Fax: 407-383-9412	E-mail: sun@digital.net
Surface Measurement Systems	24040 Camino del Avion, Monarch Beach, CA 92629	Phone: 949-495-1897 Fax: 949-495-7795	E-mail: domingue@smsna.com Web Site: http://www.smsna.com
Terra Universal	700 N Harbor Blvd. Anaheim, CA 92805	Phone: 714-526-0100 Fax: 714-992-2179	E-mail: info@terrauni.com Web Site: http://www.terrauni.com
Thomas Scientific	99 High Hill Road Swedesboro, NJ 08085	Phone: 856-467-2000 Fax: 800-345-5232; 856-467-3087	E-mail: value@thomassci.com Web Site: http://www.thomassci.com
VTI Corp	7650 W 26th Avenue Hialeah, FL 33016	Phone: 305-828-4700 Fax: 305-828-0299	E-mail: vti@vticorp.com Web Site: http://www.vticorp.com
VWR Scientific Products	1310 Goshen Pkwy. West Chester, PA 19380	Phone: 800-932-5000	Web Site: http://www.vwrsp.com

the storage conditions, the package description, and several other factors. The stability protocol is a detailed plan for the conduct of stability studies to generate stability data and their analysis using statistical principles where applicable, support of retest of drug substance, and for the expiration dating of drug products at labeled storage conditions. For drug products, acceptable labeled storage conditions include controlled room temperature, refrigerator temperature, or freezer temperature. In addition, stability protocols should be used to support an extension of the retest period or the expiration dating period through annual reports.

A stability protocol for product registrations should also include a commitment that labeled storage temperature stability data for the first three post-production batches will be generated to justify confirmation and extension of the expiration dating period. The commitment also includes monitoring of the stability of annual batches after the first three production batches. The commitment actually constitutes an agreement with the Food and Drug Administration

Table 2 Suppliers and Manufacturers of Photostability Test Chambers

Name	Address	Phone/fax	E-mail/web site
Caron Products & Services, Inc.	PO Box 715, Products Lane Marietta, Ohio 45750	Phone: 740-373-6809 Fax: 740-374-3760	E-mail: caron@frognet.net Web Site: http://www.caronproducts.com
Environmental Growth Chambers	510 E Washington Street Chagrin Falls, OH 44022-4448	Phone: 800-321-6854 Fax: 440-247-8710	E-mail: sales@egc.com Web Site: http://www.egc.com
Environmental Specialties Inc.	4412 Tryon Road Raleigh, NC 27606-4218	Phone: 919-829-9300 Fax: 919-833-9476	E-mail: esi@esionline.com Web Site: http://www.esionline.com
Lunaire	2121 Reach Road Williamsport, PA 17701	Phone: 570-326-1770 Fax: 570-326-7304	E-mail: marketing@lunaire.com Web Site: http://www.lunaire.com
Powers Scientific Refrigeration & Incubators Inc.	P.O. Box 268 Pipersville, PA 18947	Phone: 215-230-7100 Fax: 215-230-7200	E-mail: psioff@voicenet.com Web Site: http://www.powersscientific.com
Southern New England Ultraviolet	550-29 E Main Street Branford, CT 06450	Phone: 203-483-5810 Fax: 203-481-8589	Web Site: http://www.ultravioletlamps.com

indicating that the plan outlined in the product registration document will be strictly complied with. The commitment also requires submission of stability data at periodic intervals and/or annually as specified in the application. Finally, the commitment mandates that any batches that are found outside of the specifications approved with the product will be withdrawn from the market. If the deviation from approved specifications is observed to be a single occurrence and is not expected to affect the safety and efficacy of the product, regulatory guidance allows notification to and discussion with the approving regulatory division with detailed justification for distribution of the product.

3.3. Standard Operating Procedures

Standard operating procedures (SOPs), as the name suggests, are written procedures describing operational procedures. Current good manufacturing practices (cGMPs) mandate that every organization develops a set of SOPs to describe their operations. It is almost impossible to develop any stability management function without having a comprehensive set of SOPs. Standard operating procedures cover all aspects of operations. In the United States, numerous citations (Form 483 Inspection Reports) are issued every year to pharmaceutical companies to point out operational deficiencies, which must be addressed. Frequently, the regulatory citations are the cause of recommendations by FDA field inspectors to withhold approval of product registrations. These citations often are issued as a result of Pre-Approval Inspections. Many of those citations deal with two key principles, one with lack of standard operating procedure, the other with inconsistencies with practice and SOPs. Later deviation dealing with differences between practice and what is contained in the SOP is much more severe than lack of SOPs. We do not wish to condone errors. However, if one were to choose to err, it is better to not have an SOP than to have one and not follow it. While specific SOPs to address specific activity are required, the SOPs should be written in general flexible form without including rigid specific references that may make them too restrictive and require frequent revision. The SOPs should be periodically reviewed and revised as needed.

This chapter is not intended to cover all aspects of operating procedure applicable to every organization. However, a list of significant SOPs that must be developed can be constructed. The following list is based on numerous form 483 citations that have been issued during the past few years. It should be used as an instructive list that is more likely to result in the issuance of regulatory citations, if not followed. The list is not intended to be an exhaustive but merely covers significant issues that must be addressed in SOPs.

There should be a written procedure establishing a time limit for samples, after removal from the stability cabinets, awaiting laboratory analysis. The written program must list the acceptable procedure and practice. It should document a rationale for acceptance of data generated. An assurance of the integrity of the sample should be supported by the procedure.

The SOP should address how "out of specification" results are handled. When, how, and why retesting is performed must be addressed. This applies to in-house as well as outside contract laboratories. How an investigation is carried out? Who signs the investigation report? Under what circumstances are the results classified as "inconclusive" immediate retesting is allowed? How are follow-up investigations done to determine the reason for failure? Who is responsible for performing and evaluating such investigations?

Whenever an investigation of an "out of specification" result leads to a lab-related issue, the procedure should call for extending the investigation to determine if other lots or other product analyses could have been likewise affected, by the equipment or analyst.

There should be a procedure to evaluate stability trends. This pertains to systematic downward or upward trends in any particular test result. For example, it is not uncommon to see a downward shift in in-vitro dissolution test results.

There should be a written procedure establishing the steps for inactivating (closing) stability projects in the stability system scientific software, whenever applicable. The procedure should address the criteria for closing or inactivating projects and a detailed explanation for documenting the reasons for these activities. Stability projects or stability data in the software system should never be deleted. Any changes for data or projects should be made in the form of amendments through clear documentation stating the reason and justification for doing so.

A procedure should exist, which outlines how the periodic review of temperature/relative humidity charts from stability rooms and cabinets or continuous monitoring and surveillance of stability rooms and cabinets is conducted. It should address the procedure for handling normal reports as well as how the discrepancies, deficiencies, deviations, or malfunctions in the system are detected. Who is responsible for investigating deviations? There should be a procedure for documenting that stability samples are not adversely affected. The procedure should describe how corrective actions are undertaken.

Stability procedures dealing with analytical methodology should address how and when significant changes in analytical test results are investigated. For example, and adequate investigation needs to be undertaken when an additional peak shows on a chromatogram compared to previous results. The investigation should determine and document the cause of the additional peak or other significant change.

Procedures for preventive maintenance, including calibration of stability rooms, cabinets, and laboratory equipment, instrumentation, balances, monitoring equipment, etc. should list the frequency of testing at periodic intervals and documentation of results. Records for calibration should be maintained for specified periods. In addition, the calibration tests, which are performed by a vendor, should be described in a written procedure, contract, or report.

Documentation of label preparation, review, and approval should be addressed. It should address the retention of label specimens with batch records. Further details regarding the identity of the dates, the persons performing the operations, the equipment used, and any control procedures employed, and yield determination should be included in the procedures.

A procedure for the operation and control of the computer software used for input and output of stability data, including verification of the data, should be developed. The procedure should address who is authorized to flag erroneous data and enter corrected data whenever errors are discovered. The procedure should clearly state that data should never be deleted.

There should be a procedure available to ensure proper handling of stability samples whose due date falls on Saturdays, Sundays, and known holidays. The procedure should clearly delineate how and when the samples are actually pulled. In general, the practice of pulling samples *after* their actual test date rather than *before the test date* can be defended better on scientific grounds.

3.4. Training of Personnel

There should be a written procedure that describes a training program for laboratory personnel to assure that the employees are trained in and remain familiar with laboratory control procedures and current good manufacturing practice regulations. In addition, the procedures and criteria used to qualify or certify employees in specific analytical methods should be described in the written programs.

Regulations require that "each person engaged in the manufacture, processing, packing, or holding of a drug product should have education, training, and experience, or any combination thereof, to enable that person to perform the assigned functions. Training shall be in the particular operations that the employees performs." Training must be conducted by qualified individuals on a continuing basis and with sufficient frequency to assure that employees remain familiar with particular assignments. Maintenance of training records is an equally important function. Often, many employers are able to provide adequate training but fail to maintain appropriate records of training.

3.5. Selection of Batches and Samples

During the discovery phase, stability evaluation of almost every batch of drug substance used in the conduct of preclinical safety testing and/or for clinical trials is carried out. The purpose of this is to establish a database and a history of critical stability characteristics of drug substance. These data provide a foundation for justifying a retest period for drug substance. Because of the paucity of data, initially a shorter retest period is commonly employed. A longer retest period can be justified as more stability information is generated. The data is essential for the establishment of stability trends and also for setting of specifications for the drug substance. These data serve as the basis of supporting information with the registration applications.

As the product continues through further clinical testing and reaches the marketable stage, stability testing for at least three batches of drug substance at normal storage and at accelerated conditions must be available with the product registration application. Generally a minimum of twelve months long-term stability data is sufficient for filing purposes. Selection of three batches provides a means for evaluation of statistical variation in the process development and demonstrates the consistency of the results. These three batches must be made on a pilot plant scale. They must use the same synthetic or biological process for manufacture that is intended for manufacture on a large-scale commercial production. The synthetic procedure must represent the evolutionary development from early preclinical and clinical testing stages. The packaging for drug substance for long-term stability testing must simulate the intended drug substance storage. It is very common to employ smaller representative samples of the same material composition and to simulate a comparable drug-to-surface-area ratio as would be expected on large-scale storage. The overall quality of the three-batches must be comparable to or better than the quality of the batches used in early stages of development. If the three batches of drug substances included in the product registration application do not represent the full-scale manufacture, then the first three production batches of drug substance post approval must be placed on the same long-term stability evaluation protocol as was used earlier.

Similarly, during the early clinical testing (IND) stage, stability evaluation of several batches of drug product used in the conduct of clinical testing is carried out. The purpose of the data generation is to establish a stability database and identify critical stability characteristics of the drug product. The goal is to identify the expiration dating period for clinical formulations. During product development stages, formulation development continues through further refinement in terms of strengths, size and shape where applicable, and scale-up to support larger clinical trials. Since these stability data may be a multitude of packages and may represent minor changes in the formulations, they serve as the supportive data for product registration purposes.

The primary stability data for submission with the product registration application must be generated on at least three batches of drug product packaged in the container/closure system intended for marketing. Primary stability studies are intended to show that drug product stored in the proposed container/closure for marketing will remain within specifications if stored under storage conditions that support the proposed shelf life. Guidelines suggest the submission of at least 12 months labeled storage temperature data at the filing time. In addition, at least 6 months accelerated stability data should be included with the application. Two of the three batches of drug product should be at pilot scale. Generally that is not much of a problem for products indicated for chronic therapeutic uses where clinical trials require large-scale manufacture of clinical supplies. The third batch may be from a smaller lot (for example, 25,000 to 50,000 tablets or capsules in case of solid dosage forms). Preferably these three batches of drug product should be manufactured from three different batches of drug substance. Drug substance for the manufacture of product must be from the synthetic process intended for commercial production. The process used for the manufacture of drug product must be representative of what is intended for large-scale manufacture. It should provide acceptable continuity with batches manufactured during early development stages. The overall quality of the product should be the same quality intended for marketing, and it should meet all the specifications included in the registration document.

3.6. Stability Storage Test Conditions and Stability Rooms/Cabinets/Chambers

Selection of stability storage test conditions for conducting long-term stability studies is based on an analysis on the effects of climatic conditions existing in the region where the drug product distribution is expected to occur. For multinational pharmaceutical companies, stability testing can become unwieldy, if a separate study were to be conducted for each country. Fortunately, the ICH guidelines have made the task easier by adopting the climatic zones concept. The concept of mean kinetic temperature in any region of the world is nicely defined in terms of climatic zones (Grimm, 1985, 1986; Haynes, 1971). Based on the analysis, the world is divided into four climatic zones, to which the individual countries are assigned. The four climatic zones are defined as I. Temperate climate, II. Subtropical and Mediterranean climates, III. Hot, dry climate, and IV. Hot, humid climate. The climate zone II covers the European Union, Japan, and the United State.

The ICH Guidelines and the recent FDA draft guidance recommend the long-term stability storage at 25°C ± 2°C 60% RH ± 5% for most stable products and for many solid dosage forms. The length of the stability studies and the long-term storage condition should be sufficient to cover shipment, storage, and subsequent use (for example, reconstitution or dilution as recommended in the label). For products whose long-term storage can be used at 25°C ± 2°C 60% RH ± 5%, accelerated conditions at 40°C ± 2°C 75% RH ± 5% is recommended in the guidelines. Long-term stability study generally should be scheduled for up to 60 months at 25°C ± 2°C 60% RH ± 5%. A minimum of 12 months stability data at the long-term stability storage (25°C ± 2°C 60% RH ± 5% in this case) should be submitted with the registration application. A minimum of 6 months data at accelerated condition (40°C ± 2°C 75% RH ± 5%) should be available along with the long-term stability data. The regulatory guidance suggests, "when significant change occurs due to accelerated testing, additional testing at an intermediate condition (for example, 30°C ± 2°C/60% RH ± 5%) should be conducted" (FDA, 1998b). *Significant change* at the accelerated condition is defined as

1. A 5 percent potency loss from the initial assay value of a batch
2. Any specified degradant exceeding its specification limit
3. The product exceeding its pH limit
4. Dissolution exceeding the specification limits for 12 capsules or 12 tablets (USP Stage 2)
5. Failure to meet specification for appearance and physical properties (e.g., color, phase separation, resuspendibility, delivery per actuation, caking, hardness)

Whenever significant change occur at 40°C + 2°C 75% RH ± 5%, a minimum of 6 months' data from an ongoing 1-year study at 30°C 60% RH should be included with the initial registration application with the provision that same significant change criteria apply.

The draft guidance (FDA, 1998b) includes recommendations regarding continued occurrence of significant change during 12 months' exposure at 30°C ± 2°C 60% RH ± 5% as well as for several other situations.

For products, which are stable at refrigerated conditions, long-term stability storage at 5°C ± 3°C should be conducted. Monitoring of relative humidity but not its control is recommended at the refrigerated storage temperature. Generally, accelerated conditions at 15°C higher than the long-term stability storage conditions can be used as the accelerated conditions. Following the principle, accelerated conditions of 25°C ± 2°C 60% RH ± 5% are appropriate for refrigerated storage product. Likewise, when the product is extremely sensitive to temperature and requires freezer storage, long-term storage at − 15°C ± 5°C with accelerated testing at refrigerated temperature condition of 5°C ± 3°C is recommended.

Draft guidance includes several special situations. For example, for liquids in glass bottles, vials, or sealed glass ampoules, which provide an impermeable barrier to water loss, the following factors should be taken into consideration.

Accelerated condition, 40°C and ambient humidity is an acceptable alternative to 40°C ± 2°C 75% RH ± 5%.

Intermediate condition, 30°C and ambient humidity is an acceptable alternative to 30°C ± 2°C 60% RH ± 5%.

Long-term condition, 25°C and ambient humidity is an acceptable alternative to 25°C ± 2°C 60% RH ± 5%.

One of the complications with these special situation is that it requires stability rooms and/or cabinets at several identical conditions with minor differences. Often these can not be justified economically, as they add to the total expense not only for acquisition but also for continuous monitoring, maintenance, and record keeping. As an aid to the readers interested in pursuing the establishment of stability rooms and/or cabinets, Table 1 provides a list of equipment suppliers and/or manufacturers in alphabetical order. The list is not intended to be exhaustive. The reader is cautioned that inclusion in this list does not constitute an endorsement of these suppliers in any way. These are provided as an aid only. Likewise exclusion of other important suppliers in the list does not mean that they are not reliable and thus should not be considered. Every effort has been made to check the accuracy of the list contents. Please convey any errors to the author.

3.7. Thermal Cycling

Most drug products, especially solid dosage forms (tablets, capsules), oral powders, and lyophilized cakes, are not affected by short-term exposure to variations in temperature conditions encountered during shipping and distribution. However, most heterogeneous systems, for example ointments, creams, suspensions, emulsions, lotions, inhalation aerosols, and suppositories may be adversely affected by variations in extreme temperature fluctuations. They may undergo phase separation, precipitation, crystallization, changes in viscosity, creaming, sedimentation, aggregation, and so on. These types of drug products should be tested under cycling temperature conditions to simulate shipping and distribution conditions. These studies are generally performed on the packaged product during stress testing of the product development phase. Testing parameters for those studies should include not only the chemical analysis but also physical changes and homogeneity. Draft guidance (FDA, 1998b) provides several recommendations for thermal cycling studies.

For products that may be exposed to above-freezing temperature variations, a temperature cycling study consisting of three cycles of 2 days or refrigerated storage (2–8°C) followed by 2 days under accelerated conditions (40°C) should be performed.

For products that may be exposed to subfreezing temperature variations, a temperature cycling study consisting of three cycles of 2 days at freezer temperature (−10 to −20°C) followed by 2 days under accelerated conditions (40°C) is recommended.

For frozen drug products, the recommended cycling study should include an evaluation of effects due to thawing under a hot water bath or in a microwave unless the drug in the product is known to degrade at high temperatures.

For inhalation aerosols, the recommended cycling condition consisting of three to four cycles of 6 h per day, between subfreezing temperatures (−10 to −20°C) and accelerated conditions (40°C 75%–85% RH) for a period of up to 6 weeks should be followed.

Specific drug product characteristics may require an alternative cycling condition study as long as it is scientifically justified.

3.8. Photostability Testing

The ICH Harmonized Tripartite Guideline as well as the draft guidance (ICH, 1996; FDA, 1998b) state that light testing should be an integral part of stress testing. During the development phase, the intrinsic photostability characteristics of new drug substances and products should be determined. The purpose of such study is to demonstrate that light exposure does not result in unacceptable change and to determine whether precautionary measures in manufacturing, labelling, and packaging are needed to overcome changes due to light exposure. Generally, photostability testing is carried out on a single batch of material selected.

Photostability information is generally required for submission in registration applications for new molecular entities and drug products made from the entity. Both the ICH Guideline and the draft guidance recommend a systematic approach to photostability testing to cover studies such as

Tests on the drug substance
Tests on the exposed drug product outside of the immediate pack, and if necessary
Tests on the drug product in the immediate pack, and if necessary,
Tests on the drug product in the marketing pack.

The Decision Flow Chart for Photostability Testing of Drug Products included in the ICH guideline suggests the extent of drug product testing by assessing whether acceptable change has occurred at the end of the light exposure testing. Acceptable change is defined as change within limits justified by the applicant.

Generally, the photostability studies are carried out by exposing the samples to specific light sources as defined in the guideline. The guideline offers two options for light sources for testing. The first is "any light source that is designed to produce an output similar to the D65/ID65 emission standard such as an artificial daylight fluorescent lamp combining visible and ultraviolet (UV) outputs, xenon, or metal halide lamp. D65 is the internationally recognized standard for outdoor daylight. ID65 is the equivalent indoor indirect daylight standard." For option 2, the same sample should be exposed to both a cool white fluorescent and a near-ultraviolet fluorescent lamp meeting the criteria specified.

Since temperature changes occur when the samples are exposed to light sources, protected samples (for example samples wrapped in aluminum foil) are also exposed side by side with the authentic sample so that any contribution due to thermal exposure can be determined in the dark control sample.

For the drug substance, photostability testing is divided into two parts, forced stress testing and confirmatory testing. Generally, forced stress testing is done by exposing the drug alone and/or in solutions/suspensions to evaluate the photosensitivity, to validate the analytical method, and to elucidate degradation pathways. Confirmatory studies should then be undertaken to generate the information necessary for handling, packaging, and labeling. For confirmatory studies the guideline suggests the exposure of samples to light providing an overall illumination of not less than 1.2 million lux hours and an integrated near ultraviolet energy of not less than 200 watt hours/square meter to allow direct comparisons between the drug substance and the drug product.

The guideline recommends the conducting of studies on drug products to be carried out in a sequential manner starting with testing the fully exposed product and then progressing as necessary to the product in the immediate pack and then in the marketing pack. Testing should continue until the results demonstrate that the drug product is adequately protected from exposure to light. Normally, only one batch of drug product is tested during the development phase, and the photostability characteristics should be confirmed on a single batch of drug product selected. It should be noted that photostability testing is stress testing designed to determine the intrinsic photostability characteristics of new drug substances and drug products. No correlation has been developed or established that can equate the within-specification result to the labeled expiration dating period. Detailed description of the presentation of samples, the analysis of samples, and the judgment of results is covered in the guideline and the draft guidance. Details concerning an actinometric procedure for monitoring exposure to a near-UV fluorescent lamp and other actinometric procedures and light sources are included in the guidelines. The reader is also referred to several references on the subject (Drew et al., 1998a and 1998b; Tonnesen, 1996; Sager et al., 1998; Yoshioka et al., 1994). The draft guidance also provide an account of acceptable/unacceptable photostability change and photostability labeling considerations.

As an aid to the reader interested in pursuing the establishment of photostability testing, Table 2 provides a list of equipment suppliers and/or manufacturers in alphabetical order. The list is not intended to be exhaustive. The reader is cautioned that inclusion in this list does not constitute an endorsement of these suppliers in any way. These are provided as an aid only. Likewise exclusion of other important suppliers in the list does not mean that they are not reliable and thus should not be considered. Every effort has been made to check the accuracy of the list contents. Please convey any errors to the author.

3.9. Container/Closure Systems

The draft guidance recommends that long-term stability data should be generated for the drug product in all types of sizes of immediate containers and closures proposed for marketing, promotion, or bulk storage. Good science dictates that rigorous studies be undertaken to test any possible interaction between drug product and the immediate container and/or closure. This includes consideration of leachables or extractables from the container/closure into the drug product environment or surface. Similarly, absorption or adsorption of product components into the container/closure should be investigated for liquid or semisolid products. The studies must employ meaningful and sensitive assay methodology to detect this type of interaction.

Most oral solutions, sterile solutions in small volume container systems, and sterile solutions in large volume parenteral container/closure system should be tested in upright and inverted and/or on-the-side positions at both long-term and accelerated conditions to ensure that no adverse effect from any interactions is produced. A direct comparison of the data from upright and inverted or on-the-side positions demonstrates whether any extractable and/or absorption/adsorption has occurred or is likely to occur.

Regulatory guidelines do allow for conducting bracketed stability studies for intermediate containers when the drug product is packaged in the smallest and the largest containers, provided that all containers and closures share the same composition and design.

3.10. Testing Frequency

Testing frequency for both drug substance and drug product for product registration should be sufficient to establish stability characteristics. The ICH Guidelines and draft guidance require that testing should be every 3 months over the first year, every 6 months over the second year, and annually thereafter. For drug products packaged in identical container/closures of different sizes and/or identical formulations of different strengths, matrixing and/or bracketing can be used with valid justification. During early evaluation phase, it is not uncommon to use more frequent, for example every 2 weeks during the first 3 months. For accelerated stability conditions, testing is generally done for 6 months. The guidance recommends testing at 0, 2, 4, and 6 months so that at least four stability time-points are generated.

3.11. Stability Indicating Analytical Methodology and Validation

A critical component of stability testing is the development of stability indicating assay methodology. Analytical methodology should be developed and validated to assure that the method can accurately and quantitatively analyze the major component(s), impurities, degradation products, and other selected components in the drug substance and drug products. Development and validation of the method should ensure that the method is able to analyze the component(s) in the presence of other components. Methodology development should address key analytical and statistical parameters, such as accuracy, precision, detection and quantitation limits, linearity, range of analytical limits, robustness, reproducibility, repeatability, ruggedness, and system suitability testing. The standard operating procedures must describe the procedure in sufficient detail, the steps necessary for performing the test, including the sample preparation, reference standards, preparation of reagents, instrumentation used, calibration, calculations, and so on to ensure the validity of data. A partial revalidation of assay methodology should be undertaken any time there is a significant change in the product formulation or in the process that may affect the validity of the original assay.

3.12. Stability of Reconstituted Products

Drug products that require reconstitution prior to dispensing to the patient or the consumer require special consideration with regard to stability testing. Section 211.166 of 21 CFR states that "testing of drug products for reconstitution at the time of dispensing (as directed in the labeling) as well as after they are reconstituted" shall be included in the written program for stability testing. Examples of reconstituted products include lyophilized products, powders or granules for reconstitution, dilution of active product with additives, admixture of drug product with other components, and so on. For reconstituted products, two distinct stability periods are in operation. The first period covers up to its expiry dating followed through long-term and accelerated stability testing prior to reconstitution. The second period consists of usually short-term stability after reconstitution. Stability study after reconstitution as directed on the label should be conducted by appropriate sampling and testing to cover a period beyond that specified on the label. For example, if the reconstituted product were recommended to be used within 1 week under refrigeration, then an appropriate stability study could cover testing at periodic intervals up to 10 days or 2 weeks under refrigeration. In addition, an accelerated condition stability of reconstituted product at 25°C 60% RH for 1 to 3 days may be appropriate.

3.13. Dosage Form Considerations

The draft guidance provides specific recommendations for evaluation of stability attributes specific to many dosage forms. In general, assay, appearance, and degradation products should be evaluated for all dosage forms. It should be noted that not all tests need to be done for all dosage forms, nor it is intended to be an exhaustive list. Selection of attributes for stability evaluation should be made on the basis of stability behavior of new drug product observed during development process. Finally, it is not expected that every listed test be performed at every time point. Not all the dosage forms included in the draft guidance are discussed below. For the complete list, the reader should refer to the draft guidance. The draft guidance recommends that

Tablets should be evaluated for appearance, color, odor, assay, degradation products, dissolution, moisture, and friability.

Hard gelatin capsules should be evaluated for appearance (including brittleness), color, odor of contents, assay, degradation products, dissolution, moisture, and microbial limits.

Soft gelatin capsules should be tested for appearance color, odor of contents, assay, degradation products, dissolution, microbial limits, pH, leakage, and pellicle formation. In addition, precipitation and cloudiness in the fill medium should be examined.

Emulsions should be evaluated for appearance (including phase separation, color, odor, assay, degradation products, pH, viscosity, microbial limits, preservative content, and mean size and distribution of dispersed phase globules.

Oral solutions should be tested for appearance (including precipitate formation, clarity), color, odor, assay, degradation products, pH, microbial limits, and preservative content.

Oral suspensions should be evaluated for appearance, color, odor, assay, degradation products, pH, microbial limits, preservative content, redispersibility, rheological properties, and mean size and distribution of particles.

Oral powders for reconstitution should be evaluated for appearance, color, moisture, and reconstitution time. After reconstitution, testing for oral solutions or suspensions should be applied.

Metered-dose inhalations and nasal sprays should be evaluated for appearance (including content, container, valve, and its components), color, taste, assay, degradation products, assay for cosolvents, dose content uniformity, labeled number of medicated actuations per container, aerodynamic particle size distribution, microscopic evaluation, water content, leak rate, microbial limits, valve delivery (shot weight), and extractables/leachables from plastic and elastomeric components. Samples should be stored upright and inverted or on the side.

Inhalation solutions should be tested for appearance, color, assay, degradation products, pH, sterility, particulate matter, preservative and antioxidant content (if present), net contents, weight loss, and extractables/leachables from plastic, elastomeric, and other packaging components.

Topical, ophthalmic, and otic preparations include ointments, creams, lotions, pastes, gels, solutions, and nonmetered aerosols for application to the skin. Topical preparations should be evaluated for appearance, clarity, color, homogeneity, odor, pH, resuspendibility (for lotions), consistency, viscosity, particle size distribution (for suspensions), assay, degradation products, preservatives, and antioxidant content, microbial limits/sterility, and weight loss.

Small volume parenterals (SVPs). Included in this category is a wide range of injection products such as drug injection, drug for injection, drug injectable suspension, drug for injected suspension, and drug injectable emulsion. Stability evaluation of drug injection should include appearance, color, assay, preservative content (if present), degradation products, particulate matter, pH, sterility, and pyrogenicity. Drug for injection should include evaluation of appearance, clarity, color, reconstitution time, and residual moisture content. After reconstitution, stability should be monitored for attributes listed in drug injection.

Large volume parenterals (LVPs) should be evaluated for appearance, color, assay, preservative content (if present), degradation products, particulate matter, pH, sterility, pyrogenicity, clarity and volume.

3.14. Bracketing

A lot of products are marketed in more than one type of product configuration. For example, tablets are packaged in multiple sizes, which may include a very small size for promotional use, several intermediate sizes for pharmacy or patient use, and large sizes for institutional usage. A similar situation may exist with liquid products, semisolids, and others. Other examples may include tablets in several strengths packaged in identical container/closure systems. The use of bracketing in those situations to reduce the full stability evaluation is important in stability studies. A detailed account of bracketing designs and other complex designs are included in recent publications (Fairweather et al., 1995; Nordbrock, 1992).

Bracketing may be defined as the design of a stability testing schedule so that at any time point only the stability samples on the extremes, for example, of container size and/or dosage strengths, are tested. The key assumption for the bracketing design is that the stability of the intermediate size samples is represented by that of the extremes. A bracketing design may be used for primary stability studies to be included in the registration application, stability batches as committed after approval of the product, and annual batches or batches included in supplemental changes. Bracketing design should generally not be used for clinical formulations during the development phase. Protocols for bracketing designs should be endorsed by the FDA prior to initiation of primary stability studies (FDA, 1998b).

Draft guidance provides a detailed discussion of bracketing design. Bracketing design is applicable to most dosage forms, including immediate release solid dosage forms and modified release oral solids, liquids, semisolids, and injectable products. It may not be readily applicable to certain types of dosage forms, such as metered dose inhalers, dry powder inhalers, and transdermal products, unless convincing scientific arguments can be justified. The Food and Drug Administration encourages and welcomes consultation on the subject.

The simplest situation for bracketing design is where the same drug product is packaged in different size/fill containers of the same composition. Where the range of fill/size of the same strength drug product is to be evaluated, bracketing may be applicable if the compositions of the containers and closures are identical throughout the range. Bracketing design is applicable without further justification, where the range of size or fill varies while other factors are constant. Bracketing design is also applicable, where both the container size and the fill vary, provided good justification is included and supported by data. Justification should include discussion of the surface-area-to-volume ratio, the dead-space-to-volume ratio, the container wall thickness, and the geometry of the closures. As long as the intermediate sizes are adequately bracketed by extreme sizes, the bracketing design should be justifiable.

Bracketing design is suitable and applicable to identical or closely related formulations packaged in different size identical container/closure systems. Examples include tablets compressed to different weights from the same granulation or powder mix. Similar examples include a range of different size capsules filled to different plug weights from the same granulation. Wherever more than one variable is involved, such as different amounts of active ingredient while the amount of each excipient or the total weight of the dosage form remains constant, bracketing design may not be applicable unless justified by supportive stability data from development studies. Bracketing designs are generally not applicable whenever multiple variables exist, for example significantly different formulation strengths made from different excipients, colorants, flavors, and so on. In selected cases where multiple variables are involved matrixing may be considered.

3.15. Matrixing

The ICH Guidelines define matrixing as the statistical design of a stability schedule so that only a fraction of the total number of stability samples are tested at any specified sampling point. At a subsequent sampling point, different sets of samples of the total number are tested. The design assumes that the stability of the samples tested represents the stability of all samples. The reduced stability testing in the matrixing design offers an excellent alternative for monitoring stability where multiple factors for the same product, such as different strengths of the same formulation, different batches, different sizes of packages, and different fills are involved. Other factors, such as different container and closure systems, different strengths of closely related formulations, different orientations of containers during storage (upright and inverted or on the side), multiple drug substance manufacturing sites, multiple drug product manufacturing sites, may be matrixed with appropriate scientific justification. Supportive stability data from the development is most crucial for justifying complex matrix designs. For example, for primary formulations that show excellent stability to moisture, oxidative exposure, and light exposure are less likely to be affected by differences in package configurations versus those that show marginal stability against these variables. It may be possible to justify matrixing for different packages, such as glass bottles, blister packages, and high-density polyethylene bottles for such a formulation with excellent stability behavior. On scientific grounds, it may be more difficult to justify matrixing significantly different formulations with different excipients, unless the drug substance has been found to be highly stable and almost unaffected by most environmental variables.

The draft guidance provides a detailed discussion of matrixing design. Matrixing design is applicable to most dosage forms, including immediate release solid dosage forms, modified release oral solids, liquids, semisolids, and injectable products. It may not be readily applicable to certain types of dosage forms such as metered dose inhalers, dry powder inhalers, and transdermal products unless convincing scientific arguments can be made. The regulatory guidelines require that in every case all batches be tested initially and at the end of the long-term testing. The Food and Drug Administration encourages and welcomes consultation on the subject. (FDA, 1998b).

Factors affecting the applicability of matrixing design for primary stability studies are the stability of drug substance, drug product formulations during development stages, and the variability in the stability data, other than analytical method variation. Stability data variability for matrixing design purposes refers to batch-to-batch variability, strength-to-strength, size-to-size, and across different variables, such as batch versus strength, strength versus size. In general, the better the stability behavior of the drug substance and formulations, the better the applicability of matrixing when accompanied by less variability in the data. Likewise, the opposite is true when the matrixing design should not be applied in situations with poor stability behavior and larger variability. Moderate stability and moderate variability means matrixing may be applied with appropriate justification. Similarly historical data from primary stability determines whether the matrixing design may be applied to post-approval commitment batches or supplemental batches.

The draft guidance suggests that for product registration applications, a matrixing design should always include the initial and final test points, as well as two additional points through the first 12 months. All samples including those that are scheduled to be tested should be placed on the stability program. The protocol should be followed as planned without deviation once the study has begun. The size of the matrixing design is expressed as a fraction of the total number of the sample to be tested in the full corresponding stability protocol. The size depends on the amount and quality of supportive data as stated earlier and the number of variables to be matrixed.

For example, for a product with excellent stability available in three batches, with three different strengths, and packaged in three packages/fill sizes, the number of factors to be tested in a full stability protocol is $3 \times 3 \times 3 \times 27$. For such a product, the size of the matrixing design could be as small as half the original full stability protocol. Thus fractional $\frac{1}{2}$ means that only half as many samples as the full stability protocol will be tested in the matrixing design. One of the key elements of matrixing design is that it should be a well-balanced design. If possible, an estimate of the probability that stability outcomes from the matrixed study would be the same for a given factor or across factors should be provided with the protocol. Several examples with different fractional matrixing designs are included in the draft guidance (FDA, 1998b). A detailed account of matrixing designs and other complex analysis is provided in recent articles (Fairweather et al., 1995; Nordbrock, 1992).

3.16. Statistical Analysis and Evaluation of Data

The presentation and evaluation of the stability information should be based on a systematic approach. Evaluation of stability should cover all chemical assay data and, as necessary, physical, chemical, biological, and microbiological quality characteristics, including particular properties of the dosage form (for example, in-vitro dissolution rate for oral solid dose forms).

The goal in the design of the stability study is to help establish a shelf-life and label storage conditions that can be applied to all future batches of the dosage form manufactured and packaged under identical conditions. This is accomplished by testing a minimum of three batches of the drug product under a variety of stability conditions. Generally, the better the stability of the product, the better is the confidence that a future batch will remain within acceptable specification until the expiration date. Likewise, the greater the degree of variability of individual batches, the greater will it influence the confidence that a future batch will remain within acceptable specification until the expiration date.

Generally, an expiration period for drug products and a retest period for drug substances should be determined on the basis of statistical analysis of the observed long-term stability data. The draft guidance allows limited extrapolation beyond the observed range to extend the expiration dating or retest period, provided it is supported by the statistical analysis of the real-time data, satisfactory accelerated data, and other supportive data.

For the purpose of the determination of expiration date, the initial value for the batch is critical. The initial value is in terms of a percentage of the label rather than a percentage of the initial average of the results. The time during which a batch can be expected to remain within specifications depends not only on the rate of physical, chemical, and microbiological changes but also on the initial average value for the batch.

The expiration dating period for an individual batch is based on the observed pattern of change in the quantitative attributes (for example assay, degradation products) under study and its precision. The key is the quantitative nature of the attribute.

A commonly acceptable statistical approach for analyzing an attribute that is expected to decrease with time is to determine the time at which the 95% one-sided lower confidence limit (also known as the 95% lower confidence bound) intersects the lower specification limit. For most degradation pathways, since the overall observed degradation is less than 10% one cannot distinguish between a zero-order rate of degradation and a higher-order rate of degradation. Consequently, the rate of degradation can safely be assumed to be linear. For drug products, usually the 95% lower confidence bound at 90% of labeled claim is assumed. For the appearance of degradation products, which are expected to increase with time, the 95% one-sided upper limit for the mean should be used.

When the 95% one-sided lower limit for assay and the 95% one-sided upper limit for degradation products are used, both the average assay values and the degradation product values are likely to remain within specifications with 95% confidence.

The expiration dating period using the point at which the fitted least-squares line intersects the specification limit should not be used. It is as likely to provide an overestimate of the expiration dating period as it is likely to underestimate it in which case the batch average can be expected to remain within specifications at expiration if the fitted least-squares line is used with a confidence level of only 50%.

If analysis shows that the batch-to-batch variability is small, that is, the relationship between assay or degradation products and time is essentially the same from batch to batch, the stability data should be combined into one overall estimate. Combining the stability data from batches should be based on batch similarity. The similarity of estimated curves among the batches tested should be assessed by applying statistical tests of equality of the slopes and of zero time intercepts. A detailed account of these principles is covered in the draft guidance.

A useful concept is the release limits of drug products. The release limits of dosage forms are defined as the bounds on the potency at which an individual lot can be released for marketing which will ensure that it remains within registered limits throughout its shelf life. A statistically based method is used for calculating release limits for any type of dosage form and any parameter for which the rate of change with time is predictably uniform and linear. When the mean release assay result for a specific product lot is at or within the calculated release limit bounds, assurance is provided at the specified confidence level that the average assay results obtained at any subsequent time within the expiry dating period will remain within registered specifications (Allen et al., 1991).

Often the data may show so little degradation and so little variability that it is apparent from looking at the data that the requested shelf life can easily be supported. Under the circumstances, it is normally unnecessary to go through the formal statistical analysis, one merely provides a full justification for the omission.

Any evaluation should consider not only the assay but also the levels of degradation products and appropriate attributes. Where appropriate, attention should be paid to reviewing the adequacy of the mass balance, different stability behavior, and degradation performance.

4. MANAGEMENT OF STABILITY INFORMATION USING COMPUTERS

Development of a stability information management system is the pharmaceutical industry requires careful consideration of organizational structure, its needs, and the desire to integrate many factors involved in stability data collection, storage, retrieval, analysis, computation, validation, and so on. The easy availability of personal computers and of access to the internet have made the development of such a system much closer to reality than what would have been possible just a few years ago. Prior to the explosion of personal computers, only mainframe or minicomputer systems were available for information management systems. These required a lot of resources for the development of stability information systems because of complex operating systems and a whole host of computational languages. Personal computer expansion has made the development of specialized systems such as stability systems possible. Because of the unique nature and the limited market, only a very few commercially available systems exist for managing stability information; they are specifically designed to serve the needs of the pharmaceutical industry.

From regulatory perspectives, the management of a stability system or other laboratory data acquisitions systems requires careful consideration of the whole process of collection of the data. The FDA guidance related to the subject provides some insight.

"It is important, for computerized and non-computerized systems to define the universe of data that will be collected, the procedures to collect, and the means to verify its accuracy. Equally important are the procedures to audit data and the program and the process for correcting errors. Several issues must be addressed when evaluating computerized laboratory systems. These include data collection, processing, data integrity, and security. Guidance on security and authenticity issues for computerized systems covers the following items:

Provision must be made so that only authorized individuals can make data entries.
Data entries may not be deleted.
The data must be made as tamperproof as possible.
The standard operating procedures must describe the procedures for ensuring the validity of the data.

One basic aspect of validation of laboratory computerized acquisition requires a comparison of data from the laboratory system(s) with the same data electronically processed through the system and emanating on a printer. Periodic data comparisons would be sufficient only when such comparisons have been made over a sufficient period of time to assure that the computerized system produces consistent and valid results (FDA, 1993)."

As an aid to the reader interested in pursuing the acquisition or development of stability information systems, Table 3 provides a list of developers of computerized stability systems and laboratory information management systems in alphabetical order. Only three of the suppliers from the list are known to offer stability information systems (Lycoming Analytical, Metrics, and ScienTek Software). The list is not intended to be exhaustive. The reader is cautioned that inclusion in this list does not constitute an endorsement of these suppliers in any way. These are provided as an aid only. Likewise exclusion of other important suppliers from the list does not mean that they are not reliable and thus should not be considered. Every effort has been made to check the accuracy of the list contents. Please convey any errors to the author.

A well-designed stability system: should provide the flexibility in customizing the system to suit the organization's needs; should be user friendly; should follow intuitive and logical operational procedure; and should be easy to validate. It should reduce clerical needs; should reduce turnaround time; should provide dependable storage of data; and should reduce errors. The system should reduce costs and improve capabilities; should improve efficiency; should provide easy retrieval; should serve global needs; and should allow extensive search capability.

The system should allow inclusion of attributes, that the organization may want or require to investigate specific questions or for managing the information, in a searchable field. For example, if we are interested in determining all stability batches of a product containing a specific batch of active ingredient, or all drug product batches in a specific package, unless the information is included in a searchable field, it may not be possible to retrieve the information.

The stability system should offer the following capabilities for the management of stability operations. Not all the capabilities may be needed or useful for an organization.

User-friendly system
Initiation of stability of new batch
Stability protocol modification
Inactivation of stability
Entry, verification, and approval of data

Table 3 Suppliers and Developers of Stability System Software and Laboratory Information Management Systems

Name	Address	Phone/fax	E-mail/web site
Artel Inc.	25 Bradley Dr Westbrook, ME 04092	Phone: 207-854-0860 Fax: 207-854-0867	E-mail: post@artel-usa.com Web Site: http://www.artel-usa.com
Beckman Coulter	90 Boroline Rd Allendale, NJ 07401	Phone: 201-818-8900 Fax: 201-818-9740	E-mail: labsales@beckmanlab.com Web Site: http://www.beckmancoulter.com
Compex USA Inc.	2625 Butterfield Rd Suite 307E Oak Brook, IL 60523	Phone: 630-368-0905 Fax: 630-368-1086	Web Site: http://www.compex.be
Lycoming Analytical Laboratories (CF Technical Development, Inc.)	2687 Euclid Avenue Duboistown, PA 17702	Phone: 570-323-5001 Fax: 570-323-0009	
Metrics Inc.	P.O. Box 4035 Greenville, NC 27836	Phone: 252-752-3800 Fax: 252-758-8522	E-mail: marketing@metricsinc.com Web Site http://www.metricsinc.com
PE Informatics	3822 N First St San Jose, CA 95134	Phone: 408-577-2200 Fax: 408-894-9307	Web Site http://www.peinformatics.com
Perkin-Elmer Corp	761 Main Ave Wilton, CT 06859	Phone: 203-762-4000 Fax: 203-762-4228	E-mail: info@perkin-elmer.com Web Site: http://www.perkin-elmer.com
Pharm-Eco Laboratories	128 Spring St Lexington, MA 02421	Phone: 781-861-9303 Fax: 781-861-9386	E-mail: boast@pharmeco.com Web Site: http://www.pharmeco.com
ScienTek Software, Inc.	P.O. Box 323 Tustin, CA 92681-0323	Phone: 714-832-7435 Fax: 714-832-7435	
Thermo LabSystems	100 Cummings Center Suite 407J Beverly, MA 01915	Phone: 978-524-1400 Fax: 978-524-1244	E-mail: sales@labsystems.com Web Site http://www.labsystems.com

Automated error detection
Storage of stability data
Retrieval of stability data
Printing of data in multiple formats
Tabulation of data
Graphical presentation of data
Maintenance of schedules
Generation of test assays scheduled

Statistical evaluation of the data
Linear and multiple regression
Expiry data evaluation
Shelf life projections
Analysis for pooling of batches
Multivariable search capability
Projections for current and future workload
Maintenance of active batches
Maintenance of inactive batches
Archival of data
Pending scheduled tests
Classification of studies by type
Printing of labels
Electronic reading of samples/data (e.g. bar codes)
Provision for multiple levels of authorizations
Maintenance of audit trails
Amendments for data
Validation capabilities
Inventory management
Back-up security provision

The following parameters should be considered as an essential list of attributes for initiation of stability studies in the software system. It should be noted that many of these items are part of the information for submission to the regulatory agencies, while others are for management of the stability function within the organization.

Drug substance or drug product name
Date stability initiated
Purpose of stability study
Safety handling precautions
Stability study number or code
Formulation/product number or code
Batch number or lot number
Storage conditions
Location of samples
Tests to be performed
Internal test specifications or target
Registration test specification
Analytical method to be used
Laboratory performing the test
Testing frequency
Number of replicates
Duration of the study
Cost per test
Internal accounting charges
Dosage form and strength
Labeled strength
Theoretical strength (including overage)
Batch type and size

Date of manufacture
Site of manufacture
Drug substance batch number
Drug substance manufacturer
Drug substance manufacturing site
Container description or composition
Closure description or composition
Other packaging components description or composition
Packaging site
Date of packaging
Person(s) responsible for the study

Should all organizations consider developing or acquiring the computer system for stability information? There is no simple answer. It depends on the needs of the organization, the volume of stability data generated, the computational demands, and the need for dissemination of information. Development or acquisition, establishment and implementation of computerized stability systems are fairly expensive propositions. Not every organization can justify the cost associated with the development, training, validation, maintenance, upgrades, and operational needs.

In reality, stability data is not much different from other data, such as inventory, payroll, or employee information data. It does, however, require careful consideration of compliance with good laboratory practices and/or good manufacturing practices regulations, especially if the systems are used as a primary database for making decisions. One can apply general principles involved in the development of information systems to stability information systems.

The terms data and information are often used interchangeably, but they refer to two distinct concepts. Since we are dealing with stability data and information, let us examine the concepts from stability perspectives. Data are objective representations of events or concepts. Results of a chemical assay, pH measurements, and in-vitro dissolution results are all examples of data. Information on the other hand is subjective and relative. While data and information are separate concepts they are distinctly related; information is produced from data. The computation of an expiry date, either from the graphical representation of stability data or from a rigorous statistical analysis of the data, is an example of information. In simple terms, data are the raw material from which the finished product, information, is produced. Although data are the key ingredient used to produce information, not all data provide relevant and timely information. The sheer volume of the data can be a burden to an organization. Collection of the data just for the sake of collecting is a most nonproductive use of resources. Data collection must add value to the organization. The only way to reduce this burden of unnecessary information is through implementing an effective and efficient user-oriented information system.

The purpose of information is to increase the knowledge level of the recipient. In simple terms, the information should provide a clear picture of what has happened. Information serves another very important function for the person who receives it. That function is to reduce the variety of choices and the uncertainty related to these choices. Information is critical in the decision-making process.

A single stability result for a 6 months time point for a given batch may not be meaningful, but when it is viewed in conjunction with initial results for the same batch, it becomes useful information whether any significant loss from initial result has occurred or not.

4.1. Data Operations

How are the data converted into information? Several specific data operations are involved in the conversion process. Any one or combination of these operations can producc information from data. The data operations are:

Capturing: This operation refers to recording of data from an event. For example, a reading from a chart paper, a pH meter, or a liquid chromatograph are examples of capturing data.

Verifying: This pertains to validating data to insure that it was captured and recorded correctly. This is critical operational step. This assures the integrity of the data. Example includes the review by another person to verify the accuracy of the data. In computer data entry, the verification often means rekeying in the data in order to match the earlier entry.

Classifying: This operation places data elements into specific categories, which provide meaning for the user. Examples of this could be classification of stability data for solid dosage forms, sterile products, and topical dosage forms. One may classify the solids data by breaking up further into tablets, soft gelatin capsules, hard filled capsules, etc.

Arranging (sorting): This operation places data elements in a specified or predetermined sequence. For example, arranging the stability data for a given lot by age of the samples, or by temperatures.

Summarizing: This operation combines or aggregates data elements in either a mathematical or a logical sense. For example, the stability data could be summarized for a product type. Within a product type, one could summarize by different formulations. Another way is to summarize the data by divisional responsibility, e.g., production versus research.

Calculating: As the name implies, it performs the arithmetic and or logical calculations, for example calculation of percent of label or theory from raw data, to determine if the assay results are below or above a certain specified limit, or to determine if it meets a test or not.

Storing: This operation simply places data into some storage media such as a file drawer, a computer disk, microfilm, or magnetic tape from where it can be retrieved.

Retrieving: This operation means searching out and gaining access to specific data elements from the storage medium.

Reproducing: This operation duplicates data from one medium to another. A common example would be a back-up copy of a magnetic tape or disk for further processing or security purposes or for a disaster plan.

Communicating: This involves transferring data from one place to another. The examples could be the transfer of data onto a computer monitor or to a printer. The ultimate aim of all information systems is to communicate or disseminate information to the final user.

4.2. Data Processing Operations

These data operations are involved in any management of data. They are independent of the data processing method, whether it is done manually or through the use of a computer. The decision to use one of these methods depends on the economic considerations, the processing requirements, and the performance factors related to the data processing method. Data processing requirements are based on

Volume of the data: Volume simply means the number of data units that must be processed to achieve an information goal. For a stability information system, each data unit directly comes from each stability result. The total volume in any organization will depend on the number of stability batches, and the average number of tests performed per batch in a given period.

Complexity: Complexity refers to the number of intricate and interrelated data operations that must be performed to achieve an information goal. In the case of a stability system, once the results are entered, they must be checked for any systematic types of errors, added to the proper batch, inform the user, keep track of results in the file, identify any results that have not been entered, and so on.

Processing time constraints: Time constraints can be defined as the amount of time allowed or acceptable between when the data are available and when the information is required. In today's information age, all of us want more and more information and we want it sooner rather than later.

Computational demands: These are a unique combination of both volume and complexity. For a stability system, the computational requirements could be the calculation of shelf-life based on a zero order or a first order, or the application of the Arrhenius relation to multitemperature stability data.

4.3. Performance Factors

So far we have identified the processing requirements for a stability information system. Development of the system requires careful consideration of many performance factors as follows.

Initial investment: This is just the initial expense of acquiring any materials or hardware and software required for processing.

Setup: This is the expense required to capture initial data for subsequent processing.

Conversion: The one-time expense of processing data with the new computer method.

Skilled personnel requirements: This simply means the education and training of the individuals involved in operating and using the system.

Variable cost: The cost of a data unit based on the total volume. In general, computers allow greater efficiency in this area. The greater the volume of data that needs to be handled, the cheaper it becomes per unit cost.

Modularity: This represents the ability to increase or decrease processing capability to match the requirements for processing. For example, if we need one data entry computer terminal to process 5,000 records a day, two computer terminals will provide for data entry up to 10,000 records per day.

Flexibility: The ability to change the processing procedure to satisfy new or changing requirements. No system remains static for long. Well-designed systems provide excellent flexibility to modify the system to meet changing needs of the organization.

Versatility: The ability to perform many different tasks. This is where the computer excels as it can handle multiple tasks.

Processing speed: This represents the time required to convert inputs to outputs. This is not a problem as the processing speeds of today's computers continue to improve in an exponential manner.

Computational power: The ability to perform complex mathematical operations. This is very inefficient in the manual system. This is excellent for the computer.

Automatic error detection: The ability to identify processing errors. The human mind is capable of detecting errors, but we have a fatigue problem. On the other hand, the computer never gets tired. This capability in the computer system provides excellent dividends in terms of continuous monitoring.

Decision-making power: The ability to chose among various alternatives in order to continue processing. Again the speed with which the manual system would work in this task is slow. The computer has high capability of making decisions along the way if the system is well designed.

System degradation: The level to which the processing system is degraded because of breakdown or unavailability of a component. When the computer breaks down, the system is completely down.

The design and developmental consideration of a computer system for stability information management involves the following major steps.

SYSTEMS ANALYSIS. Here we need to focus on current capabilities of present systems and determine what the new systems should do. In this phase, we have to look at users and any problems faced by the users. What do they need? Are there any problems with the system as it is being used? Sometimes the users feel that they need the information because the predecessors have done so. Often, the users do not know what can be provided. Systems scope covers what capability the system is supposed provide. Should it include only the scheduling of stability? Should it provide computations?

Finally the requirements for the system are formulated through the system design phase.

GENERAL SYSTEMS DESIGN. In systems analysis the focus was with what the system is doing, and what the new system should be doing to meet the requirements. In the system design phase one has to concentrate on how the system is developed to meet these requirements. In the broad design phase questions like How it should be done? What if? Why not? need to be addressed. A rough sketch or an outline of the system is developed. The input and output of the system need to be closely looked at. One needs to determine whether there is more than one way to get what is expected from the information system.

SYSTEMS EVALUATION AND JUSTIFICATION. In this phase one has to evaluate the impact of the proposed system on the employees. The system should be such that it assists the user, rather than requiring the user to change significantly the way he or she operates. One should not expect the user to be a genius in order to use the system. The system should serve the needs of the users and not vice versa. Next, the evaluation of cost-effectiveness should be undertaken. What are the direct costs, the indirect costs? One has to determine the fixed costs, which have to be borne independent of continuing needs as well as operational costs depending on the complexities and the volume of data. Again, one has to consider various alternatives just as in the system design.

DETAILED SYSTEMS DESIGN. This is the most critical step in the design phase, representing the fine-tuning of the broad design. This is the stage at which the detailed specifications of the systems are developed. One has to look at not only what the input would be but also how the input would be prepared, what types of forms are needed for input, and what different types of output are needed. What type of data entry method is most suitable, and who would enter the data? It is at this stage that one should keep asking the users for more and more information in trying to refine the system. During this phase every aspect of the system is clearly spelled out.

Often people overlook the importance of a well-designed form for input or output of the data. The input forms should be designed with simplicity in mind. Complicated forms only cause confusion. The form should be designed so that it serves the needs of the person who will record the data on it. Secondly, it should assist the person in data processing. Often the forms are designed by and for data processing people, completely ignoring the users.

Today's computer printers have come a long way from the fixed font type of printer of earlier years. They allow printing of information in a nice layout with exceptional clarity if enough thought process has gone into designing the input and output forms. Physical layout as well as order of items on the forms should follow reasonable logic that serves the user's needs.

SYSTEM VALIDATION AND IMPLEMENTATION. This is the final phase of system design. During this phase, one should familiarize the users about the system by proper training and education. The more the users know the intricacies of the system, the more they will be able to utilize all the capabilities. Furthermore, the users are the ones who will continually come up with a list of improvements in the system. Before final implementation, extensive validation of the system needs to be undertaken. Next comes the testing of all of the individual components or programs of the system. All of the inputs should be used to produce the desired output. One has to check the output for accuracy and reliability. Validation of computerized systems requires a lot of resources. If not properly planned, validation can consume more resources than that required for the development or implementation (Budihandojo, 1997, 1998; Samways et al. 1996; Schoenauer et al., 1993).

Finally the system requires continuous maintenance. No system can survive without proper maintenance. Maintenance does not only mean to keep the system going but also to keep the system up to date to serve the changing needs of the user. There is nothing more frustrating than using a system designed using outmoded standards.

REFERENCES

Allen, P. V., Dukes, G. R., Gerger, M. E. Determination of release limits: general methodology. Pharm. Res. 8(9):1210–1213, 1991.

Budihandojo, Rory. Computerized system validation: a concept approach in the preparation of a validation plan document. Pharmaceutical Technology 21(2):70, 1997.

Budihandojo, Rory. Computerized system validation: preparation of a design specifications document. Pharmaceutical Technology 22(3):150, 1998.

Drew, H. D., Thornton, L. K., Juhl, W. E., Brower, J. F. An FDA/PhRMA interlaboratory study of the International Conference on Harmonization's proposed photostability testing procedures and guidelines. Pharmacopeial Forum 24(3):6317, 1998a.

Drew, H. D., Brower, J. F., Juhl, W. E., Thornton, L. K. Quinine photochemistry: a proposed chemical actinometer system to monitor UV-A exposure in photostability studies of pharmaceutical drug substances and drug products. Pharmacopeial Forum 24(3):6334, 1998b.

Fairweather, W. R., Lin, T.-Y. D., Kelly, R. Regulatory design and analysis aspects of complex stability studies. J. Pharm. Sci. 84:1322–1326, 1995.

Food and Drug Administration (FDA). FDA guide to inspections of pharmaceutical quality control laboratories. July 1993.

FDA. Content and format of investigational new drug applications (INDs) for phase 1 studies of drugs, including well-characterized, therapeutic, biotechnology-derived products, guidance for industry. Center for Drug Evaluations Research (CDER), Center for Biologics Evaluation and Research (CBER), November 1995.

FDA. SUPAC-IR: immediate release solid dosage forms: scale-up and post-approval changes, chemistry, manufacturing, and controls, in vitro dissolution testing, in vivo bioequivalence documentation. CDER, November 1995b.

FDA. SUPAC-IR/MR: immediate and modified release solid dosage forms, manufacturing equipment addendum, draft guidance. April 1998a.

FDA. Guidance for industry: stability testing of drug substances and drug products, draft guidance. June 1998b.

Grimm, W. Drugs Made in Germany 28:196–202, 1985.

Grimm, W. Drugs Made in Germany 29:39–47, 1986.

Haynes, J. D. The world with virtual temperatures for product stability testing. J. Pharm. Sci. 60:927–929, 1971.

International Conference on Harmonization (ICH). Q1A stability testing for new drug substances and products. September 1994.

ICH. Q1B photostability testing of new drug substances and products. November 1996.

Nordbrock, E. Statistical comparison of stability study designs. J. Biopharm. Stat. 2(1):91–113, 1992.

Sager, N., Baum, R. G., Wolters, R. J., Layloff, T. Photostability testing of pharmaceutical products. Pharmacopeial Forum 24(3):6331, 1998.

Samways, K., Sattler, L., Wherry, R. Computerized system validation: one company's approach to assigning tasks and responsibilities. Pharmaceutical Technology 20(8):45, 1996.

Schoenauer, Ciaran M., Wherry, Robert J., et al. Computer system validation—staying current: security in computerized systems PMA's computer systems validation committee. Pharmaceutical Technology 17(5):48, 1993.

Tonnesen, H. H., ed. Photostability of Drugs and Drug Formulations. London: Taylor and Francis, 1996.

Yoshioka, S., et al. Quinine actinometry as a method for calibrating ultraviolet radiation intensity in light-stability testing of pharmaceuticals. Drug Development and Industrial Pharmacy 20(13):2049–2062, 1994.

16

Stability of Polypeptides and Proteins

MARY D. DIBIASE

Biogen, Cambridge, Massachusetts

MARY K. KOTTKE

Cubist Pharmaceuticals, Inc., Cambridge, Massachusetts

1. INTRODUCTION

Several differences exist between conventional small molecular weight compounds and polypeptides, many of which impact their predicted stability profile. Polypeptides consist of a regularly repeating backbone with distinctive side chains that interact with each other to contribute to the three-dimensional structure of the protein (see Fig. 1). Typically, small molecular weight compounds are either linear or cyclical in nature, and their size prohibits extensive intramolecular bonding. Thus the majority of conventional small molecular weight compounds do not exhibit higher level structures found in polypeptide molecules.

It is possible to make replacements of similar amino acids within a polypeptide with little impact to the biological activity and structural stability of the molecule.

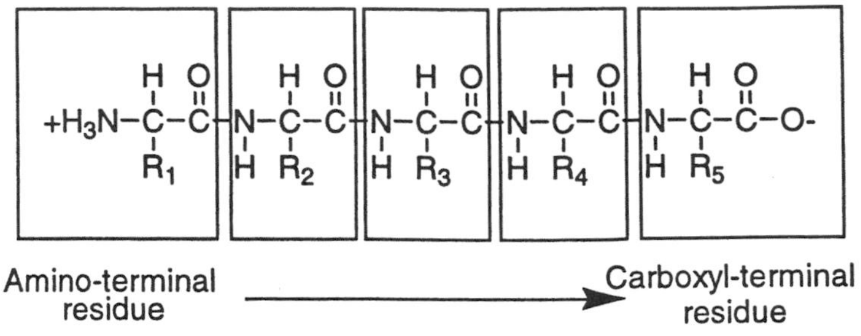

Amino-terminal residue → Carboxyl-terminal residue

Fig. 1 Schematic representation of a pentapeptide.

Therefore chemical changes can be made to the compound without loss of activity. This is in distinct contradiction to small molecular weight compounds.

Loss of chemical integrity or conformational structure of a polypeptide can affect the molecule's activity at several levels. As with conventional small molecular weight compounds, potency can be decreased or entirely lost due to degradation of the molecule. In addition, protein degradation can result in acquired immunogenicity, altered pharmacokinetics, and protein self-association. This chapter will discuss various methods that cause protein degradation and will provide several methods for detecting these changes.

2. PROTEIN BACKGROUND

2.1. Formation of Peptide Bonds

Polypeptides are generated by the condensation of two amino acid residues. As illustrated in Fig. 2, the peptide bond is formed between the α-carboxyl group of one amino acid and the α–amino group of another. It is noted that hydrolysis of the peptide bond is favored over its synthesis. Therefore the formation of peptide bonds requires input energy.

2.2. Common Amino Acids

Provided in Table 1 is a list of the 20 naturally occurring amino acids that make up most polypeptides and proteins. These residues can be classified further based on their side chain functionality. With the exception of glycine, all amino acids contain a chiral carbon and are thus restricted in their conformation.

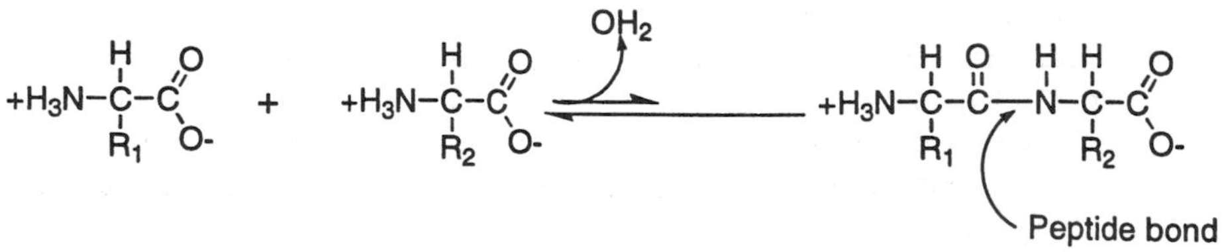

Peptide bond

Fig. 2 Formation of peptide bond.

Table 1 Twenty Naturally Occuring Amino Acids with Corresponding Abbreviation and Mode(s) of Degradation

Residue	Abbreviation		Potential mode(s) of degradation
	3-Letter	1-Letter	
Alanine	Ala	A	Relatively stable
Arginine	Arg	R	Relatively stable
Asparagine	Asn	N	Deamidation, racemization isomerization
Aspartic acid	Asp	D	Hydrolysis, racemization, isomerization
Cysteine	Cys	C	Oxidation, β-elimination, racemization, disulfide exchange
Glutamic acid	Glu	E	Relatively stable
Glutamine	Gln	Q	Deamidation, racemization, isomerization
Glycine	Gly	G	N-terminal location promotes diketopiperazine formation
Histidine	His	H	Oxidation
Isoleucine	Ile	I	Relatively stable
Leucine	Leu	L	Relatively stable
Lysine	Lys	K	Relatively stable
Methionine	Met	M	Oxidation
Phenylalanine	Phe	F	Relatively stable
Proline	Pro	P	N-terminal location promotes diketopiperazine formation
Serine	Ser	S	β-elimination, racemization
Threonine	Thr	T	β-elimination, racemization
Tryptophan	Trp	W	Oxidation
Tyrosine	Tyr	Y	Oxidation
Valine	Val	V	Relatively stable

2.2.1. Hydrophobic Amino Acids

Hydrophobic amino acids contain either aliphatic or aromatic side chains (see Figs. 3 and 4). Most hydrophobic residues are located in the interior of the polypeptide, thereby making them relatively unreactive. Although this group of amino acids is primarily hydrophobic in nature, the nitrogen in the indole ring of tryptophan and the phenolic group in tyrosine can H-bond with other residues. In addition, the proximity of the methyl groups in valine and isoleucine to the main chain restricts polypeptide conformation in regions where these residues are present.

2.2.2. Neutral-Polar Amino Acids

Neutral-polar amino acids contain polar groups that are not readily ionizable (see Fig. 5). Asparaginyl and glutaminyl residues have the propensity to deamidate to aspartic and glutamic acid, respectively. In addition, when centrally located, asparagine and glutamine often H-bond with amides. Hydroxyl groups present in the side chain of serine and threonine allow them to hydrogen bond with the main

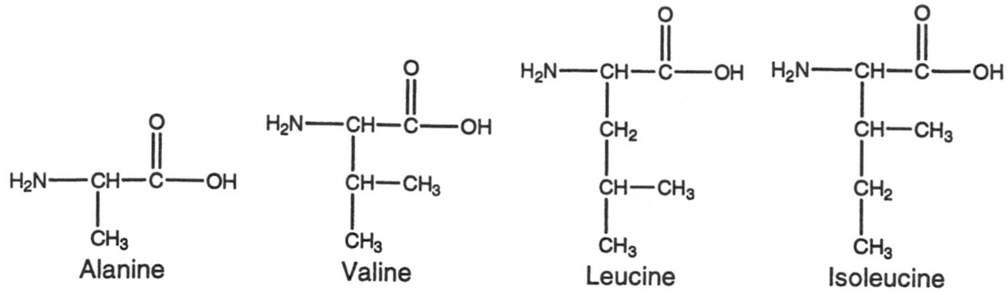

Alanine Valine Leucine Isoleucine

Fig. 3 Hydrophobic-aliphatic amino acids.

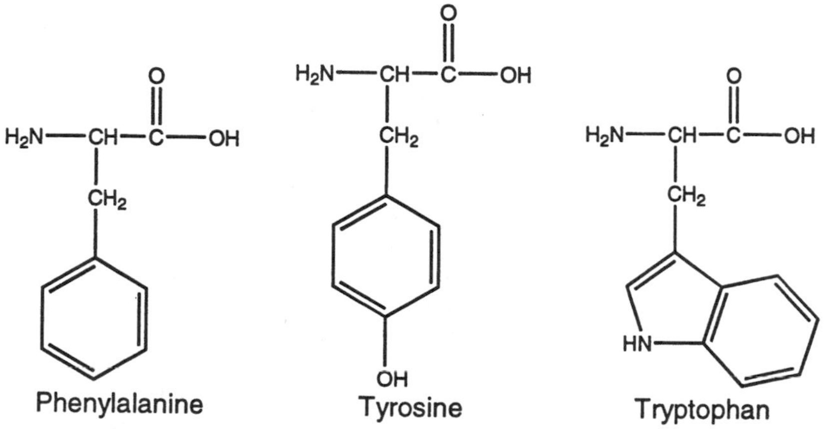

Phenylalanine Tyrosine Tryptophan

Fig. 4 Hydrophobic-aromatic amino acids.

chain thus providing the opportunity for β-elimination and racemization. Methionine residues are primarily located within the interior region of polypeptides and are highly susceptible to oxidation. Tryptophan and tyrosine also have the potential to oxidize, though to a lesser extent than methionine. The sulfhydryl groups within each of the cysteine residues react together to form disulfide bridges.

2.2.3. Acidic Amino Acids

Acidic amino acids contain side chains that are negatively charged at physiological pH (see Fig. 6). Acidic amino acids are typically located at the polypeptide surface and are frequently the active site for enzymatic reactions. Aspartyl residues are often susceptible to hydrolysis, racemization, and isomerization while glutamyl residues are relatively stable.

2.2.4. Basic Amino Acids

Basic amino acids either possess a neutral (histidine) or positive (lysine, arginine) charge at physiological pH (see Fig. 7). Histidine is often involved in enzymatic reactions and can be prone to oxidation. When located in the interior portion of a polypeptide, basic amino acids (lysine, arginine) can be involved in electrostatic interactions with acidic residues (i.e., aspartic acid, glutamic acid).

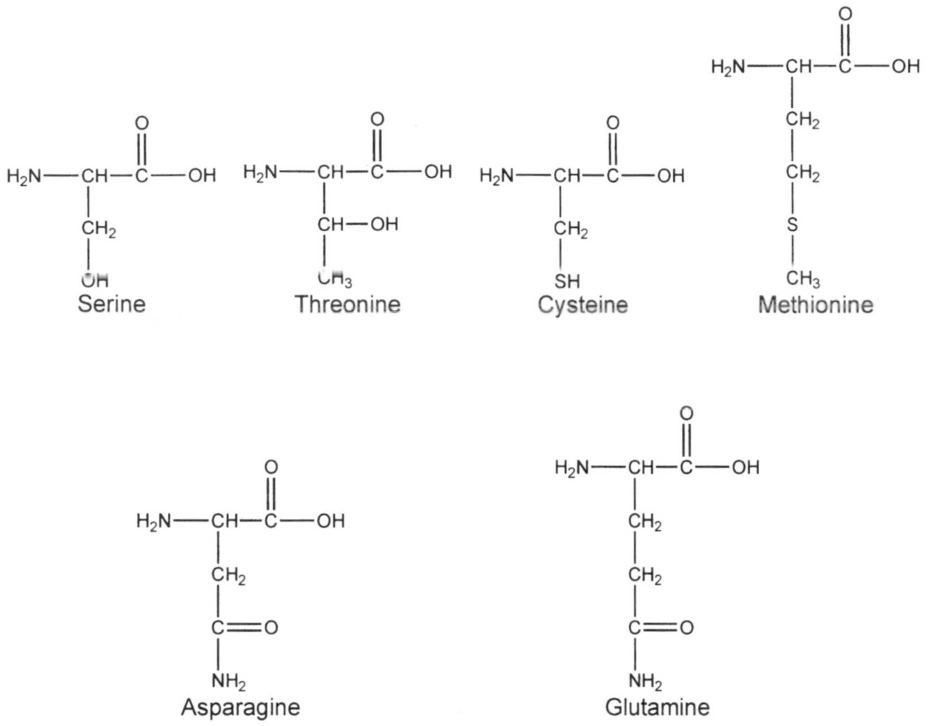

Serine Threonine Cysteine Methionine

Asparagine Glutamine

Fig. 5 Neutral-polar amino acids.

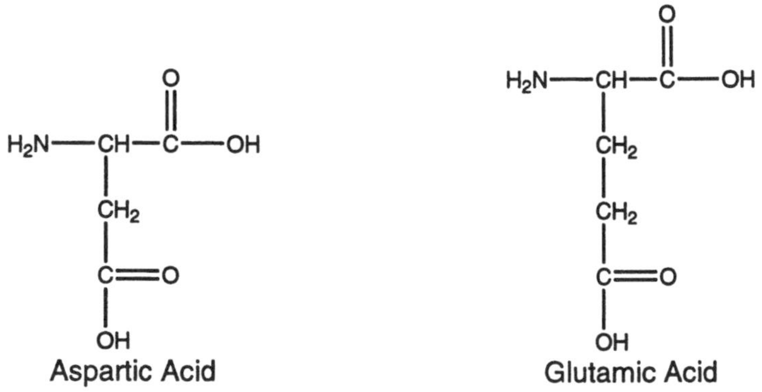

Aspartic Acid Glutamic Acid

Fig. 6 Acidic amino acids.

2.2.5. Conformation Influencing Amino Acids

Glycine and proline (see Fig. 8) are two residues that play a major role in a protein's conformation. These two residues, however, behave quite differently from one another. Due to its achiral nature, glycine is able to adopt conformations forbidden by other residues. In addition, glycine is the smallest of the 20 amino acid residues. Thus, glycine is highly flexible and is often located at turns in the protein. Proline, on the other hand, is very rigid because its side chain nitrogen is covalently linked

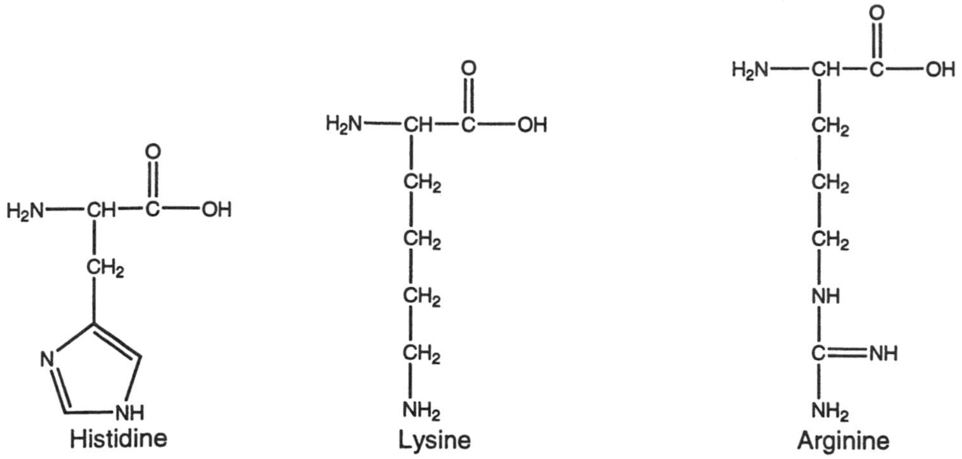

Histidine Lysine Arginine

Fig. 7 Basic amino acids.

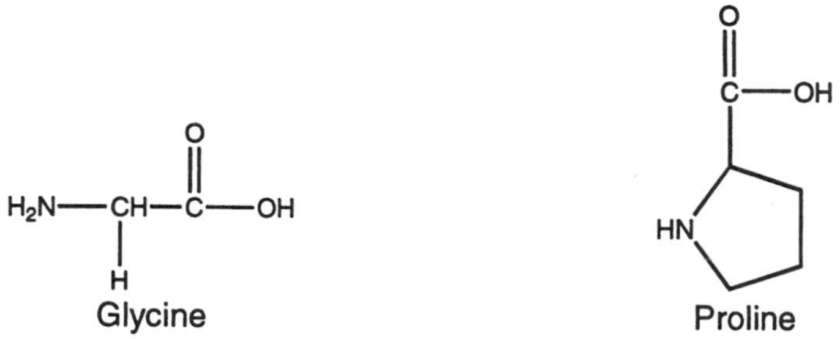

Glycine Proline

Fig. 8 Conformation influencing amino acids.

to the main chain of the polypeptide. Therefore regions containing proline residues are restricted in their conformation. Interestingly, the presence of either glycine or proline in an N-terminal position promotes diketopiperazine formation.

2.3. Protein Structure

As noted in the introduction, the primary factor that differentiates polypeptides and proteins from conventional small molecular weight compounds is the varying levels of structure that polypeptides and proteins exhibit. Changes at any level of the protein structure can impact the molecule's activity. Protein chains fold spontaneously into their native (i.e., biologically active) conformation, most often with the hydrophobic residues interior to the protein and the charged residues on the surface. Protein denaturation occurs when a polypeptide loses its higher level structure and often results in loss of biological activity.

2.3.1. Primary

The primary structure (Fig. 9) of a polypeptide is composed of its amino acid sequence and the location of any disulfide bridges. Thus all of the covalent connections within a polypeptide contribute to its primary structure.

a. primary

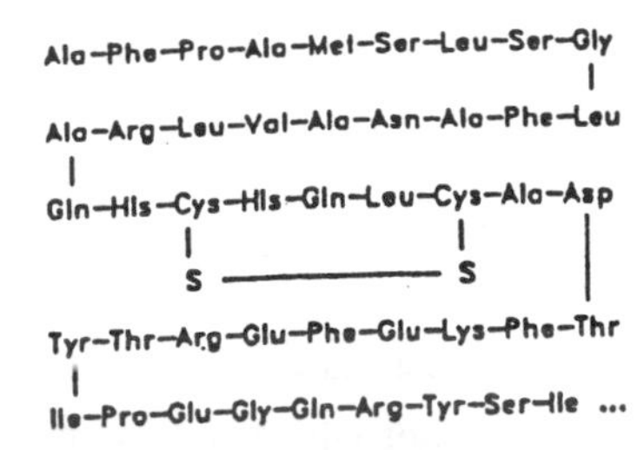

b. secondary

c. tertiary

d. quaternary

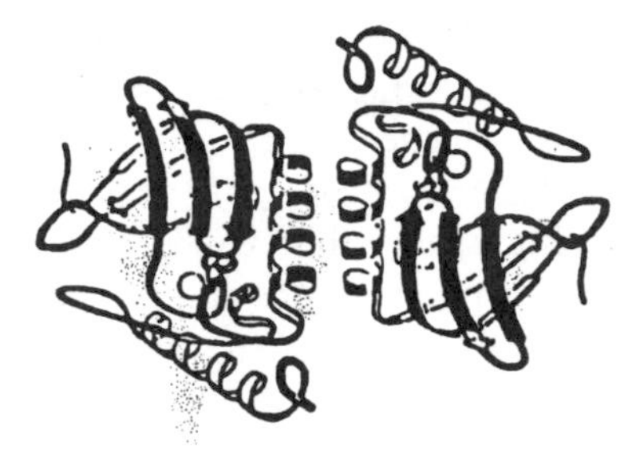

Fig. 9 Protein structure: (a) primary; (b) secondary; (c) tertiary; (d) quaternary.

2.3.2. Secondary

Secondary structure refers to the localized folding or shape of a polypeptide. Amino acid sequence (*i.e.*, primary structure), residue alignment, hydrogen bonding, and disulfide bridges all contribute to the secondary structure of a polypeptide. Therefore secondary structure results from the interaction between adjacent residues, or those within close proximity of each other. Examples of periodic secondary structure include the α-helix (as depicted in Fig. 9), the β-pleated sheet, and the β-turn.

2.3.3. Tertiary

A polypeptide's tertiary structure imparts three-dimensionality on the molecule. It is derived from the spatial arrangement of nonadjacent residues. In other words, tertiary structure is the result of the interactions between the residues' side chains and their resultant orientation.

2.3.4. Quaternary

Proteins containing more than one polypeptide chain, or subunit, exhibit quaternary structure. The quaternary structure of a protein is a result of noncovalent binding of its subunits. The manner in which the subunits are aligned with each other contribute to its quaternary structure.

3. PROTEIN DEGRADATION

Although polypeptide and protein stability have been separated into chemical and physical modes, there is obviously a significant relationship between conformational stability and the chemical integrity of each molecule. This relationship is of particular importance to the understanding of the mechanisms of protein inactivation. Perturbation of secondary or tertiary structure can lead to exposure of previously buried amino acids, facilitating their chemical reactivity; alternatively, chemical changes can lead to loss of conformational stability.

Numerous excipients incorporated into pharmaceutical products can also affect protein stability. For example, reducing sugars can react with protein amino groups to result in the formation of a Schiff's base and subsequent Maillard reaction. In addition, several surfactants, such as Tween 20, Tween 80, and Poloxamer polymers, can cause the oxidation of susceptible amino acids due to residual peroxides present in these materials that often increase on storage. Highly purified forms of some of these surfactants are now being made available with minimal peroxide levels, and some are also being produced with antioxidants, such as butylated hydroxytoluene (BHT). Antioxidants themselves should not be incorporated into formulations of proteins that contain disulfide bonds, since they may contain reducing agents that will destroy this bond. As many of the chemical reactions involving polypeptides and proteins are catalyzed or facilitated by metal ions, attention should also be made to levels of these in formulated products.

3.1. Chemical Instability

Within the context of this chapter, chemical instability refers to the formation or destruction of covalent bonds within a polypeptide or protein molecule. These changes alter the primary structure of the protein and can also impact higher levels of its structure. Common causes of chemical instability of polypeptide molecules include deamidation, oxidation, and cystine destruction/disulfide exchange.

3.1.1. Deamidation

Nonenzymatic deamidation of asparagine residues occurs in many proteins and peptides and is a major source of spontaneous degradation and loss of amino acid sequence homogeneity. While glutamine residues can also undergo deamidation, they are much less labile to this reaction than asparagine. Asparagine residues deamidate to aspartate or isoaspartate, changing positive charge to negative. Although the reaction is generally associated with conditions of neutral to basic pH, mechanistic studies have shown that deamidation can also occur in acidic pH.

The neutral-to-basic pH deamidation reaction is intramolecular and occurs via a five-membered ring succinimide intermediate (see Fig. 10) by attack of the side chain amide on the backbone of the peptide carbonyl. The succinimide ring is labile to base and readily hydrolyzes in two possible ways, yielding either aspartic acid in a normal α-linked form and/or isoaspartyl characterized by an atypical bond between the β-carbonyl of aspartate and the α-nitrogen of the neighboring amino acid. Although, in general, proteins are less susceptible to deamidation under acidic conditions, deamidation by direction hydrolysis, rather than via imide formation, can occur as pH continues to decrease.

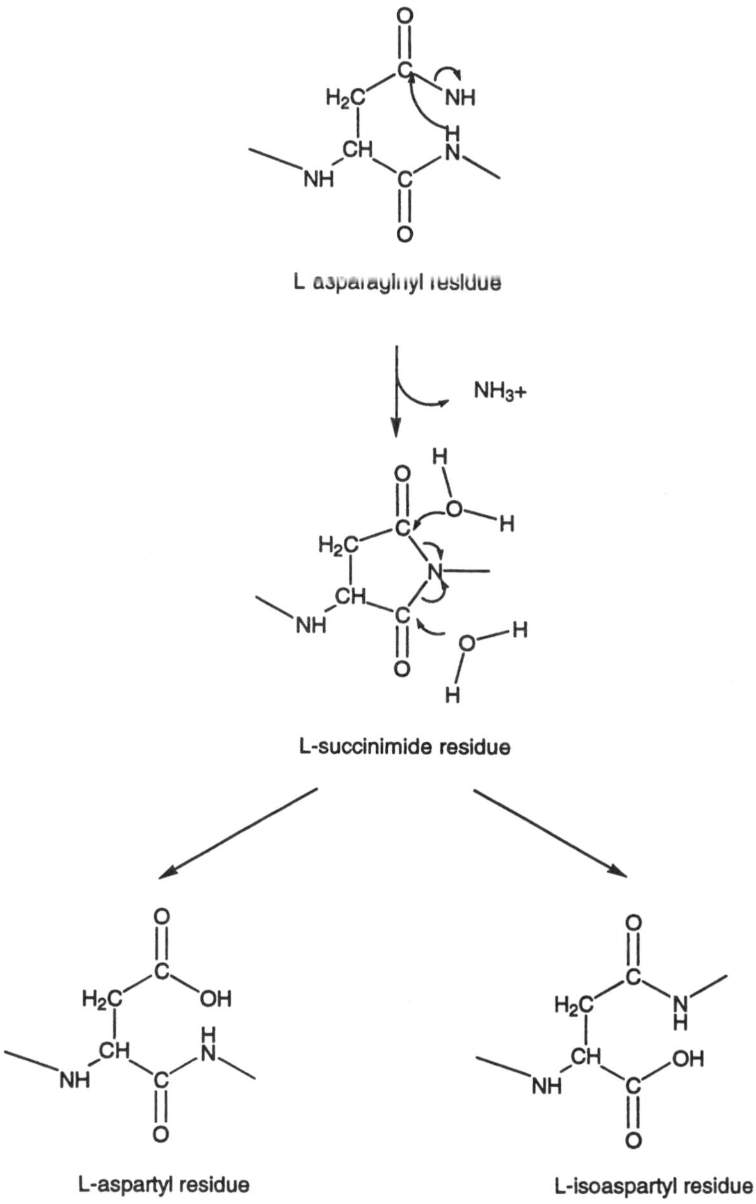

Fig. 10 Deamidation of L-asparagyl residue by succinimide formation.

Susceptibility of asparagine or glutamine to deamidation is highly sequence and conformation dependent. Studies with model peptides show that asparagine, proceeded by glycine and serine residues, is particularly labile due to the decreased steric hindrance of succinimide formation by the C-terminal residue (1). In contrast, bulky residues such as proline reduce the reactivity of asparaginyl residues. Therefore, although conformationally rigid regions of a protein molecule may inhibit deamidation at labile asparagine residues, the presence of a sensitive aspara-

Table 2 Several Polypeptides and Proteins that are Subject to Deamidation

Protein/peptide	Affected residues	Reference
Insulin	$Asn_{B3}Gln$	Brange et al. 1992 (2); Berson and Yalow 1996 (3)
Insulin	Asn_{A21} (C-terminus)	Strickley and Anderson 1996 (4)
Adrenocorticotropin hormone (ACTH)	$Asn_{25}Gly$	Bhatt et al. 1990 (5)
Human growth hormone (hGH)	$Asn_{149}Asp$	Becker et al. 1987 (6); Johnson et al. 1989 (7)
Interleukin-1α (IL-1α)	$Asn_{36}Asp$	Wingfield et al. 1987 (8)
Human growth hormone releasing factor (hGHRF)	$Asn_{8}Ser$	Friedman et al. 1991 (9)
CD4	$Asn_{52}Asp$	Teshima et al. 1991 (10)
Tissue plasminogen Activator (tPA)	$Asn_{37}Ser$, $Asn_{58}Gly$, $Asn_{177}Ser$	Paranandi et al. 1994 (11)
Hirudin	$Asn_{33}Gly$, $Asn_{53}Gly$	Tuong et al. 1992 (12); Bischoff et al. 1993 (13)

gine sequence in a particularly flexible region (e.g., flanked by glycine or serine) may enhance the susceptibility of this site to deamidation.

As noted in Table 2, deamidation has been documented in a number of proteins and can impact significantly the molecule's activity. Deamidation can make proteins more susceptible to proteases and denaturation. Increased susceptibility to proteases induced by deamidation can reduce the in vivo half-life of the molecule. An additional problem posed by deamidation of asparaginyl residues is the formation of isoaspartate. Isoaspartate residues can dramatically affect the activity and conformation of a protein (14) and may also increase the immunogenicity of certain proteins (15). Note that isoaspartate can also arise from direct isomerization of aspartate. This reaction, which is generally more difficult to detect as no change in charge occurs, is optimal at pH 5 (16,17).

Deamidation in pharmaceutical products can be detected by charge or molecular weight changes or by direct monitoring of the formation of succinimide or isoaspartic residues. Methods used to identify deamidation include peptide maps, capillary electrophoresis, isoelectric focusing, and enzyme catalyzed radiolabeling of the isoaspartyl sites (n.b., commercial kits are available).

Formulation approaches to minimize deamidation generally involve maintaining conditions of low pH where deamidation is less likely to occur. However, desialylation of glycosylated proteins is prevalent at low pH. Therefore when dealing with proteins that are subject to both deamidation and desialylation, optimization studies to balance these conflicting objectives should be performed. As deamidation may also be affected by buffer composition, studies evaluating polypeptide stability in varying buffers may also provide valuable information.

3.1.2. Oxidation

Methionine, cysteine, tryptophan, and tyrosine residues are all susceptible to oxidation. Cysteine and methionine oxidation, however, are more common and have been studied in greater depth.

Methionine oxidation, induced by exposure to air, intense fluorescent light, or residual peroxide content, results in conversion of the thioether to sulfoxide as shown in Fig. 11. Harsher oxidative conditions result in the formation of sulfone. Methionine oxidation can occur during fermentation or cell culture and purification of recombinant proteins (18), so great care should be taken when modifying such processes in a product where susceptible methionine residues are present. It is noteworthy, however, that several new approaches to remove oxidized material during purification are being developed (19).

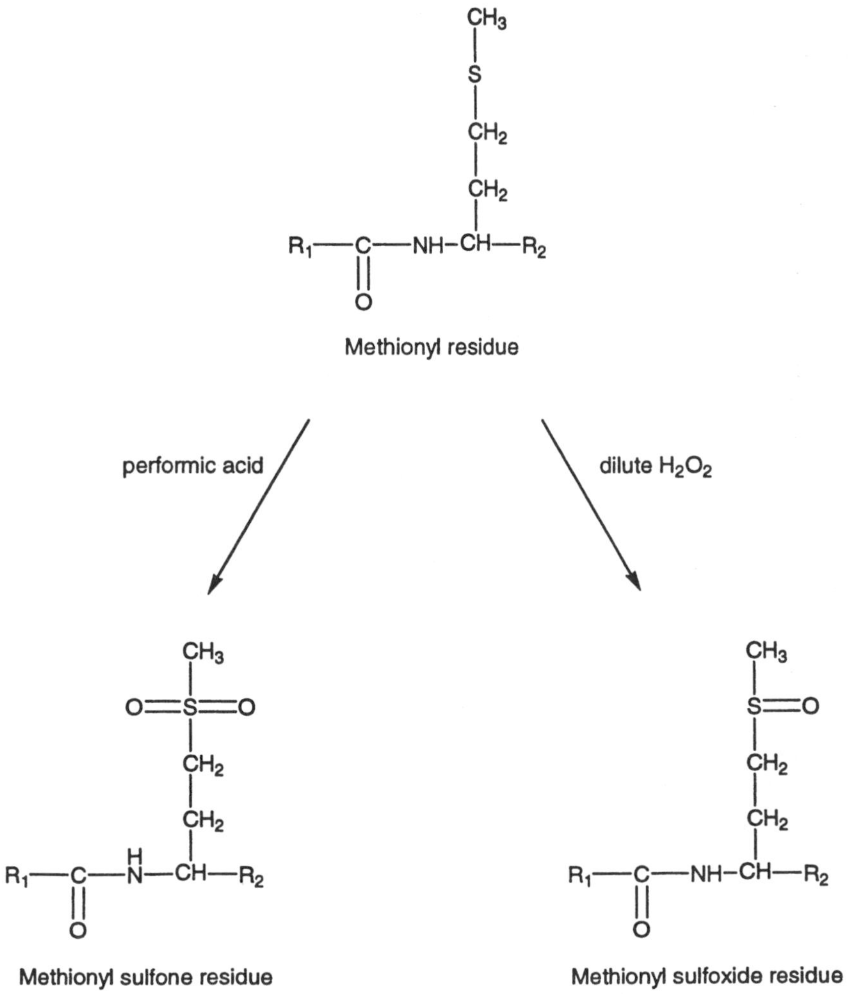

CH_3
S
CH_2
CH_2
R_1—C(=O)—NH—CH—R_2

Methionyl residue

performic acid

dilute H_2O_2

CH_3
O=S=O
CH_2
CH_2
R_1—C(=O)—N(H)—CH—R_2

CH_3
S=O
CH_2
CH_2
R_1—C(=O)—NH—CH—R_2

Methionyl sulfone residue

Methionyl sulfoxide residue

Fig. 11 Methionine oxidation under mild (dilute H_2O_2) and strong (performic acid) conditions.

Methionine oxidation is also of particular concern during the storage of both liquid and lyophilized formulated products (20–23). The susceptibility of methionine to oxidation has been shown to be dependent on the relative exposure of the residue (24,25). Alteration of pH, ionic strength, and solvent polarity can change both the rate and the extent of methionine oxidation.

Although the impact of oxidative modifications on proteins is variable, studies with recombinant human leptin show that the four methionine residues in the molecule are not equally susceptible to oxidation and that the resulting oxidative products each have a different magnitude of effect on maintaining leptin's structural and biological activity (26). Selective oxidation of another protein, recombinant human granulocyte colony stimulating factor (rhG-CSF), by hydrogen peroxide generated four different modified forms of the molecule containing one to four modified methionine residues. All oxidized forms of G-CSF retained a conformational structure similar to that of the native molecule, but only one retained all of its biological function. In addition, site-directed mutagenesis with leucine replacement of methionine produced increased stability and retained the biological activity of the molecule (24). Thus chemical modification to G-CSF was made without impact to its activity.

Antithrombin provides another interesting example of the impact of methionine oxidation on biological activity. Antithrombin contains four methionine residues; two adjacent residues (Met_{314} and Met_{315}) near the reactive site loop cleaved by thrombin, and two exposed residues (Met_{17} and Met_{20}) that border on the heparin-binding region. In forced oxidation studies with hydrogen peroxide, the methionine residues at positions 314 and 315 were found to be highly susceptible to oxidation, but their oxidation did not affect the molecule's thrombin-inhibitory activity or heparin binding. Significant oxidation of methionine residues at positions 17 and 20 only occurred at higher peroxide concentrations, at which point heparin affinity was decreased. Structural studies indicate that highly oxidized antithrombin is less able to undergo the complete conformational change induced by heparin (27). Additional studies evaluating the oxidation of methionine residues are provided in Table 3.

Table 3 Several Polypeptides and Proteins that Are Subject to Methionyl Oxidation

Protein/peptide	Reference
Antistasin	Przysiecki et al. 1992 (18)
Cystatin C	Berti et al. 1997 (19)
Recombinant human monoclonal antibody HER2 (rhuMAbHER2)	Lam et al. 1997 (21)
Granulocyte colony-stimulating factor (G-CSF)	Lu et al. 1999 (25)
Leptin	Liu et al. 1998 (26)
Antithrombin	Van Patten et al. 1999 (27)
Epidermal growth factor (EGF)	George-Nascimento et al. 1988 (28)
Adrenocorticotropin hormone (ACTH)	Antonini and Spoto 1988 (29)

Peptide maps are convenient approaches for detecting methionine oxidation, with the peptide containing the sulfoxide being identified by mass spectroscopy (MS). Reversed-phase HPLC (RP-HPLC) is often successful at separating oxidized forms and is also used to identify oxidized products by coelution of hydrogen peroxide stressed molecules.

Approaches to minimize the formation of oxidative products are based on the addition of antioxidants to the formulation and the manipulation of environmental conditions (i.e., pH or temperature adjustment). Antioxidants such as methionine, sodium thiosulfate, catalase, or platinum can prevent methionine oxidation, presumably by acting as free radicals or oxygen scavengers (21). Oxidation of cysteine residues is promoted at both neutral and elevated pH and can result in aggregation or disulfide scrambling (see Sec. 3.1.3). Thus maintaining low pH may reduce the incidence of methionine oxidation.

A recent study reports the identification of tryptophan oxidation products (hydroxytryptophan, N-formylkynurenine, and kynurenine) in intact bovine alpha-crystallin protein (30). Oxidation of tryptophan residues has also been reported to occur in luteinizing hormone-releasing factor (LRF), somatostatin, valine-gramicidin A, and ACTH (31,32).

3.1.3. Cystine Destruction and Thiol–Disulfide Exchange

Cysteine residues contain thiol groups that can be oxidized to form disulfide bonds (cystine residues). These bonds are naturally occurring cross-links that covalently connect protein or polypeptide chains either intra- or intermolecularly. Disulfides can be formed either by direct oxidation of cysteine or by thiol–disulfide interchange and are often catalyzed by trace copper and iron, the relative stability of a reduced cysteine being dependent on the redox potential of the protein's environment. Each pH unit change towards the acidic has been reported to decrease the oxidative reactivity of thiols by approximately an order of magnitude (33).

Disulfide bonds generally aid in stabilizing folded forms of proteins, but can be subject to destruction under certain conditions such as reducing potential, heat, and alkaline pH. Destruction of disulfide bonds can occur at elevated pH by β-elimination (see Fig. 12). For example, at pH 9 and 50°C, formation of atypical cross-links in hirudin resulted in the molecule's loss of activity (34). Disulfide scrambling has also been documented in human insulin-like growth factor (IGF) (23) where formation of intermolecular disulfides resulted in dimerization. Another protein in which atypical disulfide bonds are formed is acidic fibroblast growth factor (aFGF). aFGF normally contains three reduced cysteine residues and no disulfide bonds. However, copper is able to catalyze the formation of an intramolecular disulfide thereby causing inactivation of the aFGF molecule (35).

The presence of atypical disulfide bonding can often be suggested by running reduced and nonreduced sodium dodecyl sulfate polyacrylamide gel electrophoresis (SDS-PAGE); and Ellman's reagent is widely used for the determination of free thiols (36). Methods used to study reactions between free thiols and disulfide bonds in more detail include various proteolytic digestions, peptide mapping, partial reduction, and assignment of disulfides by N-terminal sequencing and matrix-assisted laser desorption ionization (MALDI) MS. The use of the majority of these methods is documented in a recent study on the determination of the disulfide structure of tumor necrosis factor binding protein (37).

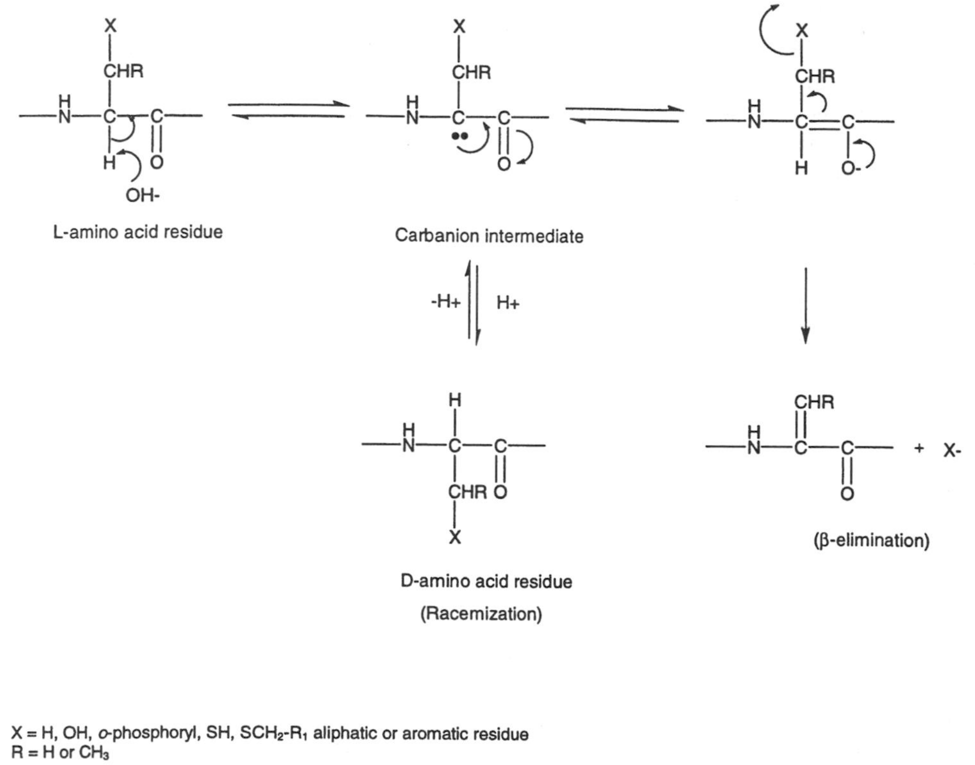

X = H, OH, *o*-phosphoryl, SH, SCH_2-R_1 aliphatic or aromatic residue
R = H or CH_3

Fig. 12 *β*-Elimination and racemization in alkaline media.

Formulation approaches to minimize the destruction of disulfides and thiol–disulfide exchange include targeting acidic pH formulations and avoiding any potential reducing agents (n.b., a number of antioxidant excipients have significant reducing potential). Lyophilization of therapeutic proteins obviously reduces the rate of many of these types of reactions, but it is possible to facilitate thiol–disulfide exchange by structural modification due to dehydration of the molecule by lyophilization. One alternative approach to resolving potential instability caused by free thiols in the presence of disulfides is the site-specific substitution of noncritical cysteines with other residues. This has been done with a human interferon (IFN) beta analogue, IFN beta-1b, which contains one disulfide bond and a serine in place of the free thiol (38).

3.1.4. Other Covalent Bond Modifications in Polypeptides and Proteins

A number of other less common or less documented bond cleavage reactions also occur in peptides and proteins and are discussed below.

3.1.4.1. Hydrolysis at Aspartic Acid Residues Under Acidic Conditions

The mechanism of hydrolysis of the N-terminal or C-terminal peptide bond of aspartic acid is shown in Fig. 13 (39). This hydrolysis, documented primarily at acidic pH, has been shown to occur at a rate at least 100 times greater than any

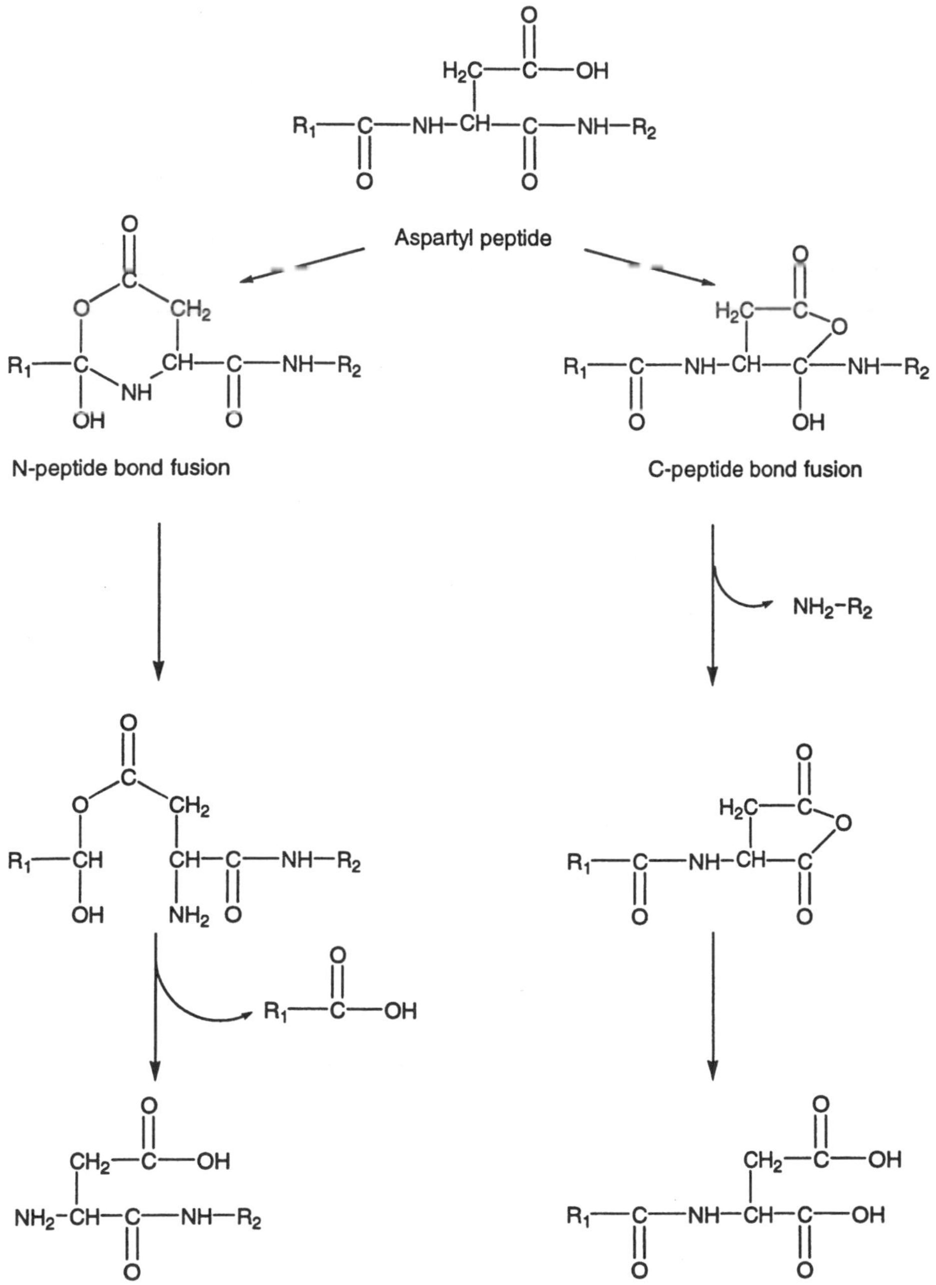

Aspartyl peptide

N-peptide bond fusion

C-peptide bond fusion

NH_2-R_2

R_1—C(=O)—OH

Fig. 13 Hydrolysis of aspartyl peptide in dilute acid.

other amino acid (40), and the Asp-Pro bond is known to be particularly labile, possibly due to either an enhanced α/β isomerization of the aspartic acid residue or the more basic nature of the proline nitrogen (41). Cleavage of an Asp-Ser bond has been reported on storage of recombinant human epidermal growth factor (rhEGF) under the accelerated conditions of pH 3 and 45°C (42).

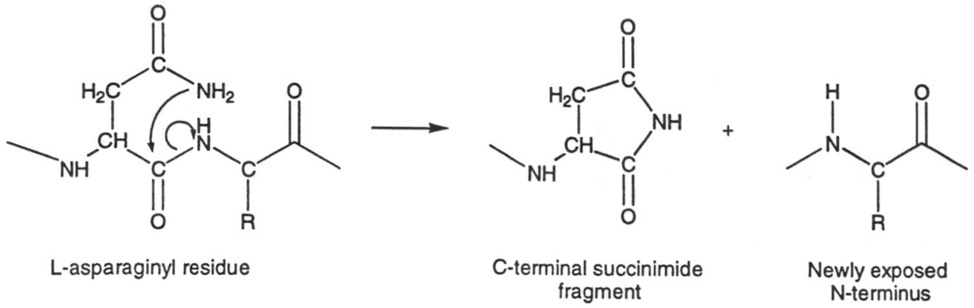

L-asparaginyl residue

C-terminal succinimide fragment

Newly exposed N-terminus

Fig. 14 C-Terminal succinimide formation of asparginyl residue.

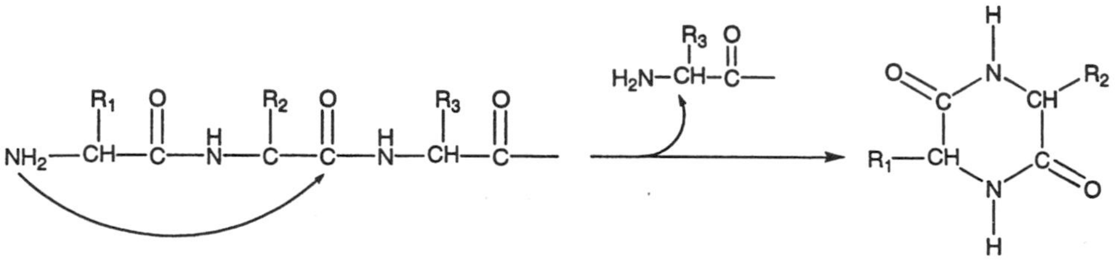

Fig. 15 Diketopiperazine formation.

3.1.4.2. C-Terminal Succinimide Formation at Asparagine Residues

Succinimide formation as occurs during deamidation reactions can also result in the spontaneous cleavage of the peptide bond at asparagine residues. In this case, the side chain amide nitrogen attacks the peptide bond to form a C-terminal succinimide residue and a newly formed amino acid terminus (43). See Fig. 14.

3.1.4.3. Diketopiperazine Formation

Rearrangement of the N-terminal dipeptide can result in the formation and cleavage of a cyclic diketopiperazine at high pH (see Fig. 15). Proline and glycine at the N-terminus (44) promote this base-catalyzed reaction. The kinetics of diketopiperazine formation has been studied using model peptides. In one study, degradation due to the formation of diketopiperazine followed pseudo–first order kinetics and showed a significant dependence on pH (45).

3.1.4.4. Enzymatic Proteolysis and Autolysis

Contaminating proteases present during fermentation, cell culture, or purification can result in the cleavage of recombinant proteins. Optimization of processing conditions and addition of carefully selected protease inhibitors can minimize this occurrence. Storage of purified proteases may also lead to peptide bond cleavage; tissue plasminogen activator (tPA), a serine protease, contains an activation cleavage site at Arg_{275}. Even after autolysis, the two chains of the clipped protein remain held together by a disulfide bond between residues 264 and 395. Resolution of the clipping by site-directed mutagenesis of the labile protein was attempted by replacing the Arg_{275} residue with other amino acids. The resulting one-chain forms of tPA, however, exhibited diminished activity in the absence of a fibrin cofactor (46).

3.1.4.5. Alternative Covalent Aggregation

Although the most common form of covalent aggregation is due to the formation of disulfide bonds, aggregation can also occur via free amino, carbonyl, carboxyl, or hydroxyl functional groups with ester or amide linkages, for example. Such aggregates have been documented for insulin (47,48) and were also found during the analysis of relaxin (49).

3.1.5. Deglycosylation and Desialylation

There are a number of glycosylated therapeutic proteins that have sugar and sialic acid molecules covalently attached to the peptide backbone. Runkel et al. report that IFN-beta produced in mammalian cell culture as a glycosylated protein, identical to the human molecule, has greater stability to aggregation than the corresponding protein produced by bacterial fermentation in the nonglycosylated form (50). Generally, bonds attaching sugars to proteins are stable, and any modifications in carbohydrate structure are due to process changes during the manufacture of the product. Desialylation can however occur at acidic pH on storage. Modifications to carbohydrate structure, or changes in the glycosylation of a therapeutic protein, can result in significant changes in the pharmacokinetics of the molecule. Differing sialic acid content has been shown to be responsible for variability in bioactivity of highly purified human pituitary luteinizing hormone isoforms (51).

Changes in glycosylation can be detected by various gel methods including fluorophore-assisted carbohydrate electrophoresis (FACE) and MS. Changes in sialic acid content can be detected by measurement of free sialic acid. New methods to quantitate oligosaccharides are being developed. For example, high resolution of normal phase high-pH anion-exchange chromatography and normal phase HPLC methods combined with MS can provide effective technology for analyzing a wide repertoire of oligosaccharides structures (52).

3.1.6. Photodegradation of Proteins

Both ionizing and nonionizing radiation can cause protein inactivation. Tryptophan, tyrosine, and cysteine are particularly susceptible to nonionizing radiation, such as ultraviolet (UV) light. The absorption of photons leads to photoionization and the formation of photodegradation products through either direct interaction with amino acid or indirectly via various sensitizing agents, such as oxygen. Commonly observed photodegradation products in an aerated, neutral-pH, aqueous protein solution include S–S bond fission, conversion of tyrosine to DOPA, 3-(4-hydroxyphenyl)lactic acid, and dityrosine, as well as fragmentation by-products and the conversion of tryptophan residues to kynurenine and N-formyl-kynurenine (53). The position of a sensitive residue within the three-dimensional structure of the protein also influences its reactivity towards photolysis. It is also important to take into account potential damage to the protein during analysis using circular dichroism (CD), UV, or fluorescent measurements where incident radiation is being used.

3.2. Physical Instability

Physical instabilities, for the purpose of this chapter, have been categorized as those reactions that do not involve the formation or destruction of covalent bonds. Aggregation of proteins, discussed primarily as a physical instability, can also occur as the result of intermolecular disulfide bond formation or transamidation as discussed previously.

Conformational changes (i.e., secondary and tertiary structure changes) in proteins can occur as the result of exposure to structural perturbants or environmental stress. Often, stability studies of proteins involve changes in temperature and pH, both of which can cause structural modifications that can be detected by standard spectroscopic techniques such as CD, fluorescence and infrared (IR) spectroscopy.

3.2.1. Aggregation and Precipitation

The term protein aggregation covers an extremely wide range of reactions and outcomes, some of which have been extensively studied for specific proteins whereas others remain challenging to study and define. The monomer form of a protein is generally the desirable physiologically active species (with the exception of multimeric species), but there are few proteins that are not susceptible to some form of either soluble or insoluble, reversible or irreversible aggregation.

The most common mechanism of protein aggregation is believed to involve protein denaturation and noncovalent association via hydrophobic interfaces. Denaturation is usually induced at gas–liquid, liquid–liquid (54) (as during microencapsulation processes) or container–liquid interfaces, although heat, pH variations, solvents, salts, and excipients may also contribute.

Thermal stress studies are often used to evaluate the rank ability of a formulation to stabilize a protein. Chan et al. showed, via differential scanning calorimetry (DSC), that calcium chloride and sugars stabilized recombinant human deoxyribonuclease (rhDNAse) against thermal denaturation while divalent cations, urea, and guanidine hydrochloride destabilize the protein (55). A recent study on recombinant human Flt3 ligand aggregation elucidates the importance of thermal reversibility as it pertains to the minimization of the ligand aggregation and its potential role for determining solution conditions that can achieve the greatest long-term stability. Evidence that unfolding preceded the formation of the aggregate was provided by far-UV CD (56).

Other noncovalent mechanisms of aggregation include ionic complexation, salting out, charge neutralization close to the isoelectric pH, and limiting solubility of the molecule. These types of aggregates, unlike hydrophobic aggregates, should be reversible, for example by dilution.

Along with accelerated studies, a number of other pharmaceutical development–associated activities have been found to induce denaturation. Studies on the stress of spray drying and nebulization on proteins have been published (57–60), and investigations into the effect of freeze–thaw cycling (61,62) are also documented. Formulation and filling activities along with shipping of liquid formulations often result in significant exposure of proteins to gas–liquid interfaces. Surfactants such as Tween 80, Pluronic F-68, and Brij 35 have been shown to stabilize interface-induced aggregation of human growth hormone (hGH) (63) but do not stabilize hGH against thermal stress in DSC studies. In fact, higher concentrations of the surfactants were found to destabilize the molecule (64). Formulation with sugars is also known to stabilize a number of proteins against aggregation (65,66).

Soluble aggregates remain in solution often as dimer or low molecular weight multimers. In general, these types of aggregates can be detected by high performance size exclusion chromatography (HP-SEC) and have been found by this method in many proteins including hGH, insulin, interleukin-2 (IL-2), antitrypsin-α_1, IFN-γ, basic fibroblast growth factor, and IFN-β (67). Conventional HP-SEC methods can also be modified to provide more accurate separation. For example, guanidine hydrochloride is often added to the mobile phase in order to differentiate native and denatured protein. In general, low levels of soluble aggregates in pharmaceutical products can be tolerated as long as the product remains stable and the soluble aggregates do not progress to insoluble forms (68). There is concern that increasing levels of soluble aggregate may be responsible for changes in immunogenicity of therapeutic proteins.

Insoluble aggregates usually manifest as visible particulates, although on occasion the level and size of the particles is so small that visual identification can be difficult. Insoluble aggregates have been documented for insulin and IL-1β (69,70). Insoluble aggregates can be evaluated by a number of solid-state techniques such as Fourier transform infrared (FTIR) and Raman and electron spin resonance spectroscopy. Light scattering techniques such as simple UV absorption at 320 and 360 nm can be useful for simple comparative evaluations. Protein concentration determination using UV absorbance is often used with a background subtraction to take into account low levels of insoluble material.

3.2.2. Adsorption

Adsorption onto various surfaces has been exploited for benefit in many disciplines such as analytical and separation technology. There are many documented studies of adsorption in the analytical and biomaterials literature. The propensity for proteins to adsorb is generally a problem to be overcome in the formulation and container closure selection of pharmaceutical products. Care must also be taken to evaluate parenteral administration sets for compatibility to assure that the targeted dose is delivered. Generally, adsorption is more of an issue for low-dose therapeutics where significant portions of a dose may be lost to container/closure or administration set surfaces.

In general, addition of surfactant molecules to formulations is known to reduce adsorption losses. Calcitonin has been shown to adsorb to glass with adsorption isotherms of the Langmuir and Freundlich type, depending on pH. The addition of nonionic surfactants, such as Pluronic F68 and Tween 80, to calcitonin solutions demonstrated inhibition of adsorption and reduction of adsorption rate (71).

The adsorption of proteins to liquid–gas interfaces, as well as solid–liquid interfaces, has recently become an area of interest using new techniques such as x-ray or neutron reflection. Studies have included evaluation of the adsorbed amount, the total thickness of the adsorbed layer, and the effect of pH and excipients on these parameters (72–74). Atomic force microscopy in a force-spectroscopy mode has been used to investigate the kinetics of the adsorption process of fibrinogen mol-

Table 4 Useful Methods for Detecting Protein Degradation/Inactivation

Degradation/ inactivation	Method	Peptide/ Protein	Reference
Deamidation	Urea-IEF	aFGF	Volkin et al. 1995 (78)
	RP-HPLC	Hirudin	Bischoff et al 1992 (13)
	RP-HPLC	Pramlintide	Hekman et al. 1998 (79)
	RP-HPLC	Insulin	Ladisch and Kohlmann 1992 (80)
Methionine oxidation	RP-HPLC	G-CSF	Reubsaet et al. 1998 (22)
	RP-HPLC	IGF-1	Fransson 1997 (23)
	RP-HPLC	ACTH	Antonini and Spoto 1988 (29)
Aggregation	SDS-PAGE	IGF-1	Fransson 1997 (23)
	SDS-PAGE	human Flt3 ligand	Remmele et al. 1999 (56)
	SDS-PAGE	IFN-β	Geigert et al. 1988 (81)
	SDS-PAGE	HGH	Mockridge et al. 1998 (82)
	HP-SEC	bFGF	Shahrokh et al. 1994 (67)
	HP-SEC	rbST	Chang et al. 1996 (83)
	HP-SEC	sCT)	Windisch et al. 1997 (84)
	UV Absorption	bFGF	Shahrokh et al. 1994 (67)
	UV Absorption	Insulin	Sluzky et al. 1991 (85)
	RP-HPLC	bFGF	Shahrokh et al. 1994 (67)
	AUC	vWF-A	Williams et al. 1999 (86)
	AUC	IL-2	Advant et al. 1995 (87)
	AUC	MIP-1α	Varley et al. 1997 (88)
	ELISA	IFN-α	Braun and Alsenz 1997 (89)
	Light scattering	Insulin	Bohidar 1998 (47)
	Light scattering (dynamic)	Insulin	Dathe et al. 1990 (90)
	Light scattering (quasi-elastic)	Insulin	Sluzky et al. 1991 (85)
	Viscosity	Insulin	Bohidar 1998 (47)
	PFG NMR	Insulin	Lin and Larive 1995 (91)
Denaturation	CD	α-crystallin	Doss-Pepe et al. 1988 (92)
	CD	Insulin	Vecchio et al. 1996 (93)
	CD	IFN-γ	Beldarrain et al. 1999 (94)
	far UV CD	human Flt3 ligand	Remmele et al. 1999 (56)
	DSC	tPA	Radek and Castellino 1988 (95)
	DSC	IFN-γ	Beldarrain et al. 1999 (94)
	DSC	IL-1 receptor	Remmele et al. 1998 (96)
	FTIR	Insulin	Vecchio et al. 1996 (93)
	FTIR	IL-2	Prestrelski et al. 1995 (97)
	CZE	Bovine β-lactoglobulin	Rochu et al. 1999 (98)
	CZE	Insulin, hGH	Nielsen et al. 1989 (99)
Adsorption	X-ray reflection	Urease	Gidalevitz et al. 1999 (72)
	Neutron reflection	Bovine serum albumin (bSA)	Lu et al. 1999 (73)
	Neutron reflection	Beta-casein	Puff et al. 1998 (74)
	Force-spectroscopy mode force spectroscopy	Fibrinogen	Hemmerle et al. 1999 (75)

Table 4 *(continued)*

Peptide/Protein Abbreviations	*Analytic Method Abbreviations*
ACTH – adrenocorticotropin hormone	AUC – analytical ultracentrifugation
aFGF – acid fibroblast growth factor	CD – circular dichroism
bFGF – basic fibroblast growth factor	CZE – capillary zone electrophoresis
G-CSF – granulocyte colony-stimulating factor	DSC – differential scanning calorimetry
IFN-interferon	ELISA – enzyme-linked immunosorbent assay
IGF-1 – insulin-like growth factor 1	FTIR – fourier transform infrared
IL – interleukin	HP-SEC – high performance size exclusion chromatography
MIP-1α – macrophage inflammatory protein-1α	IEF – isoelectric focusing
rbST – recombinant bovine somatotropin	PFG NMR – pulsed-field gradient nuclear magnetic resonance spectroscopy
sCT – salmon calcitonin	RP-HPLC – reversed-phase high performance liquid chromatography
vWF-A – type A domain of von Willebrand Factor	SDS-PAGE – sodium dodecyl sulfate polyacrylamide gel electrophoresis
	UV – ultra-violet

ecules onto silica (75), and various efforts at modeling adsorption have also been investigated (76,77).

4. ANALYTICAL METHODS

Although analytical methods used to detect and elucidate mechanisms of polypeptide degradation have been touched upon in relevant sections of this chapter, Table 4 provides a fairly comprehensive list of these methods. Typically, the evaluation of the stability of polypeptides will require the use of several methods (e.g., one method sensitive to chemical changes and one method sensitive to conformational modifications).

5. CONCLUSIONS

In conclusion, polypeptides and proteins are unique in the higher levels of structure that they exhibit. Loss of higher level structure typically results in loss of biological activity with or without loss to the molecule's chemical composition.

REFERENCES

1. T Geiger, S Clarke. Deamidation, isomerization and racemization at asparaginyl and aspartyl residues in peptides: succinimide-linked reactions that contribute to protein degradation. J Biol Chem 262:785–794, 1987.
2. J Brange, S Havelund, P Hougaard. Chemical stability of insulin 2: formation of higher molecular weight transforming products during storage of pharmaceutical insulin preparations. Pharm Res 9(6):727, 1992.
3. SA Berson, RS Yalow. Deamidation of insulin during storage in frozen state. Diabetes 15(12):875–879, 1966.
4. RG Strickley, BD Anderson. Solid-state stability of human insulin. I. Mechanism and the effect of water on the kinetics of degradation in lyophiles from pH 2–5 solutions. Pharm Res 13(8):1142–1153, 1996.
5. NP Bhatt, K Patel, RT Borchardt. Chemical pathways of peptide degradation I. Pharm Res 7(6):593, 1990.

6. GW Becker, RR Bowsher, WC Mackeller, ML Poor, PM Tackitt, RM Riggin. Chemical, physical and biological characterization of a dimeric form of biosynthetic Human Growth Hormone. Biotech Appl Biochem 9:478, 1987.
7. BA Johnson, JM Shirokawa, WS Hancock, MW Spellman, LJ Basa, DW Aswad. Formation of isoaspartate at two distinct sites during in vitro aging of human growth hormone. J Biol Chem 264:14262–14271, 1989.
8. PT Wingfield, RJ Mattaliano, HF MacDonald S Craig, GM Glore, AM Groneneborn, U Schemeissner. Recombinant-derived interleukin-1α stabilized against specific deamidation. Protein Eng 1:413, 1987.
9. AR Friedman, AK Ichhpurani, DM Brown, RM Hillman, LF Krabill, RA Martin, HA Zurcher-Neely, DM Guido. Degradation of growth hormone releasing factor analogs in neutral aqueous solution is related to deamidation of asparagine residues. Int J Pept Protein Res 37(1):14–20, 1991.
10. G Teshima, J Porter, K Kim, V Ling, A Guzzetta. Deamidation of soluble CD4 at Asparagine-52 results in reduced binding capacity for the HIV-1 envelope glycoprotein gp120. Biochemistry 30:3916, 1991.
11. MV Paranandi, AW Guzzetta, WS Hancock, DW Aswad. Deamidation and isoaspartate formation during in vitro aging of recombinant tissue plasminogen activator. J Biol Chem 269(1):243–253, 1994.
12. A Tuong, M Maftouh, C Ponthus, O Whitechurch, C Roitsch, C Picard. Characterization of the deamidated forms of recombinant hirudin. Biochemistry 31(35):8291–8299, 1992.
13. R Bischoff, P Lepage, M Jaquinod, G Cauet, M Acker-Klein, D Clesse, M Laporte, A Bayol, A Van Dorsselaer, C Roitsch. Sequence-specific deamidation: isolation and biochemical characterization of succinimide intermediates of recombinant hirudin. Biochemistry 32(2):725–734, 1993.
14. G Teshima, WS Hancock, E Canova-Davis. In: DW Aswad, ed. Deamidation and Isoaspartate Formation in Peptides and Proteins. Boca Raton, FL: CRC Press, 1994.
15. DT Liu. Deamidation: a source of microheterogeneity in pharmaceutical proteins. Trends Biotechnol 10(10):364–369, 1992.
16. C Oliyai, RT Borchardt. Chemical pathways of peptide degradation. VI. Effect of the primary sequence on the pathways of degradation of aspartyl residues in model hexapeptides. Pharm Res 11(5):751, 1994.
17. TV Brennan, S Clarke, In: DW Aswad, ed. Deamidation and Isoaspartate Formation in Peptides and Proteins. Boca Raton, FL: CRC Press, 1994.
18. CT Przysiecki, JG Joyce, PM Keller, HZ Markus, CE Carty, A Hagopian, MK Sardana, CT Dunwiddie, RW Ellis, WJ Miller, ED Lehman. Characterization of recombinant antistasin secreted by Saccharomyces cerevisiae. Protein Expr Purif 3(3):185–195, 1992.
19. PJ Berti, I Ekiel, P Lindahl, M Abrahamson, AC Storer. Affinity purification and elimination of methionine oxidation in recombinant human cystatin C. Protein Expr Purif 11(1):111–118, 1997.
20. MJ Pikal, KM Dellerman, ML Roy, RM Riggin. The effects of formulation variables on the stability of freeze dried human growth hormone. Pharm Res 8(4):427–436, 1991.
21. XM Lam, JY Yang, JL Cleland. Antioxidants for prevention of methionine oxidation in recombinant monoclonal antibody HER2. J Pharm Sci 86(11):1250–1255, 1997.
22. JL Reubsaet, JH Beijnen, A Bulu, E Hop, SD Scholten, J Teeuwsen, WJ Underberg. Oxidation of recombinant methionyl human granulocyte colony stimulating factor. J Pharm Biomed Anal 17(2):283–289, 1998.
23. JR Fransson. Oxidation of human insulin-like growth factor I in formulation studies. 3. Factorial experiments of the effects of ferric ions, EDTA, and visible light on methionine oxidation and covalent aggregation in aqueous solution. J Pharm Sci 86(9):1046–1050, 1997.

24. HS Lu, PR Fausset, LO Narhi, T Horan, K Shinagawa, G Shimamoto, TC Boone. Chemical modification and site-directed mutagenesis of methionine residues in recombinant human granulocyte colony-stimulating factor: effect on stability and biological activity. Arch Biochem Biophys 362(1):1–11, 1999.
25. T Kornfelt, E Persson, L Palm. Oxidation of methonine residues in coagulation factor VIIa. Arch Biochem Biophys 363(1):43–54, 1999.
26. JL Liu, KV Lu, T Eris, V Katta, KR Westcott, LO Narhi, HS Lu. In vitro methionine oxidation of recombinant human leptin. Pharm Res 15(4):632–640, 1998.
27. SM Van Patten, E Hanson, R Bernasconi, K Zhang, P Manavalan, ES Cole, JM McPherson, T Edmunds. Oxidation of methionine residues in antithrombin. Effects on biological activity and heparin binding. J Biol Chem 274(15):10268–10276, 1999.
28. C George-Nascimento, A Gyenes, SM Halloran, J Merryweather, P Valenzuela, KS Steimer, FR Masiarz, A Randolph. Characterization of recombinant human epidermal growth factor produced in yeast. Biochemistry 27(2):797–802, 1988.
29. G Antonini, G Spoto. Reversed phase high performance liquid chromatography of adrenocorticotropin 1-39 and its fragments in the native and oxidized forms. Biochem Int 16(6):1009–1017, 1988.
30. EL Finley, J Dillon, RK Crouch, KL Schey. Identification of tryptophan oxidation products in bovine alpha-crystallin. Protein Sci 7(11):2391–2397, 1998.
31. WE Savige, A Fontana. Oxidation of tryptophan to oxindolylalanine by dimethyl sulfoxide-hydrochloric acid. Selective modification of tryptophan containing peptides. Int J Pept Protein Res 15(3):285–297, 1980.
32. E Canova-Davis, J Ramachandran. Chemical modification of the tryptophan residue in adrenocorticotropin. Biochemistry 5(4):921–927, 1976.
33. TE Creighton. Protein Structure: A Practical Approach. IRL Oxford, 1989, pp. 155–167.
34. JY Chang. Stability of hirudin, a thrombin-specific inhibitor. The structure of alkaline-inactivated hirudin. J Biol Chem 266(17):10839–10843, 1991.
35. DL Linemeyer, JG Menke, LJ Kelly, J Disalvo, D Soderman, MT Schaeffer, S Ortega, G Gimenex-Gallego, KA Thomas. Disulfide bonds are neither required, present, nor compatible with full activity of human recombinant acidic fibroblast growth factor. Growth Factors 3(4):287–298, 1990.
36. D Gergel, AI Cederbaum. Inhibition of the catalytic activity of alcohol dehydrogenase by nitric oxide is associated with S nitrosylation and the release of zinc. Biochemistry 35(50):16186–16194, 1996.
37. MD Jones, J Hunt, JL Liu, SD Patterson, T Kohno, HS Lu. Determination of tumor necrosis factor binding protein disulfide structure: deviation of the fourth domain structure from the TNFR/NGFR family cysteine-rich region signature. Biochemistry 36(48):14914–14923, 1997.
38. LL Betaseron. Dev Biol Stand 96:97–104, 1998.
39. MC Manning, P Kamlesh, RT Borchardt. Stability of protein pharmaceuticals. Pharm Res 6(11):903–918, 1989.
40. J Schultz. Meth Enzymol 11:255–263, 1967.
41. D Piszkiewicz, M Landon, El Smith. Anomalous cleavage of aspartyl-proline peptide bonds during amino acid sequence determinations. Biochem Biophys Res Comm 40(5):1173–1178, 1970.
42. C George-Nascimento, J Lowsenson, M Borissenko, M Calderon, A Medina-Selby, J Kuo, S Clarke, A Randolph. Replacement of a labile aspartyl residue increase the stability of human epidermal growth factor. Biochemistry 29(41):9584–9591, 1990.
43. S Clarke, RC Stephenson, JD Lowenson. Liability of asparagine and aspartic acid residues in proteins and peptides. In: TJ Ahern, MC Manning, eds. Stability of Protein Pharmaceuticals: Part A. Chemical and Physical Pathways of Protein Degradation. New York: Plenum Press. 1992, pp. 1–29.

44. Y-CJ-S2, MA Hanson. Parenteral formulations of proteins and peptides: stability and stabilizers. J Parent Sci Tech 42(2S):S3–S26, 1988.
45. C Goolcharran, RT Borchardt. Kinetics of diketopoperazine formation using model peptides. J Pharm Sci 87(3):283–288, 1998.
46. DL Higgins, MC Lamb, SL Young, DB Powers, S Anderson. The effect of the one-chain to two-chain conversion in tissue plasminogen activator: characterization of mutations at position 275. Throm Res 57(4):527–539, 1990.
47. HB Bohidar. Light scattering and viscosity study of heat aggregation of insulin. Biopolymers 45(1):1–8, 1998.
48. J Brange, L Langkjoer. Insulin structure and stability. Pharm biotechnol 5:315–350, 1993.
49. S Li, TH Nguyen, C Schoneich, RT Borchardt. Aggregation and precipitation of human relaxin induced by metal-catalyzed oxidation. Biochemistry 34(17):5762–5772, 1995.
50. L Runkel, W Meier, RB Pepinsky, M Karpusas, A Whitty, K Kimball, M Brickelmaier, C Muldowney, W Jones, SE Goelz. Structural and functional differences between glycosylated and nonglycosylated forms of human interferon-beta (IFN-beta). Pharm Res 15(4):641–649, 1998.
51. PG Burgon, PG Stanton, K Pettersson, DM Robertson. Effect of desialylation of highly purified isoforms of human luteinizing hormone on their bioactivity in vitro, radioreceptor activity and immunoactivity. Reprod Fertil Dev 9(5):501–508, 1997.
52. KR Anumula, ST Dhume. High resolution and high sensitivity methods for oligosaccharide mapping and characterization by normal phase performance liquid chromatography following derivatization with highly fluorescent anthranilic acid Glycobiology 8(7):685–694, 1998.
53. DB Volkin, H Mach, CR Middaugh. Degradative covalent reactions important to protein stability. Mol Biotechnol 8(2):105–122, 1997.
54. H Shah. Stabilization of proteins against methylene chloride/water interface-induced denaturation and aggregation. J Controlled Release 58(2):143–151, 1999.
55. HK Chan, KL Au-Yeung, I Gonda. Effects of additives on heat denaturation of rhDNASE in solutions. Pharm Res 13(5):756–761, 1996.
56. RL Remmele, SD Bhat, DH Phan, WR Gombotz. Minimization of recombinant human Flt3 ligand aggregation at the Tm plateau: a matter of thermal reversibility. Biochemistry 38(16):5241–5247, 1999.
57. JD Andya, YF Maa, HR Costantino, PA Nguyen, N Dasovich, TD Sweeny, CC Hsu, SJ Shire. The effect of formulation excipients on protein stability and aerosol performance of spray-dried powders of a recombinant humanized anti-IgE monoclonal anitbody. Pharm Res 16(3):350–358, 1999.
58. ST Tzannis, SJ Prestrelski. Activity-stability considerations of trypsinogen during spray-drying: effects of sucrose. J Pharm Sci 88(3):351–359, 1999.
59. AY Ip, T Arakawa, H Silvers, CM Ransone, RW Niven. Stability of recombinant consensus interferon to air-jet and ultrasonic nebulization. J Pharm Sci 84(10):1210–1214, 1995.
60. YF Maa, PA Nguyen, SW Hsu. Spray-drying of air-liquid interface sensitive recombinant human growth hormone. J Pharm Sci 87(2):152–159, 1998.
61. F Jameel, R Bogner, F Mauri, D Kalonia. Investigation of physiological changes to L-asparaginase during freeze-thaw cycling. J Pharm Pharmacol 49(5):472–477, 1997.
62. L Kreilgaard, LS Jones, TW Randolph, S Frokjaer, JM Flink, MC Manning, JF Carpenter. Effect of Tween 20 on freeze-thawing and agitation-induced aggregation of recombinant human factor XIII. J Pharm Sci 87(12):1597–1603, 1998.
63. M Katakam, AK Banga. Use of poloxamer polymer to stabilize recombinant human growth hormone against various processing stresses. Pharm Dev Technol 2(2):143–149, 1997.

64. M Katakam, LN Bell, AK Banga. Effects of surfactants on the physical stability of recombinant human growth hormone. J Pharm Sci 84(6):713–716, 1995.
65. M Katakam, AK Banga. Aggregation of insulin and its prevention by carbohydrate excipients. PDA J Pharm Sci Technol 49(4):160–165, 1995.
66. M Katakam, AK Banga. Aggregation of proteins and its prevention by carbohydrate excipients: albumins and gamma-globulin. J Pharm Pharmacol 47(2):103–107, 1995.
67. Z Shahrokh, PR Stratton, GA Eberlein, YJ Wang. Approaches to analysis of aggregates and demonstrating mass balance in pharmaceutical protein (Basic Fibroblast Growth Factor) formulations. J Pharm Sci 83(12):1645–1650, 1994.
68. BL Chen, T Arakawa, CF Morris, WC Kenney, CM Wells, CG Pitt. Aggregation pathway of recombinant human keratinocyte growth factor and its stabilization. Pharm Res 11(1):1581–1587, 1994.
69. LC Gu, EA Erdos, H-S Chiang, T Calderwodd, K Tsaie, GC Visor, J Duffy, W-C Hsu, L Foster. Stability of interleukin-1β (IL-1β) in aqueous solution: analytical methods, kinetic products and solution formulation implications. Pharm Res 8(4):485–490, 1991.
70. J Schrier, RA Kenley, R Williams, RJ Corcoran, Y Kim, RP Nothey, D Augusta, M Huberty. Degradation pathways for recombinant human macrophage colony-stimulating factor in aqueous solution. Pharm Res 10(7):993–944, 1993.
71. SL Law, CL Shih. Adsorption of calcitonin to glass. Drug Dev Ind Pharm 25(2):253–256, 1999.
72. D Gidalevitz, Z Huang, SA Rice. Urease and hexadecylamine-urease films at the air–water interface: an x-ray reflection and grazing incidence x-ray diffraction study. Biophys J 76(5):2792–2802, 1999.
73. JR Lu, TJ Su, RK Thomas. Structural conformation of bovine serum albumin layers at the air–water interface studied by neutron reflection. J Colloid Interface Sci 213(2):426–437, 1999.
74. N Puff, A Cagna, V Aguie-Beghin, R Douillard. Effect of ethanol on the structure and properties of beta-casein adsorption layers at the air/buffer interface. J Colloid Interface Sci 208(2):405–414, 1998.
75. J. Hemmerle, SM Altmann, M Maaloum, JK Horber, L Heinrich, JC Vogel, P Schaaf. Direct observation of the anchoring process during the adsorption of fibrinogen on a solid surface by force-spectroscopy mode force spectroscopy. Proc Natl Acad Sci USA 96(12):6705–6710, 1999.
76. C Ruggerio, M Mantelli, A Curtis, S Zhang, P Rolfe. Computer modeling of the adsorption of proteins on solid surfaces under the influence of double layer and van der Waals energy. Med Biol Eng Comput 37(1):119–124, 1999.
77. WK Lee, J McGuire, MK Bothwell. A mechanistic approach to modeling single protein adsorption at solid water interfaces. J Colloid Interface Sci 213(1):265–267, 1999.
78. DB Volkin, AM Verticelli, MW Bruner, KE Marfia, PK Tsai, MK Sardana, CR Middaugh. Deamidation of polyanion-stabilized acidic fibroblast growth factor. J Pharm Sci 84(1):7–11, 1995.
79. C Hekman, W DeMond, T Dixit, S Mauch, M Nuechterlein, A Stepanenko, JD Williams, M Ye. Isolation and identification of peptide degradation products of heat stressed pramlinitide injection drug product. Pharm Res 15(4):650–659, 1998.
80. MR Ladisch, KL Kohlmann. Recombinant human insulin. Biotechnol Prog 8(6):469–478, 1992.
81. J Geigert, BM Panschar, S Fong, HN Huston, DE Wong, DY Wong, C Taforo, M Pemberton. The long-term stability of recombinant (serine-17) human interferon-beta. J Interferon Res 8(4):539–547, 1988.
82. JW Mockridge, R Aston, DJ Morrell, AT Holder. Cross-linked growth hormone dimers have enhanced biological activity. Eur J Endocrinol 138(4):449–459, 1998.

83. JP Chang, TH Ferguson, PA Record, DA Dickson, DE Kiehl, AS Kennington. Improved potency assay for recombinant bovine somatotropin by high-performance size-exclusion chromatography. J Chromatogr A 736(1–2):97–104, 1996.
84. V Windisch, F DeLuccia, L Duhau, F Herman, JJ Mencel, SY Tang, M Vuilhorgne. Degradation pathways of salmon calcitonin in aqueous solution. J Pharm Sci 86(3):359–364, 1997.
85. V Sluzky, JA Tamada, AM Klibanov, R Langer. Kinetics of insulin aggregation in aqueous solutions upon agitation in the presence of hydrophobic surfaces. Proc Natl Acad Sci USA 88(21):9377–9381, 1991.
86. SC Williams, J Hinshelwood, SJ Perkins, RB Sim. Production and functional activity of a recombinant von Willebrand factor-A domain from human complement factor B. Biochem J 342(t3):625–632, 1999.
87. SJ Advant, EH Braswell, CV Kumar, DS Kalonia. The effect of pH and temperature on the self-association of recombinant human interleukin-2 as studied by equilibrium sedimentation. Pharm Res 12(5):637–641, 1995.
88. PG Varley, AJ Brown, HC Dawkes, NR Burns. A case study and use of sedimentation equilibrium analytical ultracentrifugation as a tool for biopharmaceutical development. Eur Biophys J 25(5–6):437–443, 1997.
89. A Brunn, J Alsenz. Development and use of enzyme-linked immunosorbent assays (ELISA) for the detection of protein aggregates in interferon-alpha (IFN-alpha) formulations. Pharm Res 14(10):1394–1400, 1997.
90. M Dathe, K Gast, D Zirwer, H Welfle, B Mehlis. Insulin aggregation in solution. Int J Pept Protein Res 36(4):344–349, 1990.
91. M Lin, CK Larive. Detection of insulin aggregates with pulsed-field gradient nuclear magnetic resonance spectroscopy. Anal Biochem 229(2):214–220, 1995.
92. EW Doss-Pepe, EL Carew, JF Koretz. Studies of the denaturation patterns of bovine alpha-crystallin using an ionic denaturant, guanidine hydrochloride and a non-ionic denaturant, urea. Exp Eye Res 67(6):657–679, 1998.
93. G Vecchio, A Bossi, P Pasta, G Carrea. Fourier-transform infrared conformational study of bovine insulin in surfactant solutions. Int J Pept Protein Res 48(2):113–117, 1996.
94. A Beldarrain, JL Lopez-Lacomba, G Furrazola, D Barberia, M Cortijo. Thermal denaturation of human gamma-interferon. A calorimetric and spectroscopic study. Biochemistry 38(24):7865–7873, 1999.
95. JT Radek, FJ Castellino. A differential scanning calorimetric investigation of the domains of recombinant tissue plasminogen activator. Arch Biochem Biophys 267(2):776–786, 1988.
96. RL Remmele Jr, NS Nightlinger, S Srinivasan, WR Gombotz. Interleukin-1 receptor (IL-1R) liquid formulation development using differential scanning calorimetry. Pharm Res 15(2):200–208, 1998.
97. SJ Prestrelski, KA Pikal, T Arakawa. Optimization of lyophilization conditions for recombinant human interleukin-2 by dried-state conformational analysis using Fourier-transform infrared spectroscopy. Pharm Res 12(9):1250–1259, 1995.
98. D Rochu, G Ducret, F Ribes, S Vanin, P Masson. Capillary zone electrophoresis with optimized temperature control for studying thermal denaturation of proteins at various pH. Electrophoresis 20(7):1586–1594, 1999.
99. RG Nielsen, GS Sittampalam, EC Rickard. Capillary zone electrophoresis of insulin and growth hormone. Anal Biochem 177(1):20–26, 1989.

17

Regulatory Aspects of Stability Testing in Europe

BRIAN R. MATTHEWS

Alcon Laboratories (U.K.) Limited, Hemel Hempstead, Hertfordshire, England

---

Reprinted from Drug Development and Industrial Pharmacy, 25(7): 831–856. Marcel Dekker, Inc., 1999

1. INTRODUCTION

The stability of pharmaceutical ingredients and the products containing them depends on (a) the chemical and physical properties of the materials concerned (including the excipients and container systems used for formulated products) and (b) environmental factors such as temperature, humidity, and light and their effect on the substances in the product. Stability data on the drug substance and the formulated product are required in connection with applications for marketing authrorizations for pharmaceutical products containing new active substances, for products containing known active substances (including reformulations), for addiitonal strenghts or dosage forms, for new containers, and for amendments or variations to existing marketing authorizations, including new sources for active ingredients, extensions to the shelf life, and so on. In this chapter, the general regulatory requirements for stability testing of pharmaceuticals for registration in the European Community (European Union) are discussed with respect to the sources of information on stability requirements, the design of stability studies for new drug substances and finished products (drug products) containing new and known (pharmacopoeial or other) active ingredients, the presentation of stability data, and conclusions.

In the context of pharmaceutical products, stability should be considered not in the dictionary definition sense of "fixed" or "not likely to change," but in the sense of "controlled, documented and acceptable change." The types of change that have to be taken into account include chemical, physical, pharmacotechnical, microbiological, toxicological, as well as changes to bio-availability and other clinically significant changes. An important question then becomes: What degree of change is acceptable?

This question needs to be addressed in the marketing authorization application. In arriving at an answer that is mutually acceptable to the pharmaceutical manufacturer and the regulatory authorities, due account will need to be taken of the body of available evidence and theoretical extrapolations based on known physical and chemical laws and statistical models. However, many of the final proposals are often based on subjective, rather than objective, assessment of data, and the basis for decisions should therefore be discussed in the application.

In addition to the need for these data, at the time of marketing authorization application it should be kept in mind that there are also requirements for stability data generation in connection with other parts of the regulatory process. This includes data generation during the conduct of safety studies under good laboratory practice conditions (e.g., the stability of the drug in the medium being used for administration in animals and the stability of the radiolabel used in tracer studies), for clinical trial approvals in many countries (in which both the investigational product and any active comparator may need to be investigated for stability), as well as for preclinical and clinical studies (for which the stability of the analyte in biological specimens may need to be investigated in additional to the stability of the product under test).

In addition, such studies may be needed to meet good manufacturing practice requirements (including the retention of reference samples and starting materials) after the product has been approved for marketing, to undertake stability studies on manufactured batches, and to meet the requirements of the rapid alert system with respect to products that have stability problems during distribution. Medicine inspectors may also request to see data being generated at the request of assessor, including ongoing stability investigations.

Some of the types of change in pharmaceutical products that may need to be considered (depending on the dosage form concerned) are

Physical changes: appearance, consistency, product uniformity, clarity of solution, absence of particulates, color, odor, or taste, hardness, friability, disintegration, dissolution, sedimentation and resuspendability, weight change, moisture content, particle shape and size, pH, package integrity

Chemical changes: degradation product formation, loss of potency (active ingredient), loss of excipients (antimicrobial preservatives, antioxidants)

Microbial changes: proliferation of microorganisms in nonsterile products, maintenance of sterility, preservative efficacy changes

In designing a stability program, due account should be taken of the nature of the ingredients and the finished product, the intended clinical uses, and any special factors.

The discussion in this chapter centers on consideration of the requirements that apply to pharmaceutical products containing chemical active ingredients intended for use by humans. However, stability data are required for all types of products subjected to a regulatory procedure under the terms of the EC Pharmaceutical

Directives, including homoeopathic products, herbal products, immunological products, biological products, products derived from biotechnology, and veterinary medicinal products. There are a number of specific sources of information on the data requirements for these classes of products, but these are not discussed here.

In addition, a summary of World Health Organization (WHO) stability study recommendations is included, as is a glossary of stability-related terms.

2. REASONS FOR STABILITY STUDIES

Stability data are produced to establish the storage conditions and retest interval of the active ingredient and the storage conditions and shelf life for the manufactured products. Part of the information can also be used to justify overages included in products for stability reasons. When sufficient data are not available to support fully a desired shelf life, it is open to the manufacturer to base an application on the available information, but to offer an ongoing commitment to generate additional data and to advise the regulatory authority should any value then fall outside the agreed specifications.

In its recently adopted stability guidelines, the Committee for Proprietary Medicinal Products (CPMP) indicates that the objective of stability testing is "to provide evidence on how the quality of an active substance or medicinal product varies with time under the influence of a variety of environmental factors such as temperature, humidity and light, and enables recommended storage conditions, re-test periods and shelf-lives to be established" (1).

3. PHARMACEUTICAL EXPERT REPORTS: DISCUSSION OF STABILITY DATA

The pharmaceutical Expert Report is an important tool in the regulatory submission in Europe and can be used to present a rationale and justification for the proposals made in the application, not least concerning any unusual or controversial aspects of stability data or shelf life proposals. In the case of a request for a shelf life based on limited experimental data, the Expert should put forward an opinion on the reasonableness of that request.

In addition, the pharmaceutical Expert Report should provide a critical assessment of all other relevant data, as well as a summary of the data in the main file, and could include a discussion of the points below.

For active ingredients,

- a summary of the results and an analysis of the data;
- the variability of the stability profiles of different batches of drug substance;
- the most appropriate storage conditions for the active ingredient;
- the proposed duration of storage before retesting;
- the significance of the degradation products that can form on storage with cross references to the relevant information in the pharmacotoxicological Expert Report; and
- the reasonableness of the specification proposed for the active ingredient in relation to the available batch analysis, the analytical methods used and their validation, and the available stability data.

For finished products,

- a discussion of the basis for the relevance of the data included in the marketing authorization application other than those relating to the intended marketing formulation and proposed marketing packaging;
- discussion of data relating to the potential effects of manufacturing procedures (such as heating, exposure to ionizing radiation, etc.) on the stability of the active ingredient or the finished product;
- a summary of the results of the stability trials and an analysis of the data, particularly the assay results for the active ingredient and the content of significant degradation products;
- commentary on any inconsistencies in the data;
- a discussion of the variations in stabiity found between batches of the finished product;
- an explanation of the calculations used to estimate the shelf life from the stability test data;
- a justification for any excesses in the quantities of active ingredients (and excipients if appropriate) in the manufactured products indicating whether they are there becasue of stability reasons or to cover manufacturing losses;
- discussion of the proposed finished product specification, particularly a justification for the release specification limits (with cross references to appropriate sections of the pharmacotoxicological data or Expert Report); and
- a justification for the proposed finished product storage conditions and shelf life in the light of the supporting data, particularly for the shelf life of a reconstituted or diluted product, and a commentary on the specific instructions for the preparation of the material that will be administered (particularly products administered by infusion).

There should also be an interplay between the stability section and other parts of the marketing authorization application. For example, the pharmaceutical Expert Report may need to discuss a number of items included in the development pharmaceutics section of the application in the context of the stability data, for example:

- compatibility of the active ingredients and the excipients used in the formulated product and compatibility of the excipients with each other;
- physical parameters such as the effect of pH within the specified range (including potential effects on active ingredients, antimicrobial efficacy, etc.) and any long-term effects observed in the stability studies;
- dissolution performance of conventional release products—the data in the development pharmaceutics and stability sections can be discussed by the pharmaceutical Expert when considering whether a dissolution test should form part of the routine finished product test specification;
- a discussion of the potential for interaction, particularly between semiliquid and liquid manufactured products (including aerosols) and their container/closure systems (sorption or leaching) and other products (such as administration sets) with which the products must be used in clinical practice, and proposals for or justification of any proposed statements for inclusion in the summary of product characteristics or patient information leaflet; and

- discussion of the potential impurities that could be present in the active ingredient and the product, including synthetic by-products; impurities from starting materials or isomerization; residual solvents, reagents, or catalysts and degradation products and the adequacy of the analytical methods used in the stability studies with respect to their specificity, and so on. The availability (or otherwise) of synthesized reference materials for potential degradation products and/or other impurities may need to be discussed.

When a product contains a novel pharmaceutical excipient, full data may be required as for a new active ingredient. In such cases, stability data (including tests relevant to the intended performance of the novel material) will also be required for the excipient, as well as for the product containing it. The pharmaceutical Expert should discuss the data and their adequacy.

4. POSTMARKETING CHANGES

Even after an application has been approved and the product has been placed on the market, consideration has to be given to the effects of changes to the manufacturing formula or manufacturing processes in the light of production experience. Bioequivalence of the product and stability changes may be affected. Appropriate studies will therefore need to be undertaken, and applications for variations will need to be submitted as necessary. The current data requirements for such applications are discussed below.

5. DRUG–DEVICE COMBINATIONS

For Europe, in the case of medical devices containing an integral substance that in other circumstances would be a medicinal product, stability data will be required as part of a file that will have to be submitted to a "notified body" and that will be referred in part to a pharmaceutical competent authority for consideration of matters relating to the "pharmaceutical" material. In the case of medical devices supplied with a pharmaceutical as a single nonreusable unit, the pharmaceutical regulatory regime will apply.

In both cases, the type and extent of data that the regulatory agency examining the dossier will expect are likely to be similar. However, for some types of medical device, specific advice on stability studies may be available in the form of European Committee for Standardization (CEN) or International Organization for Standardization (ISO) standards (for example, for contact lens care products).

6. SOURCES OF INFORMATION

The primary sources of regulatory requirements in the European Community are the pharmaceutical Regulations and Directives. In addition, there is guidance on how to meet those requirements in a series of advisory guidelines. The relevant documents can now be accessed through the Internet.

There are two primary sites of relevance. The first is EudraLex at http://dg3.eudra.org/eudralex, which has all of the relevant documents for human medicines that form the Rules Governing Medicinal Products in the European

Union in portable document format (PDF), but some only in read-only format. Second, the CPMP and International Conference on Harmonization (ICH) draft and adopted guidelines on the European Agency for the Evaluation of Medicinal Products (EMEA)/CPMP home page at www.eudra.org, which has PDF copies of the documents (which are printable). Hard copies of the EudraLex documents can be purchased from agents of the EU's Office for Official Publications. The CPMP adopted and draft guidelines may be purchased from a number of commercial sources, including the Medicines Control Agency (MCA) EuroDirect service and professional associations such as the Regulatory Affairs Professionals Society (RAPS) and Drug Information Association (DIA), as well as obtained from trade associations.

As indicated above, in the case of data requirements in the stability area, there are a number of documents that need to be studied, with specific information in the annex to Directive 91/507/EEC (which is actually a replacement for the annex of Directive 75/318/EEC), as well as various guidelines on stability testing, including the primary guideline for requirements applying to new drug substances and products containing them that was adopted through the ICH process, several secondary ICH guidelines, CPMP guidance on stability issues, some relevant references included in the body of other guidelines (CPMP or ICH), the requirements built into the Variations Regulation, and so on. A list of relevant documents is included at the end of this chapter. Each of the main sources of information is discussed below.

6.1. Directives

Articles 4 and 4a of Directive 65/65/EEC require the submission of results of physicochemical, biological, or microbiological tests, pharmacological and toxicological tests, and clinical trials. Data on stability and storage need to be included in the application for a marketing authorization to support the information included in the section of the Summary of Product Characteristics on shelf life and precautions for storage (Article 4a of Directive 65/65/EEC). When reliance is placed on a European Drug Master File (DMF) rather than inclusion of data in the application itself, it should be borne in mind that stability data on the active ingredient will need to be included in the DMF.

Directive 91/507/EEC introduced a formal requirement for all analytical methods used in connection with a marketing authorization application to have been validated and for the validation data to be included with the dossier. This includes the analytical methods used in the development pharmaceutics and stability sections of the file. The number of replicates of individual determinations undertaken for validated methods should be related to the results of the validation studies. Revalidation of methods may be required when, for example, significant changes to the method of manufacture of the active ingredient or changes to the composition of the finished product have occurred.

6.2. Guidelines

Several guidelines should be taken into account by those undertaking stability tests for submission in a European marketing authorization application. The European Community's CPMP adopted a guideline on stability testing in July 1988, and this came into effect in 1989. For new active ingredients and products containing them, this in effect has been replaced by the guideline adopted through the ICH process, and it is expected that all applications submitted since 1998 will include studies that have been conducted according to the ICH-derived guideline. A revision of the old CPMP stability guideline (the new one relating to requirements for stability data required for existing active ingredients and products containing them) was adopted in April 1998, with an operational date of October 1998. At the same time, a guideline was adopted on the data requirements for a type II variation to a marketing authorization.

In addition to the main ICH stability guideline, recent documents include a note for guidance on photostability testing and another for new dosage forms (of new active substances). More details of the guideline content is given below.

7. DATA REQUIREMENTS

As indicated above, all analytical methods should be adequately validated. For stability studies, it is important that the analytical methods be capable of detecting and quantitating degradation products and interaction products (active ingredient/excipient and/or active ingredient/container components) in the finished product and should also be capable of separating these from substances unique to the active ingredient (e.g., synthetic impurities that are not degradation products). The basic requirements for information to be included in a marketing authorization application on stability of the active ingredients and the manufactured product are summarized below.

For active ingredients,

- information on the batches tested (including date and place of manufacture, batch size, and use of batches);
- details of the test methodology, including normal, stress, and accelerated conditions, as appropriate;
- analytical test procedures and their validation data, with particular reference to the assay and the determination of degradation products;
- analytical results, including initial results;
- conclusions;
- a proposed retest interval and storage conditions for the ingredient.

For dosage forms,

- proposed specification to apply at the end of the shelf life (the release specification being included elsewhere in the dossier), including the acceptable limits for impurity or degradation product content following storage under the recommended conditions;
- batches tested and the packaging used, including details of the composition (if not those of the proposed marketing formula), packaging, batch manufacturing date and site, batch size, and use of batch;
- study design and methods for real-time and accelerated conditions;
- characteristics studied, including physical, microbiological, chemical, chromatographic, and the like characteristics of the product, as well as characteristics of the product such as potency and purity, and including potential interactions between the container and closure system and the pro-

duct, as well as characteristics such as preservative efficacy and sterility in appropriate cases;

- evaluation of the test procedures, including description and validation, and consideration of potential interference in the test results;
- test results, including initial results, and information on degradation products;
- discussion of the available information regarding compatibility and stability with administration devices such as infusion sets and syringes, for example;
- conclusions;
- proposed shelf life and storage conditions, including the shelf life after reconstitution or after first opening of the product, and simulated in-use stability data.

Information should also be provided in the application regarding any ongoing stability studies.

Physical characteristics such as crystal form, moisture content, and particle size may need to be considered for the formulated product. These may have an effect on bioavailability, content uniformity, suspension properties, and possibly patient comfort (e.g., for ophthalmic products). Critical characteristics may need to be controlled in the finished product specification.

When a chiral substance is used in a pharmaceutical product, both the raw material and the finished product containing it should be investigated for the absence of unacceptable changes in the stereochemical purity or ratio during the proposed retest interval and shelf life.

For some dosage forms, particular aspects of stability testing may need to be given particular attention. Examples include the formation of particulates on storage in products intended for intravenous infusion; the stability of intravenous additives in the diluents proposed for use; the stability characteristics of aerosol inhalation products (especially when these have been reformulated to replace chlorofluorocarbon propellants). Products in plastic containers (especially for semisolid/liquid products, including aerosol inhalation, parenteral, and ophthalmic products) will require special consideration, particularly with respect to protection of the product from external factors, extraction of materials from the polymer, or release of materials from polymers (including antioxidants, mono- and oligomers, plasticizers, catalysts, processing aids, mineral compounds such as calcium, barium, or tin, etc.); and relating to the rate of moisture vapor loss and the tightness of closures, for example.

7.1. Impurities in New Active Substances

The ICH guideline on impurities in new active substances requires characterization of recurring impurities and degradation products found at or above an apparent level of 0.1% (e.g., calculated with reference to the response factor of the drug). When it is not possible to identify an impurity, a summary of the studies undertaken to try to make such identification should be included in the dossier. Data may also be reported regarding studies undertaken on impurities present below 0.1%, although this is not normally required unless the potential impurities are expected to be unusually potent or toxic or to have pharmacological activity. However, analytical values between 0.05% and 0.09% should not be rounded up when reporting study results.

7.2. Impurities in New Medicinal Products

A second ICH guideline addresses impurities in new medicinal products. The degradation products of the active ingredient or reaction products between the active ingredient, and excipients or container components, but not impurities arising from excipients, or extraneous contaminants (which should not be present and which should be the subject of good manufacturing practice controls), observed during stability trials under the recommended storage conditions should be reported, identified, and/or qualified when suggested thresholds are exceeded.

The suggested thresholds are as follows:

Maximum daily dose	Threshold for reporting
≤1 g	0.1%
>1 g	0.05%

Regarding the threshold for reporting, the degradation test procedure can be an important support tool for monitoring the manufacturing quality, as well as for use in setting a shelf life for the product, and the reporting level should be set below the level of the identification threshold.

Maximum daily dose	Threshold for identification
< 1 mg	1.0% or 5 μg total daily intake, whichever is lowest
1–10 mg	0.5% or 20 μg total daily intake, whichever is lowest
>10 mg–2 g	0.2% or 2 mg total daily intake, whichever is lowest
>2 g	0.1%

Maximum daily dose	Threshold for qualification
< 10 mg	1.0% or 50 μg total daily intake, whichever is lowest
10–100 mg	0.5% or 200 μg total daily intake, whichever is lowest
>100 mg–2 g	0.2% or 2 mg total daily intake, whichever is lowest
>2 g	0.1%

Higher thresholds may be proposed (with justification) if the target reporting threshold cannot be achieved. When it is not possible to identify particular degradation products, a summary of the studies undertaken should be included in the application. However, these requirements will affect the design of stability studies, particularly the choice of analytical method.

Identification should also be considered for particularly potent, toxic, or pharmacologically significant degradation products below the stated levels. The specification for the finished product should include limits for degradation products known to occur under the stated storage conditions and should take into account the qualification of those substances. The basis for the proposals should be data on the product intended to be marketed. A rationale should be given for the inclusion and exclusion of impurities from such specifications.

7.3. Stability Testing of New Drug Substances and Products

The guideline called Stability Testing of New Active Substances and Medicinal Products, was adopted by the CPMP in December 1993 after the usual ICH development process and may be considered as the "parent" ICH guideline. It is expected to be used as the basis for stability trials reported for all applications for new drugs and products containing them from January 1, 1998, particularly for the provision of data on the required number of batches with the use of long-term testing conditions of 25°C and 60% relative humidity (RH) for 12 months and accelerated testing at 40°C and 75% RH for 6 months with testing under controlled humidity conditions, but it also was accepted for earlier submissions. (For abridged applications, the CPMP guideline on stability testing of existing active substances and related finished products was adopted in April 1998 and had an operational date of October 1998. It was based closely on the parent ICH guideline.)

Sampling and specific test requirements for particular dosage forms or packaging etc. are not covered by the ICH guidelines.

The ICH guideline was developed taking into account the concept of climatic zone and conditions (Table 1). The ICH regions are encompassed by climatic zones I and II.

7.3.1. General

Stability studies (conditions and duration) should be of sufficient duration to cover storage, shipment, and subsequent use (including reconstitution or dilution of the finished product if appropriate). The use of the same storage conditions for the drug substance and the finished product facilitates comparisons of the data generated, although other justified storage conditions can be included.

The recommended storage conditions are as follows:

Test	Conditions	Minimum period
Long term	25°C ± 2°C 60% ± 5% RH or t°C ± 2°C and appropriate RH	12 months; extended to cover retest interval or shelf life
Intermediate (example)	30°C ± 2°C 60% ± 5% RH	6 months of data from 12-month study
Accelerated	40°C ± 2°C 75% ± 5% RH or (t + 15)°C ± 2°C and appropriate RH	6 months

Table 1 International Climatic Zones and Conditions (CPMP Stability Guideline)

Climatic condition	Zone I: Temperate (e.g., Northern Europe, United Kingdom, Canada, Russia)	Zone II: Mediterranean (subtropical) (e.g., United States, Southern Europe, Japan)	Zone III: Hot/dry or hot/moderate RH (e.g., Iran, Iraq, Sudan)	Zone IV: Very hot/humid (e.g., Brazil, Ghana, Indonesia, Nicaragua, Philippines)
Mean annual temperature	< 20.5°C	20.5°C–24°C	> 24°C	> 24°C
Kinetic mean temperature	21°C	26°C	31°C	31°C
Mean annual relative humidity (%)	45%	60%	40%	70%

It would normally be expected that data would be generated at 25°C and at 40°C, but alternative justified conditions will be accepted, for example, in the case of products with normal storage conditions that would be lower than 25°C (t°C), in which case the accelerated condition would be $(t+15)$°C. The intermediate condition would be used when "significant change" (i.e., failure to meet the proposed or agreed specification for an active substance) or (*a*) a 5% loss of potency from the initial batch analysis of the finished product, (*b*) the level of a specified degradation product exceeded its specification limit, (*c*) the product exceeded its pH limits, (*d*) the dissolution performance fell outside the specification, or (*e*) there was a failure to meet specifications for appearance or physical properties (color, phase separation, resuspendability, delivery per actuation, caking, hardness, etc.) occurred during accelerated testing or when a product was not suitable for testing at that temperature (e.g., an eye ointment designed to melt below 40°C). The data at 30°C and/or 40°C can be used to evaluate the effect of short-term excursions outside the labeled storage conditions (e.g., during shipping).

Other storage conditions may be used if justified. Special consideration should be given to products that change physically or chemically at lower storage conditions, such as suspensions and emulsions (which may cream or sediment) or creams, oils, or semisolid preparations that may increase in viscosity. Conditions of high relative humidity are particularly applicable to solid dosage forms or other products in packs designed to offer a permanent barrier to moisture vapor loss; aqueous solutions, suspensions, and the like packaged in semipermeable containers may be tested under low relative humidity conditions (e.g., 10–20% RH). It is understood that, at least for the United States and Europe, the long-term test conditions for semipermeable containers should be 25°C ± 2°C at 40% RH ± 5% RH, and accelerated conditions should be 40°C ± 2°C at 10–20% RH, although this is due to be reconsidered during the revision of the ICH guideline.

The testing intervals should be sufficient to establish the stability characteristics of the drug substance and the dosage form. The suggested sampling frequency is every 3 months for the first year and then every 6 months for the second year and annually thereafter.

The long-term studies should be undertaken in packaging that is intended for storage and distribution or material that simulates that packaging for drug substances. Containers and closure systems used in primary stability studies of finished products should be those intended for marketing.

Matrixing or bracketing studies may be applied if justified. This is discussed in a separate section of this chapter (page 597).

Stability tests should be designed to cover those features of the active ingredient or finished product that are susceptible to change during storage. The information collected should cover any factor that is likely to affect quality, safety, or efficacy, including chemical, physical, and microbiological characteristics.

The limits of acceptability in the proposed specifications should be derived from the stability profiles obtained. The specification will need to include individual and total upper limits for impurities and degradation products. Justification for the proposed levels will need to include reference to the levels of the impurities and degradation products that were present in the materials used in preclinical and clinical studies (i.e., whether or not they have been "validated").

7.3.2. Drug Substance

In addition to the accelerated and long-term studies, stress tests of the drug substance should also be used. These should be designed to determine the intrinsic degradation pathways and to identify likely degradation products and show the suitability of the analytical procedures.

Three batches of drug substance, prepared at not less than pilot manufacturing scale utilizing the intended synthetic route to be used for commercial material, should be used to evaluate stability. The data obtained are used to establish a retest interval. The batches tested should also be representative of the quality of the material used in preclinical and clinical studies, as well as the intended manufacturing-scale material. Laboratory-scale batches can be used for supporting data. The first three production batches should be placed in a long-term study if these data are not included in the submission; the study should use the same stability protocol reported in the dossier.

For quantitative characteristics expected to decrease with time, it is suggested that the time at which the 95% one-sided confidence limit for the mean degradation curve intersects the acceptable lower specification limit be used to establish this interval. Data may be combined if the batch-to-batch variability is small and if satisfactory results are first obtained from statistical tests applied to the regression lines and zero time intercepts for the individual batches (e.g., if the p value for the level of significance of rejection is more than 0.25). Depending on the degradation relationship, the data may be transformed (e.g., using linear, quadratic, or cubic functions) for linear regression analysis using arithmetic or logarithmic scales; statistical tests should be applied for the goodness of fit of the data to the assumed degradation plot. When it is not possible to combine data, the retest interval may be established according to the least stable batch.

When there is little degradation or little variability, formal statistical analysis of the data is not usually required if visual examination is sufficient.

Limited extrapolation of real-time data (especially when supported by accelerated data) may be allowed. This should be justified by reference to the mechanism for degradation, the goodness of fit of the statistical model, the batch sizes concerned, and available supporting data. The justification should take into account assay results and levels of degradation products and any other relevant characteristics.

The storage conditions for inclusion in the labeling should be in accordance with local regulatory requirements based on a consideration of the stability data. Specific warnings should be included (e.g., "Do not freeze"). Terms such as "ambient conditions" and "room temperature" should be avoided.

7.3.3. Finished Product

The information on the stability of the drug substance and data on the clinical trial formulation should be taken into account in designing stability studies for the dosage form of the finished product. The chosen design should be justified.

The stability studies of the finished product should be conducted on at least three batches of the formulation proposed for marketing; the batches should be made by a process that meaningfully simulates the intended commercial process (and ideally uses separate lots of the active ingredient) in the intended commercial container system. Two of the batches should be at least of pilot-scale manufacture (i.e., representative of and simulating the full-scale production process, such as a minimum of scale or 100,000 tablets or capsules). A third batch may be smaller (e.g., 25,000–50,000 tablets or capsules). The same quality specification should apply as that proposed for the commercial product. If appropriate data are not submitted with the application, the first three production batches should be placed on accelerated and long-term stability tests using the same protocol as that reported in the application.

Data on laboratory scale batches are not normally acceptable as primary data. They may be used to support primary data, as may information on products with related, but not identical, formulations and/or different packaging.

The presentation and evaluation of the data should be systematic and should cover necessary physical, pharmacotechnical (e.g., dissolution), chemical, biological, and microbiological attributes. The degree of variability of the data from the tested batches will affect the confidence that future production batches will meet the specification to the end of their shelf life.

For quantitative characteristics expected to decrease with time, it is suggested that the time at which the 95% one-sided confidence limit for the mean degradation curve intersects the acceptable lower specification limit be used to establish this interval. Data may be combined if the batch-to-batch variability is small and if satisfactory results are first obtained from statistical tests applied to the regression lines and zero time intercepts for the individual batches (e.g., if the p value for the level of significance of rejection is more than 0.25). Depending on the degradation relationship, the data may be transformed (e.g., using linear, quadratic, or cubic functions) for linear regression analysis using arithmetic or logarithmic scales; statistical tests should be applied for the goodness of fit of the data to the assumed degradation plot. When it is not possible to combine data, the retest interval may be established according to the least stable batch.

When there is little degradation or little variability, formal statistical analysis of the data is not usually required if visual examination is sufficient.

Limited extrapolation of real-time data (especially when supported by accelerated data) may be allowed. This should be justified by reference to the mechanism for degradation, the goodness of fit of the statistical model, the batch sizes concerned, and available supporting data. The justification should take into account assay results and levels of degradation products and any other relevant characteristics.

The storage conditions for inclusion in the labeling should be in accordance with local regulatory requirements based on a consideration of the stability data. Specific warnings should be included (e.g., "Do not freeze"). Terms such as "ambient conditions" and "room temperature" should be avoided. However, see the discussion of specific European requirements for labeling and leaflets, later.

7.4. International Conference on Harmonization Guideline: Stability Testing: Requirements for New Dosage Forms

The ICH guideline called Stability Testing: Requirements for New Dosage Forms (an annex to the parent ICH stability guideline) was adopted by the CPMP in December 1996 and came into operation for studies commencing after January 1998.

It is addressed to the manufacturer of the originally approved product and applies to applications for new dosage forms (i.e., a different pharmaceutical product type, including products intended for different routes of administration, e.g., oral to parenteral; and different functional types, e.g., immediate-release tablets to modified-release tablets; and different dosage forms for the same route of administration, e.g., capsules to tablets or solutions to suspensions) containing the same active ingredient as the approved product. Such products should, in principle, follow the main guideline, but it might be possible in justified cases to provide a limited amount of data at the time of submission, such as 6 months of accelerated and 6 months of long-term data from ongoing studies.

7.5. International Conference on Harmonization Guideline: Photostability Testing of New Drug Substances and Products

The ICH document called Photostability Testing of New Active Substances and Medicinal Products was adopted by the CPMP in December 1996 and came into effect for studies commencing after January 1998. It is an annex to the parent ICH stability guideline.

The guideline is intended to be taken into account when designing light challenge tests and stress tests for marketing authorization applications for new active substances and products containing them, but it may also be used in connection with variation applications for formulation or packaging amendments. However, photostability testing is not required as a part of the stability test program for marketed products. The guideline does not address in-use photostability of products (but see below) or extend to products not covered by the parent stability guideline.

Photostability tests are normally undertaken on a single representative batch of the material under consideration, although this may need to be extended if certain changes are made to the product (e.g., to the formulation or the packaging) or depending on the data initially reported concerning photostability.

The approach to the test should be systematic and involve (*a*) tests on the drug substance; (*b*) tests on the formulated product outside its immediate pack; (*c*) tests on the drug product in the immediate pack if there are signs of photolability in this condition, and (*d*) tests on the finished product in the marketing pack if there are signs of instability in this configuration. The applicant should define and justify the limits of acceptable change, and this will determine the extent of photostability testing required. Any resulting labeling requirements are those required by the relevant regulatory authority.

The light sources to be used for photostability studies and described in the guideline are (*a*) one producing an output similar to the D65 (outdoor daylight, ISO 10977: 1993)/ID65 (indoor indirect daylight) emission standard (such as an artificial daylight fluorescent lamp combining ultraviolet [UV] and visible outputs, xenon, or metal halide lamps) with a window glass filter for a device having a significant output below 320 nm or (*b*) exposure of the same sample to both a cool white fluorescent lamp designed to produce an output similar to that specified in ISO 10977: 1993 and a near-UV fluorescent lamp having a spectral distribution from 320 nm to 400 nm with maximum energy emission between 350 nm and 370 nm and a significant proportion of output in bands at 320–360 nm and at 360–400 nm.

Containers for test samples should be chemically inert and transparent to light. Companies should either maintain an appropriate temperature control to minimize the potential for local effects due to temperature or include a dark control in the same environment side by side with the test samples (unless otherwise justified). Precautions should be taken to minimize changes in physical state of the samples (e.g., sublimation, evaporation, or melting) and to ensure minimal interference with the exposure of the test samples. Possible interaction between samples and containers or protective materials should be considered and eliminated as much as possible.

The study design for confirmatory testing (see below) should include an overall illumination of not less than 1.2 million lux hours and an integrated near-UV energy exposure of not less than 200 watt hours m^{-2} for drug and formulated product. Samples may be exposed side by side with a validated actinometric system (such as quinine for the near-UV region; the guideline includes two options for calibrated quinine actinometers) or for an appropriate period of time using calibrated radiometers/lux meters.

The results should be evaluated to determine whether acceptable levels of change are induced by light exposure, taking into account other stability data obtained for the drug and the product concerned. Special labeling and packaging may then be identified to ensure that the product will remain within the proposed specifications for the shelf life of the product.

7.5.1. Drug Substance Testing

The test program for drug substance testing should be in two phases: forced degradation and confirmatory testing. In the forced degradation study, the drug may be tested as is or in simple solution or suspension. A variety of (justified) exposure conditions may be applied depending on the photosensitivity of the drug and the intensity of the light sources, but exposure may need to be limited if extensive decomposition occurs and may be terminated if the material is found to be photostable.

These studies are used to establish the overall light sensitivity of the material and may also be used for elucidation of the degradation pathways and validation studies. Under the conditions used for this phase of the study, degradation products may be formed that are unlikely to be formed under the confirmatory test conditions. However, the information may be useful in the development and validation of analytical procedures. If it is demonstrated that these degradation products are not formed in the confirmatory studies, they do not need to be studied further. The initial test is normally conducted on a single batch of material; if the results are equivocal, up to an additional two batches may need to be examined.

Confirmatory tests should be carried out under the conditions listed above to provide necessary information concerning the precautions needed for handling the drug and for its packaging and labeling. The confirmatory test is normally carried out on a single batch of material unless the results are equivocal, which requires an additional two batches to be examined.

Testing samples should be placed in chemically inert and transparent containers. Direct challenge studies of solid drugs may involve a suitable quantity of the substance being placed as a layer not more than 3 mm deep in a glass or suitable plastic dish, protected with a suitable transparent cover if necessary.

Samples should be examined for changes in physical properties (appearance, color of solution, clarity of solution, etc.), assayed, and examined for degradation products. The analytical methods should have been validated for potential photodegradation products and should be capable of resolving and detecting such degradation products, especially those appearing in the confirmatory tests. Sampling of solids should ensure that representative portions are tested. Other materials should also be homogeneous. Protected samples used as dark controls should be analyzed at the same time.

The results from the forced degradation stress tests cannot be expected to establish qualitative or quantitative limits for change but should be used to develop and validate test methods for use in the confirmatory studies. The confirmatory studies should identify the extent of change that occurs (and the manufacturer then has to decide whether this is acceptable) and will help to identify precautions necessary for the manufacture, formulation, and packaging of the finished product. Results from other formal stability studies should also be taken into account to confirm that the substance will remain within justified limits at the time of use.

7.5.2. Finished Product

Studies of formulated product should follow the sequence indicated above: fully exposed product, product in its primary container (unless the container is impermeable to light, e.g., a metal tube or can), and, if necessary, product in its full marketing packaging. In-use photostability testing may be considered for products such as infusion fluids and dermal products, depending on the instructions for use. Normally, only one batch of product is tested unless the results obtained are equivocal, in which case an additional two batches should be examined.

Samples should be placed in suitable containers, taking into account the physical characteristics of the product and the need to ensure minimal changes of the physical state while ensuring minimal interference with the irradiation of the test samples. The use of cooling or the placement of samples in sealed containers may be justified. Interactions between containers and the like and product should be considered, and any such materials should be eliminated when they are not essential for the conduct of the test.

Product tested outside its primary pack should be placed in chemically inert and transparent containers. Direct challenge studies of solid products may involve a suitable quantity of the material being placed as a layer not more than 3 mm deep in a glass or suitable plastic dish and protection with a suitable transparent cover if necessary. Tablets, capsules, or other forms should be placed in a single layer. If direct exposure is not practical (e.g., due to oxidation), the sample should be placed in a suitable inert container (e.g., made of quartz). Product tested in the primary container should be placed to provide the most uniform exposure (e.g., horizontal or transverse placement), although special adjustments may be needed in the case of large-volume containers.

Samples should be tested after exposure using representative samples (e.g., of 20 capsules or tablets) or other samples (or solutions), ensuring homogeneous distribution. Physical properties (appearance, clarity and color of solution, disintegration or dissolution of capsules, etc.) and chemical properties (assays for the active ingredient and degradation products) should be undertaken. Exposed and dark control samples should be tested at the same time.

7.6. Committee for Proprietary Medicinal Products Guideline for Reduced Stability Testing: Bracketing and Matrixing

The CPMP guideline called Reduced Stability Testing Plan—Bracketing and Matrixing, Annex to Note for Guidance on Stability Testing: Stability Testing of New Drug Substances and Products, is an annex to the parent ICH stability guideline and was issued for comment in September 1996 and adopted by the CPMP in October 1997. The concept of reduced testing plans is acceptable only for closely related substances and products such as the same dosage form (comparable specifications), same (or very closely similar) formulation, and the same product using different batches of the same active ingredient. Batch-to-batch variability should be small.

A manufacturer who has used matrix or bracketing studies in developing a product should justify the chosen protocol and show that the proposed shelf life for each product strength and pack is supported using stability data on both the active ingredient and the formulated products.

It is normally expected that three batches of each strength, of each container type, and of each pack size are stability tested at 25°C and 60% RH for 12 months and at 40°C and 75% RH for 6 months (or alternative conditions) as per the main ICH guideline. However, by the use of appropriate bracketing or matrixing studies, the number of samples analyzed at each test point may be substantially reduced yet produce adequate information to justify the proposed shelf life of products concerned. The manufacture and start of analysis may be staggered.

Bracketing or matrixing would normally be acceptable in the following cases:

- strength changes with no or small changes in the proportion of ingredients or achieved by varying the amount of active ingredient and one or two major excipients
- container size changes with the same contact materials
- change to an equivalent closure system
- change to a manufacturing site within the same company
- change of batch size

Matrixing would normally be acceptable in the following cases:

- strength changes associated with significant changes to the proportions of ingredients or changes in one or two minor components
- changes to fill volumes in containers
- introduction of a closure system of nonequivalent performance
- changes to the manufacturing process
- the introduction of manufacturing by a different company

7.6.1. Bracketing

The example in Table 2 is given in the guideline for a product line having three strengths of product with the same relative composition of excipients, two types of packaging, and three pack sizes for one of these types of packaging.

Table 2 Example of Bracketing Design

	Dosage strength/active substance lot (A, B, C)								
	50 mg			75 mg			100 mg		
Pack type	A	B	C	A	B	C	A	B	C
Blister	X	X	X	(X)	(X)	(X)	X	X	X
HDPE/15	X	X	X	(X)	(X)	(X)	X	X	X
HDPE/100	(X)	(X)	(X)	(X)	(X)	(X)	(X)	(X)	(X)
HDPE/500	X	X	X	(X)	(X)	(X)	X	X	X

(X) = sample not tested.

7.6.2. Matrixing

In the matrixing type of protocol, all product strengths, packaging materials, pack sizes, and the like are included and tested to some extent. None of the product variants are excluded from the test program. Examples of the sampling schedules for the first and second time pulls is included in the guideline and given in Tables 3 and 4.

Other statistically based designs may also be used. In all cases, the manufacturer will need to ensure that the distribution of all variables is even during the period of the protocol. Consultation with regulatory agencies may be advisable before employing complex matrix designs.

Initial and final time points should include all samples. The statistical basis of a low testing matrix such as a 1/3 design is better when the protocol covers several strengths and/or pack sizes rather than a product with a single strength and pack size, for example.

The degree of reduction in the testing may be affected by factors such as the stability of the active ingredient and the physical stability of the dosage form. With a very stable drug in a conventional tablet formulation, a 1/3 matrix could be acceptable, but if evidence of significant changes is found during the real-time arm of the stability trial, the testing frequency will need to be increased to (for example) a 2/3 matrix for the remaining part of the study.

Table 3 Matrix Design (First Time Point)

	Dosage strength/active substance lot (A, B, C)								
	50 mg			75 mg			100 mg		
Pack type	A	B	C	A	B	C	A	B	C
Blister	1	1	(1)	(1)	1	1	1	(1)	1
HDPE/1	(1)	1	1	1	(1)	1	1	1	(1)
HDPE/2	1	(1)	1	1	1	(1)	(1)	1	1

First time point: for the 1/3 testing matrix, 1 = sample not tested, (1) = sample tested; for the 2/3 testing matrix, 1 = sample tested, (1) = sample not tested.

Table 4 Matrix Design (Second Time Point)

	Dosage strength/active substance lot (A, B, C)								
	50 mg			75 mg			100 mg		
Pack type	A	B	C	A	B	C	A	B	C
Blister	1	1	(1)	(1)	1	1	1	(1)	1
	(2)	2	2	2	(2)	2	2	2	(2)
HDPE/1	(1)	1	1	1	(1)	1	1	1	(1)
	2	(2)	2	2	2	(2)	(2)	2	2
HDPE/2	1	(1)	1	1	1	(1)	(1)	1	1
	2	2	(2)	(2)	2	2	2	(2)	2

First and second time points.
Test/no test convention as in note to Table 3.

7.7. Committee for Proprietary Medicinal Products Guideline: Maximum Shelf Life for Sterile Products After First Opening or Following Reconstitution

The CPMP draft guideline on maximum shelf life for sterile products after first opening or following reconstitution was issued for comment in June 1996; a second revised version was released for comment in June 1997; and a considerably revised third version was adopted in January 1998, with an operational date of July 1998. The guideline applies to all sterile products for human use except radiopharmaceuticals and extemporaneously prepared or modified preparations. It is emphasized that the end user is responsible for maintaining the quality of the product to be administered to patients, but that the manufacturer (authorization holder) should conduct appropriate studies and provide relevant information in the product information. Reference is also made to the European Pharmacopoeial recommendations regarding storage times and conditions of storage for particular categories of sterile products once they have been opened.

The guideline suggests wording to be used for unpreserved sterile products (including injections and intravenous infusions) and for aqueous preserved products and nonaqueous products. The proposals include references to storage time (in hours or days, as appropriate) and conditions (temperature). For injectable products not containing a preservative, it is suggested that the storage time and conditions should not normally be longer than 24 h at 2° to 8°C unless reconstitution or dilution had taken place in controlled and validated aseptic conditions. For preserved sterile products, it is indicated that usage periods should not normally be more than 28 days.

7.8. Committee for Proprietary Medicinal Products Stability Guideline: Abridged Applications (1989 Version)

In the CPMP stability guideline for abridged applications (1989 version), it is indicated that information on the stability of the active ingredient may be determined by experimental studies or by evidence from the scientific literature for known substances (although comparative accelerated studies may need to be considered, for example, when there is a significant change in the route of synthesis from one approved by the regulatory authorities or when there is a significant change to the production method). When multiple sources of the active ingredient are requested, stability data relating to each may be required. Data on a minimum of two batches will be required when such studies are needed.

Studies of the finished product (normally at least three batches unless the product is stable, i.e., no significant degradation products are found, or when the active ingredient has already been used in licensed products) will be required to establish the shelf life specification, the shelf life, and recommended storage conditions. Specific studies will also be required when the product is labile once the container has been opened.

Real-time studies will be required. These should be carried out under a variety of controlled conditions: The properties of the product at temperatures between 20°C and 30°C should be able to be evaluated. The mean kinetic temperature for Europe should be taken as 25°C. A variety of other storage conditions should also be considered, such as below −15°C, from 2°C to 8°C (refrigeration), and freeze–thaw cycling. High humidity may be relevant (i.e., >75% RH), as may combinations of elevated temperature and humidity (e.g., 40°C/75% RH) and exposure to natural or defined artificial light conditions.

The summarized (tabular or graphical) data should be presented for each batch of product. The results of ongoing studies should be provided as they become available, including the results from the first two or three production batches. These results should be discussed in the Expert Report.

No labeling statement is required if the product is stable at up to 30° unless special storage requirements apply (e.g., do not freeze, do not refrigerate, etc.). In other cases, a recommended temperature storage range should be stated (in °C).

Data will also be required for certain variations, such as when there are major changes to the composition of the product, to packaging materials, or to the method of preparation of the product. Results from comparative accelerated and long-term stability studies would normally be required prior to authorization.

This guideline has been replaced by a newer one, "Stability Testing of Existing Active Substances and Related Finished Products" (discussed next).

7.9. Committee for Proprietory Medicinal Products Guideline: Stability Testing of Existing Active Substances and Related Finished Products

The CPMP guideline regarding Stability Testing of Existing Active Substances and Related Finished Products was released for comment in March 1997 and was adopted by the CPMP in April 1998. It came into effect in October 1998.

It is intended that this guideline apply to chemically active substances that have been previously approved in the EU and to products containing them, but not to radiopharmaceuticals or biological/biotechnological products. The test conditions in the guideline refer to the parent ICH guideline on stability testing of new active substances and products containing them. A systematic approach to the presentation and interpretation of data is encouraged.

7.9.1. Active Ingredients

Data from stress testing and formal studies may be expected. The extent of data requirements will depend on the status of the active ingredient concerning the availability of relevant (i.e., European) monographs, published data, and the like.

When a pharmacopoeial monograph is available and includes data under the headings "purity test" and/or "transparency statement" for named degradation products or when additional information on the degradation pathways is available in the published scientific literature, no stress testing will be required. Such data should be generated in other circumstances.

When a pharmacopoeial monograph is available that includes suitable degradation product limits (but no definition of a retest interval), it would be open to the applicant to state that the material complies with the monograph immediately prior to use in the manufacture of the finished product. In this case, no stability data would be required, provided that the suitability of the pharmacopoeial monograph has been established for the particular named ingredient source. Alternatively, the retest interval could be set based on results from long-term stability tests.

When data are required, the following general suggestions are included in the guideline concerning the selection of batches for testing, although alternative test conditions can be proposed and justified:

Option 1: Two industrial-scale batches made using the same synthetic route and in the same packaging used for storage and distribution or one batch simulating that packaging at the time of submission (accelerated and long-term study data) and a further batch (placed in a long-term test) to the same protocol after approval. Accelerated test conditions are 40°C ± 2°C, 75% RH ± 5% RH; 6 months of data. Long-term test conditions are 25°C ± 2°C, 60% RH ± 5% RH; 6 months of data.

Option 2: At least three pilot-scale batches using the same synthetic route at the time of submission (6 months of accelerated and 12 months of long-term study data) and extended to cover the retest period and in the appropriate packaging, with the first three production-scale batches manufactured postapproval placed in a long-term test to the same protocol, with results being submitted as they become available.

Results from studies should be submitted to the regulatory authorities when available.

In both cases, the testing intervals should be according to the main ICH guideline. Intermediate testing conditions can be used when significant changes occur under accelerated conditions. For temperature-sensitive materials, long-term testing should be performed at 5°C ± 3°C with the relative humidity conditions reported, and accelerated studies should be undertaken at 25°C ± 2°C and 60% RH ± 5% RH for 12 and 6 months. The usual requirements for validated analytical methods, collection of data on features susceptible to change and having a potential impact on quality safety or efficacy and so on apply.

The specification should include limits of acceptability. This should include individual and total upper limits for impurities, including degradation products. The limits should be based on safety and efficacy considerations.

Labeling statements for active ingredients should indicate appropriate conditions based on the stability evaluation, but should avoid terms such as "ambient conditions" or "room temperature."

7.9.2. Finished Product

At the time the application for authorization is submitted, data will normally be expected from accelerated (6 months of data at 40°C ± 2°C/60% RH ± 5% RH) and long-term studies of the proposed formulation and dosage form intended for marketing. For conventional dosage forms such as immediate-release solid dosage forms and solutions and when the active substances are known to be stable (i.e., remains within initial specifications after 2 years at 25°C/60% RH and after 6 months at 40°C/75% RH), long-term data from at least two pilot-scale batches for 6 months at 25°C ± 2°C/60% RH ± 5% RH will normally be accepted. In other cases (e.g., prolonged-release dosage forms or when the active substance is known to be unstable), stability data are likely to be required for three batches, two at pilot scale and a third smaller batch, for 12 months at the same conditions. Other storage conditions may be justified, for example, if there is "significant change" under the normal conditions. The permitted conditions in special cases (e.g., products stored under refrigerated conditions or liquid products in semipermeable containers) follow the advice given in the parent ICH guideline. The usual pattern of test sampling points should be followed.

In all cases, the manufacturing method should be a simulation of that to be applied to large-scale marketing batches, and the packaging used should be the same as that used for the commercial product.

The methods should investigate preservative losses, physical property changes, and, when necessary, microbiological attributes, as well as chemical and biological stability. Preservative efficacy tests, as well as preservative assays, should be undertaken on stored samples, particularly when different limits are proposed for preservatives in the release and shelf life specifications. In the case of particle size and/or dissolution rate specifications, reference should be made to batches used to establish bioavailability or bioequivalence.

Specifications should be proposed for shelf life and for release (when the values differ from the shelf life specification values). The shelf life specifications, including specific upper limits for degradation products (based on efficacy and safety considerations), may deviate from the release limits provided that these are justified by the changes observed in storage.

The data should be analyzed statistically to establish the degree of variability between batches (as per the parent ICH guideline) unless there is so little variability that it is apparent that the requested shelf life will be granted.

In establishing the acceptable shelf life, limited extrapolation of the available real-time data may be allowed, particularly when the accelerated data are supportive. However, since this assumes that the degradation mechanism will continue to apply beyond the observed data points, it is important to consider what is known about the degradation mechanism and the goodness of fit of the data to any mathematical model and any other supportive data, including mass balance. It is indicated that the normal extrapolation is twice the length of the real-time studies (but not more than 3 years).

Based on the available data, appropriate storage and labeling statements should be proposed. These should refer, as appropriate, to conditions such as temperature (avoiding the terms "ambient conditions" and "room temperature"), light, and humidity.

7.10. Committee for Veterinary Medicinal Products Guideline: In Use Stability Testing of Veterinary Medicinal Products (Excluding Immunological Veterinary Medicinal Products)

The guideline for In Use Stability Testing of Veterinary Medicinal Products (Excluding Immunological Veterinary Medicinal Products) was adopted by the Committee for Veterinary Medicinal Products (CVMP) in March 1996 and came into force in September 1996. While it is primarily addressed to veterinary products other than immunologicals, it may also be applied to (or this type of data may be requested for) human products by regulatory agencies. The test procedure involves removal of aliquots of the (reconstituted if appropriate) product from the container at regular intervals over the proposed in-use shelf life.

The guideline is intended to offer information for the design of studies to establish the period for which a multiuse product may be used after the first dose has been removed from the container without the integrity of the product being adversely affected. Such testing is required for all parenteral products in multidose containers, and particularly so if the headspace in the vial is replaced by an inert atmosphere during manufacture. Multidose oral or topical dosage forms should be tested if a specific stability-related problem has been identified once the container has been opened. Oral products may also need to be tested (e.g., when the active ingredient was subjected to oxidative degradation and was packed in a well-filled oxygen-impermeable container).

At least two batches of product (of stated batch number and batch size; production-scale lots are preferred, although representative development-scale batches will be accepted) in the proposed marketing container–closure system should be tested. At least one of those batches should be approaching the end of its shelf life in the case of products that have chemical characteristics that will alter significantly during the shelf life (e.g., antioxidant levels). When different package sizes will be marketed, the one offering the greatest demand on the system should be used.

The test should simulate as much as possible the intended use of the product. At appropriate intervals, aliquots of the product should be withdrawn by the usual technique, with the product being stored under the recommended conditions throughout the test period. In some cases, it may be appropriate to remove a significant proportion of the contents at the start of the test only and then to reexamine the product up to the end of its proposed in-use shelf life; or, it may be more relevant to remove aliquots on a daily basis if this is more representative of the normal use pattern for the product. More than one test may be necessary if a product may be used in different ways. In any case, the rationale for the design adopted should be explained.

Chemical (active ingredient assays, antimicrobial and chemical preservative content, degradation products), physical (pH, color, clarity, closure integrity, particulate contamination, particle size), and microbiological properties (total viable count, antimicrobial preservative efficacy, single or repeat challenge) of the product should be monitored during the study. Test procedures should be described and validated. The results should be summarized and tabulated. Conclusions should be made, drawing attention to any anomalous results.

The in-use shelf life and any advice on discarding the product should be included in the summary of product characteristics. Suitable labeling should be applied—including a space in which the user can write in the date when the first dose of product is withdrawn from the container. The in-use shelf life should also be stated on the label if space permits.

7.11. Committee for Proprietary Medicinal Products and Committee for Veterinary Medicinal Products Guideline: Inclusion of Antioxidants and Antimicrobial Preservatives in Medicinal Products

The CPMP and CVMP guideline for the inclusion of antioxidants and antimicrobial preservatives in medicinal products was released for comment in July 1996 and adopted in July 1997, and had an operational date of January 1998. With respect to stability studies, it is suggested that the guideline on stability of new dosage forms should be followed. Antioxidant and antimicrobial preservative levels should be quantified periodically throughout the shelf life of the finished product, and preservative efficacy should be established using the European Pharmacopoeial methodology both at the end of shelf life and at the lower end of the proposed preservative content.

Products intended for use on more than one occasion should also be tested for preservative efficacy under simulated in-use conditions. It might also be necessary to investigate preservative efficacy following storage of opened or used containers during the proposed in-use shelf-life.

7.12. Draft Committee for Proprietary Medicinal Products Guideline: Dry Powder Inhalers

The CPMP guideline concerning dry powder inhalers was released for comment in October 1997 and adopted in June 1998, and had an operational date of December 1998. It is suggested that dose, dose delivery, and particle characteristics of these products need to be shown over the proposed shelf life and over the proposed period of use. Active ingredient assays and determinations of related substances will be expected. The particular significance of exposure to moisture is emphasized in the guideline, with suggested testing conditions of 25°C/60% RH and 40°C/75% RH. No special storage instructions relating to moisture exposure are likely to be required if the product complies with its shelf life specification after exposure to such conditions. The exception to this is when an overwrap is used for the product, but the unwrapped product was unable to withstand the suggested stability test conditions for a period of, say, 3 to 6 months (to stimulate normal use of the product).

7.13. Committee for Proprietary Medicinal Products Guideline: Declaration of Storage Conditions for Medicinal Products in the Product Particulars

The CPMP guideline concerning declaration of storage conditions for medicinal products in the product particulars was circulated for comment in June 1997, adopted in January 1998, and came into effect in June 1998. The guideline defines, based on the stability results, the types of statement to be included in the summary of product characteristics and the product labeling and leaflet. It is intended that the wording in the guideline be used verbatim, although additional or alternative storage statements will be allowed when necessary to protect the product, but these statements will need to be supported with data. Any alternative storage recommendation must be achievable in practice. The use of the terms "room temperature" and "ambient conditions" is stated to be unacceptable.

When a product is stable at 25°C/60% RH and 40°C/75% RH, no labeling statement is required, but "Do not refrigerate or freeze" may be added. When products are stable at 25°C/60% RH and 30°C/60% RH, the labeling should state "Do not store above 30°C" and may state "Do not refrigerate or freeze." Products stable at 25°C/60% RH should state "Do not store above 25°C" and may state "Do not refrigerate or freeze." For products stable at 5°C, the label should state "Store at 2°C–8°C" and may state "Do not freeze." Products that are stable below 0°C should state "Store in a freezer" and state a justified temperature.

With regard to statements relating to protection from moisture or light, it is indicated that the inclusion of warning statements should not be an alternative to the correct choice of container in the first place. However, the following statements may be allowed: regarding moisture sensitivity, "Keep the container tightly closed" (for plastic bottles, etc.) or "Store in the original package" (for blisters); and regarding light sensitivity, "Store in the original container" or "Keep container in the outer carton."

7.14. European Union Requirements for Variations (Changes) Related to Stability

In the European Union, Commission Regulation 541/95 introduced a procedure for classification and assessment of applications to vary marketing authorizations, including those related to stability. This was amended by Commission Regulation 1146/98, which came into effect in June 1998. These measures introduced the concept of classifying variations into three classes: minor, or type I (with a "notification" procedure); major, or type II (with an assessment and approval procedure); and those with changes that were so significant that they required a new marketing authorization application. The simplified type I procedure could be applied in cases for which (a) the requested changes were supported by data generated according to the originally approved protocol and that showed that the originally agreed specifications were still met, and (b) that the maximum requested shelf life was not more than 5 years, and that consequential changes to the product information formed, part of the application for a variation. Examples of such variations are discussed below.

Changes that do not meet these requirements will be subjected to the type II variation procedure, which includes a longer processing time and requires positive approval prior to implementation.

The main types of application with stability-related changes that are identified in the regulations as being type I (and the conditions that have to be met) are the following:

20. *Extensions to the shelf life as foreseen at the time of authorization.* Conditions to be fulfilled: Stability studies have been done according to the protocol that was approved at the time of the issue of the marketing authorization; the studies must show that the agreed end of shelf life specifications are still met; the shelf life does not exceed 5 years.

20A. *Extension to the shelf life/retest period of the active substance.* Condition to be fulfilled: Stability studies must show that the agreed specifications are still met.

21. *Change in the shelf life after first opening.* Condition to be fulfilled: Studies must show that the agreed end of shelf life specifications are still met.

22. *Change in the shelf life after reconstitution.* Condition to be fulfilled: Studies must show that the agreed end of shelf life specifications are still met for the reconstituted product.

23. *Change in storage conditions.* Condition to be fulfilled: Stability studies have been done according to the protocol that was approved at the time of issue of the marketing authorization. The studies must show that the agreed end of shelf life specifications are still met.

Stability data will be expected to form part of the supporting package for variations for excipient replacement (provided that the replacement is for a material with comparable characteristics, except for vaccines and immunological products, for which a new application may be required), for coating changes on solid dosage forms (for which dissolution characteristics should be unchanged), and for changes to the container and closure system. See below for further discussion.

Other variations concerned directly with shelf life and in-use life of pharmaceuticals are likely to be type II variations, such as those that do not comply with the conditions to be fulfilled listed above.

The CPMP adopted a guideline on stability studies for type II variations in April 1998, and this had an operational date of October 1998. It is indicated that the guidance was intended to provide general information, but that it was intended to allow sufficient flexibility for different practical situations that may be encountered (although alternative approaches should be scientifically justified).

The guideline points out that stability studies should be continued to the approved retest interval for active substances and shelf life for products, and that the regulatory authorities should be advised for any problems that are identified. The specific examples covered in the document include changes to the manufacturing process for the active substance, a change in composition of the finished product, or a change in the immediate packaging of the finished product.

In all cases, the relevant available information on the stability of the active substance (stability profile, including stress test results, supportive data, and primary accelerated and long-term testing data) and finished product (primary data from accelerated and long-term studies and any supportive data) should be taken into account. Appropriate additional studies will then need to be reported.

The data requirements at the time of submission of the variation will depend on the particular circumstances. The applicant should investigate whether the proposed change(s) have a potential impact on the quality characteristics of the active substance and/or the product. Examples are discussed below. In addition, the first three production batches made post approval should be placed in long-term and accelerated stability studies using the same protocols as used for the submitted data. Data should be submitted to the authorities when available. Normal extrapolation of the data will be allowed (e.g., up to two times the real-time data up to a limit of 3 years).

7.14.1. Changes to the Manufacturing Process for the Active Substance

7.14.1.1. Modification(s) to One or More Steps of the Same Route of Synthesis

When the physical characteristics, impurity profile, or other quality characteristics of the active substance are adversely affected by the change, there will be a need to provide 3 months of comparative accelerated and long-term data on one batch of product at not less than pilot scale for stable drugs. In other cases, 6 months of data will be required from three pilot-scale batches.

When the change is to the final stage of synthesis (e.g., new solvent or new crystallization method), additional accelerated and long-term data (3 months, two pilot-scale batches) may be needed for solid dosage forms, for which the physical characteristics may have an impact on the stability of the product.

7.14.1.2. Change of Synthetic Route

Comparative accelerated and long-term data will be required on three batches of at least pilot-scale manufacture for 6 months.

When there is a change of active substance specification that might affect the finished product (e.g., physical characteristics, degradation product levels) data may also be required on the finished product using at least two pilot-scale batches and reporting 3 months of data at the time of submission.

7.14.2. Changes to the Composition of the Finished Product

For conventional dosage forms such as conventional release solid dosage forms and solutions in which the active substance is stable, there will be a requirement for 6 months of data to be supplied from two pilot-scale batches from long-term and accelerated conditions. For critical dosage forms and less stable active substances, 6 months of data from three batches of product under long-term and accelerated conditions will be required.

7.14.3. Changes to the Immediate Packaging of the Finished Product

When a less protective package is introduced or there is a risk of interaction (e.g., with liquid or semiliquid products), 6 months of data will be required from long-term and accelerated studies of three pilot-scale batches of the product.

7.15. World Health Organization Stability Guidelines

While the WHO guidelines for stability testing are not directly applicable in the ICH regions, a summary of their requirements is included in the Appendix to this chapter. It seems probable that some account will be taken of these during the revision of the main ICH guideline.

8. DISCUSSION

The European regulatory agencies have adopted the ICH guidelines for stability testing, although the practical implementation may differ in detail from that applied in the other ICH regions. However, there is a wide range of other relevant guidelines—generated either by the ICH or by the CPMP—that add to the statutory requirements included in the EU Directives and Regulations affecting pharmaceutical products, particularly Directives 65/65/EEC and 75/318/EEC as amended by 91/507/EEC.

This paper has attempted to bring together and summarize many of the requirements and proposals included in guidance documents that affect the types of stability data required for new chemical active substances and products containing them and for existing chemical active substances and products containing them.

GLOSSARY

The terms in this listing are taken from ICH guidelines unless otherwise indicated.

Accelerated testing
Studies designed to increase the rate of chemical degradation or physical change of an active substance or drug product by using exaggerated storage conditions as part of the formal, definitive, storage programme.

These data, in addition to the long term stability studies, may also be used to assess longer term chemical effects at non-accelerated conditions and to evaluate the impact of short-term excursions outside the label storage conditions such as might occur during shipping. Results from accelerated testing studies are not always predictive of physical change.
WHO definition similar: Accelerated stability testing.

Active substance; Active ingredient; Drug substance; Medicinal substance
The unformulated drug substances which may be subsequently formulated with excipients to produce the drug product.

Bracketing
The design of a stability schedule so that at any time point only the samples on the extremes, for example of container size and/or dosage strengths, are tested. The design assumes that the stability of the intermediate condition samples are represented by those at the extremes.
[Where a range of dosage strengths is to be tested, bracketing designs may be particularly applicable if the strengths are very closely related in composition (eg for a tablet range made with different compression weights of a similar basic granulation, or a capsule range made by filling different plug fill weights of the same basic composition into different size capsule shells). Where a range of sizes of inter-

mediate containers are to be evaluated, bracketing designs may be applicable if the material of composition of the container and the type of closure are the same throughout the range.]

Climatic zones
The concept of dividing the world into four zones based on defining the prevalent annual climatic conditions.
[See Table 1.]
WHO definition similar.

Confirmatory studies [Photostability testing]
Those [studies] undertaken to establish photostability characteristics under standardised conditions. These studies are used to identify precautionary measures needed in manufacturing or formulation and whether light resistant packaging and/or special labelling is needed to mitigate exposure to light.

Dosage form; Preparation
A pharmaceutical product type, for example tablet, capsule, solution, cream, etc, that contains a drug ingredient generally, but not necessarily, in association with excipients.

Drug product; Finished product
The dosage form in the final immediate packaging intended for marketing.

Excipient
Anything other than the drug substance in the dosage form.

Expiry; Expiration date
The date placed on the container/labels of a drug product designating the time during which a batch of the product is expected to remain within the approved shelf-life specification if stored under defined conditions, and after which it must not be used.
WHO definition:
The expiry date given on the individual container (usually on the label) of a drug product, designates the date up to and including which the product is expected to remain within specifications, if stored correctly. It is established for every batch by adding the shelf life to the manufacturing date.

Forced degradation studies
Those studies undertaken to degrade the sample deliberately. These studies, which may be undertaken in the development phase normally on the drug substance, are used to evaluate the overall photosensitivity of the material for method development purposes and/or degradation pathway elucidation.

Formal (systematic) studies
Stability evaluation of the physical, chemical, biological, and microbiological characteristics of a drug product and a drug substance, covering the expected duration of the shelf life and re-test period, which are claimed in the submission and which will appear in the labeling.

Immediate (primary) pack
That constituent of the packaging that is in direct contact with the drug substance or drug product, and includes an appropriate label.

Marketing pack
The combination of immediate pack and other secondary packaging such as a carton.

Mass balance; Material balance
The process of adding together the assay values and levels of degradation products to see how closely these add up to 100 per cent of the initial value, with due consideration of the margin of analytical precision.

This concept is a useful scientific guide for evaluating data but it is not achievable in all circumstances. The focus may instead be on assuring the specificity of the assay, the completeness of the investigation of routes of degradation, and the use, if necessary, of identification degradants as indicators of the extent of degradation via particular mechanisms.

Matrixing
The statistical design of a stability schedule so that only a fraction of the total number of samples are tested at any specified sampling point. At a subsequent sampling point, different sets of samples of the total number would be tested. The design assumes that the stability of samples tested represents the stability of all samples. [The differences in the samples for the same drug product should be identified as, for example, covering different batches, different strengths, different sizes of the same container and closure and possibly in some cases different container/closure systems.

Matrixing can cover reduced testing when more than one variable is being evaluated. Thus the design of the matrix will be dictated by the factors needing be covered and evaluated. This potential complexity precludes inclusion of specific details and examples, and it may be desirable to discuss design in advance with the regulatory authority, where this is possible. In every case it is essential that all batches are tested initially and at the end of the long-term testing.]

Mean kinetic temperature
When establishing the mean values of the temperature, the formula of J D Haynes (*J Pharm Sci,* **60**, 927–929, 1971) can be used to calculate the mean kinetic temperature. It is higher than the arithmetic mean temperature and takes into account the Arrhenius equation from which Hayes derived his formula.
WHO definition:
The single test temperature corresponding to the effects on chemical reaction kinetics of a temperature-time distribution of each of the four world climatic zones and according to the formula developed by Hayes J D, Journal of Pharmaceutical Sciences, *1971, 60: 927–929. It is a higher value than that of the arithmetic mean temperature.*

New molecular entity; New drug substance
A substance which has not previously been registered as a new drug substance with the national or regional authority concerned.

Pilot plant scale
The manufacture of either drug substance or drug product by a procedure fully representative of and stimulating that to be applied on a full manufacturing scale. For oral solid dosage forms this is generally taken to be at a minimum scale of one tenth of full production or 100000 tablets or capsules, whichever is the larger.

Primary stability data
Data on the drug substance stored in the proposed packaging under storage conditions that support the proposed re-test date.

Data on the drug product stored in the proposed container-closure for marketing under storage conditions that support the proposed shelf life.

WHO definition:
Real time (long term) stability studies
Evaluation of experiments for physical, chemical, biological, biopharmaceutical and microbiological characteristics of a drug, during and beyond the expected time of shelf-life and storage of samples at expected storage conditions in the intended market. The results used to establish shelf-life, to confirm projected shelf-life and recommend storage conditions.

Re-test date
The date when samples of the drug substance should be re-examined to ensure that material is still suitable for use.

Re-test period
The period of time during which the drug substance can be considered to remain within the specification and therefore acceptable for use in the manufacture of a given drug product, provided that it has been stored under the defined conditions; after this period, the batch should be re-tested for compliance with specification and then used immediately.

Shelf-life: Expiration dating period
The time interval that a drug product is expected to remain within the approved shelf-life specification provided that it is stored under the conditions defined on the label in the proposed containers and closure.
WHO definition:
The period of time during which a pharmaceutical product is expected, if stored correctly, to comply with the specification ['shelf life specification: ie requirements to be met throughout the shelf-life of the drug product (should not be confused with 'release specification')] as determined by stability studies on a number of batches of the product. The shelf-life is used to establish the expiry date of each batch.

Specification—check/shelf life
The combination of physical, chemical, biological and microbiological test requirements that drug substance must meet up to its re-test date or a drug product must meet throughout its shelf life.

Specification—release
The combination of physical, chemical, biological and microbiological test requirements that determine a drug product is suitable for release at the time of its manufacture.

WHO definition:
Stability
The ability of a pharmaceutical product to retain its properties within specified limits throughout its shelf-life. The following aspects of stability are to be considered: chemical, physical, microbiological and biopharmaceutical.

WHO definition:
Stability tests
Stability tests are series of tests designed to obtain information on the stability of a pharmaceutical product in order to define its shelf-life and utilisation period under specified packaging and storage conditions.

Storage conditions tolerances
The acceptable variation in temperature and relative humidity of storage facilities. The equipment must be capable of controlling temperature to within a range of ±2°C and relative humidity to ±5%. The actual temperatures and humidities should be monitored during stability storage. Short term spikes due to opening of doors of the storage facility are accepted as unavoidable. The effect of excursions due to equipment failure should be addressed by the applicant and reported if judged to impact stability results. Excursions that exceed these ranges (ie ±2°C and/or ±5% RH) for more than 24 hours should be described in the study report and their impact assessed.

Stress testing (drug product)
Light testing should be an integral part of stress testing. Special test conditions for specific products (eg metered dose inhalations and creams and emulsions) may require additional stress testing.

Stress testing (drug substance)
These studies are undertaken to elucidate intrinsic stability characteristics. Such testing is part of the development strategy and is normally carried out under more severe conditions than those used for accelerated tests.

Stress testing is conducted to provide data on forced decomposition products and decomposition mechanisms for the drug substance. The severe conditions that may be encountered during distribution and be covered by stress testing of definitive batches of the drug substance.

These studies should establish the inherent stability characteristics of the molecule, such as the degradation pathways, and lead to identification of degradation products and hence support the suitability of the proposed analytical procedures. The detailed nature of the studies will depend on the individual drug substance and type of drug product.

This testing is likely to be carried out on a single batch of material and to include the effect of temperature in 10°C increments above the accelerated temperature test conditions (e.g., 50°C, 60°C, etc); humidity where appropriate (e.g., 75% or greater); oxidation and photolysis on the drug substance plus its susceptibility to hydrolysis across a wide range of pH values when in solution or suspension.

Results from these studies will form an integral part of the information provided to the regulatory authorities.

Light testing should be an integral part of stress testing. It is recognised that some degradation pathways can be complex and that under forcing conditions decomposition products may be observed which are unlikely to be formed under accelerated and long term testing. This information may be useful in developing and validating suitable analytical methods, but it may not always be necessary to examine specifically for all degradation products, if it has been demonstrated that in practice these are not found.

Supporting stability data
Data other than primary stability data, such as stability data on early synthetic route batches of drug substance, small scale batches of materials, investigational formulations not proposed for marketing, related formulations, product presented in containers and/or closures other than those proposed for marketing, information regarding test results on containers, and other scientific rationale that support the analytical procedures, the proposed re-test period of shelf life and storage conditions.
WHO definition:
Supplementary data, such as stability data on small scale batches, related formulations, products presented in containers other than those proposed for marketing and other scientific rationale that support the analytical procedures, the proposed re-test period or shelf-life and storage conditions.

WHO definition:
Utilisation period
A period of time during which a reconstituted preparation or the finished dosage form in an opened multidose container can be used.

APPENDIX

Provisional World Health Organization Guidelines on Stability Testing of Pharmaceutical Products Containing Well-Established Drug Substances in Conventional Dosage Forms

A provisional guideline on the stability testing of well-established drug substances in conventional dosage forms was adopted at a meeting of the Expert Committee on Specifications for Pharmaceutical Preparations in November/December 1994.

For products of this type, data are often available in the literature on the decomposition processes and degradation of active substances (e.g., in the unpublished report. WHO/PHARM/86.529) together with adequate analytical methods. In such cases, attention may be concentrated on the stability of the dosage form. Stability of formulated products should be adequately investigated during development of the product, including consideration of the possibility of interactions between the drug substance and excipients and container system components. Due consideration should be given to transport and storage conditions and to the climatic conditions of the places where the product is to be marketed. Appropriate storage recommendations may be needed to ensure compliance with the shelf life proposed for the product as indicated by the labeled expiry date.

The WHO guideline includes definitions of accelerated stability testing, batch, climatic zones, expiry date, mean kinetic temperature, real-time (long-term) stability studies, shelf life, stability, stability tests, supporting stability data, and utilization period.

The purposes of stability testing are identified in Table A1.

The importance of climatic zone is emphasized in the WHO guideline, and additional data are included in the document compared with the CPMP guideline. A summary of the information is included in Table A2.

Table A1 Uses of Stability Studies

Phase/purpose	Accelerated studies	Real-time studies
Development: selection of adequate formulations and container–closure systems	X	
Development and dossier: determination of shelf life and storage conditions	X	X
Dossier: substantiation of claimed shelf life		X
Quality assurance, quality control: verification that no changes have been introduced into the formulation or manufacturing process that can adversely affect the stability of the product	X	X

Table A2 Climatic Conditions Data (WHO)

Climatic zone		Measured data			
		Open Air	Storage room	Calculated data	Derived data
I	Temperature	10.9°C	18.7°C	20.0°C	21°C
	Relative humidity	75%	45%	42%	45%
II	Temperature	17.0°C	18.7°C	21.6°C	25°C
	Relative humidity	70%	52%	52%	60%
III	Temperature	24.4°C	26.0°C	26.4°C	30°C
	Relative humidity	39%	54%	35%	35%
IV	Temperature	26.5°C	28.4°C	26.7°C	30°C
	Relative humidity	77%	70%	76%	70%

Note: References to the derivation of these figures are included in the WHO document.

The recommendation in the WHO document is for products intended for a global market to be tested under the climatic zone IV conditions, that is, real-time studies under conditions as close as possible to those that will be encountered during storage in the distribution system (with a minimum of 12 months of data being submitted with the application) and accelerated testing at 40°C ± 2°C and 75% RH ± 5% RH for 6 months, although other conditions may be used (e.g., 3 months at 45°C–50°C and 75% RH).

In the case of zone II conditions, accelerated conditions of 40°C ± 2°C and 75% RH ± 5% RH for 3 months or 6 months are recommended in the case of a less stable drug substance or for those products for which a limited amount of data is available. When these conditions are not appropriate, alternative conditions may be used, provided that accelerated studies are undertaken at not less than 15°C above the expected long-term storage temperature and relevant humidity conditions.

In addition to the indicated temperature and humidity conditions, it is also suggested that stability studies for liquid products should consider the use of low temperatures (e.g., below 0°C, −10°C to −20°C) freeze–thaw cycles, and refrigeration conditions (2°C to 8°C). Light exposure may also be relevant for some products. Publicly available information on stability of drugs should be taken into account when designing studies.

Intermediate test conditions (e.g., 30°C ± 2°C and 60% RH ± 5% RH) may be used if significant changes occur (e.g., 5% loss of initial potency value; any degradation product exceeding its specification limits; dissolution test failure; fail to meet requirements for physical properties or appearance) in the accelerated studies, with 6 months of data from a 1 year study being submitted with the application.

Three batches of product should be tested except when the active ingredient is known to be fairly stable. The batches should be representative of the manufacturing process and be made at pilot scale or full scale. When possible, different batches of active ingredient should be used to manufacture the stability lots. In addition, production batches should also be tested (e.g., for stable formulations, one batch every other year; for products for which the stability profile has been established, one batch every 3 to 5 years unless a major changes has been made to the product, e.g., to the formula or to the method of manufacture). Details of the batches used in the stability trials should be stated: batch number, date of manufacture, batch size, packaging, and so on.

The suggesting sampling program for new products is (a) real-time studies, 0, 6, 12, 24, 36 months (or more); (b) accelerated studies, 0, 1, 2, 3, (6) months (or more); except when the active ingredient is less stable or when limited data are available (when sampling every 3 months would be appropriate) in the first year, followed by sampling every 6 months in the second year and annually thereafter. For ongoing studies to support a provisional shelf life, the study may be based on sampling every 6 months; for highly stable formulations, annual testing may be sufficient.

Analytical methods should be validated, and assays should be stability indicative. Methods used to quantify degradation products or related substances should be specific and of adequate sensitivity. Suitable methods should be applied to ensure that excipients remain effective and unchanged during the proposed shelf life. The use of a checklist to identify stability characteristics is suggested in the guideline.

A stable product is one that shows no significant degradation or changes in its physical, chemical, microbiological, and biological properties, with the product remaining within its specification.

It is suggested that the stability results should be presented in a tabular or graphical format, with details of the initial and other data points for each batch of product. A standard format is suggested as an example (see Table A3).

The study report should include information on the study design, the results, and the conclusions. The stability evaluation and the derived recommended storage conditions and shelf life relate to a particular formulation and method of production. Some extrapolation from the real-time data—when supported by accelerated data—may be appropriate.

Table A3 Suggested Format for Stability Test Summary Sheet (WHO)

Accelerated/real-time studies

Name of drug product: ______________________________

Manufacturer: ______________________________

Address: ______________________________

Active ingredient (INN): ______________________________

Dosage form: ______________________________

Packaging: ______________________________

Batch number *Date of manufacture* *Expiry date*

1 ______________________________

2 ______________________________

3 ______________________________

Shelf life: ______ years ______ months

Batch size *Type of batch (experimental, pilot plant, production)*

1 ______________________________

2 ______________________________

3 ______________________________

Samples tested (per batch): ______________________________

Storage/test conditions

Temperature ______ °C ______ Humidity ______ % RH

Light ____________ (lux hours) Pressure ____________ (bar)

RESULTS

1. Chemical findings: ______________________________
2. Microbiological and biological findings: ______________________________
3. Physical findings: ______________________________
4. Conclusions: ______________________________

____

Responsible officer: ______________________________ Date: ____________

____

A tentative shelf life of 24 months may be proposed based on

a stable active ingredient;
no significant changes in a controlled stability study;
similar formulations having been assigned a 24-month shelf life;
a claimed shelf life of not more than twice the period for which real-time data are submitted; and
an undertaking on the part of the manufacturer to continue real-time stability studies until the proposed shelf life has been covered.

The labeling that may be accepted may be based on

storage under "normal conditions," defined by WHO as "Storage in dry, well-ventilated premises at temperatures of 15°C–25°C or, depending on climatic conditions, up to 30°C. Extraneous odours, contamination, and intense light have to be excluded," although alternative normal conditions may be defined locally having regard to prevalent conditions;
storage between 2°C and 8°C under refrigeration, no freezing;
store below 8°C under refrigeration;
store in a freezer at −5°C to −20°C;
store below −18°C in a deep freezer.

Additional statements such as "protect from light" or "store in a dry place" may be used (but not to cover up stability problems).

Appropriate information may also be given for use and storage periods after the product is opened, diluted, or the like.

Relevant European Union Legislative Requirements and Guidelines Relating to Stability Testing

The documents listed here can be found in *The Rules Governing Medicinal Products in the European Union* (published by the Commission of the European Communities) or are available through the following Internet addresses: http:/dg3.eudra.org/eudralex or http:/www.eudra.org (CPMP home page at the EMEA site).

Legislation

Commission Regulation 541/95 of March 10, 1995, concerning the examination of variations of the terms of a marketing authorization granted by a competent authority of a member state, as amended by Commission Regulation 1146/98
Council Directive 65/65/EEC of January 26, 1965, on the approximation of provisions laid down by law, regulation, or administrative action relating to medicinal products
Council Directive 75/318/EEC of May 20, 1975, on the approximation of the laws of the member states relating to analytical, pharmacotoxicological, and clinical standards, and protocols in respect of the testing of medicinal products

Quality Guidelines

Development Pharmaceutics
Validation of Analytical Procedures: Methodology
Stability Testing of New Active Substances and Medicinal Products
Stability Testing of Active Ingredients and Finished Products
Stability Testing: Requirements for New Dosage Forms
Photostability Testing of New Active Substances and Medicinal Products
Inclusion of Antioxidants and Antimicrobial Preservatives in Medicinal Products
Impurities in New Active Substances
Impurities in New Medicinal Products
Reduced Stability Testing Plan—Bracketing and Matrixing, Annex to Note for Guidance on Stability Testing: Stability Testing of New Drug Substances and Products
Declaration of Storage Conditions for Medicinal Products in the Product Particulars, Annex to Note for Guidance on Stability Testing of New Active Substances; Annex to Note for Guidance on Stability Testing of Existing Active Substances and Related Finished Products
Maximum Shelf Life for Sterile Products for Human Use After First Opening or Following Reconstitution
Stability Testing of Existing Active Substances and Related Finished Products
Stability Testing for a Type II Variation to a Marketing Authorization
Dry Powder Inhalers

REFERENCE

1. Committee for Proprietary Medicinal Products, European Commission. *The Rules Governing Medicinal Products in the European Union, Vol. 3, Guidelines: Medicinal Products for Human Use, Part A, Quality and Biotechnology*. Luxembourg. European Commission, 1998, p. 129.

18

Regulatory and Scientific Aspects of Stability Testing: Present and Possible Future Trends

C. T. RHODES

University of Rhode Island, Kingston, Rhode Island

I. INDUSTRY TRENDS

1.1. Mergers and Takeovers

During the 1990s, there was a substantial increase in the number of pharmaceutical companies, involved in mergers or takeovers. As a result, a number of very large pharmaceutical companies, transnational in nature, have emerged. These companies have tremendous revenues but also tremendous costs; in order to improve their bottom lines, they often give rigorous evaluation to all aspects of the operations, merging some units from different geographical locations, centralizing some activities, and contracting out to outside organizations some operations that previously were totally in-house. Stability testing is not exempt from these winds of change.

1.2. Specialization

At the same time as the emergence of the giant transnational pharmaceutical companies, small companies concentrating on some specialty niche that they are able to exploit rapidly because of their corporate maneuverability are becoming of increasing importance. Often, such companies grow rapidly or are taken over by a larger company. Also, unfortunately, not all succeed and many fall by the wayside.

1.3. Contracting Out

As recently as the 1960s, there were many moderate-size companies that had a corporate, full-service approach. The company took complete responsibility for all aspects of the research, development, and regulatory approval of its products. Nowadays, there are many companies that contract out a variety of activities: manufacture of clinical trials materials, supervision of clinical trials, design of formulations to be marketed, preparation of regulatory submission documents, and so on. It is now quite common for pharmaceutical companies to contract out part or all of their stability work. It is claimed that there can be substantial financial advantage to this practice. However, there are potential problems with this practice, and it is essential, if a contract house is selected to carry out stability work, that the sponsor carefully monitor the work.

The selection of a contract house for stability work should be a careful, thoughtful process. Questions to be addressed include: What is the reputation of the contract house in the pharmaceutical industry and with regulators? How long has the company been providing services in the stability area? What type of facilities and equipment does the company possess? What is the education and training of the staff? Thus it is important to carry out a site visit to explore these and other pertinent questions.

The inspection of the contract house should be conducted using predetermined audit procedures. It is normally desirable that the audit team include at least two people, one of whom might be a consultant.

The audit team should visit the contract house when work is in progress, since as well as examining various documents it is highly desirable that there should be an opportunity to observe chemists at work in the laboratory, as well as other persons involved in storage control, data management, etc. Obviously, particular attention should be given to the validation of stability-indicating assays.

When it is decided to use the services of a contract house for stability testing, the contract that defines the relationship should, as far as possible, deal with all eventualities. For example, if stability data generated at the contract house is for some reason perceived to be of questionable value by a regulatory agency, such as the FDA, the contract house should, without additional charge, attend meetings at the regulatory agency and explain how the data was generated and interpreted.

Even when a contract house has, after careful inspection, been selected to perform stability work, its performance should be the subject of a continuing monitoring process. Those interested in using contract houses for stability testing may find useful the Register of Contract Support Organizations (1), published by the Drug Information Association.

1.4. Globalization

The results of the mergers and takeovers have been to produce a significant number of transnational pharmaceutical companies that have research, development, manufacturing, and marketing units in various parts of the world. In order to take advantage of economy of scale, there is an obvious trend for such companies to try to standardize formulation design, package selection, and laboratory procedures at all their facilities throughout the world. This is not always possible. For example, a product for distribution in a Climate Zone 4 country may require a special, heavy-duty tropical pack that is not necessary in North America, Western Europe, or Japan. Also, not all pharmaceutical excipients are universally acceptable throughout the world; thus, it may not be possible to have a formulation and pack that are identical in all areas. However, such matters as standard operating procedures design and implementation can be unified to a very considerable extent.

1.5. More Effective Cost Control

As a result of mergers or takeovers, the megapharmaceutical companies so formed are often under considerable pressure to cut costs and hence improve profitability. The size of such companies often means that a single company budget is quite insufficient as an instrument to control costs in esoteric areas, such as stability testing. There is, therefore, a tendency to develop budgets for the many individual cost centers which, together, comprise the company. In Chapter 1 of this book, I expressed my opinion about the value of stability budgets. In some companies, this activity has progressed further so that the stability costs associated with any of the company's marketed products can be quantified. This type of data can be most useful and can provide material assistance for decisions about whether any given product of low sales is really a loss maker or a profit generator.

One money-saving procedure adopted in some companies is to accelerate the use of automated and robotic equipment for stability testing. If an adequate period is allowed for the introduction and validation of this type of equipment, this approach can well be financially attractive, although of course initially it does require some significant capital expenditure on equipment. However, the reduction in personnel costs will often amortize the equipment capital costs in a surprisingly short period of time.

Centralization of all R&D-type stability testing in just one world-wide central laboratory can also, in some situations, reduce costs.

Perhaps the most attractive method of controlling costs in the testing of market product retained samples will be an increased use of bracketing and even more sophisticated, fractional factorial designs.

2. HARMONIZATION

2.1. The ICH Process

The International Conference on Harmonization of Technical Requirements of Pharmaceuticals for Human Use (ICH) consists of representatives of the three pre-eminent pharmaceutical areas of the world, the European Union, the United States of America, and Japan. The stated objectives of the harmonization process are "the more economical use of human, animal and material resources and the elimination of unnecessary delay in the development and availability of new medicines while maintaining safeguards on quality, safety and efficacy and regulatory obligations to protect public health." The ICH Steering Committee developed a five-step procedure for the development of Guidelines on a number of topics concerning research, development, and quality control of pharmaceutical products. The Guidelines most likely to be of interest to those involved in stability testing are Q1A Stability Testing of New Drugs and Products, Q2A Validation of Analytical

Methods, Definitions and Terminology, Q3A Impurities in New Drug Substances, Q5C Quality of Biotechnology Products Stability of Products, Q1B Photostability Testing, Q1C Stability Testing New Dosage Forms, and Q2B Validation of Analytical Methods Methodology.

Although much of the ICH material is new, those expert in exegesis can often tease out from the Guidelines the individual contributions of the Europeans, Japanese, and Americans.

2.2. Other Harmonization Activities

Pharmacopoeias such as U.S.P., J.P., B.P., F.P., and E.P. are also involved in activities that ultimately will, it is hoped, result in considerable standardization of monographs and test methods. Such harmonization will probably take a very considerable period of time but will eventually be of great value.

3. POSSIBLE INCREASED REGULATORY CONCERN ABOUT THE STABILITY OF BOTANICALS, NUTRACEUTICALS, AND OTCs

At present, there is a rather obvious difference in the level of attention concerning stability issues given to prescription drugs as compared to botanicals, nutraceuticals (2), and OTCs (over-the-counter products). There is reason to believe that this difference will increasingly become a source of friction: but eventually, "harmonization" of stability standards may prevail.

4. MORE CONCERN ABOUT THE STABILITY OF DRUGS IN CHANNELS OF DISTRIBUTION

As has already been indicated in Section 3.6 of Chapter 1, our present stability testing procedures might perhaps legitimately be criticized for giving insufficient attention to the stability of "real" product in the channel of distribution, as opposed to "ideal" product kept in pampered luxury in a manufacturer's retained stability storage area.

However, it would be quite invalid to conclude that pharmaceutical companies have not given attention to this important aspect of stability testing. Even as long ago as the 1950s, there were some companies that routinely obtained samples of their own products from the marketplace and evaluated their quality.

4.1. Market Challenge Tests

During the process of developing a new pharmaceutical formulation, it is relatively common for pharmaceutical scientists to challenge products by tests specifically designed to simulate some aspect of market stress. For example, for compressed tablets a drop fragility test may be more predictive of physical stability during distribution than the more common (crushing strength) hardness testing. There is not, at present any official drop fragility test. However, one version of this test involves dropping fifty tablets, one at a time, through a 50 cm glass tube onto a steel base plate, with passing results being obtained if no more than one tablet shows any sign of chipping.

Freeze/thaw tests, such as are commonly used for polypeptides and proteins (3) are another form of market challenge tests commonly used to simulate market challenge.

4.2. Test Distributed Samples

The practice of sending samples of a pharmaceutical product out through a small, selected number of the channels of distribution and then retrieving the samples to the manufacturers' laboratories so that the effects of market place stress can be evaluated has been used by some companies for many years. Obviously, it is not normally feasible to test all channels of distribution, so there is a possibility of both alpha and beta sampling errors. However, many experienced pharmaceutical scientists believe this type of study to be useful. Of course, it must be kept in mind that the results of this type of study may well be, to some extent at least, dependent on the season of the year. Thus, whereas a product left on a loading dock in January may be subjected to an external temperature of −20°C, the temperature on the same dock in July may be 45°C.

A level of additional sophistication can be added to test distribution by using either temperature/color change labels or temperature/time computer chip recorders. The labels are fabricated so that they change color (often to red) at some predetermined temperature deemed to be critical for the product (for example, 40°C). The computer chip will record a temperature/time profile for the whole distribution sequence, the information is downloaded, and hard copy is obtained when the chip is returned to the manufacturer's laboratory.

4.3. The USP Mailbox Study

Recognizing that mail order supply of drug products has become an increasingly important element in drug distribution, the USP Subcommittee on Packaging Storage and Distribution implementation in 1996 a study of temperature and humidity conditions that may be experienced in main transit (4). Mailing from Rockville, MD, was performed between June and August, with destinations in various parts of the United States. The rather worrying results of this survey showed that only 84% of the packages recorded temperatures within USP specifications. The remainder were subjected to temperatures above 30°C, and 21.6% experienced excessive heat (over 40°C).

4.4. Labile Products

It is generally recognized that some pharmaceutical products are much more liable to degrade than others. Products that are unusually sensitive to temperature have been designated as labile. The USP test for lability is based on showing that the product fails to comply with monograph specifications either after two weeks storage at 50°C, in the market container, at any humidity, or four weeks at 40°C in the market container, preferably at 75% relative humidity, or four weeks at 25°C, with 60% relative humidity in an open dish.

There has been some misunderstanding concerning this definition. It does not mean that all three tests have to be performed in order for a product to be designated as labile. What is meant is that if any item fails any one of the three above tests, it may be designated as "labile" (5).

It has been proposed that products which have been designated as labile should be shipped with a time temperature indicator that discloses a history of exposure equivalent to 24 hours at a temperature of 40°C. It has been suggested that Labile Preparations should have a Beyond-Use date of 120 days (i.e., once dispensed or repackaged) the product should be used within four months.

Kommanboyina and Rhodes have reviewed recent trends in stability testing (6) and reported data concerning the effect of temperature "spikes" on assigned shelf lives (7). Data has also been presented concerning stability of compounded products (8).

It appears likely that in the near future more attention will be given to the status of all pharmaceutical products (manufactured or compounded) during their transit through the channels of distribution and their quality when they are supplied to the final user. In particular, it seems likely that the stability of repackaged material and also products which have been shipped by mail will be subjected to increasing scrutiny.

Unfortunately, at present there is a paucity of the reliable temperature/time data that might allow us to make firm conclusions about how serious the problem of degradation of some products in distribution may be.

One question that has recently received attention is, How substantial or frequent can excursions up to 30°C be for controlled room temperature products before the assigned shelf life becomes invalidated? Kommanaboyina and Rhodes have conducted a computer study (7), the results of which indicate that for products with a shelf life of more than 12 months at 30°C for as long as 2 months may be tolerable if the baseline MKT is 22.5°C. Even for products that only have a shelf life of 1 month, it is highly unlikely that an otherwise normal product would be adversely affected by a 30°C spike of 7 hours or less.

4.5. Complaints

Another source of information about how pharmaceutical products are tolerating the stress of storage and distribution after leaving the locus of production is the company complaints file. Of course, for many pharmaceutical products, the majority of complaints probably do not involve any aspect of stability. However, there is often a significant proportion of complaints which do involve, at least peripherally, some stability aspects. Broadly speaking, stability problems that are the cause of complaints may be attributed to formulation or processing (including pack selection) inadequacies by the producer or inappropriate storage or handling at some point (or points) during distribution. If it is found that the stability problem is due to limitations in the design, processing, or quality control evaluation of the product at the site of manufacture, then obviously the manufacturer must take appropriate remedial action. When it is found that the problem has been caused by maltreatment during the distribution, it is not always easy to determine who or what was responsible. Even when causality is established, it is often difficult to take appropriate remedial action. Of course, as with any complaints, all stability complaints investigations must be documented (9).

4.6. Returned Product

Many pharmaceutical companies have a policy that if product is still on the market at (or near) the end of the shelf life period, it can be returned, with full credit being given to the hospital, community pharmacy, or other entity that returned the product. Such returned product is then destroyed and the residue disposed of in an appropriate manner. However, before the returned product is destroyed, there are strong arguments in favor of a selective evaluation of some of the material. It is not recommended that every unit of returned product necessarily be subjected to a comprehensive physicochemical evaluation. In many situations it may be sufficient merely to inspect the returned product. Does the pack still have a clean, fresh appearance? Is the label still readily legible? Is the label still adhering fully to the container? Are all seals (tamper-evident, child-resistant) still intact? Is the back-off torque of the cap still satisfactory? Does the product look, feel, and smell the same as newly manufactured material? Are there any signs of physical damage (e.g., tablet breakage)? When thought appropriate, specific physicochemical tests can be applied to some samples.

There is probably no specific regulatory requirement that manufacturers evaluate returned product. However, the company that is dedicated to protecting its reputation as a producer of high-quality products has a strong incentive to do so. Since returned product was, until shortly before its return, available for supply to patients, there is good reason to conclude that its quality, when returned, is not significantly different from what it was when in the marketplace. Thus although it is probably true that in many instances returned goods are not representative of the average quality of the product on the market, it is still prudent for attention to be given to monitoring the quality of returned product.

5. NEW APPROACHES TO STABILITY TESTING

The analytical and statistical methods used in the execution and interpretation of stability studies, as well as regulatory policies, are highly unlikely to remain static. In this section of the chapter, the author considers some of the possible changes in stability testing that may become increasingly important in the future. It certainly cannot be guaranteed that the concepts discussed in this section will all become of importance in all areas of the world, nor can we be assured that there will not be other important developments. Prophesying scientific and regulatory developments is not an exact discipline!

5.1. Constant Date Testing Method

Carstensen and Rhodes (10) proposed in 1993 the use of a stability protocol which they called the Constant Date method (to distinguish it from the conventional Constant Interval method). This method has attracted the attention of pharmaceutical scientists and regulators, and it appears likely that its use will become increasingly common.

The conventional Constant interval method is based on the concept that the retained stability samples should be tested at fixed intervals during shelf life. Thus, if a product has a 5-year shelf life, it would normally be tested at 0, 3, 6, 9, 12,

18, 24, 36, 48, and 60 months. Once the product becomes successful with say 80 batches being manufactured per year, there will be, on average, about one stability sample requiring evaluation every other week. Setting up the assay procedure for, say, just one or two stability samples is both tedious and expensive. In the past, some companies tested some samples early and some later than the theoretical date. Unfortunately, some FDA investigators have adopted a rather legalistic approach to the times when retained stability samples should be tested. In some cases, 483's (Notice of Adverse Findings) have been issued by FDA personnel even though the testing evaluation date was less than a week late. Thus there has been increasing concern within industry to discover a procedure that would be acceptable to the FDA but that would not require the wasteful expense of setting up an HPLC, running standards and controls, etc., all for just one or two retained stability samples.

The Constant Date Method proposed by Carstensen and Rhodes is probably applicable to a large majority of pharmaceutical products. Using this approach, the stability testing of retained samples for any individual product is simply restricted to four specified dates each year (e.g., the first working Tuesday or February, May, August, and November). Thus some samples may be tested up to about 6 weeks "early" and some up to 6 weeks "late," but the totality of data obtained is the same (or, in some instances, somewhat greater) than would be obtained by the Constant Interval Method. Obviously, the use of the Constant Date Method must be specified in any required regulatory approval document. This method is not only more cost effective, it is also likely to improve the extent to which trend analysis of stability data is completed in a timely fashion.

5.2. Marketplace Sampling

In Ref. 10, Carstensen and Rhodes not only made the quite moderate proposal concerning the Constant Date protocol but also made a more radical proposal for the increased use of marketplace sample testing as a major element in the stability testing of at least some pharmaceutical products. In essence, what Carstensen and Rhodes proposed was that if a pharmaceutical has been on the market for a significant period of time (perhaps 5 years) in substantially the same form without any major stability problems, then consideration should be given to the possibility of discontinuing, partially or completely, retained sample stability testing, relying instead on an official testing program of material sampled from the marketplace. Obviously, any marketplace testing program for the evaluation of stability must have built into it the recognition that any defect that may be discerned in marketplace samples may be due to either or both of two factors: (a) a defect in the formulation design or manufacturing process, or (b) inappropriate storage or handling subsequent to the time at which the product was released onto the market by the manufacturer. Under category (b) we would, of course, include any adverse effects on stability caused by a repackaging operation.

Although the author is not aware of any formal official implementation of the marketplace testing of stability as proposed in Ref. 10, there is undoubtedly a higher level of awareness of the importance of marketplace stress as a factor of importance governing the quality of products supplied to patients.

5.3. Use of the Near IR

Near infrared spectroscopy has considerable potential in the stability testing of pharmaceutical products (11). Within the past decade or so, the pharmaceutical potential of the near IR for quantitative purposes has become increasingly evident. Now that the calibration techniques for the quantitative near IR have become well established and rugged, relatively inexpensive equipment is generally available. The use of this most flexible and widely applicable method of analysis is predicted to become increasingly common in the pharmaceutical industry.

In those instances where the near IR spectrum of parent drug is significantly different from that of degradation products, there may be a possibility of using near IR to quantitate the rate of decrease of parent drug and/or the rate of increase of a specific degradation product. Since the near IR can be used for nondestructive sample evaluation, this method may well prove to have considerable advantages in reducing costs of stability testing. Also, this method may allow an improved precision in stability testing.

When one examines the scatter seen in typical stability plots of percentage of label claim as a function of time, it is apparent that there are several factors that contribute to the finding that experimental data is not all on the mean regression line and thus the 90% confidence envelope is relatively wide. Obviously, error derived from the limitations of the analytical method will, in many cases, be of major importance. However, for many pharmaceutical products, especially tablets, allowable content uniformity variation will play a major role in defining the variability of retained sample stability data and hence the width of the 90% confidence envelope. For many USP tablets, the allowable content uniformity variation is defined by a standard deviation of 6%. Thus even if the analytical method had zero error and there was no other factor causing scatter on the stability plot, we would still expect to see substantial divergence between the loci of individual experimental points and the mean regression line simply because different sample tablets were, at time 0, characterized by percentage label claim values that varied perhaps from 93 to 107%.

Kommanboyina et al. (12) have proposed the use of near IR for the potency stability testing of retained samples using the assay of the same individually identified tablets for the whole testing protocol. This method removes the variance in stability data caused by content uniformity factors. As a result, it is possible that, in some cases, the improved precision of the analytical data may allow for a legitimate extension in the shelf life.

5.4. Extensive Use of Fractional Factorial Designs

Both ICH and FDA guidelines provide specific authorization of some reduction in stability testing by use of bracketing designs. However, this is probably only a rather timid beginning to a much more extensive use of well-authenticated mathematical methods that can substantially reduce the amount of stability testing with minimal effects on the level of stability assurance provided by the experimental results (13).

6. INTERACTING WITH REGULATORY AGENCIES

6.1. Inspections

Some people in the pharmaceutical industry tend to regard visits by investigators from regulatory agencies as similar to visits to a dentist—time wasting at best, remarkably unpleasant at worst. Although this attitude is perhaps understandable on some occasions, it is highly desirable that, whenever possible, we approach visits to our facilities by investigators from regulatory agencies in a positive manner. Our own company self-audit and other self-inspection procedures should be so comprehensive and rigorous in nature that we should be able to contemplate visits by investigators from regulatory agencies with a quite high degree of confidence and even serenity. We should be prepared to respond to any reasonable comment or question. Of course, unfortunately, there may be occasions when an investigator from a regulatory agency raises what we perceive to be a quite unreasonable issue. In the event of such an occurrence, we still must keep our cool, respond in a factual, logical, professional, nonadversarial manner. This is not always easy, since it would be difficult to put all investigators from regulatory agencies into the category of saints. Indeed, at some times we might be tempted to use an epithet of quite the reverse meaning for certain investigators. However, it is the clear responsibility of persons from the pharmaceutical industry to avoid, whenever possible, all personal animosity and to be cool, calm, and collected.

It has been my experience in dealing with FDA personnel that they rarely have forked tails. Most are hard-working individuals, simply trying to do their jobs. If we approach a visit by an investigator from a regulatory agency in a well-prepared, positive frame of mind, rather than a negative, adversarial one, we have started out in the right direction.

The company should, of course, have a well-designed, unambiguous SOP (standard operating procedure) for inspections by personnel from regulatory agencies. This should cover all aspects of the type of visit, including such matters as continuous escort by a person of the same gender throughout the visit and how to respond to questions during the visit or written reports left at the end of the visit. In the past, it was quite common for investigators to direct their questions mainly, if not exclusively, to the company escort or some senior company employee. It now seems to be increasingly common for some investigators to ask questions of any company employee. It is therefore important that all employees should be clearly instructed in how to respond to questions from an investigator from a regulatory agency. The Golden Rules are:

Answer honestly.
Respond only to the specific points raised in the question.
If you don't know the answer, say so.

If, at the end of an inspection, a 483 (or some comparable document) is distributed, a well-prepared written response should be issued as soon as possible.

Hynes (14) has recently edited a book on FDA Pre-Approval Inspections.

6.2. Submissions

Data pertaining to stability is often an important and quite substantial part of a regulatory submission document, such as an NDA or ANDA. It is suggested that some of the more important points to consider for a market approval submission document are as follows.

The document should be well prepared, easy to read, of simple linguistic style, and free of ambiguity and error.

Those preparing the document should be fully conversant with all the latest information about the policies and procedures of the regulatory agency to which the application is submitted.

Often particularly true for stability data is the old adage that a picture can do the work of a thousand words. Thus appropriate figures can very often supplement tabular data and give the reader a rapid insight into the overall thrust of the results.

If there are any marginal, aberrant, or unusual data, this occurrence should be commented on. It is not appropriate simply to hope that the plethora of data will so overwhelm the reader that they will not notice a problem. This tactic might work on some occasions, but it is more likely simply to cause annoyance and delay.

6.3. Meetings

There are occasions in stability work when there is no substitute for a face-to-face meeting with appropriate officials at regulatory agencies. The decision to seek such a meeting should not be taken lightly, and the topic or topics for the meeting should be considered very carefully.

When it is decided that a meeting is necessary, it is important to plan very carefully so that the maximum value can be obtained. In general, we should only request a meeting at a regulatory agency concerning stability matters when

1. The problem is an atypical one not covered by Guidelines or other documents
2. There is a serious and continuing personnel problem with an individual regulatory agency employee
3. There is a substantial chance of legal action involving the regulatory agency

It is imperative that the company clearly inform the regulatory agency in advance of the meeting of the topic or topics for discussion at the meeting, and the names and roles of the persons whom the company plans to bring to the meeting. It is often helpful to provide some succinct background data on the agenda items.

At the meeting, the company shall present a lucid account of the problem an outline how it proposes to solve the problem. It is normally inappropriate and imprudent to ask regulators what to do. During the meeting, some member of the company team, which should be kept as small as possible, should take notes. Subsequent to the meeting, a brief letter should be sent to the senior official of the regulatory agency who was present at the meeting summarizing what the company believes was the consensus reached at the meeting.

7. THE NEW FDA GUIDANCE

7.1. Interaction of Science and Regulation in Stability Testing

Pharmaceutical scientists involved in stability testing obviously must be knowledgeable of the principles of chemistry and pharmaceutics that underlie an understanding of the degradation of drug products. Equally true, of course, is the statement that those of us who work for the pharmaceutical industry must be aware of, and comply with, all relevant regulations and other official directives.

Because the pharmaceutical industry is pervasively regulated, there is a danger that some pharmaceutical scientists, working in stability testing, may become so overawed by official guidance that they relinquish all scientific judgment of their own and simply try to follow as slavishly as possible any whisper from a regulatory agency. This is most undesirable. There must never be any occasion when a regulatory guidance become an excuse for not using good scientific judgement. (The comments on guidelines provided by Mr. Dean in his Packaging chapter are most appropriate to this topic.)

7.2. The Functions of a Guidance

A guidance should be designed and used for exactly what the name implies. It should be a document that clearly defines the general principles that should be followed in the subject area to which it is addressed. It should provide sufficient detail so that the scientist to whom it is directed should be able to understand and apply the basic principles that underlie its structure. However, it is not possible for the authors of a guidance—although they may be most knowledgeable and experienced and possessed of considerable insight—to prepare a document that can reasonably be expected to deal with all possible problems that may arise in the field of interest. Thus, recognizing that it is not possible to forecast in advance all conceivable situations that may develop, there must be some wiggle-room in the interpretation and application of any guidance. The question then arises as to how much detail is it prudent for a guidance to provide. If we give too little information we run the risk of having more than one possible interpretation of how the guidance should be applied in a particular situation. If, however, we provide a plethora of detail there is a risk that our guidance may become a straightjacket, inhibiting scientific judgment and making the resolution of abnormal problems more difficult than need be.

There appears to be a significant difference in the approach used in the preparation of a guidance, between those that originate in the European Union, and those that are developed in the United States. The tradition in the United States is to develop guidances that are comprehensive and quite detailed. In Europe, by contrast, guidances tend to be substantially shorter and give more emphasis to general principles than to detailed instructions.

7.3. Development of the FDA Stability Guidance

In 1987 FDA issued the *Guideline for Submitting Documentation for the Stability of Human Drugs and Biologicals*. This document was of great significance in the evolution of the regulation of the stability testing of pharmaceuticals. For a number of years, industry waited expectantly for a new guideline. There were delays caused, I believe, primarily by the negotiations conducted under the auspices of the ICH.

In June 1998 the new, eagerly-awaited Draft Guidance *Stability Testing of Drug Substances and Products* was issued by the FDA. The comment period stated in the June 1998 Draft Guidance was ninety days. Thus it was expected that the period for comment would end on 31 August 1998. However, there was considerable dismay, expressed by many in the pharmaceutical industry, that the comment period was only ninety days for a document comprising 3,349 lines.

The FDA yielded to the arguments put to it concerning the period allowed for public comment and extended the deadline to 31 December 1999. As a result of this and other delays the date for the guidance to become official showed progressive slippage. Thus in June 1998 there were FDA watchers who were relatively confident that the new FDA guidance would become official in June 1999. At the time of writing (July 1999) it now seems improbable that the guidance will become official in its final form before June 2000. Enclosed as an appendix to this book is the June 1999 FDA Draft Guidance. The reader should be aware that the Agency has clearly labeled the document a draft—not for implementation. However, it is this writer's opinion, based on discussions with knowledgeable persons in the pharmaceutical area, that truly radical changes in the guidance are not likely.

In October 1999 ICH will meet in Washington DC. It is believed that, although there are some thorny issues awaiting resolution, it is highly probable that agreement will ultimately be reached. However, further delay to the issuance of the final version of the FDA guidance is not impossible.

8. COMMENTS ON SOME SPECIFIC ASPECTS OF THE GUIDANCE

Since the guidance is generally very clearly written, it would be redundant to give a line-by-line exegesis. In this section attention is only directed to what may be considered to be the more salient features of the guidance.

8.1. Introduction

The introduction makes it very clear that the FDA guidance is based on, and fully compatible with, ICH Guidelines concerning stability testing of pharmaceuticals. The guidance gives copious reference to ICH policies.

The FDA points out that applicants may chose procedures other than those specified in the guidance, but applicants are encouraged to discuss such action with the FDA before departing from the procedures given in the guidance.

8.2. Long-Term/Accelerated Testing Conditions

Table 1 of the guidance specifies temperature and relative humidity values to be used for long-term/accelerated testing. It will be seen that the temperatures defined are ±2°C and relative humidity values (RH) are ±5%. With presently available storage chambers, temperature control to ±0.5% is often quite readily obtained. However, it is not always easy to maintain RH values to within the required range.

8.3. Switching to ICH Storage Conditions

Manufacturers of pharmaceutical products, approved by the FDA for marketing in the United States are placed in somewhat of a dilemma by the relatively recent acceptance by the relevant U.S. regulatory authorities (FDA and USP) of ICH defined storage conditions. Such manufacturers will, in many instances, still be marketing products for which the on-going stability testing commitment was based on non-ICH conditions.

The guidance provides two possible plans, either of which would be acceptable to the FDA. Clearly, Plan B, which involves a side-by-side comparison of stability under both the original and the new ICH storage conditions is more expensive. However, it is also the more reliable method in terms of obtaining data relevant to the effect of the change in storage conditions. Thus, if there is any doubt about which option to select it may well be prudent to select Plan B.

8.4. Stability Testing for Abbreviated New Drug Applications

Those pharmaceutical scientists employed by generic pharmaceutical companies are likely to examine with particular attention the section on ANDA-type products. A number of pharmaceutical companies active in the development and manufacture of generic pharmaceutical products have expressed considerable concern when the ICH stability documents were published that there was a noteworthy absence of specific advice on stability studies for generic products. The FDA guidance does directly address this topic. Allowance is made depending on the availability of significant information and the complexity of the product, for a reduction in the amount of stability data required for an ANDA as compared to an NDA.

The FDA classifies dosage forms that are the subject of ANDAs as "simple" or "not simple." For simple dosage forms the FDA allows a tentative expiration date of up to 24 months based on satisfactory accreted test data at 0, 1, 2, and 3 months (for products to be stored at controlled room temperature, 25°C, the accreted conditions are normally 40°C and 75% RH). However, for ANDA products that do not meet the requirements that would allow them to be classified as "simple," the requirements are less liberal. This has caused considerable concern to some generic companies.

8.5. Clinical Trials Stability (INDs)

The guidance provides advice on the amount of stability data needed for IND clinical trails product. In general, the required data is quite limited.

8.6. Sampling Times

The guidance makes it clear that the normal sampling schedule for retained stability samples is every three months during the first year, every six months during the second year, and annually thereafter. The guidance makes it clear that freezing of samples for the convenience of scheduling analysis is not acceptable.

8.7. Bracketing

Bracketing is given quite extensive treatment. The treatment is helpful and provides most useful advice including, for example, the statement that bracketing should not be used for clinical trials materials.

8.8. Matrixing

The matrixing section in the guidance is of great importance. It is the present author's opinion that of all the sections of the guidance the discussion of matrixing is likely to be the one that many readers will examine more than once. Matrixing offers the potential of allowing a very considerable reducing in the costs associated with stability testing. Of course, the scientific principles that underlie matrixing are well established. Fractional factorial design has been used by pharmaceutical and other scientists for many years. The ICH guidelines do allow matrixing. However, because the official recognition of matrixing is so recent, there is considerable uncertainty to what extent the technique may be applied. That is where the FDA guidance is so useful. It provides considerable specific information on how matrixing may be applied.

Although the FDA guidance provides considerable detail, it is prudent to consult with the FDA before any matrixing stability design is put into operation. Although there is probably going to be considerable flexibility in acceptable matrixing design, the FDA has indicated what factors should not be matrixed. In general a matrixing design is more likely to be accepted for products with good stability and low variability rather than for those with poor stability and large variability.

8.9. Thermal Cycling

The FDA guidance makes specific recommendations about thermal cycling tests. Not all pharmaceutical scientists will find the recommendations given in the guidance to be in line with current practice. Many of us would recommend a 24 hour cycling period with maximum and minimum temperatures selected as may be appropriate for the product. For many products a −4 to 40°C or +4 to 40°C range might be chosen. Also the practice of extending the rest to 2 or 3 weeks (i.e. 14 or 21 cycles) does seem to have merit.

8.10. Biotechnology Drug Products

The FDA guidance contains a most useful section on stability testing of biotechnology drug products. This section, which may well have been prepared originally by separate authors from those involved in the other sections of the document, gives useful, and in many cases quite detailed, advice on many aspects of the stability testing of macromolecules. The guidance frankly addresses the several significant problems that face those of us who work in this field. Stability studies of proteins can be both fascinating and frustrating. The complex nature and hypersensitivity of many biotechnology drugs make demands on pharmaceutical scientists that are often very difficult to respond to effectively. Fortunately the unit of the FDA that deals with stability issues for many macromelecular drugs, the Center of Biologics Evaluation and Research, is staffed by many knowledgeable scientist/regulators, who often display considerable flexibility in their review of stability data for biotechnology drugs.

As the guidance points out, primary data in support of the shelf life assignment for biotechnology drug must be based on long-term, real-time conditions. Also, even for well characterized proteins it is highly unlikely that any one analytical technique will be sufficient to characterize fully the stability of these products.

The guidance points out that for biotechnology drugs, purity is often a relative rather than an absolute term. Also, the guidance reminds us that it is highly desirable to use appropriate reference materials in our stability studies. Of course, if we are developing a new chemical entity that has not previously been used as a drug, it is highly unlikely that there will be a national or international reference standard available. It is thus extremely useful that the potential sponsor of a regulatory application should develop an in-house reference standard as early as possible.

Since many proteins are hypersensitive to temperature stress it may not be feasible to conduct stability studies at temperatures other than that at which the product is to be stored. Also, since many biotechnology drug products are packaged in glass, humidity stress studies may not be necessary.

8.11. Guidance on Stability Studies for Specific Dosage Forms

The FDA guidance provides specific advice on what particular attributes should normally be measured for different dosage forms. The systems to which attention is directed include

- Tablets
- Capsules
- Emulsions
- Oral solutions and suspensions
- Oral powders for reconstitutions
- Metered-dose inhalations and nasal aerosols
- Inhalation solutions and powders
- Nasal sprays: solutions and suspensions
- Topical, ophthalmic, and otic preparations
- Transdermals
- Suppositories
- Small-volume parenterals
- Large-volume parenterals
- Drug additives
- Implantable subdermal, vaginal, and intrauterine devices that deliver drug products

The information provided in each of these sections is most useful and should be carefully examined by all those concerned with products that fall into any of the above categories.

8.12. Stability Testing for Post-Approval Changes

FDA has directed a good deal of time and energy to consideration of the amount of new stability data that should be generated when post-approval changes are made. The changes include those that involve the manufacturing process for the drug substance, manufacturing site, testing laboratory, process, site or equipment used in the fabrication of the drug delivery system or change in batch size.

It seems to the author of this chapter that regulatory authorities in Europe and Japan have probably given much less attention than the FDA to the implications for stability testing that derive from post-approval changes. The FDA proposals in this area have been the subject of lively—in some instances quite vehement—criticism. However, even those who are presently unable to accept the rationale for all aspects of these proposals should be willing to commend the FDA for trying to clarify the situation in this important area.

REFERENCES

1. Pharmaceutical Contract Support Organizations Register, Published by the Drug Information Association, Ambler, PA, 1996.
2. MK Kottke. Drug Development and Industrial Pharmacy 24:1177–1196, 1998.
3. L Underhill, RC Rhodes III, CT Rhodes. Drug Development and Industrial Pharmacy 25:67–70, 1999.
4. CC Okeke, LC Bailey, T Medwick, LT Grady. P.F. 23:4155–4182, 1997.
5. LC Bailey, T Medwick. P.F. 24:5643–5644, 1998.
6. B Kommanboyina, CT Rhodes. Drug Development and Industrial Pharmacy 25:857–867, 1999.
7. B Kommanboyina, CT Rhodes. Drug Development and Industrial Pharmacy, 25:1301–1306.
8. B Kommanboyina, RF Lindauer, CT Rhodes, TL Grady. International Journal of Compounding, accepted for publication.
9. CT Rhodes. Clin. Res. and Drug Regulatory Affairs 10:65–69, 1993.
10. JT Carstensen, CT Rhodes. Drug Development and Industrial Pharmacy 10:177–185, 1993.
11. KM Morisseau, CT Rhodes. Pharm. Tech. (Tableting Yearbook) 6–12, 1997.
12. KM Morisseau, CT Rhodes. Encyclopedia of Pharmaceutical Technology 19:357–370, 2000. Published by Marcel Dekker, Inc.
13. GA Lewis, D Mathieu, R Pham-Tam-Luu, eds. Pharmaceutical Experimental Design. New York: Marcel Dekker, 1999.
14. MD Hynes, III, ed. Preparing for FDA Pre-Approval Inspections. New York: Marcel Dekker, 1999.

Guidance for Industry

Stability Testing of Drug Substances and Drug Products

DRAFT GUIDANCE

This guidance document is being distributed for comment purposes only.

Comments and suggestions regarding this draft document should be submitted within 90 days of publication in the *Federal Register* of the notice announcing the availability of the draft guidance. Submit comments to Dockets Management Branch (HFA-305), Food and Drug Administration, 12420 Parklawn Dr., rm. 1-23, Rockville, MD 20857. All comments should be identified with the docket number listed in the notice of availability that publishes in the *Federal Register*.

Copies of this draft guidance are available from the Office of Training and Communications, Division of Communications Management, Drug Information Branch, HFD-210, 5600 Fishers Lane, Rockville, MD 20857 (Phone 301-827-4573) or from the Internet at http://www.fda.gov/cder/guidance/index.htm.

Copies also are available from the Office of Communication, Training and Manufacturers Assistance, HFM-40, CBER, FDA, 1401 Rockville Pike, Rockville, MD 20852-1448, or from the Internet at http://www.fda.gov/cber/guidelines.htm. Copies also may be obtained by fax from 1-888-CBERFAX or 301-827-3844 or by mail from the Voice Information System at 800-835-4709 or 301-827-1800.

For questions on the content of the draft document, contact Kenneth Furnkranz (301) 827-5848.

U.S. Department of Health and Human Services
Food and Drug Administration
Center for Drug Evaluation and Research (CDER)
Center for Biologics Evaluation and Research (CBER)
June 1998

J: \GUIDANC 1707DFT9.WPD
5 27 98

DRAFT - Not for Implementation

TABLE OF CONTENTS

DRAFT - Not for Implementation

LIST OF TABLES

DRAFT - Not for Implementation

Draft - Not for Implementation

GUIDANCE FOR INDUSTRY[1]

Stability Testing of Drug Substances and Drug Products

(Due to the length and complexity of this draft document, please identify specific comments by line number.)

1 **I. INTRODUCTION**

2 The guidance is intended to be a comprehensive document that provides information on all
3 aspects of stability data generation and use. It references and incorporates substantial text from
4 the following International Conference on Harmonization (ICH) guidance:

5 • *ICH Harmonized Tripartite Guideline for Stability Testing of New Drug Substances and*
6 *Products, September 23, 1994* [ICH Q1A]
7 • *ICH Guideline for Stability Testing of New Dosage Forms* [ICH Q1C]
8 • *ICH Guideline for Photostability Testing of New Drug Substances and Products* [ICH Q1B]
9 • *ICH Guideline for Stability Testing of Biotechnological/Biological Products* [ICH Q5C].

10 Where text from one of these documents has been incorporated in this guidance, it has been
11 denoted by the use of a reference in square brackets in the beginning of a particular section or at
12 the end of an individual paragraph.

13 The purpose of stability testing is to provide evidence on how the quality of a drug substance or
14 drug product varies with time under the influence of a variety of environmental factors such as
15 temperature, humidity, and light. Stability testing permits the establishment of recommended
16 storage conditions, retest periods, and shelf lives. [ICH Q1A]

17 This guidance provides recommendations regarding the design, conduct and use of stability
18 studies that should be performed to support:

19 • Investigational new drug applications (INDs) (21 CFR 312.23(a)(7)),

---

[1] This guidance has been prepared by the Stability Technical Committee of the Chemistry Manufacturing Controls Coordinating Committee (CMCCC) of the Center for Drug Evaluation and Research (CDER) at the Food and Drug Administration with input from the Center for Biologics Evaluation and Research (CBER). This guidance document represents the Agency's current thinking on stability testing of drug substances and products. It does not create or confer any rights for or on any person, and does not operate to bind FDA or the public. An alternative approach may be used if such an approach satisfies the requirements of the applicable statute, regulations, or both.

Draft - Not for Implementation

20 • New drug applications (NDAs) for both new molecular entities (NMEs) and non-NMEs,
21 • New dosage forms (21 CFR 314.50(d)(1)),
22 • Abbreviated new drug applications (ANDAs) (21 CFR 314.92 - 314.99),
23 • Supplements and annual reports (21 CFR 314.70, and 601.12),
24 • Biologics license application (BLAs) and product license applications (PLAs) (21 CFR 601.2).

25 The principle established in ICH Q1A — that information on stability generated in any one of the
26 three areas of the EU, Japan, and the USA would be mutually acceptable in both of the other two
27 areas — is incorporated in this guidance document. In fact, much of the text of the guidance on
28 drug substances and drug products (Sections II.A. and II.B.) is incorporated directly from the
29 ICH Q1A text.

30 This guidance is intended to replace the *Guideline For Submitting Documentation for the*
31 *Stability of Human Drugs and Biologics*, published in February 1987. It applies to all drug
32 substances and products submitted to the Center for Drug Evaluation and Research (CDER).
33 This guidance also applies to biological products that are included in the scope of the ICH Q5C
34 Stability Annex, *Stability Testing of Biotechnology Drug Products* (July 1996) and all other
35 products submitted to the Center for Biologics Evaluation and Research (CBER).

36 The guidance provides recommendations for the design of stability studies for drug substances
37 and drug products that should result in a statistically acceptable level of confidence for the
38 established retest or expiration dating period for each type of application. The applicant is
39 responsible for confirming the originally established retest and expiration dating periods by
40 continual assessment of stability properties (21 CFR 211.166). Continuing confirmation of these
41 dating periods should be an important consideration in the applicant's stability program.

42 The choice of test conditions defined in this guidance is based on an analysis of the effects of
43 climatic conditions in the EU, Japan, and the USA. The mean kinetic temperature in any region
44 of the world can be derived from climatic data (Grimm, W., Drugs Made in Germany,
45 28:196-202, 1985, and 29:39-47, 1986). [ICH Q1A]

46 The recommendations in this guidance are effective upon publication of the final guidance and
47 should be followed in preparing new applications, resubmissions, and supplements. This guidance
48 represents FDA's current thinking on how the stability section of drug and biologics applications
49 should be prepared. An applicant may choose to use alternative procedures. If an applicant
50 chooses to depart from the recommendations set forth in this guidance, the applicant is
51 encouraged to discuss the matter with FDA prior to initiating studies that may later be determined
52 to be unacceptable.

53 FDA recognizes that the time necessary for applicants to establish new procedures, install, and
54 commission the new temperature and relative humidity-controlled rooms/cabinets, carry out
55 appropriate stability studies on batches of product, and submit the information in an application
56 may prevent some applicants from generating data consistent with the recommendations in the
57 guidance for some time. However, since this guidance represents FDA's current thinking and
58 recommendations regarding stability, submission of data not conforming with this guidance is

Draft - Not for Implementation

59 possible with justification. Applications withdrawn prior to publication of this guidance should
60 not normally have to include stability data in conformance with the guidance upon resubmission.
61 However, if new stability studies are conducted to support the submission, such studies should be
62 conducted as recommended in the guidance.

63 A comprehensive glossary has been included, which contains definitions of the major terms and
64 the origin of the definitions (e.g., ICH, CFR, USP) where applicable.

65 **II. STABILITY TESTING FOR NEW DRUG APPLICATIONS**

66 **A. Drug Substance**

67 Information on the stability of a drug substance under defined storage conditions is an integral
68 part of the systematic approach to stability evaluation. Stress testing helps to determine the
69 intrinsic stability characteristics of a molecule by establishing degradation pathways to identify the
70 likely degradation products and to validate the stability indicating power of the analytical
71 procedures used.

72 Stress testing is conducted to provide data on forced decomposition products and decomposition
73 mechanisms for the drug substance. The severe conditions that may be encountered during
74 distribution can be covered by stress testing of definitive batches of the drug substance. These
75 studies should establish the inherent stability characteristics of the molecule, such as the
76 degradation pathways, and lead to identification of degradation products and hence support the
77 suitability of the proposed analytical procedures. The detailed nature of the studies will depend on
78 the individual drug substance and type of drug product.

79 This testing is likely to be carried out on a single batch of a drug substance. Testing should
80 include the effects of temperatures in 10°C increments above the accelerated temperature test
81 condition (e.g., 50°C, 60°C) and humidity, where appropriate (e.g., 75 percent or greater). In
82 addition, oxidation and photolysis on the drug substance plus its susceptibility to hydrolysis across
83 a wide range of pH values when in solution or suspension should be evaluated. Results from
84 these studies will form an integral part of the information provided to regulatory authorities.
85 Light testing should be an integral part of stress testing. The standard test conditions for
86 photostability are discussed in the ICH Q1B guidance.

87 It is recognized that some degradation pathways can be complex and that under forced conditions,
88 decomposition products may be observed that are unlikely to be formed under accelerated or
89 long-term testing. This information may be useful in developing and validating suitable analytical
90 methods, but it may not always be necessary to examine specifically for all degradation products if
91 it has been demonstrated that in practice these are not formed.

92 Primary stability studies are intended to show that a drug substance will remain within
93 specifications during the retest period if stored under recommended storage conditions. [ICH
94 Q1A].

Draft - Not for Implementation

95 1. Selection of Batches

96 Stability information from accelerated and long-term testing should be provided on at least three
97 batches. Long-term testing should cover a minimum of 12 months' duration on at least three
98 batches at the time of submission. The batches manufactured to a minimum of pilot plant scale
99 should be by the same synthetic route and use a method of manufacture and procedure that
100 simulates the final process to be used on a manufacturing scale. The overall quality of the batches
101 of drug substance placed on stability should be representative of both the quality of the material
102 used in preclinical and clinical studies and the quality of material to be made on a manufacturing
103 scale. Supporting information may be provided using stability data on batches of drug substance
104 made on a laboratory scale. [ICH Q1A]

105 The first three production batches[2] of drug substance manufactured post approval, if not
106 submitted in the original drug application, should be placed on long-term stability studies post
107 approval, using the same stability protocol as in the approved drug application. [ICH Q1A]

108 2. Test Procedures and Test Criteria
109
110 The testing should cover those features susceptible to change during storage and likely to
111 influence quality, safety and/or efficacy. Stability information should cover as necessary the
112 physical, chemical, biological, and microbiological test characteristics. Validated
113 stability-indicating test methods should be applied. The extent of replication will depend on the
114 results of validation studies. [ICH Q1A]

115 3. Specifications
116
117 Limits of acceptability should be derived from the quality profile of the material as used in the
118 preclinical and clinical batches. Specifications will need to include individual and total upper
119 limits for impurities and degradation products, the justification for which should be influenced by
120 the levels observed in material used in preclinical studies and clinical trials. [ICH Q1A]

121 4. Storage Conditions

122 The length of the studies and the storage conditions should be sufficient to cover storage,
123 shipment, and subsequent use. Application of the same storage conditions applied to the drug
124 product will facilitate comparative review and assessment. Other storage conditions are allowable
125 if justified. In particular, temperature-sensitive drug substances should be stored under an
126 alternative lower temperature condition, which will then become the designated long-term testing
127 storage temperature. The 6-month accelerated testing should then be carried out at a temperature
128 at least 15°C above this designated long-term storage temperature (together with the appropriate

---

[2] The terms *production batch* and *manufacturing scale production batch* are used interchangeably throughout this guidance to mean a batch of drug substance or drug product manufactured at the scale typically encountered in a facility intended for marketing production.

Draft - Not for Implementation

129 relative humidity conditions for that temperature). The designated long-term testing conditions
130 will be reflected in the labeling and retest date. [ICH Q1A]

131 Where *significant change* occurs during 6 months of storage under conditions of accelerated
132 testing at 40°C ± 2°C/75% RH± 5%, additional testing at an intermediate condition (such as
133 30°C± 2°C/60% RH± 5%) should be conducted for a drug substance to be used in the
134 manufacture of a dosage form tested for long-term at 25°C ± 2°C/60% RH ± 5% and this
135 information should be included in the drug application.[3] The initial drug application should
136 include at the intermediate storage condition a minimum of 6 months of data from a 12-month
137 study. [ICH Q1A]

138 *Significant change* at 40°C/75% RH or 30°C/60% RH is defined as failure to meet the
139 specifications.[ICH Q1A] If any parameter fails *significant change* criteria during the
140 accelerated stability study, testing of all parameters during the intermediate stability study should
141 be performed.

142 If stability samples have been put into the intermediate condition, but have not been tested, these
143 samples may be tested as soon as the accelerated study shows significant change in the drug
144 substance. Alternatively, the study at the intermediate condition would be started from the initial
145 time point.

146 Where a *significant change* occurs during 12 months of storage at 30°C/60%RH, it may not be
147 appropriate to label the drug substance for controlled room temperature (CRT) storage with the
148 proposed retest period even if the stability data from the full long-term studies at 25°C/60%RH
149 appear satisfactory. In such cases, alternate approaches, such as qualifying higher acceptance
150 criteria for a degradant, shorter retest period, refrigerator temperature storage, or more protective
151 container and/or closure, should be considered during drug development.

152 The long-term testing should be continued for a sufficient period of time beyond 12 months to
153 cover all appropriate retest periods, and the further accumulated data can be submitted to the
154 FDA during the assessment period of the drug application. [ICH Q1A]

155 The data (from accelerated testing and/or from testing at an intermediate storage condition) may
156 be used to evaluate the impact of short-term excursions outside the label storage conditions such
157 as might occur during shipping. [ICH Q1A]

158 5. Testing Frequency

---

[3]The equipment must be capable of controlling temperature to a range of ± 2 °C and relative humidity to ± 5% RH. The actual temperatures and humidities should be monitored during stability storage. Short-term spikes due to opening of doors of the storage facility are accepted as unavoidable. The effect of excursions due to equipment failure should be addressed by the applicant and reported if judged to impact stability results. Excursions that exceed these ranges (i.e., ± 2 °C and/or ± 5% RH) for more than 24 hours should be described and their impact assessed in the study report.

Draft - Not for Implementation

159 Frequency of testing should be sufficient to establish the stability characteristics of the drug
160 substance. Testing under the defined long-term conditions will normally be every 3 months over
161 the first year, every 6 months over the second year, and then annually. [ICH Q1A]

162 6. Packaging /Containers

163 The containers to be used in the long-term, real-time stability evaluation should be the same as or
164 simulate the actual packaging used for storage and distribution. [ICH Q1A]

165 7. Evaluation

166 The design of the stability study is to establish a retest period applicable to all future batches of
167 the bulk drug substance manufactured under similar circumstances, based on testing a minimum of
168 three batches of the drug substance and evaluating the stability information (covering as necessary
169 the physical, chemical, and microbiological test characteristics). The degree of variability of
170 individual batches affects the confidence that a future production batch will remain within
171 specifications until the retest date. [ICH Q1A]
172
173 An acceptable approach for quantitative characteristics that are expected to decrease with time is
174 to determine the time at which the 95 percent one-sided confidence limit for the mean degradation
175 curve intersects the acceptable lower specification limit. If analysis shows that the batch to batch
176 variability is small, it is advantageous to combine the data into one overall estimate, and this can
177 be done by first applying appropriate statistical tests (for example, p values for level of
178 significance of rejection of more than 0.25) to the slopes of the regression lines and zero time
179 intercepts for the individual batches. If it is inappropriate to combine data from several batches,
180 the overall retest period may depend on the minimum time a batch may be expected to remain
181 within acceptable and justified limits. [ICH Q1A]

182 The nature of any degradation relationship will determine the need for transformation of the data
183 for linear regression analysis. Usually the relationship can be represented by a linear, quadratic, or
184 cubic function on an arithmetic or logarithmic scale. Statistical methods should be employed to
185 test the goodness of fit of the data on all batches and combined batches (where appropriate) to the
186 assumed degradation line or curve. [ICH Q1A]

187 The data may show so little degradation and so little variability that it is apparent from looking at
188 the data that the requested retest period will be granted. Under the circumstances, it is normally
189 unnecessary to go through the formal statistical analysis; providing a full justification for the
190 omission is usually sufficient. [ICH Q1A]

191 Limited extrapolation may be undertaken of the real-time data beyond the observed range to
192 extend retest period at approval time, particularly where the accelerated data support this.
193 However, this assumes that the same degradation relationship will continue to apply beyond the
194 observed data, and hence the use of extrapolation must be justified in each application in terms of
195 what is known about such factors as the mechanism of degradation, the goodness of fit of any
196 mathematical model, batch size, and existence of supportive data. Any evaluation should cover

Draft - Not for Implementation

197 not only the assay, but the levels of degradation products and other appropriate attributes. [ICH
198 Q1A]

199 8. Statements/Labeling

200 A storage temperature range may be used in accordance with relevant national/regional
201 requirements. The range should be based on the stability evaluation of the drug substance.
202 Where applicable, specific requirements should be stated, particularly for drug substances that
203 cannot tolerate freezing. The use of terms such as *ambient conditions* or *room temperature* is
204 unacceptable. [ICH Q1A]

205 A retest period should be derived from the stability information. [ICH Q1A]

206 **B. Drug Product**
207
208 1. General

209 The design of the stability protocol for the drug product should be based on the knowledge
210 obtained on the behavior, properties, and stability of the drug substance and the experience gained
211 from clinical formulation studies. The changes likely to occur upon storage and the rationale for
212 the selection of drug product parameters to be monitored should be stated. [ICH Q1A]

213 2. Selection of Batches
214
215 Stability information from accelerated and long-term testing is to be provided on three batches of
216 the same formulation of the dosage form in the container and closure proposed for marketing.
217 Two of the three batches should be at least pilot scale. The third batch may be smaller (e.g.,
218 25,000 to 50,000 tablets or capsules for solid oral dosage forms). The long-term testing should
219 cover at least 12 months' duration at the time of submission. The manufacturing process to be
220 used should meaningfully simulate that to be applied to large-scale production batches for
221 marketing. The process should provide product of the same quality intended for marketing, and
222 meeting the same quality specification to be applied for release of material. Where possible,
223 batches of the finished product should be manufactured using identifiably different batches of the
224 drug substance. [ICH Q1A]

225 Data on laboratory-scale batches are not acceptable as primary stability information. Data on
226 associated formulations or packaging may be submitted as supportive information. The first three
227 production batches manufactured post approval, if not submitted in the original application,
228 should be placed on accelerated and long-term stability studies using the same stability protocols
229 as in the approved drug application. [ICH Q1A]

230 3. Test Procedures and Test Criteria

231 The test parameters should cover those features susceptible to change during storage and likely
232 to influence quality, safety and/or efficacy. Analytical test procedures should be fully validated
233 and the assays should be stability-indicating. The need for replication will depend on the results

234 of validation studies. [ICH Q1A]

235 The range of testing should cover not only chemical and biological stability, but also loss of
236 preservative, physical properties and characteristics, organoleptic properties, and where required,
237 microbiological attributes. Preservative efficacy testing and assays on stored samples should be
238 carried out to determine the content and efficacy of antimicrobial preservatives. [ICH Q1A]

239 4. Specifications

240 Where applicable, limits of acceptance should relate to the release limits to be derived from
241 consideration of all the available stability information. The shelf-life specifications could allow
242 acceptable and justifiable deviations from the release specifications based on the stability
243 evaluation and the changes observed on storage. They need to include specific upper limits for
244 degradation products, the justification for which should be influenced by the levels observed in
245 material used in preclinical studies and clinical trials. The justification for the limits proposed for
246 certain other tests, such as particle size and/or dissolution rate, will require reference to the results
247 observed for the batch(es) used in bioavailability and/or clinical studies. Any differences between
248 the release and shelf-life specifications for antimicrobial preservatives content should be supported
249 by preservative efficacy testing. [ICH Q1A]

250 5. Storage Test Conditions

251 The length of the studies and the storage conditions should be sufficient to cover storage,
252 shipment and subsequent use (e.g., reconstitution or dilution as recommended in the labeling). See
253 Table 1 below for recommended accelerated and long-term storage conditions and minimum
254 times. Assurance that long-term testing will continue to cover the expected shelf life should be
255 provided. [ICH Q1A]

256 Other storage conditions are allowable if justified. Heat-sensitive drug products should be stored
257 under an alternative lower temperature condition, which will eventually become the designated
258 long-term storage temperature. Special consideration may need to be given to products that
259 change physically or even chemically at lower storage temperatures (e.g., suspensions or
260 emulsions which may sediment, or cream, oils and semi-solid preparations, which may show an
261 increased viscosity). Where a lower temperature condition is used, the 6-month accelerated
262 testing should be carried out at a temperature at least 15°C above its designated long-term
263 storage temperature (together with appropriate relative humidity conditions for that temperature).
264 For example, for a product to be stored long-term under refrigerated conditions, accelerated
265 testing should be conducted at 25°C ± 2°C/60% RH ± 5%. The designated long-term testing
266 conditions will be reflected in the labeling and expiration date. [ICH Q1A]

267 Storage under conditions of high relative humidity applies particularly to solid dosage forms. For
268 drug products such as solutions and suspensions contained in packs designed to provide a
269 permanent barrier to water loss, specific storage under conditions of high relative humidity is not
270 necessary but the same range of temperatures should be applied. Low relative humidity (e.g., 10 -
271 20% RH) can adversely affect products packed in semi-permeable containers (e.g., solutions in

Draft - Not for Implementation

272 plastic bags, nose drops in small plastic containers), and consideration should be given to
273 appropriate testing under such conditions. [ICH Q1A]

274 **Table 1: Long-Term/Accelerated Testing Conditions**

	Conditions	Minimum time period at submission
275 Long-term testing	25°C ± 2°C/60% RH ± 5%	12 Months
276 Accelerated Testing	40°C ± 2°C/75% RH ± 5%	6 Months

277 Where *significant change* occurs due to accelerated testing, additional testing at an intermediate
278 condition (e.g., 30°C± 2°C/60% RH ± 5%) should be conducted. *Significant change* at the
279 accelerated conditions is defined as:

280 1. A 5 percent potency loss from the initial assay value of a batch.
281 2. Any specified degradant exceeding its specification limit.
282 3. The product exceeding its pH limits.
283 4. Dissolution exceeding the specification limits for 12 capsules or tablets (USP Stage 2).
284 5. Failure to meet specifications for appearance and physical properties (e.g., color, phase
285 separation, resuspendability, delivery per actuation, caking, hardness) [ICH Q1A].

286 Should significant change occur at 40°C/75% RH, the initial application should include a
287 minimum of 6 months' data from an ongoing 1-year study at 30°C/60 percent RH; the same
288 significant change criteria shall then apply. [ICH Q1A]

289 If any parameter fails *significant change* criteria during the accelerated stability study, testing of
290 all parameters during the intermediate stability study should be performed.

291 If stability samples have been put into the intermediate condition, but have not been tested, testing
292 these samples may begin as soon as the accelerated study shows significant change in the drug
293 product. Alternatively, the study at the intermediate condition would be started from the initial
294 time point.

295 Where a *significant change* occurs during 12 months of storage at 30°C/60%RH, it may not be
296 appropriate to label the drug product for CRT storage with the proposed expiration dating period
297 even if the stability data from the full long-term studies at 25°C/60%RH appear satisfactory. In
298 such cases, alternate approaches, such as qualifying higher acceptance criteria for a degradant,
299 shorter expiration dating period, refrigerator temperature storage, more protective container
300 and/or closure, modification to the formulation and/or manufacturing process, should be
301 considered during drug development. If CRT storage is ultimately justified, it may be necessary
302 to add to the product labeling a cautionary statement against prolonged exposure at or above
303 30°C.

Draft - Not for Implementation

304 The long-term testing will be continued for a sufficient period of time beyond 12 months to cover
305 shelf life at appropriate test periods. The further accumulated data should be submitted to the
306 FDA during the assessment period of the drug application. [ICH Q1A]

307 The first three production batches manufactured post approval, if not submitted in the original
308 application, should be placed on accelerated and long-term stability studies using the same
309 stability protocol as in the approved drug application. [ICH Q1A] A minimum of 4 test stations
310 (e.g., 0, 2, 4, and 6 months) are recommended for the 6-month accelerated stability study.

311 6. Stability Storage Conditions not Defined in ICH Q1A

312 The stability sample storage conditions for most dosage forms (e.g., solid oral dosage forms,
313 solids for reconstitution, dry and lyophilized powders in glass vials) are defined in Section V.E. of
314 the ICH Q1A Guidance and in Section II.B.5 of this guidance. However, the stability storage
315 conditions are not indicated in ICH Q1A for certain other drug products including those packaged
316 in semi-permeable containers (except for accelerated studies), products intended to be stored
317 under refrigerator or freezer temperatures, or certain studies on metered dose inhalations (MDIs)
318 and dry powder inhalers (DPIs). Further information about these products and containers is
319 provided in this section.

320 a. Stability Storage Conditions for Drug Products in Semi-Permeable and Permeable
321 Containers

322 For large volume parenterals (LVPs), small volume parenterals (SVPs), ophthalmics, otics, and
323 nasal sprays packaged in semi-permeable containers, such as plastic bags, semi-rigid plastic
324 containers, ampules, vials and bottles with or without droppers/applicators, which may be
325 susceptible to water loss, the recommended stability storage conditions are:

326 • Accelerated condition: 40°C ± 2°C/15% RH ± 5% (hereafter referred to as 40°C/15%
327 RH)[ICH Q1A];
328 • Intermediate condition: 30°C ± 2°C/40% RH ± 5% (hereafter referred to as 30°C/40% RH);
329 • Long-term condition: 25°C ± 2°C/40% RH ± 5%

330 For liquids in glass bottles, vials, or sealed glass ampules, which provide an impermeable barrier
331 to water loss,

332 • Accelerated condition: 40°C/ambient humidity is an acceptable alternative to 40°C/75% RH;
333 • Intermediate condition: 30°C/ambient humidity is an acceptable alternative to 30°C/60% RH;
334 • Long-term condition: 25°C/ambient humidity is an acceptable alternative to 25°C/60% RH.

335 b. Stability Storage Conditions for Drug Products Intended to be Stored at
336 Refrigerator Temperature

337 • Accelerated conditions: 25°C/60% RH, with ambient humidity an acceptable alternative for
338 aqueous products that would not be affected by humidity conditions;
339 • Long-term conditions: 5°C ± 3°C, with monitoring, but not control of, humidity.

Draft - Not for Implementation

340 c. Stability Storage Conditions for Drug Products Intended to be Stored at Freezer
341 Temperature

342 • Accelerated conditions: 5°C ± 3°C/ambient humidity;
343 • Long-term conditions: -15°C ± 5°C.

344 d. Stability Storage Conditions for Some Inhalation Products

345 Additional storage conditions may apply to inhalation powders and suspension inhalation aerosols
346 when significant change in aerodynamic particle size distribution or in dose content uniformity
347 occurs at accelerated conditions (40C/75%RH). (The Agency currently is developing a draft
348 guidance to address chemistry, manufacturing, and controls documentation for MDIs and DPIs.)

349 7. Testing Frequency

350 Frequency of testing should be sufficient to establish the stability characteristics of the drug
351 product. Testing will normally be every 3 months over the first year, every 6 months over the
352 second year, and then annually. Matrixing or bracketing can be used, if justified. [ICH Q1A] A
353 minimum of 4 test stations (e.g., 0, 2, 4, and 6 months) are recommended for the 6-month
354 accelerated stability study.
355
356 8. Application of ICH Stability Study Storage Conditions to Approved Applications

357 Although the ICH Guidance for *Stability Testing of New Drug Substances and Products* applies
358 only to new molecular entities and associated drug products, applicants may wish to voluntarily
359 switch to the ICH-recommended storage conditions as defined in ICH Q1A and Sections II.A.4.
360 and II.B.5. of this guidance or other FDA-recommended conditions as described in Section II.B.6.
361 of this guidance, as appropriate, for previously approved drug or biologic products. Applicants
362 are not required to make such a switch for either annual stability batches or batches intended to
363 support supplemental changes. Although the following discussions refer only to the ICH
364 conditions, the same recommendations can be applied when a switch to other FDA-recommended
365 conditions is contemplated.

366 Two plans are presented to assist applicants who desire to switch their approved drug products to
367 the ICH-recommended storage conditions. Under each plan, recommendations will be made on
368 how to initiate a switch to the ICH storage testing conditions, select batches, collect data, report
369 results, and proceed if products fail the approved specifications under the ICH conditions.

370 **a. Plan A: Using the ICH Storage Testing Conditions for an Approved Stability**
371 **Protocol**

372 This plan may be most suitable for drug products that have been confirmed to be stable when
373 exposed to the controlled level of humidity on a long-term basis. Only one set of conditions (i.e.,
374 the ICH conditions) and one set of testing for each of the three verification batches, as defined
375 below, are necessary under this plan.

Draft - Not for Implementation

376 i. Drug Products with an Approved Stability Protocol
377
378 Applicants who have previously performed drug product stability studies under an approved
379 protocol at 25°C, 30°C, or 25-30°C without humidity controls may switch over to the ICH
380 long-term conditions, as defined in V.E. of the ICH Q1A guidance and incorporated in Section
381 II.B. of this guidance, for all of their annual stability studies. A revised stability protocol may be
382 submitted in the annual report, reflecting changes in temperature and humidity to conform with
383 those recommended by the ICH. Any other changes to the stability protocol should be submitted
384 as a prior-approval supplement. Once adopted through an annual report, the ICH conditions
385 should be used to generate stability data for subsequent supplemental changes. Alternatively, the
386 applicant may report the ICH switch in a supplement, which requires stability data, if the
387 supplement occurs before the next scheduled annual report. Data from the first three consecutive
388 annual batches after the switch can be used to verify the previously approved expiration dating
389 period. However, if the applicant wishes to verify product stability under the ICH conditions over
390 a shorter time span, three production batches within one year, instead of three consecutive annual
391 batches, may be studied.

392 ii. Products Without an Approved Stability Protocol

393 Applicants who have previously performed stability studies on a drug product without an
394 approved protocol are required to submit an appropriate protocol under a prior-approval
395 supplement under 21 CFR 314.70(b) or (g) or 601.12(b) (see Section V regarding an Approved
396 Stability Protocol). Upon approval of the protocol, applicants may initiate stability studies on all
397 annual batches under the ICH long-term conditions. Data from the first three consecutive annual
398 batches after the switch can be used to verify the current, or to establish a new, expiration dating
399 period. However, if the applicant wishes to verify product stability under the ICH conditions over
400 a shorter time span, three production batches within one year, instead of three consecutive annual
401 batches, may be studied

402 iii. Stability Data for Supplemental Changes

403 Stability data submitted in support of supplemental changes for an existing drug product may be
404 generated with samples stored at the ICH-recommended accelerated testing conditions, and
405 long-term testing conditions, and, if applicable, intermediate conditions, as described in V.E. of
406 the ICH Q1A guidance (Section II.B. of this guidance) or Section III.B of this guidance.

407 iv. Other Considerations

408 For a moisture-sensitive product, the applicant may wish to explore the possibility of improving
409 the container/closure before embarking on the switch-over to the ICH condition.

410 Although 30°C/60% RH is an acceptable alternative to 25°C/60% RH for long-term studies,
411 these conditions should not be used as the basis for a labeling statement such as "Store at 30°C"
412 or "Store at 15-30°C" to gain marketing advantage.

413 With respect to ongoing stability studies, applicants may carry them to completion under the

Draft - Not for Implementation

414 previously approved conditions or may, for practical or economic reasons, choose to make an
415 immediate switch to ICH conditions and report the change in the next annual report.

416 v. Data Submission to FDA

417 *Satisfactory data*:

418 If the stability data generated on the first three annual batches after the switch to the
419 ICH-recommended long-term testing conditions using an approved protocol, as defined
420 above, support the previously approved expiration dating period under the non-ICH
421 conditions, the data can be submitted in the next annual report, and the current expiration
422 dating period can be retained.

423 *Unsatisfactory data*:

424 If the stability data under the ICH conditions fall outside the specifications established for the
425 previously approved expiration dating period, the applicant should perform an investigation to
426 determine the probable cause of the failure in accordance with CGMP regulations under 21
427 CFR 211.192. Additionally, the applicant should submit an NDA Field-Alert Report in
428 accordance with 21 CFR 314.81(b)(1)(ii) or an error and accident report for a biological
429 product under 21 CFR 600.14. A recall of the corresponding product in the market place may
430 also be necessary. If it is determined that the ICH storage conditions, particularly the added
431 humidity, is the cause for the stability failure, the applicant may shorten the expiration dating
432 period in a changes-being-effected supplement while retaining the ICH storage condition.
433 Subsequently, if justified, the applicant may request an approval for a revision of the product
434 specifications and for reinstating the previously approved expiration dating period under the
435 non-ICH conditions through a prior-approval supplement. Other measures (e.g., more
436 protective container/closure or product reformulation) may be considered through a
437 prior-approval supplement.

438 Alternatively, the applicant may, after careful consideration of all aspects, request for a return
439 to the previous storage conditions in a changes-being-effected supplement if justification,
440 including all related data and investigational results, is provided.

441 **b. Plan B: Using the ICH Conditions under an Alternate Protocol**

442 An alternative to Plan A is to conduct two side-by-side studies by simultaneously placing samples
443 from the same batch of drug product under the ICH conditions as well as the previously approved
444 storage condition. The protocol containing the ICH storage conditions is considered an
445 alternative to the approved protocol. This plan may prove to be particularly useful if the drug
446 product is believed to be moisture-sensitive.
447
448 i. Products with an Approved Stability Protocol
449
450 Applicants may initiate stability studies under the ICH-recommended long-term testing conditions,
451 in addition to the previously approved conditions at 25°C, 30°C, or 25-30°C without humidity

Draft - Not for Implementation

452 controls, for three consecutive annual batches. Data from these annual batches under the ICH
453 conditions should be used to verify the current expiration dating period. However, if the applicant
454 wishes to verify the ICH conditions over a shorter time span, three production batches within one
455 year or less may be selected, instead of three consecutive annual batches.

456 ii. Products without an Approved Stability Protocol

457 Applicants who have previously performed stability studies on a drug product without an
458 approved protocol should submit an appropriate protocol as a prior-approval supplement. This
459 protocol should contain 25°C/ambient humidity as the primary long-term storage testing
460 conditions and the ICH long-term conditions, as the alternative, as well as the IC-recommended
461 accelerated testing conditions. Upon approval of the protocol, applicants may initiate stability
462 studies on three consecutive annual batches at both 25°C/ambient humidity and 25°C/60% RH or
463 25°C/40% RH. Data from these annual batches under the ICH conditions can be used to verify
464 the current, or to establish a new, expiration dating period.

465 iii. Other Considerations

466 Same as in Plan A.

467 iv. Protocol Revisions
468
469 *Products with an approved stability protocol*:

470 Applicants who have an approved stability protocol may submit the alternate stability protocol
471 in the annual report, reflecting the temperature and humidity as recommended by the ICH.
472 Other changes to the stability protocol generally should be submitted in a prior-approval
473 supplement, unless the changes are to comply with the current compendium.

474 Once adopted as an alternate protocol through an annual report, the ICH conditions can be
475 used, in parallel with the previously approved conditions, to generate stability data for
476 subsequent supplemental changes. Alternatively, the applicant may report the alternative ICH
477 conditions in a supplement, which requires stability data, if the supplement occurs before the
478 next scheduled annual report.

479 If the complete stability data generated on the first three annual batches under the ICH
480 long-term conditions using an approved alternate protocol (as defined above) support the
481 previously approved expiration dating period under the non-ICH conditions, the alternate
482 stability protocol can be adopted as the primary stability protocol through an annual report.

483 *Products without an approved stability protocol*:

484 For applications that do not contain an approved stability protocol as defined above, a new or
485 revised stability protocol may be submitted in a prior-approval supplement marked *expedited*
486 *review requested*. This protocol should encompass 25°C/ambient humidity as the primary
487 long-term storage conditions and the ICH long-term conditions, as the alternate, as well as

Draft - Not for Implementation

488 accelerated stability storage conditions, as defined by the ICH Guidance and above, and other
489 recommendations described in this guidance. Upon approval of the protocol, stability studies
490 may be initiated on annual batches and batches intended to support supplemental changes.

491 v. Stability Data for Supplemental Changes

492 Applicants may provide stability data in support of postapproval supplemental changes with
493 samples stored at the ICH-recommended accelerated testing conditions and long-term testing
494 conditions, both previously approved and ICH, as well as, if applicable, intermediate conditions.
495 See Change in Stability Protocol (Section IX.J.) for the recommended filing mechanism.

496 vi. Data Submission

497 *Satisfactory data*:

498 If the complete stability data generated on the first three annual batches under the ICH
499 long-term conditions using an approved alternate protocol support the previously approved
500 expiration dating period under the non-ICH conditions, the data can be submitted in the
501 annual report and the current expiration dating period can be retained.

502 *Unsatisfactory data*

503 If the stability data under the ICH conditions fall outside the acceptance criteria while data
504 from the parallel study under the previously approved conditions or 25°C/ambient humidity,
505 whichever applies, are satisfactory during the previously approved expiration dating period,
506 and the added humidity is determined to be the cause for the stability failure, the product will
507 still be considered to be in compliance with the regulatory specifications approved in the
508 application. If the applicant decides to adopt the ICH conditions, a changes-being-effected
509 supplement with shortened expiration dating period or a prior-approval supplement with
510 revised product specifications may be submitted where justified. Other measures (e.g., more
511 protective container/closure or product reformulation) may be considered through a
512 prior-approval supplement.

513 Alternatively, after careful consideration of all aspects, the applicant may decide not to pursue
514 the switch-over to the ICH conditions for the product. The applicant may eliminate the
515 alternate stability protocol in the next annual report if a full explanation, including all related
516 data and investigational results, is provided.

517 In the case where the product fails to meet the specifications under the non-ICH conditions,
518 irrespective of whether it also fails under the ICH conditions, a thorough investigation in
519 accordance with CGMP should be performed and appropriate corrective actions should be
520 taken, including a Field-Alert Report and recall of the affected product from the market place
521 if warranted.

522 9. Packaging Materials [ICH Q1A]

Draft - Not for Implementation

523 The testing should be carried out in the final packaging proposed for marketing. Additional
524 testing of the unprotected drug product can form a useful part of the stress testing and package
525 evaluation, as can studies carried out in other related packaging materials in supporting the
526 definitive pack(s).

527 10. Evaluation [ICH Q1A]

528 A systematic approach should be adopted in the presentation of the evaluation of the stability
529 information, which should cover, as necessary, physical, chemical, biological and microbiological
530 quality characteristics, including particular properties of the dosage form (for example, dissolution
531 rate for oral solid dose forms).

532 The design of the stability study is to establish a shelf-life and label storage instructions applicable
533 to all future batches of the dosage form manufactured and packed under similar circumstances
534 based on testing a minimum of three batches of the drug product. The degree of variability of
535 individual batches affects the confidence that a future production batch will remain within
536 specifications until the expiration date.

537 An acceptable approach for quantitative characteristics that are expected to decrease with time is
538 to determine the time at which the 95 percent one-sided confidence limit for the mean degradation
539 curve intersects the acceptable lower specification limit. If analysis shows that the batch-to-batch
540 variability is small, it may be advantageous to combine the data into one overall estimate by first
541 applying appropriate statistical tests (e.g., p values for level of significance of rejection of more
542 than 0.25) to the slopes of the regression lines and zero time intercepts for the individual batches.
543 If combining data from several batches is inappropriate, the overall retest period may depend on
544 the minimum time a batch may be expected to remain within acceptable and justified limits.

545 The nature of the degradation relationship will determine the need for transformation of the data
546 for linear regression analysis. Usually the relationship can be represented by a linear, quadratic, or
547 cubic function of an arithmetic or logarithmic scale. Statistical methods should be employed to
548 test the goodness of fit of the data on all batches and combined batches (where appropriate) to the
549 assumed degradation line or curve.

550 Where the data show so little degradation and so little variability that it is apparent from looking
551 at the data that the requested shelf life will be granted, it is normally unnecessary to go through
552 the formal statistical analysis; but a justification for the omission should be provided.

553 Limited extrapolation may be taken of the real-time data beyond the observed range to extend
554 expiration dating at approval time, particularly where the accelerated data support this. However,
555 this assumes that the same degradation relationship will continue to apply beyond the observed
556 data, and hence the use of extrapolation must be justified in each application in terms of what is
557 known about such factors as the mechanism of degradation, the goodness of fit of any
558 mathematical model, batch size, and existence of supportive data.

559 Any evaluation should cover not only the assay, but also the levels of degradation products and

Draft - Not for Implementation

560 appropriate attributes. Where appropriate, attention should be paid to reviewing the adequacy of
561 the mass balance, different stability, and degradation performance.

562 The stability of the drug product after reconstituting or diluting according to labeling should be
563 addressed to provide appropriate and supportive information.

564 See Section VIII.N. for additional information on drug products which are reconstituted or
565 diluted.

566 11. Statements/Labeling

567 A storage temperature range may be used in accordance with FDA regulations. The range should
568 be based on the stability evaluation of the drug product. Where applicable, specific requirements
569 should be stated, particularly for drug products that cannot tolerate freezing.

570 The use of terms such as ***ambient conditions*** or ***room temperature*** is unacceptable.

571 There should be a direct linkage between the label statement and the demonstrated stability
572 characteristics of the drug product.

573 A single set of uniform storage statements (USSs) for NDAs, ANDAs, PLAs and BLAs is
574 recommended to avoid different labeling storage statements for products stored under controlled
575 room temperature conditions. The storage statements and storage conditions provided in this
576 section of the guidance are intended to be standardized and harmonized with the CRT definition
577 in the USP and the recommendations in the ICH guidance.

578 a. Room Temperature Storage Statements
579
580 i. Liquid Dosage Forms in Semi-Permeable Containers

581 The recommended storage statement for LVPs, SVPs, ophthalmics, otics and nasal sprays
582 packaged in semi-permeable containers, such as plastic bags, semi-rigid plastic containers,
583 ampules, vials and bottles with or without droppers/applicators, that may be susceptible to water
584 loss but have been demonstrated to be stable at 25°C ± 2°C/40% or 60% RH ± 5% (or 30°C ±
585 2°C/40% or 60% RH ± 5%); at 25°C/NMT 40% or 30°C/NMT 40% RH; or 30°C, 25-30°C, or
586 25°C without humidity controls, is:

587 **Store at 25°C (77°F); excursions permitted to 15-30°C (59-86°F)**
588 [see USP Controlled Room Temperature]

589 For sterile water for injection (WFI) and LVP solutions of inorganic salts packaged in
590 semi-permeable containers (e.g., plastic bags) the following statement may be used on the
591 immediate container labels:

Draft - Not for Implementation

592 **Store at 25°C (77°F); excursions permitted to 15-30°C (59-86°F)**
593 [see USP Controlled Room Temperature]
594 (see insert for further information)

595 and the following statement may be used in the "How Supplied" section of the package insert:

596 **Store at 25°C (77°F); excursions permitted to 15-30°C (59-86°F)**
597 [see USP Controlled Room Temperature]
598 **Brief exposure to temperatures up to 40°C/104°F may be tolerated provided the**
599 **mean kinetic temperature does not exceed 25°C (77°F).**
600 **However, such exposure should be minimized.**

601 LVP solutions packaged in a semi-permeable container (e.g., a plastic bag) and containing simple
602 organic salts (e.g.,acetate, citrate, gluconate, and lactate, and dextrose 10 percent or less) may be
603 labeled as above, provided there are adequate stability data (at least 3 months' at 40°C ±
604 2°C/15% RH± 5% or 40°C/NMT 20% RH) to support such labeling.
605
606 ii. All Other Dosage Forms

607 For all other dosage forms (e.g., solid oral dosage forms, dry powders, aqueous liquid, semi-solid
608 and suspension dosage forms) that have been demonstrated to be stable at the ICH-recommended
609 conditions (25°C ± 2°C/60% RH ± 5%, or 30°C/60% RH ± 5%) or at non-ICH conditions, such
610 as 30°C, 25-30°C, or 25°C without humidity controls and intended to be stored at room
611 temperature, the recommended labeling statement is:

612 **Store at 25°C (77°F); excursions permitted to 15-30°C (59-86°F)**
613 [see USP Controlled Room Temperature]

614 iii. Where Space on the Immediate Container is Limited

615 Where an abbreviated labeling statement is necessary because space on the immediate container is
616 limited, either of the following statements is acceptable provided the full labeling statement, as
617 shown above, appears on the outer carton and in the package insert:

618 **Store at 25°C (77°F); excursions 15-30°C (59-86°F)**
619 **Store at 25°C (77°F) (see insert)**

620 b. Refrigerator Storage Statement

621 For a drug product demonstrated to be stable at 5°C ± 3°C, 2-5°C, or 2-8°C with or without

Draft - Not for Implementation

622 humidity control and which is intended to be stored at refrigerator temperature, the recommended
623 storage statement for labeling may be one of the following:

624 **Store in a refrigerator, 2-8°C (36-46°F)**
625 **Store refrigerated, 2-8°C (36-46°F)**

626
627 Where an abbreviated labeling statement is necessary because space on the immediate container is
628 limited, the following statement is acceptable, provided one of the full labeling statements, as
629 shown above, appears on the outer container and in the package insert:

630 **Refrigerate (see insert)**

631 c. Room Temperature and/or Refrigerator Storage Statement

632 For a drug product demonstrated to be stable both at 25°C ± 2°C/60% RH ± 5% and at
633 refrigerator temperature, either/or both of the room temperature and refrigerator labeling
634 statements, as described above, are acceptable, depending on the storage conditions intended for
635 the product. A statement such as "store at 2-25°C" is not recommended.

636 d. Additional Cautionary Statements

637 If warranted, additional cautionary statements to protect a drug product from excessive heat,
638 light, humidity, freezing, and other damaging conditions, should be included on the container label
639 and the package insert. If the space on the container label is too limited to display all the
640 recommended statements in detail, a reference to the package insert for further information (e.g.,
641 *see insert*) is recommended. The uniform storage statements and stability conditions are
642 summarized in Tables 2 and 3, respectively.

Draft - Not for Implementation

643 **Table 2: Summary of Uniform Storage Statements in Drug Product Labeling**

		Recommended Storage Statement in Drug Product Labeling	
		Full	Abbreviated
Intended storage conditions for drug product	Room Temperature	Store at 25°C (77/F) excursions permitted to 15-30°C (59-86°F) [see USP Controlled Room Temperature]	Store at 25/C (77°F) excursions 15-30°C (59-86°F) or Store at 25°C (77°F) (see insert)
	Refrigerator Temperature	Store in a refrigerator, 2-8°C (36-46°F) or Store refrigerated	Refrigerate (see insert)

644
645
646
647
648

649 **Table 3: Conditions under which Product has been Shown to be Stable**
650 **to Apply Uniform Storage Statements**
651

Intended storage conditions for drug product	Room Temperature		Refrigerator Temperature
Type of product	LVP in a plastic bag[a] or Aqueous Solution in a LDPE bottle or prefilled syringe	All other types	All drug products (as appropriate)
Conditions under which product has been shown to be stable	25°C ± 2°C/60% RH ±5% 30°C ± 2°C/40% RH ± 5% 25°C/NMT 40% RH 30°C/NMT 40% RH or 25°C, 30°C or 25-30°C and ambient humidity	25°C ± 2°C /60% RH ± 5% 30°C/60% RH ± 5% or 25°C, 30°C or 25-30°C and ambient humidity	5°C ± 3°C 2-5°C or 2-8°C

652
653
654
655
656
657
658
659

660 [a] See Section II.B.11.a. for additional information on sterile water for injection and LVPs containing inorganic salts
661 or simple organic salts.

Draft - Not for Implementation

662 e. Other Considerations

663 The applicant may wish to include the definition of USP CRT in its entirety in the package insert
664 to provide easy reference.

665 f. Implementation of the USSs in Labeling for New Product Applications

666 The recommended storage statements in labeling should be adopted for new or pending NDA,
667 ANDA, BLA or PLA products. For applications approved prior to the publication of the
668 guidance, the recommended storage statements should be adopted through the annual report
669 mechanism at the next printing opportunity if desired, but within three years of the date of the
670 final guidance. With respect to room temperature storage statements for already approved
671 products, new stability studies under the ICH conditions are not required to adopt the
672 recommended room temperature labeling statements, provided the products have been
673 demonstrated to be stable through expiry under one of the following controlled temperatures:
674 30°C, 25-30°C, 25°C and at ambient humidity.

675 **C. New Dosage Forms [ICH Q1C]**

676 A new dosage form is defined as a drug product that is a different pharmaceutical product type,
677 but contains the same active substance as included in an existing drug product approved by the
678 FDA.

679 New dosage forms include products of different administration route (e.g., oral, when the original
680 new drug product was a parenteral), new specific functionality/delivery system (e.g., modified
681 release tablet, when the original new drug product was an immediate release tablet, and different
682 dosage forms of the same administration route (e.g., capsule to tablet, solution to suspension).

683 Stability protocols for new dosage forms should follow the guidance in the ICH Q1A in principle.
684 However, a reduced stability database at submission time (e.g., 6 months' accelerated and 6
685 months' long-term data from ongoing studies) may be acceptable in certain justified cases.

686 **D. Other NDAs**

687 Stability protocols for new combination products or new formulations (which require clinical data
688 for approval) should follow the guidance in the ICH Q1A in principle. However, a reduced
689 stability database at submission time (e.g., 6 months' accelerated and 6 months' data from
690 ongoing studies at the long-term condition) may be acceptable in certain justified cases, such as
691 when there is a significant body of information on the stability of the drug product and the dosage
692 form.

693 III. STABILITY TESTING FOR ABBREVIATED NEW DRUG APPLICATIONS

694 Much of the general information provided in this guidance is applicable to abbreviated new drugs
695 (ANDAs). However, depending upon the availability of significant information on, and the

Draft - Not for Implementation

696 complexity of, these drug products/dosage forms, the amount of information necessary to support
697 these applications may vary from that proposed for NDAs. This section is intended to provide
698 specific recommendations on abbreviated applications.

699 **A. Drug Substance Stability Data Submission**

700 For drug products submitted under an ANDA, including antibiotics, supporting information may
701 be provided directly to the drug product ANDA or by reference to an appropriately referenced
702 drug master file (DMF). Publications may be provided or referenced as supportive information.
703 For ANDA bulk drug substances, stability data should be generated on a minimum of one pilot-
704 scale batch. All batches should be made using equipment of the same design and operating
705 principle as the manufacturing-scale production equipment with the exception of capacity. For
706 ANDA bulk drug substances produced by fermentation, stability data should be provided on three
707 production batches, at least two of which should be generated from different starter cultures.
708
709 **B. Drug Substance Testing**

710 A program for stability assessment may include storage at accelerated, long-term, and, if
711 applicable, intermediate stability study storage conditions (refer to IV.G. of the ICH Q1A
712 Guidance and Section II.A. of this guidance). Stability samples should be stored in the bulk
713 storage container equivalent (e.g., same composition and type of container, closure and liner, but
714 smaller in size).

715 If not previously generated or available by reference, stress testing studies should be conducted to
716 establish the inherent stability characteristics of the drug substance, and support the suitability of
717 the proposed analytical procedures. The detailed nature of the studies will depend on the
718 individual drug substance, type of drug product and available supporting information. Any
719 necessary testing may be carried out as described in Section II.A.

720 **C. Drug Product**

721 Original ANDAs should contain stability data generated under the long-term and accelerated
722 stability storage conditions delineated in V.E. of the ICH Q1A guidance (Section II.B. of this
723 guidance). The data package for ANDAs (e.g., number of batches, length of studies needed at
724 submission and at approval, and accelerated, intermediate and long-term stability data) should be
725 based on several factors, including the complexity of the dosage form, the existence of a
726 significant body of information for the dosage form, and the existence of an approved application
727 for a particular dosage form.

728 **D. ANDA Data Package Recommendations**

729 For *Simple Dosage Forms* the following stability data package is recommended:

730 • Accelerated stability data at 0, 1, 2, and 3 months. A tentative expiration dating period of up
731 to 24 months will be granted based on satisfactory accelerated stability data unless not
732 supported by the available long-term stability data.

Draft - Not for Implementation

733 • Long-term stability data (available data at the time of original filing and subsequent
734 amendments).
735 • A minimum of one batch; pilot scale.
736 • Additional stability studies (12 months at the intermediate conditions, or long-term data
737 through the proposed expiration date) if ***significant change*** is seen after 3 months during the
738 accelerated stability study. The tentative expiration dating period will be determined based on
739 the available data from the additional study.

740 **E. Exceptions to the ANDA Data Package Recommendations**

741 The following may be considered exceptions to the general ANDA recommendations:

742 • Complex dosage forms, such as modified-release products, transdermal patches, metered-dose
743 inhalers.
744 • Drug products without a significant body of information.
745 • New dosage forms submitted through the ANDA suitability petition process (Q1C
746 applications).
747 • Other exceptions may exist and should be discussed with the Office of Generic Drugs.

748 An ANDA that is determined to be one of the above categories should contain a modified ICH
749 Q1A stability data package, including:

750 • 3-month accelerated stability studies.
751 • Long-term stability studies (available data at the time of original filing and subsequent
752 amendments). The expiration dating period for complex dosage forms will be determined
753 based on available long-term stability data submitted in the application.
754 • A minimum of three batches manufactured in accordance with the ICH Q1A batch size
755 recommendations (refer to V.B. of the ICH Q1A guidance and Section II.B. of this guidance).
756 • Additional stability studies (12 months at the intermediate conditions or long-term stability
757 testing through the proposed expiration date) if significant change is seen after 3 months
758 during the accelerated stability studies (the tentative expiration dating period will be
759 determined based on the available data from the additional studies).

760 **F. Data Package for Approval**

761 Full-term stability testing of the primary stability batch(es) is suggested. However, in the absence
762 of full-term stability data for the drug product, adequate accelerated stability data combined with
763 available long-term data can be used as the basis for granting a tentative expiration dating period.
764 The batch(es) used for stability testing should comply fully with the proposed specifications for
765 the product and be packaged in the market package, and the release assay should be within
766 reasonable variation (taking into account inherent assay variability) from the labeled strength or
767 theoretical strength of the reference listed drug. If formulated with an overage, the overage
768 should be justified as necessary to match that of the reference listed drug.

769 Other supportive stability data may be submitted on drug product batches that may or may not
770 meet the above criteria. Data on relevant research batches, investigational formulations, alternate

Draft - Not for Implementation

771 container/closure systems, or from other related studies may also be submitted to support the
772 stability of the drug product. The supportive stability data should be clearly identified.

773 **G. Stability Study Acceptance**

774 If the results are satisfactory, a tentative expiration dating period of up to 24 months at the labeled
775 storage conditions may be granted. Where data from accelerated studies are used to project a
776 tentative expiration dating period that is beyond a date supported by actual long-term studies on
777 production batches, the application should include a commitment to conduct long-term stability
778 studies on the first three production batches and annual batches until the tentative expiration
779 dating period is verified, or the appropriate expiration dating period is determined. Extension of
780 the tentative expiration dating period should be based on data generated on at least three
781 production batches tested according to the approved protocol outlined in the stability
782 commitment. Reporting of the data should follow Section VI. of this guidance.

783 ANDAs withdrawn prior to publication of this guidance should not normally have to include
784 stability data in conformance with the guidance upon resubmission if the original application was
785 withdrawn due to non-stability related issues. However, if new stability studies are conducted to
786 support the submission, such studies should be conducted as recommended in the guidance.

787 **IV. STABILITY TESTING FOR INVESTIGATIONAL NEW DRUG APPLICATIONS**

788 Much of the following information is taken from the guidance for industry, *Content and Format*
789 *of Investigational New Drug Applications (INDs) for Phase 1 Studies of Drugs, Including*
790 *Well-Characterized, Therapeutic Biotechnology-derived Products* (November 1995).

791 The regulation at 312.23(a)(7) emphasizes the graded nature of manufacturing and controls
792 information. Although in each phase of the investigation, sufficient information should be
793 submitted to ensure the proper identification, quality, purity, and strength of the investigational
794 drug, the amount of information needed to achieve that assurance will vary with the phase of the
795 investigation, the proposed duration of the investigation, the dosage form, and the amount of
796 information otherwise available. Therefore, although stability data are required in all phases of
797 the IND to demonstrate that the new drug substance and drug product are within acceptable
798 chemical and physical limits for the planned duration of the proposed clinical investigation, if very
799 short-term tests are proposed, the supporting stability data can be correspondingly very limited.

800 It is recognized that modifications to the method of preparation of the new drug substance and
801 dosage form, and even changes in the dosage form itself, are likely as the investigation progresses.
802 In an initial phase 1 CMC submission, the emphasis should generally be placed on providing
803 information that will allow evaluation of the safety of subjects in the proposed study. The
804 identification of a safety concern or insufficient data to make an evaluation of safety are the only
805 reasons for placing a trial on clinical hold based on the CMC section.

806 **A. Phase 1**

Draft - Not for Implementation

807 Information to support the stability of the drug substance during the toxicologic studies and the
808 proposed clinical study(ies) should include the following: a brief description of the stability study
809 and the test methods used to monitor the stability of the drug substance and preliminary tabular
810 data based on representative material. Neither detailed stability data nor the stability protocol
811 need to be submitted.

812 Information to support the stability of the drug product during the toxicologic studies and the
813 proposed clinical study(ies) should include the following: a brief description of the stability study
814 and the test methods used to monitor the stability of the drug product packaged in the proposed
815 container/closure system and storage conditions and preliminary tabular data based on
816 representative material. Neither detailed stability data nor the stability protocol need to be
817 submitted.

818 When significant decomposition during storage cannot be prevented, the clinical trial batch of
819 drug product should be retested prior to the initiation of the trial and information should be
820 submitted to show that it will remain stable during the course of the trial. This information should
821 be based on the limited stability data available when the trial starts. Impurities that increase
822 during storage may be qualified by reference to prior human or animal data.

823 **B. Phase 2**

824 Development of drug product formulations during phase 2 should be based in part on the
825 accumulating stability information gained from studies of the drug substance and its formulations.

826 The objectives of stability testing during phases 1 and 2 are to evaluate the stability of the
827 investigational formulations used in the initial clinical trials, to obtain the additional information
828 needed to develop a final formulation, and to select the most appropriate container and closure
829 (e.g., compatibility studies of potential interactive effects between the drug substance(s) and other
830 components of the system). This information should be summarized and submitted to the IND
831 during phase 2. Stability studies on these formulations should be well underway by the end of
832 Phase 2. At this point the stability protocol for study of both the drug substance and drug
833 product should be defined, so that stability data generated during phase 3 studies will be
834 appropriate for submission in the drug application.

835 **C. Phase 3**

836 In stability testing during phase 3 IND studies, the emphasis should be on testing final
837 formulations in their proposed market packaging and manufacturing site based on the
838 recommendations and objectives of this guidance. It is recommended that the final stability
839 protocol be well defined prior to the initiation of phase 3 IND studies. In this regard,
840 consideration should be given to establish appropriate linkage between the preclinical and clinical
841 batches of the drug substance and drug product and those of the primary stability batches in
842 support of the proposed expiration dating period. Factors to be considered may include, for
843 example, source, quality and purity of various components of the drug product, manufacturing
844 process of and facility for the drug substance and the drug product, and use of same containers

Draft - Not for Implementation

845 and closures.

846 V. APPROVED STABILITY PROTOCOL

847 A. Stability Protocol

848 An *approved stability protocol* is a detailed plan described in an approved application that is used
849 to generate and analyze stability data to support the retest period for a drug substance or the
850 expiration dating period for a drug product. It also may be used in developing similar data to
851 support an extension of that retest or expiration dating period via annual reports under 21 CFR
852 314.70(d)(5). If needed, consultation with FDA is encouraged prior to the implementation of the
853 stability protocol.

854 To ensure that the identity, strength, quality, and purity of a drug product are maintained
855 throughout its expiration dating period, stability studies should include the drug product packaged
856 in the proposed containers and closures for marketing as well as for physician and/or promotional
857 samples. The stability protocol may also include an assessment of the drug product in bulk
858 containers to support short-term storage prior to packaging in the market container.

859 The stability protocol should include methodology for each parameter assessed during the
860 stability evaluation of the drug substance and the drug product. The protocol should also address
861 analyses and approaches for the evaluation of results and the determination of the expiration
862 dating period, or retest period. The stability-indicating methodology should be validated by the
863 manufacturer and described in sufficient detail to permit validation and/or verification by FDA
864 laboratories.
865
866 The stability protocol for both the drug substance and the drug product should be designed in a
867 manner to allow storage under specifically defined conditions. For the drug product, the protocol
868 should support a labeling storage statement at CRT, refrigerator temperature, or freezer
869 temperature. See Sections II.B.5 and 6.

870 A properly designed stability protocol should include the following information:

871 • Technical grade and manufacturer of drug substance and excipients
872 • Type, size, and number of batches
873 • Type, size, and source of containers and closures
874 • Test parameters
875 • Test methods
876 • Acceptance criteria
877 • Test time points
878 • Test storage conditions
879 • Container storage orientations
880 • Sampling plan
881 • Statistical analysis approaches and evaluations
882 • Data presentation
883 • Retest or expiration dating period (proposed or approved)

Draft - Not for Implementation

884 Stability commitment

885 The use of alternative designs, such as bracketing and matrixing, may be appropriate (see Sections
886 VII.G. and H.).

887 At the time of a drug application approval, the applicant has probably not yet manufactured the
888 subject drug product repeatedly on a production scale or accrued full long-term data. The
889 expiration dating period granted in the original application is based on acceptable accelerated
890 data, statistical analysis of available long-term data, and other supportive data for an NDA, or on
891 acceptable accelerated data for an ANDA. It is often derived from pilot-scale batches of a drug
892 product or from less than full long-term stability data. An expiration dating period assigned in this
893 manner is considered tentative until confirmed with full long-term stability data from at least three
894 production batches reported through annual reports. The stability protocol approved in the
895 application is then crucial for the confirmation purpose.

896 **B. Stability Commitment**

897 A stability commitment is acceptable when there are sufficient supporting data to predict a
898 favorable outcome with a high degree of confidence, such as when an application is approved
899 with stability data available from pilot-plant batches, when a supplement is approved with data
900 that do not cover the full expiration dating period, or as a condition of approval. This
901 commitment constitutes an agreement to:

902 1. Conduct and/or complete the necessary studies on the first three production batches and
903 annual batches thereafter of each drug product, container, and closure according to the
904 approved stability protocol through the expiration dating period.

905 2. Submit stability study results at the time intervals and in the format specified by the FDA,
906 including the annual batches.

907 3. Withdraw from the market any batches found to fall outside the approved specifications for
908 the drug product. If the applicant has evidence that the deviation is a single occurrence that
909 does not affect the safety and efficacy of the drug product, the applicant should immediately
910 discuss it with the appropriate chemistry team and provide justification for the continued
911 distribution of that batch . The change or deterioration in the distributed drug or biological
912 product must be reported under 21 CFR 314.81(b)(1)(ii) or 21 CFR 601.14, respectively.

913 For postapproval changes, items 2 and 3 remain the same and item 1 becomes:

914 1. Conduct and/or complete the necessary studies on the appropriate number of batches. The
915 amount of stability data supplied will depend on the nature of the change being made.
916 Applicants may determine the appropriate data package by consulting the PostApproval
917 Changes section of this guidance (Section IX.) and in consultation with the appropriate
918 chemistry review team.

919 The approved stability protocol should be revised as necessary to reflect updates to USP

920 monographs or the current state-of-the-art regarding the type of parameters monitored,
921 acceptance criteria of such parameters, and the test methodology used to assess such parameters.
922 However, other modifications are discouraged until the expiration dating period granted at the
923 time of approval has been confirmed by long-term data from production batches. Once a
924 sufficient database is established from several production batches to confirm the originally
925 approved expiration dating period, it may be appropriate to modify the stability protocol. See
926 Section IX.J.

927 VI. REPORTING STABILITY DATA

928 A. General

929 Stability data should be included in the application (NDA, ANDA, BLA, PLA, IND, supplement)
930 they are intended to support. The extent of stability data expected at the time of submission is
931 discussed at length throughout this guidance. Additional data from ongoing studies and regular
932 annual batches should be included in the application's annual report.

933 Annual reports should include new or updated stability data generated in accordance with the
934 approved stability protocol. These data may include accelerated and long-term studies for each
935 product to satisfy the standard stability commitment made in the original or supplemental
936 application, including the annual batch(es), and to support postapproval changes. The data should
937 be presented in an organized, comprehensive, and cumulative format.

938 B. Content of Stability Reports

939 It is suggested that stability reports include the following information and data to facilitate
940 decisions concerning drug product stability:

941 1. General Product Information

942 • Name, source, manufacturing sites, and date of manufacture of drug substance and drug or
943 biological product.
944 • Dosage form and strength, including formulation. (The application should provide a table of
945 specific formulations under study. When more than one formulation has been studied, the
946 formulation number is acceptable.)
947 • Composition, type, source, size, and adequate description of container and closure. Stuffers,
948 seals, and desiccants should also be identified.

949 2. Specifications and Test Methodology Information

950 • Physical, chemical, and microbiological attributes and regulatory specifications (or specific
951 references to NDA, BLA, PLA, or USP).
952 • Test methodology used (or specific reference to IND, ANDA, NDA, BLA, PLA prior
953 submissions, or USP) for each sample tested.
954 • Information on accuracy, precision, and suitability of the methodology (cited by reference to

Draft - Not for Implementation

955 appropriate sections).
956 • Where applicable, a description of the potency test(s) for measuring biological activity,
957 including specifications for potency determination.

958 3. Study Design and Study Conditions

959 • Description of the sampling plan, including:
960 • Batches and number selected.
961 • Container and closures and number selected.
962 • Number of dosage units selected and whether tests were conducted on individual units or
963 on composites of individual units.
964 • Sampling time points.
965 • Testing of drug or biological products for reconstitution at the time of reconstitution (as
966 directed on the labeling) as well as through their recommended use periods.
967 • Expected duration of the study.
968 • Conditions of storage of the product under study (e.g., temperature, humidity, light, container
969 orientation).

970 4. Stability Data/Information

971 • Batch number (research, pilot, production) and associated manufacturing date.
972 • For antibiotic drug products, the age of the bulk active drug substance(s) used in
973 manufacturing the batch.
974 • Analytical data, source of each data point, and date of analysis (e.g., batch, container,
975 composite, etc). Pooled estimates may be submitted if individual data points are provided.
976 • Individual data as well as mean and standard deviation should be reported.
977 • Tabulated data by storage condition.
978 • Summary of information on previous formulations during product development. This
979 summary may be referenced (if previously submitted) and should include other containers and
980 closures investigated.

981 5. Data Analysis

982 The following data analysis of quantitative parameters should be provided:

983 • Evaluation of data, plots, and/or graphics.
984 • Documentation of appropriate statistical methods and formulas used.
985 • Results of statistical analysis and estimated expiration dating period.
986 • Results of statistical tests used in arriving at microbiological potency estimates.

987 6. Conclusions

988 • Proposed expiration dating period and its justification.
989 • Regulatory specifications (establishment of acceptable minimum potency at the time of initial
990 release for full expiration dating period to be warranted).

991 **C. Formatting Stability Reports**

Draft - Not for Implementation

992 Submitted information should be cumulative and in tabular form. Examples are provided on the
993 following list and in Table 4.

994 **Summary Of Stability Studies For Drug Product X**

995	Study Number	Container Composition/Supplier
996	Drug Product Batch #/Control #*, **	Closure Composition/Supplier
997	Formulation Code/No	Seal/Supplier
998	Dosage and Strength	Mfg/Site/Date
999	Batch Type and Size	Packager/Site/Date
1000	Storage Conditions	Location of Data in Application
1001	Drug Substance Mfg/Site/Batch#	Specs Failures
1002	Length of Study	Reporting Period

1003 *Batches Used in Clinical Studies and Biostudies (Specify)
1004 **Batches of Different Formulation

Draft - Not for Implementation

1005 **Table 4: Model Stability Data Presentation**

1006 Summary of Stability Studies for Drug Product X

1007 Product Name
1008 Study Number
1009 Formulation Code/Number
1010 Dosage and Strength
1011 Drug Product Batch Number/Control Number[a,b]
1012 Batch Type and Size
1013 Drug Product Manufacturer/Site/Date
1014 Drug Substance Manufacturer/Site/Batch Number
1015 Container Composition/Supplier
1016 Closure Composition/Supplier
1017 Seal/Supplier
1018 Packager/Site/Date
1019 Sampling Plan
1020 Specifications and Test Methods
1021 Storage Conditions
1022 Length of Study
1023 Reporting Period
1024 Location of Data in Application
1025 Summary of Data
1026 Data Analysis
1027 Conclusions

1028 [a] Batches used in clinical studies and biostudies (specify).
1029 [b] Batches of different formulations.

Draft - Not for Implementation

1030 **Table 4: (cont.)**

1031 Stability Raw Data for Drug Product X, Batch Y

1032	Product Name/Strength	Study Number	Purpose of Study
1033	Batch Number	Batch Size	Date Study Started
1034	Date Manufactured	Manufacturer/Site	Container/Size/Supplier
1035	Date Packaged	Packager/Site	Closure Supplier
1036	Storage Condition	Storage Orientation	Seal Supplier

1037 Drug Substance Manufacturer/Site/Batch Number
1038 Approved/Proposed Expiration Dating Period

	Attributes	Method	Specification	Time (Months)							
1039		SOP #	(Low/High)	0	3	6	9	12	18	24	etc.
1040	Appearance										
1041	Assay										
1042 1043	Degradation Product A										
1044 1045	Degradation Product B										
1046 1047	Degradation Product C										
1048	etc.										

Draft - Not for Implementation

1049 VII. SPECIFIC STABILITY TOPICS

1050 A. Mean Kinetic Temperature

1051 1. Introduction

1052 Section 501(a)(2)(B) of the Federal Food, Drug, and Cosmetic Act states that a drug shall be
1053 deemed to be adulterated if the facilities or controls used for holding drugs do not conform to or
1054 are not operated or administered in conformity with good manufacturing practice to assure that
1055 such drugs meet the requirements of the Act as to safety, and have the identity and strength, and
1056 meet the quality and purity characteristics, which they purport or are represented to possess. This
1057 applies to all persons engaged in manufacture and holding, i.e., storage, of drugs.
1058
1059 Current good manufacturing practices (CGMP) regulations applicable to drug manufacturers (21
1060 CFR 211.142) state that written procedures describing the warehousing of drug products shall be
1061 established and followed. These regulations also state that such procedures shall include
1062 instructions for the storage of drug products under appropriate conditions of temperature,
1063 humidity, and light so the identity, strength, quality, and purity of the drug products are not
1064 affected.

1065 The regulation governing state licensing of wholesale prescription drug distributors (21 CFR
1066 205.50 (c)) states that all prescription drugs shall be stored at appropriate temperatures and under
1067 appropriate conditions in accordance with requirements, if any, in the labeling of such drugs, or
1068 with requirements in the current edition of an official compendium, such as the USP/NF. The
1069 regulation also states that if no storage requirements are established for a prescription drug, the
1070 drug may be held at CRT, as defined in an official compendium, to help ensure that its identity,
1071 strength, quality and purity are not adversely affected (21 CFR 205.50 (c)(1)).

1072 Mean kinetic temperature (MKT)[4] is defined as the isothermal temperature that corresponds to
1073 the kinetic effects of a time-temperature distribution. The Haynes formula can be used to
1074 calculate the MKT. It is higher than the arithmetic mean temperature and takes into account the
1075 Arrhenius equation from which Haynes derived his formula. Thus, MKT is the single calculated
1076 temperature that simulates the nonisothermal effects of storage temperature variations. This
1077 section of the guidance explains how to calculate MKT. It also recommends a course of action
1078 should a facility containing products that are labeled for CRT storage fail to maintain the drugs at
1079 appropriate temperature conditions as defined in this guidance. Because MKT is intended to
1080 provide guidance on temperature control of drug storage facilities and is not correlated to any
1081 specific lot of drug product in the storage facility, an MKT in excess of 25°C does not, on its
1082 own, infer that CGMPs have been violated.

---

[4] J.D. Haynes,"Worldwide Virtual Temperatures for Product Stability Testing", *J. Pharm. Sci.*, Vol. 60, No. 6, 927 (June 1971).

Draft - Not for Implementation

1083 2. Calculation

1084 There are a variety of ways to approximate a MKT. The FDA recommends that, for
1085 manufacturers, repackagers, and warehouses, all data points obtained be inserted directly into the
1086 MKT equation. A minimum of weekly high and low readings is recommended, and more rigorous
1087 approximations using daily highs and lows or even more frequent temperature readings would be
1088 acceptable. Storage temperatures may be obtained using automated recording devices, chart
1089 recorders, or a high-low thermometer.

1090 The temperature readings (minimum of 104 weekly high and low readings) would be inserted into
1091 the MKT equation to calculate a yearly MKT. The yearly MKT for the preceding twelve months
1092 should be calculated every month. At times when no drugs are stored in a facility, those intervals
1093 should not be used in MKT calculations. The MKT equation is shown below:

1094
$$T_k = \frac{\frac{-\Delta H}{R}}{\ln\left(\frac{e^{-\frac{\Delta H}{RT_{1H}}} + e^{-\frac{\Delta H}{RT_{1L}}} + \cdots + e^{-\frac{\Delta H}{RT_{nH}}} + e^{-\frac{\Delta H}{RT_{nL}}}}{2n}\right)}$$

1095 Where:

1096 T_k= the mean kinetic temperature in °K
1097 ΔH= the heat of activation, 83.144 kJ•mole^{-1}
1098 R = the universal gas constant, 8.3144 x 10^{-3} kJ•mole^{-1}•°K^{-1}
1099 T_{1H}= the high temperature in °K during the 1st week
1100 T_{1L}= the low temperature in °K during the 1st week
1101 T_{nH}= the high temperature in °K during the nth week
1102 T_{nL}= the low temperature in °K during the nth week
1103 n = the total number of weeks (i.e. 52)
1104 T = absolute temperature in °K
1105 °K = °C (Celsius) + 273.2
1106 °K = [(°F (Fahrenheit) -32)•0.555] + 273.2

1107 Note that 83.144 kJoules/mol is an average value based upon many common organic reactions.
1108 Since $\Delta H/R$ = 10,000°K, the above equation can be simplified as:
1109

1110
$$T_k = \frac{-10{,}000}{\ln\left(\frac{e^{-\frac{10{,}000}{T_{1H}}} + e^{-\frac{10{,}000}{T_{1L}}} + \cdots + e^{-\frac{10{,}000}{T_{nH}}} + e^{-\frac{10{,}000}{T_{nL}}}}{2n}\right)}$$

1111

Draft - Not for Implementation

1112 3. Application

1113 Any time the yearly MKT of a facility approaches 25°C, the occurrence should be documented,
1114 the cause for such an occurrence should be investigated, and corrective actions should be taken to
1115 ensure that the facility is maintained within the established conditions for drug product storage.
1116 FDA recognizes that, when the yearly MKT of a facility begins to exceed 25°C, it may not
1117 necessarily have an impact on products that have been stored for less than one year at the time,
1118 but should be a warning that the facility itself may not be under adequate control

1119 In addition, whenever the recorded temperature (as opposed to the calculated MKT) exceeds the
1120 allowable excursions of 15-30°C in a facility that contains drugs labeled for storage at CRT, the
1121 occurrence should be documented. The cause for such an occurrence should be investigated, and
1122 corrective actions taken to ensure that the facility is maintained within the established conditions
1123 for drug product storage. The FDA recognizes that brief spikes outside of 15-30°C may, in fact,
1124 be expected from time to time in the real world and may not necessarily have an impact on
1125 product quality. However, depending on the duration and extent of such an exposure and the
1126 dosage form, it may be necessary to determine if the product quality has been adversely affected.

1127 **B. Container/Closure**

1128 Stability data should be developed for the drug product in each type of immediate container and
1129 closure proposed for marketing, promotion, or bulk storage. The possibility of interaction
1130 between the drug and the container and closure and the potential introduction of extractables into
1131 the drug product formulations during storage should be assessed during container/closure
1132 qualification studies using sensitive and quantitative procedures. These studies are recommended
1133 even if the container and closure meet compendial suitability tests, such as those outlined in the
1134 USP for plastic containers and elastomeric or plastic closures. A draft guidance is available on
1135 this topic entitled *Submission of Documentation in Drug Applications for Container Closure*
1136 *Systems Used for the Packaging of Human Drugs and Biologics* (June 1997).

1137 1. Container and Closure Size

1138 Stability data for a given strength may be bracketed by obtaining data for the smallest and the
1139 largest container and closure to be commercially marketed, provided that the intermediate
1140 container and closure is of comparable composition and design (Section VII.G.).

1141 Physician and/or promotional samples that are in different containers and closures or sizes from
1142 the marketed package should be included in the stability studies. Samples in similar container
1143 closure systems may be included in bracketing or matrixing studies (Section VII.H.).
1144 For solid oral dosage forms packaged in large containers (i.e., those not intended for direct
1145 distribution to the patient) full stability studies should be performed if further packaging by health
1146 institutions or contract packagers is anticipated. Samples for stability testing at different time
1147 points may be taken from the same container. Stability data also may be necessary when the
1148 finished dosage form is stored in interim bulk containers prior to filling into the marketed package.
1149 If the dosage form is stored in bulk containers for over 30 days, real-time stability data under

Draft - Not for Implementation

1150 specified storage conditions should be generated to demonstrate comparable stability to the
1151 dosage form in the marketed package. Interim storage of the dosage form in bulk containers
1152 should generally not exceed six months. The computation of the expiration dating period of the
1153 final marketed product should begin within 30 days of the date of production (see Glossary) of the
1154 dosage form, as defined in the section on Computation of Expiration Dating Period (Section
1155 VII.F.1.), irrespective of the packaging date. If the dosage form is shipped in bulk containers
1156 prior to final packaging, a simulated study may be important to demonstrate that adverse shipping
1157 and/or climatic conditions do not affect its stability.

1158 2. Container Orientations

1159 Solutions (i.e., oral, SVPs, LVPs, oral and nasal inhalations, and topical preparations), dispersed
1160 systems (oral, MDIs, injectables), and semi-solid drug products (topical, ophthalmics, and otics)
1161 should be stored in both the upright and either inverted or on-the-side positions until contact with
1162 the container/closure system has been shown not to impact on drug product quality. The
1163 comparison between upright and inverted or on-the-side position is important to determine
1164 whether contact of the drug product (or solvent) with the closure results in extraction of chemical
1165 substances from the closure components or adsorption and absorption of product components
1166 into the container/closure. The evaluation should include the set of test parameters that are listed
1167 in Considerations for Specific Dosage Forms (Section VIII.). Upright versus inverted/on-the-side
1168 stability studies should be performed during the preapproval and postapproval verification stages
1169 of the stability program. Once it has been demonstrated that the product in maximum contact
1170 with the primary pack does not have a significantly greater impact on drug product quality than
1171 the upright orientation, stability studies may be continued only in the most stressful orientation,
1172 which is generally the inverted or on-the-side position.

1173 3. Extractables and Adsorption/Absorption of Drug Product Components

1174 Specific extractables testing on a drug product is not recommended. Inverted versus upright
1175 stability testing during preapproval and postapproval verification is usually adequate. Extensive
1176 testing for extractables should be performed as part of the qualification of the container/closure
1177 components, labels, adhesives, colorants and ink (see previously cited packaging guidance for
1178 additional information). Such testing should demonstrate that the levels of extractables found
1179 during extraction studies, which are generally performed with various solvents, elevated
1180 temperatures and prolonged extraction times, are at levels determined to be acceptable, and that
1181 those levels will not be approached during the shelf life of the drug product.

1182 Loss of the active drug substance or critical excipients of the drug product by interaction with the
1183 container/closure components or components of the drug delivery device is generally evaluated as
1184 part of the stability protocol. This is usually accomplished by assaying those critical drug product
1185 components, as well as monitoring various critical parameters (e.g., pH, preservative
1186 effectiveness). Excessive loss of a component or change in a parameter will result in the failure of
1187 the drug product to meet applicable specifications.

1188 **C. Microbiological Control and Quality**

Draft - Not for Implementation

1189 1. Preservatives Effectiveness

1190 Both sterile and nonsterile drug products may contain preservative systems to control bacteria and
1191 fungi that may be inadvertently introduced during manufacturing. Acceptance criteria should be
1192 provided as part of the drug product specifications for the chemical content of preservatives at the
1193 time of product release and/or through the product shelf life.

1194 The minimum acceptable limit for the content of preservatives in a drug product should be
1195 demonstrated as microbiologically effective by performing a microbial challenge assay of the drug
1196 formulated with an amount of preservative less than the minimum amount specified as acceptable.
1197 This approach provides a margin of safety within the limit and a margin of error for the assays.
1198 Additionally, compatibility of the preservative system with the container, closure, formulation and
1199 devices (e.g., pumps, injection pens) should be demonstrated over the contact period. Multiple
1200 use container systems, for example, containers that are used after the closure is replaced with an
1201 applicator or dropper and large bottles containing syrups or suspensions should be tested for the
1202 microbiological effectiveness of the preservatives system following simulated uses, including
1203 breaches of the container system as permitted in the labeling. USP "Antimicrobial
1204 Preservatives-Effectiveness"<51> provides a microbial challenge assay.

1205 For the purpose of approval of drug applications, stability data on pilot-scale batches should
1206 include results from microbial challenge studies performed on the drug product at appropriate
1207 intervals. Generally, microbial challenge studies conducted initially, annually, and at the end of
1208 the expiration dating period are adequate. Chemical assays of preservative content(s) should be
1209 performed at all test points.

1210 For postapproval testing, the first three production batches should be tested with a microbial
1211 challenge assay at the start and the end of the stability period and at one point in the middle of the
1212 stability period if the test period equals or exceeds two years. The first three production batches
1213 should be assayed for the chemical content of the preservatives at all appropriate test points.
1214 Upon demonstration of chemical content commensurate with microbial effectiveness in the first
1215 three production batches, chemical assays may be adequate to demonstrate the maintenance of the
1216 specified concentrations of preservatives for subsequent annual batches placed into stability
1217 testing.

1218 2. Microbiological Limits for Nonsterile Drug Products

1219 Nonsterile drug products that have specified microbial limits for drug product release should be
1220 tested for conformance to the specified limits at appropriate, defined time points during stability
1221 studies. The USP provides microbiological test methods for microbial limits and guidance
1222 concerning microbiological attributes of nonsterile drug products.

1223 3. Sterility Assurance for Sterile Drug Products

1224 The stability studies for sterile drug products should include data from a sterility test of each batch
1225 at the beginning of the test period. Additional testing is recommended to demonstrate

Draft - Not for Implementation

1226 maintenance of the integrity of the microbial barrier provided by the container and closure system.
1227 These tests should be performed annually and at expiry.

1228 Integrity of the microbial barrier should be assessed using an appropriately sensitive and
1229 adequately validated container and closure integrity test. The sensitivity of this test should be
1230 established and documented to show the amount of leakage necessary to detect a failed barrier in
1231 a container and closure system. The number of samples to be tested should be similar to the
1232 sampling requirement provided in current USP "Sterility Tests" <71>. The samples that pass
1233 container and closure integrity testing may be used for other stability testing for that specific time
1234 point, but should not be returned to storage for future stability testing. Container and closure
1235 integrity tests do not replace the current USP "Sterility Tests" <71> or 21 CFR 610.12 for
1236 product release.

1237 4. Pyrogens and Bacterial Endotoxins

1238 Drug products with specified limits for pyrogens or bacterial endotoxins should be tested at the
1239 time of release and at appropriate intervals during the stability period. For most parenteral
1240 products, testing at the beginning and the end of the stability test period may be adequate. Sterile
1241 dosage forms containing dry materials (powder filled or lyophilized products) and solutions
1242 packaged in sealed glass ampoules may need no additional testing beyond the initial time point.
1243 Products containing liquids in glass containers with flexible seals or in plastic containers should be
1244 tested no less than at the beginning and the end of the stability test period. For test procedures
1245 and specifications, refer to the FDA *Guideline on Validation of the Limulus Amoebocyte Lysate*
1246 *Test as an End-Product Endotoxin Test for Human and Animal Parenteral Drugs, Biological*
1247 *Products, and Medical Devices*, the USP "Bacterial Endotoxins Test" <85>, and the USP
1248 "Pyrogen Test" <151>.

1249 **D. Stability Sampling Considerations**

1250 The design of a stability study is intended to establish, based on testing a limited number of
1251 batches of a drug product, an expiration dating period applicable to all future batches of the drug
1252 product manufactured under similar circumstances. This approach assumes that inferences drawn
1253 from this small group of tested batches extend to all future batches. Therefore, tested batches
1254 should be representative in all respects such as formulation, manufacturing site, container and
1255 closure, manufacturing process, source and quality of bulk material of the population of all
1256 production batches and conform with all quality specifications of the drug product.

1257 The design of a stability study should take into consideration the variability of individual dosage
1258 units, of containers within a batch, and of batches to ensure that the resulting data for each dosage
1259 unit or container are truly representative of the batch as a whole and to quantify the variability
1260 from batch to batch. The degree of variability affects the confidence that a future batch would
1261 remain within specifications until its expiration date.

1262 1. Batch Sampling

Draft - Not for Implementation

1263 Batches selected for stability studies should optimally constitute a random sample from the
1264 population of production batches. In practice, the batches tested to establish the expiration dating
1265 period are often made at a pilot plant that may only simulate full-scale production. Future
1266 changes in the production process may thus render the initial stability study conclusions obsolete.

1267 At least three batches, preferably more, should be tested to allow an estimate of batch-to-batch
1268 variability and to test the hypothesis that a single expiration dating period for all batches is
1269 justifiable. Testing of less than three batches does not permit a reliable estimate of batch to batch
1270 variability unless a significant body of information is available on the dosage form and/or drug
1271 product. Although data from more batches will result in a more precise estimate, practical
1272 considerations prevent collection of extensive amounts of data. When a significant body of
1273 information is not available, testing at least three batches represents a compromise between
1274 statistical and practical considerations.

1275 2. Container, Closure, and Drug Product Sampling

1276 Selection of containers, such as bottles, packages, and vials, from the batch chosen for inclusion in
1277 the stability study should ensure that the samples represent the batch as a whole. This can be
1278 accomplished by taking a random sample of containers from the finished batch, by using a
1279 stratification plan whereby at a random starting point every nth container is taken from the filling
1280 or packaging line (*n* is chosen such that the sample is spread over the whole batch), or by some
1281 other plan designed to ensure an unbiased selection.

1282 Generally, samples to be assayed at a given sampling time should be taken from previously
1283 unopened containers. For this reason, at least as many containers should be sampled as the
1284 number of sampling times in the stability study.

1285 For products packaged in containers intended for dispensing by a pharmacy to multiple patients,
1286 or intended for repackaging or packaged in unit-of-use containers, samples may be taken from
1287 previously opened containers. More than one container should be sampled during the stability
1288 study. The sampling protocol should be submitted in the drug application.

1289 Dosage units should be sampled from a given container randomly, with each dosage unit having
1290 an equal chance of being included in the sample. If the individual units entered the container
1291 randomly, then samples may be taken from units at the opening of the container. However,
1292 because dosage units near the cap of large containers may have different stability properties than
1293 dosage units in other parts of the container, dosage units should be sampled from all parts of the
1294 container. For dosage units sampled in this fashion, the location within the container from which
1295 the samples were taken should be documented and this information included with the test results.

1296 Unless the product is being tested for homogeneity, composites may be assayed instead of
1297 individual units. If more than one container is sampled at a given sampling time, an equal number
1298 of units from each container may be combined into the composite. If composites are used, their
1299 makeup should be described in the stability study report. The same type of composite should be
1300 used throughout the stability study. For example, if 20-tablet composites are tested initially, then

Draft - Not for Implementation

1301 20-tablet composites should be used throughout. If a larger sample at a given sampling time is
1302 desired, replicated 20-tablet composites should be assayed rather than a single assay of a
1303 composite made from more than 20 tablets. An average of these composite values may be used
1304 for the release assay. However, the individual assay values should be reported as well. Although
1305 other release and stability tests may be performed on these samples (e.g., impurities, preservatives
1306 effectiveness), the results of these tests do not need to be subjected to top/middle/bottom
1307 comparisons.

1308 Semisolid drug products in sizes that are intended for multiple uses should be tested for
1309 homogeneity. Homogeneity testing may be bracketed by container and/or fill size, with testing
1310 done only on the smallest and largest marketed package sizes of each strength. Stability protocols
1311 should provide for increased testing in the event of homogeneity failures, or following a change in
1312 packaging materials or procedures, for example, with a change to a new sealant, or a change in
1313 tube crimping procedures. Where the largest marketed size is more than 20 times the smallest,
1314 homogeneity testing of an intermediate size is recommended.

1315 Semisolid drug products in sizes that are intended for single use need not be tested for
1316 homogeneity.

1317 3. Sampling Time

1318 The sample time points should be chosen so that any degradation can be adequately profiled (i.e.,
1319 at a sufficient frequency to determine with reasonable assurance the nature of the degradation
1320 curve). Usually, the relationship can be adequately represented by a linear, quadratic, or cubic
1321 function on an arithmetic or a logarithmic scale.

1322 Stability testing for long-term studies generally should be performed at three-month intervals
1323 during the first year, six-month intervals during the second, and yearly thereafter. For drug
1324 products predicted to degrade more rapidly, for example, certain radiopharmaceuticals, the
1325 intervals between sampling times should be shortened. Stability testing for accelerated studies
1326 generally should be performed at a minimum of four time points, including the initial sampling
1327 time.

1328 Freezing samples after sampling for the convenience of scheduling analysis is not an acceptable
1329 practice because it may cause delay in finding and responding to out-of-specification test results,
1330 or may adversely affect the stability of a product that does not withstand freezing.

1331 The degradation curve is estimated most precisely, in terms of the width of the confidence limit
1332 about the mean curve (Figure 1, Section VII.E.2.), around the average of the sampling times
1333 included in the study. Therefore, testing an increased number of replicates at the later sampling
1334 times, particularly the latest sampling time, is encouraged because this will increase the average
1335 sampling time toward the desired expiration dating period.

1336 4. Annual Stability Batches

1337 After the expiration dating period has been verified with three production batches, a testing

Draft - Not for Implementation

1338 program for an approved drug product should be implemented to confirm on-going stability. For
1339 every approved application, at least one batch of every strength in every approved
1340 container/closure system, such as bottles or blisters, should be added to the stability program
1341 annually in all subsequent years. If the manufacturing interval is greater than one year, the next
1342 batch of drug product released should be added to the stability program. Bracketing and
1343 matrixing can be used to optimize testing efficiency.

1344 The recommendations in this section do not apply to compressed medical gases, blood, or blood
1345 products.

1346 **E. Statistical Considerations and Evaluation**

1347 1. Data Analysis and Interpretation for Long-term Studies

1348 A stability protocol should describe not only how the stability study is to be designed and carried
1349 out, but also the statistical method to be used in analyzing the data. This section describes an
1350 acceptable statistical approach to the analysis of stability data and the specific features of the
1351 stability study that are pertinent to the analysis. Generally, an expiration dating or retest period
1352 should be determined based on statistical analysis of observed long-term data.. Limited
1353 extrapolation of the real-time data beyond the observed range to extend the expiration dating or
1354 retest period at approval time may be considered if it is supported by the statistical analysis of
1355 real-time data, satisfactory accelerated data, and other nonprimary stability data.

1356 The methods described in this section are used to establish with a high degree of confidence an
1357 expiration dating period during which average drug product attributes such as assay and
1358 degradation products of the batch will remain within specifications. This expiration dating period
1359 should be applicable to all future batches produced by the same manufacturing process for the
1360 drug product.

1361 If an applicant chooses an expiration dating period to ensure that the characteristics of a large
1362 proportion of the individual dosage units are within specifications, different statistical methods
1363 than those proposed below should be considered.[5] In this setting, testing of individual units,
1364 rather than composites, may be important.

1365 Applicants wishing to use a statistical procedure other than those discussed in this guidance
1366 should consult with the chemistry review team prior to the initiation of the stability study and data
1367 analysis.

1368 2. Expiration Dating Period for an Individual Batch

1369 The time during which a batch may be expected to remain within specifications depends not only
1370 on the rate of physical, chemical or microbiological changes, but also on the initial average value

[5] R.G. Easterling, J. Am. Stat. Assoc., "Discrimination Intervals for Percentiles in Regression", 64, 1031-41, 1969.

Draft - Not for Implementation

1371 for the batch. Thus, information on the initial value for the batch is relevant to the determination
1372 of the allowable expiration dating period and should be included in the stability study report.
1373 Percentage of label claim, not percentage of initial average value, is the variable of interest.

1374 The expiration dating period for an individual batch is based on the observed pattern of change in
1375 the quantitative attributes (e.g., assay, degradation products) under study and the precision by
1376 which it is estimated.

1377 An acceptable approach for analyzing an attribute that is expected to decrease with time is to
1378 determine the time at which the 95 percent one-sided lower confidence limit, also known as the 95
1379 percent lower confidence bound, for the estimated curve intersects the acceptable lower
1380 specification limit. In the example shown in Figure 1 where the estimated curve is assumed to be
1381 linear based on 24 months of real time data and the lower specification limit is assumed to be 90
1382 percent of label claim, an expiration dating period of 24 months could be granted. When
1383 analyzing an attribute that is expected to increase with time, the 95 percent one-sided upper
1384 confidence limit for the mean is recommended.

1385 When analyzing an attribute with both an upper and a lower specification limit, special cases may
1386 lead to application of a two-sided 95 percent confidence limit. For example, although chemical
1387 degradation of the active ingredient in a solution product would cause a decrease in the assayed

Figure 1: Statistical Analysis of Long-Term Stability Data

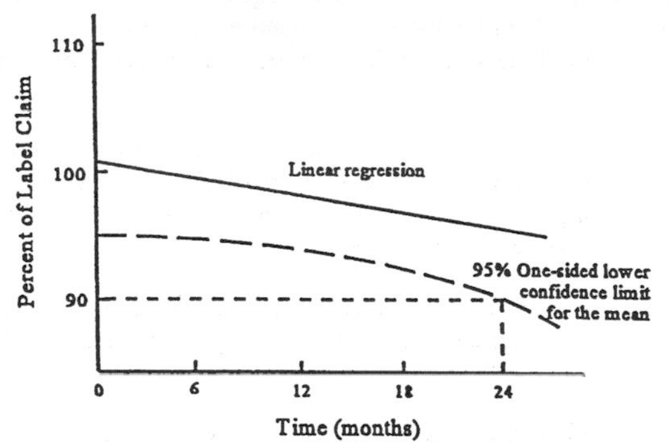

1388 concentration, evaporation of the solvent in the product (through the container/closure) would
1389 result in an increase in the concentration. Because both possibilities should be taken into account,
1390 two-sided confidence limits would be appropriate. If both mechanisms were acting, the
1391 concentration might decrease initially and then increase. In this case, the degradation pattern
1392 would not be linear, and more complicated statistical approaches should be considered.

1393 If the approach presented in this section is used, average parameters such as assay and

Draft - Not for Implementation

1394 degradation products of the dosage units in the batch can be expected to remain within
1395 specifications to the end of the expiration dating period at a confidence level of 95 percent. The
1396 expiration dating period should not be determined using the point at which the fitted least-squares
1397 line intersects the appropriate specification limit. This approach is as likely to overestimate the
1398 expiration dating period as to underestimate it, in which case the batch average can be expected to
1399 remain within specifications at expiration if the fitted least-squares line is used with a confidence
1400 level of only 50 percent.

1401 The statistical assumptions underlying the procedures described above, such as the assumption
1402 that the variability of the individual units from the batch average remains constant over the several
1403 sampling times, are well known and have been discussed in numerous statistical texts. The above
1404 procedures will remain valid even when these assumptions are violated to some degree. If severe
1405 violation of the assumptions in the data is noted, an alternate approach may be necessary to
1406 accomplish the objective of determining an expiration dating period with a high degree of
1407 confidence.

1408 3. Expiration Dating Period for All Batches

1409 If batch-to-batch variability is small, that is, the relationship between the parameter of interest
1410 such as assay or degradation products and time is essentially the same from batch to batch,
1411 stability data should be combined into one overall estimate. Combining the data should be
1412 supported by preliminary testing of batch similarity.[6] The similarity of the estimated curves
1413 among the batches tested should be assessed by applying statistical tests of the equality of slopes
1414 and of zero time intercepts. The level o significance of the tests, expressed in the p-value, should
1415 be chosen so that the decision to combine the data is made only if there is strong evidence in favor
1416 of combining. A p-value of 0.25 for preliminary statistical tests has been recommended.[7] If the
1417 tests for equality of slopes and for equality of intercepts do not result in rejection at a level of
1418 significance of 0.25, the data from the batches could be pooled. If these tests resulted in p-values
1419 less than 0.25, a judgment should be made as to whether pooling could be permitted. The
1420 appropriate FDA chemistry review team should be consulted regarding this determination.

1421 If the preliminary statistical test rejects the hypothesis of batch similarity because of unequal initial
1422 intercept values, it may still be possible to establish that the lines are parallel (i.e., that the slopes
1423 are equal). If so, the data may be combined for the purpose of estimating the common slope. The
1424 individual expiration dating period for each batch in the stability study may then be determined by
1425 considering the initial values and the common slope using appropriate statistical methodology. If
1426 data from several batches are combined, as many batches as feasible should be combined because
1427 confidence limits about the estimated curve will become narrower as the number of batches
1428 increases, usually resulting in a longer expiration dating period. If it is inappropriate to combine

---

[6] K.K. Lin, T-Y.D. Lin, and R.E. Kelley, "Stability of Drugs: Room Temperature Tests", in *Statistics in the Pharmaceutical Industry*, ed. C.R. Buncher and J-Y. Tsay, pp. 419-444, Marcel Dekker, Inc.: New York, 1994.

[7] T. A. Bancroft, "Analysis and Inference for Incompletely Specified Models Involving the Use of Preliminary Test(s) of Significance," *Biometrics*, 20(3), 427-442, 1964.

1429 data from several batches, the overall expiration dating period will depend on the minimum time a
1430 batch may be expected to remain within acceptable limits.

1431 4. Precautions in Extrapolation Beyond Actual Data

1432 The statistical methods for determining an expiration dating period beyond the observed range of
1433 time points are the same as for determining an expiration dating period within the observed range.
1434 The a priori correctness of the assumed pattern of change as a function of time is crucial in the
1435 case of extrapolation beyond the observed range. When estimating a line or curve of change
1436 within the observed range of data, the data themselves provide a check on the correctness of the
1437 assumed relationship, and statistical methods may be applied to test the goodness of fit of the data
1438 to the line or curve. No such internal check is available beyond the range of observed data. For
1439 example, if it has been assumed that the relationship between log assay and time is a straight line
1440 when, in fact, it is a curve, it may be that within the range of the observed data, the true curve is
1441 close enough to a straight line that no serious error is made by approximating the relationship as a
1442 straight line. However, beyond the observed data points, the true curve may diverge from a
1443 straight line enough to have a significant effect on the estimated expiration dating period.

1444 For extrapolation beyond the observed range to be valid, the assumed change must continue to
1445 apply through the estimated expiration dating period. Thus, an expiration dating period granted
1446 on the basis of extrapolation should always be verified by actual stability data as soon as these
1447 data become available.

1448 **F. Expiration Dating Period/Retest Period**

1449 1. Computation of Expiration Date

1450 The computation of the expiration dating period of the drug product should begin no later than
1451 the time of quality control release of that batch, and the date of release should generally not
1452 exceed 30 days from the production date, regardless of the packaging date. The data generated in
1453 support of the assigned expiration dating period should be from long-term studies under the
1454 storage conditions recommended in the labeling. If the expiration date includes only a month and
1455 year, the product should meet specifications through the last day of the month.

1456 In general, proper statistical analysis of long-term stability data collected, as recommended in
1457 Section VII.E. and exemplified in Figure 1, should support at least a one-year expiration dating
1458 period. Exceptions do exist, for example, with short half-life radioactive drug products.

1459 If the production batch contains reprocessed material, the expiration dating period should be
1460 computed from the date of manufacture of the oldest reprocessed material used.

1461 a. Extension of Expiration Dating Period

1462 An extension of the expiration dating period based on full long-term stability data obtained from
1463 at least three production batches in accordance with a protocol approved in the application may
1464 be described in an annual report (21 CFR 314.70(d)(5). The expiration dating period may be

Draft - Not for Implementation

1465 extended in an annual report only if the criteria set forth in the approved stability protocol are met
1466 in obtaining and analyzing data, including statistical analysis, if appropriate.

1467 Alternatively, if the stability study on at least three pilot-scale batches is continued after the
1468 NDA/BLA approval, it is feasible to extend the tentative expiration dating period based on full
1469 long-term data obtained from these batches in accordance with the approved protocol, including
1470 statistical analysis if appropriate, through a prior approval supplement. However, the expiration
1471 dating period thus derived remains tentative until confirmed with full long-term data from at least
1472 three production batches.

1473 Unless a new stability protocol has been adopted via a prior approval supplement before the
1474 change is made, stability protocols included in drug applications prior to the 1985 revisions to the
1475 NDA regulations (50 FR 7452) may not support the extension of expiration dating periods
1476 through annual reports. If the data are obtained under a *new* or *revised* stability protocol, a prior
1477 approval supplement under 21 CFR 314.70(b) or (g) or 21 CFR 601.12 should be submitted to
1478 extend the expiration dating period.

1479 b. Shortening of Expiration Dating Period

1480 When warranted, a previously approved expiration dating period may be shortened via a changes-
1481 being-effected supplement (21 CFR 314.70(c)(1) or 21 CFR 601.12). The supplemental
1482 application should provide pertinent information and the data that led to the shortening of the
1483 expiration dating period. The expiration dating period should be shortened to the nearest
1484 available real-time long-term test point where the product meets acceptance criteria. The
1485 expiration dating period thus derived should be applied to all subsequent production batches and
1486 may not be extended until the cause for the shortening is fully investigated, the problem is
1487 resolved, and satisfactory stability data become available on at least three new production batches
1488 to cover the desired expiration dating period and are submitted in a changes-being-effected
1489 supplement.

1490 2. Retest Period for Drug Substance

1491 A retest period for a drug substance may be established based on the available data from
1492 long-term stability studies and, as such, can be longer than 24 months if supported by data. A
1493 retest date should be placed on the storage container and on the shipping container for a bulk
1494 drug substance. A drug substance batch may be used without retest during an approved retest
1495 period. However, beyond the approved retest period, any remaining portion of the batch should
1496 be retested immediately before use. Retest of different portions of the same batch for use at
1497 different times as needed is acceptable, provided that the batch has been stored under the defined
1498 conditions, the test methods are validated and stability-indicating, and all stability-related
1499 attributes are tested and test results are satisfactory.

1500 Satisfactory retest results on a drug substance batch after the retest date do not mean that the
1501 retest period can be extended for that batch or any other batch. The purpose of retest is to qualify
1502 a specific batch of a drug substance for use in the manufacture of a drug product, rather than to

1503 recertify the drug substance with a new retest date. To extend the retest period, full long-term
1504 data from a formal stability study on three production batches using a protocol approved in an
1505 application or found acceptable in a DMF should be provided.

1506 Similar to the extension of an expiration dating period for a drug product, a retest period for a
1507 drug substance may be extended beyond what was approved in the original application. This can
1508 be achieved through an annual report based on full long-term stability data (i.e., covering the
1509 desired retest period on three production batches using an approved stability protocol).

1510 In a case where testing reveals a limited shelf-life for a drug substance, it may be inappropriate to
1511 use a retest date. An expiration dating period, rather than a retest period, should be established
1512 for a drug substance with a limited shelf-life (e.g., some antibiotics, biological substances).

1513 3. Holding Times for Drug Product Intermediates

1514 Intermediates such as blends, triturates, cores, extended-release beads or pellets may be held for
1515 up to 30 days from their date of production without being retested prior to use. An intermediate
1516 that is held for longer than 30 days should be monitored for stability under controlled, long-term
1517 storage conditions for the length of the holding period. In addition, the first production batch of
1518 the finished drug product manufactured with such an intermediate should be monitored on
1519 long-term stability. When previous testing of an intermediate or the related drug product batches
1520 suggests that an intermediate may not be stable for 30 days, the holding time should be kept to a
1521 minimum and qualified by appropriate stability testing.

1522 The frequency of testing of an intermediate on stability is related to the length of the holding time.
1523 Where practical, testing should be done at a minimum of three time points after the initial testing
1524 of an intermediate. At a minimum, all critical parameters should be evaluated at release of an
1525 intermediate and immediately prior to its use in the manufacture of the finished drug product.

1526 In the event that the holding time for an intermediate has not been qualified by appropriate
1527 stability evaluations, the expiration date assigned to the related finished drug product batch should
1528 be computed from the quality control release date of the intermediate if this date does not exceed
1529 30 days from the date of production of the intermediate. If the holding time has been qualified by
1530 appropriate stability studies, the expiration date assigned to the related finished drug product can
1531 be computed from its quality control release date if this release date does not exceed 30 days from
1532 the date that the intermediate is introduced into the manufacture of the finished drug product.

1533 **G. Bracketing**

1534 1. General

1535 The use of reduced stability testing, such as a bracketing design, may be a suitable alternative to a
1536 full testing program where the drug is available in multiple sizes or strengths. This section
1537 discusses the types of products and submissions to which a bracketing design is applicable and the
1538 types of factors that can be bracketed. Applicants are advised to consult with the FDA when

Draft - Not for Implementation

1539 questions arise.

1540 2. Applicability

1541 The factors that may be bracketed in a stability study are outlined in ICH Q1A and described in
1542 further detail below. The types of drug products and the types of submissions to which
1543 bracketing design can be applied are also discussed.

1544 a. Types of drug product

1545 Bracketing design is applicable to most types of drug products, including immediate- and
1546 modified-release oral solids, liquids, semi-solids, injectables. Certain types of drug products, such
1547 as metered-dosed inhalers (MDIs), dry powder inhalers (DPIs) and transdermal delivery systems
1548 (TDSs), may not be amenable to, or may need additional justification for, bracketing design.

1549 b. Factors

1550 Where a range of container/fill sizes for a drug product of the same strength is to be evaluated,
1551 bracketing design may be applicable if the material and composition of the container and the type
1552 of closure are the same throughout the range. In a case where either the container size or fill size
1553 varies while the other remains the same, bracketing design may be applicable without justification.
1554 In a case where both container size and fill size vary, bracketing design is applicable if appropriate
1555 justification is provided. Such justification should demonstrate that the various aspects (surface
1556 area/volume ratio, dead-space/volume ratio, container wall thickness, closure geometry) of the
1557 intermediate sizes will be adequately bracketed by the extreme sizes selected.

1558 Where a range of dosage strengths for a drug product in the same container/closure (with
1559 identical material and size) is to be tested, bracketing design may be applicable if the formulation
1560 is identical or very closely related in components/composition. Examples for the former include a
1561 tablet range made with different compression weights of a common granulation, or a capsule
1562 range made by filling different plug fill weights of the same composition into different size capsule
1563 shells. The phrase *very closely related formulation* means a range of strengths with a similar, but
1564 not identical, basic composition such that the ratio of active ingredient to excipients remains
1565 relatively constant throughout the range (e.g., addition or deletion of a colorant or flavoring).

1566 In the case where the amount of active ingredient changes while the amount of each excipient or
1567 the total weight of the dosage unit remains constant, bracketing may not be applicable unless
1568 justified. Such justification may include a demonstration of comparable stability profile among the
1569 different strengths based on data obtained from clinical/development batches, primary stability
1570 batches, and/or production batches in support of primary stability batches, commitment batches,
1571 and/or annual batches and batches for postapproval changes, respectively. With this approach,
1572 the formulations should be identical or very closely related, and the container/closure system
1573 should be the same between the supportive batches and the batches for which the bracketing

Draft - Not for Implementation

1574 design is intended.

1575 If the formulation is significantly different among the different strengths (e.g., addition or deletion
1576 of an excipient, except colorant or flavoring), bracketing is generally not applicable.

1577 Due to the complexity in product formulation, applicants are advised to consult the appropriate
1578 chemistry review team in advance when questions arise in the above situations or where
1579 justification is needed for bracketing design.

1580 In the case where the strength and the container and/or fill size of a drug product both vary,
1581 bracketing design may be applicable if justified.

1582 c. Types of submissions

1583 A bracketing design may be used for primary stability batches in an original application,
1584 postapproval commitment batches, annual batches, or batches intended to support supplemental
1585 changes. Bracketing design should not be applied to clinical batches during the IND stages when
1586 the product is still under development. Where additional justification is needed for applying a
1587 bracketing design, product stability should be demonstrated using supportive data obtained from
1588 clinical/development or NDA batches, commitment batches, or production batches. Before a
1589 bracketing protocol is applied to primary stability batches to support an application, the protocol
1590 should be endorsed by Agency chemistry staff via an IND amendment, an end-of-phase 2 meeting,
1591 or prior to submission of an ANDA. Bracketing protocols to be applied to postapproval
1592 commitment batches and annual batches, if proposed, will be approved as part of the original
1593 application.

1594 A bracketing design that is not contained in the approved protocol in the application is subject to
1595 supplemental approval (21 CFR 314.70(b)(2)(ix)) (601.12). If the new bracketing design is used
1596 to generate stability data to support two different chemistry, manufacturing or controls changes,
1597 the two proposed changes could be combined into one prior-approval supplement even though
1598 the latter may otherwise qualify for a changes-being-effected supplement or annual report under
1599 314.70 (c) or (d) or 601.12, or relevant SUPAC guidances. Alternatively, the applicant may
1600 consult the appropriate Agency review staff through general correspondence regarding the
1601 acceptability of the new bracketing design prior to the initiation of the stability studies, and
1602 subsequently submit the data to support the proposed change through the appropriate filing
1603 mechanism.

1604 3. Design

1605 A bracketing protocol should always include the extremes of the intended commercial sizes and/or
1606 strengths. Physician samples or bulk pharmacy packs intended to be repackaged should be
1607 excluded from the bracketing protocol for commercial sizes, but could be studied under their own
1608 bracketing protocols, if applicable. Where a large number, for example four or more, of

Draft - Not for Implementation

1609 sizes/strengths is involved, the inclusion of the one batch each of the intermediates or three
1610 batches of the middle size/strength in the bracketing design is recommended. Where the ultimate
1611 commercial sizes/strengths differ from those bracketed in the original application, a commitment
1612 should be made to place the first production batches of the appropriate extremes on the stability
1613 study postapproval. Such differences should, however, be justified. Where additional justification
1614 for the bracketing design is needed in the original application, one or more of the first production
1615 batches of the intermediate(s) should be placed on the postapproval long-term stability study.

1616 An example of bracketing design is presented in Table 5, where both strengths and container/fill
1617 sizes are bracketed in one protocol and "X" denotes the combination of strength and container/fill
1618 size to be placed on stability study. In this hypothetical situation, the capsule dosage form is
1619 available in three different strengths made from a common granulation and packaged in three
1620 different sizes of HDPE bottles with different fills: 30 counts, C1; 100 counts, C2; and 200
1621 counts, C3. The surface area/volume ratio, dead space/volume ratio, container wall thickness,
1622 and closure performance characteristics are assumed to be proportional among the three
1623 container/fill sizes for each strength of the capsules.

1624 **Table 5: Bracketing Example**

1625 Batch	1									2									3								
1626 Strength	100 mg			200 mg			300 mg			100 mg			200 mg			300 mg			100 mg			200 mg			300 mg		
1627 Container/ 1628 Closure	C1	C2	C3	C1	C2	C3	C1	C2	C3	C1	C2	C3	C1	C2	C3	C1	C2	C3	C1	C2	C3	C1	C2	C3	C1	C2	C3
1629 Sample on 1630 Stability	X		X				X		X	X		X				X		X	X		X				X		X

1631 4. Data evaluation

1632 The stability data obtained under a bracketing protocol should be subjected to the same type of
1633 statistical analysis described in Section VII.E. The same principle and procedure on poolability
1634 should be applied (i.e., testing data from different batches for similarity before combining them
1635 into one overall estimate). If the statistical assessments of the extremes are found to be dissimilar,
1636 the intermediate sizes/strengths should be considered to be no more stable than the least stable

1637 extreme.[8]

1638 **H. Matrixing**

1639 1. General

1640 The use of reduced stability testing, such as a matrixing design, may be a suitable alternative to a
1641 full testing program where multiple factors involved in the product are being evaluated. The
1642 principle behind matrixing is described in ICH Q1A. This section provides further guidance on
1643 when it is appropriate to use matrixing and how to design such a study. Consultation with FDA is
1644 encouraged before the design is implemented.

1645 2. Applicability

1646 The types of drug products and the types of submissions to which matrixing design can be applied
1647 are the same as described for bracketing above. The factors that can be matrixed with or without
1648 justification and those that should not be matrixed are discussed below. Additionally, data
1649 variability and product stability, as demonstrated through previous supportive batches, should be
1650 considered when determining if matrixing can be applied to the batches of interest.

1651 a. Types of drug product

1652 Matrixing design is applicable to most types of drug products, including immediate- and modified-
1653 release oral solids, liquids, semisolids, injectables. Certain types of drug products such as MDIs,
1654 DPIs, and TDSs may not be amenable to, or may need additional justification for, matrixing
1655 design.

1656 b. Factors

1657 Some of the factors that can be matrixed include batches, strengths with identical formulation,
1658 container sizes, fill sizes, and intermediate time points. With justification, additional factors that
1659 can be matrixed include strengths with closely related formulation, container and closure
1660 suppliers, container and closure systems, orientations of container during storage, drug substance
1661 manufacturing sites, and drug product manufacturing sites. For example, to justify matrixing
1662 across HDPE bottles and blister packs, a tablet dosage form could be shown not to be sensitive to
1663 moisture, oxygen, or light (through stressed studies, including open-dish experiments) and that it
1664 is so stable that the protective nature of the container/closure system made little or no difference
1665 in the product stability (through supportive data). Alternatively, it could be demonstrated, if

[8] For additional information on bracketing studies, see W.R. Fairweather, T.-Y. D. Lin, and R. Kelly, "Regulatory, Design, and Analysis Aspects of Complex Stability Studies," *J. Pharm. Sci.*, 84, 1322-1326, 1995.

Draft - Not for Implementation

1666 appropriate, that there is no difference in the protective nature between the two distinctively
1667 different container/closure systems. The justification is needed to ensure that the matrixing
1668 protocol would lead to a successful prediction of the expiration dating period when two otherwise
1669 different container/closure systems are studied together.

1670 Factors that should not be matrixed include initial and final time points, attributes (test
1671 parameters), dosage forms, strengths with different formulations (i.e., different excipients or
1672 different active/excipient ratios, and storage conditions).

1673 c. Data variability and Product Stability

1674 The applicability of matrixing design to primary stability batches depends on the product stability
1675 and data variability demonstrated through clinical or developmental batches. Data variability
1676 refers to the variability of supportive stability data within a given factor (i.e., batch-to-batch,
1677 strength-to-strength, size-to-size) and across different factors (e.g., batch vs strength, strength vs
1678 size). It is assumed that there is very little variability in the analytical methods used in the testing
1679 of stability samples. Matrixing design is applicable if these supportive data indicate that the
1680 product exhibits excellent stability with very small variability. Where the product displays
1681 moderate stability with moderate variability in the supportive data, matrixing design is applicable
1682 with additional justification. Conversely, if supportive data suggest poor product stability with
1683 large variability, matrixing design is not applicable. Similarly, whether or not matrixing design can
1684 be applied to postapproval commitment batches or supplemental changes will depend on the
1685 cumulative stability data on developmental batches, primary batches, and/or production batches,
1686 as appropriate.

1687 Table 6 illustrates the range of situations under which matrixing design is applicable, applicable if
1688 justified, generally not applicable, and not applicable. The table is intended, in a qualitative
1689 manner, to serve as a general guide for sponsors when determining if matrixing design is
1690 appropriate for a drug product with respect to the likelihood that such a design would result in a
1691 successful prediction of the expiration dating period. It does not seek to quantitatively define the
1692 different degrees of product stability or data variability.

Draft - Not for Implementation

1693 **Table 6: Applicability of Matrixing Design**

Data Variability[b]	Product Stability[a]		
	Excellent	Moderate	Poor
Very Small	Applicable	Applicable	Applicable if justified
Moderate	Applicable	Applicable if justified	Generally not applicable
Large	Applicable if justified	Generally not applicable	Not Applicable

1699 [a] In general, *moderate* and *excellent* stability mean little or no change in product test results for a period of 2-3 years
1700 and 4-5 years, respectively, as indicated by supportive data. *Poor stability* means measurable changes in test results
1701 within 1 year.

1702 [b] Variability in supportive stability data within a given factor or across different factors.

1703 d. Types of submission

1704 Same as Section VII.G.1.c.

1705 3. Design

1706 a. General

1707 For original applications, a matrixing design should always include the initial and final time points,
1708 as well as at least two additional time points through the first 12 months, that is at least three time
1709 points including the initial and 12-month time points. This approach is especially important if the
1710 original application contains less than full long-term data at the time of submission.

1711 Although matrixing should not be performed across attributes, different matrixing designs for
1712 different attributes may be suitable where different testing frequencies can be justified. Likewise,
1713 each storage condition should be treated separately under its own matrixing design, if applicable.
1714 Care must be taken to ensure that there are at least three time points, including initial and end
1715 points, for each combination of factors under an accelerated condition. If bracketing is justified,
1716 the matrixing design should be developed afterward.

1717 All samples should be placed on stability including those that are not to be tested under the
1718 matrixing design. Once the study begins, the protocol should be followed without deviation. The

Draft - Not for Implementation

1719 only exception is that, if necessary, it is acceptable to revert back to full stability testing during the
1720 study. But once reverted, the full testing should be carried out through expiry.

1721 b. Size of matrixing design

1722 The appropriate size of a matrix is generally related to the number of combinations of factors and
1723 the amount of supportive data available (Table 7). The size of a matrixing design is expressed as
1724 a fraction of the total number of samples to be tested in the corresponding full stability protocol
1725 For a product available in 3 batches, 3 strengths, and 3 container/fill sizes, the number of
1726 combinations of factors to be tested in a full design is 3x3x3 or 27. Similarly, if there are 3
1727 batches with one strength and no other factors, the number of combinations of factors is
1728 expressed as 3x1. The larger the number of combinations of factors to be tested *and* the greater
1729 the amount of available supportive data, the smaller the size of matrixing design that may be
1730 justified. The phrase *substantial amount of supportive data* means that a sufficient length of
1731 stability data are available on a considerable number of clinical/development batches, primary
1732 stability batches, and/or production batches to justify the use of matrixing design on primary
1733 stability batches, commitment batches, and/or annual batches and batches for postapproval
1734 changes. The formulations used in a matrixing design should be identical or very closely related,
1735 and the container/closure system should be the same between the supportive batches and the
1736 batches for which the matrixing design is intended. The size of matrixing design shown in the
1737 table takes into account all possible combinations of factors and time points. For example, where
1738 there are 3x3x3 combinations of factors and a substantial amount of supportive data are available,
1739 the size of the matrixing design could be as small as one half of that of a full testing protocol.
1740 Thus, *fractional ½* means that only one half of the total number of samples in the corresponding
1741 full protocol will be tested under the matrixing design. Refer to Examples 2 and 3 below for two
1742 designs with an overall size of 5/12 and ½, respectively.

Draft - Not for Implementation

Table 7: Size of Matrixing Design[a]

Number of Combinations of Factors[a]	Amount of Supportive Data[c] Available		
	Substantial	Moderate	Little or none
Large (e.g., 3x3x3 or greater)	Fractional (e.g., ½)	Fractional (e.g., 5/8)	Full (i.e., no matrixing[d])
Moderate (e.g., 3x2)	Fractional (e.g., 5/8)	Fractional (e.g., 3/4)	Full (i.e., no matrixing)
Very small (e.g., 3x1)	Fractional (e.g., ¾)	Full (i.e., no matrixing)	Full (i.e., no matrixing)

[a] Expressed as a fraction of the total number of samples to be tested in the corresponding full design.

[b] Excluding time points.

[c] Cumulative stability data obtained from clinical/development batches, primary stability batches, and/or production batches, as appropriate, to form the basis to support the stability profile of the product.

[d] *No matrixing* means that matrixing is not suitable.

c. Statistical Considerations

The design should be well balanced. An estimate of the probability that stability outcomes from the matrixed study would be the same for a given factor or across different factors should be provided if available.[9]

d. Examples

Matrixing Example #1. Complete design with five-sevenths' time points (overall size: five-sevenths of full testing protocol)

[9] For additional information on matrixing studies see W.R. Fairweather, T.-Y. D. Lin and R. Kelly, "Regulatory, Design, and Analysis Aspects of Complex Stability Studies," *J. Pharm. Sci.*, 84, 1322-1326, 1995.

Draft - Not for Implementation

1766 The following example (Table 8) involves a complete design of 3x3x3 combinations of
1767 factors with five-sevenths' time points for a capsule dosage form available in 3 strengths of a
1768 common granulation and packaged in 3 container/closure systems and/or sizes: C1, HDPE
1769 bottle; 30 counts; C2, HDPE bottle, 100 counts; and blister-pack. A 24-month expiration
1770 dating period is proposed. While stability samples for all 27 combinations of factors will be
1771 tested, they will be tested only at five-sevenths of the usual time points; thus the overall size
1772 of design is 5/7 of the corresponding full testing protocol.

1773 **Table 8: Matrixing Example #1**

1774–1781

Batch		1									2									3								
Strength		100 mg			200 mg			300 mg			100 mg			200 mg			300 mg			100 mg			200 mg			300 mg		
Container/ Closure		C1	C2	C3	C1	C2	C3	C1	C2	C3	C1	C2	C3	C1	C2	C3	C1	C2	C3	C1	C2	C3	C1	C2	C3	C1	C2	C3
Schedule		T1	T2	T3	T2	T3	T1	T3	T1	T2	T2	T3	T1	T3	T1	T2	T1	T2	T3	T3	T1	T2	T1	T2	T3	T2	T3	T1
Time Points (mo)	0	x	x	x	x	x	x	x	x	x	x	x	x	x	x	x	x	x	x	x	x	x	x	x	x	x	x	x
	3	x					x		x				x		x		x				x		x					x
	6		x		x					x	x					x		x				x		x		x		
	9	x		x		x	x	x	x			x	x	x	x		x		x	x	x		x		x		x	x
	12		x	x	x	x		x		x	x	x		x		x		x	x	x		x		x	x	x	x	
	18	x	x	x	x	x	x	x	x	x	x	x	x	x	x	x	x	x	x	x	x	x	x	x	x	x	x	x
	24	x	x	x	x	x	x	x	x	x	x	x	x	x	x	x	x	x	x	x	x	x	x	x	x	x	x	x

Draft - Not for Implementation

1782 Matrixing Example #2. Two-thirds fractional design with five-eighths time points (overall
1783 size: five-twelfths of full testing protocol)

1784 The following example (Table 9) involves a two-thirds fractional design of 3x3x3
1785 combinations of factors with five-eighths time points for a capsule dosage form which is
1786 available in 3 strengths of a common granulation and packaged in 3 container/closure
1787 systems and/or sizes: C1, HDPE bottle; 30 counts; C2, HDPE bottle, 100 counts; and C3,
1788 HDPE bottle, 200 counts. A 36-month expiration dating period is proposed. The overall
1789 size of this design can be referred to as 2/3 (of 27 combinations of factors) x 5/8 (of 8 time
1790 points), or 5/12 (of 216 samples in a full testing protocol).

1791 **Table 9: Matrixing Example #2**

1792–1799

Batch		1									2									3								
Strength		100 mg			200 mg			300 mg			100 mg			200 mg			300 mg			100 mg			200 mg			300 mg		
Container/ Closure		C1	C2	C3	C1	C2	C3	C1	C2	C3	C1	C2	C3	C1	C2	C3	C1	C2	C3	C1	C2	C3	C1	C2	C3	C1	C2	C3
Schedule		T1	T2		T2		T1		T1	T2	T2		T1		T1	T2	T1	T2			T1	T2	T1	T2		T2		T1
Time Points (mo)	0	x	x		x		x		x	x	x		x		x	x	x	x			x	x	x	x		x		x
	3	x					x		x				x		x		x				x		x					x
	6		x		x					x	x					x		x				x		x		x		
	9	x					x		x				x		x		x				x		x					x
	12		x		x					x	x					x		x				x		x		x		
	18	x					x		x				x		x		x				x		x					x
	24		x		x					x	x					x		x				x		x		x		
	36	x	x		x		x		x	x	x		x		x	x	x	x			x	x	x	x		x		x

Draft - Not for Implementation

1800 Matrixing Example #3. Bracketing design and three-fourths Matrix (overall size: one-half
1801 of full testing protocol)

1802 The following example (Table 10) illustrates how bracketing (of one factor) and matrixing
1803 (with three-fourths time points) can be combined in one protocol. The description of the
1804 drug product is as shown in Example 2. The overall size of this design is 2/3 X 3/4, or ½ of
1805 that of a full testing protocol.

1806 **Table 10: Matrixing Example #3**

1807 Batch		1									2									3								
1808 Strength		100 mg			200 mg			300 mg			100 mg			200 mg			300 mg			100 mg			200 mg			300 mg		
1809 1810 Container/ Closure		C1	C2	C3	C1	C2	C3	C1	C2	C3	C1	C2	C3	C1	C2	C3	C1	C2	C3	C1	C2	C3	C1	C2	C3	C1	C2	C3
1811 Schedule		T1	T2	T3				T3	T1	T2	T2	T3	T1				T1	T2	T3	T3	T1	T2				T2	T3	T1
1812 1813 1814 Time Points (mo)	0	x	x	x				x	x	x	x	x	x				x	x	x	x	x	x				x	x	x
	3	x	x						x	x	x		x				x	x			x	x				x		x
	6		x	x				x		x	x	x						x	x	x		x				x	x	
	9	x		x				x	x			x	x				x		x	x	x						x	x
	12	x	x	x				x	x	x	x	x	x				x	x	x	x	x	x				x	x	x
	18	x							x				x				x				x							x
	24		x	x				x		x	x	x						x	x	x		x				x	x	
	36	x	x	x				x	x	x	x	x	x				x	x	x	x	x	x				x	x	x

1815

Draft - Not for Implementation

1816 4. Data Evaluation

1817 The stability data obtained under a matrixing protocol should be subjected to the same type
1818 of statistical analysis with the same vigor and for the same aspects as outlined in Section
1819 VII.E. The same principle and procedure on poolability (i.e., testing data from different
1820 batches for similarity before combining them into one overall estimate, as described in
1821 Section VII.E.1) should be applied.

1822

1823 I. Site-Specific Stability Data For Drug and Biologic Applications

1824 1. Purpose

1825 At the time of NDA submission, at least 12 months of long-term data and 6 months of
1826 accelerated data should be available on three batches of the drug substance (all of which
1827 should be at least pilot scale) and three batches of the drug product (two of which should be
1828 at least pilot scale); reference is made to the drug substance and drug product sections of the
1829 ICH Q1A Guidance and to Sections II.A and II.B. of this guidance, respectively. Because
1830 the ICH Guidance did not address where the stability batches should be made, this section
1831 provides recommendations on site-specific stability data: the number and size of drug
1832 substance and drug product stability batches made at the intended manufacturing-scale
1833 production sites and the length of stability data on these batches, for an original NDA,
1834 ANDA, BLA or PLA application. Applicants are advised to consult with the respective
1835 chemistry review team when questions arise.

1836 2. Original NDAs, BLAs, or PLAs

1837 In principle, primary stability batches should be made at the intended commercial site. If the
1838 primary stability batches are not made at the intended commercial site, stability data from
1839 the drug substance/product batches manufactured at that site (i.e., site-specific batches)
1840 should be included in the original submission to demonstrate that the product made at each
1841 site is equivalent. If at the time of application submission, there are 12 months of long-term
1842 data and 6 months of accelerated data on three primary stability batches made at other than
1843 the intended commercial site, a reduced number of site-specific batches with shorter
1844 duration of data than the primary batches may be acceptable. In addition, these site-specific
1845 batches may be of pilot scale.

1846 A drug substance should be adequately characterized (i.e., results of chemical, physical,
1847 and, when applicable, biological testing). Material produced at different sites should be of
1848 comparable quality. In general, three to six months of stability data on one to three site-
1849 specific drug substance batches, depending on the availability of sufficient primary stability
1850 data from another site, should be provided at the time of application submission. Table 11
1851 depicts the site-specific stability data recommended for the drug substance in an original
1852 application.

Draft - Not for Implementation

1853 1854 **Table 11: Site-Specific Stability Data for a Drug Substance in an Original Application**

Scenario[a]	Site-Specific Stability Data Recommended at Time of Submission[b]	Stability Commitment[c]
Sufficient primary stability data are available for the drug substance	3 months of accelerated (from a 6-month study) and long-term data on 1 site-specific batch.	First 3 drug substance production batches on long-term and accelerated stability studies.
Sufficient primary stability data are not available for the drug substance	3 months of accelerated (from a 6-month study) and long-term data on 3 site-specific batches.	First 3 drug substance production batches on long-term and accelerated stability studies.

1862 1863 1864 [a] The phrase ***sufficient primary stability data*** means that, at the time of submission, there are 6 months of accelerated data and at least 12 months of long-term data on three primary stability batches made at a different pilot or production site from the intended site.

1865 1866 [b] Additional long-term stability data and, if applicable, accelerated data, should be submitted for review as soon as they become available prior to the approval.

1867 1868 1869 [c] A commitment should be provided in the application to place the first three production batches at each site on long-term and accelerated stability studies and annual batches thereafter on long-term studies using the approved protocol and to report the resulting data in annual reports.

1870 The complexity of the drug product dosage form is a critical factor in determining the
1871 number of site-specific batches for an original application. The quality and/or stability of a
1872 simple dosage form is less likely to vary due to a different manufacturing site than that of a
1873 complex dosage form. Three site-specific batches are needed for a complex dosage form to
1874 provide an independent and statistically meaningful stability profile for the product made at
1875 that site. One site-specific batch may be sufficient to verify the stability profile of a simple
1876 dosage form. Table 12, below, illustrates the site-specific stability data recommended for
1877 drug products in an original application:

Draft - Not for Implementation

1878 **Table 12: Site-Specific Stability Data for a Drug Product**
1879 **in an Original NDA, BLA, or PLA**

1880	Scenario[a]	Site-Specific Stability Data Recommended at Time of Submission[b]	Stability Commitment[c]
1881 1882 1883	Simple dosage form where sufficient primary stability data are available	3 months of accelerated (from a 6-month study) and long-term data on 1 site-specific batch.	First 3 production batches on long-term and accelerated stability studies.
1884 1885 1886	Complex dosage form where sufficient primary stability data are available	3 months of accelerated (from a 6-month study) and long-term data on 3 site-specific batches.	First 3 production batches on long-term and accelerated stability studies.
1887 1888 1889	Any dosage form where sufficient primary stability data are not available	6 months of accelerated and 12 months of long-term data on 3 site-specific batches.	First 3 production batches on long-term and accelerated stability studies.

1890 1891 1892 [a] The phrase *sufficient primary stability data* means that, at the time of submission, there are 6 months of accelerated data and at least 12 months of long-term data on three primary stability batches made at a different pilot or production site from the intended site.

1893 1894 [b] Additional long-term stability data and, if applicable, accelerated data should be submitted for review as soon as they become available prior to the approval.

1895 1896 1897 [c] A commitment should be provided in the application to place the first 3 production batches at each site on long-term and accelerated stability studies and annual batches thereafter on long-term studies using the approved protocol and to report the resulting data in annual reports.

1898 Other factors, such as lack of experience at the new site in a particular dosage form, or
1899 difference in the environmental conditions between the sites, can potentially affect the
1900 quality and/or stability of a drug product. Therefore, one site-specific batch may not be
1901 sufficient in these cases. More than one site-specific batch may be needed for a drug
1902 substance/product that is intrinsically unstable.

1903 Although one site-specific batch may be sufficient under certain situations, the data so
1904 generated, particularly if limited to accelerated studies, may not be amenable to statistical
1905 analysis for the establishment of a retest period or expiration dating period. Instead, the
1906 single site-specific batch may only serve to verify the stability profile of a drug
1907 substance/product that has been established based on primary stability batches at a pilot
1908 plant.

1909 In general, site-specific drug product batches should be made with identifiable site-specific
1910 drug substance batches both for original applications, wherever possible, and for
1911 postapproval stability commitment.

1912 Although pilot and commercial facilities may or may not be located on the same campus or

Draft - Not for Implementation

1913 within the same geographical area, they will generally employ similar processes and
1914 equipment of the same design and operating principles. If different processes and/or
1915 equipment are used, more site-specific batches and/or longer duration of data are
1916 recommended. If the pilot plant where the primary stability batches are made is located at
1917 the intended commercial site (i.e., on the same campus as the intended manufacturing-scale
1918 production facility) the site-specific stability recommendations are met (provided the
1919 processes and equipment are the same) and no additional data will be needed. A
1920 commitment should be made to place the first three production batches on accelerated and
1921 long-term stability studies. If more than one manufacturing-scale production site is
1922 proposed for an original NDA, BLA or PLA, the recommendations above would be
1923 applicable to each site.

1924 3. Site-Specific Data Package Recommendations for ANDAs

1925 For ANDAs, the primary batch(es) to support the application are usually manufactured in
1926 the production facility. If the primary stability batch(es) are not made at the intended
1927 commercial site, stability data should be generated, as outlined in Table 13, on the drug
1928 product manufactured at that site, i.e. site-specific batches, and the data should be included
1929 in the original submission to demonstrate that the product made at each site is equivalent.

1930 If the pilot plant where the primary stability batches are made is located at the intended
1931 commercial site (i.e., on the same campus as the intended commercial facility), the
1932 site-specific stability recommendations are met and no additional data will be needed. A
1933 commitment should be made to place the first three production batches and annual batches
1934 thereafter on long-term stability studies.

1935 For complex dosage forms as described in the previous section, a reduced number of
1936 site-specific batches may be justified if accelerated and long-term data are available at the
1937 time of application submission on batches made at a different pilot or commercial site from
1938 the intended commercial facility.
1939

Draft - Not for Implementation

1940 **Table 13: Site-Specific Stability Data for a Drug Product in an Original ANDA**

Scenario	Site-Specific Stability Data Recommended at Time of Submission[a]	Stability Commitment[b]
Simple Dosage Form	3 months of accelerated and available long-term data on 1 site-specific batch.	First 3 production batches on long-term stability studies.
Complex Dosage Form	3 months of accelerated and available long-term data on 3 site-specific batches.	First 3 production batches on long-term stability studies.

1941
1942
1943
1944
1945

1946 [a] Drug substance batches used to produce site-specific drug product batches should be clearly identified. Additional
1947 long-term stability data should be submitted for review as soon as they become available prior to approval.

1948 [b] A commitment should be provided in the application to place the first three production batches at each site on long-
1949 term stability studies and annual batches thereafter on long-term studies using the approved protocol and to report
1950 the resulting data in annual reports.

1951 **J. Photostability**

1952 1. General

1953 *The ICH Harmonized Tripartite Guideline on Stability Testing of New Drug Substances and*
1954 *Products* (hereafter referred to as the *parent guidance*) notes that light testing should be an
1955 integral part of stress testing.

1956 The ICH Q1B guidance *Photostability Testing of New Drug Substances and Products* primarily
1957 addresses the generation of photostability information for new molecular entities and associated
1958 drug products and the use of the data in determining whether precautionary measures in
1959 manufacturing, labeling, or packaging are needed to mitigate exposure to light. Q1B does not
1960 specifically address other photostability studies that may be needed to support, for example, the
1961 photostability of a product under in-use conditions or the photostability of analytical samples.
1962 Because data are generated on a directly exposed drug substance alone and/or in simple solutions
1963 and drug products when studies are conducted as described in the Q1B guidance, knowledge of
1964 photostability characteristics may be useful in determining when additional studies may be needed
1965 or in providing justification for not performing additional studies. For example, if a product has
1966 been determined to photodegrade upon direct exposure but is adequately protected by packaging,
1967 an in-use study may be needed to support the use of the product (e.g., a parenteral drug that is
1968 infused over a period of time). The test conditions for in-use studies will vary depending on the
1969 product and use but should depend on and relate to the directions for use of the particular product.

1970 Photostability studies are usually conducted only in conjunction with the first approval of a new
1971 molecular entity. Under some circumstances, photostability studies should be repeated if certain
1972 postapproval or supplemental changes, such as changes in formulation or packaging, are made to

Draft - Not for Implementation

1973 the product, or if a new dosage form is proposed. Whether these studies should be repeated
1974 depends on the photostability characteristics determined at the time of initial filing and the type of
1975 changes made. For example, if initial studies demonstrate that an active moiety in a simple solution
1976 degrades upon exposure to light and the tablet drug product is stable, a subsequent filing
1977 requesting approval of a liquid dosage form may warrant additional studies to characterize the
1978 photostability characteristics of the new dosage form.

1979 Photostability studies need not be conducted for products that duplicate a commercially available
1980 listed drug product provided that the packaging (immediate container/closure and market pack)
1981 and labeling storage statements regarding light duplicate those of the reference listed drug. If
1982 deviations in packaging or labeling statements are made, additional studies may be recommended.
1983 The decision as to whether additional studies should be conducted will be made on a case-by-case
1984 basis by the chemistry review team.

1985 The intrinsic photostability characteristics of new drug substances and products should be
1986 evaluated to demonstrate that, as appropriate, light exposure does not result in unacceptable
1987 change. Normally, photostability testing is carried out on a single batch of material selected as
1988 described in the section Selection of Batches, in the parent guidance. Under some circumstances,
1989 these studies should be repeated if certain variations and changes are made to the product (e.g.,
1990 formulation, packaging). Whether these studies should be repeated depends on the photostability
1991 characteristics determined at the time of initial filing and the type of variation and/or change made.
1992 [ICH Q1B]

1993 A systematic approach to photostability testing is recommended covering, as appropriate, studies
1994 such as:

1995 • Tests on the drug substance;
1996 • Tests on the exposed drug product outside of the immediate pack; and if necessary,
1997 • Tests on the drug product in the immediate pack; and if necessary,
1998 • Tests on the drug product in the marketing pack.[ICH Q1B]

1999 The extent of drug product testing should be established by assessing whether or not acceptable
2000 change has occurred at the end of the light exposure testing as described in Figure 2, the Decision
2001 Flow Chart for Photostability Testing of Drug Products. Acceptable change is change within limits
2002 justified by the applicant. [ICH Q1B]

2003 The formal labeling requirements for photolabile drug substances and drug products are established
2004 by national/regional requirements. [ICH Q1B]

2005 2. Light Sources
2006
2007 The light sources described below may be used for photostability testing. The applicant should
2008 either maintain an appropriate control of temperature to minimize the effect of localized
2009 temperature changes or include a dark control in the same environment unless otherwise justified.
2010 For both options 1 and 2, a pharmaceutical manufacturer/applicant can rely on the spectral
2011 distribution specification of the light source manufacturer. [ICH Q1B]

Draft - Not for Implementation

2012 **Option 1**

2013 Any light source that is designed to produce an output similar to the D65/ID65 emission standard
2014 such as an artificial daylight fluorescent lamp combining visible and ultraviolet (UV) outputs,
2015 xenon, or metal halide lamp. D65 is the internationally recognized standard for outdoor daylight as
2016 defined in ISO 10977 (1993). ID65 is the equivalent indoor indirect daylight standard. For a light
2017 source emitting significant radiation below 320 nanometers (nm), an appropriate filter(s) may be
2018 fitted to eliminate such radiation. [ICHQ1B]

2019 **Option 2**

2020 For option 2 the same sample should be exposed to both the cool white fluorescent and near
2021 ultraviolet lamp.

2022 • A cool white fluorescent lamp designed to produce an output similar to that specified in ISO
2023 10977 (1993); and

2024 • A near UV fluorescent lamp having a spectral distribution from 320 nm to 400 nm with a
2025 maximum energy emission between 350 nm and 370 nm; a significant proportion of UV should
2026 be in both bands of 320 to 360 nm and 360 to 400 nm. [ICH Q1B]
2027
2028 3. Procedure [ICH Q1B]

2029 For confirmatory studies, samples should be exposed to light providing an overall illumination of
2030 not less than 1.2 million lux hours and an integrated near ultraviolet energy of not less than 200
2031 watt hours/square meter to allow direct comparisons to be made between the drug substance and
2032 drug product.

2033 Samples may be exposed side-by-side with a validated chemical actinometric system to ensure the
2034 specified light exposure is obtained, or for the appropriate duration of time when conditions have
2035 been monitored using calibrated radiometers/lux meters. An example of an actinometric procedure
2036 is provided in the Annex.

2037 If protected samples (e.g., wrapped in aluminum foil) are used as dark controls to evaluate the
2038 contribution of thermally induced change to the total observed change, these should be placed
2039 alongside the authentic sample. [ICH Q1B]

Draft - Not for Implementation

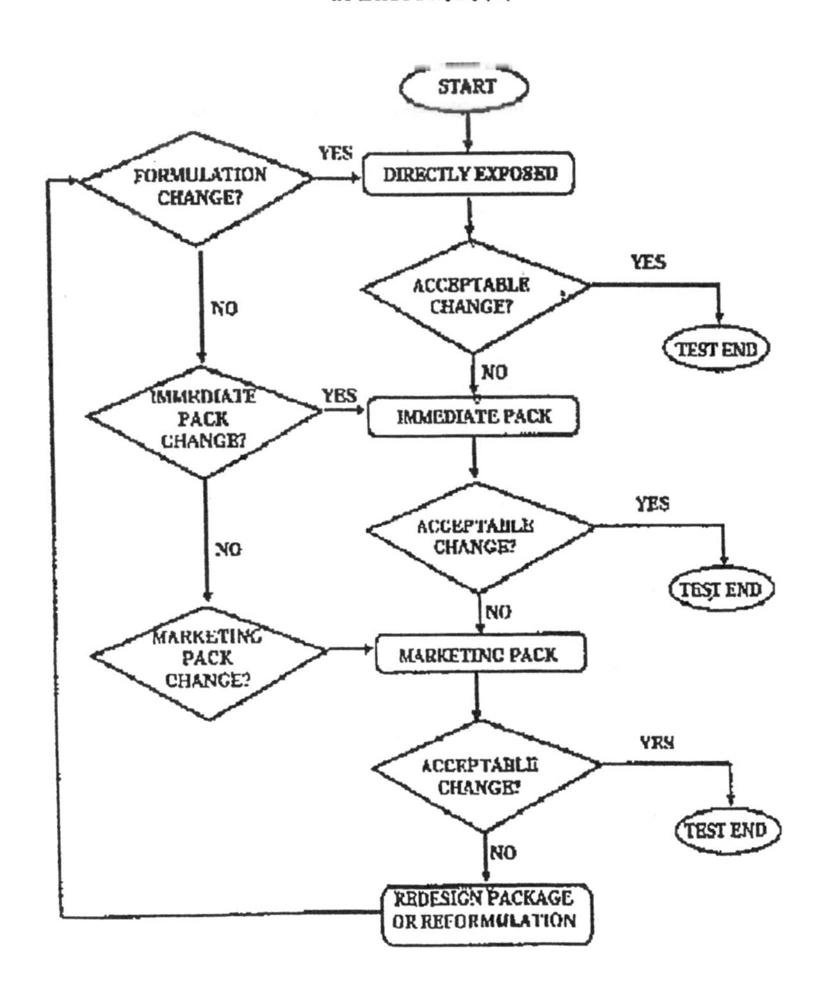

Draft - Not for Implementation

2040 4. Drug Substance [ICH Q1B]

2041 For drug substances, photostability testing should consist of two parts: Forced degradation testing
2042 and confirmatory testing.

2043 The purpose of forced degradation testing studies is to evaluate the overall photosensitivity of the
2044 material for method development purposes and/or degradation pathway elucidation. This testing
2045 may involve the drug substance alone and/or in simple solutions/suspensions to validate the
2046 analytical procedures. In these studies, the samples should be in chemically inert and transparent
2047 containers. In these forced degradation studies, a variety of exposure conditions may be used,
2048 depending on the photosensitivity of the drug substance involved and the intensity of the light
2049 sources used. For development and validation purposes, it is appropriate to limit exposure and end
2050 the studies if extensive decomposition occurs. For photostable materials, studies may be
2051 terminated after an appropriate exposure level has been used. The design of these experiments is
2052 left to the applicant's discretion although the exposure levels used should be justified.

2053 Under forcing conditions, decomposition products may be observed that are unlikely to be formed
2054 under the conditions used for confirmatory studies. This information may be useful in developing
2055 and validating suitable analytical methods. If in practice it has been demonstrated they are not
2056 formed in the confirmatory studies, these degradation products need not be examined further.

2057 Confirmatory studies should then be undertaken to provide the information necessary for handling,
2058 packaging, and labeling (see Section VIII.J.3., Procedure, and 4.a., Presentation of Samples, for
2059 information on the design of these studies).

2060 Normally, only one batch of drug substance is tested during the development phase, and then the
2061 photostability characteristics should be confirmed on a single batch selected as described in the
2062 parent guidance if the drug is clearly photostable or photolabile. If the results of the confirmatory
2063 study are equivocal, testing of up to two additional batches should be conducted. Samples should
2064 be selected as described in the parent guidance.

2065 a. Presentation of Samples [ICH Q1B]

2066 Care should be taken to ensure that the physical characteristics of the samples under test are taken
2067 into account, and efforts should be made, such as cooling and/or placing the samples in sealed
2068 containers, to ensure that the effects of the changes in physical states such as sublimation,
2069 evaporation, or melting are minimized. All such precautions should be chosen to provide minimal
2070 interference with the exposure of samples under test. Possible interactions between the samples
2071 and any material used for containers or for general protection of the sample should also be
2072 considered and eliminated wherever not relevant to the test being carried out.

2073 As a direct challenge for samples of solid drug substances, an appropriate amount of sample should
2074 be taken and placed in a suitable glass or plastic dish and protected with a suitable transparent
2075 cover if considered necessary. Solid drug substances should be spread across the container to give

Draft - Not for Implementation

2076 a thickness of typically not more than 3 millimeters. Drug substances that are liquids should be
2077 exposed in chemically inert and transparent containers.

2078 b. Analysis of Samples

2079 At the end of the exposure period, the samples should be examined for any changes in physical
2080 properties (e.g., appearance, clarity or color of solution) and for assay and degradants by a method
2081 suitably validated for products likely to arise from photochemical degradation processes.

2082 Where solid drug substance samples are involved, sampling should ensure that a representative
2083 portion is used in individual tests. Similar sampling considerations, such as homogenization of the
2084 entire sample, apply to other materials that may not be homogeneous after exposure. The analysis
2085 of the exposed sample should be performed concomitantly with that of any protected samples used
2086 as dark control if these are used in the test.

2087 c. Judgment of Results

2088 The forced degradation studies should be designed to provide suitable information to develop and
2089 validate test methods for the confirmatory studies. These test methods should be capable of
2090 resolving and detecting photolytic degradants that appear during the confirmatory studies. When
2091 evaluating the results of these studies, it is important to recognize that they form part of the stress
2092 testing and are not therefore designed to establish qualitative or quantitative limits for change.

2093 The confirmatory studies should identify precautionary measures needed in manufacturing or in
2094 formulation of the drug product and if light resistant packaging is needed. When evaluating the
2095 results of confirmatory studies to determine whether change due to exposure to light is acceptable,
2096 it is important to consider the results from other formal stability studies to ensure that the drug will
2097 be within justified limits at time of use (see the relevant ICH stability and impurity guidance).

2098 5. Drug Product [ICH Q1B]
2099
2100 Normally, the studies on drug products should be carried out in a sequential manner starting with
2101 testing the fully exposed product then progressing as necessary to the product in the immediate
2102 pack and then in the marketing pack. Testing should progress until the results demonstrate that
2103 the drug product is adequately protected from exposure to light. The drug product should be
2104 exposed to the light conditions described under the procedure in Section VII.J.3.

2105 Normally, only one batch of drug product is tested during the development phase, and then the
2106 photostability characteristics should be confirmed on a single batch selected as described in the
2107 parent guidance if the product is clearly photostable or photolabile. If the results of the
2108 confirmatory study are equivocal, testing of up to two additional batches should be conducted.

2109 For some products where it has been demonstrated that the immediate pack is completely
2110 impenetrable to light, such as aluminum tubes or cans, testing should normally only be conducted
2111 on directly exposed drug product.

Draft - Not for Implementation

2112 It may be appropriate to test certain products, such as infusion liquids or dermal creams, to
2113 support their photostability in-use. The extent of this testing should depend on and relate to the
2114 directions for use, and is left to the applicant's discretion.

2115 The analytical procedures used should be suitably validated.

2116 a. Presentation of Samples

2117 Care should be taken to ensure that the physical characteristics of the samples under test are taken
2118 into account, and efforts, such as cooling and/or placing the samples in sealed containers, should
2119 be made to ensure that the effects of the changes in physical states are minimized, such as
2120 sublimation, evaporation, or melting. All such precautions should be chosen to provide minimal
2121 interference with the irradiation of samples under test. Possible interactions between the samples
2122 and any material used for containers or for general protection of the sample should also be
2123 considered and eliminated wherever not relevant to the test being carried out.

2124 Where practicable when testing samples of the drug product outside of the primary pack, these
2125 should be presented in a way similar to the conditions mentioned for the drug substance. The
2126 samples should be positioned to provide maximum area of exposure to the light source. For
2127 example, tablets and capsules should be spread in a single layer.

2128 If direct exposure is not practical (e.g., due to oxidation of a product), the sample should be placed
2129 in a suitable protective inert transparent container (e.g., quartz).

2130 If testing of the drug product in the immediate container or as marketed is needed, the samples
2131 should be placed horizontally or transversely with respect to the light source, whichever provides
2132 for the most uniform exposure of the samples. Some adjustment of testing conditions may have to
2133 be made when testing large volume containers (e.g., dispensing packs).

2134 b. Analysis of Samples

2135 At the end of the exposure period, the samples should be examined for any changes in physical
2136 properties (e.g., appearance, clarity, or color of solution, dissolution/disintegration for dosage
2137 forms such as capsules) and for assay and degradants by a method suitably validated for products
2138 likely to arise from photochemical degradation processes.

2139 When powder samples are involved, sampling should ensure that a representative portion is used in
2140 individual tests. For solid oral dosage form products, testing should be conducted on an
2141 appropriately sized composite of, for example, 20 tablets or capsules. Similar sampling
2142 considerations, such as homogenization or solubilization of the entire sample, apply to other
2143 materials that may not be homogeneous after exposure (e.g., creams, ointments, suspensions). The
2144 analysis of the exposed sample should be performed concomitantly with that of any protected
2145 samples used as dark controls if these are used in the test.

2146 c. Judgment of Results

Draft - Not for Implementation

2147 Depending on the extent of change, special labeling or packaging may be needed to mitigate
2148 exposure to light. When evaluating the results of photostability studies to determine whether
2149 change due to exposure to light is acceptable, it is important to consider the results obtained from
2150 other formal stability studies to ensure that the product will be within proposed specifications
2151 during the shelf life (see the relevant ICH stability and impurity guidance).

2152 6. Quinine Chemical Actinometry [ICH Q1B]

2153 The following provides details of an actinometric procedure for monitoring exposure to a near UV
2154 fluorescent lamp (based on work done by FDA/National Institute of Standards and Technolo
2155 study). For other light sources/actinometric systems, the same approach may be used, but ea
2156 actinometric system should be calibrated for the light source used.

2157 Prepare a sufficient quantity of a 2 percent weight/volume aqueous solution of quinine
2158 monohydrochloride dihydrate (if necessary, dissolve by heating).
2159
2160 **Option 1**

2161 Put 10 milliliters (mL) of the solution into a 20 mL colorless ampoule (see drawing, below),
2162 hermetically, and use this as the sample. Separately, put 10 mL of the solution into a 20 mL
2163 colorless ampoule (see note 1), seal it hermetically, wrap in aluminum foil to protect comple
2164 from light, and use this as the control. Expose the sample and control to the light source fo
2165 appropriate number of hours. After exposure, determine the absorbances of the sample (AT
2166 the control (AO) at 400 nm using a 1 centimeter (cm) path length. Calculate the change in
2167 absorbance units (AU): A = AT - AO. The length of exposure should be sufficient to ensu
2168 change in absorbance of at least 0.9 AU. *Note: Shape and Dimensions (See Japanese Indu*
2169 *Standard (JIS) R3512 (1974) for ampoule specifications).*[10]

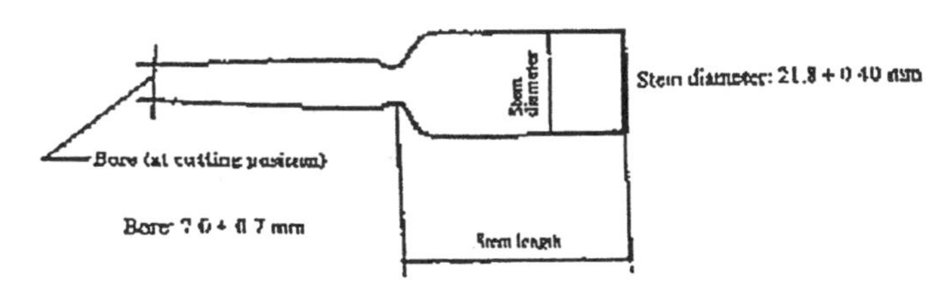

Stem length 80.0 = 1.2 mm

[10] Yoshioka, S., "Quinine Actinometry as a Method for Calibrating Ultraviolet Radiation Intensity in Light-stability Testing of Pharmaceuticals," *Drug Development and Industrial Pharmacy*, 20(13):2049-2062, 1994

Draft - Not for Implementation

2170 **Option 2**

2171 Fill a 1 cm quartz cell and use this as the sample. Separately fill a 1 cm quartz cell, wrap in
2172 aluminum foil to protect completely from light, and use this as the control. Expose the sample and
2173 control to the light source for an appropriate number of hours. After exposure, determine the
2174 absorbances of the sample (AT) and the control (AO) at 400 nm. Calculate the change in
2175 absorbance,) A = AT - AO. The length of exposure should be sufficient to ensure a change in
2176 absorbance of at least 0.3.

2177 Alternative packaging configurations may be used if appropriately validated. Alternative validated
2178 chemical actinometers may be used.

2179 7. Acceptable/Unacceptable Photostability Change

2180 The extent of the drug product photostability testing depends on the change that has occurred at
2181 the end of each test tier described in Figure 2, above, the Decision Flow Chart for Photostability
2182 Testing of Drug Products. Test results that are outside the proposed acceptance criteria for the
2183 product would not be considered acceptable change. This is a stress test designed to determine the
2184 intrinsic photostability characteristics of new drug substances and products, and no correlation has
2185 been developed to equate a within specification result to an expiration dating period. The
2186 acceptability of any observed changes should be justified in the application. It may be important to
2187 consider other degradative processes (e.g., thermal) when justifying a photostability change as
2188 acceptable because the processes may be independent and additive. For example, a 5 percent loss
2189 in potency due to photodegradation may be considered acceptable if that is the only type of
2190 degradation observed. If the product is also expected to degrade 5 percent over the shelf-life due
2191 to thermal degradation, the photodegradation may then be considered unacceptable based on the
2192 potential additive effect of the changes. In this case, precautions should be taken to mitigate the
2193 product's exposure to light.

2194 Under the intense light exposure conditions included in the Q1B guidance, certain colors in solid
2195 dosage forms may fade. Quantitative analysis of the color change is not recommended as these
2196 changes are not likely to occur under actual storage conditions. In the absence of change in other
2197 parameters such as assay, these color changes may be acceptable.
2198
2199 8. Photostability Labeling Considerations
2200
2201 The data generated using the procedure described in the ICH Q1A guidance is useful in
2202 determining when special handling or storage statements regarding exposure to light should be
2203 included in the product labeling (21 CFR 201.57(k)(4)). The labeling guidance provided below
2204 pertains only to products as packaged for distribution. Instructions and stability statements that
2205 may be needed to address in-use conditions pursuant to 21 CFR 201.57(j) are not covered.

2206 **Change after direct exposure:** If changes that are observed when the product is directly exposed

Draft - Not for Implementation

2207 under the light conditions described in the Q1B guidance are acceptable, no labeling storage
2208 statement regarding light is needed.

2209 **Change after exposure in the immediate container/closure:** If changes observed when the
2210 product is directly exposed are unacceptable, but are acceptable when the product is tested in the
2211 immediate container/closure under the conditions described in the Q1B guidance, the inclusion of a
2212 labeling storage statement regarding light would depend on the likelihood of the product being
2213 removed from the immediate package during the distribution process.

2214 • For those products that are unlikely to be removed from the immediate container, such as
2215 creams or ointments in tubes dispensed directly to the patient, and ophthalmic products, the
2216 use of a labeling storage statement regarding light is optional.

2217 • For products that may be removed from the immediate pack, such as pharmacy bulk packs, a
2218 light storage statement should be included such as "PROTECT FROM LIGHT. Dispense in a
2219 light-resistant container."

2220 **Change after exposure in the market pack:** If changes that are observed are acceptable only
2221 when the product in the market pack is exposed under the conditions described in the Q1B
2222 guidance, labeling storage statements regarding light should be included.
2223
2224 Examples of typical storage statements are, for single-dose and multiple-dose products
2225 respectively, "PROTECT FROM LIGHT. Retain in carton until time of use." and "PROTECT
2226 FROM LIGHT. Retain in carton until contents are used."

2227 **K. Degradation Products**

2228 When degradation products are detected upon storage, the following information about them
2229 should be submitted:

2230 • Procedure for isolation and purification
2231 • Identity and chemical structures
2232 • Degradation pathways
2233 • Physical and chemical properties
2234 • Detection and quantitation levels
2235 • Acceptance Criteria (individual and total)
2236 • Test methods
2237 • Validation data
2238 • Biological effect and pharmacological actions, including toxicity studies, at the concentrations
2239 likely to be encountered (cross-reference to any available information is acceptable)

2240 If racemization of the drug substance in the dosage form is possible, the information described
2241 above also should be provided.

2242 **L. Thermal Cycling**

2243 A study of the effects of temperature variation, particularly if appropriate for the shipping and

Draft - Not for Implementation

2244 storage conditions of certain drug products, should be considered. Drug products susceptible to
2245 phase separation, loss of viscosity, precipitation, and aggregation should be evaluated under such
2246 thermal conditions. As part of the stress testing, the packaged drug product should be cycled
2247 through temperature conditions that simulate the changes likely to be encountered once the drug
2248 product is in distribution.

2249 • A temperature cycling study for drug products that may be exposed to temperature variations
2250 above freezing may consist of three cycles of two days at refrigerated temperature (2-8°C)
2251 followed by two days under accelerated storage conditions (40°C).

2252 • A temperature cycling study for drug products that may be exposed to sub-freezing
2253 temperatures may consist of three cycles of two days at freezer temperature (-10° to -20°C)
2254 followed by two days under accelerated storage conditions (40°C).

2255 • For inhalation aerosols, the recommended cycle study consists of three or four six-hour cycles
2256 per day, between subfreezing temperature and 40°C (75-85 percent RH) for a period of up to
2257 six weeks.

2258 • For frozen drug products, the recommended cycle study should include an evaluation of effects
2259 due to accelerated thawing in a microwave or a hot water bath unless contraindicated in the
2260 labeling.

2261 • Alternatives to these conditions may be acceptable with appropriate justification.

2262 **M. Stability Testing in Foreign Laboratory Facilities**

2263 Stability testing (as well as finished product release testing) performed in any foreign or domestic
2264 facility may be used as the basis for approval of an application. This includes all NDAs, ANDAs,
2265 and related CMC supplements. A satisfactory inspection of the laboratory(ies) that will perform
2266 the testing will be necessary.[11]

2267 Applicants should consider the effects of bulk packaging, shipping, and holding of dosage forms
2268 and subsequent market packaging, and distribution of the finished drug product, and be aware of
2269 the effect of such operations on product quality. Time frames should be established to encompass
2270 the date of production, date of quality control release of the dosage form, bulk packaging,
2271 shipping, and market packaging, and initiation and performance of the stability studies on the drug
2272 product should be established, controlled, and strictly followed. Maximum time frames for each
2273 operation should be established and substantiated by the applicant.

[11] This statement replaces a previous position, established via a CDER Office of Generic Drugs guidance, which recommended that finished product and stability testing be conducted at a United States laboratory for drug products manufactured in foreign facilities and shipped in bulk containers to the United States for packaging into immediate containers for marketing.

Draft - Not for Implementation

2274 **N. Stability Testing of Biotechnology Drug Products**

2275 1. General [ICH Q5C]

2276 The ICH harmonized tripartite guidance entitled Q1A *Stability Testing of New Drug Substances*
2277 *and Products* issued by ICH on October 27, 1993, applies in general to biotechnological/biological
2278 products. However, biotechnological/biological products have distinguishing characteristics to
2279 which consideration should be given in any well-defined testing program designed to confirm their
2280 stability during the intended storage period. For such products in which the active components are
2281 typically proteins and/or polypeptides, maintenance of molecular conformation and, hence, of
2282 biological activity, is dependent on noncovalent as well as covalent forces. The products are
2283 particularly sensitive to environmental factors such as temperature changes, oxidation, light, ionic
2284 content, and shear. To ensure maintenance of biological activity and to avoid degradation,
2285 stringent conditions for their storage are usually necessary.

2286 The evaluation of stability may necessitate complex analytical methodologies. Assays for
2287 biological activity, where applicable, should be part of the pivotal stability studies. Appropriate
2288 physicochemical, biochemical, and immunochemical methods for the analysis of the molecular
2289 entity and the quantitative detection of degradation products should also be part of the stability
2290 program whenever purity and molecular characteristics of the product permit use of these
2291 methodologies.

2292 With these concerns in mind, the applicant should develop the proper supporting stability data for a
2293 biotechnological/biological product and consider many external conditions that can affect the
2294 product's potency, purity, and quality. Primary data to support a requested storage period for
2295 either drug substance or drug product should be based on long-term, real-time, real-condition
2296 stability studies. Thus, the development of a proper long-term stability program becomes critical to
2297 the successful development of a commercial product. The purpose of this document is to give
2298 guidance to applicants regarding the type of stability studies that should be provided in support of
2299 marketing applications. It is understood that during the review and evaluation process, continuing
2300 updates of initial stability data may occur.

2301 2. Scope [ICH Q5C]

2302 The guidance in this section applies to well-characterized proteins and polypeptides, their
2303 derivatives and products of which they are components and which are isolated from tissues, body
2304 fluids, cell cultures, or produced using recombinant deoxyribonucleic acid (r-DNA) technology.
2305 Thus, the section covers the generation and submission of stability data for products such as
2306 cytokines (interferons, interleukins, colony-stimulating factors, tumor necrosis factors),
2307 erythropoietins, plasminogen activators, blood plasma factors, growth hormones and growth
2308 factors, insulins, monoclonal antibodies, and vaccines consisting of well-characterized proteins or
2309 polypeptides. In addition, the guidance outlined in the following sections may apply to other types
2310 of products, such as conventional vaccines, after consultation with the product review office. The
2311 section does not cover antibiotics, allergenic extracts, heparins, vitamins, whole blood, or cellular
2312 blood components.

Draft - Not for Implementation

2313 3. Terminology [ICH Q5C]

2314 For the basic terms used in this section, the reader is referred to the Glossary. However, because
2315 manufacturers of biotechnological/biological products sometimes use traditional terminology,
2316 traditional terms are specified in parentheses to assist the reader.

2317 4. Selection of Batches [ICH Q5C]

2318 a. Drug Substance (Bulk Material)

2319 Where bulk material is to be stored after manufacture, but before formulation and final
2320 manufacturing, stability data should be provided on at least three batches for which manufacture
2321 and storage are representative of the manufacturing scale of production. A minimum of six
2322 months' stability data at the time of submission should be submitted in cases where storage periods
2323 greater than six months are requested. For drug substances with storage periods of less than six
2324 months, the minimum amount of stability data in the initial submission should be determined on a
2325 case-by-case basis. Data from pilot-scale batches of drug substance produced at a reduced scale of
2326 fermentation and purification may be provided at the time the application is submitted to the
2327 Agency with a commitment to place the first three manufacturing scale batches into the long-term
2328 stability program after approval.

2329 The quality of the batches of drug substance placed into the stability program should be
2330 representative of the quality of the material used in preclinical and clinical studies and of the quality
2331 of the material to be made at manufacturing scale. In addition, the drug substance (bulk material)
2332 made at pilot-scale should be produced by a process and stored under conditions representative of
2333 that used for the manufacturing scale. The drug substance entered into the stability program should
2334 be stored in containers that properly represent the actual holding containers used during
2335 manufacture. Containers of reduced size may be acceptable for drug substance stability testing
2336 provided that they are constructed of the same material and use the same type of container/closure
2337 system that is intended to be used during manufacture.

2338 b. Intermediates

2339 During manufacture of biotechnological/biological products, the quality and control of certain
2340 intermediates may be critical to the production of the final product. In general, the manufacturer
2341 should identify intermediates and generate in-house data and process limits that ensure their
2342 stability within the bounds of the developed process. Although the use of pilot-scale data is
2343 permissible, the manufacturer should establish the suitability of such data using the manufacturing-
2344 scale process.

2345 c. Drug Product (Final Container Product)

2346 Stability information should be provided on at least three batches of final container product
2347 representative of that which will be used at manufacturing scale. Where possible, batches of final
2348 container product included in stability testing should be derived from different batches of bulk
2349 material. A minimum of six months' data at the time of submission should be submitted in cases

Draft - Not for Implementation

2350 where storage periods greater than six months are requested. For drug products with storage
2351 periods of less than six months, the minimum amount of stability data in the initial submission
2352 should be determined on a case-by-case basis. Product expiration dating should be based upon the
2353 actual data submitted in support of the application. Because dating is based upon the
2354 real-time/real-temperature data submitted for review, continuing updates of initial stability data
2355 should occur during the review and evaluation process. The quality of the final container product
2356 placed on stability studies should be representative of the quality of the material used in the
2357 preclinical and clinical studies. Data from pilot-scale batches of drug product may be provided at
2358 the time the application is submitted to the Agency with a commitment to place the first three
2359 manufacturing scale batches into the long-term stability program after approval. Where pilot-plant
2360 scale batches were submitted to establish the dating for a product and, in the event that the product
2361 produced at manufacturing scale does not meet those long-term stability specifications throughout
2362 the dating period or is not representative of the material used in preclinical and clinical studies, the
2363 applicant should notify the appropriate FDA reviewing office to determine a suitable course of
2364 action.

2365 d. Sample Selection

2366 Where one product is distributed in batches differing in fill volume (e.g., 1 milliliter (mL), 2 mL, or
2367 10 mL), unitage (e.g., 10 units, 20 units, or 50 units), or mass (e.g., 1 milligram (mg), 2 mg, or 5
2368 mg), samples to be entered into the stability program may be selected on the basis of a matrix
2369 system and/or by bracketing.

2370 Matrixing — the statistical design of a stability study in which different fractions of samples are
2371 tested at different sampling points — should only be applied when appropriate documentation is
2372 provided that confirms that the stability of the samples tested represents the stability of all samples.
2373 The differences in the samples for the same drug product should be identified as, for example,
2374 covering different batches, different strengths, different sizes of the same closure, and, possibly, in
2375 some cases, different container/closure systems. Matrixing should not be applied to samples with
2376 differences that may affect stability, such as different strengths and different containers/closures,
2377 where it cannot be confirmed that the products respond similarly under storage conditions,

2378 Where the same strength and exact container/closure system is used for three or more fill contents,
2379 the manufacturer may elect to place only the smallest and largest container size into the stability
2380 program (i.e., bracketing). The design of a protocol that incorporates bracketing assumes that the
2381 stability of the intermediate condition samples are represented by those at the extremes. In certain
2382 cases, data may be needed to demonstrate that all samples are properly represented by data
2383 collected for the extremes.

2384 5. Stability-Indicating Profile [ICH Q5C]

2385 On the whole, there is no single stability-indicating assay or parameter that profiles the stability
2386 characteristics of a biotechnological/biological product. Consequently, the manufacturer should
2387 propose a stability-indicating profile that provides assurance that changes in the identity, purity,
2388 and potency of the product will be detected.

Draft - Not for Implementation

2389 At the time of submission, applicants should have validated the methods that comprise the
2390 stability-indicating profile, and the data should be available for review. The determination of which
2391 tests should be included will be product-specific. The items emphasized in the following
2392 subsections are not intended to be all-inclusive, but represent product characteristics that should
2393 typically be documented to demonstrate product stability adequately.

2394 a. Protocol

2395 The marketing application should include a detailed protocol for the assessment of the stability of
2396 both drug substance and drug product in support of the proposed storage conditions and
2397 expiration dating periods. The protocol should include all necessary information that demonstrates
2398 the stability of the biotechnological/biological product throughout the proposed expiration dating
2399 period including, for example, well-defined specifications and test intervals. The statistical methods
2400 that should be used are described in the ICH Q1A guidance on stability.

2401 b. Potency

2402 When the intended use of a product is linked to a definable and measurable biological activity,
2403 testing for potency should be part of the stability studies. For the purpose of stability testing of the
2404 products described in this guidance, potency is the specific ability or capacity of a product to
2405 achieve its intended effect. It is based on the measurement of some attribute of the product and is
2406 determined by a suitable in vivo or in vitro quantitative method. In general, potencies of
2407 biotechnological/biological products tested by different laboratories can be compared in a
2408 meaningful way only if expressed in relation to that of an appropriate reference material. For that
2409 purpose, a reference material calibrated directly or indirectly against the corresponding national or
2410 international reference material should be included in the assay.

2411 Potency studies should be performed at appropriate intervals as defined in the stability protocol
2412 and the results should be reported in units of biological activity calibrated, whenever possible,
2413 against nationally or internationally recognized standards. Where no national or international
2414 reference standards exist, the assay results may be reported in in-house derived units using a
2415 characterized reference material.

2416 In some biotechnological/biological products, potency is dependent upon the conjugation of the
2417 active ingredient(s) to a second moiety or binding to an adjuvant. Dissociation of the active
2418 ingredient(s) from the carrier used in conjugates or adjuvants should be examined in
2419 real-time/real-temperature studies (including conditions encountered during shipment). The
2420 assessment of the stability of such products may be difficult because, in some cases, in vitro tests
2421 for biological activity and physicochemical characterization are impractical or provide inaccurate
2422 results. Appropriate strategies (e.g., testing the product before conjugation/binding, assessing the
2423 release of the active compound from the second moiety, in vivo assays) or the use of an
2424 appropriate surrogate test should be considered to overcome the inadequacies of in vitro testing.

2425 c. Purity and Molecular Characterization

Draft - Not for Implementation

2426 For the purpose of stability testing of the products described in this guidance, purity is a relative
2427 term. Because of the effect of glycosylation, deamidation, or other heterogeneities, the absolute
2428 purity of a biotechnological/biological product is extremely difficult to determine. Thus, the purity
2429 of a biotechnological/biological product should be typically assessed by more than one method and
2430 the purity value derived is method-dependent. For the purpose of stability testing, tests for purity
2431 should focus on methods for determination of degradation products.

2432 The degree of purity, as well as the individual and total amounts of degradation products of the
2433 biotechnological/biological product entered into the stability studies, should be reported and
2434 documented whenever possible. Limits of acceptable degradation should be derived from the
2435 analytical profiles of batches of the drug substance and drug product used in the preclinical and
2436 clinical studies.

2437 The use of relevant physicochemical, biochemical, and immunochemical analytical methodologies
2438 should permit a comprehensive characterization of the drug substance and/or drug product (e.g.,
2439 molecular size, charge, hydrophobicity) and the accurate detection of degradation changes that
2440 may result from deamidation, oxidation, sulfoxidation, aggregation, or fragmentation during
2441 storage. As examples, methods that may contribute to this include electrophoresis (SDS-PAGE,
2442 immunoelectrophoresis, Western blot, isoelectrofocusing), high-resolution chromatography (e.g.,
2443 reversed-phase chromatography, gel filtration, ion exchange, affinity chromatography), and peptide
2444 mapping.

2445 Wherever significant qualitative or quantitative changes indicative of degradation product
2446 formation are detected during long-term, accelerated, and/or stress stability studies, consideration
2447 should be given to potential hazards and to the need for characterization and quantification of
2448 degradation products within the long-term stability program. Acceptable limits should be proposed
2449 and justified, taking into account the levels observed in material used in preclinical and clinical
2450 studies.

2451 For substances that cannot be properly characterized or products for which an exact analysis of the
2452 purity cannot be determined through routine analytical methods, the applicant should propose and
2453 justify alternative testing procedures.

2454 d. Other Product Characteristics

2455 The following product characteristics, though not specifically relating to
2456 biotechnological/biological products should be monitored and reported for the drug product in its
2457 final container:

2458 • Visual appearance of the product (color and opacity for solutions/suspensions; color, texture,
2459 and dissolution time for powders), visible particulates in solutions or after the reconstitution of
2460 powders or lyophilized cakes, pH, and moisture level of powders and lyophilized products.

2461 • Sterility testing or alternatives (e.g., container/closure integrity testing) should be performed at
2462 a minimum initially and at the end of the proposed shelf life.

Draft - Not for Implementation

2463 • Additives (e.g., stabilizers, preservatives) or excipients may degrade during the dating period
2464 of the drug product. If there is any indication during preliminary stability studies that reaction
2465 or degradation of such materials adversely affect the quality of the drug product, these items
2466 may need to be monitored during the stability program.

2467 • The container/closure has the potential to affect the product adversely and should be carefully
2468 evaluated (see below).

2469 6. Storage Conditions [ICH Q5C]

2470 a. Temperature

2471 Because most finished biotechnological/biological products need precisely defined storage
2472 temperatures, the storage conditions for the real-time/real-temperature stability studies may be
2473 confined to the proposed storage temperature.

2474 b. Humidity

2475 Biotechnological/biological products are generally distributed in containers protecting them against
2476 humidity. Therefore, where it can be demonstrated that the proposed containers (and conditions of
2477 storage) afford sufficient protection against high and low humidity, stability tests at different
2478 relative humidities can usually be omitted. Where humidity-protecting containers are not used,
2479 appropriate stability data should be provided.

2480 c. Accelerated and Stress Conditions

2481 As previously noted, the expiration dating should be based on real-time/real-temperature data.
2482 However, it is strongly recommended that studies be conducted on the drug substance and drug
2483 product under accelerated and stress conditions. Studies under accelerated conditions may provide
2484 useful support data for establishing the expiration date, provide product stability information or
2485 future product development (e.g., preliminary assessment of proposed manufacturing changes such
2486 as change in formulation, scale-up), assist in validation of analytical methods for the stability
2487 program, or generate information that may help elucidate the degradation profile of the drug
2488 substance or drug product. Studies under stress conditions may be useful in determining whether
2489 accidental exposures to conditions other than those proposed (e.g., during transportation) are
2490 deleterious to the product and also for evaluating which specific test parameters may be the best
2491 indicators of product stability. Studies of the exposure of the drug substance or drug product to
2492 extreme conditions may help to reveal patterns of degradation; if so, such changes should be
2493 monitored under proposed storage conditions. Although the OCH Q1A guidance on stability
2494 describes the conditions of the accelerated and stress study, the applicant should note that those
2495 conditions may not be appropriate for biotechnological/biological products. Conditions should be
2496 carefully selected on a case-by-case basis.

2497 d. Light

Draft - Not for Implementation

2498 Applicants should consult the FDA on a case-by-case basis to determine guidance for testing.
2499
2500 e. Container/Closure

2501 Changes in the quality of the product may occur due to the interactions between the formulated
2502 biotechnological/biological product and container/closure. Where the lack of interactions cannot be
2503 excluded in liquid products (other than sealed ampules), stability studies should include samples
2504 maintained in the inverted or horizontal position (i.e., in contact with the closure) as well as in the
2505 upright position, to determine the effects of the closure on product quality. Data should be
2506 supplied for all different container/closure combinations that will be marketed.

2507 In addition to the standard data necessary for a conventional single-use vial, the applicant should
2508 demonstrate that the closure used with a multiple-dose vial is capable of withstanding the
2509 conditions of repeated insertions and withdrawals so that the product retains its full potency,
2510 purity, and quality for the maximum period specified in the instructions-for-use on containers,
2511 packages, and/or package inserts. Such labeling should be in accordance with FDA requirements.

2512 f. Stability after Reconstitution of Freeze-Dried Product

2513 The stability of freeze-dried products after their reconstitution should be demonstrated for the
2514 conditions and the maximum storage period specified on containers, packages, and/or package
2515 inserts. Such labeling should be in accordance with FDA requirements.

2516 7. Testing Frequency [ICH Q5C]

2517 The shelf lives of biotechnological/biological products may vary from days to several years. Thus,
2518 it is difficult to draft uniform guidances regarding the stability study duration and testing frequency
2519 that would be applicable to all types of biotechnological/biological products. With only a few
2520 exceptions, however, the shelf lives for existing products and potential future products will be
2521 within the range of 0.5 to 5 years. Therefore, the guidance is based upon expected shelf lives in
2522 that range. This takes into account the fact that degradation of biotechnological/biological
2523 products may not be governed by the same factors during different intervals of a long storage
2524 period.

2525 When shelf lives of one year or less are proposed, the real-time stability studies should be
2526 conducted monthly for the first three months and at three month intervals thereafter. For products
2527 with proposed shelf lives of greater than one year, the studies should be conducted every three
2528 months during the first year of storage, every six months during the second year, and annually
2529 thereafter.

2530 While the testing intervals listed above may be appropriate in the preapproval or prelicense stage,
2531 reduced testing may be appropriate after approval or licensing where data are available that
2532 demonstrate adequate stability. Where data exist that indicate the stability of a product is not
2533 compromised, the applicant is encouraged to submit a protocol that supports elimination of
2534 specific test intervals (e.g., nine-month testing) for postapproval/postlicensing, long-term studies.

Draft - Not for Implementation

2535 8. Specifications [ICH Q5C]

2536 Although biotechnological/biological products may be subject to significant losses of activity,
2537 physicochemical changes, or degradation during storage, international and national regulations
2538 have provided little guidance with respect to distinct release and end of shelf life specifications.
2539 Recommendations for maximum acceptable losses of activity, limits for physicochemical changes,
2540 or degradation during the proposed shelf life have not been developed for individual types or
2541 groups of biotechnological/biological products but are considered on a case-by-case basis. Each
2542 product should retain its specifications within established limits for safety, purity, and potency
2543 throughout its proposed shelf life. These specifications and limits should be derived from all
2544 available information using the appropriate statistical methods. The use of different specifications
2545 for release and expiration should be supported by sufficient data to demonstrate that the clinical
2546 performance is not affected, as discussed in the OCH Q1A guidance on stability.

2547 9. Labeling [ICH Q5C]

2548 For most biotechnological/biological drug substances and drug products, precisely defined storage
2549 temperatures are recommended. Specific recommendations should be stated, particularly for drug
2550 substances and drug products that cannot tolerate freezing. These conditions, and where
2551 appropriate, recommendations for protection against light and/or humidity, should appear on
2552 containers, packages, and/or package inserts. Such labeling should be in accordance with section
2553 II.B.11 of this document.

2554 VIII. CONSIDERATIONS FOR SPECIFIC DOSAGE FORMS

2555 The following list of parameters for each dosage form is presented as a guide for the types of tests
2556 to be included in a stability study. In general, appearance, assay, and degradation products should
2557 be evaluated for all dosage forms.

2558 The list of tests presented for each dosage form is not intended to be exhaustive, nor is it expected
2559 that every listed test be included in the design of a stability protocol for a particular drug product
2560 (for example, a test for odor should be performed only when necessary and with consideration for
2561 analyst safety). Furthermore, it is not expected that every listed test be performed at each time
2562 point.

2563 **A. Tablets**

2564 Tablets should be evaluated for appearance, color, odor, assay, degradation products, dissolution,
2565 moisture, and friability.

2566 **B. Capsules**

2567 Hard gelatin capsules should be evaluated for appearance (including brittleness), color, odor of
2568 contents, assay, degradation products, dissolution, moisture, and microbial limits.

Draft - Not for Implementation

2569 Testing of soft gelatin capsules should include appearance, color, odor of content, assay,
2570 degradation products, dissolution, microbial limits, pH, leakage, and pellicle formation. In
2571 addition, the fill medium should be examined for precipitation and cloudiness.

2572 **C. Emulsions**

2573 An evaluation should include appearance (including phase separation), color, odor, assay,
2574 degradation products, pH, viscosity, microbial limits, preservative content, and mean size and
2575 distribution of dispersed phase globules.

2576 **D. Oral Solutions and Suspensions**

2577 The evaluation should include appearance (including formation of precipitate, clarity for solutions),
2578 color, odor, assay, degradation products, pH, preservative content, and microbial limits.

2579 Additionally, for suspensions, redispersibility, rheological properties, and mean size and
2580 distribution of particles should be considered. After storage, samples of suspensions should be
2581 prepared for assay according to the recommended labeling (e.g., shake well before using).

2582 **E. Oral Powders for Reconstitution**

2583 Oral powders should be evaluated for appearance, odor, color, moisture, and reconstitution time.

2584 Reconstituted products (solutions and suspensions) should be evaluated as described in VIII.D.
2585 above, after preparation according to the recommended labeling, through the maximum intended
2586 use period.

2587 **F. Metered-Dose Inhalations and Nasal Aerosols**

2588 Metered-dose inhalations and nasal aerosols should be evaluated for appearance (including
2589 content, container, valve and its components), color, taste, assay, degradation products, assay for
2590 co-solvent (if applicable), dose content uniformity, labeled number of medication actuations per
2591 container meeting dose content uniformity, aerodynamic particle size distribution, microscopic
2592 evaluation, water content, leak rate, microbial limits, valve delivery (shot weight), and
2593 extractables/leachables from plastic and elastomeric components. Samples should be stored in
2594 upright and inverted/on-the-side orientations.

2595 For suspension-type aerosols, the appearance of the valve components and container's contents
2596 should be evaluated microscopically for large particles and changes in morphology of the drug
2597 surface particles, extent of agglomerates, crystal growth, as well as foreign particulate matter.
2598 These particles lead to clogged valves or non-reproducible delivery of a dose. Corrosion of the
2599 inside of the container or deterioration of the gaskets may adversely affect the performance of the
2600 drug product.

2601 A stress temperature cycling study should be performed under the extremes of high and low
2602 temperatures expected to be encountered during shipping and handling to evaluate the effects of

Draft - Not for Implementation

2603 temperature changes on the quality and performance of the drug product. Such a study may
2604 consist of three or four six-hour cycles per day, between subfreezing temperature and 40°C (75-85
2605 percent RH), for a period of up to six weeks.

2606 Because the inhalant drug products are intended for use in the respiratory system, confirmation
2607 that initial release specifications are maintained should be provided to ensure the absence of
2608 pathogenic organisms (e.g., Staphylococcus aureus, Pseudomonas aeruginosa, Escherichia coli,
2609 and Salmonella species) and that the total aerobic count and total mold and yeast count per
2610 canister are not exceeded.

2611 **G. Inhalation Solutions and Powders**

2612 The evaluation of inhalation solutions and solutions for inhalation should include appearance,
2613 color, assay, degradation products, pH, sterility, particulate matter, preservative and antioxidant
2614 content (if present), net contents (fill weight/volume), weight loss, and extractables/leachables
2615 from plastic, elastomeric and other packaging components.

2616 The evaluation of inhalation powders should include appearance, color, assay, degradation
2617 products, aerodynamic particle size distribution of the emitted dose, microscopic evaluation,
2618 microbial limit, moisture content, foreign particulates, content uniformity of the emitted dose, and
2619 number of medication doses per device meeting content uniformity of the emitted dose (device
2620 metered products).

2621 **H. Nasal Sprays: Solutions and Suspensions**

2622 The stability evaluation of nasal solutions and suspensions equipped with a metering pump should
2623 include appearance, color, clarity, assay, degradation products, preservative and antioxidant
2624 content, microbial limits, pH, particulate matter, unit spray medication content uniformity, number
2625 of actuations meeting unit spray content uniformity per container, droplet and/or particle size
2626 distribution, weight loss, pump delivery, microscopic evaluation (for suspensions), foreign
2627 particulate matter, and extractables/leachables from plastic and elastomeric components of the
2628 container, closure, and pump.

2629 **I. Topical, Ophthalmic and Otic Preparations**

2630 Included in this broad category are ointments, creams, lotions, pastes, gels, solutions, and
2631 nonmetered aerosols for application to the skin.

2632 Topical preparations should be evaluated for appearance, clarity, color, homogeneity, odor, pH,
2633 resuspendability (for lotions), consistency, viscosity, particle size distribution (for suspensions,
2634 when feasible), assay, degradation products, preservative and antioxidant content (if present),
2635 microbial limits/sterility, and weight loss (when appropriate).

2636 Appropriate stability data should be provided for products supplied in closed-end tubes to support
2637 the maximum anticipated use period, during patient use, once the tube seal is punctured allowing
2638 product contact with the cap/cap liner. Ointments, pastes, gels, and creams in large containers,

Draft - Not for Implementation

2639 including tubes, should be assayed by sampling at the surface, top, middle, and bottom of the
2640 container. In addition, tubes should be sampled near the crimp (see also Section VII.D.2.).

2641 Evaluation of ophthalmic or otic products (e.g., creams, ointments, solutions, and suspensions)
2642 should include the following additional attributes: sterility, particulate matter, and extractables.

2643 Evaluation of nonmetered topical aerosols should include: appearance, assay, degradation
2644 products, pressure, weight loss, net weight dispensed, delivery rate, microbial limits, spray pattern,
2645 water content, and particle size distribution (for suspensions).

2646 **J. Transdermals**

2647 Stability studies for devices applied directly to the skin for the purpose of continuously infusing a
2648 drug substance into the dermis through the epidermis should be examined for appearance, assay,
2649 degradation products, leakage, microbial limit/sterility, peel and adhesive forces, and the drug
2650 release rate.

2651 **K. Suppositories**

2652 Suppositories should be evaluated for appearance, color, assay, degradation products, particle size,
2653 softening range, appearance, dissolution (at 37°C,) and microbial limits.

2654 **L. Small Volume Parenterals (SVPs)**

2655 SVPs include a wide range of injection products such as *Drug Injection*, *Drug for Injection*, *Drug*
2656 *Injectable Suspension*, *Drug for Injectable Suspension*, and *Drug Injectable Emulsion*.

2657 Evaluation of *Drug Injection* products should include appearance, color, assay, preservative
2658 content (if present), degradation products, particulate matter, pH, sterility, and pyrogenicity.

2659 Stability studies for *Drug for Injection* products should include monitoring for appearance, clarity,
2660 color, reconstitution time, and residual moisture content. The stability of *Drug for Injection*
2661 products should also be evaluated after reconstitution according to the recommended labeling.
2662 Specific parameters to be examined at appropriate intervals throughout the maximum intended use
2663 period of the reconstituted drug product, stored under condition(s) recommended in labeling,
2664 should include appearance, clarity, odor, color, pH, assay (potency), preservative (if present),
2665 degradation products/aggregates, sterility, pyrogenicity, and particulate matter.

2666 The stability studies for *Drug Injectable Suspension* and *Drug for Injectable Suspension* products
2667 should also include particle size distribution, redispersibility, and rheological properties in addition
2668 to the parameters cited above for *Drug Injection* and *Drug for Injection* products.

2669 The stability studies for *Drug Injectable Emulsion* products should include, in addition to the
2670 parameters cited above for *Drug Injection*, phase separation, viscosity, and mean size and
2671 distribution of dispersed phase globules.

Draft - Not for Implementation

2672 The functionality and integrity of parenterals in prefilled syringe delivery systems should be
2673 ensured through the expiration dating period with regard to factors, such as the applied extrusion
2674 force, syringeability, pressure rating, and leakage.

2675 Continued assurance of sterility for all sterile products can be assessed by a variety of means,
2676 including evaluation of the container and closure integrity by appropriate challenge test(s), and/or
2677 sterility testing as described in Section VII.C. Stability studies should evaluate product stability
2678 following exposure to at least the maximum specified process lethality (e.g., F_0, Mrads).

2679 Inclusion of testing for extractables/leachables in the stability protocol may be appropriate in
2680 situations where other qualification tests have not provided sufficient information or assurance
2681 concerning the levels of extractables/leachables from plastics and elastomeric components.

2682 Interaction of administration sets and dispensing devices with parenteral drug products, where
2683 warranted, should also be considered through appropriate use test protocols to assure that
2684 absorption and adsorption during dwell time do not occur.

2685 **M. Large Volume Parenterals (LVPs)**

2686 Evaluation of LVPs should include appearance, color, assay, preservative content (if present),
2687 degradation products, particulate matter, pH, sterility, pyrogenicity, clarity, and volume.

2688 Continued assurance of sterility for all sterile products may be assessed by a variety of means,
2689 including evaluation of the container and closure integrity by appropriate challenge test(s) and/or
2690 sterility testing as described in Section VII.C. Stability studies should include evaluation of
2691 product stability following exposure to at least the maximum specified process lethality (e.g., F_0,
2692 Mrads).

2693 Interaction of administration sets and dispensing devices with this type of dosage form should also
2694 be considered through appropriate use test protocols to ensure that absorption and adsorption
2695 during dwell time do not occur.

2696 **N. Drug Additives**

2697 For any drug product or diluent that is intended for use as an additive to another drug product, the
2698 potential for incompatibility exists. In such cases, the drug product labeled to be administered by
2699 addition to another drug product (e.g., parenterals, inhalation solutions), should be evaluated for
2700 stability and compatibility in admixture with the other drug products or with diluents both in
2701 upright and inverted/on-the-side orientations, if warranted.

2702 A stability protocol should provide for appropriate tests to be conducted at 0-, 6-to-8-, and
2703 24-hour time points, or as appropriate over the intended use period at the recommended
2704 storage/use temperature(s). Tests should include appearance, color, clarity, assay, degradation
2705 products, pH, particulate matter, interaction with the container/closure/device, and sterility.
2706 Appropriate supporting data may be provided in lieu of an evaluation of photodegradation.

Draft - Not for Implementation

2707 The compatibility and the stability of the drug products should be confirmed in all diluents and
2708 containers and closures as well as in the presence of all other drug products indicated for
2709 admixture in the labeling. Compatibility studies should be conducted on at least the lowest and
2710 highest concentrations of the drug product in each diluent as specified in the labeling. The stability
2711 and compatibility studies should be performed on at least three batches of the drug product.
2712 Compatibility studies should be repeated if the drug product or any of the recommended diluents
2713 or other drug products for admixture are reformulated.

2714 Testing for extractables/leachables on stability studies may be appropriate in situations where other
2715 qualification tests have not provided sufficient information or assurance concerning the levels of
2716 extractables/leachables from plastics and elastomeric components. Interaction of administration
2717 sets and dispensing devices with parenteral drug products, where warranted, should also be
2718 considered through appropriate use test protocols to ensure that absorption and adsorption during
2719 dwell time do not occur.

2720 **O. Implantable Subdermal, Vaginal and Intrauterine Devices that Deliver Drug**
2721 **Products**

2722 A device containing a drug substance reservoir or matrix from which drug substance diffuses
2723 should be tested for total drug substance content, degradation products, extractables, in vitro drug
2724 release rate, and as appropriate, microbial burden or sterility. The stability protocol should include
2725 studies at 37°C or 40°C over a sufficient period of time to simulate the in vivo use of the drug
2726 delivery device.

2727 Stability testing for intrauterine devices (IUDs) should include the following tests: deflection of
2728 horizontal arms or other parts of the frame if it is not a T-shaped device (frame memory), tensile
2729 strength of the withdrawal string, and integrity of the package (i.e., seal strength of the pouch),
2730 and sterility of the device.

2731 IX. STABILITY TESTING FOR POSTAPPROVAL CHANGES

2732 **A. General**

2733 Due to the great variety of changes that may be encountered after a drug application is approved, it
2734 is impossible to address stability requirements for all changes in an exhaustive manner in this
2735 guidance. Some more common examples of changes to an approved drug application for which
2736 supportive stability data should be submitted are listed below. All changes should be accompanied
2737 by the standard stability commitment to conduct and/or complete long-term stability studies on the
2738 first 1 or 3 batches of the drug substance and/or drug product and annual batches thereafter, in
2739 accordance with the approved stability protocol. The accumulated stability data should be
2740 submitted in the subsequent annual reports. Unless otherwise noted, if the data give no reason to
2741 believe that the proposed change will alter the stability of the drug product, the previously
2742 approved expiration dating period can be used.

Draft - Not for Implementation

2743 Historically, all postapproval changes were considered together and required extensive stability
2744 documentation. With the publication of the SUPAC-IR guidance, this approach was changed and
2745 the likelihood of a specific CMC change affecting a drug product's performance was considered in
2746 creating a multitiered system for evaluating postapproval changes. That system is used in this
2747 guidance. With a higher level change, more stability data will be expected to support that change.
2748 Thus, five stability data package types have been defined, as explained in Table 14.

2749 **Table 14: Stability Data Packages to Support Postapproval Changes**

2750 2751	Stability Data Package	Stability Data at Time of Submission	Stability Commitment
2752	Type 0	None	None beyond the regular annual batches
2753	Type 1	None	First (1) production batch and annual batches thereafter on long-term stability studies.
2754	Type 2	3 months of comparative accelerated data and available long-term data on 1 batch[a] of drug product with the proposed change.	First (1) production batch[b] and annual batches thereafter on long-term stability studies[c].
2755	Type 3	3 months of comparative accelerated data and available long-term data on 1 batch[a] of drug product with the proposed change.	First 3 production batches[b] and annual batches thereafter on long-term stability studies.[c]
2756	Type 4	3 months of comparative accelerated data and available long-term data on 3 batches[a] of drug product with the proposed change.	First 3 production batches[b] and annual batches thereafter on long-term stability studies.[c]

2757 [a] Pilot scale batches acceptable.

2758 [b] If not submitted in the supplement.

2759 [c] Using the approved stability protocol and reporting data in annual reports.

Draft - Not for Implementation

2760 The following sections address a number of possible postapproval changes and contain summary
2761 tables with examples of the different levels of change, the stability data package type and, wherever
2762 possible, the filing documentation (AR = annual report; CBE = changes-being-effected
2763 supplement; PA = prior approval supplement) recommended to support each change. The
2764 information presented here is not intended to be exhaustive. Where a specific issue is not covered,
2765 consultation with FDA staff is recommended.

2766 **B. Change in Manufacturing Process of the Drug Substance**

2767 A change in the manufacturing process of the drug substance at the approved manufacturing site
2768 should be supported by the submission of sufficient data to show that such a change does not
2769 compromise the quality, purity, or stability of the drug substance and the resulting drug product.
2770 Because chemical stability of a substance is an intrinsic property, changes made in the preparation
2771 of that substance should not affect its stability, provided the isolated substance remains of
2772 comparable quality for attributes such as particle size distribution, polymorphic form, impurity
2773 profile, and other physiochemical properties. Special concerns for biological products may exist if
2774 changes are made in the manufacturing process of a drug substance that may not exist in a
2775 chemically synthesized drug substance.
2776
2777 Specific submission and stability issues will be addressed in detail in a separate forthcoming
2778 guidance dealing with postapproval changes for drug substances.

2779 **C. Change in Manufacturing Site**

2780 Site changes consist of changes in the location of the site of manufacture, packaging operations,
2781 and/or analytical testing laboratory both of company-owned as well as contract manufacturing
2782 facilities. The stability data package and filing mechanisms indicated below apply to site changes
2783 only. If other changes occur concurrently, the most extensive data package associated with the
2784 individual changes should be submitted.

2785 When a change to a new manufacturer or manufacturing site for any portion of the manufacturing
2786 process of a drug substance or drug product is made, sufficient data to show that such a change
2787 does not alter the characteristics or compromise the quality, purity, or stability of the drug
2788 substance or drug product may be necessary. The data should include a side-by-side comparison
2789 of all attributes to demonstrate comparability and equivalency of the drug substance or drug
2790 product manufactured at the two facilities. New manufacturing locations should have a
2791 satisfactory CGMP inspection.

2792 1. Site Change for the Drug Substance

2793 For a change limited to an alternate manufacturing site for the drug substance using similar
2794 equipment and manufacturing process, stability data on the drug substance may not always be
2795 necessary because, for essentially pure drug substances, stability is an intrinsic property of the
2796 material. Biotechnology and biologic products may be an exception (see 21 CFR 601.12 and
2797 314.70 (g)). In general, such a change can be made in a CBE supplement as allowed under 21

2798 CFR 314.70(c)(3). The standard stability commitment should be made to conduct long-term
2799 stability studies in accordance with the approved stability protocol on the first production batch of
2800 drug product produced from a production batch of drug substance manufactured at the new site.
2801 Ordinarily, the approved expiration dating period for the drug product may be retained if the drug
2802 substance is shown to be of comparable quality (e.g., particle size distribution, polymorphic form,
2803 impurity profile, and other physiochemical properties). If the drug substance is not of comparable
2804 quality, then more extensive stability data on the drug product manufactured from the drug
2805 substance will be needed.

2806 Specific submission and stability issues pertaining to manufacturing site changes for a drug
2807 substance or its intermediates in the drug substance manufacturing process will be addressed in a
2808 separate forthcoming guidance on postapproval changes for the drug substance.

2809 2. Site Change for the Drug Product

2810 For a move of the manufacturing site within an existing facility or a move to a new facility on the
2811 same campus using similar equipment and manufacturing processes, submission of stability data on
2812 the drug product in the new facility prior to implementation is generally not necessary (Table 15).

2813 For a move to a different campus using similar equipment and manufacturing processes, stability
2814 data on the drug product in the new facility should be submitted in a supplemental application.
2815 Three months of accelerated and available long-term stability data on one to three batches of drug
2816 product manufactured in the new site is recommended, depending on the complexity of the dosage
2817 form and the existence of a significant body of information (Table 15). A commitment should be
2818 made to conduct long-term stability studies on the first or first three production batch(es) of the
2819 drug product, depending on the dosage form and the existence of a significant body of information,
2820 manufactured at the new site in accordance with the approved stability protocol. If the stability
2821 data are satisfactory, the existing expiration dating period may be used.

2822 Table 15 reflects the guidance provided in existing SUPAC documents that address the stability
2823 recommendations for the various levels of site change. The stability data package type and filing
2824 mechanisms are as indicated in the table. Note that SUPAC guidances and Table 14 currently do
2825 not apply to biotechnology/biological products (see 21 CFR 314.70(g) and 601.12).

2826 3. Change in Packaging Site for Solid Oral Dosage Form Drug Products

2827 A stand-alone packaging operation site change for solid oral dosage form drug products using
2828 container(s)/closure(s) in the approved application should be submitted as a CBE supplement. No
2829 up-front stability data are necessary. The facility should have a current and satisfactory CGMP
2830 compliance profile for the type of packaging operation under consideration before submitting the
2831 supplement. The supplement should also contain a commitment to place the first production batch
2832 and annual batches thereafter on long-term stability studies using the approved protocol in the
2833 application and to submit the resulting data in annual reports.

2834 A packaging site change for other than solid oral dosage form drug products is considered a
2835 manufacturing site change and the data package that should be submitted for approval is indicated

Draft - Not for Implementation

2836 in Section IX.C.2.

2837 4. Change in Testing Laboratory

2838 An analytical testing laboratory site change may be submitted as a CBE supplement under certain
2839 circumstances (see *PAC-ATLS: Postapproval Changes, Analytical Testing Laboratory Sites*, CMC
2840 10, April 1998). No stability data are required.

2841 **Table 15: Stability Data to Support Postapproval**
2842 **Drug Product Manufacturing Site Changes**[a]

Level of Change	Definition/Examples	Filing Documentation	Stability Data Package	
1	a. Manufacturing site change within a facility with the same equipment, SOPs, environmental conditions, controls, personnel (e.g., remodeling an existing building, add-on to an existing facility).	AR	Type 0	
	b. Packaging site change for solid oral dosage form drug products.	CBE	Type 1	
	c. Test laboratory site change to a new location.	CBE	Type 0	
2	Change within a contiguous campus, or between facilities in adjacent city blocks, with the same equipment, SOPs, environmental conditions, controls, personnel:			
	a. *Immediate release solid oral and semisolid dosage forms*	CBE	Type 1	
	b. *Modified release dosage forms*	CBE	Type 2	
3	Manufacturing site change to a different facility with the same equipment, SOPs, environmental conditions, and controls:		SBI[b]	No SBI[b]
	a. *Immediate Release Solid Oral Dosage Forms*	CBE	Type 2	Type 3
	b. *Semisolid Dosage Forms*	CBE	Type 3	Type 3
	c. *Modified Release Dosage Forms*	PA	Type 3	Type 4

2848 [a] Note that metered dose inhalers and dry powder inhalers, transdermal patches, and sterile aqueous solutions are the
2849 subjects of forthcoming guidances and, except for changes in testing laboratory, are not covered in this table. In
2850 addition, this table does not apply to biotechnology/biological products.

2851 [b] Significant body of information.

2852

Draft - Not for Implementation

2853 **D. Change in Formulation of the Drug Product**

2854 Historically, all changes in drug product formulation were grouped together and required extensive
2855 stability documentation, usually submitted as a prior-approval supplement. An exception was the
2856 deletion of a color from a product that could be reported in an annual report without supporting
2857 stability data (21 CFR 314.70(d)(4)). Excipients play a critical role in certain complex dosage
2858 forms, including semisolid and modified release drug products. Table 16 provides information on
2859 stability recommendations to support postapproval formulation changes.[12]

[12] Please refer to the following guidance for industry: SUPAC-IR (November 1995), *SUPAC-SS (May 1997)*, and *SUPAC-MR (September 1997)* for more detailed information on formulation changes for those specific dosage forms.

Draft - Not for Implementation

2860 **Table 16: Stability Data to Support Postapproval Formulation Changes**[a]

Line	Level of Change	Definition/Examples	Filing Documentation	Stability Data Package
2861 2862				
2863	1	a. *All Dosage Forms*: Deletion or partial deletion of an ingredient intended to affect the color, taste or fragrance of the drug product. b. *Immediate Release Solid Oral and Semisolid Dosage Forms*: The total additive effect of all excipient changes does not exceed 5%, with individual changes within the limits specified in SUPAC-IR and -SS.[b] c. *Semisolid Dosage Forms*: Change in supplier of a structure-forming excipient which is primarily a single chemical entity (purity 95%). d. *Modified Release Dosage Forms*: See SUPAC-MR guidance document for specific information on what excipient quantity changes constitute a level 1 change.	AR	Type 1
2864	2	a. *Immediate Release Solid Oral and Semisolid Dosage Forms*: The total additive effect of all excipient changes is >5-10% with individual changes within the limits specified in SUPAC-IR and -SS.[b]	PA	Type 2
		b. *Semisolid Dosage Forms*: Change in supplier or grade of a structure forming excipient not covered under level 1. c. *Semisolid Dosage Forms*: Change in the particle size distribution of active drug substance, if the drug is in suspension.	CBE	
		d. *Modified Release Dosage Forms*: Change in the technical grade and/or specifications of a nonrelease controlling excipient. e. *Modified Release Dosage Forms*: See SUPAC-MR Guidance document for specific information on what release controlling excipient quantity changes constitute a level 2 change.	PA	see SUPAC-MR
2865	3	a. *All Dosage Forms*: Any qualitative or quantitative change in excipient beyond the ranges noted in the level 2 change. b. *Semisolid Dosage Forms*: Change in the crystalline form of the drug substance, if the drug is in suspension.	PA	SBI[c]: Type 2 / Type3/4 No SBI[c]: Type 3/4 / Type 4

2866 2867 [a] Note that metered dose inhalers and dry powder inhalers, transdermal patches, and sterile aqueous solutions are the subjects of forthcoming guidances and are not covered in this table.

2868 2869 2870 [b] Allowable changes in the composition are based on the approved target composition and not on previous Level 1 or level 2 changes in the composition. Changes in diluent (q.s. excipient) due to component and composition changes in excipients are allowed and are excluded from the 10% change limit.

2871 [c] Significant body of information.

Draft - Not for Implementation

2872 **E. Addition of a New Strength for the Drug Product**

2873 The addition of a new strength for an approved drug product will generally require the submission
2874 of a prior-approval supplement. Demonstration of equivalent stability between the approved drug
2875 product and the new strength will allow extension of the approved drug product expiration dating
2876 to the new strength. Depending on issues specific to the drug product (e.g., dosage form)
2877 availability of a significant body of information for the approved dosage form, a Type 2, 3, or 4
2878 stability data package may be appropriate as shown in Table 17. New strengths intermediate to
2879 those of an approved drug product may be supported by bracketing/matrixing studies (See Section
2880 VII.G. and VII.H.).

2881 **Table 17: Stability Data to Support Addition of a New Strength for a Drug Product[a]**

2882
2883

2884–2888

2889–2893

Definition of Change	Examples	Filing Documentation	Stability Data Package
New strength of identical qualitative and quantitative composition[b]	a. Addition of a score to an immediate release tablet.	PA	Type 1
	b. Change in the fill of an immediate release hard gelatin capsule.	PA	Type 2
	c. Change in the fill of a hard gelatin capsule containing modified release encapsulated beads.	PA	Type 2
	d. Change in the size of an immediate release tablet or capsule.	PA	Type 3
New strength involving a change in the drug substance to excipient(s) ratio	a. *Simple solutions*	PA	Type 2
	b. *Immediate release solid oral dosage forms*	PA	Type 3
	c. *Semisolid and modified release oral dosage forms*	PA	Type 4

2894 [a] Note that metered dose inhalers and dry powder inhalers, transdermal patches, and sterile aqueous solutions are the
2895 subjects of forthcoming guidances and are not covered in this table.

2896 [b] No change in drug substance to excipient(s) ratio from the approved drug product.

Draft - Not for Implementation

2897 **F. Change in Manufacturing Process and/or Equipment for the Drug Product**

2898 A change limited to the manufacturing process of the drug product, such as a change in the type of
2899 equipment used, can be supported by the submission of sufficient data to show that such a change
2900 does not alter the characteristics or compromise the stability of the drug product. For information
2901 on determining when equipment is considered to be of the same design and operating principle,
2902 refer to the Supac-IR/MR draft manufacturing equipment addendum (April 1998). In general,
2903 stability data on the drug product demonstrating comparability with and equivalency to the
2904 previously approved drug product should be submitted. The submission types and stability data
2905 packages shown in Table 18 apply to immediate release solid oral dosage forms and semisolid
2906 dosage forms and incorporate the criteria provided by those SUPAC documents. Because
2907 additional data may be appropriate for more complex dosage forms, the chemistry review team
2908 should be consulted. The standard stability commitment to conduct and/or complete the stability
2909 studies on the first three production batches produced by the revised manufacturing process in
2910 accordance with the approved stability protocol is necessary. If the data are found acceptable, the
2911 approved expiration dating period may be retained.

2912 Submissions for approval of a change of manufacturing site for any portion of the manufacturing
2913 process for the drug product are addressed in Section IX.C.

Draft - Not for Implementation

2914 **Table 18: Stability Data to Support Manufacturing Process Changes[a]**

2915
2916
2917
2918

Level of Chang e	Definition/Examples	Filing Documentation	Stability Data Package
	Process: Changes in processing parameters such as mixing times, operating speeds within application/validation ranges.	AR	Type 0
1	Equipment: Change from nonautomated to automated or mechanical equipment; or Change to alternative equipment of the same design and operating principles.	AR	Type 1
2	Process: Changes in processing parameters such as mixing times, operating speeds outside of application/validation ranges: a. *Immediate release solid oral dosage forms* b. *Semisolid dosage forms* c. *Modified release dosage forms*	 CBE CBE CBE	SBI[b] No SBI[b] Type 1 Type 1 Type 2 Type 4 Type 2 Type 2
	Equipment: Changes to equipment of different design and/or operating principles: a. *Immediate release solid oral dosage forms* b. *Semisolid dosage forms* c. *Modified release dosage forms*	 PA CBE PA	SBI[b] No SBI[b] Type 2 Type 3/4 Type 2 Type 4 Type 3 Type 4
3	Process: Changes in type of process used in the manufacture of the product, such as a change from wet granulation to direct compression of dry powder: a. *Immediate release solid oral dosage forms* b. *Modified release dosage forms*	 PA PA	SBI[b] No SBI[b] Type 2 Type 3/4 Type 4 Type 4

2919 2920 2921

2922
2923
2924 [a] Note that metered dose inhalers and dry powder inhalers, transdermal patches, and sterile aqueous solutions are the subjects of forthcoming guidances and are not covered in this table. In addition, this table does not apply to biotechnology/biological products.

2925 [b] Significant body of information.

Draft - Not for Implementation

2926 **G. Change in Batch Size of the Drug Product**

2927 A key question in considering an increase in batch size beyond the production batch size approved
2928 in the application is whether the change involves a change in equipment or its mode of operation,
2929 or other manufacturing parameters described for the approved batch size. If no equipment change
2930 is planned, then the next concern is the size of the change relative to the approved batch size, with
2931 larger changes expected to present a greater risk of stability problems in the drug product. Table
2932 19 presents the recommended stability data packages for a variety of batch size situations not
2933 involving equipment or mode of operation changes.

2934 If an equipment change is part of the batch size change, please refer to Change in Manufacturing
2935 Process of the Drug Product (Section IX.F.).

2936 **Table 19: Stability Data to Support Postapproval Batch Size Changes[a]**

	Level of Change	Definition/Examples	Filing Documentation	Stability Data Package
2937 2938				
2939	1	*Solid oral dosage forms* (i.e., tablets, capsules, powders for reconstitution), *semisolid dosage forms, and oral solutions*: A change in batch size up to and including a factor of ten times the size of the pivotal clinical trial/biobatch.	AR	Type 1
2940	2	*Solid oral dosage forms* (i.e., tablets, capsules, powders for reconstitution), *semisolid dosage forms, and oral solutions*: A change in batch size beyond a factor of ten times the size of the pivotal clinical trial/biobatch.	CBE	Type 2

2941 [a] Note that metered dose inhalers and dry powder inhalers, transdermal patches, and sterile aqueous solutions are the
2942 subjects of forthcoming guidances and are not covered in this table.

2943

Draft - Not for Implementation

2944 **H. Reprocessing of a Drug Product**

2945 Stability data submitted in support of reprocessing of a specific batch of a drug product should
2946 take into account the nature of the reprocessing procedure and any specific impact that might have
2947 upon the existing stability profile of the drug. The expiration dating period for a reprocessed batch
2948 should not exceed that of the parent batch, and the expiration date should be calculated from the
2949 original date of manufacture of the oldest batch.

2950 The acceptability of reprocessing of a specific batch of a drug product will depend on the nature of
2951 the reprocessing procedure, which can range from repackaging a batch when packing equipment
2952 malfunctions to regrinding and recompressing tablets. The appropriate chemistry review team
2953 should be contacted to determine whether or not the reprocessing procedure is acceptable. Any
2954 batch of the drug product that is reprocessed should be placed on accelerated and long-term
2955 stability studies using the approved protocol to generate a Type 2 stability data package.

2956 **I. Change in Container and Closure of the Drug Product**

2957 The stability data packages for changes in container and closure of a drug product vary (Table 20).
2958 The first factor used in determining the stability data package recommendation is whether or not
2959 the protective properties of the container/closure system are affected by the proposed change.
2960 Protective properties of the container/closure system include, but are not limited to, moisture
2961 permeability, oxygen permeability, and light transmission. Changes that may affect these
2962 properties should be supported by a greater amount of data to support the change. The second
2963 factor is the nature of the dosage form itself. A solid dosage form will generally be less affected by
2964 a container change than a liquid dosage form. Because considerably more information will be
2965 needed to document a container/closure change than just stability data, applicants are encouraged
2966 to consult with the appropriate chemistry review team to determine the appropriate filing
2967 mechanisms. Please refer to the guidance for industry: *Submission of Documentation in Drug*
2968 *Applications for Container Closure Systems Used for the Packaging of Human Drugs and*
2969 *Biologics* for qualification and quality control information requested for container closure
2970 systems.[13] Table 20 below describes what type of stability data should be supplied for some of the
2971 most common post-approval changes to container/closure systems for solid and liquid oral drug
2972 products.

---

[13] A forthcoming guidance will deal more extensively with postapproval packaging changes for all dosage forms.

Draft - Not for Implementation

2973 **Table 20: Stability Data to Support Postapproval Container/Closure**
2974 **Changes for Solid and Liquid Oral Drug Products[a]**

2975

2976
2977
2978
2979
2980
2981

Type of change	Definition	Examples	Stability Data Package
Changes that do not affect the protective properties of the container/closure system	1. Closure changes	Adding or changing a child-resistant feature to a packaging system or changing from a metal to a plastic screw cap, while the inner seal remains unchanged.	Type 0
	2. Changing the secondary packaging	Changing a carton.	Type 0
	3. Removal of non-drug product material	Removing: a. an insert. b. a filler.	 Type 0 Type 1
	4. Changing shape of container/closure	(Without changing the size)	Type 0
	5. Changing size of container/closure	a. Within the approved range of sizes. b. Outside the approved range of sizes.	Type 0 Type 2
Changes that may affect the protective properties of the container/closure system	1. Adding or changing a component to increase protection within the same system.	a. Adding, or changing to, a heat-induction seal: i. For a solid oral drug product. ii. For a liquid oral drug product. b. Adding or changing a desiccant or a filler. c. Adding an overwrap or carton.	 Type 1 Type 2 Type 2 Type 2
	2. Changing the manufacturer or formulation of a container/closure component, including bottle or blister resin, cap liner, seal laminate, desiccant, filler, etc., within the same system.	a. Using an approved or compendial container or closure equivalency protocol for: i. a solid oral drug product. ii. a liquid oral drug product. b. Without an approved or compendial container or closure equivalency protocol.	 Type 1 Type 1 Type 2
	3. Changing to a different container and closure system	For any solid or liquid oral drug product.	SBI[b]: Type 3 No SBI[b]: Type 4

2982
2983
2984
2985
2986
2987

2988 [a] In certain situations, e.g., for particularly sensitive drug products, additional stability requirements may apply. Note
2989 that Metered Dose Inhalers and Dry Powder Inhalers, Transdermal Patches, and Sterile Aqueous Solutions are the
2990 subject of a forthcoming guidance and are not covered in this table.

2991 [b] Significant body of information.

2992 **J. Changes in the Stability Protocol**

2993 In general, modification of the approved stability protocol is discouraged until the expiration dating
2994 period granted at the time of approval has been confirmed by long-term data from production
2995 batches. However, changes in analytical methods provide increased assurance in product identity,
2996 strength, quality, and purity, or to comply with USP monographs, may be appropriate prior to the
2997 confirmation of the expiration dating period.

2998 Certain parameters may be reduced in test frequency or omitted from the stability protocol for annual
2999 batches on a case-by-case basis through a prior-approval supplement. A justification for such a
3000 reduction or omission should be adequately provided.

3001 If justified, test frequency for all parameters may be reduced for *annual batches* based on
3002 accumulated stability data. Such a modification to the approved stability protocol should be
3003 submitted as a prior-approval supplement. The justification may include a demonstrated history of
3004 satisfactory product stability, which may in turn include, but not be limited to, full long-term stability
3005 data from at least three production batches. The reduced testing protocol should include a minimum
3006 of four data points, including the initial time point, and the expiry and two points in between. For
3007 example, drug products with an expiration dating period of less than 18 months should be tested at
3008 quarterly intervals; products with an expiration dating period of 18 but not more than 30 months
3009 should be tested semiannually; and products with an expiration dating period of 36 months or longer
3010 should be tested annually. It should be noted, however, that the reduced testing protocol applies
3011 only to annual batches and does not apply to batches used to support a postapproval change that
3012 requires long-term stability data at submission and/or as a commitment. Furthermore, whenever
3013 product stability failures occur, the original full protocol should be reinstated for annual batches until
3014 problems are corrected.

3015 A bracketing or matrixing design, if proposed for annual batches or to support a supplemental
3016 change, should be submitted as a prior-approval supplement (see Sections VII.G. and H.). It is
3017 acceptable to submit these modifications to the protocol, along with data generated therefrom to
3018 support a supplemental change, in one combined prior-approval supplement. However, the applicant
3019 is encouraged to consult with the appropriate FDA chemistry review team before initiating such
3020 studies.

Draft - Not for Implementation

3021 **BIBLIOGRAPHY**

3022 Bancroft, T. A., "Analysis and Inference for Incompletely Specified Models Involving the Use of
3023 Preliminary Test(s) of Significance," *Biometrics*, 20(3), 427-442, 1964.

3024 Easterling, R.G.,"Discrimination Intervals for Percentiles in Regression," *J. Am. Stat. Assoc.*, 64,
3025 1031-41, 1969.

3026 Fairweather, W.R., T.-Y. D. Lin, and R. Kelly, "Regulatory, Design, and Analysis Aspects of
3027 Complex Stability Studies," *J. Pharm. Sci.*, 84, 1322-1326, 1995.

3028 Food and Drug Administration (FDA), *PAC-ATLS: Post approval Changes: Analytical Testing*
3029 *Laboratory Sites*, Center for Drug Evaluation and Research (CDER), April 1998.

3030 FDA, *SUPAC-MR, Modified Release Solid Oral Dosage Forms, Scale-Up and Post-Approval*
3031 *Changes: Chemistry, Manufacturing and Controls, In Vitro Dissolution Testing, In Vivo*
3032 *Bioequivalence Documentation*, CDER, September 1997.

3033 FDA, *Submission of Documentation in Drug Applications for Container Closure Systems used for*
3034 *the Packaging of Human Drugs and Biologics*, CDER/CBER (Center for Biologics Evaluation and
3035 Research), draft guidance, July 1997.

3036 FDA, *SUPAC-SS, Non-Sterile Semi-Solid Dosage Forms. Scale-Up and Post-Approval Changes:*
3037 *Chemistry, Manufacturing and Controls, In Vitro Release Testing, In Vivo Bioequivalence*
3038 *Documentation*, CDER, May 1997.

3039 FDA , *SUPAC-IR/MR: Immediate and Modified Release Solid Oral Dosage Forms, Manufacturing*
3040 *Equipment Addendum*, draft guidance, April 1998.

3041 FDA , *SUPAC-IR, Immediate Release Solid Oral Dosage Form: Scale-Up and Post-Approval*
3042 *Changes: Chemistry, Manufacturing and Controls, In Vitro Dissolution Testing, In Vivo*
3043 *Bioequivalence Documentation* Center for Drug Evaluation and Research (CDER), November 1995.

3044 Haynes, J.D., "Worldwide Virtual Temperatures for Product Stability Testing," *J. Pharm. Sci.*, Vol.
3045 60, No. 6, 927 (June 1971).

3046 International Conference on Harmonisation (ICH), *Q1A Stability Testing for New Drug Substances*
3047 *and Products*, September 1994.

3048 ICH, *Q5C Quality of Biotechnological Products: Stability Testing of Biotechnological/Biological*
3049 *Products*, July 1996.

3050 ICH, *Q1C Stability Testing for New Dosage Forms*, November 1996.

3051 ICH, *Q1B Photostability Testing of New Drug Substances and Products*, May 1997.

Draft - Not for Implementation

3052 Lin, K.K., T-Y.D. Lin, and R.E. Kelley, "Stability of Drugs: Room Temperature Tests", in *Statistics*
3053 *in the Pharmaceutical Industry*, ed. C.R. Buncher and J-Y. Tsay, p 419-444, Marcel Dekker, Inc.,:
3054 New York 1994.

3055 U.S. Department of Health and Human Services, Public Health Service, FDA, *Guideline on*
3056 *Validation of the Limulus Amebocyte Lysate Test as an End-Product Endotoxin Test for Human and*
3057 *Animal Parenteral Drugs, Biological Products and Medical Devices,* December 1987.

3058 Yoshioka, S. et al., "Quinine Actinometry as a Method for Calibrating Ultraviolet Radiation Intensity
3059 in Light-stability Testing of Pharmaceuticals," *Drug Development and Industrial Pharmacy,*
3060 20(13):2049-2062, 1994.

Draft - Not for Implementation

3061 GLOSSARY

3062 **Accelerated Testing [ICH Q1A]**

3063 Studies designed to increase the rate of chemical degradation or physical change of an active drug
3064 substance and drug product by using exaggerated storage conditions as part of the formal, definitive,
3065 stability protocol. These data, in addition to long-term stability data, may also be used to assess
3066 longer term chemical effects at nonaccelerated conditions and to evaluate the impact of short-term
3067 excursions outside the label storage conditions such as might occur during shipping. Results from
3068 accelerated testing studies are not always predictive of physical changes.

3069 **Acceptance Criteria [21 CFR 210.3]**

3070 Product specifications and acceptance/rejection criteria, such as acceptable quality level and
3071 unacceptable quality level, with an associated sampling plan, that are necessary for making a decision
3072 to accept or reject a lot or batch (or any other convenient subgroups of manufactured units).

3073 **Active Substance; Active Ingredient; Drug Substance; Medicinal Substance** [ICH Q1A]

3074 The unformulated drug substance which may be subsequently formulated with excipients to produce
3075 the drug product.

3076 **Approved Stability Protocol**

3077 The detailed study plan described in an approved application to evaluate the physical, chemical,
3078 biological, and microbiological characteristics of a drug substance and a drug product as a function
3079 of time. The approved protocol is applied to generate and analyze acceptable stability data in
3080 support of the expiration dating period. It may also be used in developing similar data to support an
3081 extension of that expiration dating period, and other changes to the application. It should be
3082 designed in accordance with the objectives of this guidance.

3083 **Batch** [21 CFR 210.3(b)(2)]

3084 A specific quantity of a drug material that is intended to have uniform character and quality, within
3085 specified limits, and is produced according to a single manufacturing order during the same cycle of
3086 manufacture.

3087 **Bracketing** ICH Q1A]

3088 The design of a stability schedule so that at any time point only the samples on the extremes, for
3089 example, of container size and/or dosage strengths, are tested. The design assumes that the stability
3090 of the intermediate condition samples is represented by those at the extremes.
3091

Draft - Not for Implementation

3092 **Climatic Zones** [ICH Q1A]

3093 The concept of dividing the world into four zones based on defining the prevalent annual climatic
3094 conditions.

3095 **Complex Dosage Form**

3096 A complex dosage form is one where quality and/or stability is more likely to be affected by changes
3097 because the release mechanism, delivery system, and manufacturing process are more complicated
3098 and thus more susceptible to variability.

3099 Examples of complex dosage forms include modified-release dosage forms, metered-dose inhalers,
3100 transdermal patches, liposome preparations. Due to the diversity of currently marketed dosage forms
3101 and the ever-increasing complexity of new delivery systems, it is impossible to clearly identify simple
3102 vs. complex dosage forms in an exhaustive manner. Applicants are advised to consult with the
3103 appropriate FDA chemistry review team when questions arise.

3104 **Conjugated Product** [ICH Q5C]

3105 A conjugated product is made up of an active ingredient (e.g., peptide, carbohydrate) bound
3106 covalently or noncovalently to a carrier (e.g., protein, peptide, inorganic mineral) with the objective
3107 of improving the efficacy or stability of the product.

3108 **Confirmatory Studies** [ICH Q1B]

3109 Those studies undertaken to establish photostability characteristics under standardized conditions.
3110 These studies are used to identify precautionary measures needed in manufacturing or formulation
3111 and whether light-resistant packaging and/or special labeling is needed to mitigate exposure to light.
3112 For the confirmatory studies, the batch(es) should be selected according to batch selection for
3113 long-term and accelerated testing which is described in the parent guidance.

3114 **Controlled Room Temperature (CRT)** [USP]

3115 A temperature maintained thermostatically that encompasses the usual and customary working
3116 environment of 20°C to 25°C (68°F to 77°F) that results in a mean kinetic temperature (MKT)
3117 calculated to be not more than 25°C and that allows for excursions between 15°C and 30°C (59°F
3118 to 86°F) that are experienced in pharmacies, hospitals and warehouses.

3119 **Date of Production**

3120 The date that the first step of manufacture is performed which involves the combining of an active
3121 ingredient, antioxidant, or preservative, with other ingredients in the production of a dosage form.
3122 For drug products consisting of a single ingredient filled into a container, the date of the production
3123 is the initial date of the filling operation. For a biological product subject to licensure see the
3124 definition of date of manufacture in 21 CFR 610.50.

:ntation

bulk material) brought about over time.
)ed in this guidance, such changes could
substance or drug product,
dation, oxidation, aggregation, proteolysis).
ıon products may be active.

ontainer product) that is not
sule, solution, cream, that contains a drug
dditives to the drug
with excipients.

:nded for marketing.

anufacturing process that is
critical to the successful
b)]
nediate will be quantifiable
ıacological activity or other direct effect in the
on of the manufacturing
of disease or to affect the structure or any
rial that may undergo
urther processing.

ge form.
cal characteristics of a drug
life and retest period,

product designating the time during which a batch
proved shelf-life specification if stored under
: used.
r and quality within
process, it is a specific
ssures its having uniform

ıiner/closure which have been transferred into the

1B]
duct production for

le deliberately. These studies, which may be
y on the drug substances, are used to evaluate the
:thod development purposes and/or degradation

a carton.

:approval stability protocol which embraces the
ducts to see how closely

3155 principles of these guidances.

3156 **Immediate (Primary) Pack** [ICH Q1B]

3157 That constituent of the packaging that is in direct contact with the drug
3158 and includes any appropriate label.

3159 **Impurity**

3160 Any entity of the drug substance (bulk material) or drug product (final c
3161 the chemical entity defined as the drug substance, an excipient, or other
3162 product.

3163 **Intermediate** [ICH Q5C]

3164 For biotechnological/biological products, a material produced during a m
3165 not the drug substance or the drug product but for which manufacture is
3166 production of the drug substance or the drug product. Generally, an intern
3167 and specifications will be established to determine the successful completi
3168 step before continuation of the manufacturing process. This includes mate
3169 further molecular modification or be held for an extended period before f

3170 **Long-Term (Real-Time) Testing** [ICH Q1A]

3171 Stability evaluation of the physical, chemical, biological, and microbiolog
3172 product and a drug substance, covering the expected duration of the shelf
3173 which are claimed in the submission and will appear on the labeling.

3174 **Lot** [21 CFR 210.3(b)(10)]

3175 A batch, or a specific identified portion of a batch, having uniform charact
3176 specified limits; or, in the case of a drug product produced by continuous p
3177 identified amount produced in a unit of time or quantity in a manner that as
3178 character and quality within specific limits.

3179 **Manufacturing-Scale Production** [ICH Q5C]

3180 Manufacture at the scale typically encountered in a facility intended for pro
3181 marketing.

3182 **Marketing Pack** [ICH Q1B]

3183 The combination of immediate pack and other secondary packaging such as

3184 **Mass Balance (Material Balance)** [ICH Q1A]

3185 The process of adding together the assay value and levels of degradation pr

Draft - Not for Implementation

3155 principles of these guidances.

3156 **Immediate (Primary) Pack** [ICH Q1B]

3157 That constituent of the packaging that is in direct contact with the drug substance or drug product,
3158 and includes any appropriate label.

3159 **Impurity**

3160 Any entity of the drug substance (bulk material) or drug product (final container product) that is not
3161 the chemical entity defined as the drug substance, an excipient, or other additives to the drug
3162 product.

3163 **Intermediate** [ICH Q5C]

3164 For biotechnological/biological products, a material produced during a manufacturing process that is
3165 not the drug substance or the drug product but for which manufacture is critical to the successful
3166 production of the drug substance or the drug product. Generally, an intermediate will be quantifiable
3167 and specifications will be established to determine the successful completion of the manufacturing
3168 step before continuation of the manufacturing process. This includes material that may undergo
3169 further molecular modification or be held for an extended period before further processing.

3170 **Long-Term (Real-Time) Testing** [ICH Q1A]

3171 Stability evaluation of the physical, chemical, biological, and microbiological characteristics of a drug
3172 product and a drug substance, covering the expected duration of the shelf life and retest period,
3173 which are claimed in the submission and will appear on the labeling.

3174 **Lot** [21 CFR 210.3(b)(10)]

3175 A batch, or a specific identified portion of a batch, having uniform character and quality within
3176 specified limits; or, in the case of a drug product produced by continuous process, it is a specific
3177 identified amount produced in a unit of time or quantity in a manner that assures its having uniform
3178 character and quality within specific limits.

3179 **Manufacturing-Scale Production** [ICH Q5C]

3180 Manufacture at the scale typically encountered in a facility intended for product production for
3181 marketing.

3182 **Marketing Pack** [ICH Q1B]

3183 The combination of immediate pack and other secondary packaging such as a carton.

3184 **Mass Balance (Material Balance)** [ICH Q1A]

3185 The process of adding together the assay value and levels of degradation products to see how closely

Draft - Not for Implementation

3125 **Degradation Product** [ICH Q5C]

3126 A molecule resulting from a change in the drug substance bulk material) brought about over time.
3127 For the purpose of stability testing of the products described in this guidance, such changes could
3128 occur as a result of processing or storage (e.g., by deamidation, oxidation, aggregation, proteolysis).
3129 For biotechnological/biological products, some degradation products may be active.

3130 **Dosage Form; Preparation** [ICH Q1A]

3131 A pharmaceutical product type, for example tablet, capsule, solution, cream, that contains a drug
3132 substance, generally, but not necessarily, in association with excipients.

3133 **Drug Product; Finished Product** [ICH Q1A]

3134 The dosage form in the final immediate packaging intended for marketing.

3135 **Drug Substance; Active Substance** [21 CFR 312.3(b)]

3136 An active ingredient that is intended to furnish pharmacological activity or other direct effect in the
3137 diagnosis, cure, mitigation, treatment, or prevention of disease or to affect the structure or any
3138 function of the human body.

3139 **Excipient** [ICH Q1A]

3140 Anything other than the drug substance in the dosage form.

3141 **Expiry/Expiration Date** [ICH Q1A]

3142 The date placed on the container/labels of a drug product designating the time during which a batch
3143 of the product is expected to remain within the approved shelf-life specification if stored under
3144 defined conditions, and after which it must not be used.

3145 **Extractables/Leachables**

3146 Materials or components derived from the container/closure which have been transferred into the
3147 contained drug substance or drug product.

3148 **Forced Degradation Testing Studies** [ICH Q1B]

3149 Those studies undertaken to degrade the sample deliberately. These studies, which may be
3150 undertaken in the development phase normally on the drug substances, are used to evaluate the
3151 overall photosensitivity of the material for method development purposes and/or degradation
3152 pathway elucidation.

3153 **Formal (Systematic) Studies** [ICH Q1A]

3154 Formal studies are those undertaken to a preapproval stability protocol which embraces the

Draft - Not for Implementation

3186 these add up to 100 per cent of the initial value, with due consideration of the margin of analytical
3187 precision.

3188 This concept is a useful scientific guide for evaluating data but it is not achievable in all
3189 circumstances. The focus may instead be on assuring the specificity of the assay, the completeness of
3190 the investigation of routes of degradation, and the use, if necessary, of identified degradants as
3191 indicators of the extent of degradation via particular mechanisms.

3192 **Matrixing** [ICH Q1A]

3193 The statistical design of a stability schedule so that only a fraction of the total number of samples are
3194 tested at any specified sampling point. At a subsequent sampling point, different sets of samples of
3195 the total number would be tested. The design assumes that the stability of the samples tested
3196 represents the stability of all samples. The differences in the samples for the same drug product
3197 should be identified as, for example, covering different batches, different strengths, different sizes of
3198 the same container and closure, and, possibly, in some cases different containers/closure systems.

3199 Matrixing can cover reduced testing when more than one variable is being evaluated. Thus the
3200 design of the matrix will be dictated by the factors needing to be covered and evaluated. This
3201 potential complexity precludes inclusion of specific details and examples, and it may be desirable to
3202 discuss design in advance with the FDA chemistry review team where this is possible. In every case,
3203 it is essential that all batches are tested initially and at the end of the long-term testing period.

3204 **Mean Kinetic Temperature** [ICH Q1A]

3205 Mean kinetic temperature (MKT) [14] is defined as the isothermal temperature that corresponds to the
3206 kinetic effects of a time-temperature distribution.

3207 **Modified Release Dosage Forms** [SUPAC-MR]

3208 Dosage forms whose drug-release characteristics of time course and/or location are chosen to
3209 accomplish therapeutic or convenience objectives not offered by conventional dosage forms such as a
3210 solution or an immediate release dosage form. Modified release solid oral dosage forms include both
3211 delayed and extended release drug products.

3212 **New Dosage Form** [ICH Q1C]

3213 A drug product which is a different pharmaceutical product type, but contains the same active
3214 substance as included in the existing drug product approved by the pertinent regulatory authority.

3215 **New Molecular Entity; New Active Substance** [ICH Q1A]

3216 A substance which has not previously been registered as a new drug substance with the national or

[14] J.D. Haynes, "Worldwide Virtual Temperatures for Product Stability Testing", *J. Pharm. Sci.*, Vol. 60, No. 6, 927 (June 1971).

Draft - Not for Implementation

3217 regional authority concerned.

3218 **Pilot-Plant Scale**

3219 The manufacture of either drug substance or drug product by a procedure fully representative of and
3220 simulating that to be applied on a full manufacturing scale.

3221 For oral solid dosage forms this is generally taken to be at a minimum scale of one tenth that of full
3222 production or 100,000 tablets or capsules, whichever is the larger. [Q1A]

3223 For biotechnology products, the methods of cell expansion, harvest, and product purification should
3224 be identical except for the scale of production.
3225 [ICH Q5C]

3226 **Primary Stability Data** [ICH Q1A]

3227 Data on the drug substance stored in the proposed packaging under storage conditions that support
3228 the proposed retest date.

3229 Data on the drug product stored in the proposed container/closure for marketing under storage
3230 conditions that support the proposed shelf life.

3231 **Production Batch**

3232 A batch of a drug substance or drug product manufactured at the scale typically encountered in a
3233 facility intended for marketing production.

3234 **Random Sample**

3235 A selection of units chosen from a larger population of such units so that the probability of inclusion
3236 of any given unit in the sample is defined. In a simple random sample, each unit has equal chance of
3237 being included. Random samples are usually chosen with the aid of tables of random numbers found
3238 in many statistical texts.

3239 **Reference Listed Drug** [21 CFR 314.3]

3240 The listed drug identified by FDA as the drug product upon which an applicant relies in seeking
3241 approval of its abbreviated application.

3242 **Retest Date** [ICH Q1A]

3243 The date when samples of the drug substance should be reexamined to ensure that the material is still
3244 suitable for use.

3245 **Retest Period** [ICH Q1A]

Draft - Not for Implementation

3246 The time interval during which the drug substance can be considered to remain within the
3247 specifications and therefore acceptable for use in the manufacture of a given drug product, provided
3248 that it has been stored under the defined conditions; after this period the batch should be retested for
3249 compliance with specifications and then used immediately.

3250 **Semi-Permeable Container**

3251 A container which permits the passage of a solvent, such as water contained therein, but prevents the
3252 passage of the dissolved substance or solute, thus resulting in an increased concentration of the latter
3253 over time. It may also permit the ingress of foreign volatile materials. The transport of the solvent,
3254 its vapor, or other volatile material occurs through the container by dissolution into one surface,
3255 diffusion through the bulk of the material, and desorption from the other surface, all caused by a
3256 partial-pressure gradient. Examples of semi-permeable containers include plastic bags or semi-rigid
3257 LDPE for LVPs, and LDPE ampoules, vials, or bottles for inhalation or ophthalmic solutions.

3258 **Semisolid Dosage Forms** [SUPAC-SS]

3259 Semi-solid dosage forms include non-sterile and semi-solid preparations, e.g., creams, gels and
3260 ointments, intended for all topical routes of administration.

3261 **Shelf Life; Expiration Dating Period** [ICH Q1A]

3262 The time interval that a drug product is expected to remain within the approved shelf-life
3263 specification provided that it is stored under the conditions defined on the label in the proposed
3264 containers and closure.

3265 **Significant Body of Information** [SUPAC-IR/MR]
3266
3267 Immediate Release Solid Oral Dosage Forms

3268 A significant body of information on the stability of the drug product is likely to exist after five years
3269 of commercial experience for new molecular entities, or three years of commercial experience for
3270 new dosage forms.

3271 Modified Release Solid Oral Dosage Forms
3272
3273 A significant body of information should include, for "Modified Release Solid Oral Dosage Forms," a
3274 product-specific body of information. This product-specific body of information is likely to exist
3275 after five years of commercial experience for the original complex dosage form drug product, or
3276 three years of commercial experience for any subsequent complex dosage form drug product.

3277 **Significant Change** [ICH Q1A]

3278 Significant change for a drug product at the accelerated stability condition and the intermediate
3279 stability condition is defined as:

3280 1. A 5 percent potency loss from the initial assay value of a batch;

Draft - Not for Implementation

3281 2. Any specified degradant exceeding its specification limit;
3282 3. The product exceeding its pH limits;
3283 4. Dissolution exceeding the specification limits for 12 capsules or tablets;
3284 5. Failure to meet specifications for appearance and physical properties, e.g., color, phase
3285 separation, resuspendibility, delivery per actuation, caking, hardness.

3286 **Simple Dosage Form**

3287 A dosage form whose quality and/or stability is less likely to be affected by the manufacturing site
3288 because the release mechanism, delivery system, and manufacturing process are less complicated and
3289 less susceptible to variability.

3290 Examples of simple dosage forms include immediate-release solid oral dosage forms, e.g., tablets,
3291 capsules, semi-solid dosage forms, and oral and parenteral solutions. Due to the diversity of
3292 currently marketed dosage forms and the ever-increasing complexity of new delivery systems, it is
3293 impossible to clearly identify simple vs. complex dosage forms in an exhaustive manner. Applicants
3294 are advised to consult with the appropriate FDA chemistry review team when questions arise.

3295 **Site-Specific Batches**

3296 Batches of drug substance or drug product made at the intended manufacturing scale production site
3297 from which stability data are generated to support the approval of that site, as well as to support the
3298 proposed retest period or expiration dating period, respectively, in an application. The site-specific
3299 batch(es) of the drug product should be made from identifiable site-specific batch(es) of the drug
3300 substance whenever possible.

3301 **Specification-Check/Shelf-life** [ICH Q1A]

3302 The combination of physical, chemical, biological and microbiological test requirements that a drug
3303 substance must meet up to its retest date or a drug product must meet throughout its shelf life.

3304 **Specification-Release** [ICH Q1A]

3305 The combination of physical, chemical, biological and microbiological test requirements that
3306 determine that a drug product is suitable for release at the time of its manufacture.

3307 **Stability**

3308 The capacity of a drug substance or a drug product to remain within specifications established to
3309 ensure its identity, strength, quality, and purity throughout the retest period or expiration dating
3310 period, as appropriate.

3311 **Stability Commitment**

3312 A statement by an applicant to conduct and/or complete prescribed studies on production batches of
3313 a drug product after approval of an application.

Draft - Not for Implementation

3314 **Stability-Indicating Methodology**

3315 Validated quantitative analytical methods that can detect the changes with time in the chemical,
3316 physical, or microbiological properties of the drug substance and drug product, and that are specific
3317 so that the contents of active ingredient, degradation products, and other components of interest can
3318 be accurately measured without interference.

3319 **Stability Profile**

3320 The physical, chemical, biological, and microbiological behavior of a drug substance or drug product
3321 as a function of time when stored under the conditions of the Approved Stability Protocol.

3322 **Storage Conditions Tolerances** [ICH Q1A]

3323 The acceptable variation in temperature and relative humidity of stability storage.

3324 **Strength** [21 CFR 210.3(b)(16)]

3325 The concentration of the drug substance (for example weight/weight, weight/volume, or unit
3326 dose/volume basis), and/or the potency, that is, the therapeutic activity of the drug product as
3327 indicated by appropriate laboratory test or by adequately developed and controlled clinical data
3328 (expressed for example, in terms of units by reference to a standard).

3329 **Stress Testing - Drug Substance** [ICH Q1A]
3330
3331 Studies undertaken to elucidate intrinsic stability characteristics. Such testing is part of the
3332 development strategy and is normally carried out under more severe conditions than those used for
3333 accelerated tests.

3334 **Stress Testing - Drug Product** [ICH Q1A]

3335 Light testing should be an integral part of stress testing.

3336 Special test conditions for specific products (e.g., metered dose inhalations and creams and
3337 emulsions) may require additional stress studies.

3338 **Supporting Stability Data** [ICH Q1A]

3339 Data other than the primary stability data, such as stability data on early synthetic route batches of
3340 drug substance, small scale batches of materials, investigational formulations not proposed for
3341 marketing, related formulations, product presented in containers and/or closures other than those
3342 proposed for marketing, information regarding test results on containers, and other scientific
3343 rationale that support to the analytical procedures, the proposed retest period or shelf life and
3344 storage conditions.

3345 **Tentative Expiration Dating Period**

Draft - Not for Implementation

3346 A provisional expiration dating period which is based on acceptable accelerated data, statistical
3347 analysis of available long-term data, and other supportive data for an NDA product, or on acceptable
3348 accelerated data for an ANDA product, but not on full long-term stability data from at least three
3349 production batches.

Author Index

Subject Index